FARADAY DISCUSSIONS
NO. 100 1995

Atmospheric Chemistry: Measurements, Mechanisms and Models

The Faraday Division
The Royal Society of Chemistry
London

Organising Committee

Prof. I. W. M. Smith (*Co-Chairman*)
Dr. J. R. Sodeau (*Co-Chairman*)
Dr. R. A. Cox
Dr. J. A. Pile
Dr. J. M. C. Plane
Prof. F. Taylor

ISBN: 0-85404-873-1

ISSN: 0301-7249

Typeset by Santype International Ltd., Netherhampton Road, Salisbury, Wiltshire and printed and bound in Great Britain by Whitstable Litho Printers Ltd.

A General Discussion

on

Atmospheric Chemistry: Measurements, Mechanisms and Models

19th, 20th and 21st April, 1995

A General Discussion on Atmospheric Chemistry was held at the University of East Anglia, Norwich on 19th, 20th and 21st April, 1995

Contents

Faraday Discuss., 1995, **100**, 1–21

Introductory Lecture

Overview of Tropospheric Chemistry: Developments during the Past Quarter Century and a Look Ahead

Paul J. Crutzen
Max-Planck-Institute for Chemistry, P.O. Box 3060, 55020 Mainz, Germany

An overview is given of the main processes that affect tropospheric chemistry, in particular regarding the roles that are played by ozone, hydroxyl radicals, methane, carbon monoxide, NO and NO_2. Model calculations are presented of the current concentrations of O_3 and OH, as well as of changes compared with pre-industrial times. Based on these calculations, proposals are presented for atmospheric budgets of CH_4 and CO.

The important roles of the tropics and subtropics in atmospheric chemistry are emphasized and the large changes that are expected in the developing world due to expanding agricultural and industrial activities are described. Finally, a personal view is presented of some research topics which deserve future attention.

1. Introduction

Over the past quarter of a century much has been learned about the crucial role of ozone in the chemistry of the stratosphere and troposphere and for the Earth's climate. This progress in knowledge has gone hand in hand with the realization that human activities are increasingly affecting the concentrations of ozone in both regions of the atmosphere. Major and unforeseen depletions of stratospheric ozone are now observed during springtime at high latitudes, especially over the Antarctic continent. At the same time, near and downwind of industrial centres, large increases in tropospheric ozone concentrations are commonplace and an overall increase in the ozone concentration in the background troposphere in the northern hemisphere has likewise been noted. High tropospheric ozone concentrations are also common during the dry season in the tropics and subtropics in connection with agricultural biomass burning. Stratospheric ozone loss due to the emission of industrial chlorofluorocarbon gases has been recognized as a truly global environmental problem and, as a consequence, international agreements to phase out their production have been put in place. Although the discovery of the Antarctic 'ozone hole' came as a total surprise, a few years of intense research activities were sufficient to identify clearly the catalytic role of minute quantities of ozone-destroying chlorine and bromine radicals of anthropogenic origin as the cause for the unexpectedly large ozone depletions. A further discovery was the importance of reactions on low-temperature particulate surfaces for the conversion of the reservoir molecules HCl and $ClONO_2$, which do not react with ozone, to Cl_2 and further to the active chlorine compounds ($Cl + ClO + 2Cl_2O_2$) that rapidly destroy ozone by catalytic reactions. These reactions were of minor significance under natural conditions, but they have gained major importance as a result of a sixfold increase in the stratospheric chlorine loading over natural levels due to the release of the $CFCl_3$ and CF_2Cl_2, and other organic chlorine- and bromine-containing industrial gases over the past decades.

Although there remain several interesting, unanswered questions regarding stratospheric chemistry and dynamics, especially concerning microphysics, phase changes and chemical composition of various types of particulate matter, they are being addressed by major research efforts. It is my opinion that the main gaps in present knowledge of atmospheric chemistry may well involve the troposphere. For that reason the following discussion will be concerned with tropospheric photochemistry. In this paper, a brief overview is given of the main reactions in tropospheric chemistry, some model results concerning past and present distributions of key chemical constituents are presented and, finally, a personal preference is presented of some major research topics.

2. Photochemistry of the Background Troposphere

Only *ca.* 10% of all atmospheric ozone is located in the troposphere, with typical volume mixing ratios in the range of 20–100 ppbv (1 ppbv = 10^{-9} by volume). Until the early 1970s it was believed that this ozone originated from the stratosphere, was chemically inert in the troposphere and was only destroyed at the earth's surface. We now know that this picture is far from true and that tropospheric ozone is of fundamental importance for the composition of the earth's atmosphere, owing to its crucial role in the production of OH radicals.[1,2] This is achieved by ozone's absorption of solar UV radiation of wavelengths shorter than *ca.* 325 nm, generating electronically excited oxygen atoms, $O(^1D)$,

$$O_3 + h\nu \rightarrow O(^1D) + O_2 \ (\lambda < 325 \text{ nm}) \quad (1)$$

which possess enough energy to react with H_2O to produce two OH radicals

$$O(^1D) + H_2O \rightarrow 2OH \quad (2)$$

for each ozone molecule that is destroyed by reactions (1) and (2). Reaction with OH initiates the oxidation and removal of most trace gases from the atmosphere. Hydroxyl radicals, therefore, determine the oxidation efficiency (also called the oxidation capacity or oxidation power) of the atmosphere and their chemical reactivity is of the greatest importance for the composition of the atmosphere. The chemical significance of OH radicals depends on the fact that only a few of the gases that are produced at the earth's surface can be removed by rainfall and none by reaction with O_2, although oxygen is 10^{13} times more abundant than OH. It is highly remarkable that the primary formation of OH is accomplished by the same so-called UV-B radiation of wavelengths shorter than 325 nm that can be harmful to life on Earth. Thus, although too much UV-B radiation and ozone are detrimental to the biosphere, the same factors are also essential for relieving the atmosphere of most compounds of natural and anthropogenic origin. The residence times of many atmospheric gases are largely determined by their reactivity towards OH radicals. With a global average OH concentration of about 10^6 molecule cm^{-3}, the residence time of CH_4 is *ca.* 9 years and that of CO on the average 1–2 months.[3] These compounds are the most abundant, photochemically active carbon-containing gases in the background troposphere, in which *ca.* 60% of the OH radicals react with CO, and 40% with CH_4 and its oxidation products. Consequently, a clear understanding of tropospheric chemistry and assessments of the future impact of human activities requires detailed knowledge of the photochemical reactions which affect ozone, carbon monoxide and methane. The concentrations of ozone and hydroxyl radicals are also strongly affected by reactions involving NO and NO_2 as catalysts.[4,5] For instance, in the presence of sufficiently large concentrations of nitric oxide, the oxidation of CO to CO_2 leads to formation of ozone through the reaction chain:

$$CO + OH \rightarrow H + CO_2 \tag{3}$$

$$H + O_2 + M \rightarrow HO_2 + M \tag{4}$$

$$HO_2 + NO \rightarrow OH + NO_2 \tag{5}$$

$$NO_2 + h\nu \rightarrow NO + O\ (\leqslant 420\ \text{nm}) \tag{6}$$

$$O + O_2 + M \rightarrow O_3 + M \tag{7}$$

$$CO + 2O_2 \rightarrow CO_2 + O_3 \tag{C1}$$

with OH, HO_2, NO and NO_2 serving as catalysts.

On the contrary, in NO-poor environments, CO oxidation leads to ozone destruction:

$$CO + OH \rightarrow H + CO_2 \tag{3}$$

$$H + O_2 + M \rightarrow HO_2 + M \tag{4}$$

$$HO_2 + O_3 \rightarrow OH + 2O_2 \tag{8}$$

$$CO + O_3 \rightarrow CO_2 + O_2 \tag{C2}$$

The second reaction sequence is more important than the first one whenever the ratio of the concentrations of NO to O_3 is less than 2×10^{-4}, the ratio between the rate coefficients of reactions (8) to (5). With global average ozone volume mixing ratios increasing from *ca.* 25 ppbv at the earth's surface to 100 ppbv at the tropopause, the break-even point between the reactions (C2) and (C1) is attained at NO volume mixing ratios of 5×10^{-12} (5 pptv) in the lower and 20 pptv in the upper troposphere. Although these are very low concentrations, they may nevertheless not be exceeded in extensive regions of the troposphere[6–9] owing to small emissions of NO in the oceanic atmosphere and rapid conversion, within days, of NO_x($NO + NO_2$) to highly water-soluble nitric acid by reaction with OH

$$NO + O_3 \rightarrow NO_2 + O_2 \tag{9}$$

$$NO_2 + OH(+M) \rightarrow HNO_3(+M) \tag{10}$$

during daytime or during night-time *via* N_2O_5 formation

$$NO_2 + O_3 \rightarrow NO_3 + O_2 \tag{11}$$

$$NO_3 + NO_2 = N_2O_5 \tag{12}$$

followed by the reaction of N_2O_5 with H_2O on cloud droplets and water-containing, *e.g.* sulfate, aerosol

$$N_2O_5 + H_2O \rightarrow 2HNO_3 \tag{13}$$

HNO_3 formation by heterogeneous reactions is especially important during dark and cold winter periods, when N_2O_5 formation is favoured.

Because of the very short lifetime of NO_x in the troposphere, photochemical ozone production is largely concentrated near the industrialized region at mid-latitudes in the Northern Hemisphere, the subtropical and tropical continents which are affected by biomass burning during the dry season, and the continental tropics and mid-latitude during summer as a consequence of NO formation by lightning. Elsewhere measured background mixing ratios of NO of less than 10 pptv are common in the oceanic marine boundary layer and free troposphere.[6–9] In these regions efficient ozone destruction takes place by reactions (1), (2) and (8). Very low ozone mixing ratios, <10 ppbv, have consequently been observed on various occasions in the tropical marine boundary layer[10–14] and during March 1993 also over the entire tropospheric column in a large tract of the Pacific ocean, covering a distance of *ca.* 2000 km between Fiji and Christmas Island.[15]

Besides reacting with NO and O_3 *via* reactions (5) and (8), HO_2 can also react with itself, leading to the production of hydrogen peroxide, which can photodissociate, resulting in the oxidation of CO to CO_2 without any effect on ozone

$$CO + OH \rightarrow H + CO_2 (2\times) \quad (3)$$

$$H + O_2 + M \rightarrow HO_2 + M (2\times) \quad (4)$$

$$HO_2 + HO_2 \rightarrow H_2O_2 + O_2 \quad (14)$$

$$H_2O_2 + h\nu \rightarrow 2OH \ (\leqslant 350 \text{ nm}) \quad (15)$$

$$2CO + O_2 \rightarrow 2CO_2 \quad (C3)$$

H_2O_2 plays an important role in the aqueous-phase oxidation of SO_2:[16–18]

$$SO_2 \cdot H_2O = H^+ + HSO_3^- \quad (16)$$

$$HSO_3^- + H_2O_2 \rightarrow HSO_4^- + H_2O \quad (17)$$

$$HSO_4^- = H^+ + SO_4^{2-} \quad (18)$$

Because highly water-soluble H_2O_2, which is formed in reaction (14) from two HO_2 radicals, can be effectively removed from the atmosphere by uptake in clouds, rainout is an important sink for HO_2 and, through prevention of reaction (15), also of OH radicals. Furthermore, H_2O_2 is involved in the catalytic reaction pair:

$$HO_2 + HO_2 \rightarrow H_2O_2 + O_2 \quad (14)$$

$$H_2O_2 + OH \rightarrow HO_2 + H_2O \quad (19)$$

$$OH + HO_2 \rightarrow H_2O + O_2 \quad (C4)$$

which removes both OH and HO_2. The oxidation of CO in NO-poor environments, therefore, generally leads to loss of HO_x ($OH + HO_2$) radicals; in NO-rich environments ozone is formed.[4,5] However, in a cloudy atmosphere, O_3 formation may be substantially suppressed[19,20] *via* reactions which will be discussed later. An increase in atmospheric CO by anthropogenic activities alone leads to a decrease in OH concentrations *via* reaction (3). As they do also cause NO emissions, however, the loss of OH may be partially or fully compensated by reaction (5), causing greater conversion of HO_2 to OH and more ozone production, in turn leading to more OH by reactions (1) and (2). The net result of increased anthropogenic emissions of CO, CH_4 and NO on OH concentrations is thus not straightforward and requires the use of three-dimensional (3D), global models of atmospheric chemistry, an example of which will be shown later in this paper.

By affecting the atmospheric concentrations of OH and O_3,[4,5] the oxidation of CH_4 is also of great importance for tropospheric photochemistry. Following the initial attack of OH on CH_4, the prevailing concentration of NO determines which oxidation pathway dominates. In NO-rich environments, rapid formation of ozone and formaldehyde occurs:

$$CH_4 + OH \rightarrow CH_3 + H_2O \tag{20}$$

$$CH_3 + O_2 + M \rightarrow CH_3O_2 + M \tag{21}$$

$$CH_3O_2 + NO \rightarrow CH_3O + NO_2 \tag{22}$$

$$CH_3O + O_2 \rightarrow CH_2O + HO_2 \tag{23}$$

$$HO_2 + NO \rightarrow OH + NO_2 \tag{5}$$

$$NO_2 + h\nu \rightarrow NO + O\ (<420\ \text{nm})\ (2\times) \tag{6}$$

$$O + O_2 + M \rightarrow O_3 + M\ (2\times) \tag{7}$$

$$CH_4 + 4O_2 \rightarrow CH_2O + H_2O + 2O_3. \tag{C5}$$

In NO-poor environments, CH_4 is oxidized to CH_2O by either of the following pathways:

$$CH_4 + OH \rightarrow CH_3 + H_2O \tag{20}$$

$$CH_3 + O_2 + M \rightarrow CH_3O_2 + M \tag{21}$$

$$CH_3O_2 + HO_2 \rightarrow CH_3O_2H + O_2 \tag{24}$$

$$CH_3O_2H + h\nu \rightarrow CH_3O + OH\ (<330\ \text{nm}) \tag{25}$$

$$CH_3O + O_2 \rightarrow CH_2O + HO_2 \tag{23}$$

$$CH_4 + O_2 \rightarrow CH_2O + H_2O \tag{C6}$$

or

$$CH_4 + OH \rightarrow CH_3 + H_2O \tag{20}$$

$$CH_3 + O_2 + M \rightarrow CH_3O_2 + M \tag{21}$$

$$CH_3O_2 + HO_2 \rightarrow CH_3O_2H + O_2 \tag{24}$$

$$CH_3O_2H + OH \rightarrow CH_2O + H_2O + OH \tag{26a}$$

$$CH_4 + OH + HO_2 \rightarrow CH_2O + 2H_2O \tag{C7}$$

Analogously to reaction chain (C4), the catalytic HO_x-destroying pair of reactions

$$CH_3O_2 + HO_2 \rightarrow CH_3O_2H + O_2 \tag{24}$$

$$CH_3O_2H + OH \rightarrow CH_3O_2 + H_2O \tag{26b}$$

$$OH + HO_2 \rightarrow H_2O + O_2 \tag{C8}$$

is also important. The oxidation pathways of methane in NO-poor environment are, therefore, more complex than in NO-rich environments, but it is clear that OH and HO_2 radicals are being lost.

Three reaction pathways lead to the gas phase oxidation of CH_2O to CO:

$$CH_2O + h\nu \rightarrow H + CHO \; (\leqslant 350 \text{ nm}) \qquad (27a)$$

$$H + O_2 + M \rightarrow HO_2 + M \qquad (4)$$

$$CHO + O_2 \rightarrow CO + HO_2 \qquad (28)$$

$$CH_2O + 2O_2 \rightarrow CO + 2HO_2 \qquad (C9a)$$

or

$$CH_2O + OH \rightarrow CHO + H_2O \qquad (29)$$

$$CHO + O_2 \rightarrow CO + HO_2 \qquad (28)$$

$$CH_2O + OH + O_2 \rightarrow CO + H_2O + HO_2 \qquad (C9b)$$

or directly *via*

$$CH_2O + h\nu \rightarrow CO + H_2 \; (\leqslant 350 \text{ nm}) \qquad (27b)$$

The HO_2 radicals formed in reaction cycles (C9*a*) and (C9*b*) can react further *via* reactions (5), (8), (14), or (24), leading to formation or destruction of O_3, or formation of H_2O_2, depending on the concentrations of NO and O_3. The overall net production of HO_2 radicals in the gas phase is *ca.* 0.8 per CH_2O molecule that is oxidized to CO.

The chemistry of CH_2O is not restricted to the gas phase. CH_2O is very soluble in water, leading to $CH_2(OH)_2$, and a sequence of aqueous-phase reactions[16–20]

$$CH_2(OH)_2 + OH + O_2 \rightarrow HCOOH + H_2O + HO_2 \qquad (30)$$

$$HCOOH = H^+ + HCOO^- \qquad (31)$$

$$H^+ + HCOO^- + OH + O_2 \rightarrow CO_2 + HO_2 + H_2O \qquad (32)$$

$$CH_2O + 2OH + 2O_2 \rightarrow CO_2 + 2HO_2 + H_2O \qquad (C10)$$

producing formic acid and CO_2 but not CO, thus significantly reducing atmospheric CO production, despite the fact that air parcels on average spend only *ca.* 15% of the time in clouds.[19,20]

Clouds can significantly affect the photochemistry of the atmosphere in other ways through (i) dissolution and scavenging of water-soluble substances, such as H_2O_2, SO_2 and various acids (*e.g.* HNO_3, H_2SO_4, HCl, HCOOH, CH_3COOH); (ii) conversion of NO_x to HNO_3 *via* reactions (11)–(13); (iii) uptake of HO_2 in water droplets, leading to aqueous-phase ozone destruction

$$HO_2 = H^+ + O_2^- \qquad (33)$$

$$O_2^- + O_3 \rightarrow O_2 + O_3^- \qquad (34)$$

$$O_3^- + H^+ \rightarrow OH + O_2 \qquad (35)$$

$$HO_2 + O_3 \rightarrow OH + 2O_2 \qquad (C11)$$

while at the same time largely preventing O_3 formation *via* reaction (5)–(7), as NO, which is only sparingly water-soluble, remains in the gas phase.[19,20]

The oxidation sequences leading from CH_4 to CO, and from there to CO_2, thus have a strong effect on the concentrations of both hydroxyl radicals and ozone in the background troposphere. In clouds they will lead to a lowering of OH and HO_2 radicals. In cloud-free regions, depending on the ambient concentrations of NO, the oxidation of one molecule of CH_4 to one molecule of CO_2 yields the following net results:
(a) in NO-poor environments, *i.e.* especially in the unpolluted background troposphere, a net loss of about two HO_x radicals, and (b) in NO-rich environments: a net gain of 0.5 HO_x radicals and almost four ozone molecules.

The amount of methane in the atmosphere has been increasing for a considerable time, in agreement with the growth of various anthropogenic activities, which are summarized in Table 1. Long-term observations indicated an average yearly increase close to 1%[21–23] until about the mid-1980s. This increase has, however, levelled off during the past decade.[24–26] Analyses of air trapped in ice cores have shown that the atmospheric methane content during the present interglacial until about 200 years ago was two to three times lower than at present.[27–31] The observed increases in methane have, therefore, probably caused substantial increases in ozone concentrations in NO-rich environments and decreases in hydroxyl radical concentrations in NO-poor environments. During this century, the production of ozone at northern hemisphere mid-latitudes has also been strongly enhanced by concomitant increases in NO emissions from industrial activities (Table 1), and probably also by increases by *ca.* 0.85% per year in the CO concentrations of the free troposphere as deduced from solar spectra taken at the Jungfraujoch in Switzerland between 1950 and 1985.[32] The observed rise in CO at mid-latitudes in the northern hemisphere may have led to a decrease of OH through reaction (3). However, as noted before, increasing nitric oxide emissions, leading to enhanced ozone formation, as well as enhanced HO_2 to OH conversion by reaction (5) have counteracted this. Unfortunately, no reliable CO determinations have been performed

Table 1 Estimated sources and sinks of CO, CH_4, reactive biogenic hydrocarbons (*e.g.* C_5H_8) and NO, their average atmospheric residence times and ranges of volume mixing ratios

gas	sources/10^{14} g a^{-1} or 10^{12} g N a^{-1}	sinks and atmospheric increase/10^{14} g a^{-1} or 10^{12} g N a^{-1}	atmospheric lifetimes (volume mixing ratios)
CO	25: total 4 ± 1: fossil-fuel burning 7 ± 4: biomass burning 1: vegetation 9: methane oxidation 4: C_5H_8 *etc.* oxidation	25: reaction with OH	2 months ($40–200 \times 10^{-9}$)
CH_4	6.3: total 0.8 ± 0.2: rice fields 2.6 ± 0.3: natural wetlands 0.7: ruminants 0.4 ± 0.1: animal wastes 0.15 ± 0.05: termites 0.35 ± 0.05: biomass burning 0.65 ± 0.15: gas leakage, oil production 0.35: coal mining 0.4 ± 0.1: landfills	5.3: reaction with OH (troposphere) 0.4: soils 0.3: stratospheric loss 0.3: increase	8 years ($1.7–1.8 \times 10^{-6}$)
C_5H_8 *etc.*	6–10: forests		<1 day ($0–10 \times 10^{-9}$)
NO:	44 ± 10: total 10 ± 5: soils 5 ± 3: lightning 0.5: flux from stratosphere 24: fossil fuel combustion 4 ± 2: biomass burning 0.6: aircraft	20 ± 5: 'dry' deposition 18 ± 5: OH (model) 6 ± 4: N_2O_5 on aerosol	few days ($0–50 \times 10^{-9}$)

on ice cores and long-term CO monitoring has so far only been performed at two measurement sites in the southern hemisphere, Cape Grim/Tasmania and Cape Town/South Africa. For these sites no significant CO trends could so far be deduced from the records. In recent years CO measurements by two research groups have shown surprisingly large annual downward trends by 1.4% in the northern hemisphere and 5.2% in the southern hemisphere between 1987 and 1992,[33] or by 6.1% and 7%, respectively, between June 1990 and 1993.[34] Over the past two years, however, the downward trends at the measurements sites have again reversed by similar rates (P. Novelli, personal communication). The reasons for these very rapid changes are not known. Later in this paper we will return to this issue.

3. Observed and Model Calculated Global Ozone Distributions

Global models have been developed to calculate the worldwide distributions of O_3, OH and other chemical constituents.[35–40] As an example of such model calculations, we will next study some of the results that were obtained with the 3D, global transport photochemistry model, MOGUNTIA, which has been developed at the Max-Planck-Institute for Chemistry in Mainz, Germany and which is run with 10° horizontal and 100 hPa vertical resolution.[35] This model is based on monthly averaged winds and parametrizations that describe transport by subgrid-scale processes and convection.[41] The latter is of particular importance for vertical transport of short-lived gases which would have no chance to reach the upper troposphere if only mean and eddy motion would be considered. The model takes into account the 'background chemistry' of the atmosphere involving reactions in the O_3–NO_x–HO_x–CH_4–CO system as outlined before in this paper. Heterogeneous reactions in clouds and on particulate matter were also taken into account.[19,20,42] CH_4 surface concentrations were prescribed based on observations both for the year 1985 and for the pre-industrial atmosphere, in the latter case provided by ice-core data.[27–31] For NO, emissions were based on the estimates given in Table 1, disregarding industrial inputs and assuming 10% of the current biomass burning emissions for the pre-industrial simulations. In the case of CO, simulations of the current atmosphere were based on the limited amount of surface CO observations that were available for the year 1985. For the pre-industrial atmosphere, CO fluxes were assigned at the lower boundary based on fluxes derived for the contemporary atmosphere reduced by the industrial and 90% of the biomass burning emissions.[35] With additional upper boundary conditions, describing the downward flux of NO_y and O_3 from the stratosphere, as well as CO and CH_4 destruction in the stratosphere, the model calculates the distributions of O_3 and OH, as well as the other intermediate compounds in the background photochemical system for both the industrial and pre-industrial atmospheres. The calculated distributions of OH are used to estimate the sources and sinks of CH_4 and CO.

The calculated seasonal and zonal average meridional ozone volume mixing ratio distributions for the middle of the past decade (Fig. 1) show some of the main features that are also present in the observations,[43] such as larger concentrations in the northern hemisphere and minimum ozone concentrations in the tropics [as a consequence of maximum ozone destruction by reactions (1) + (2) and (9) owing to the high levels of photochemically active UV radiation that can penetrate through the thin tropical ozone shield]. The model calculates too much ozone in the tropics, however. This discrepancy may have several causes: (i) too large an influx of ozone from the stratosphere to the upper tropical troposphere from the extratropical lower stratosphere, (ii) too much NO production from lightning, leading to too much ozone production by reaction cycles (C1) and (C5), (iii) insufficient consideration of the effects of aqueous and heterogeneous reactions (33)–(35) and (13).[19,20]

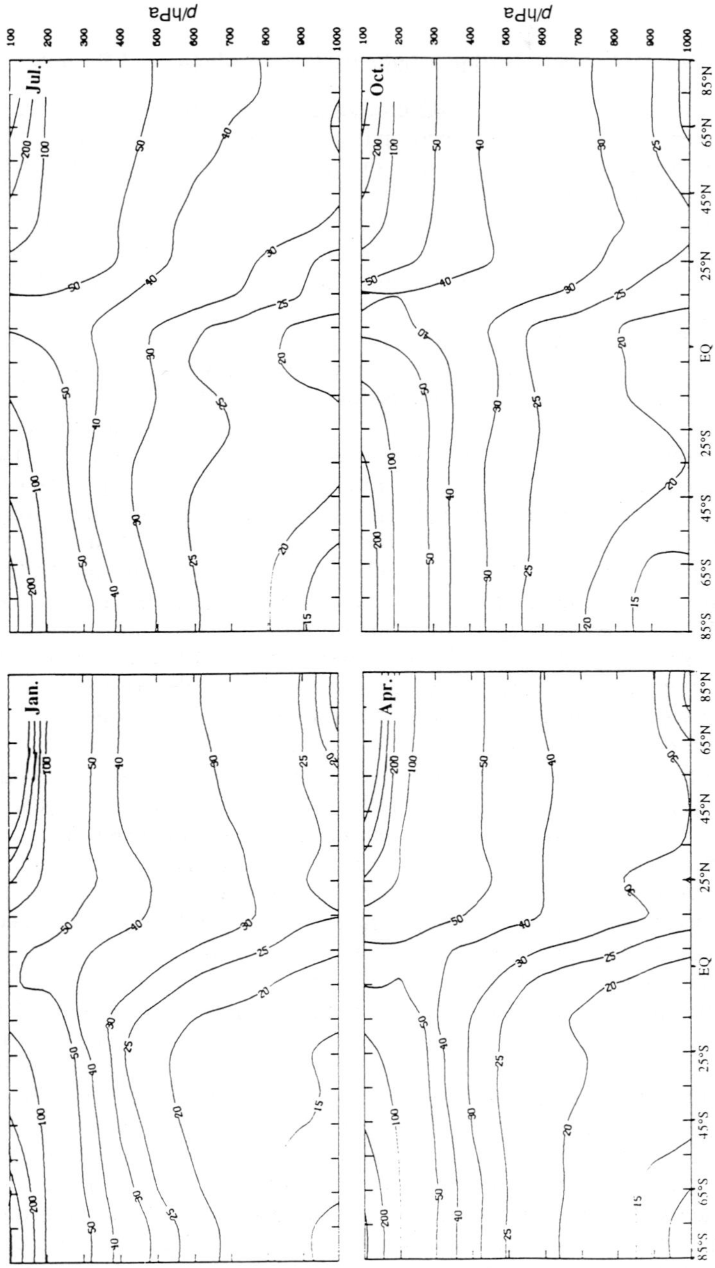

Fig. 1 Calculated meridional cross sections of zonal average ozone volume mixing ratios in ppbv (10^{-9}) for the months of January, April, July and October; 1985 conditions

Unfortunately, despite the critical importance of ozone in atmospheric chemistry, measurements of tropospheric ozone are too limited to allow a critical assessment of the model's performance. This applies especially for the tropics and subtropics in which regions the atmospheric oxidation efficiency dominates, owing OH concentrations.

4. O_3 and OH Concentrations: Industrial *vs.* Pre-industrial Periods

With the large growth in tropospheric NO emissions due to biomass burning, and especially fossil-fuel combustion, and the accompanying enhancements in the atmospheric CH_4 and CO emissions, tropospheric ozone should have increased during the past few centuries, especially in the northern hemisphere. This is indeed reflected in the model calculations. Comparing Fig. 1 and 2 we note almost a doubling of surface ozone concentrations in the northern hemisphere since pre-industrial times. In the southern hemisphere and the tropics, the calculated ozone concentration also increased during the same period by up to *ca.* 50%.

A substantial growth in tropospheric ozone concentrations in the northern hemisphere has indeed been established by observations. According to analyses by Volz and Kley,[44] reliable surface ozone measurements had been made by the end of the 19th century by Levy and co-workers at Montsouris on the outskirts of Paris. These showed, for all months of the year, ozone volume mixing ratios below 10 ppbv for air coming from the unpolluted rural sector, *ca.* four times lower than currently observed background values at mid-latitudes in the northern hemisphere. The more extensive measurements made in the 1920s to 40s in Western Europe, including many optical observations, indicate since then up to a doubling of summertime tropospheric ozone concentrations at 3–4 km altitude.[45] At Barrow, Alaska; Mauna Loa, Hawaii; and Fichtelberg and Hohenpeissenberg, Germany, surface ozone concentrations increased by *ca.* 1% yearly between 1966 and 1986.[46] Analyses of background ozone measurements for the last 20 years present a somewhat complicated picture[47,48] with large increases of a few per cent per year in Western Europe and Japan, but smaller or insignificant trends elsewhere.

We also present in Fig. 3 and 4 the seasonal distributions of zonal average OH concentrations calculated for the industrial (1985) and pre-industrial periods. As a test for the validity of the calculated OH fields, comparison between calculated and observed methylchloroform (CH_3CCl_3) concentrations at the five stations at which such measurements are carried out is particularly valuable. Taking into account recent corrections in the absolute calibrations of the CH_3CCl_3 measurements[3] and loss in the oceans and stratosphere, we conclude that the calculated OH concentrations for 1985 in Fig. 3 are on average too low by *ca.* 20%. Considering the many approximations and simplifications that are introduced in the model, this is remarkably good agreement. Correcting for the 20% discrepancy, the average global tropospheric OH concentration is *ca.* 10^6 molecule cm^{-3}. The calculated OH distributions can then be used to estimate the spatial distribution of the atmospheric removal of CH_4 and CO by reaction with OH, thus yielding important information on the required sources of these gases at the earth's surface. In this way, the total worldwide destruction of CH_4 by reaction with OH is estimated to have been equal to *ca.* 530 Mt in 1985. Adding uptake by soils, and destruction in the stratosphere as additional sinks for CH_4, and accounting for the atmospheric growth rate, the total source of CH_4 in 1985 should have been *ca.* 630 Mt (see Table 1 for detailed information). This analysis clearly shows the power of modelling, which puts a limit on the sum of the individual CH_4 sources, each of which are still quite uncertain. It is interesting to note that 'authoritative' budget estimates of CH_4 about 30 years ago, when the role of OH in atmospheric removal processes was still unknown, contained much larger CH_4 sources than those given above.

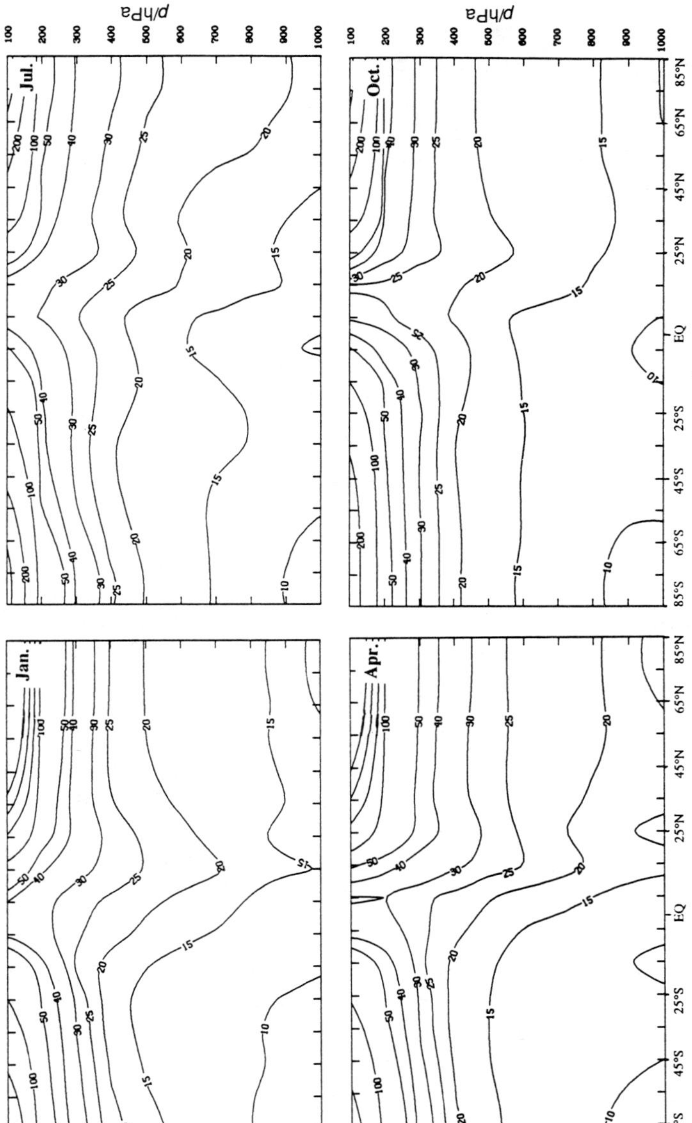

Fig. 2 As in Fig. 1, but for pre-industrial conditions

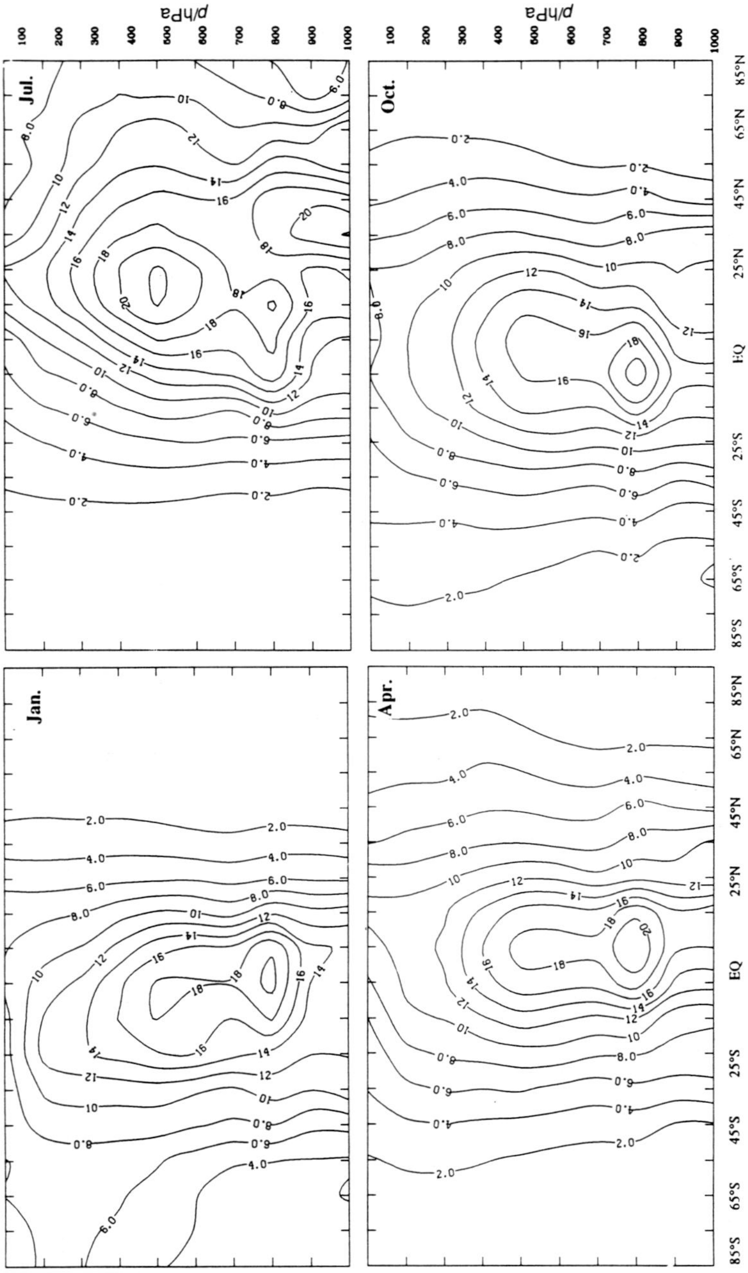

Fig. 3 Calculated meridional cross-sections of zonal and diurnal (24 h) average OH concentrations (in 10^6 molecule cm^{-3}) for industrial conditions

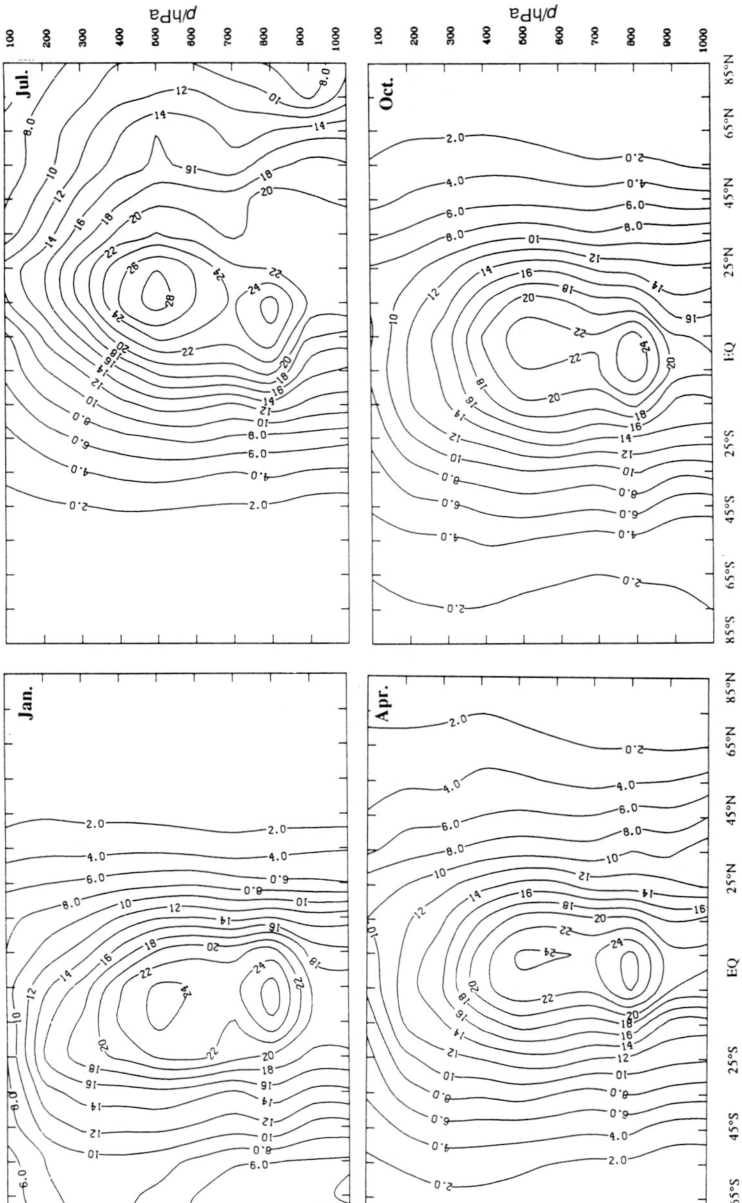

Fig. 4 As in Fig. 3, but for pre-industrial conditions

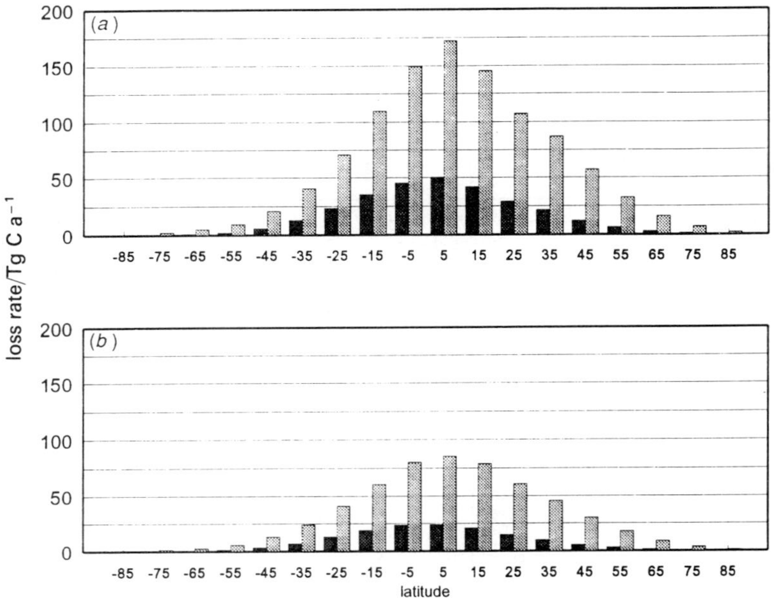

Fig. 5 Calculated annual CH_4 (black bars) and CO (grey bars) loss rates in Tg C per year in 10° latitude bins for (*a*) the industrial and (*b*) the pre-industrial periods

The high OH concentrations in the tropics lead to pronounced maxima in the atmospheric destruction rates of CH_4 and CO in these regions, for both the industrial and pre-industrial periods (see Fig. 5), demonstrating the great importance of the tropics in atmospheric chemistry. Calculations of CH_4 and CO losses in the pre-industrial period may be taken to represent the natural contributions to atmospheric methane, although of course these are uncertain, since there is no way to check the results as in the case for the contemporary atmosphere. It is, however, interesting to note, comparing Fig. 3 and 4, that the model calculations show a decrease in global average OH concentrations since pre-industrial times of *ca.* 25%. The information obtained by the model calculations and analyses compiled by WMO[49] together with some personal assessments of the available data have been combined in Table 1 to provide some estimates of individual sources of CH_4 and CO.

5. Sources and Sinks of Background Tropospheric Ozone

The term in the global, tropospheric ozone budget that can be estimated with the greatest confidence is the ozone loss by reactions (1) and (2), which on average amounts to *ca.* 15×10^{10} molecule cm^{-2} s^{-1}. The uncertainty is due mainly to incomplete knowledge of the ozone distributions and their correlations with those of water vapour in the tropics and subtropics. The estimate takes into account recent revisions in the quantum yields of $O(^1D)$ formation by reaction.[50,51] The ozone loss at the ground, although quite uncertain, may be of similar magnitude.[52] In addition, *ca.* 7×10^{10} molecule cm^{-2} s^{-1} of ozone is lost *via* reaction (9) and various other minor reactions. The estimated global average downward transport of stratospheric ozone by meteorological processes of *ca.* 4×10^{10} molecule cm^{-2} s^{-1}, which was assumed in this model,[43] is not, therefore, sufficient to balance the calculated chemical O_3 losses. This conclusion even holds if we accept the higher estimates of the stratospheric source that can be found in the literature (for a discussion see ref. 43). It is thus clear that significant amounts of ozone must also be produced in environments containing sufficient NO. On the other hand, if all oxidation of CO and CH_4, with average tropospheric destruction rates by reaction with OH of *ca.* 5×10^{11} and 1.5×10^{11} molecule cm^{-2} s^{-1}, respectively, were to occur in NO-rich environments, yielding one ozone molecule for each CO that is oxidized to CO_2, and about three for each CH_4 molecule that is oxidized to CO, then the global average net ozone production rate would be close to 10^{12} molecule cm^{-2} s^{-1}, far larger than the atmospheric plus surface loss reactions. Consequently, a substantial fraction of the background troposphere must contain so little NO_x that ozone is not produced but destroyed. This analysis shows that the ozone production potential is utilized in only *ca.* 30% of the troposphere, and that the availability of NO is the main limiting factor. As a large fraction of NO is produced by anthropogenic activities (see Table 1), it is not surprising that substantial increases in tropospheric ozone have occurred during the industrial era, especially in the northern hemisphere. In future, increasing NO emissions by industrial activities, and maybe also from agricultural biomass burning, have the potential to increase ozone concentrations considerably in the developing world of the tropics and subtropics.

In Table 2 we summarize the model-calculated terms in the ozone budget of the troposphere for both the northern and southern hemispheres. The results clearly show the great importance of tropospheric photochemical processes, mostly in the northern, but also in the southern hemisphere. The high ozone production is caused by anthropogenic activities, as can be seen by comparison of the budget analyses for the industrial and pre-industrial atmospheres. However, surprisingly, it appears that, even in the pre-industrial era, photochemical reactions were significant contributors to the tropospheric ozone budget, mainly owing to reactions in the tropics stimulated by natural NO emissions coming from lightning and soils. It should be remembered that the budget analyses presented in Table 2 consider only CH_4 and CO as 'fuels' for the ozone-formation processes. The contributions coming from the oxidation of more reactive hydrocarbons may be significant, but very difficult to estimate, because of major uncertainties in natural hydrocarbon and NO production rates, as well as reaction mechanisms. Other major uncertainties are concerned with the escape of NO_x and reactive hydrocarbons from the boundary layer and interactions of chemical products with the aqueous and ice phase in the thunderstorms in which much NO may be produced by lightning. This would in principle lead to a situation with considerable potential for O_3 formation.[53] However, observations show that this does not really occur. Research is needed to find out why.

Extrapolation of estimates of photochemical O_3 production due to anthropogenic NO_x and mixed natural and anthropogenic reactive hydrocarbon emissions,[54,55] during the summer season over the US, indicates the possibility of an average summertime

Table 2 Global and hemispheric sources and sinks of tropospheric ozone (in 10^{12} mol a^{-1})

sources/sinks	global	northern hemisphere	southern hemisphere
$HO_2 + NO$	+50/+22	+30/+12	+20/+10
$CH_3O_2 + NO$	+17/+8	+10/+4	+7/+4
$O(^1D) + H_2O$	−37/−22	−21/−12	−16/−10
$O_3 + HO_x$	−22/−10	−14/−7	−8/−3
stratospheric	+10/+10	+7/+7	+3/+3
surface destruction	−18/−8	−12/−4	−6/−4

First numbers for the contemporary (1985) atmosphere, second numbers for the pre-industrial atmosphere.

ozone production flux of 10^{11} molecule cm^{-2} s^{-1} for the northern hemisphere, about 40% of the global, annual mean production flux calculated to come from CH_4 and CO oxidation (Table 2). This is clearly a significant contribution. For the other seasons, the ozone production rates are probably much lower owing to smaller hydrocarbon emissions from forests and relatively greater removal of NO_x by the heterogeneous reaction (13). In fact, simultaneous measurements of CO and O_3 at a station on the Atlantic coast of Ireland show positive correlations from June to August and negative correlations from December to February.[56] Averaged over the year, the present contribution of 'photochemical smog ozone' from the US to the free troposphere may be 5–10% of that coming from CH_4 and CO oxidation. However, our model may overestimate the calculated ozone production from reaction cycles (C1) and (C5), as the coarse grid spacing (10° longitude by 10° latitude) that is used in the MOGUNTIA model artificially enhances the dispersion of the shortlived NO_x compounds, causing apparently higher ozone production efficiencies.[54] Clearly, for the summer months, photochemical ozone production in the polluted boundary layers of the US and other industrial regions may well be significant contributors to ozone concentrations at the mid-latitudes of the northern hemisphere. The same may be said about ozone formation due to biomass burning in the tropics and subtropics during the dry season; this is clearly observed even from the southern hemisphere which is less affected by anthropogenic and natural NO emissions than the northern hemisphere.[57]

6. Some Topics for Future Research

Despite the fundamental progress that has been made over the past two or three decades, much research will be needed in the coming decades to fill major gaps in our knowledge of tropospheric chemistry. In closing this paper I will try to indicate some of those research areas which I consider to be of greatest interest.

Tropospheric Ozone Observations

Despite the great importance of tropospheric ozone in atmospheric chemistry, there are still major uncertainties concerning its budget and global concentration distribution. Everywhere, but especially in the tropics and the subtropics, there is a severe lack of data on background tropospheric ozone concentrations. Considering the enormous role of tropical ozone for the oxidation efficiency of the atmosphere, the already recognized large anthropogenic impact on ozone through biomass burning, and the expected major agricultural and industrial expansion of human activities in this part of the world, this knowledge gap is very serious and must be remedied as soon as possible. At this stage it is fair to say that current global photochemical transport models cannot be satisfactorily tested, owing to the severe scarcity of ozone observations, especially in the tropics and subtropics. Of critical importance in the effort to obtain data from the tropics and subtropics in the short and the long term will be the active participation of scientists from the developing countries. Under the auspices of the International Global Atmospheric Chemistry (IGAC) Programme, a core project of the International Geosphere Biosphere Programme (IGBP), a major research and scientific training effort, including the launching of ozone sondes 1–3 times per week at 24 tropical and subtropical stations over a two-year period, has thus been prepared for submission to the Global Environment Fund (GEF) of the United Nations Environmental Programme (UNEP). At a subset of the stations, it is proposed to continue long-term ozone measurements beyond the planned ITOY period in order to document future trends. At these stations it would be of great value if the ozone measurements were complemented with ground-based observations of *e.g.* CO, CH_4 and chemical constituents in precipitation.

Long-term Monitoring of Atmospheric Properties

Two major findings have demonstrated the extreme value of long-term monitoring of important atmospheric properties. One example was the discovery of the rapid depletion of stratospheric ozone over Antarctica during the spring.[58] Had these measurements not been continued since the end of 1950s, following the conclusion of the International Geophysical Year, the discovery of the major, and totally unforseen, impact of human activities on stratospheric ozone probably would have been postponed by many years with substantially increased risks for the environment. Another example is the detection of unexpected breaks in the trends of CH_4 and CO since about the beginning of the present decade. Most surprising were the recent changes in the trends of CO for which Khalil and Rasmussen[33] derived a downward trend in surface concentrations by $1.4 \pm 0.9\%$ per year in the northern hemisphere and by $5.2 \pm 0.7\%$ per year in the southern hemisphere between 1987 and 1992. Even larger downward trends, $6.1 \pm 1\%$ per year in the northern hemisphere and $7 \pm 0.6\%$ per year in the southern hemisphere, were reported for the period between June 1990 and June 1993 by Novelli *et al.*[34] Most recently it appears that these trends have again reversed (P. Novelli personal communication). The reasons for this surprising behaviour are not known. They may consist of a combination of the following factors: (i) variable annual emissions especially from biomass burning, (ii) higher concentrations of OH radicals, maybe owing to loss of statospheric ozone, triggered by an increase in the aerosol loading of the stratosphere owing to the Pinatubo volcanic eruption in June 1991,[59] (ii) a dynamically forced global redistribution of CO and a bias of the measuring sites, mostly in marine locations, to regions which experienced lower CO concentrations, (iv) reduced CO formation from the oxidation of natural hydrocarbons emitted by tropical forests, or, most likely, (v) other, yet unknown, factors.

At this stage we can only conclude that the causes for the surprisingly rapid CO trend changes are not known, the main reason being incomplete global coverage of the CO measurement network. The same applies for CH_4.

Concentrated Measurement Campaign

Comprehensive field programmes that have been conducted in the past with detailed observations of all factors that influence the photochemistry of the troposphere, will also be much needed in the future, especially in various regions of the marine and continental tropics and subtropics during the dry and the wet seasons in order to find out whether we understand the major processes that determine the distribution of ozone and other photochemically active compounds. Applications of comprehensive chemical transport models should be an important part of these activities. Among the factors to be studied we mention especially: (i) improved quantification of the extratropical, stratospheric influx of ozone; (ii) distributions, sources and sinks of NO_x and NO_y, including organic nitrates; quantification of natural NO emissions from lightning and soils; (iii) distributions of CO and reactive hydrocarbons.

Cloud Transport

The role of clouds as transporters of boundary-layer chemical constituents (such as CO, reactive hydrocarbons and their oxidation products) to the middle and upper troposphere (and possibly into the lower stratosphere in the tropics) should be better understood and quantified, so that they can be parametrized for inclusion in large-scale photochemical models of the atmosphere.[53,54,60] Similarly, the production of NO by lightning and its vertical redistribution by convective storms should also be much better quantified, both for marine and continental conditions. Current uncertainties in the amount of NO production by lightning are at least a factor of four.[61,62]

Chemical Interactions with Hydrometeors

The chemical interactions of constituents emanating from the boundary layer with liquid and solid hydrometeors in the clouds will be of special importance. There is, for instance, the question of why high ozone formation has not been noted around the most convective regions in the continental tropics, in which forest-derived reactive hydrocarbons, such as isoprene (C_5H_8), and their oxidation products can be rapidly lifted to the middle and upper troposphere and mixed with lightning produced NO to provide favourable conditions for photochemical ozone formation. Could it be that the expected ozone formation is prevented by chemical interactions of the hydrocarbon reaction products and NO_x with the hydrometeors? Could significant ozone destruction take place *via* reactions (33)–(35) in cloud water and/or on the surface of ice particles which may be partially covered by water? Such questions regarding potential loss of ozone by reaction with hydrometeors may be especially relevant in connection with observations of a record low O_3 volume mixing ratio of less than 10 ppbv over the entire tropospheric column in March 1993 in an extended region between Fiji and Christmas Island during the CEPEX field programme.[15] Although such low ozone volume mixing ratios had been noted on several occasions in the tropical marine boundary layer, and can be explained by the ozone-destroying action of reactions (1) and (2) in the lower troposphere, it should be ascertained whether this mechanism alone is sufficient for such drastic ozone depletion over such a large volume as observed during CEPEX.

Photolysis Rates in Cloudy Atmosphere

Regarding the photochemistry taking place in cloudy atmospheric conditions, recent observations of unexpectedly high absorption of solar radiation during the same field programme,[63] and since then confirmed by ground-based and satellite observations,[64] point to the possibility that multiple scattering between and inside clouds[65] may lead to strongly enhanced photolysis rates and photochemical activity, leading *e.g.* to much higher O_3 destruction and OH production rates by reactions (1) and (2) than thought so far. The influence of clouds on the photochemically active UV radiation field is a new and very important research topic which should be pursued by measurements and the development of appropriate radiative transfer models.

Biogenic Sources of Hydrocarbons, CO and NO

The continental biosphere is an especially large source of hydrocarbons, carbon monoxide and NO.[66,67] Quantification of these sources in terms of geophysical (*e.g.* temperature, humidity, light levels) and biogeochemical (soil physical and chemical properties, land use) parameters is much needed for inclusion in atmospheric models. The hydrocarbon oxidation mechanisms[68] in the atmosphere should also be better understood, so that formation of ozone, carbon monoxide, partially oxidized gaseous hydrocarbons, and organic aerosols can be better quantified. The formation of organic aerosols from hydrocarbon precursors and their capability for serving as cloud condensation nuclei are related, potentially important, subjects which have not been studied in any depth so far.[69]

Effect of Biomass Burning on Tropical Ozone

It is now well known that tropical ozone concentrations are much higher during the dry than during the wet season,[70] probably owing to reactive hydrocarbon and NO emissions by fires. Although satellite observations clearly indicate high continental O_3 and CO concentrations in the troposphere during the dry season and spillover to mid-latitudes in the southern hemisphere,[57] a good quantitative analysis of the sources and sinks of ozone produced as a consequence of biomass burning is not available.

Potential Role of Halogen Radicals in Ozone Destruction

There are strong observational indications that tropospheric ozone can be destroyed by reactions additional to those discussed so far. Surface ozone observations during polar sunrise in the Arctic have frequently shown the occurrence of unmeasurably low ozone concentrations, coinciding with high 'filterable Br'.[71] Further measurements identified BrO as one of the active Br compounds,[72] which, as is well known from stratospheric measurements, may rapidly attack ozone by a series of catalytic reactions, such as

$$2(\mathrm{Br} + \mathrm{O_3} \rightarrow \mathrm{BrO} + \mathrm{O_2}) + (\mathrm{BrO} + \mathrm{BrO} \rightarrow 2\mathrm{Br} + \mathrm{O_2}) = 2\mathrm{O_3} \rightarrow 3\mathrm{O_2}$$

$$(\mathrm{Br} + \mathrm{O_3} \rightarrow \mathrm{BrO} + \mathrm{O_2}) + (\mathrm{Cl} + \mathrm{O_3} \rightarrow \mathrm{ClO} + \mathrm{O_2}) + (\mathrm{ClO} + \mathrm{BrO} \rightarrow \mathrm{Br} + \mathrm{Cl} + \mathrm{O_2}) = 2\mathrm{O_3} \rightarrow 3\mathrm{O_2}$$

$$(\mathrm{Br} + \mathrm{O_3} \rightarrow \mathrm{BrO} + \mathrm{O_2}) + (\mathrm{BrO} + \mathrm{HO_2} \rightarrow \mathrm{HOBr} + \mathrm{O_2} + (\mathrm{HOBr} + h\nu \rightarrow \mathrm{OH} + \mathrm{Br}) + (\mathrm{OH} + \mathrm{CO} + \mathrm{O_2} \rightarrow \mathrm{HO_2} + \mathrm{CO_2}) = \mathrm{CO} + \mathrm{O_3} \rightarrow \mathrm{CO_2} + \mathrm{O_2}$$

The source of reactive bromine has not been established. One possibility is that it originates from marine-biological organobromine compounds. However, it is also possible that reactive Br and Cl compounds are released from seasalt through reactions of Br^- and Cl^- with NO_3 and N_2O_5,[74,75] originally emanating from 'Arctic haze' NO_x pollution sources. It is definitely interesting to explore whether such ozone-depleting reactions may also take place elsewhere in the marine boundary layer, when polluted air masses are transported over the sea under stable meteorological conditions.[73]

Heterogeneous Reactions on Aerosol Particles

The issue of interactions between gases and atmospheric aerosol is largely unexplored and has not been extensively considered in tropospheric chemistry models. Examples are interactions of DMS-derived sulfur compounds with seasalt in the marine boundary layer and reactions of SO_2, H_2SO_4, NO_x, N_2O_5 and HNO_3 on soil dust particles which remove these compounds from the gas phase. In the case of industrial SO_2, the neglect of such heterogeneous reactions may well have led to overestimations of the climatic cooling effects of anthropogenic aerosols,[76] as any incorporation of sulfur in soil dust or seasalt will prevent the formation of new sunlight backscattering sulfate particles.

Importance of Ozone for the Earth's Climate

Ozone is a significant greenhouse gas with an IR absorption band in the atmospheric window region, centred at 9.6 μm. Although the amount of tropospheric ozone is only *ca.* 10% of that of the stratosphere, the effective longwave optical depth of tropospheric ozone is larger. Of most importance would be any changes that might take place in the ozone concentrations in the tropopause regions as a result of human activities,[78] in particular those due to NO emissions from expanding fleets of civil aircraft flying in the stratosphere and upper troposphere. This is a topic of intense research and international assessment.[49] The importance of ozone for the earth's climate is, however, not limited to its role as a greenhouse gas. We should also consider the strong coupling between the temperature and ozone concentration profiles in the stratosphere: increasing O_3 concentrations with altitude leads to increased temperatures, these in turn leading to reduced vertical mixing, thus promoting the concentration of ozone in the lower stratosphere.

Changes in chemical and radiative conditions in the lower stratosphere may, therefore, create feedbacks which should be well understood, including their potential impact on tropopause height, high-altitude cloud characteristics and possibly other properties of the hydrological cycle. Recent observations of increasing trends of water vapour concentrations in the lower stratosphere over Boulder[79] emphasize this point.

References

1 H. Levy, *Science*, 1971, **173**, 141.
2 A. M. Thompson, *Science*, 1992, **256**, 1157.
3 R. G. Prinn, R. F. Weiss, B. R. Miller, J. Huang, F. N. Alyea, D. M. Cunnola, P. J. Fraser, D. E. Hartley and P. G. Simmonds, *Science*, submitted.
4 P. J. Crutzen, *Pure Appl. Geophys.*, 1973, **106–108**, 1385.
5 P. J. Crutzen, *Tellus*, 1974, **26**, 47.
6 M. McFarland, D. Kley, J. M. Drummond, A. L. Schmeltekopf and R. H. Winkler, *Geophys. Res. Lett.*, 1979, **6**, 605.
7 A. L. Torres and A. M. Thompson, *J. Geophys. Res.*, 1993, **98**, 16949.
8 W. L. Chameides, D. D. Davis, J. Bradshaw, M. Rodgers, S. Sandholm and D. B. Bai, *J. Geophys. Res.*, 1987, **92**, (D2), 2131.
9 B. A. Ridley, M. A. Carroll and G. L. Gregory, *J. Geophys. Res.*, 1987, **92**, 2025.
10 S. C. Liu, M. McFarland, D. Kley, O. Zafiriou and B. Huebert, *J. Geophys. Res.*, 1983, **88**, 1349.
11 F. Routhier, R. Dennett, D. D. Davis, A. Wartburg, P. Haagenson and A. C. Delany, *J. Geophys. Res.*, 1980, **85**, 7307.
12 S. R. Piotrowicz, H. F. Bezdek, G. R. Harvey, M. Springer-Young and K. J. Hanson, *J. Geophys. Res.*, 1991, **96** (D10), 18679.
13 J. E. Johnson, R. H. Gammon, J. Larson, T. S. Bates, S. J. Oltmans and J. C. Farmer, *J. Geophys. Res.*, 1990, **95** (D8), 11847.
14 A. M. Thompson, J. E. Johnson, A. L. Torres, T. S. Bates, K. C. Kelly, E. Atlas, J. P. Greenberg, N. M. Donahue, S. A. Yuon, E. S. Saltzman, B. G. Heikes, B. W. Mosher, A. A. Shashkov and V. I. Yegorov, *J. Geophys. Res.*, 1993, **98**, 16955.
15 D. Kley, H. G. J. Smit, H. Vömel, S. Oltmans, H. Grassl, V. Ramanathan and P. J. Crutzen, *Science*, 1995, submitted.
16 W. L. Chameides and D. D. Davis, *J. Geophys. Res.*, 1982, **87**, 4863.
17 T. E. Graedel and C. J. Wechsler, *Rev. Geophys. Space Phys.*, 1981, **19**, 505.
18 S. A. Penkett, B. M. R. Jones, K. A. Brice and A. E. J. Eggleton, *Atmos. Environ.*, 1979, **13**, 123.
19 J. Lelieveld and P. J. Crutzen, *Nature (London)*, 1990, **343**, 227.
20 J. Lelieveld and P. J. Crutzen, *J. Atmos. Chem.*, 1991, **12**, 229.
21 R. A. Rasmussen and M. A. K. Khalil, *J. Geophys. Res.*, 1981, **86**, 9826.
22 D. R. Blake, E. W. Mayer, S. C. Tyler, D. C. Montague, Y. Makide and F. S. Rowland, *Geophys. Res. Lett.*, 1982, **9**, 477.
23 C. P. Rinsland, J. S. Levine and T. Miles, *Nature (London)*, 1985, **318**, 245.
24 M. A. K. Khalil and R. A. Rasmussen, *Chemosphere*, 1993, **26**, 803.
25 L. P. Steele, E. J. Dlugokencky, P. M. Lang, P. P. Tans, R. C. Martin and K. A. Masarie, *Nature (London)*, 1992, **358**, 313.
26 E. J. Dlugokencky, L. P. Steele, P. M. Lang and K. A. Masarie, *J. Geophys. Res.*, 1994, **99D**, 17021.
27 R. A. Rasmussen and M. A. K. Khalil, *J. Geophys. Res.*, 1981, **86**, 9826.
28 H. Craig and C. C. Chou, *Geophys. Res. Lett.*, 1982, **9**, 1221.
29 D. M. Etheridge, G. I. Pearman and P. J. Fraser, *Tellus, Ser. B*, 1992, **44**, 282.
30 T. Nazakawa, T. Machida, M. Tanaka, Y. Fujii, S. Aoki and O. Watanabe, *Geophys. Res. Lett.*, 1993, **20**, 943.
31 J. Chappelaz, T. Blunier, D. Raynaud, J. M. Barnola, J. Schwander and B. Stauffer, *Nature (London)*, 1993, **366**, 443.
32 R. Zander, Ph. Demoulin, D. H. Ehhalt, U. Schmidt and C. P. Rinsland, *J. Geophys. Res.*, 1989, **94**, 11021.
33 M. A. K. Khalil and R. A. Rasmussen, *Nature (London)*, 1994, **370**, 639.
34 P. C. Novelli, K. A. Masarie, P. P. Tans and P. M. Lang, *Science*, 1994, **263**, 1587.
35 P. J. Crutzen and P. H. Zimmermann, *Tellus, Ser. B*, 1991, **43**, 136.
36 I. Fung, J. John, J. Lerner, E. Matthews, M. Prather, L. P. Steele and P. J. Fraser, *J. Geophys. Res.*, 1991, **96**, 13003.
37 P. S. Kasibhatla, H. Levy II and W. J. Moxim, *J. Geophys. Res.*, 1993, **98**, 7165.
38 J-F. Müller and G. Brasseur, *J. Geophys. Res.*, in the press.
39 J. E. Penner, C. S. Atherton, J. Dignon, S. J. Ghan, J. J. Walton and S. Hameed, *J. Geophys. Res.*, 1991, **96**, 959.

40 C. M. Spivakovsky, R. Yevitch, J. A. Logan, S. C. Wofsy, M. B. McElroy and M. J. Prather, *J. Geophys. Res.*, 1990, **95**, 18441.
41 J. Feichter and P. J. Crutzen, *Tellus, Ser. B*, 1990, **42**, 100.
42 F. J. Dentener and P. J. Crutzen, *J. Geophys. Res.*, 1993, **98**, 7149.
43 P. J. Crutzen, in *Composition, Chemistry and Climate of the Atmosphere*, ed. H. Singh, Van Nostrand, New York, 1995, pp. 349–393.
44 A. Volz and D. Kley, *Nature (London)*, 1988, **332**, 240.
45 P. J. Crutzen, in *Tropospheric Ozone*, ed. I.S.A. Isaksen, Reidel, Dordrecht, 1988, pp. 3–32.
46 R. D. Bojkov, in *Tropospheric Ozone*, ed. I.S.A. Isaksen, Reidel, Dordrecht, 1988, pp. 83–96.
47 J. A. Logan, *J. Geophys. Res.*, 1985, **90**, 10463.
48 J. A. Logan, *J. Geophys. Res.*, 1994, **99**, 25553.
49 World Meteorological Organization, *Scientific Assessment of Ozone Depletion*: 1994, 1995, Geneva.
50 H. A. Michelsen, R. J. Salawitch, P. O. Wennberg and J. G. Anderson, *Geophys. Res. Lett.*, 1994, **21**, 2227.
51 M. Müller, A. Kraus and A. Hofzumahaus, *Geophys. Res. Lett.*, 1995, **22**, 679.
52 I. E. Galbally and C. R. Roy, *Q. J. R. Meteorol. Soc.*, 1980, **106**, 599.
53 R. R. Dickerson, G. J. Huffmann, W. T. Luke, L. J. Nunnermacker, K. E., Pickering, A. C. D. Leslie, W. G. N. Slinn, T. J. Kelly, P. H. Daum, A. C. Delany, J. P. Greenberg, P. R. Zimmerman, J. F. Boatman, J. D. Ray and D. H. Stedman, *Science*, 1987, **235**, 460.
54 S. C. Liu, M. Trainer, F. C. Fehsenfeld, D. D. Parrish, E. J. Williams, D. W. Fahey, G. Hübler and P. C. Murphy, *J. Geophys. Res.*, 1987, **92**, 4191.
55 D. J. Jacob, J. A. Logan, G. M. Gardner, R. M. Yevich, C. M. Spivakovsky, S. C. Wofsy, S. Sillman and M. J. Prather, *J. Geophys. Res.*, 1993, **98**, (D8), 14817.
56 B. G. Doddridge, WMO GAW Report No. 98, 1994, Geneva.
57 J. Fishman, C. E. Watson, J. C. Larsen and J. A. Logan, 1990, *J. Geophys. Res.*, **95** (D4), 3599.
58 J. C. Farman, B. G. Gardiner and J. D. Shanklin, *Nature (London)*, 1985, **315**, 207.
59 S. Madronich and C. Granier, *Geophys. Res. Lett.*, 1992, **19**, 465.
60 J. Lelieveld and P. J. Crutzen, *Science*, 1994, **264**, 1759.
61 C. Price and D. Rind. *Mon. Weather Rev.*, 1994, **122**, 1930.
62 M. G. Lawrence, W. L. Chameides, P. S. Kasibhatla, H. Levy II and W. Moxim, in *Handbook of Atmospheric Electrodynamics*, ed. Hans Volland, CRC Press, Boca Raton, 1995, vol. I, pp. 197–211.
63 V. Ramanathan, B. Subasilar, G. J. Zhang, W. Contant, R. D. Cess, J. T. Kiehl, H. Grassl and L. Shi, *Science*, 1995, **267**, 499.
64 R. D. Cess, M. H. Zhang, P. Minnis, L. Corsetti, E. G. Dutton, B. W. Forgan, D. P. Garber, W. L. Gates, J. J. Hack, E. F. Harrison, X. Jing, J. T. Kiehl, C. N. Long, J-J. Morcrette, G. L. Potter, V. Ramanathan, B. Subasilar, C. H. Whitlock, D. F. Young and Y. Zhou, *Science*, 1995, **267**, 496.
65 F. Erle, K. Pfeilsticker and U. Platt, *Geophys. Res. Lett.*, 1995, submitted.
66 F. C. Fehsenfeld, J. Calvert, R. Fall, P. Goldan, A. B. Guenther, C. N. Hewitt, B. Lamb, S. Liu, M. Trainer, H. Westberg and P. Zimmerman, *Global Biogeochem. Cycles*, 1992, **6**, 389.
67 H. B. Singh and P. B. Zimmerman, *Adv. Environ. Sci. Technol.*, 1992, **24**, 177.
68 R. Atkinson, *Atmos. Environ. A*, 1990, **24**, 1.
69 T. Novakov and J. Penner, *Nature (London)*, 1993, **365**, 823.
70 P. J. Crutzen and M. O. Andreae, *Science*, 1990, **250**, 1669.
71 L. A. Barrie, J. W. Bottenheim, R. C. Schnell, P. J. Crutzen and R. A. Rasmussen, *Nature (London)*, 1988, **334**, 138.
72 M. Hausmann and U. Platt, *J. Geophys. Res.*, 1994, **99D**, 25399.
73 R. Sander and P. J. Crutzen, *J. Geophys. Res.*, submitted.
74 B. J. Finlayson-Pitts, M. J. Ezell and J. N. Pitts Jr., *Nature (London)*, 1989, **337**, 241.
75 C. Zetsch and W. Behnke, *Ber. Bunsen-Ges. Phys. Chem.*, 1992, **96**, 488.
76 R. L. Charlson, J. Langner, H. Rodhe, C. B. Leovy and S. G. Warren, *Tellus, Ser. B*, 1991, **43**, 152.
77 V. Ramanathan and R. E. Dickinson, *J. Atmos. Sci.*, 1979, **36**, 1084.
78 A. A. Lacis, D. J. Wuebbles and J. A. Logan, *J. Geophys. Res.*, 1990, **95**, 9971.
79 S. J. Oltmans and D. J. Hofmann, *Nature (London)*, 1995, **379**, 146.

Paper 5/03728J; *Received* 12*th June*, 1995

Faraday Discuss., 1995, **100**, 23–37

Reactions of Alkoxyl Radicals in the Atmosphere

Roger Atkinson, Eric S. C. Kwok, Janet Arey and Sara M. Aschmann
Statewide Air Pollution Research Center, University of California, Riverside, CA 92521, USA

The products of the gas-phase reactions of the OH radical with *n*-pentane and [$^2H_{12}$]pentane have been studied at 296 ± 2 K and 740 Torr total pressure of N_2–O_2 diluent gas in the presence of NO. Direct air sampling, atmospheric-pressure ionisation tandem mass spectrometry showed the formation of the previously predicted C_5-hydroxycarbonyl product, together with pentanones. The effect of varying the O_2 concentration on the C_5-hydroxycarbonyl and pentanones was consistent with the hydroxycarbonyl being the product formed after isomerisation of the pentan-2-oxyl radical. Gas chromatographic analyses of pentan-2-one and pentan-3-one and pentan-2-yl and pentan-3-yl nitrates allowed rate constants for the decomposition and isomerisation reactions of the pentan-2-oxyl and pentan-3-oxyl radicals, relative to their reactions with O_2, to be determined.

Introduction

Alkoxyl (RO˙) radicals and hydroxy- and nitrato-substituted alkoxyl radicals are key intermediates in the tropospheric degradation reactions of alkanes and alkenes.[1,2] In the troposphere, alkoxyl radical reactions with NO and NO_2 are minor[2] and alkoxyl radicals react with O_2, decompose unimolecularly, or isomerise by a 1,5-H shift through a six-member transition state (Scheme I).[1,2]

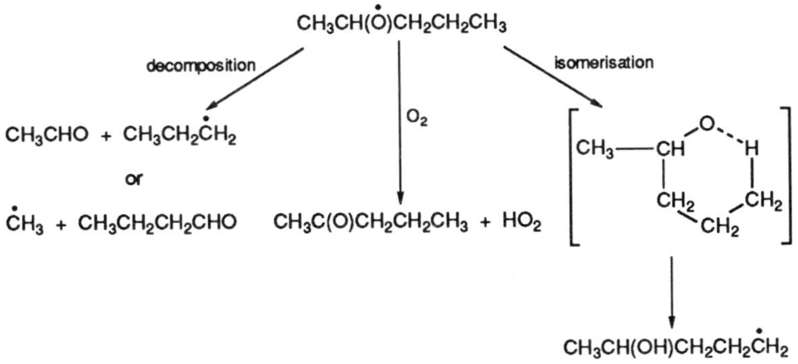

Scheme I

A major uncertainty in the alkane and alkene degradation reactions in the troposphere, which affects the products predicted to be formed, involves the relative importance of these three reaction pathways of the alkoxyl radical intermediates.[1,2]

Absolute rate constants are available only for the reactions of the CH_3O˙, C_2H_5O˙ and $(CH_3)_2CHO$˙ radicals with O_2,[2] and it is not clear whether or not the rate constants for the $C_2H_5O^{\bullet}$ and $(CH_3)_2CHO^{\bullet}$ radicals can be taken as representative of those for the $\geqslant C_2$ primary and secondary alkoxyl radicals, respectively.[2] Rate constants for the alkoxyl radical decomposition reactions have generally been measured relative to those for reaction of the alkoxyl radicals with O_2 or NO.[2–4]

$$RO^{\bullet} + NO \xrightarrow{M} RONO \tag{1}$$

Furthermore, apart from the studies of Dóbé *et al.*,[5] in which the 4-hydroxypentan-1-yl radical formed after isomerisation of the pentan-2-oxyl radical was trapped by reaction with methyl radicals, and Eberhard *et al.*,[6] in which 2-hydroxyhexan-5-one formed from the hexan-2-oxyl radical was detected and quantified as the 2,4-dinitrophenylhydrazone derivative, the occurrence of the isomerisation reaction of alkoxyl radicals formed from the alkanes has been inferred from the low yields of the products expected from the competing decomposition and O_2 reactions.[7–10] While an empirical method for assessing the relative importance of alkoxyl radical decomposition *vs.* reaction with O_2 has been proposed,[1,2] the lack of knowledge of the alkoxyl radical isomerisation rate constants precludes reliable assessments of the relative importance of the three alkoxyl radical reaction pathways.

In this work, we have investigated the products of the gas-phase reactions of the OH radical with pentane and $[^2H_{12}]$pentane in the presence of NO. Products of the pentane reaction were quantified by gas chromatography, and the reactions of pentane and $[^2H_{12}]$pentane were studied using a direct air sampling, atmospheric-pressure ionisation tandem mass spectrometer. The data obtained show the formation of the δ-hydroxycarbonyl(s) expected from the alkoxyl radical isomerisation reaction. Additionally, the relative importance of the various reactions of the pentan-2-oxyl and pentan-3-oxyl radicals, formed subsequent to H-atom abstraction by the OH radical from the 2- and 3-positions in pentane,[2] has been determined.

Experimental

Experiments were carried out in *ca.* 6500 and 7900 l all-Teflon chambers, with the *ca.* 6500 l all-Teflon chamber being interfaced to a PE SCIEX API III MS/MS direct air sampling, atmospheric-pressure ionisation triple quadrupole mass spectrometer. Both chambers are equipped with two parallel banks of blacklamps for irradiation, and all experiments were carried out at 296 ± 2 K and 740 Torr total pressure of purified air, or N_2–O_2 mixtures, at *ca.* 5% relative humidity. In both chambers, hydroxyl radicals were generated by the photolysis of methyl nitrite (CH_3ONO) in air at wavelengths >300 nm.[11]

$$CH_3ONO + h\nu \rightarrow CH_3O^{\bullet} + NO \tag{2}$$

$$CH_3O^{\bullet} + O_2 \rightarrow HCHO + HO_2 \tag{3}$$

$$HO_2 + NO \rightarrow OH + NO_2 \tag{4}$$

NO was added to the reactant mixtures to suppress the formation of O_3, and hence of NO_3 radicals.[11]

Analyses by API MS/MS

The ion source and instrument design of the PE SCIEX API III MS/MS atmospheric-pressure ionisation triple quadrupole mass spectrometer have been described previously.[12,13] The atmospheric-pressure ionisation source is a point-to-plane design, with the point electrode being a discharge needle and the plane being part of the MS atmospheric-pressure-to-vacuum interface flange. Sample introduction from the Teflon reaction chamber into the ion source is through a *ca.* 25 mm diameter, *ca.* 75 cm length Pyrex tube, at flow rates of *ca.* 22 l min^{-1} provided by a suction pump. Under positive APCI (atmospheric-pressure chemical ionisation) conditions, protonated hydrates $[H_3O^+(H_2O)_n]$ (n = *ca.* 3 to 6 at 298 K and *ca.* 5% relative humidity)[14] generated by a corona discharge in the chamber diluent gas are responsible for the protonation of analytes

$$H_3O^+(H_2O)_n + M \rightarrow MH^+(H_2O)_m + (n - m + 1)H_2O \quad (5)$$

where M is the neutral analyte of interest. Ions are drawn by an electric potential from the ion source through the sampling orifice into the mass-analysing first quadrupole (Q1) or third quadrupole (Q3). Neutral molecules and particles are prevented from entering the orifice by a counterflow of ultra-pure nitrogen ('curtain') gas. The protonated molecular ($[M + H]^+$) ions mass-analysed are the result of the declustering action of the curtain gas on the hydrated ions. The sensitivity of API-MS to the analyte M is determined by the stability of $MH^+(H_2O)_m$ and the water vapour concentration in the samples.[15] The lack of an $[M + H]^+$ signal from pentane and $[^2H_{12}]$pentane in the APCI mass spectra of the CH_3ONO–NO–pentane (and $[^2H_{12}]$pentane)–air reaction mixtures suggests that hydrates of $[C_5H_{12} + H]^+$ and $[C_5D_{12} + H]^+$ were not stable under our instrumental conditions.

The mass spectrometer was operated in the positive-ion mode throughout these experiments. Both MS (scanning) and MS/MS with collision activated dissociation were used to analyse the CH_3ONO–NO–pentane (or $[^2H_{12}]$pentane)–O_2–N_2 mixtures. In the MS mode, mass spectra (each the sum of 10 scans) of the reactants and the reacted mixtures were obtained by the first quadrupole (Q1), with the second and third quadrupoles (Q2 and Q3, respectively) being operated in the 'total-ion' mode (rf-only mode). Each scan was acquired over the range 10–360 u using a step size of 0.2 u and a dwell time of 10 ms. The curtain gas flow (CG), orifice voltage (OR), and rod-offset voltage of the rf-quadrupole (R0) of the API-MS used during sample analyses were adjusted to minimise the fragmentation of ions in the vacuum side of the orifice.[16] It was not possible to eliminate fragmentation prior to Q1, but the system could be optimised to favour protonated molecular ions $[M + H]^+$ and/or protonated dimer ions $[M_2 + H]^+$. The channel electron multiple (CEM) was operated in pulse-counting mode and optimised by the criterion of a 40% increase in signal intensity per 200 V increment in the CEM voltage. The quadrupole power supplies were adjusted so that unit resolution (50% valley definition) and 0.7 u peak width (at 70% peak height) were achieved. The protonated dimer ion of diisopropylmethylphosphonate (DIMP) was used to optimise the instrument parameters CG, OR, R0 and CEM. Mass calibration and quadrupole power supplies optimisation were performed on ions at 29, 181 and 361 u, corresponding to $[N_2H]^+$, $[DIMP + H]^+$, and $[(DIMP)_2 + H]^+$, respectively.

The pentanones were quantified by external calibration of the integrated signal of the ion peak at 87 u ($[M + H]^+$) to that of a standard of pentan-3-one, assuming identical molar responses for pentan-2-one and pentan-3-one. Calibration of hydroxycarbonyl(s) was carried out by measuring the relative response of the $[M + H]^+$ ions of 4-hydroxy-4-methylpentan-2-one *vs.* hexan-3-one, and assuming the same relative response holds for C_5-hydroxycarbonyl(s) *vs.* pentan-2-one and pentan-3-one. The molar response of 4-hydroxy-4-methylpentan-2-one was *ca.* 0.05 of that of hexan-3-one. These calibrations should be viewed as semi-quantitiative.

Two MS/MS modes of the API-MS were employed in this work. In the 'daughter ion' scan mode, the precursor ion was selected by Q1 and after collision-activated dissociation (CAD) in Q2, the fragment ions were scanned by Q3. In the 'parent ion' mode, the fragment ions were selected by Q3 and matched to their precursors by scanning Q1. Ultra-pure argon was used as the collision gas in an open rf-only Q2 collision cell. The

CAD spectrum of an analyte (sum of 60–200 scans) was acquired using a step size of 0.2 u and a dwell-time of 10 ms. The lower and upper limits of the mass range used for acquiring 'parent and daughter ion' spectra, respectively, was determined by the molecular weight of the ion of interest. Optimisation of the MS/MS instrumental parameters was performed by multiple reaction monitoring.[17]

For the API-MS experiments, the initial reactant concentrations (in molecule cm^{-3}) were: CH_3ONO, NO, pentane and [$^2H_{12}$]pentane (2.4–4.8) $\times 10^{13}$ each, with the initial concentrations of CH_3ONO, NO and pentane (or [$^2H_{12}$]pentane) being equal in each experiment. Irradiations were carried out at 100% of the maximum light intensity for 5 min. In addition to experiments carried out in air (21% O_2), OH radical reactions with pentane were carried out in N_2–O_2 mixtures with low and high O_2 concentrations (4 and 80% O_2, respectively) to investigate the reaction mechanism (see, for example, Scheme I). To check on any effects of changing the O_2 concentration on the API-MS analyses, the relative responses of 4-hydroxy-4-methylpentan-2-one and hexan-3-one were measured at 4, 21 and 80% O_2.

Because the API-MS instrument as used in these experiments is insensitive to alkanes, pentane and [$^2H_{12}$]pentane were analysed during the experiments by gas chromatography with flame ionisation detection (GC-FID), as described below. For the API-MS calibrations described above, 4-hydroxy-4-methylpentan-2-one, pentan-3-one and hexan-3-one were also analysed by GC-FID, using the sampling procedure and GC conditions described below for pentan-2-one and pentan-3-one and pentan-2-yl and pentan-3-yl nitrate.

Analyses by GC-FID

Experiments with GC-FID analyses of pentane and selected products were carried out in a 7900 l all-Teflon chamber, with initial reactant concentrations (in molecule cm^{-3}) of: CH_3ONO, 2.1×10^{14}; NO, $(1.6–1.9) \times 10^{14}$; and pentane, $(4.29–4.39) \times 10^{13}$. Irradiations were carried out at the maximum light intensity for 3–15 min. Experiments were carried out in air [21% O_2 (155 Torr O_2)] and in 'high-O_2' diluent gas [74–85% O_2 (590 ± 40 Torr O_2)].

Pentane and selected products were analysed during the experiments by GC-FID. For the analyses of pentane, gas samples were collected from the chamber in 100 cm^3 all-glass, gas-tight syringes and introduced *via* a 1 cm^3 gas-sampling loop onto a 30 m DB-5 megabore column in a Hewlett Packard (HP) 5890 GC, initially held at -25 °C and then temperature-programmed to 200 °C at 8 °C min^{-1}. For the analyses of the pentan-2-one and pentan-3-one and pentan-2-yl and pentan-3-yl nitrate products, 100 cm^3 gas samples were collected from the chamber onto Tenax-TA solid adsorbent, with subsequent thermal desorption at 225 °C onto a 30 m DB-5.625 megabore column in an HP 5710 GC, initially held at 0 °C and then temperature-programmed to 200 °C at 8 °C min^{-1}. Gas samples were also collected onto Tenax-TA solid adsorbent for subsequent thermal desorption and analysis by combined gas chromatography–mass spectrometry (GC–MS), using a 60 m DB-5MS fused silica capillary column, temperature-programmed from -80 °C at 10 °C min^{-1} to 250 °C, in an HP 5890 GC interfaced to an HP 5970 mass-selective detector operated in the scanning mode. The NO concentrations and initial NO_2 concentrations were monitored during these experiments using a Thermo Environmental Instruments, Inc., Model 42 NO-NO_2-NO_x analyser.

Chemicals

The chemicals used, and their stated purities, were hexan-3-one (98%), 4-hydroxy-4-methylpentan-2-one (99%), and pentane (99+%), Aldrich Chemical Company;

$[^2H_{12}]$pentane (98% D) and $[^2H_{12}]$pentanol (98%), Cambridge Isotope Laboratories; pentan-2-one and pentan-3-one, MCB; pentan-2-yl and pentan-3-yl nitrate, Fluorchem, Inc.; and NO ($\geqslant$99.0%), Matheson Gas Products. Methyl nitrite was prepared and stored as described previously.[11]

Results

In the presence of NO, the OH radical-initiated reactions of pentane and $[^2H_{12}]$pentane lead to the formation of pentanoxyl radicals by the series of reactions[2]

$$OH + RH \rightarrow H_2O + R^{\cdot} \quad (6)$$

$$R^{\cdot} + O_2 \xrightarrow{M} RO_2^{\cdot} \quad (7)$$

$$RO_2^{\cdot} + NO \xrightarrow{M} RONO_2 \quad (8a)$$

$$RO_2^{\cdot} + NO \rightarrow RO^{\cdot} + NO_2 \quad (8b)$$

The expected products then arise from reaction (8*a*) and from the reactions of the alkoxyl radicals (as shown, for example, in Scheme I for the pentan-2-oxyl radical).

Analyses by API MS/MS

Fig. 1(*a*) shows the positive APCI mass spectrum of an irradiated CH_3ONO–NO–pentane–air mixture after *ca.* 20% reaction of the initial pentane. As noted above, no signal from pentane was observed. The C_5-carbonyls (primarily pentan-2-one and pentan-3-one) formed in the reaction appeared as a peak at 87 u, $[M + H]^+$. As shown in Fig. 2(*c*) for an authentic standard of pentan-2-one, C_5-carbonyls gave a

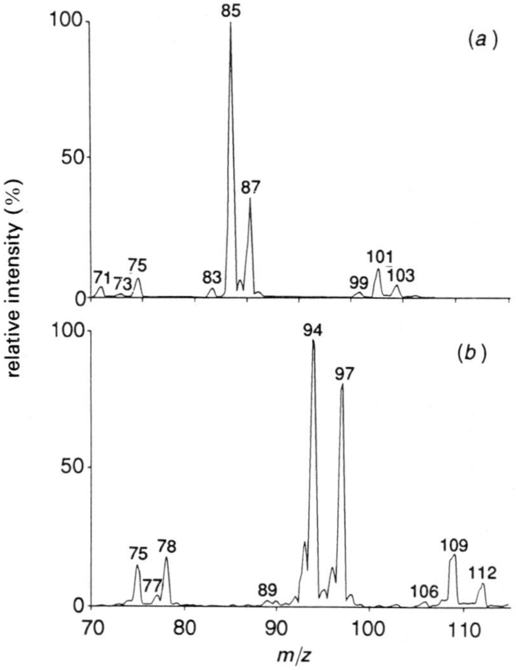

Fig. 1 Positive APCI mass spectra of irradiated (*a*) CH_3ONO–NO–pentane–air and (*b*) CH_3ONO–NO–$[^2H_{12}]$pentane–air mixtures

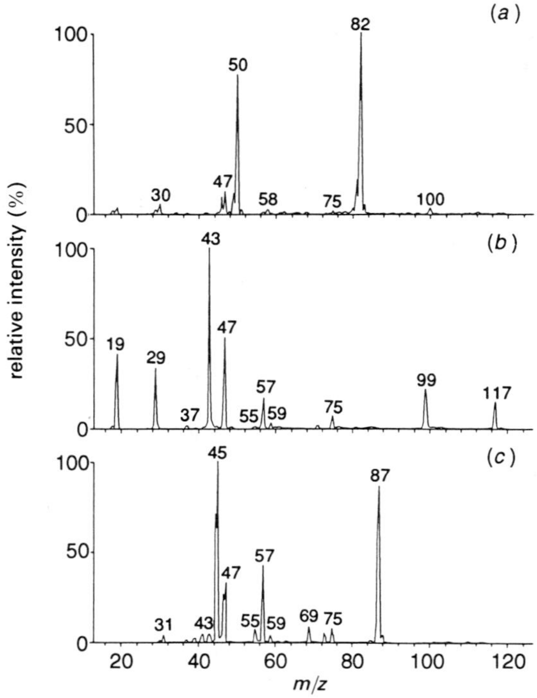

Fig. 2 Positive APCI mass spectra of authentic standards of (*a*) [$^2H_{12}$]pentanol, (*b*) 4-hydroxy-4-methylpentan-2-one and (*c*) pentan-2-one

strong protonated molecular ion under our instrumental conditions. The assignment of the 87 u peak in the reaction to pentanones was confirmed by matching its CAD spectrum with those of authentic standards (the pentan-2-one and pentan-3-one could not be distinguished by their CAD spectra). The GC-FID analyses discussed below show that both pentan-2-one and pentan-3-one were formed.

Product ion peaks of irradiated CH_3ONO–NO–pentane–air mixtures were also observed at 103, 101, and 85 u [Fig. 1(*a*)]. The molecular weight of the expected δ-hydroxycarbonyl 1-hydroxypentan-4-one formed from the pentan-2-oxyl radical by reactions (9)–(13) is 102 u, suggesting that the 103 u peak was the $[M + H]^+$ ion of this compound.

$$CH_3CH(\dot{O})CH_2CH_2CH_3 \xrightarrow{\text{isom}} CH_3CH(OH)CH_2CH_2\dot{C}H_2 \tag{9}$$

$$CH_3CH(OH)CH_2CH_2\dot{C}H_2 + O_2 \xrightarrow{M} CH_3CH(OH)CH_2CH_2CH_2\dot{O}_2 \tag{10}$$

$$CH_3CH(OH)CH_2CH_2CH_2\dot{O}_2 + NO \longrightarrow CH_3CH(OH)CH_2CH_2CH_2\dot{O} + NO_2 \tag{11}$$

$$CH_3CH(OH)CH_2CH_2CH_2\dot{O} \xrightarrow{\text{isom}} CH_3\dot{C}(OH)CH_2CH_2CH_2OH \tag{12}$$

$$CH_3\dot{C}(OH)CH_2CH_2CH_2OH + O_2 \longrightarrow CH_3C(O)CH_2CH_2CH_2OH + HO_2 \tag{13}$$

As shown in Fig. 2(*b*), the spectrum of an authentic sample of 4-hydroxy-4-methylpentan-2-one included ions at 117 and 99 u, corresponding to $[M + H]^+$ and

$[M + H—H_2O]^+$. The 85 u peak in the pentane reaction was, therefore, attributed to loss of water from the protonated molecular ion of the δ-hydroxycarbonyl. A 'parent ion' scan showed that the 101 u peak also originated from this compound and this $[M—H]^+$ ion was attributed to loss of H_2 from the protonated molecular ion. As described below, the ion peak at 103 u may also include a contribution from the $[(N_2)_3H_3O]^+$ cluster observed in the background spectrum without added reactants.

To confirm the identification of the δ-hydroxycarbonyl, the mass spectrometer was deliberately tuned to induce dimer formation, and the presence of an ion peak at 205 u, corresponding to the protonated δ-hydroxycarbonyl dimer, provided additional evidence for the occurrence of this pentan-2-oxyl radical isomerisation product. Under these experimental conditions, corresponding fragment ions to the δ-hydroxycarbonyl peaks at 101 and 85 u were also observed at 203 u ($[M_2—H]^+$) and 187 u ($[M_2 + H—H_2O]^+$), respectively. Furthermore, an ion peak which could correspond to the protonated dimer of pentanone and δ-hydroxycarbonyl was observed at 189 u, as was an ion peak at 171 u (corresponding to the protonated heterodimer less water).

Nitrogen clusters $[(N_2)_n(H_3O)]^+$ and $[(N_2)_mH]^+$ (where $n \geqslant 0$ and $m \geqslant 1$) and water clusters $[(H_2O)_n(H_3O)]^+$ (where $n \geqslant 0$) may be formed at the APCI interface,[18,19] resulting in peaks at 19, 29, 37, 47, 55, 57, 73, 75, 85 and 103 u, and certain of these peaks are observed in the spectra of the background air and in the CH_3ONO–NO–pentane–air reaction mixtures. Because they also appear in the CAD spectra of protonated pentanone, δ-hydroxycarbonyl and their corresponding dimers, the 19, 29, 47, 75 and 85 u peaks are attributed mainly to ion fragments derived from these products. Because both fragment ions and $[M + H]^+$ molecular ions are present at odd masses, distinguishing the presence of low-molecular-weight products, for example, propanal ($[M + H]^+ = 59$) and acetaldehyde ($[M + H]^+ = 45$), is difficult.

As shown in Fig. 1(*b*), the reaction of the OH radical with $[^2H_{12}]$pentane produced product peaks at 94, 97 and 112 u, with the 97 u peak corresponding to the protonated $[^2H_{10}]$pentanone ($[M + H]^+$). The protonated, fully deuteriated $[^2H_{10}]\delta$-hydroxycarbonyl ion peak would occur at 113 u. However, the expected rapid exchange of the D atom of the —OD group with water vapour in the chamber to form the —OH group[20] leads to the protonated $[^2H_9]\delta$-hydroxycarbonyl peak at 112 u (for example, $\{[CD_3C(O)CD_2CD_2CD_2OH] + H\}^+$). Such D/H exchange in the —OD group under our experimental conditions was demonstrated for fully deuteriated $[^2H_{12}]$pentanol, with the $[M + H]^+$ ion appearing at 100 u [Fig. 2(*a*)] instead of 101 u if D/H exchange did not occur. The peak at 112 u [Fig. 1(*b*)] is therefore assigned to a protonated $[^2H_9]\delta$-hydroxycarbonyl, and this assignment is further supported by the occurrence of the $[M + H—H_2O]^+$ peak at 94 u [corresponding to $\{CD_3C(O)CD_2CD_2CD_2]^+\}$. Analogous to the situation for the pentane reaction, $[M + H—H_2O]^+$ was the major ion from the δ-hydroxycarbonyl observed in the spectrum from irradiated CH_3ONO–NO–$[^2H_{12}]$pentane–air reaction mixtures [Fig. 1(*a*)], and an $[M—D]^+$ peak was also observed at 109 u $\{$corresponding to $[CD_3C(O)CD_2CD_2CDOH]^+\}$. Thus, despite optimising the instrument to preserve the molecular ions of interest, the majority of the δ-hydroxycarbonyl molecules preferentially fragmented by loss of a water molecule after their protonation.

Fig. 3 shows the mass spectra obtained from the reaction of the OH radical with pentane at low (*ca.* 4%), normal (21%) and high (*ca.* 80%) O_2 levels, with the percentage of the pentane reacted being $23 \pm 3\%$. As seen in Fig. 3, the ratio of the peak at 87 u (pentanones) *vs.* that at 85 u (C_5-hydroxycarbonyl) increased with increasing O_2 concentration (note that varying the O_2 concentration did not change the relative responses of hexan-3-one and 4-hydroxy-4-methylpentan-2-one in a standard mixture). From the peak areas of the hydroxycarbonyl and pentanone signals, and assuming that the response for the C_5-hydroxycarbonyl relative to the pentanones was identical to that of 4-hydroxy-4-methylpentan-2-one relative to hexan-3-one (*ca.* 0.05), the measured

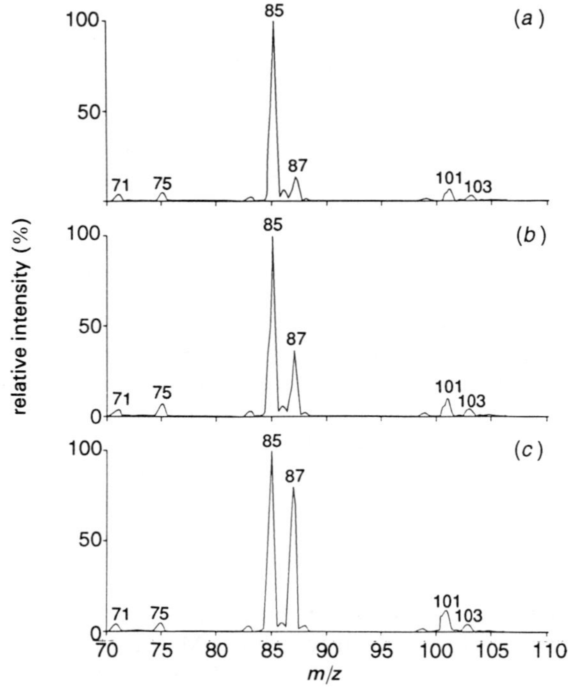

Fig. 3 Positive APCI mass spectra of irradiated CH_3ONO–NO–pentane–air mixtures in N_2–O_2 diluent gas at 740 Torr total pressure and *ca.* 5% relative humidity, with (*a*) 4, (*b*) 21 and (*c*) 80% O_2

C_5-hydroxycarbonyl/pentanones yield ratios were 4.6, 2.5 and 1.1 at 4, 21 and 80% O_2, respectively. These yield ratios do not take into account reactions of the OH radical with the C_5-hydroxycarbonyl and pentanone products. Because the C_5-hydroxycarbonyl is calculated to be more reactive than the pentanones towards the OH radical,[2,21,22] the corrected C_5-hydroxycarbonyl/pentanones formation yield ratios are estimated to be *ca.* 6.4, *ca.* 3.5 and *ca.* 1.5 at 4, 21 and 80% O_2, respectively.

Clearly, the observed change in the C_5-hydroxycarbonyl and pentanones distribution with varying O_2 concentration (Fig. 3) is consistent with the expected reactions of the pentan-2-oxyl radical with O_2, decomposition, and isomerisation, and of the pentan-3-oxyl radical with O_2 and decomposition (the pentan-3-oxyl radical cannot undergo isomerisation *via* a six-member transition state)[1,2,7] [see also the Discussion below].

GC-FID Analyses

GC-MS and GC-FID analyses of irradiated CH_3ONO–NO–pentane–air mixtures showed the formation of pentan-2-one and pentan-3-one and pentan-2-yl and pentan-3-yl nitrate. (The pentyl nitrates did not give a detectable positive APCI signal at the concentrations present in the APCI-MS/MS experiments with the instrumental conditions employed.) Since these products also react with the OH radical, their secondary reactions must be considered in determining their formation yields. Secondary reactions were taken into account as described by Atkinson *et al.*,[23] using rate constants for the OH radical reactions (in 10^{-12} cm^3 $molecule^{-1}$ s^{-1}) of:[2,22] pentane, 3.91; pentan-2-one, 5.07; pentan-3-one, 2.09; pentan-2-yl nitrate, 1.85 and pentan-3-yl nitrate, 1.12. The multiplicative correction factors accounting for secondary reactions with the OH radical, which increase with the extent of reaction and with the rate constant ratio k(OH + product)/k(OH + pentane),[23] were $\leqslant 1.30$ for pentan-2-one, $\leqslant 1.12$ for pentan-3-one, $\leqslant 1.12$ for pentan-2-yl nitrate and $\leqslant 1.06$ for pentan-3-yl nitrate.

Plots of the amounts of pentan-2-one and pentan-3-one formed, corrected to take into account secondary reactions with the OH radical, against the amounts of pentane reacted are shown in Fig. 4, and the formation yields obtained by least-squares analyses of these data are given in Table 1. The relative yields of pentan-2-one and pentan-3-one at 296 ± 2 K and 1 atm of air measured in this work, with (pentan-2-one)/(pentan-2-one + pentan-3-one) = 0.18 ± 0.04, are in good agreement with the ratios reported by Carter *et al.*,[7] of 0.16–0.21. Fig. 4 and Table 1 show that, as expected from Scheme I, the pentan-2-one and pentan-3-one formation yields increase with increasing O_2 concentration, with the effect being most marked for pentan-2-one. In contrast, the pentan-2-yl and pentan-3-yl nitrate yields were independent of the O_2 concentrations (Fig. 5), as

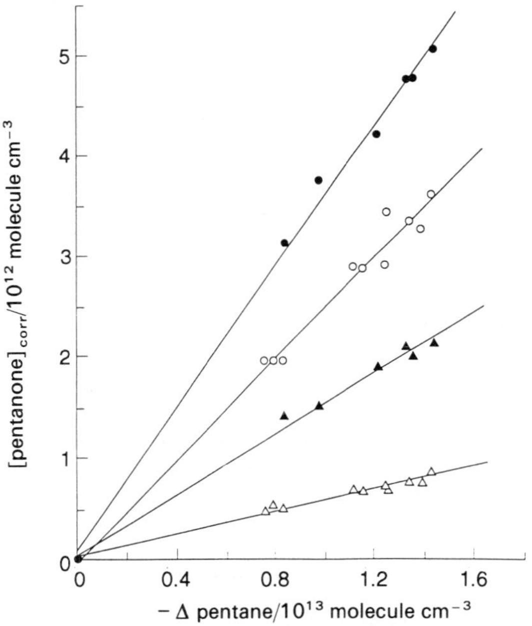

Fig. 4 Amounts of (▲, △) pentan-2-one and (●, ○) pentan-3-one formed, corrected for reaction with the OH radical, *vs.* the amount of pentane reacted at 296 ± 2 K and 740 Torr total pressure. Open symbols 155, and filled symbols 590 ± 40 Torr O_2.

Table 1 Measured formation yields at 296 ± 2 K and 740 Torr total pressure

	formation yields[a]			
O_2 pressure/Torr	pentan-2-one	pentan-3-one	pentan-2-yl nitrate	pentan-3-yl nitrate
155	0.054 ± 0.008	0.250 ± 0.030	0.061 ± 0.012	0.047 ± 0.006
590 ± 40	0.149 ± 0.017	0.353 ± 0.032	0.064 ± 0.009	0.043 ± 0.004
			0.060 ± 0.010[b]	0.044 ± 0.006[b]

[a] Indicated errors are the two least-squares standard deviations combined with the estimated overall uncertainties in the GC-FID response factors of ±5% for pentane and the products.
[b] Combined data at both O_2 concentrations.

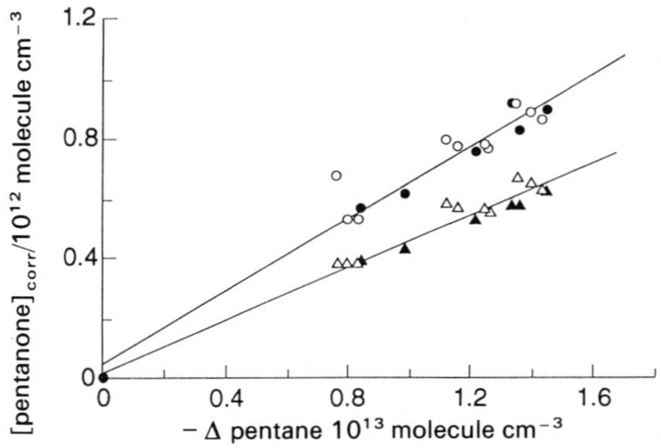

Fig. 5 Amounts of (○, ●) pentan-2-yl and (△, ▲) pentan-3-yl nitrate formed, corrected for reaction with the OH radical, *vs.* the amount of pentane reacted at 296 ± 2 K and 740 Torr total pressure. Open symbols 155 and filled symbols 590 ± 40 Torr O_2.

expected since these nitrates are formed from the reactions of the pentan-2-yl and pentan-3-yl peroxy radicals with NO [reaction (8*a*)] in competition with the pathway forming the alkoxyl radical plus NO_2 [reaction (8*b*)].[2] The pentyl nitrate yields are dependent on the temperature (increasing with decreasing temperature) and total pressure (increasing with increasing total pressure),[2,24,25] but should be independent of the O_2 concentration providing that O_2 and N_2 have similar third-body efficiencies in reaction (8*a*).

The formation yields of the pentan-2-yl and pentan-3-yl nitrates obtained from least-squares analyses of the entire data set are 0.060 ± 0.010 for pentan-2-yl nitrate and 0.044 ± 0.006 for pentan-3-yl nitrate (Table 1). These formation yields can be compared with our previous yields at room temperature (299–300 K) and atmospheric pressure (735–740 Torr) of 0.071 ± 0.009 and 0.046 ± 0.006, respectively,[23] and 0.074 ± 0.002 and 0.052 ± 0.004, respectively,[24] where the indicated errors in the Atkinson *et al.*[23,24] data are two least-squares standard deviations. The agreement is reasonably good.

Discussion

The expected reactions of the pentan-2-oxyl and pentan-3-oxyl radicals formed after H-atom abstraction from the 2- and 3-positions in pentane are reaction with O_2 [reaction (14*a*)], decomposition [reaction (14*b*)], and isomerisation [reaction (14*c*)] for the pentan-2-oxyl radical, and O_2 reaction [reaction (15*a*)] and decomposition [reaction (15*b*)] for the pentan-3-oxyl radical (isomerisation *via* a six-member transition state cannot occur).[2]

$$CH_3CH(\dot{O})CH_2CH_2CH_3 \xrightarrow{O_2} CH_3C(O)CH_2CH_2CH_3 + HO_2 \quad (14a)$$

$$CH_3CH(\dot{O})CH_2CH_2CH_3 \longrightarrow CH_3CHO + CH_3CH_2\dot{C}H_2 \quad (14b)$$

$$CH_3CH(\dot{O})CH_2CH_2CH_3 \longrightarrow CH_3CH(OH)CH_2CH_2\dot{C}H_2 \quad (14c)$$

$$CH_3CH_2CH(\dot{O})CH_2CH_3 \xrightarrow{O_2} CH_3CH_2C(O)CH_2CH_3 + HO_2 \quad (15a)$$

$$CH_3CH_2CH(\dot{O})CH_2CH_3 \longrightarrow CH_3CH_2CHO + CH_3\dot{C}H_2 \quad (15b)$$

The pentan-2-one and pentan-3-one formation yields should therefore exhibit a dependence on the O_2 concentration, as observed (Fig. 4 and Table 1). Plots of $(\text{yield})^{-1}$ *vs.* $[O_2]^{-1}$ for pentan-2-one and pentan-3-one should be straight lines, with intercepts of (yield of pentan-2-oxyl or pentan-3-oxyl radical from pentane)$^{-1}$ and slope/intercept ratios of $[(k_{decomp} + k_{isom})/k_{O_2}]$ (where $k_{isom} \ll k_{decomp}$ for the pentan-3-oxyl radical).[26] Such plots of $(\text{yield})^{-1}$ *vs.* $[O_2]^{-1}$ are shown in Fig. 6, and the intercepts lead to fractional yields of the pentanoxyl radicals from the reaction of the OH radical with pentane in the presence of NO of 0.40 for the pentan-2-oxyl radical and 0.42 for the pentan-3-oxyl radical.

Combining the pentan-2-one and pentan-3-one yields extrapolated to $[O_2]^{-1} = 0$ with the measured pentan-2-yl and pentan-3-yl nitrate yields implies a yield of the pentan-2-yl radical from pentane of 0.46, and of the pentan-3-yl radical from pentane of 0.46. Since the calculated formation yield of the pentan-1-yl radical from pentane is 0.095,[21] then a complete accounting of the pentanyl radicals formed from pentane is obtained. The formation yields of pentan-2-yl and pentan-3-yl radicals derived from this product study, of 0.46 each, are somewhat different from (though in agreement within the experimental uncertainties) the formation yields of 0.55 and 0.36, respectively, calculated by the Atkinson estimation technique.[21]

The values of $[(k_{decomp} + k_{isom})/k_{O_2}]$ obtained from the slope/intercept ratios in Fig. 6 are 3.3×10^{19} molecule cm^{-3} for the pentan-2-oxyl radical and 3.3×10^{18} molecule cm^{-3} for the pentan-3-oxyl radical, both at 296 ± 2 K and 740 Torr total pressure and with estimated overall uncertainties of a factor of *ca.* 2. Literature data for the rate constant ratios k_{isom}/k_{O_2} and k_{decomp}/k_{O_2} for the butan-1-oxyl,[8–10] butan-2-oxyl,[8,9] and hexan-2-oxyl and hexan-3-oxyl[6] radicals at room temperature are given in Table 2. Isomerisation by a 1,5-H shift *via* a six-member transition state cannot occur for the butan-2-oxyl or pentan-3-oxyl radicals. Furthermore, isomerisations by a 1,4-H shift *via* a five-member transition state are calculated to be 3–4 orders of magnitude slower at 296 K than the 1,5-H shift isomerisations,[27] and to be slower than the decomposition reactions of the butan-2-oxyl and pentan-3-oxyl radicals.[2,26] The rate constant ratio $(k_{decomp} + k_{isom})/k_{O_2}$ obtained from the data for pentan-3-one formation shown in Fig. 6 is therefore the rate constant ratio k_{decomp}/k_{O_2} for the pentan-3-oxyl radical (Table 2). The decomposition rate constant for the pentan-2-oxyl radical is expected to be a factor of *ca.* 2 lower than that for the pentan-3-oxyl radical (owing to the degeneracy factor of

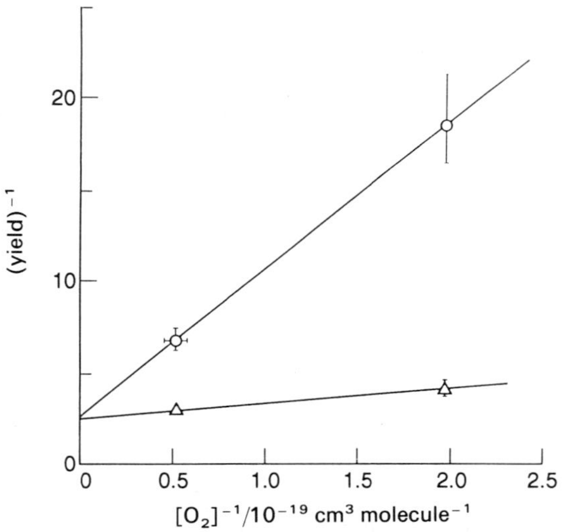

Fig. 6 $(\text{Yield})^{-1}$ for the formation of ○, pentan-2-one and △, pentan-3-one *vs.* $[O_2]^{-1}$

Table 2 k_{isom}/k_{O_2} and k_{decomp}/k_{O_2} for the pentan-2-oxyl and pentan-3-oxyl radicals at 296 ± 2 K and atmospheric pressure, together with room-temperature literature data for other alkoxyl radicals formed from alkanes

alkoxyl radical	k_{isomer}/k_{O_2}[a] molecule cm^{-3}	k_{decomp}/k_{O_2}[a] molecule cm^{-3}	ref.
butan-1-oxyl	1.6×10^{19} (303 K)		8
	$(1.5 \pm 0.5) \times 10^{19}$ (296 ± 2 K)		9
	$(1.9 \pm 0.3) \times 10^{19}$ (298 ± 2 K)		10
butan-2-oxyl		3.1×10^{18} (303 K)	8
		$(2.6 \pm 0.35) \times 10^{18}$ (296 ± 2 K)	9
pentan-2-oxyl	3.1×10^{19}[b]		this work
pentan-3-oxyl		3.3×10^{18}[b]	this work
hexan-2-oxyl	$(1.7 \pm 1.5) \times 10^{20}$[c] (297 ± 3 K)	$(3.2 \pm 1.8) \times 10^{19}$[c] (297 ± 3 K)	6
	$(5.9 \pm 4.2) \times 10^{20}$[d] (297 ± 3 K)	$(3.7 \pm 2.7) \times 10^{18}$[d] (297 ± 3 K)	6
hexan-3-oxyl	$(2.3 \pm 0.9) \times 10^{19}$[c] (297 ± 3 K)	$(5.3 \pm 2.0) \times 10^{18}$[c] (297 ± 3 K)	6
	$(5.4 \pm 3.2) \times 10^{19}$[d] (297 ± 3 K)	$(1.1 \pm 0.8) \times 10^{19}$[d] (297 ± 3 K)	6

[a] Indicated errors are two least-squares standard deviations. [b] Estimated overall uncertainties are a factor of *ca.* 2. [c] Derived from cited product yields from irradiated hexan-2-yl and hexan-3-yl nitrites–NO–air mixtures.[6] [d] Derived from cited product yields from irradiated HONO–hexane–air mixtures.[6]

2 for the pentan-3-oxyl radical) and the rate constants for the reactions of the pentan-2-oxyl and pentan-3-oxyl radicals with O_2 are expected to be similar (see below), resulting in the rate constant ratio k_{isom}/k_{O_2} for the pentan-2-oxyl radical given in Table 2. These rate constant ratios allow the C_5-hydroxycarbonyl/pentanones formation yield ratios to be calculated as a function of the O_2 concentration, with values of 4.3 at 4% O_2, 1.3 at 21% O_2, and 0.6 at 80% O_2, in surprisingly good agreement with the values derived from the API-MS data of *ca.* 6.4, *ca.* 3.5, and *ca.* 1.5, respectively.

Atkinson[2] has postulated that the rate constants for the reactions of alkoxyl radicals with O_2, k_{O_2}, depend on the exothermicity of the reactions, ΔH, with

$$k_{O_2} = 1.3 \times 10^{-19} n \exp(-0.32\Delta H) \text{ cm}^3 \text{ molecule}^{-1} \text{ s}^{-1} \text{ at 298 K} \qquad \text{(I)}$$

where n is the number of abstractable H atoms in the alkoxyl radical and ΔH is in units of kcal mol^{-1}. The calculated value of ΔH for the reaction of O_2 with the butan-1-oxyl radical is -32.3 kcal mol^{-1} and those for the butan-2-oxyl, pentan-2-oxyl, pentan-3-oxyl, hexan-2-oxyl and hexan-3-oxyl radicals are all -35.2 ± 1 kcal mol^{-1}.[28–30] From eqn. (I), this leads to essentially identical rate constants k_{O_2} of $(8–10) \times 10^{-15}$ cm^3 $molecule^{-1}$ s^{-1} at 298 K for the butan-1-oxyl, butan-2-oxyl, pentan-2-oxyl, pentan-3-oxyl, hexan-2-oxyl and hexan-3-oxyl radicals, and these calculated rate constants are very similar to those measured for the ethoxyl and propan-2-oxyl radicals[2] (as expected since eqn. (I) was derived from the rate constants k_{O_2} for the methoxyl, ethoxyl, and propan-2-oxyl radicals and the heats of reaction, ΔH, for the O_2 reactions with the ethoxyl and propan-2-oxyl radicals are similar to those for the alkoxyl radicals considered above). Hence the rate constant ratios k_{decomp}/k_{O_2} and k_{isom}/k_{O_2} can be used to investigate the dependence of the isomerisation and decomposition rate constants on the structures of the alkoxyl radicals for which data are given in Table 2.

As shown in Table 3, the rate constant for the decomposition of the butan-2-oxyl radical of $k_{decomp} = 2.8 \times 10^4$ s^{-1} at *ca.* 300 K, obtained relative to the O_2 reactions,[8,9] is significantly higher than the rate constant determined relative to the reaction with NO, of 4.8×10^3 s^{-1} at 298 K,[3] using a rate constant for the reaction with NO of $k_1 = 4.2 \times 10^{-11}$ cm^3 molecule^{-1} s^{-1}.[3] The decomposition rate constant measured by Dóbé *et al.*[5] for the pentan-2-oxyl radical, of 1.0×10^4 s^{-1} at 298 K relative to $k_1 = 4.2 \times 10^{-11}$ cm^3 molecule^{-1} s^{-1}, is in reasonable agreement with the value of $k_{decomp} = 3 \times 10^4$ s^{-1} obtained here for the pentan-3-oxyl radical relative to the rate constant for the reaction with O_2, especially considering the degeneracy factor of *ca.* 2 for the pentan-3-oxyl radical (Table 3). However, Dóbé *et al.*[5] obtained a rate constant ratio of $k_{isom}/k_{decomp} = 0.5 \pm 0.1$ at 301 K, a factor of *ca.* 20–40 lower than derived from our present data (Table 3). This marked disagreement may well be due to difficulties in quantitatively trapping the 4-hydroxypentan-1-yl radical in the study of Dóbé *et al.*[5]

The rate constant ratios k_{decomp}/k_{O_2} given in Table 2 for the butan-2-oxyl[8,9] and pentan-3-oxyl radicals are similar and are *ca.* 3×10^{18} molecule cm^{-3} (with the rate constant ratios k_{decomp}/k_{O_2} for the hexan-2-oxyl and hexen-3-oxyl radicals calculated from the product yields measured by Eberhard *et al.*[6] from irradiated hexan-2-yl and hexan-3-yl nitrite–NO–air and HONO–hexane–air mixtures exhibiting significant scatter but being consistent with this value of k_{decomp}/k_{O_2}). Using the rate constant $k_{O_2} = 9 \times 10^{-15}$ cm^3 molecule^{-1} s^{-1} calculated above, this rate constant ratio k_{decomp}/k_{O_2} leads to $k_{decomp} = 3 \times 10^4$ s^{-1} at 296 K for the butan-2-oxyl and pentan-3-oxyl radicals (Table 3).

The rate constant ratios k_{isom}/k_{O_2} for the butan-1-oxyl,[8–10] pentan-2-oxyl, and hexan-3-oxyl[6] radicals, with the isomerisation reaction involving H-atom abstraction from a $-CH_3$ group, are all similar at *ca.* (2–5) $\times 10^{19}$ molecule cm^{-3} (Table 2). For the hexan-2-oxyl radical,[6] the corresponding rate constant ratio k_{isom}/k_{O_2} is *ca.* (2–6) $\times 10^{20}$ molecule cm^{-3} (Table 2). Using the rate constant $k_{O_2} = 9 \times 10^{-15}$ cm^3 molecule^{-1} s^{-1} calculated above, these rate constant ratios k_{isom}/k_{O_2} lead to isomerisation rate constants k_{isom} of *ca.* (2–5) $\times 10^5$ s^{-1} for H-atom abstraction from a $-CH_3$ group and *ca.* (2–6) $\times 10^6$ s^{-1} for H-atom abstraction from a $-CH_2-$ group, both at room temperature (Table 3).

The increase in the isomerisation rate constant for H-atom abstraction from a $-CH_2-$ group compared to a $-CH_3$ group is expected owing to the lower C—H bond dissociation energy in the $-CH_2-$ groups compared with the $-CH_3$ groups.[1,2,4,26,27] However, the measured increase in the rate constant k_{isom} for H-atom

Table 3 Reaction rates of selected alkoxyl radicals at *ca.* 298 K and 1 atm of air

alkoxyl radical	$k_{O_2}[O_2]$/s^{-1} [a]	k_{decomp}/s^{-1} [b]	k_{isom}/s^{-1} [b]
butan-1-oxyl	4.2×10^4		1.4×10^5
butan-2-oxyl	5.2×10^4	2.8×10^4	
		4.8×10^3 [c,d]	
pentan-2-oxyl	5.2×10^4	1.0×10^4 [d,e]	3.1×10^5
			5×10^3 [d,e]
pentan-3-oxyl	5.2×10^4	3.3×10^4	
hexan-2-oxyl	5.2×10^4	*ca.* (4–30) $\times 10^4$	*ca.* (2–6) $\times 10^6$
hexan-3-oxyl	5.2×10^4	*ca.* (5–11) $\times 10^4$	*ca.* (2–5) $\times 10^5$

[a] Calculated using eqn. (I); see text. [b] From the rate constant ratios k_{decomp}/k_{O_2} and k_{isom}/k_{O_2} given in Table 2, unless indicated otherwise. [c] Ref. 3. [d] Relative to the rate constant k_1 for RO$^•$ + NO → RONO; using a value of $k_1 = 4.2 \times 10^{-11}$ cm^3 molecule^{-1} s^{-1}.[3,5] [e] From Dóbé *et al.*[5]

abstraction from a $-CH_2-$ group compared with a $-CH_3$ group of a factor of *ca.* 10 is significantly lower than the predicted increase of a factor of *ca.* 100.[2,4,26,27] Interestingly, Atkinson and Aschmann[31] also observed an increase of a factor of *ca.* 10 for alkoxyl radical isomerisation by H-atom abstraction from a $-CH_2-$ group compared with a $-CH_3$ group in the alkoxyl radicals $(CH_3)_2C(OH)CH_2C(\dot{O})(CH_3)_2$ and $CH_3CH_2C(CH_3)(OH)CH_2C(\dot{O})(CH_3)_2$ formed after OH radical reaction with 2,4-dimethylpentan-2-ol and 3,5-dimethylhexan-3-ol, respectively. These experimental data then suggest that alkoxyl radical 1,5-H shift isomerisations involving H-atom abstraction from $-CH_2-$ groups in alkoxyl radicals formed from alkanes are significantly slower, compared with H-atom abstraction from $-CH_3$ groups, than previously predicted.[2,4,26,27]

Conclusions

Employing positive APCI-MS to study the OH radical-initiated reactions of pentane provided confirmation of the occurrence of alkoxyl radical isomerisation reactions, *via* a 1,5-H shift through a six-member transition state, in the atmospheric photo-oxidation of long-chain alkanes. Quantitative measurements of the pentan-2-one and pentan-3-one formation yields as a function of the O_2 concentration lead to rate constant ratios for the decomposition and isomerisation reactions of the pentan-2-oxyl and pentan-3-oxyl radicals relative to their reactions with O_2. These data for the pentanoxyl radical reactions are reasonably consistent with the analogous data for the butan-1-oxyl and butan-2-oxyl radicals and the hexan-2-oxyl and hexan-3-oxyl radicals, and the combined data set provides important information concerning the relative importance of decomposition, isomerisation and reaction with O_2 of the alkoxyl radicals formed from the alkanes.

The authors gratefully thank the US Environmental Protection Agency (Cooperative Agreement CR 821787-01-0) and the California Air Resources Board (Contract No. A032-067) for supporting this research, and the National Science Foundation (Grant No. ATM-9015361) and the University of California, Riverside, for the funds to purchase the PE SCIEX API III MS/MS instrument. While this research has been supported by the US Environmental Protection Agency and the California Air Resources Board, it has not been subjected to review by these agencies, and therefore does not necessarily reflect their views; no official endorsement should therefore be inferred.

References

1 R. Atkinson and W. P. L. Carter, *J. Atmos. Chem.*, 1991, **13**, 195.
2 R. Atkinson, *J. Phys. Chem. Ref. Data.*, 1994, Monograph **2**, 1.
3 L. Batt, *Int. Rev. Phys. Chem.*, 1987, **6**, 53.
4 R. Atkinson, *Atmos. Environ., Part A*, 1990, **24**, 1.
5 S. Dóbé, T. Bérces and F. Márta, *Int. J. Chem. Kinet.*, 1986, **18**, 329.
6 J. Eberhard, C. Müller, D. W. Stocker and J. A. Kerr, *Environ Sci. Technol.*, 1995, **29**, 232.
7 W. P. L. Carter, K. R. Darnall, A. C. Lloyd, A. M. Winer and J. N. Pitts Jr., *Chem. Phys. Lett.*, 1976, **42**, 22.
8 W. P. L. Carter, A. C. Lloyd, J. L. Sprung and J. N. Pitts Jr., *Int. J. Chem. Kinet.*, 1979, **11**, 45.
9 R. A. Cox, K. F. Patrick and S. A. Chant, *Environ. Sci. Technol.*, 1981, **15**, 587.
10 H. Niki, P. D. Maker, C. M. Savage and L. P. Breitenbach, *J. Phys. Chem.*, 1981, **85**, 2698.
11 R. Atkinson, W. P. L. Carter, A. M. Winer and J. N. Pitts Jr., *J. Air Pollut. Control Assoc.*, 1981, **31**, 1090.
12 T. Covey and D. J. Douglas, *J. Am. Soc. Mass Spectrom.*, 1993, **4**, 616.
13 Z. Karpas, G. A. Eiceman, C. S. Harden and R. G. Ewing, *J. Am. Soc. Mass Spectrom.*, 1993, **4**, 507.
14 Y. K. Lau, S. Ikuta and P. Kebarle, *J. Am. Chem. Soc.*, 1982, **104**, 1462.
15 J. Sunner, N. Gordon and P. Kebarle, *Anal. Chem.*, 1988, **60**, 1300.
16 D. J. Douglas and J. B. French, *J. Am. Soc. Mass Spectrom.*, 1992, **3**, 398.

17 P. H. Dawson and W. F. Sun, *Int. J. Mass Spectrom. Ion Proc.*, 1983/1984, **55**, 155.
18 R. K. Mitchum and W. A. Korfmacher, *Anal. Chem.*, 1983, **55**, 1485A.
19 H. Kambara and I. Kanomata, *Anal. Chem.*, 1977, **49**, 270.
20 J. R. Dunlop and F. P. Tully, *J. Phys. Chem.*, 1993, **97**, 6457.
21 R. Atkinson, *Int. J. Chem. Kinet.*, 1987, **19**, 799.
22 R. Atkinson, *J. Phys. Chem. Ref. Data.*, 1989, Monograph **1**, 1.
23 R. Atkinson, S. M. Aschmann, W. P. L. Carter, A. M. Winer and J. N. Pitts Jr., *J. Phys. Chem.*, 1982, **86**, 4563.
24 R. Atkinson, W. P. L. Carter and A. W. Winer, *J. Phys. Chem.*, 1983, **87**, 2012.
25 W. P. L. Carter and R. Atkinson, *J. Atmos. Chem.*, 1989, **8**, 165.
26 W. P. L. Carter and R. Atkinson, *J. Atmos. Chem.*, 1985, **3**, 377.
27 A. C. Baldwin, J. R. Barker, D. M. Golden and D. G. Hendry, *J. Phys. Chem.*, 1977, **81**, 2483.
28 S. W. Benson, *Thermochemical Kinetics*, Wiley, New York, 2nd edn., 1976.
29 R. Atkinson, D. L. Baulch, R. A. Cox, R. F. Hampson Jr., J. A. Kerr and J. Troe, *J. Phys. Chem. Ref. Data*, 1992, **21**, 1125.
30 National Institute of Standards and Technology Standard Reference Database 25 *Structures and Properties Database and Estimation Program*, Version 2.0, S. E. Stein, Chemical Kinetics and Thermodynamics Division, NIST, Gaithersburg, MD, 1994.
31 R. Atkinson and S. M. Aschmann, *Environ. Sci. Technol.*, 1995, **29**, 528.

Paper 5/00171D; *Received 5th January* 1995

Faraday Discuss., 1995, **100**, 39–54

Role of Adducts in the Atmospheric Oxidation of Dimethyl Sulfide

Stephen B. Barone, Andrew A. Turnipseed and A. R. Ravishankara†
National Oceanic and Atmospheric Administration, Aeronomy Laboratory, 325 Broadway, Boulder CO 80303, USA and Department of Chemistry and Biochemistry and the Cooperative Institute for Environmental Sciences, University of Colorado, Boulder CO 80309, USA

Based on the results from recent laboratory studies and field observations, we suggest that adduct formation may play a key role in the atmospheric oxidation of dimethyl sulfide (DMS). Experimental evidence for the formation of the DMS · OH and $CH_3SO_x \cdot O_2$ adducts are described and used in a simple 'box model' to calculate the ratio of [MSA]/[SO_4^{2-}] observed in the atmosphere. These calculated ratios are compared with those inferred from field measurements. Based on these comparisons, we suggest that the changes with temperature in the mechanism of the OH + DMS reaction, one of the primary initiation steps in atmospheric DMS oxidation, are unlikely to be the source of the variation in the [MSA]/[SO_4^{2-}] ratio with temperature. We suggest that the competition between the thermal decomposition of adducts such as CH_3SO_2 and their bimolecular reactions with atmospheric species is the likely cause of the variation. Finally, a few observations needed to better our understanding of the DMS oxidation are identified.

Dimethyl sulfide (DMS, CH_3SCH_3) emitted by oceanic phytoplankton, is the major natural source of sulfur in the troposhere. It is a steady source, unlike the sporadic volcanic emissions. Much effort has been focussed, during the past two decades, on understanding the atmospheric oxidation (see for example: ref. 1) of DMS *via* atmospheric field measurements, modelling studies, and laboratory (direct and chamber) investigations of the kinetics and product identities. However, many uncertainties still remain in quantifying the eventual fate of DMS in the atmosphere. This uncertainty is primarily due to a lack of understanding of the oxidation mechanism. The extent of branching to stable end-products and the variation of this branching with temperature and composition are unclear. Such information is essential for evaluating the role of DMS in the atmosphere and assessing its impact on Earth's climate.

The likely major end-products of DMS oxidation are species such as sulfur dioxide, sulfur trioxide and methanesulfonic acid (CH_3SO_3H, MSA). SO_2 and MSA have been measured in chamber studies and in the atmosphere.[2–8] The effect of DMS on the climate is critically dependent on the production of gas-phase sulfuric acid and new particles. If SO_2 and/or SO_3 is produced, new particles can be produced because both will lead to gas-phase H_2SO_4, at least part of the time. In contrast, the formation of MSA is not expected to lead to new particles. There are subtle, but important, differences even between SO_2 and SO_3 production. If SO_2 is produced in DMS oxidation, it

† Address correspondence to this author at: NOAA/ERL, R/E/AL2, 325 Broadway, Boulder, CO 80303, USA.

is expected to lead to a more diffuse distribution of H_2SO_4 while SO_3 generation should lead to immediate conversion into H_2SO_4. In the latter case, the emitted DMS will likely influence only the marine boundary layer. This expectation is based on the lifetime of SO_3 being very short, owing to reaction with water, and that of SO_2 being a few days.

In addition, if the DMS oxidation mechanism is well characterized and the variation of the yields of end-products (and intermediates) with temperature and atmospheric composition are quantified, atmospheric abundances of products will provide insight into the concentrations of oxidants and temperature. This would be valuable for predicting the future as well as for understanding the pre-industrial atmosphere. In the latter case, DMS must have been the predominant source of sulfur. Interpretation of ice-core data would also be greatly facilitated by such an understanding of the DMS oxidation mechanism.

In this paper we argue that sulfur-containing molecules and free radicals are prone to forming adducts with OH and O_2, and the thermal stability relative to their bimolecular reactions is responsible for the changes in the branching between MSA and sulfate in the atmosphere. This thesis rests on the simple fact that unimolecular decomposition reactions, important in the atmosphere, have very large activation energies and, hence, the decomposition rate can change by orders of magnitude for small changes in temperature. In contrast, most atmospheric bimolecular reactions are not very sensitive to temperature. When the product branching depends on the competition between unimolecular reactions and bimolecular reactions it can change rapidly with temperature.

To show the importance of these adducts, we have carried out a simple modelling analysis of the DMS atmospheric oxidation based on kinetic and product studies. Field measurement data have been used to define the range of possibilities for the parameters. Thus, to some extent, our approach links observations from the atmosphere and ice-cores with the results of the numerous laboratory studies *via* a proposed DMS oxidation mechanism. Surprisingly, we can draw quite a few conclusions about the DMS oxidation mechanism and identify future studies that will be most productive.

Laboratory Studies of Adduct Formation in DMS Oxidation

Laboratory studies of the reactions involved in the atmospheric oxidation have been carried out by many groups.[9–13] The rate coefficients for the initiation reactions needed to quantify the atmospheric lifetime of DMS are now reasonably well defined. However, the subsequent reactions which influence the end-product yields are not well understood. Here, we briefly discuss the approach we have taken in our laboratory to understand the mechanism of the OH–DMS reaction and the reactions of the CH_3S radical. These studies are of particular significance since both involve formation of adducts. Based on these studies, we speculate on the bond strengths of other possible adducts in DMS oxidation.

Experimental

All measurements discussed here were carried out using the technique of pulsed laser photolysis–pulsed laser-induced fluorescence (PLP–PLIF). The PLP–PLIF apparatus and the experimental procedures used to measure rate coefficients have been described in detail elsewhere.[14,15] The OH and CH_3S radicals were detected by LIF in an excess of other reagents. OH was monitored by exciting the $(A\,^2\Sigma^+, v' = 1) \leftarrow (X\,^2\Pi, v'' = 0)$ band at *ca.* 282.1 nm using a pulsed XeCl excimer laser pumped dye laser (0.10–1.2 mJ pulse^{-1} cm^{-2}, pulse width *ca.* 8 ns, 10 Hz). CH_3S radicals were detected by exciting the $(A\,^2A_1, v_3' = 1) \leftarrow (X\,^2E, v_3'' = 0)$ transition at 371.4 nm. OH radicals were typically generated *via* the photolysis of H_2O_2 at 248 nm, while CH_3S radicals (when they were reactants) were generated by the 248 nm photolysis of DMS or DMDS (CH_3SSCH_3, dimethyl disulfide). The OH (or CH_3S) temporal profile was obtained by measuring the LIF signal at various delay times between the photolysis and probe lasers (5 μs–50 ms). The detection limit defined as $S/N = 1$, where S is the time-zero signal and N is equal to twice the standard deviation in the mean value of the measured background, was typically *ca.* 4×10^8 molecules cm^{-3} for OH and 4×10^9 molecule cm^{-3} for CH_3S in 100 Torr He upon integration of 100 laser shots. All experiments were carried out by slowly flowing the gas mixture through the reactor. The concentrations of the individual stable gases were determined from pressure and flow-rate measurements or *via* UV absorption measurements.

Status of Results in the OH + DMS Oxidation

The reaction of OH with DMS is a key step for the initiation of DMS oxidation. This is also the first place in the mechanism where branching to yield different products can take place.[9–11,16] Recently, we have focussed on the rate coefficients and products of the OH reaction with DMS. The following is a short synopsis of experiments which have been recently completed in our laboratory. A complete description of the experiments and the obtained results will be published in the near future along with detailed discussion of the chemical simulations involved in obtaining most of the values reported here. Along with some recent results from other laboratories, these studies lay the framework for understanding the branching in the initiation step in the oxidation of DMS in the atmosphere.

OH + DMS, [$^2H_6$]DMS (k_1, k_2)

The OH + DMS reaction can proceed *via* several possible channels

$$OH + DMS(+M) \rightleftarrows CH_3S(OH)CH_3(+M); \quad \Delta H^\circ = -11.0 \text{ kcal mol}^{-1} \quad (1f, 1r)$$

$$OH + DMS \rightarrow CH_3SCH_2 + H_2O; \quad \Delta H^\circ = -25.5 \text{ kcal mol}^{-1} \quad (1a)$$

$$OH + DMS \rightarrow CH_3 + CH_3SOH \quad \Delta H^\circ = 0 \pm 3 \text{ kcal mol}^{-1} \quad (1b)$$

$$OH + DMS \rightarrow CH_3S + CH_3OH; \quad \Delta H^\circ = -18.8 \text{ kcal mol}^{-1} \quad (1c)$$

Initially, the overall rate coefficients for reaction (1) and the reaction of fully deuteriated DMS

$$OH + [^2H_6]DMS \rightarrow \text{products} \quad (2)$$

were measured under pseudo-first-order conditions ([DMS] $\geqslant$ 100[OH]) at 298 K in the absence of O_2. [Note that reaction (2) has the same possible product channels as reaction (1).] The values $k_1(298\ K) = (4.95 \pm 0.35) \times 10^{-12}$ and $k_2(298\ K) = (1.75 \pm 0.25) \times 10^{-12}$ are in good agreement with other recent measurements.[12,16,17]

At low temperatures ($T \leqslant 240$ K) [OH] temporal profiles were characterized by an initial rapid exponential decay followed by a slower exponential decay. The slower decay is due to channels 1*a*–1*c*, accounted for the rate coefficient k^{bi}, and is slower for [$^2H_6$]DMS than for DMS. The faster decay is due to the rapid approach to equilibrium *via* the addition pathway and is controlled by the rate coefficients for addition of OH to DMS (k_f) and the thermal decomposition of the adduct back to reactants (k_r). By fitting the temporal profiles with a non-linear least-squares fitting routine, FACSIMILE,[18] the parameters k_f, k_r and k^{bi} were determined and the equilibrium constant, K_c,

$$K_c = \frac{[DMS \cdot OH]_{eq}}{[DMS]_{eq}[OH]_{eq}} = \frac{k_f}{k_r} = \frac{K_p}{(RT)^{\Delta N}}$$

measured from the ratio of k_f and k_r. In the above equation, $\Delta N = -1$. The equilibrium constants were determined over the temperature range 217–245 K. A second-law analysis of the K_p *vs.* $1/T$ data, shown in Fig. 1, yields the following thermochemical values at 229.5 K: $\Delta_{rxn}H^\circ = -10.2 \pm 2.0$ kcal mol^{-1}; $\Delta_{rxn}S^\circ = -28.4 \pm 6.4$ cal mol^{-1} K^{-1}.

Using ΔS° ($= -31.1$ at 229.5 K) and ΔC_p° calculated from statistical mechanics, a third-law calculation yields $\Delta_{rxn}H^\circ$(298 K) $= -10.7 \pm 2.5$ kcal mol^{-1}. This leads to $\Delta_f H^\circ$[$^2H_6$]DMS·OH) of -10.3 ± 2.5 kcal mol^{-1} at 298 K.

Equilibration of OH with [$^2H_6$]DMS has also been recently observed by Wine and co-workers[19] who report $\Delta_{rxn}H^\circ = -14.2 \pm 2.6$ kcal mol^{-1} and $\Delta_{rxn}S^\circ = -42.2 \pm 10.8$ cal mol^{-1} K^{-1} obtained from a second-law analysis. However, their measured equilibrium constants are not very different from those of the present study.

There are also two contradictory *ab initio* studies of reaction (1f). Turecek[20] predicts that there is no bound DMS·OH species; however, McKee[21] predicts a DMS·OH bond strength of *ca.* 6 kcal mol^{-1} and states that this may be underestimated by *ca.* 4 kcal mol^{-1}. Our observation of the equilibration clearly supports the existence of the DMS·OH adduct and supports the calculation of McKee. The DMS·OH adduct is a bound species with a bond dissociation energy of *ca.* 10–14 kcal mol^{-1}.

It is conceivable that the adduct could fall apart to give species other than DMS and OH. For example, it may be possible to form $CH_3SOH + CH_3$ or $CH_3OH + CH_3S$. As shown later, the possibility of the formation of $CH_3OH + CH_3S$ can be neglected. Such reactions of the adduct would manifest as bimolecular reactions in our system.

Products from Reaction (1). When OH was generated at 298 K in the presence of DMS, production of CH_3S radicals was not observed. Based on the relative detection sensitivities of our apparatus for OH and CH_3S, an upper limit for the branching ratio for k_{1c} was determined to be $\Phi_{1c} = k_{1c}/k_1 < 0.04$. Neither CH_3SCH_2 or CH_3 were detected directly by this study but they were converted into CH_3S and OH, respectively, by the addition of O_2 and NO. CH_3S and OH were detected by LIF. The results of

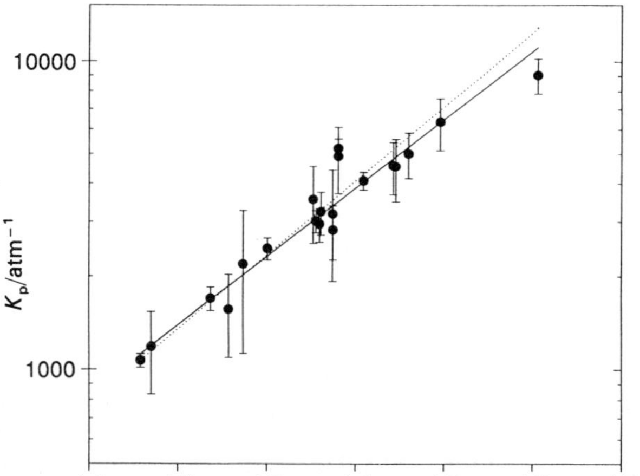

Fig. 1 K_p *vs.* $1000/T$ (van't Hoff plot) for reaction (1). K_p was measured over a range of pressures (30–100 Torr) and bath gases (He, N_2 and SF_6). The error bars shown are the 95% confidence limits in K_p and are obtained from the data analysis. The solid line drawn is the weighted least-squares fit to the data (the second-law analysis) and the dashed line represents the fit determined from a third-law analysis (see text for details).

these studies in 20 Torr of O_2 at 298 K showed that: $k_{1a}/k_1 = \Phi_{1a} = 0.86 \pm 0.26$ and $k_1^{bi} = k_{1a} + k_{1b} = (5.4 \pm 1.1) \times 10^{-12}$ cm^3 molecule^{-1} s^{-1}. The obtained value of Φ_{1a} is in excellent agreement with that measured recently by Stickel *et al.*,[22] who determined $\Phi_{1a} = 0.84 \pm 0.15$ by measuring the HOD appearance from the reaction of OD radicals with DMS. Channel (1*b*), which appears to be small, cannot be ruled out from the current experiments because OH regeneration was observed at long reaction times. However, the OH regeneration in this system could be due to the reaction

$$CH_3SCH_2O + O_2 \rightarrow CH_3SCHO + HO_2 \qquad (3)$$

followed by the conversion of HO_2 to OH *via* reaction with NO. Reaction (3) was suggested by Butkovskaya and LeBras[23] to explain their observed products in the NO_3 + DMS reaction in O_2. Our study indicates that under the experimental conditions used here (20 Torr O_2, 298 K), channel (1*a*) (the H atom abstraction) is the dominant channel and primarily leads to CH_3S formation.

O_2 + DMS · OH → *Products* (k_4)

Previous kinetic studies have shown that k_1 and k_2 increase with $[O_2]$[16,24] due to the reaction of DMS · OH with O_2,

$$DMS \cdot OH + O_2 \rightarrow products \qquad (4)$$

The rate coefficients for this reaction were determined by monitoring the OH temporal profiles in the presence of DMS (or [$^2H_6$]DMS) and O_2 and modelling the obtained profiles using the FACSIMILE program. A value of $k_4 = (9.7 \pm 3.3) \times 10^{-13}$ cm^3 molecule^{-1} s^{-1} was measured between 225 and 234 K and at 30 and 100 Torr of N_2. Thus, in the atmosphere, the only two important loss processes for the adduct are reaction with O_2 or thermal decomposition back to reactants. No other reactions of the adduct need to be considered because of the overwhelming abundance of O_2 in the atmosphere.

Our value of k_4 does not agree with the earlier reported value of Hynes *et al.*,[16] who report a value of $k_4 = 4.2 \times 10^{-12}$ cm^3 molecule^{-1} s^{-1}; however, more recent work[19] in their laboratory gives a value of $k_4 \approx 8 \times 10^{-13}$ cm^3 molecule^{-1} s^{-1} which is in good agreement with our measured value. One of the major products of this reaction was shown to be HO_2. At both 234 and 258 K, a branching ratio for the formation of HO_2 in the DMS · OH + O_2 reaction, $\Phi(HO_2) = 0.50 \pm 0.15$ was determined, independent of temperature. This is in excellent agreement with recent values from Hynes *et al.*, who also report a branching ratio for channel (4*a*) of *ca.* 50%.[19] It is assumed that the co-product of HO_2 is DMSO. In the atmosphere, it is very likely that DMSO is either deposited in the ocean and aerosols or reacts with OH. The fate of DMSO in the atmosphere is not well understood.

Reactions of the CH_3S Radical

We have shown above that, in the atmosphere, CH_3S is a major product of the abstraction part of the OH + DMS reaction. It also has to be the dominant product in the reaction of NO_3 with DMS in the atmosphere because it is known that the net result of this reaction is formation of CH_3SCH_2.[23] The CH_3S radical may also be produced by the reactions of the DMS · OH adduct. The reactions of the CH_3S radical with various atmospheric reactants such as O_3 and NO_2 have been studied in various laboratories.[9] Recently, we discovered that CH_3S adds to O_2 to form the CH_3SOO radical.[25] When CH_3S was produced *via* DMDS or DMS photolysis in the presence of O_2 at low temperatures, the CH_3S profiles were non-exponential and its concentration reached a constant value, which decreases as $[O_2]$ is increased. The rate of approach to this constant value also increases with $[O_2]$. These observations clearly show the occurrence of the equilibrium:

$$CH_3S + O_2 + M \xrightarrow{k_f} CH_3SOO + M \tag{5f}$$

$$CH_3SOO + M \xrightarrow{k_r} CH_3S + O_2 + M \tag{5r}$$

The temporal profiles of CH_3S measured at various concentrations of O_2 were analysed to obtain the equilibrium constant at each temperature. From this data the following thermodynamic quantities about CH_3SOO were calculated: $\Delta_{rxn}H°(298\ K) = -11.7 \pm 0.9$ kcal mol^{-1} and $\Delta_f H°(CH_3SOO,\ 298\ K) = 18.1 \pm 1.0$ kcal mol^{-1} using a calculated $\Delta_{rxn}S°(237\ K) = -36.8 \pm 2.6$ cal mol^{-1} K^{-1} and the calculated values of the heat capacities.

The adduct formed by this process must be $CH_3S \cdot OO$ and not CH_3SO_2 (where both oxygen atoms are bound to the sulfur) since it rapidly decomposes back to CH_3S and O_2 and does not appear to add another O_2 molecule.

Formation of the $CH_3S \cdot OO$ adduct suggests that other weakly bound adducts such as $CH_3SO(OO)$, $CH_3SO_2(OO)$, and HS(OO) could exist. Barnes *et al.*[26] and Jensen *et al.*,[27] have observed $CH_3S(O)OONO_2$ which could be formed from addition of NO_2 to $CH_3SO(OO)$. Some recent theoretical calculations also suggest that O_2 could add to HS.[28] There are also some indirect experimental observations that are consistent with HS(OO) formation.[25] Thus, it appears very likely that other S-containing free radicals could also add O_2 and that the bond strengths of these adducts would be *ca.* 10–15 kcal mol^{-1}. In particular, we assume that formation of $CH_3S(O)O_2$ and $CH_3S(O_2)O_2$ are likely.

Field Evidence for Changes in End-products with Temperature

Field measurements of DMS and its possible oxidation products in the atmosphere have cast considerable light on key aspects of the DMS oxidation mechanism. A diurnal variation in the DMS concentration in the atmosphere has been reported in several studies, all of which report a night-time maximum.[2,6,29] This observation suggests that the rate of daytime loss processes exceeds those at night if the strength of DMS emission is independent of time of day. Another observation with significant implications for the DMS oxidation mechanism lies in the $[MSA]/[SO_4^{2-}]$ ratio measured in the atmosphere and in ice-cores. (SO_4^{2-} is assumed to be formed with unit yield in the atmosphere from SO_2 or SO_3). Atmospheric measurements of the MSA/SO_4^{2-} ratio have been made by numerous investigators. The general finding is that this ratio changes with latitude and temperature. For example, Staubes and Georgii[4] report the $[MSA]/[SO_4^{2-}]$ ratio to be *ca.* 0.01 between 50°N and 30°N but increasing to *ca.* 0.25 between 53°S and 62°S and 0.44 at 70°S. This variation in the latitude must be accompanied by a temperature change in the vicinity of 300 to *ca.* 250 K. A more direct impression of the variation in the $[MSA]/[SO_4^{2-}]$ ratio with temperature can be gained from the ice-core data. For example, Legrand *et al.*[30] observed a change in the $[MSA]/[SO_4^{2-}]$ ratio from *ca.* 0.03 to *ca.* 0.18 when the surface temperature must have changed from *ca.* 210 to 200 K. Even though there has not been a very coherent measurement of the variation in the MSA/SO_4^{2-} ratio as a function of temperature, it is clear that this ratio changes rapidly with temperature. The magnitude of this change appears to be at least a factor of four or greater in going from 300 to *ca.* 250 K and another factor of *ca.* four in going from *ca.* 210 to 200 K. If both MSA and non-sea salt (nss) SO_4^{2-} are primarily produced from the oxidation of DMS, the oxidation mechanism of DMS must involve steps which reflect this temperature sensitivity.

The Role of Adducts in the DMS Oxidation Mechanism

A simplified DMS oxidation mechanism is shown in Fig. 2. A more detailed mechanism is listed in Table 1. Fig. 2 highlights various branching points in the mechanism where weakly bound adducts play a role. We will discuss how these reactions can alter the $[MSA]/[SO_4^{2-}]$ ratio and justify the assumptions in the mechanism. In this section,

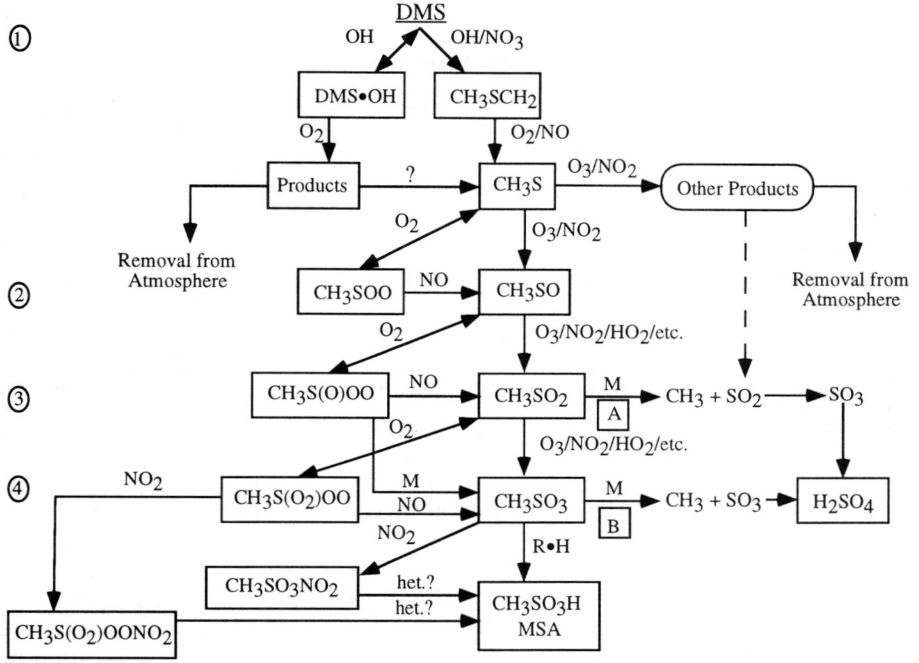

Fig. 2 Schematic representation of the atmospheric oxidation mechanism of DMS. The numbers on the left indicate branching points in the proposed mechanism where weakly bound adducts may play an important role. The letters, A and B, indicate potential unimolecular decomposition reactions important in the oxidation mechansim.

Table 1 The reactions and their rate coefficients used in the box model calculations to predict the variation of the MSA/nss sulfate ratios as a function of temperature

reaction	rate constant/cm^3 molecule^{-1} s^{-1}	ref.
$OH + DMS \rightarrow$ (abstraction)	$9.6 \times 10^{-12} \exp(-234/T)$	16
$OH + DMS \rightarrow$ (addition in 760 air)	[a]	16
$NO_3 + DMS \rightarrow CH_3SCH_2 + HNO_3$	$1.9 \times 10^{-13} \exp(-500/T)$	12
$CH_3S + O_3 \rightarrow CH_3SO + O_2$	$1.98 \times 10^{-12} \exp(290/T)$	15
$CH_3S + NO_2 \rightarrow CH_3SO + NO$	$2.06 \times 10^{-11} \exp(320/T)$	15
$CH_3SO + O_3 \rightarrow CH_3SO_2 + O_2$	6×10^{-13}	39
$CH_3SO + NO_2 \rightarrow CH_3SO_2 + NO$	8×10^{-12}	40
$CH_3SO_2 \xrightarrow{M} CH_3 + SO_2 \quad M$	$5 \times 10^{13} \exp[(17.2 \text{ kcal mol}^{-1} + RT)/RT]$	see text
$CH_3SO_2 + O_3 \rightarrow CH_3SO_3 + O_2$	3×10^{-13}	estimated
$CH_3SO_2 + NO_2 \rightarrow CH_3SO_3 + NO$	4×10^{-12}	estimated
$CH_3SO_3 \xrightarrow{M} CH_3 + SO_3 \quad M$	$5 \times 10^{13} \exp[-(22 \text{ kcal mol}^{-1} + RT)/RT]$	see text
$CH_3SO_3 + CH_2O \rightarrow CH_3SO_3H + CHO$	1.6×10^{-15}	11
$CH_3SO_3 + HO_2 \rightarrow CH_3SO_3H + O_2$	5×10^{-11}	11
$CH_3S(O)_xOO + NO \rightarrow CH_3S(O)_xOO + NO$	2.4×10^{-11}	estimated

[a] $[T \exp(-234/T) + 8.46 \times 10^{-10} \exp(7230/T) + 2.68 \times 10^{-10} \exp(7810/T)]/[1.04 \times 10^{11} T + 88.1 \exp(7460/T)]$.

CH_3S, CH_3SO, CH_3SO_2 and CH_3SO_3 are called CH_3SO_x, while the peroxy species $CH_3S(O)_xOO$ refers to CH_3SOO, $CH_3S(O)OO$, and $CH_3S(O)_2OO$.

Reactions of DMS with OH and NO_3

The results of previous studies[9,12,13] suggest that two reactions

$$NO_3 + DMS \rightarrow \text{products} \tag{6}$$

$$OH + DMS \rightarrow \text{products} \tag{1}$$

will dominate the initiation because of the large rate coefficents for these reactions and the abundances of OH and NO_3 in the troposphere. Other initiation reactions are unlikely to be important.[9] Reaction (6) has been shown to proceed through adduct formation followed by HNO_3 elimination *i.e.* equivalent to an H atom abstraction.[23,27] In contrast, the results presented in this paper combined with those of previous investigations show that reaction (1) proceeds through a complex mechanism involving H abstraction and electrophilic addition.[16] Work presented in the present study has established that H-atom abstraction primarily generates the CH_3S radical. The reaction product of the OH addition channel (DMS · OH) has been shown to react rapidly with O_2 but all the products in this reaction and their subsequent fate in the atmosphere have not fully been elucidated.[16,19] The effective overall branching between the addition and abstraction channels of reaction (1) has been characterized and is dependent on temperature.[16]

Previous investigations[31] have suggested that NO_3 + DMS will be of little significance in the oxidation of DMS in the remote marine boundary layer due to the predicted low ambient NO_3 concentrations in this region. However, if the concentrations of NO_3 exceed *ca.* six times that of OH, the rate of reaction (6) will equal that of reaction (1) at 298 K. The rate coefficients for both OH and NO_3 reactions increase with decreasing temperature. However, the rate coefficient for the OH reaction increases slightly faster than that for NO_3, such that, at 250 K for example, the NO_3 concentration has to be *ca.* 10 times that of OH for the two pathways to be equal (assuming the same number of hours of exposure to both reactants). Field measurements and modelling studies suggest that NO_3 concentrations can easily be 10 times the diurnally averaged OH value in the marine boundary layer.[32,33] As one approaches high latitudes and winter time, the $[NO_3]_{\text{night-time}}/[OH]_{\text{daytime}}$ will only get larger. Therefore, a complete quantification of the initiation step in the atmospheric oxidation of DMS requires a better knowledge of ambient NO_3 concentrations in the oceanic marine boundary layer.

As noted earlier, the diurnal cycling of DMS concentrations has been observed in the remote marine boundary layer. If this is a common phenomenon, the NO_3 concentrations in the remote boundary layer are less than the field measurements indicate or, more likely, DMS concentrations are enhanced at night. The latter possibility could be due to meteorological differences between day and night and the height of the marine boundary layer and may not require an increase in flux of DMS. This is clearly an important issue that needs resolution.

The competition between OH addition and hydrogen abstraction in reaction (1) has been suggested in several studies to be responsible for the observed temperature dependence of the $[MSA]/[SO_4^{2-}]$ ratio in the atmosphere.[2,30,34] For the $[MSA]/[SO_4^{2-}]$ ratio to increase as temperature decreases, MSA must be a product of the addition channel and SO_2 be a product of the H-atom abstraction. We use the rate coefficients for the addition *vs.* abstraction given by Wine and co-workers[16] and assume that the addition channel leads only to MSA and the subsequent steps leading to MSA from the DMS · OH adduct are insensitive to temperature. The temperature dependence of the OH + DMS reaction rate coefficient is such that the $[MSA]/[SO_4^{2-}]$ ratio will not change enough with temperature (see Fig. 3). For example, from 250 to 270 K, this ratio increases by at most 50%, while field data suggest a factor of two or more.

In addition to the above arguments, two points are worth noting: (1) The fractional change in the [MSA]/[SO_4^{2-}] ratio for a given change in T, ΔT, will depend on T and it will become quite insensitive to ΔT at low T if it depended only on the changes in the branching ratios of reaction (1). Fig. 3 shows that the temperature dependence of the [MSA]/[SO_4^{2-}] ratio tends to decrease at lower temperatures (*i.e.* the slope decreases at lower temperatures). (2) It is known from chamber studies that CH_3S does lead to MSA.[8,26,27]

Including reaction (6) decreases the temperature dependence of the [MSA]/[SO_4^{2-}] ratio even further. Fig. 3 demonstrates that the effect of NO_3 + DMS is only to lessen the importance of the branching between OH addition *cf.* abstraction. Based on these arguments, we propose that the cause of this temperature-dependent branching is unlikely to be due to changes in the OH reaction mechanism with temperature and has an entirely different origin.

The lifetime of DMS due to reactions with OH and NO_3 is of the order of a couple of days. OH exists during daytime, while NO_3 peaks at night. At night, the concentrations of species such as HO_2 and NO will be essentially zero. Therefore, DMS oxidation initiated at night will probably lead to species such as $CH_3SCH_2O_2NO_2$ and $CH_3S(O)_xOONO_2$ (see below). Unless these species are removed by deposition (wet or dry), they should be sources of CH_3S (or CH_3SO_x) radicals after sunrise and undergo further reactions. Therefore, it appears that even though NO_3 initiates DMS oxidation at night and OH during daytime, the oxidation to gas-phase MSA and sulfate takes place during daytime only.

Product-yield Branching following the Initiation Reactions

We propose that the branching in DMS oxidation arises from the oxidation of CH_3S radicals involving temperature-sensitive processes such as unimolecular decomposition and reversible adduct formation with O_2. To test this proposal we developed a simple oxidation mechanism of DMS based largely on available laboratory studies of CH_3S

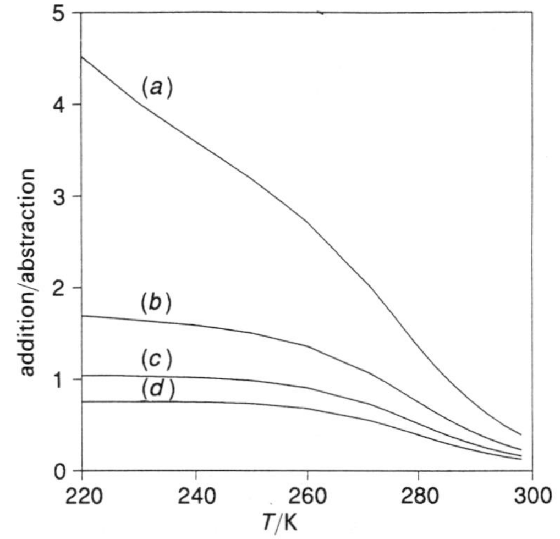

Fig. 3 The branching between MSA and SO_4^{2-} in 760 Torr air as a function of temperature for a variety of NO_3 conditions for the simple model where only the mechanism for the OH + DMS reaction changes with T. (*a*)–(d) represent ambient NO_3 concentrations of 0, 3, 6 and 9 × 10^6 molecules cm^{-3}, respectively. The diurnally averaged OH concentration is assumed to be 1 × 10^6 molecules cm^{-3}.

and its higher oxides. Where kinetic data is unavailable, we have estimated rate coefficients by analogy with other known reactions. The reactions involved in this simplified mechanism are listed in Table 1. The ambient concentrations used in our model are listed in Table 2. We have treated the initiation chemistry in the most simple fashion possible. H-atom abstraction from OH and NO_3 reactions is presumed to yield entirely CH_3S whereas the addition of OH followed by the O_2 reaction forms a reaction product different from SO_2 and MSA (probably DMSO, $DMSO_2$)

$$OH + DMS \rightarrow \rightarrow \text{products other than } SO_2 \text{ and MSA} \quad \text{(addition)} \quad (1a)$$

$$OH + DMS \rightarrow \rightarrow CH_3S \quad \text{(abstraction)} \quad (1b)$$

We realize that it is very likely that the atmospheric reactions of the DMS·OH adduct could lead to CH_3SO_x, MSA or SO_2. This occurrence will not greatly affect our arguments. When these reactions are better studied, one can easily incorporate them into our mechanism without major qualitative changes in the outcome.

Role of Higher Oxides

Our mechanism involves CH_3S and, of higher oxides, CH_3SO_x, reacting with O_3 and NO_2 in competition with their unimolecular decomposition or addition of O_2. We investigate the role of higher oxides *via* first modelling the thermal decomposition of CH_3SO_x and then the addition of O_2 to these species. The chemistry of each CH_3SO_x is treated analogously to that of CH_3S. To calculate potential unimolecular decomposition rates of the higher oxides we have assumed an A factor of 5×10^{13} s^{-1} and taken the endothermicity for the decomposition as the activation energy.

The simple mechanism given in Table 1 leads to a temperature sensitivity in the [MSA]/[SO_4^{2-}] significantly steeper than that observed in the OH initiation reaction. Fig. 4 shows this temperature dependence for the conditions outlined in Tables 1 and 2. The steep temperature-dependent branching between MSA and SO_2 is a result of the competition between the unimolecular decomposition of the CH_3SO_2 radical and the reaction of CH_3SO_2 with O_3 to oxidize sulfur further without breaking the C—S bond.

Table 2 Assumed abundance of various species in the clean marine environment

species	daytime mixing ratio	ref.
O_3	4 (–8)	41
NO_x	3 (–11)	41
CH_2O	1 (–10)	42
NO_2	2 (–11)	38
NO_3	3 (–13)	calculated
NO	1 (–11)	calculated
HO_2	2 (–12)	calculated
OH	4 (–14)	calculated

The sources of the data are given in the references shown. These are 'typical' mixing ratios during daytime and some of them have been rounded off to one significant figure. The ratios of HO_2 to OH and NO_2 to NO are kept consistent with the above values of O_3 and the known J values for NO_2.

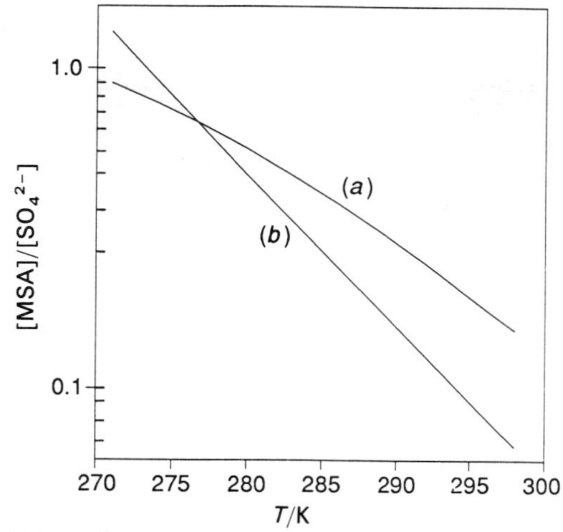

Fig. 4 Calculated [MSA]/[SO_4^{2-}] ratios from DMS oxidation as a function of temperature. (*a*) All branching due to changes in initiation reaction mechanism; (*b*) model conditions outlined in Tables 1 and 2.

The temperature dependence of the branching reflects the activation energy for the decomposition of CH_3SO_2.

We estimate the decomposition of CH_3SO_3 to be more endothermic than that of CH_3SO_2. This is because CH_3SO_2 must be oxidized to CH_3SO_3 by a species such as O_3, NO_2 or O_2/NO combination and these reactions have to be at least thermo-neutral to be important in the atmosphere. Other investigators have also estimated the C—S bond strength in CH_3SO_2 to be less than in CH_3SO_3.[11] Because of the larger endothermicity, one could expect it to contribute more to the temperature dependence of the branching between MSA and SO_4^{2-}. However, we do not expect the unimolecular decomposition of CH_3SO_3 to be a very important process in the oxidation for two reasons: (1) A large amount of the CH_3SO_2 decomposes to SO_2 before it can form CH_3SO_3 and (2) most of the CH_3SO_3 reacts with HO_2 and CH_2O. Thus, if CH_3SO_3 is formed in the atmosphere, it will very likely end up as MSA. Therefore, if our assumed rate coefficients are correct, it appears that the direct formation of SO_3 in DMS oxidation, as suggested by Bandy *et al.*, is likely to be minor.[35] Note, however, that the thermochemistry of CH_3SO_3 is very poorly defined. If the C—S bond strength in CH_3SO_3 is much smaller, say $\leqslant 20$ kcal mol^{-1}, this species could become a significant branching point in DMS oxidation. If the CH_3SO_3 radical is stable, it could react with NO_2 to form $CH_3SO_3NO_2$; this species may be hydrolysed in the particles to give MSA in the liquid phase (see below). Also, the rate coefficient for the $CH_3SO_3 + HO_2$ reaction needs to be determined to estimate this branching.

A systematic variation of the ambient NO_x concentration in our model showed a very small effect on the oxidation products of CH_3S in the atmosphere. Although the rate constants for reactions of CH_3S, CH_3SO and CH_3SO_2 with NO are large, the low NO_x levels in the remote troposphere prohibit these reactions from competing with O_3 reactions. An order of magnitude increase in the NO_x concentration used in our model results in a less than 8% change in the [MSA]/[SO_4^{2-}] ratio at 271 K. Therefore, we conclude that changes in the amount of NO_x in the boundary layer has the greatest effect on the initiation chemistry of DMS oxidation and little effect on the subsequent reactions for this proposed scheme of DMS oxidation until the NO_x level reaches > *ca.* 300 pptv.

Molecular oxygen has been observed to add reversibly to CH_3S radicals under atmospheric conditions.[25] [Reactions (5f, 5r).] The rate constants for the reactions of the

CH_3SOO radical with NO, NO_2 and O_3 have been measured.[15] In the remote atmosphere, the $CH_3S + O_3$ reaction should dominate over all other loss processes of the CH_3SOO adduct and CH_3S. To predict the effect of addition of O_2 to $CH_3S(O)_x$, we added the following reactions to the model;

$$CH_3S(O)_2 + O_2 + M \xrightarrow{k_f} CH_3S(O)_2OO + M \quad (7f)$$

$$CH_3S(O)_2OO + M \xrightarrow{k_r} CH_3S(O)_2 + O_2 + M \quad (7r)$$

$$CH_3S(O)_2OO + NO \longrightarrow CH_3SO_3 + NO_2 \quad (8)$$

Note that CH_3SO_3 would not be expected to add O_2 since the sulfur in this species is fully oxidized. Reactions (7) and (8) comprise a potentially efficient pathway for converting CH_3SO_2 to CH_3SO_3. The equilibrium constant for the reversible addition of O_2 to CH_3SO_2 would be very sensitive to the temperature and hence, this process could be another candidate in explaining the $[MSA]/[SO_4^{2-}]$ ratio observations in the field. Owing to the abundance of O_2 in the atmosphere it is likely that the equilibrium in reaction (7) will be established rapidly and reaction (8) would be the rate-limiting step in determining the overall efficiency of this mechanism.

The reaction mechanism presented up to this point is somewhat analogous to that given originally by Yin *et al.*[11] and modified by Bandy *et al.*[35] However, there are some major differences. First, we can show that NO, and not HO_2 reaction followed by photolysis of the peroxide, is responsible for converting $CH_3S(O)_xO_2$ to $CH_3S(O)_xO$. Based on simple steady-state calculations and measured abundances of NO_x,[36–38] the abundance of NO is shown to be approximately an order of magnitude greater than that of HO_2 in the marine boundary layer. The rate coefficients for the reaction of the larger peroxy radicals with NO is greater than those with HO_2 and CH_3O_2,[12,13] and, hence, the rate coefficient for the reaction of $CH_3S(O)_xOO$ with NO must be larger than that with HO_2. Therefore, the majority of the $CH_3S(O)_xOO$ must react with NO. Even if hydroperoxides are formed, since they do not usually absorb very strongly in the near-UV, they are likely to be washed out of the atmosphere.

We have estimated the rate constant for reaction (8) to be 2.4×10^{-11} cm^3 $molecule^{-1}$ s^{-1} and calculated yields for a variety of different $CH_3SO_2{-}OO$ bond strengths. Bond strengths $\leqslant 10$ kcal mol^{-1} have little effect on the outcome. However, as the bond strength is increased from 10.5 kcal mol^{-1} the addition reactions of $CH_3S(O)_x$ radicals have a pronounced effect on the absolute values of the branching to MSA and SO_2 and the temperature dependence of their ratio. Fig. 5 shows the effect of adduct bond strengths of 13 and 14 kcal mol^{-1} on the temperature dependence of the $[MSA]/[SO_4^{2-}]$ ratio. The increase in the $[MSA]/[SO_4^{2-}]$ ratio with temperature is a result of a shift toward larger equilibrium concentrations of the CH_3SO_2OO at low temperatures. This in turn increases the rate of reaction (8) relative to the unimolecular decomposition of CH_3SO_2 and, hence, adds to the overall temperature sensitivity of the $[MSA]/[SO_4^{2-}]$ ratio. Note that the incorporation of this process in the model adds to the temperature dependence in the $[MSA]/[SO_4^{2-}]$ ratio determined by the unimolecular decomposition of the CH_3SO_2 radical alone.

Aside from influencing the temperature dependence of the branching in DMS oxidation, the inclusion of the O_2 adduct formation reaction has several other consequences. First, because the $NO + CH_3S(O_2)OO$ reaction becomes the rate limiting step in the cycling of CH_3SO_2 to CH_3SO_3, NO_x will play a larger role in determining the $[MSA]/[SO_4^{2-}]$ ratio. The branching to stable end-products in DMS oxidation will be strongly NO_x dependent. Secondly, the increased efficiency in the conversion of CH_3SO_2 to CH_3SO_3 makes the unimolecular decomposition of CH_3SO_3 more important. Thus, branching to form SO_3 can become more likely. However, SO_3 formation is still likely to be minor because the rate of CH_3SO_3 decomposition is likely to be slow

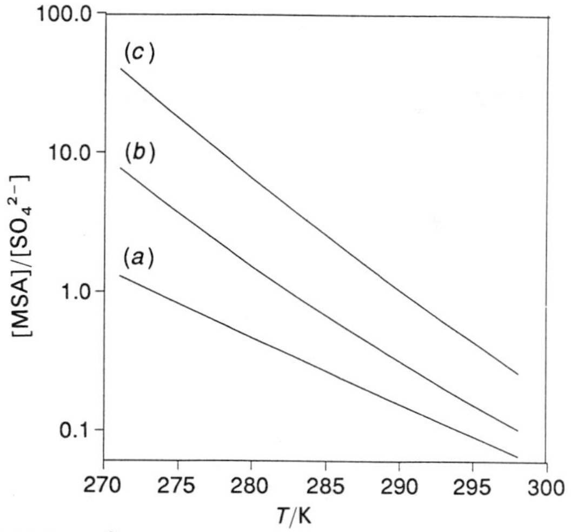

Fig. 5 Calculated $[MSA]/[SO_4^{2-}]$ ratios as a function of temperature from a DMS oxidation mechanism which includes reactions (7) and (8). (*a*), (*b*) and (*c*) are calculated from $CH_3S(O_2)\cdot OO$ adduct bond strengths of 0, 13 and 14 kcal mol^{-1}, respectively. The other rate coefficients and ambient concentrations are listed in Tables 1 and 2.

with respect to its reactions with HO_2 and CH_2O. Lastly, formation of an O_2 adduct with CH_3SO is also a potentially important pathway to form CH_3SO_2. In our simple model all CH_3SO is converted to CH_3SO_2. However, isomerization or other removal processes of this radical could play key roles in the oxidation of DMS.

The formation of nitrate from NO_2 addition to the peroxy-adducts;

$$NO_2 + CH_3S(O)_xOO \rightarrow CH_3S(O)_xOONO_2 \tag{9}$$

could be an important process in the atmosphere. Nitrates, such as these, have previously been observed in chamber studies of DMS oxidation.[26,27] This observation is an indication of the likelihood of O_2-adduct reactions in the atmospheric oxidation of DMS. The formation of these peroxynitrate radicals could be especially important if they are easily hydrolysed in cloud drops. The hydrolysis of these nitrates could prove to be an important source of MSA in the atmosphere. Also, if the nitrates are long-lived, they could play a role in the transport of sulfur to the free troposphere.

It is possible that reactions of CH_3S and CH_3SO with O_3 and of CH_3SO with NO_2 could result directly in the cleavage of the C—S bond. These channels are exothermic and would have a significant effect on the results of the mechanism developed thus far. To evaluate this possibility we have included the following reaction in our model;

$$CH_3SO + O_3 \rightarrow \rightarrow SO_2 \tag{10}$$

Note that the above reaction merely represents the cleavage of the C—S bond to lead to SO_2 formation and is not meant to be an elementary reaction. Inclusion of this chemistry significantly effects the calculated $[MSA]/[SO_4^{2-}]$ ratio. Aside from increasing the absolute magnitude of SO_2 formation, including reaction (10) decreases the temperature dependence of the SO_2 yield. Fig. 6 demonstrates the effect on the $[MSA]/[SO_4^{2-}]$ ratio of 25%, 50% and 75% channels in the reaction $CH_3SO + O_3$ to cleave the C—S bond. Reaction (10) channels significant amounts of the sulfur to end-products by bypassing the CH_3SO_2 species and, hence, avoids the temperature-dependent mechanism for SO_2 production *via* unimolecular decomposition. The effect of

$$CH_3S + O_3 \rightarrow \rightarrow SO_2 \tag{11}$$

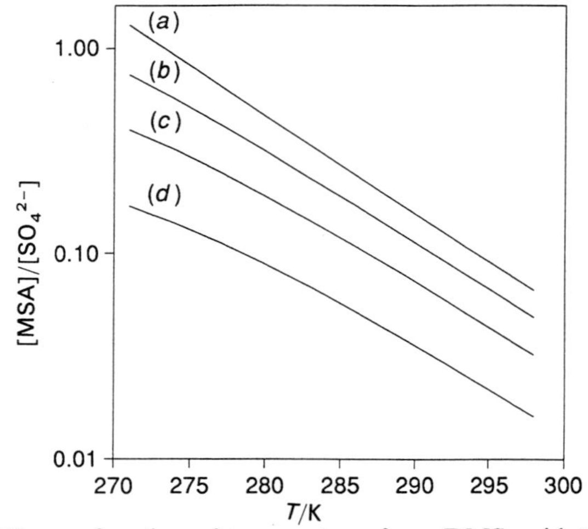

Fig. 6 $[MSA]/[SO_4^{2-}]$ as a function of temperature, from DMS oxidation, calculated with the inclusion of reaction (10). (*a*)–(*d*) reflect branching ratios of the reaction $CH_3SO + O_3$ to yield SO_2 of 0, 0.25, 0.50 and 0.75, respectively. All other relevant conditions are in Tables 1 and 2.

would be identical to reaction (10). However, the effect of;

$$CH_3SO_2 + O_3 \rightarrow CH_3 + SO_3 + O_2 \quad (12)$$

could be less than reactions (10) and (11), depending on the rate coefficient for the unimolecular decomposition of CH_3SO_2. Note that the addition of O_2 to CH_3SO, followed by the reactions of $CH_3S(O)OO$, could mediate the effects of the C—S bond cleavage reactions involving O_3.

We have focussed on the potential branching reactions of the CH_3SO_2 and CH_3SO_3 radicals, however, it is important to note that the chemistry of the preceding CH_3SO radical could also be a source of the branching between MSA and SO_4^{2-} in DMS oxidation. Branching at CH_3SO would most likely occur through a competition between O_2 addition followed by reaction with NO and reaction (10) resulting in C—S bond cleavage. C—S bond cleavage would eventually lead to SO_2 formation whereas O_2 addition could enhance the oxidation state of the sulfur without breaking the C—S bond;

$$CH_3SO + O_2 \rightarrow CH_3S(O)OO \quad (13)$$

$$CH_3S(O)OO + NO \rightarrow CH_3SO_2 + NO_2 \quad (14)$$

$$CH_3SO + O_3 \rightarrow SO_2 + CH_3 + O_2. \quad (15)$$

Role of Heterogeneous Reactions

There is a possibility that MSA is formed only *via* heterogeneous reactions of species such as $CH_3S(O)_xOONO_2$. The MSA measured in the atmosphere has always been extracted from the aerosol phase. If this is indeed the case, the $[MSA]/[SO_4^{2-}]$ ratio may be related to the heterogeneous reaction probability, *i.e.* availability of aerosols, fog, clouds *etc.* and the stability of the gas-phase species which are taken up by the condensed medium. Both these factors are enhanced at lower temperature and, hence, are consistent with the observed increases in the $[MSA]/[SO_4^{2-}]$ ratio with decreases in temperature. Measurement of gas-phase MSA abundances is essential for shedding light on this problem.

Summary of Modelling Studies

Our simple modelling analysis of DMS atmospheric oxidation has led to several basic conclusions:

(1) The temperature dependence of OH addition *vs.* H-atom abstraction in the initiation chemistry of DMS oxidation is not strong enough to rationalize completely the variations in the MSA/SO_4^{2-} ratios observed in the ice-cores and with latitude. The inclusion of NO_3 chemistry greatly reduces the overall temperature dependence of the OH addition *vs.* H-atom abstraction. An accurate knowledge of the ambient NO_3 concentrations in the marine boundary layer is the key to better understanding of the initiation chemistry.

(2) It appears that the abstraction pathway in the OH/NO_3 + DMS reaction can also lead to the formation of MSA, contrary to what has been assumed in some past studies but in accordance with many chamber studies.

(3) It appears that the major role of NO_x in DMS oxidation is limited to the initiation step, unless $CH_3SO_x \cdot O_2$ adducts are important.

(4) CH_3S oxidation *via* the formation of higher oxides that can unimolecularly decompose (cleave the C—S bond) or raise the oxidation state of sulfur without breaking the C—S bond can lead to a large temperature dependence of the branching between MSA and SO_2. This temperature dependence will be heightened if O_2 adds reversibly to some of the CH_3SO_x species. If this were to happen, NO_x could play a role in determining the branching between different endproducts.

(5) The potential for C—S bond cleavage *via* other bimolecular reactions of CH_3S or CH_3SO_x lessens the calculated temperature dependence in the [MSA]/[SO_4^{2-}] ratio. This is especially the case if it occurs before the formation of CH_3SO_2 in the oxidation mechanism.

(6) Even in the clean marine air, it is unlikely that species such as CH_3SO_2H [$CH_3S(O)$—OH or CH_3SOOH] are formed *via* the involvement of HO_2 radicals.

(7) Heterogeneous reactions of species such as $CH_3S(O)_xOONO_2$ could be key steps in MSA formation.

The simple modelling exercise highlights areas which need further effort:

(1) Measurement of the ambient levels of NO_3 and NO_x in the marine boundary layer and free troposphere is essential. The observed diurnal variation in DMS may be contradictory to the measured levels of NO_x in the marine troposphere.

(2) Determinations of the unimolecular decomposition of CH_3SO_x radicals, *via* experimental or computational investigations, are needed.

(3) The detection of peroxy and peroxy nitrate intermediates in the atmosphere and measurements of MSA as a functions of NO_x and temperature could provide valuable insight into the key aspects of the oxidation mechanism.

(4) Measurement of gas-phase MSA is needed to evaluate the role of heterogeneous processes.

(5) Development of detection techniques, such as photoionization mass spectrometry and chemical ionization mass spectrometry, for various intermediates and endproducts are needed to decipher the DMS oxidation mechanism.

This work was carried out as a part of NOAA's Climate and Global Change Programme. We thank Professor Michael McKee for assistance in calculating rotational and vibrational constants for $CH_3S(OH)CH_3$.

References

1 *Dimethylsulphide: Oceans, Atmosphere and Climate*, ed. G. Restelli and G. Angeletti, Kluwer, Dordrecht, 1992.
2 H. Berresheim, *J. Geophys. Res.* 1987, **92**, 13.
3 H. Berresheim, M. O. Andreae, G. P. Ayers, R. W. Gillett, J. T. Merrill, V. J. Davis and W. L. Chameides, *J. Atmos. Chem.*, 1990, **10**, 341.
4 R. Staubes and H-W. Georgii, *Measurements of Atmospheric and Seawater DMS Concentrations in the Atlantic, the Arctic, and the Antarctic region, Dimethylsulphide: Oceans, atmosphere and climate*, Belgirate, Italy, 1992, p.95.
5 R. W. Gillet, G. P. Ayers, J. P. Ivey and J. L. Gras, *Measurement of Dimethyl Sulfide, Sulfur Dioxide. Methane Sulfonic acid and Non-sea Salt Sulfate at the Cape Grim Baseline Station, Dimethylsulphide: Oceans, Atmosphere and Climate*, Belgirate, Italy, 1992.
6 M. O. Andreae, R. J. Ferek, K. Bermond, K. P. Byrd, R. T. Engstrom, S. Hardin, P. D. Houmere, F. LeMarrec, H. Raemdonck and R. B. Chatfield, *J. Geophys. Res.*, 1985, **90**, 12.
7 I. Barnes, K. H. Becker and N. Mihalopoulos, *FTIR Products Study of the Photolysis of CH_3SCH_3: Reactions of the CH_3S Radical, Dimethylsulphide: Oceans, Atmosphere and Climite*, Belgirate, Italy, 1992.
8 S. Hatakeyama and H. Akimoto, *J. Phys. Chem.*, 1983, **87**, 2387.
9 A. A. Turnipseed and A. R. Ravishankara, *The Atmospheric Oxidation of Dimethyl Sulfide: Elementary Steps in a Complex Mechanism, Dimethylsulphide: Oceans, Atmosphere and Climate*, Belgirate, Italy, 1992, p. 185.
10 G. S. Tyndall and A. R. Ravishankara, *Int. J. Chem. Kinet.*, 1991, **23**, 483.
11 F. Yin, D. Grosjean and J. H. Seinfeld, *J. Atmos. Chem.*, 1990, **11**, 309.
12 W. B. DeMore, S. P. Sander, D. M. Golden, R. F. Hampson, M. J. Kurylo, C. J. Howard, A. R. Ravishankara, C. E. Kolb and M. J. Molina, *Chemical Kinetics and Photochemical Data for use in Stratospheric Modeling*, Jet Propulsion Laboratory, JPL Pub. No. 92-20, 1992.
13 R. Atkinson, D. L. Baulch, R. A. Cox, R. F. Hampson, J. A. Kerr and J. Troe, *J. Phys. Chem., Ref. Data.*, 1992, **21**, 1125.
14 G. L. Vaghjiani and A. R. Ravishankara, *J. Phys. Chem.*, 1989, **93**, 1948.
15 A. A. Turnipseed, S. B. Barone and A. R. Ravishankara, *J. Phys. Chem.*, 1993, **97**, 5926.
16 A. J. Hynes, P. H. Wine and D. H. Semmes, *J. Phys. Chem.*, 1986, **90**, 4148.
17 J. P. D. Abbatt, F. F. Fenter and J. G. Anderson, *J. Phys. Chem.*, 1992, **96**, 1780.
18 A. M. Malleson, H. M. Kellet, R. G. Myhill and W. P. Sweetenham, A.E.R.E. Harwell Publication R 13729, A.E.R.E. Publications, Oxford, 1990.
19 A. J. Hynes, T. Pounds, T. McKay, J. D. Bradshaw and P. H. Wine, *12th International Symposium on Gas Kinetics*, Reading, 1992.
20 F. Turecek, *J. Phys. Chem.*, 1994, **98**, 3701.
21 M. L. McKee, *J. Phys. Chem.*, 1993, **97**, 10971.
22 R. E. Stickel, Z. Zhao and P. H. Wine, *Chem. Phys. Lett.*, 1993, **212**, 312.
23 N. I. Butkovskaya and G. LeBras, *J. Phys. Chem.*, 1994, **98**, 2582.
24 I. Barnes, V. Bastian and K. H. Becker, *Int. J. Chem. Kinet.*, 1988, **20**, 415.
25 A. A. Turnipseed, S. B. Barone and A. R. Ravishankara, *J. Phys. Chem.*, 1992, **96**, 7502.
26 I. Barnes, V. Bastian, K. H. Becker and H. Niki, *Chem. Phys. Lett.*, 1987, **140**, 451.
27 N. R. Jensen, J. Hjorth, C. Lohse, H. Skov and G. Restelli, *J. Atmos. Chem.*, 1992, **14**, 95.
28 D. Laakso, C. E. Smith, A. Goumri, J-D. Rocha and P. Marshall, *Chem. Phys. Lett.*, 1994, **227**, 377.
29 E. S. Saltzman and D. J. Cooper, *J. Atmos. Chem.*, 1988, **7**, 191.
30 M. Legrand, C. Feniet-Saigne, E. S. Saltzman, C. Germain, N. I. Barkov and V. N. Petrov, *Nature (London)*, 1991, **350**, 144.
31 R. Atkinson, J. N. Pitts and S. M. Aschmann, *J. Phys. Chem.*, 1984, **88**, 1584.
32 J. F. Noxon, *J. Geophys., Res.*, 1983, **88**, 11 017.
33 J. A. Logan, *J. Geophys. Res.*, 1983, **88**, 10 785.
34 A. J. Hynes and P. H. Wine, in *Biogenic Sulfur in the Environment* ed. E. S. Saltzmann and W. J. Cooper, American Chemical Society, Washington, DC, 1989.
35 A. R. Bandy, S. B. W. Blomquist, S. M. Chen and D. C. Thornton, *Geophys. Res. Lett.*, 1992, **11**, 1125.
36 B. A. Ridley, M. A. Carroll, D. D. Dunlap, M. Trainer, G. W. Sachse, G. L. Gregory and E. P. Condon, *J. Geophys., Res.*, 1989, **94**, 5043.
37 B. A. Ridley, S. Madronich, R. B. Chatfield, J. G. Walega, R. E. Shetter, M. A. Carroll and D. D. Montzka, *J. Geophys. Res.*, 1992, **97**, 10.
38 M. A. Carroll, B. A. Ridley, D. D. Montzka, G. Hubler, J. G. Walega, R. B. Norton, B. J. Huebert and F. E. Grahek, *J. Geophys. Res.*, 1992, **97**, 10.
39 F. Domine, A. R. Ravishankara and C. J. Howard, *J. Phys. Chem.*, 1992, **96**, 2171.
40 G. S. Tyndall and A. R. Ravishankara in *Biogenic Sulfur in the Environment*, ed., E. S. Saltzman and W. J. Cooper, American Chemical Society, Washington DC, 1989, p. 450.
41 B. A. Ridley and E. Robinson, *J. Geophys. Res.*, 1992, **97**, 10.
42 B. Heikes, *J. Geophys., Res.*, 1992, **97**, 18.

Paper 5/00197H; *Received* 9*th January*, 1995

Faraday Discuss., 1995, **100**, 55–64

Hydrofluorocarbons and Stratospheric Ozone

Timothy J. Wallington and William F. Schneider
Ford Research Laboratory, SRL-3083 Ford Motor Company Dearborn, Michigan 48121-2053, USA

Jens Sehested and Ole John Nielsen
Section for Chemical Reactivity, Environmental Science and Technology Department, Risø National Laboratory, DK-4000 Roskilde, Denmark

Recognition of the adverse environmental impact of chlorofluorocarbons (CFCs)[1] has led to an international agreement to cease their production. Hydrofluorocarbons (HFCs) are important CFC substitutes. An important question regarding HFCs is: what is their impact on stratospheric ozone? While it is well known that HFCs themselves do not react with ozone, questions have been raised regarding the possibility that species formed during the atmospheric oxidation of HFCs could deplete stratospheric ozone.

In March 1993, preliminary work was presented suggesting that CF_3O radicals formed in the degradation of many HFCs could deplete stratospheric ozone.[2] Later the same year it was shown that CF_3O radicals react only slowly with ozone and that HFCs do not destroy ozone through a catalytic cycle involving CF_3O radicals.[3–7] What was not considered at that time was the possible depletion of stratospheric ozone by other species formed in the atmospheric degradation of HFCs.

To place this concern into perspective it is instructive to contrast the magnitude of the natural ozone flux with that of HFC degradation products. The stratosphere contains *ca.* 3×10^9 t of ozone in a dynamic equilibrium. Each day 3×10^8 t of ozone are destroyed by natural processes.[8] If we assume that ozone is distributed approximately homogeneously in a region between 10 and 50 km altitude, then the volume average ozone loss rate is 4×10^6 molecule cm^{-3} s^{-1}. HFC-134a (CF_3CFH_2) is widely used now and other HFCs, such as HFC-32 (CH_2F_2), may be used increasingly in the future. By the year 2020, the annual global demand for HFC-134a and HFC-32 has been estimated to be up to 3×10^5 and 0.9×10^5 t, respectively.[9] (By comparison, the global CFC production in 1986 was 10^6 t.) A reasonable estimate of the future production of HFCs is 5×10^5 t or, assuming an average molecular weight of 100, 5×10^9 moles per year. HFC degradation is initiated by OH radical attack. Taking a typical activation energy for this reaction of 3.5 kcal mol^{-1}, and dividing the atmosphere into a tropospheric region with an average temperature of 278 K and containing 90% of the atmospheric mass and a stratospheric region with an average temperature of 224 K and containing 10% of the atmospheric mass,[10] then *ca.* 2% of HFCs will be degraded in the stratosphere. At steady state the HFC degradation flux is *ca.* 0.5 molecule cm^{-3} s^{-1}. Clearly, the HFC degradation flux is many orders of magnitude less than the natural ozone flux, so that for HFCs to impact stratospheric ozone levels requires an efficient catalytic mechanism.

Three different mechanisms can be envisioned by which HFC degradation products could impact stratospheric ozone. First, fluorine-containing radical species (*e.g.* FO_x, FCO_x or CF_3O_x) could react directly with ozone. Secondly, fluorine-containing radical species could impact ozone indirectly *via* coupling with radical (*e.g.* ClO_x, BrO_x or HO_x)

or reservoir (*e.g.* HCl or $ClONO_2$) species that are important in ozone chemistry. Thirdly, non-radical HFC degradation intermediates [(*e.g.* FNO, COF_2 or $CF_3C(O)F$] could interact directly, or indirectly, with species that are important in ozone chemistry. These possibilities are examined in turn below.

Direct Impact on Ozone *via* Reactions involving Fluorinated Radicals

To assess the potential impact of HFC degradation products on stratospheric ozone requires us to go beyond the simple estimates of HFC and ozone fluxes discussed above. A somewhat more realistic one-dimensional (1D) estimate of the natural ozone flux can be derived using the expression:[10, 11]

$$d[O_x]/dt = 2k_{NO_2+O}[NO_2][O] + 2k_{ClO+O}[ClO][O] + k_{HO_2+O_3}[HO_2][O_3] + k_{OH+O_3}[OH][O_3] + 2k_{O+O_3}[O][O_3] + k_{O+OH}[O][OH] + k_{O+HO_2}[O][HO_2]$$

where k_{X+Y} is the bimolecular rate constant for reaction of species X with Y, and [X] and [Y] are the concentrations of species X and Y. Incorporating species concentration, atmospheric density and temperature *vs.* altitude profiles from the 1D radiative convective photochemical model for mid-latitudes at equinox of Brasseur and Solomon,[10] along with kinetic data for the relevant reactions from DeMore *et al.*,[12] allows us to derive the ozone flux profile shown in Fig. 1. Because heterogeneous reactions are neglected, this approach provides only an estimate of the natural ozone flux. However, it is a reasonable basis against which any HFC-related ozone destruction can be compared. Similarly, the flux of HFC degradation products can be estimated by elaborating on the approximations above. Because the mixing time for HFCs is much less than the lifetime for oxidation, it is reasonable to assume that the HFCs are homogeneously mixed throughout the atmosphere (*i.e.* that the HFC concentrations scale with density). Taking an annual global emission rate of all HFCs of 5×10^5 t, an average molecular weight of 100, an activation energy for attack by OH radicals of 3.5 kcal mol^{-1} (as for HFC-134a, the most important HFC[12]), and the OH radical profile of Brasseur and Solomon,[10] we obtain the HFC degradation flux shown in Fig. 1. As expected from our crude estimate, the flux of HFC degradation products is four to seven orders of magnitude less than the natural ozone flux.

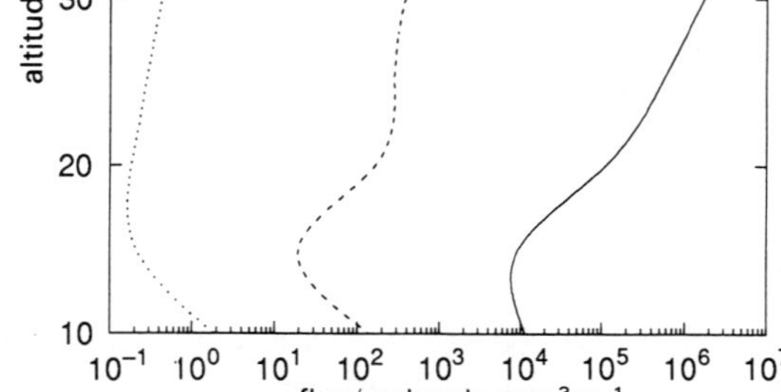

Fig. 1 Flux of HFC degradation (· · · ·), HCl (— —) and O_3 (——) as calculated using 1D model described in the text

HFCs survive for typically 1 to 40 years in the atmosphere before reacting with OH radicals. This reaction produces fluorinated alkyl radicals, whose sole fate is reaction with O_2 to form fluorinated peroxy radicals, RO_2.[13]

$$OH + RH \rightarrow R + H_2O \quad (1)$$

$$R + O_2 + M \rightarrow RO_2 + M \quad (2)$$

Different HFCs produce different peroxy radicals. For example, CH_2F_2 produces CHF_2O_2, while CF_3CFH_2 produces CF_3CFHO_2. As with other peroxy radicals, fluorinated peroxy radicals survive in the atmosphere for a few minutes before reacting with NO, NO_2 or HO_2 radicals to give directly, or indirectly, alkoxy radicals such as CHF_2O and CF_3CFHO.[14] The alkoxy radicals survive in the atmosphere for less than one minute before reacting with O_2 or decomposing through C—C bond cleavage, or both.[13] The fact that the peroxy and alkoxy radicals derived from HFCs do not react rapidly with ozone, but are lost rapidly by other atmospheric reactions, rules out any significant interaction of these species with ozone. During the subsequent oxidation reactions of all HFCs, four groups of radicals are formed that could, in principle, impact stratospheric ozone: CF_3CO_x, $x = 1$–3; CF_3O_x, $x = 1, 2$;[15] FCO_x, $x = 1$–3[16] and FO_x, $x = 0$–2.[17] These $C_xF_yO_z$ radicals could interact with stratospheric ozone directly in a catalytic process:

$$X + O_3 \rightarrow XO + O_2 \quad (3)$$

$$XO + O/O_3 \rightarrow X + O_2/2O_2$$

$$O_3 + O/O_3 \rightarrow 2O_2/3O_2 \quad (4)$$

[X = $CF_3C(O)O$, CF_3O, FC(O)O, FO, or F] as ClO_x, BrO_x, NO_x, HO_x *etc.*, are known to do.

The available experimental data concerning the atmospheric chemistry of $C_xF_yO_z$ radicals show that their direct participation in ozone destruction cycles is not important. Relevant kinetic data are gathered in Table 1. Values in parentheses are calculated lifetimes of the radicals with respect to each reaction based upon the tabulated rate constants and reactant concentrations at 20 km taken from Brasseur and Solomon.[10]

Table 1 Reaction rates and lifetimes of relevant fluorine-containing radicals

radical (X)	k_{X+O_3}[a] (lifetime)[b]	k_{X+NO}[a] (lifetime)[b]	k_{X+NO_2}[a] (lifetime)[b]	k_{X+CH_4}[a] (lifetime)[b]
Cl	$2.9 \times 10^{-11} \exp(-260/T)$ (0.024)	—	—	$1.1 \times 10^{-11} \exp(-1400/T)$ (26)
$CF_3C(O)O$[c]	—	—	—	—
FC(O)O	$<6 \times 10^{-14}$ (>3.4)	1.3×10^{-10} (7.9)	—	—
CF_3O	1.5×10^{-14} (14)	7.0×10^{-11} (15)	9×10^{-12} (62)	3.3×10^{-15} (130)
FO_2	$<3.4 \times 10^{-16}$ (605)	1.47×10^{-12} (700)	1.05×10^{-13} (5300)	$<4.1 \times 10^{-15}$ (>106)
FO	$<2 \times 10^{-16}$ (1030)	2.6×10^{-11} (40)	—	—
F[d]	$2.8 \times 10^{-11} \exp(-230/T)$ (0.021)	—	—	$8 \times 10^{-11} \exp(-400/T)$[e] (0.034)

[a] Units: cm^3 molecule^{-1} s^{-1}. Values taken from ref. 12, 19, 26–29. [b] Lifetime of species (in s) towards reaction with O_3, NO, NO_2 and CH_4 calculated using trace-gas concentrations at 20 km taken from Brasseur and Solomon.[19] [c] Lifetime towards decomposition is <1 s. [d] The rate of reaction of F atoms with O_2 is 4.4×10^{-33} cm^6 molecule^{-2} s^{-1}. The lifetime of F atoms at 20 km towards reaction with O_2 is 3.3×10^{-4} s. [e] The rate constant for the reaction of F atoms with H_2O is 1.4×10^{-11}. The lifetime of F atoms at 20 km with respect to reaction with H_2O is 0.012 s.

As discussed above the natural ozone flux is many orders of magnitude greater than that of HFC degradation products. For direct cycles involving CF_3CO_x, CF_3O_x and FCO_x to constitute a significant ozone-loss mechansim, the rate of the chain propagating reaction of the alkoxy or carboxy radical [$CF_3C(O)O$, CF_3O or $FC(O)O$] with ozone must be orders of magnitude greater than the rate of chain termination reactions. As seen in Table 1, such is not the case. $CF_3C(O)O$ radicals are lost rapidly (within 1 s) by C—C bond cleavage and are not reformed. CF_3O radicals are lost *via* reaction with CH_4 and NO. Reaction of CF_3O with CH_4 produces CF_3OH, which cannot reform CF_3O radicals chemically or photolytically.[18] Reaction of CF_3O radicals with NO gives FNO and CF_2O, neither of which can reform CF_3O. $FC(O)O$ radicals are lost by reaction with NO to give FNO and CO_2, which cannot reform $FC(O)O$ radicals. The efficient, irreversible loss mechanisms for $CF_3C(O)O$, CF_3O and $FC(O)O$ radicals, together with their relative unreactivity towards ozone, preclude any significant direct impact on the ozone layer associated with CF_3CO_x, CF_3O_x and FCO_x radicals.

Consideration of the direct impact of FO_x radicals is complicated by the rapid interconversion between F, FO, FO_2 and FNO (see Scheme 1):

$$F + O_2 + M \rightleftharpoons FO_2 + M \tag{5,–5}$$

$$FO_2 + NO \rightarrow FNO + O_2 \tag{6}$$

$$FNO + h\nu \rightarrow F + NO \tag{7}$$

$$F + O_3 \rightarrow FO + O_2 \tag{8}$$

$$FO + NO \rightarrow F + NO_2 \tag{9}$$

$$F + CH_4/H_2O \rightarrow HF + CH_3/OH \tag{10}$$

Examination of Table 1 suggests that FO_x radicals do not destroy ozone in a catalytic fashion because (i) FO and FO_2 radicals react extremely slowly with ozone and (ii) FO_x radicals are lost efficiently by reaction of F atoms with CH_4 and H_2O. The direct impact of FO_x chemistry on stratospheric ozone has been addressed quantitatively in a modelling study by Sehested *et al.*[19] and was found to be unimportant. While the efficiency of ozone destruction varies with altitude, the number of ozone molecules consumed per F atom was less than three at all altitudes.[19]

The arguments presented above rule out direct involvement of fluorinated radical species in catalytic ozone destruction cycles of any significant length.

Indirect Impact on Ozone *via* Reactions involving Fluorinated Radicals

Indirect involvement of $C_xF_yO_z$ radicals could occur either by reaction with radical (*e.g.* ClO_x, HO_x, BrO_x), or reservoir (*e.g.* $ClONO_2$ or HCl) species from the other chains. An example of such a coupling would be

$$CF_3O + ClO \rightarrow CF_3 + Cl + O_2 \tag{11}$$

A necessary condition for significant interaction of $C_xF_yO_z$ radicals with species from catalytic ozone destruction cycles is that the lifetime of these species with respect to reaction with $C_xF_yO_z$ radicals must be significant relative to that with respect to other loss processes. In this section we will show that this condition cannot be fulfilled because the concentration of $C_xF_yO_z$ radicals in the stratosphere will be too low.

We assume initially that the CF_3O_x and $FC(O)O_x$ radical fluxes can be approximated by the HFC degradation flux, or 0.1–1 molecule cm^{-3} s^{-1} each (see Fig. 1). The actual radical fluxes will be less, because neither CF_3O_x nor $FC(O)O_x$ are produced quantitatively from HFC degradation at all altitudes. Multiplication of the radical fluxes by their atmospheric lifetimes with respect to all reactions provides estimates of their steady-state concentrations. From Table 1, the atmospheric lifetimes of CF_3O and FC(O)O radicals are *ca.* 10–15 s at 20 km. These lifetimes are typical throughout the stratosphere. Hence, we derive an estimate of 1 to 15 molecule cm^{-3} for the maximum steady-state concentrations of CF_3O and FC(O)O radicals. The largest possible rate constant for reaction of CF_3O and FC(O)O radicals with any species in the stratosphere is limited by the frequency of molecular collisions and is 10^{-10} cm^3 $molecule^{-1}$ s^{-1}. Combination of the maximum possible rate constant with our estimate for the maximum steady-state concentrations provides a range of 20 to 320 years for the lower limit of the lifetime of any species in the stratosphere with respect to reaction with FC(O)O or CF_3O radicals. This lifetime is many orders of magnitude greater than those of the radical and reservoir species in the ClO_x, HO_x and BrO_x catalytic cycles, which range from a few minutes to several days. Thus, no significant interaction is possible between FC(O)O or CF_3O radicals and any catalytic cycle. Similar arguments apply for CF_3O_2 and $FC(O)O_2$ radicals, because the lifetimes of these radicals with respect to reaction with NO is *ca.* 2 min.

Calculation of the steady-state concentration of FO_x radicals is complicated by two factors. First, FO_x radicals are formed *via* photolysis of several HFC degradation products. Differences in the rates of photolysis of the various degradation products must be considered. Second, atmospheric FO_x chemistry is rather complex (see Scheme 1) and involves the equilibrium reaction (5,–5), the existence of three different radicals (F, FO and FO_2), and the formation and photolysis of FNO. To arrive at estimates for the steady-state concentrations of FO_x radicals we need to combine information on the FO_x radical lifetimes and the F atom flux from HFC degradation (labelled F_{flux} in Scheme 1).

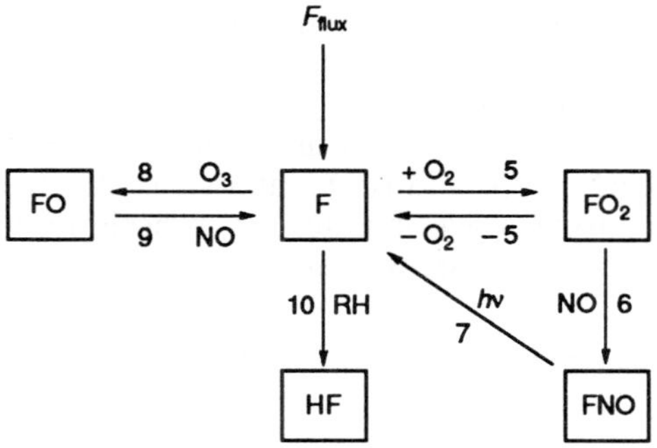

Scheme 1 Numbers refer to reactions in text

FO_x radicals are formed during the HFC degradation process following photolysis of FNO, COF_2, HC(O)F and $CF_3C(O)F$. The rate of photolysis of COF_2 depends upon the altitude and its lifetime *vs.* photolysis decreases from 64 years at 25 km to 0.6 years at 45 km.[20] Air resides for typically 3 years in the stratosphere before returning to the troposphere.[21] Hence, photolysis of COF_2 will be of negligible importance below 25 km. The rates of photolysis of HC(O)F and $CF_3C(O)F$ are expected to be similar to that of COF_2. FNO photolyses rapidly at all altitudes to give F atoms. The atmospheric lifetime of FNO is *ca.* 15 min in the daytime.[22] Below 25 km, photolysis of FNO is the sole source of F atoms and hence FO_x radicals. The only source of FNO in this region is reaction of CF_3O radicals with NO. CH_4 competes for the available CF_3O radicals and produces CF_3OH which is inert in the stratosphere.[18]

The data in Fig. 1 can be used to estimate the F-atom flux at follows. First, by analogy to HFC-134a, which is the most important HFC, the degradation of the 'HFC mix' is assumed to yield CF_3 radicals in a yield of 30% in the stratosphere.[23] At altitudes below 25 km, F_{flux} is 30% of the HFC degradation flux multiplied by that fraction of CF_3O radicals that react with NO. Secondly, at 45 km and above, we assume that all HFC degradation products photolyse to release all their F atoms. Thirdly, each HFC molecule is assumed to contain four fluorine atoms. The F-atom flux from HFC degradation at 45 km is then four times the HFC degradation flux. Fourthly, we assume that the efficiency with which F atoms are liberated at altitudes below 45 km decreases linearly to zero at 25 km. Thus, at 35 km we assume that 50% of the fluorine is liberated during the degradation and so the F-atom flux is twice (0.5 × 4) the HFC degradation flux. With these assumptions the calculated F-atom flux varies from 0.16 to 1.6 atom cm^{-3} s^{-1} over the altitude range 10 to 50 km.

To compute the FO_x radical steady-state concentrations we use the reactions given in Scheme 1. HF is the final product in atmospheric fluorine chemistry. The strength of the H—F bond precludes reaction with all but the most energetic species in the stratosphere, $O(^1D)$ atoms. While reaction of $O(^1D)$ atoms with HF proceeds at essentially the gas kinetic limit (*i.e.* on each collision) the concentration of $O(^1D)$ atoms is so low (0.1–1000 atom cm^{-3} depending on altitude[12]) that this reaction has no overall significance. The F_{flux} term is evaluated as a function of altitude as discussed above. At steady state the following relationships hold: $[F]_{ss} = F_{flux}/k_{10}$; $[FO]_{ss} = k_8[F]_{ss}/k_9$ and $[FO_2]_{ss} = k_5[F]_{ss}/(k_{-5} + k_6)$. k_5, k_{-5}, k_6, k_8, k_9 and k_{10} are the pseudo-first-order loss rates of fluorine-containing species in the corresponding reactions given above. Results are given in Fig. 2. FO_2 is the most abundant FO_x radical species, with a maximum steady-state concentration of the order of 50–500 cm^{-3}.

The upper limit for the rate constant of reaction of FO_2 radicals with any atmospheric species is 10^{-10} cm^3 $molecule^{-1}$ s^{-1}. By comparison, the largest known FO_2 radical bimolecular rate constant is 100 times less (for reaction with NO[19]). Combination of the maximum possible rate constant (10^{-10}) with the steady-state concentration provides a range of 1 to 20 years for the minimum lifetime of any species in the stratosphere with respect to reaction with FO_2 (Fig. 3). Lifetimes with respect to reaction with FO radicals will be 10 to 10 000 times longer. The lifetimes of the radical and reservoir species in the ClO_x, HO_x and BrO_x catalytic cycles due to processes unrelated to HFC chemistry range from a few minutes for radicals to a few days for reservoir species. The data in Fig. 2 and 3 show that there is no possibility of any significant

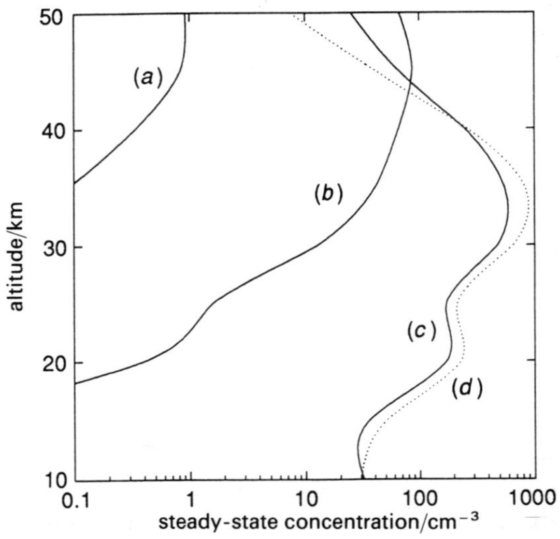

Fig. 2 Concentrations of (*a*) F, (*b*) FO and (*c*) FO_2 radicals calculated using the 1D model described in the text. For comparison, the calculated steady-state FNO concentration is shown (*d*).

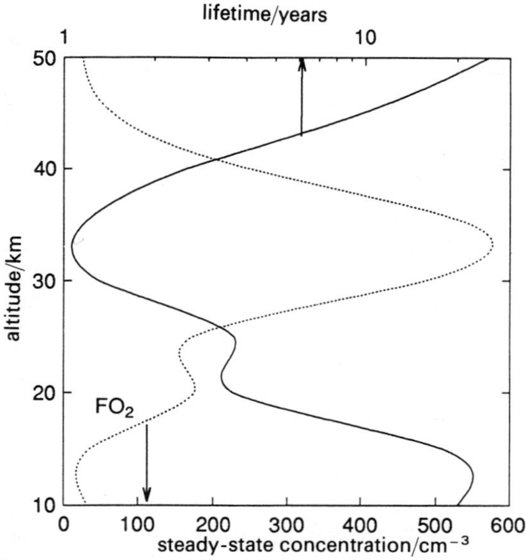

Fig. 3 Minimum lifetime of any species in the stratosphere with respect to reaction with FO_2 radicals. See text for details.

coupling of F atoms or FO radicals with radical and reservoir species from these other catalytic cycles because of the very long lifetimes for reactions of F and FO with the catalytic species (*i.e.* because of their very low concentrations).

As seen from Fig. 3, the concentration of FO_2 radicals peaks at 30 to 35 km. At this altitude the minimum lifetime of any species with respect to reaction with FO_2 radicals is *ca.* 1 year. To place this result into perspective we can consider the possibility of coupling of FO_2 radicals with the ClO_x and BrO_x cycles. HCl is the most persistent species in the ClO_x and BrO_x cycles and so is the most sensitive with respect to a potential reaction with FO_2. HCl reacts with OH radicals to give Cl atoms. Using $k(OH + HCl) = 2.6 \times 10^{-12} \exp(350/T)$ cm^3 molecule^{-1} s^{-1} [12] together with the temperature and OH radical concentration profiles discussed above, the HCl lifetime with respect to OH attack is *ca.* 13 days in the region between 30 and 35 km. By comparing this result with the minimum lifetime of 1 year with respect to reaction with FO_2, it is clear that even for the most sensitive species (HCl) the impact of FO_2 radical chemistry in the stratosphere can be, at most, of minimal (a few per cent) importance. This result is obtained assuming a gas kinetic reaction between FO_2 radicals and HCl; assuming a more realistic reaction rate would drastically reduce the potential impact. In considering the arguments presented above it is relevant to note that air in the stratosphere is recycled back into the troposphere in a time period of typically 3–5 years. Any process taking longer than 3–5 years is unlikely to be important.

Impact on Ozone *via* Reactions involving Non-radical Species

Finally, we must consider the potential impact of non-radical HFC oxidation products. The major long-lived products of the oxidation of all HFCs are FNO, HF, COF_2, HC(O)F and $CF_3C(O)F$. None of these compounds react directly with ozone. Further, none of these compounds are known to react with radical or reservoir species from other ozone-depleting catalytic cycles. As discussed above, the fluorinated radical species do not have a significant impact on stratospheric ozone. For the non-radical HFC degradation products to impact ozone they must participate in reactions that involve species that are relevant in ozone chemistry. We consider the potential importance of these interactions here.

First, as discussed above, FNO atmospheric chemistry is dominated by its rapid photolysis[22] and ultimate conversion to HF. The steady-state concentration of FNO is related to that of FO_2 radicals by the expression $[FNO]_{ss} = k_6[FO_2]_{ss}/k_7$ (see Scheme 1), where k_6 and k_7 are the pseudo-first-order loss rates of FO_2 with respect to reaction with NO, and FNO with respect to photolysis, respectively. Assuming an altitude-independent photolytic lifetime for FNO of 15 min[22] we derive the altitude profile for FNO given in Fig. 2. As seen from Fig. 2 the expected FNO concentration is similar to that of FO_2 radicals. At night, photolysis of FNO ceases and the FNO concentration could build up to equal the sum of the FO_2 radical and FNO concentrations in Fig. 2. In the preceding section it was argued that, because of their low concentration, FO_2 radicals will not couple significantly with radical or reservoir species from the ClO_x, BrO_x or HO_x cycles. The same argument applies to FNO. At the opposite extreme of reactivity, the H—F bond is too robust to be efficiently cleaved either chemically or photolytically in the stratosphere. HF is too unreactive to couple with radical or reservoir species from catalytic ozone cycles.

COF_2, HC(O)F and $CF_3C(O)F$ represent intermediate cases. Their stratospheric lifetimes are determined by photolysis and cycling into the troposphere, depending on altitude, and can be of the order of several years. These lifetimes are long enough that interaction with other ozone cycles may be possible. Interaction with species from the ClO_x, BrO_x or HO_x catalytic ozone cycles cannot be excluded on purely thermodynamic grounds either, as is possible with HF. Because their lifetimes are relatively long, the steady-state concentrations of COF_2, HC(O)F and $CF_3C(O)F$ are potentially rather large. For example, if we assume that COF_2 is formed quantitatively and essentially instantaneously from HFC degradation, and assume that the lifetime for removal of COF_2 from the stratosphere is 3 years, then at 20 km altitude the COF_2 steady-state concentration will be roughly 2×10^7 molecule cm^{-3}. We argued above that significant interaction between fluorine radical species and reservoir species from other chains could not be important because of the short lifetimes and low concentrations of the fluorine radicals. Such an argument clearly cannot be applied to the non-radical species.

We can place some limits on the potential impact of COF_2, HC(O)F and $CF_3C(O)F$, however. First, any atmospherically significant reactions involving these species must involve breaking a C—F, C—C or C—H bond. Because of the oxidizing nature of the atmosphere these bonds cannot be reformed once broken. Hence, none of these species can interact in a catalytic fashion with other ozone-depleting chains. Secondly, any reactions that merely convert these species into fluorine-containing radicals cannot be important. As discussed above, even assuming generous values for the radical fluxes, the fluorine radical species can have no significant impact on ozone. Thirdly, it follows that reactions between the non-radical fluorine species and radical species from ozone-depleting chains cannot be important, because such reactions cannot liberate new ozone-active species. Such reactions would only short circuit the ozone-depleting chains, probably diminishing their effectiveness. We conclude that the only possibly relevant reactions are between the non-radical fluorine species and reservoir species from catalytic chains, such as HCl and $ClONO_2$, with the regeneration of a catalytic ozone-depleting species.

For reactions between non-radical fluorine species and reservoir species to be significant, they must proceed at a rate at least comparable with the rate of removal of the non-radical fluorine species from the stratosphere. For example, HCl is the predominant chlorine reservoir species in the stratosphere, with a concentration of *ca.* 1×10^9 molecule cm^{-3}. For reaction with HCl to be competitive with physical removal of non-radical fluorine species (which takes about 3 years), the rate constant must be of the order of 10^{-17} cm^3 $molecule^{-1}$ s^{-1}. The reservoir species from the catalytic chains are all closed-shell molecules, as are COF_2, HC(O)F and $CF_3C(O)F$. The rate constants for reactions between two closed-shell gas-phase molecules generally lie in the range 10^{-30} to 10^{-50} cm^3 $molecule^{-1}$ s^{-1} at 298 K,[24] clearly many orders of magnitude less than 10^{-17} cm^3 $molecule^{-1}$ s^{-1}. The robustness of the bonds in the fluoro-organics COF_2, HC(O)F and $CF_3C(O)F$, as well as the low temperatures in the stratosphere, makes homogeneous gas-phase reaction between these species and reservoir species extremely unlikely. Heterogeneous reactions cannot be similarly dismissed. However, Hanson and Ravishankara have demonstrated that COF_2 is unreactive on ice surfaces,[25] and it seems reasonable to suppose that HC(O)F and $CF_3C(O)F$ will behave similarly. The available evidence indicates that interaction between non-radical fluorine species and reservoir species will not significantly affect stratospheric ozone.

As further evidence, let us suppose that some homogeneous or heterogeneous reaction between an HFC degradation product and HCl does liberate active chlorine. As mentioned above, HCl is the most persistent of reservoir species in the chlorine cycles and HCl is converted into an active form through reaction with OH radicals. The flux of chlorine atoms generated by this process can be calculated as a function of altitude (Fig. 1), and varies from 20 to 600 molecule cm^{-3} s^{-1}. This flux is two to three orders of magnitude greater than the HFC-degradation flux, depending on altitude. If one assumes that the entire HFC-degradation flux is converted to COF_2, and that all COF_2 reacts with HCl to liberate Cl atoms, then the Cl atoms will be liberated, at most, at a rate that is less than 1% faster than the background HCl + OH rate. Such a small increase in Cl-atom flux would have, at most, a marginal impact on stratospheric ozone concentrations.

Finally, one might imagine other possible minor HFC degradation products, such as $FONO_2$, FONO and FO_2NO_2, that could interact with chlorine reservoir species and be reformed by gas-phase reactions, thus constituting a cycle. However, all such reactions involve FO_x chemistry, the efficiency of which is severely limited by the irreversible formation of HF, as discussed above. Thus these processes will, at most, have very short cycle lengths and are unlikely to be significant.

Conclusion

Based upon the current understanding of the atmospheric chemistry of HFCs we can draw the following conclusions. First, HFCs are much more readily oxidized in the atmosphere than are CFCs, and the stratospheric burden of HFCs will be approximately an order of magnitude less than that of CFCs. HFC degradation products could appreciably impact stratospheric ozone only *via* catalytic cycles. Secondly, of all the known fluorine-containing radical species, only fluorine atoms themselves react at an appreciable rate with ozone, and the concentration of fluorine atoms is sharply limited by the irreversible formation of HF. Fluorine radical species do not participate in direct catalytic ozone-depletion cycles of any significant length and do not impact stratospheric ozone concentrations. Thirdly, the concentrations of all fluorine radical species in the atmosphere can be estimated and will be too low to couple significantly with radicals or reservoirs from other catalytic ozone-depletion cycles. Fluorine radical species do not indirectly impact stratospheric ozone. Fourthly, non-radical fluorine-containing species do not react with ozone. They may conceivably impact radicals or reservoirs from other catalytic ozone depletion cycles. However, this impact is unlikely to be important because it requires either unrealistically fast homogeneous gas-phase reactions between closed-shell species, or heterogeneous chemical reactions for which there is no experimental evidence. Thus, the available scientific data strongly suggest that HFCs pose no threat to stratospheric ozone.

We thank Dr. R. A. Cox (Natural Environmental Research Council, UK) and Dr. Susan Solomon and Dr. A. R. Ravishankara (National Oceanic and Atmospheric Administration, USA) for helpful discussions.

References

1 M. Molina and F. S. Rowland, *Nature (London)*, 1974, **249**, 810.
2 P. Biggs, C. E. Canosa-Mas, D. E. Shallcross, R. P. Wayne, C. Kelly and H. W. Sidebottom, *Proceedings of the STEP–HALOCSIDE/AFEAS Workshop*, University College Dublin, Ireland, March 1993, p. 177.
3 O. J. Nielsen and J. Sehested, *Chem. Phys. Lett.*, 1993, **213**, 433.
4 T. J. Wallington, M. D. Hurley and W. F. Schneider, *Chem. Phys. Lett.*, 1993, **213**, 442.
5 M. M. Maricq and J. J. Szente, *Chem. Phys. Lett.*, 1993, **213**, 449.
6 A. R. Ravishankara, A. A. Turnipseed, N. R. Jensen, S. Barone, M. Mills, C. J. Howard and S. Solomon, *Science*, 1994, **263**, 75.
7 Ch. Fockenberg, H. Saathoff and R. Zellner, *Chem. Phys. Lett.*, 1994, **218**, 21.
8 M. McFarland and J. Kaye, *Photochem. Photobiol.*, 1992, **55**, 911.
9 A. McCulloch, *Environ. Monit. Assess.*, 1994, **31**, 167.
10 G. Brasseur and S. Solomon, *Aeronomy of the Middle Atmosphere*, Reidel, Dordrecht, The Netherlands, 1986.
11 H. S. Johnston and J. Podolske, *Rev. Geophys. Space Phys.*, 1978, **16**, 491.
12 W. B. DeMore, S. P. Sander, D. M. Golden, R. F. Hampson, M. J. Kurylo, C. J. Howard, A. R. Ravishankara, C. E. Kolb and M. J. Molina, *Jet Propulsion Laboratory Publication 92-20*, Pasadena, CA, 1992.
13 T. J. Wallington, W. F. Schneider, D. R. Worsnop, O. J. Nielsen, J. Sehested, W. J. DeBruyn and J. A. Shorter, *Environ. Sci. Technol.*, 1994, **28**, 320.
14 P. D. Lightoot, R. A. Cox, J. N. Crowley, M. Destriau, G. D. Hayman, M. E. Jenkin, G. K. Moortgat and F. Zabel, *Atmos. Environ. A*, 1992, **26**, 1805.
15 M. K. W. Ko, N-D. Sze, J. M. Rodríguez, D. K. Weistenstein, C. W. Heisey, R. P. Wayne, P. Biggs, C. E. Canosa-Mas, H. W. Sidebottom and J. Treacy, *Geophys. Res. Lett.*, 1994, **21**, 101.
16 J. S. Francisco, A. N. Goldstein, Z. Li, Y. Zhao and I. H. Williams, *J. Phys. Chem.*, 1990, **94**, 4791.
17 J. S. Francisco, *J. Chem. Phys.*, 1993, **98**, 2198.
18 T. J. Wallington and W. F. Schneider, *Environ. Sci Technol.*, 1994, **28**, 1198.
19 J. Sehested, K. Sehested, O. J. Nielsen and T. J. Wallington, *J. Phys. Chem.*, 1994, **98**, 6731.
20 A. Nölle, H. Heydtmann, R. Meller, W. Schneider and G. K. Moortgat, *Geophys. Res. Lett.*, 1992, **19**, 281.
21 M. K. W. Ko, N. D. Sze and M. J. Prather, *Nature (London)*, 1994, **367**, 505.
22 T. J. Wallington, W. F. Schneider, J. J. Szente, M. M. Maricq, O. J. Nielsen and J. J. Sehested, *J. Phys. Chem.*, 1995, **99**, 984.
23 T. J. Wallington, M. D. Hurley, J. C. Ball and E. W. Kaiser, *Environ. Sci. Technol.*, 1992, **26**, 1318.
24 F. Westley, D. H. Frizzell, J. T. Herron, R. F. Hampson and W. G. Mallard, *NIST Standard Reference Database 17*, Gaithersburg, Maryland, 1993.
25 D. R. Hanson and A. R. Ravishankara, *Geophys. Res. Lett.*, 1991, **18**, 1699.
26 A. A. Turnipseed, S. B. Barone and A. R. Ravishankara, *J. Phys. Chem.*, 1994, **98**, 4594.
27 S. B. Barone, A. A. Turnipseed and A. R. Ravishankara, *J. Phys. Chem.*, 1994, **98**, 4602.
28 T. J. Wallington and J. C. Ball, *Chem. Phys. Lett.*, 1995, **234**, 187.
29 R. Zellner and H. Saathoff, paper presented at the 12th Int. Symp. on Gas Kinetics, Reading, 1992.

Paper 5/00196J; *Received 6th January*, 1995

Faraday Discuss., 1995, **100**, 65–82

Entry of Gas Molecules into Liquids

P. Davidovits and J. H. Hu†
Department of Chemistry, Merkert Chemistry Center, Boston College, Chestnut Hill, MA 02167, USA

D. R. Worsnop, M. S. Zahniser and C. E. Kolb
Center for Chemical and Environmental Physics, Aerodyne Research, Inc., Billerica, MA 01821, USA

Heterogeneous gas–liquid interactions play a fundamental role in many atmospheric processes. Pivotal for the understanding of such processes is the rate of gas uptake by the relevant liquid. Accurate experimental techniques have been developed to study uptake of gases by liquids. Gas–liquid interactions have already been investigated for more than thirty gas-phase species encompassing two classes of molecules: (1) those that do not react in the liquid but are readily solvated by it and (2) those that have low solubility in the liquid but react in the liquid rapidly to form highly soluble species. Uptake studies of non-reactive gas molecules in water have been explained by a model which treats the water surface as a narrow region of a dense gas-like state within which trace gas–solvent collisions occur. Gas accommodation occurs *via* nucleation of solvent molecules. Uptake studies for reactive gas molecules indicate that, for some species, reaction rates are much more rapid at the liquid interface than in the bulk liquid. The results of non-reactive and reactive uptake experiments are brought together here. The uptake model is briefly described, and the implication of the model for the kinetic nature of liquid surfaces is examined. Reaction-driven uptake studies which provide clear evidence for reactions at the gas/liquid interface are presented. The nature of interfacial reactions and their connection to the uptake model are examined.

Heterogeneous gas–liquid interactions play an important role in various atmospheric processes such as tropospheric ozone formation, destruction of stratospheric ozone and formation of cloud condensation nuclei.[1–3] Such interactions begin with the gas molecule striking the surface of the liquid and entering into the liquid phase. A phenomenological description of the entry of gases into liquids is straightforward. First, the gas-phase molecule is transported to the liquid surface, usually by gas-phase diffusion. The initial entry of the species into the liquid is governed by the mass accommodation coefficient, α, which is the probability that a molecule striking a liquid surface enters into the bulk liquid phase.

$$\alpha = \frac{\text{no. of molecules entering the liquid phase}}{\text{no. of molecular collisions with the surface}} \tag{1}$$

† Present address: Department of Geophysical Sciences, The University of Chicago, Chicago IL 60637, USA.

The mass accommodation coefficient determines the maximum flux, J, of gas into a liquid, which is given by

$$J = \frac{n_g c\alpha}{4} \tag{2}$$

Here n_g is the gas concentration (molecules cm^{-3}) and c is the trace gas average thermal speed.

As the species enters the liquid, a fraction may evaporate back into the gas phase owing to the limited solubility of that species in water. The equilibrium relationship between gas phase (n_g) and solvated concentrations (n_l) is determined by the Henry's law coefficient H (mol dm^{-3} atm^{-1}) as

$$n_g = HR'Tn_l \tag{3}$$

Here R' is the gas constant in units of atm dm^3 mol^{-1} K^{-1}. At equilibrium, the liquid is saturated, and the flux of molecules into the liquid is equal to the rate of desorption of these molecules out of the liquid.

Saturation can be delayed by an irreversible sink for the species in the liquid. Such a sink can be provided by a chemical reaction between the entering species A and an aqueous reactant B, that is:

$$\mathrm{A} + \mathrm{B} \xrightarrow{k_2} \text{(irreversible products)} \tag{4}$$

Here k_2 is the second-order rate coefficient for the reaction. As a result of the above reaction, a steady-state flux of molecules into the liquid can be sustained. The magnitude of this flux is a function of the reaction rate.

Because of the limitations imposed by gas-phase diffusion and liquid-phase saturation, under typical experimental conditions, the net flux of molecules into the liquid is smaller than that given by eqn. (2) and is expressed in terms of a measured uptake coefficient, γ_{meas}, as

$$J = \frac{n_g c\gamma_{meas}}{4} \tag{5}$$

In principle, the procedure for studying gas–liquid interactions is simple. The gas molecules of interest are brought into contact with the liquid and, after a controlled period of gas–liquid interaction time, the amount of gas that entered the liquid is determined. If the area of liquid in contact with the gas is known then the uptake coefficient can be determined. The non-trivial challenge in gas–liquid interaction studies is to design techniques that allow the deconvolution of the several processes involved in the overall uptake of the gas, and in that way determine the basic parameters affecting the uptake.

Over the past several years, a number of techniques have been developed for the study of gas–liquid interactions. The techniques differ in the form of gas/liquid interface, in the length of interaction time, and in the method used to detect the trace molecules.[4] In a collaborative effort between the Chemistry Department at Boston College and the Center for Chemical and Environmental Physics at Aerodyne Research, a droplet train apparatus was developed in which a monodispersed column of droplets (100 to 200 μm in diameter) interacts with the gas molecules of interest in a low-pressure flow reactor. The area of the droplets in the reaction zone is changed in a step-wise manner while the density of the gas in contact with the droplets is monitored either by optical absorption or mass spectrometry. Droplet–trace gas interaction times are controlled and can be varied in the range 2 to 15 ms.

Using this apparatus the uptake by water surfaces of more than thirty gas-phase species has been studied.[5–15] For the purposes of the present discussion, the molecules studied with the droplet apparatus can be divided into two categories; those that are non-reactive and highly soluble in water and those that are relatively less soluble but undergo irreversible aqueous reactions.

The soluble species which include H_2O_2, alcohols and acids among others, all have effective Henry's law coefficients of the order of 100 mol dm^{-3} atm^{-1} or larger. On the ms timescale of the droplet apparatus studies, the rate of gas uptake by the liquid is determined by gas-phase diffusion, Henry's law saturation and the mass accommodation coefficient (α). Experiments were conducted as a function of gas–liquid interaction time, gas density in the reaction zone and droplet temperature. Analysis of the uptake data yields resolved values for the mass accommodation coefficient (α) as a function of temperature. An examination of the results revealed patterns which led to the development of a model which describes the entry of gas molecules into liquid water.[16]

For the less soluble species which include aldehydes, halogens, N_2O_5, $ClONO_2$, O_3 and SO_2, α is not a rate-determining parameter on the timescale of the experiments. The uptake of these species is determined by their Henry's law coefficient and by their rate of irreversible aqueous reactions. In these cases the uptake data yield information about the aqueous reaction rates as a function of relevant additives and as a function of temperature. Experiments show that for some of the molecules in this category, the uptake is entirely consistent with calculations based on bulk aqueous chemistry of the species.[7,12] However, the uptake of some other species, among them SO_2(g) and acetaldehyde[8,11] as a function of pH and gas-phase halogens as a function of halide ion concentration, was not explainable by bulk-phase chemistry. Here the results indicate that, under some conditions, reactions at the gas/liquid interface may dominate the gas uptake and that chemistry at the interface is very different from that occurring either in the gas or in the liquid phase.

In this work we will bring together the results of the non-reactive and reactive uptake experiments. We briefly describe the uptake model, and then examine for the first time what the model may imply about the kinetic nature of liquid surfaces. Results will be presented which provide clear evidence for reactions at the gas/liquid interface. The nature of interfacial reactions and connections to the uptake model will be examined.

Analyses of Uptake Data

The four processes we have discussed in connection with gas uptake, namely gas-phase diffusion, mass accommodation, Henry's law saturation and liquid-phase chemical reaction, interact and, therefore, the description of the transport process involves coupled differential equations. General solutions for such equations are not available. In special cases, however, the equations can be solved.[17]

It is possible to obtain a simple approximate expression for the gas–liquid transport by decoupling the processes. In one such approach the gas–liquid transport is represented by an electrical circuit analogy shown in Fig. 1.[2,18] The dimensionless conductances representing the gas-phase diffusion (γ_{diff}), Henry's law saturation (γ_{sol}), and chemical reaction (γ_{rxn}) processes are given in eqn. (5), (6) and (7), respectively. The coefficient γ_s shown in Fig. 1 represents uptake due to reactions at the interface, and will be discussed later.

$$\frac{1}{\gamma_{diff}} = \frac{c d_f}{8 D_g} - \frac{1}{2} \tag{6}$$

$$\frac{1}{\gamma_{sol}} = \frac{\pi^{1/2}}{2} \frac{c}{4R'TH} \left(\frac{t}{D_l}\right)^{1/2} \tag{7}$$

$$\frac{1}{\gamma_{rxn}} = \frac{c}{4R'TH} \left(\frac{1}{D_l k}\right)^{1/2} \tag{8}$$

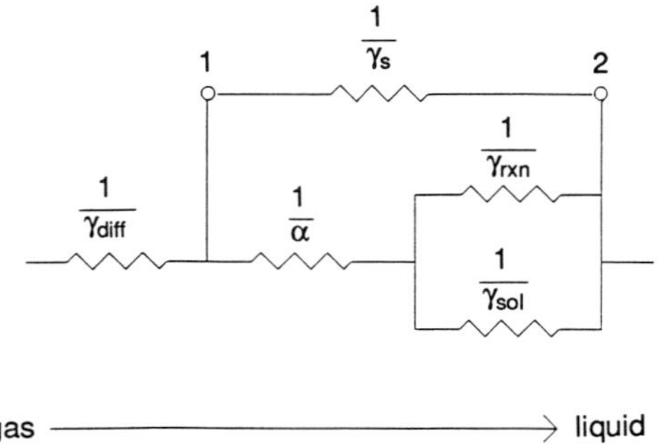

Fig. 1 Electrical circuit analogue for the gas uptake process. Explanation is found in the text.

In eqn. (6), d_f is an effective droplet diameter from the point of view of gas-phase diffusion,[6] and D_g is the gas-phase diffusion coefficient of the species. The term $(-1/2)$ in eqn. (6) accounts for distortion of the Maxwellian–Boltzmann collision rate $(n_g c/4)$ when there is net gas uptake at the surface.[19] In eqn. (7) and (8), t is the gas–liquid contact time, D_l is the liquid-phase diffusion coefficient for the species and k is the pseudo-first-order rate coefficient for aqueous reaction, *i.e.* $k = k_2[\mathrm{B}]$. Here k_2 is the bimolecular rate constant for the trace species with the liquid phase species B. The species B can be an aqueous additive at a concentration [B] or it can be liquid water itself. At high ionic concentrations the activity, a_B, of the aqueous species must be used in calculating the reaction rate. The uptake coefficient γ_{meas} is connected to the transport processes by

$$\frac{1}{\gamma_{meas}} = \frac{1}{\alpha} + \frac{1}{\gamma_{diff}} + \frac{1}{\gamma_{rxn} + \gamma_{sol}} \tag{9}$$

The uptake coefficient γ_{meas} represents the total integrated uptake during the gas–liquid contact time t (*i.e.* from 0 to t). Eqn. (9) represents the uptake under conditions when the various processes are decoupled. The coupling is strongest between the reactive and solubility processes. Therefore eqn. (9) is a good representation of the uptake process only when one of the terms, γ_{rxn} or γ_{sol}, is negligible with respect to the other. Such a condition is usually attainable in our experiments. The applicability of eqn. (9) was demonstrated by Worsnop *et al.*[6]

Uptake of Soluble Species

Formulation of the Process

The mass accommodation coefficients for soluble species obtained in our experiments were measured as a function of temperature. The temperature dependence of the coefficients allowed the determination of the Gibbs energy for the transition state between gas and the solvated state of the species. This in turn made it possible to model the process.

In Fig. 2 the mass accommodation coefficient (α) as a function of droplet temperature is shown for one such non-reactive molecule, ethanol. At droplet temperatures below 273 K the water is super-cooled but not frozen. The negative temperature dependence shown in the figure is typical of gas molecules readily solvated by water. The quality of the data for the other species studied is similar. The solid line in the figure is the best exponential fit to the data in accord with eqn. (13), discussed later. For the purposes of discussions, we show in Fig. 3 uptake coefficients for the set of alkanols studied. To simplify the presentation only the exponential fits to the data are shown. The full set of data for all species studied is found in ref. 5, 7, 10, 13, 14 and 15.

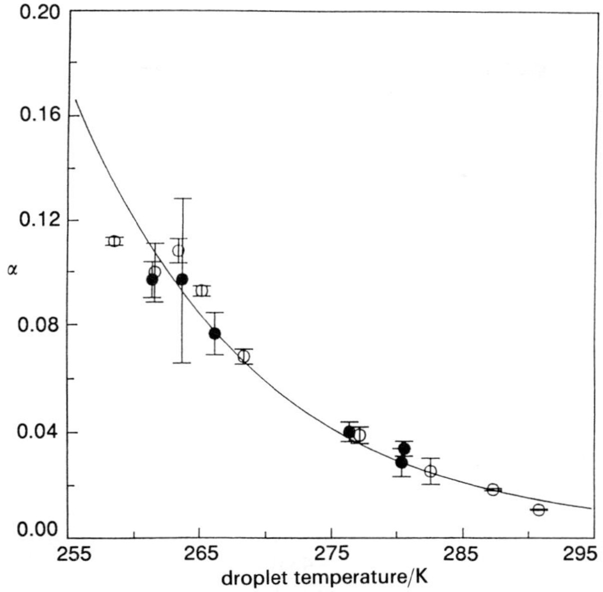

Fig. 2 Temperature dependence of α for ethanol. Solid line is an exponential fit to the data. (From ref. 10.)

Mass accommodation can be viewed as a two-step process. First, the gas molecule strikes the surface and is adsorbed. The surface species then enters the liquid. The process can be represented as

$$n_g \underset{k_{desorb}}{\overset{k_{ads}}{\rightleftarrows}} n_s \xrightarrow{k_{sol}} n_l \qquad \text{(I)}$$

Here the subscripts g, s and l represent the gas, surface and liquid state of the species.

If we assume that the gas molecule striking the liquid surface thermally accommodates with unit probability, then the adsorption rate constant k_{ads} is $k_{ads} = n_g c/4$. In the

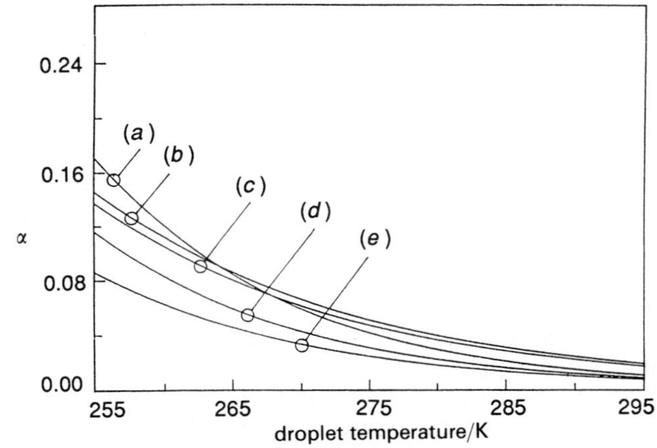

Fig. 3 Temperature dependence of α for the alkanols: (*a*) ethanol, (*b*) methanol, (*c*) *tert*-butanol, (*d*) propan-2-ol and (*e*) propan-1-ol. For clarity only the exponential fits to the data are shown. (From ref. 10).

Table 1 Measured ΔH_{obs} and ΔS_{obs} and thermodynamic solvation parameters $\Delta_{sol} H$ and $\Delta_{sol} S$

molecule	ΔH_{obs} /kcal mol^{-1}	ΔS_{obs} /cal mol^{-1} K^{-1}	$\Delta_{sol} H$ /kcal mol^{-1}	$\Delta_{sol} S$ /cal mol^{-1} K^{-1}
acetone	−12.7	−53.7	−9.84[a]	−17.6
methanol	−8.0	−34.9	−10.25[b]	−17.3
ethanol	−11.0	−46.2	−12.05[b]	−23.5
propan-1-ol	−9.2	−40.9	−13.25[b]	−28.0
propan-2-ol	−9.9	−43.0	−14[c]	−45
tert-butanol	−8.2	−35.8	−16[c]	−52
Cl-ethanol	−7.3	−32.3		
Br-ethanol	−8.4	−35.9		
I-ethanol	−8.2	−34.4		
formic acid	−7.9	−34.9		
acetic acid	−8.1	−34.9		
HCl	−8.8	−34.6		
CCl_3CO_2H	−9.9	−40.0		
CCl_2HCO_2H	−8.0	−33.0		
$CClH_2CO_2H$	−8.4	−34.4		
CF_3CO_2H	−4.5	−20.1		
CF_2ClCO_2H	−7.1	−29.4		
$DMSO_2$ $(CH_3)_2SO_2$	−10.7	−43.0		
DMSO $(CH_3)_2SO$	−5.1	−23.1		
HNO_3	−6.6	−27.6		
ethylene glycol	−5.3	−24.5		
H_2O_2	−5.5	−22.5	−13.1[d]	−15.3
MAS CH_3SO_3H	−3.5	−16.7		

[a] Calculated from data in ref. 13. [b] Ref. 20. [c] J. R. Snider and G. A. Dawson, *J. Geophys. Res.*, 1985, **90**, **D2**, 3797. [d] Calculated from results of J. A. Lind and G. L. Kok, *J. Geophys. Res.*, 1986, **91**, **D7**, 7889.

absence of liquid saturation the uptake process is expressed as

$$\alpha n_g c/4 = n_g c/4 - n_g k_{evap} \tag{10}$$

and

$$\alpha n_g c/4 = n_s k_{sol} \tag{11}$$

From these relationships we obtain

$$\frac{\alpha}{1-\alpha} = \frac{k_{sol}}{k_{desorb}} \tag{12}$$

This equation can be expressed in terms of a measured Gibbs energy $\Delta G^{\neq}_{obs} = \Delta H_{obs} - T\Delta S_{obs}$ as

$$\frac{\alpha}{1-\alpha} = \frac{k_{sol}}{k_{desorb}} = \exp\left(\frac{-\Delta G^{\neq}_{obs}}{RT}\right) \tag{13}$$

The parameter $\Delta G^{\neq}_{obs}$ can be regarded as the Gibbs energy of the transition state between gas-phase and aqueous-phase solvation. The values for ΔH_{obs} and ΔS_{obs} can be obtained from the experimental results by plotting $\ln(\alpha/1-\alpha)$ as a function of $1/T$. The slope of such a plot is $-\Delta H_{obs}/R$ and the intercept is $\Delta S_{obs}/R$.

The expression in eqn. (13) is general. The specific expression for $\Delta G^{\neq}_{obs}$ depends on the theoretical formulation of the uptake process. Therefore the parameter $\Delta G^{\neq}_{obs}$ serves as a bridge between experiment and theory.

The values for ΔH_{obs} and ΔS_{obs} obtained from the experimental results are listed in Table 1 for all the species studied. In this table we also show (where available) thermodynamic parameters $\Delta_{sol} H$ and $\Delta_{sol} S$ of solvation for the species in water at 298 K. To facilitate comparison, these thermodynamic parameters are not in the usual tabulated form. Rather, they refer to the process of transferring a mole of gas into liquid water without volume change. This point is discussed by Ben-Naim and co-workers.[20,21] In this representation the Henry's law constant H is given by $H = (R'T)^{-1} \exp(-\Delta_{sol} G/RT)$. Here R' is the gas constant ($R' = 0.082$ atm^{-1} K^{-1} mol^{-1}, to be distinguished from $R = 1.99$ cal K^{-1} mol^{-1}).

Salient Features of the Results

Several features are evident in the results. First, for all species the mass accommodation coefficient exhibits a negative temperature dependence (*i.e.* ΔH_{obs} is in all cases negative). Further, the magnitudes of ΔH_{obs} are ordered approximately as diols and DMSO < acids < haloethanols < alkanols < acetone. This arrangement seems to be inversely proportional to the expected hydrogen-bonding ability of the species. Secondly, the magnitudes of ΔH_{obs} and ΔS_{obs} do not depend on the size or shape of the molecule. This is clearly evident from a comparison of the results for methanol and *tert*-butanol. The hydrophobic component of the *tert*-butanol is significantly larger than that of the methanol. Specifically the molar volume of butanol is larger by a factor of 2.3, yet $\Delta_{sol} H$ and ΔS_{obs} are the same, within our experimental error, for the two species. Finally, there is a direct relationship between ΔH_{obs} and ΔS_{obs}; a smaller ΔH_{obs} is correlated with a smaller ΔS_{obs}. As can be seen from the data in Table 1, there is no such generally corresponding relationship in the solvation parameters. The ΔH_s value for H_2O_2 is relatively high yet the magnitude of its $\Delta_{sol} S$ is the lowest in the table. This relationship between ΔH_{obs} and ΔS_{obs} is displayed graphically in Fig. 4. The figure also shows calculated values for ΔH_{obs} and ΔS_{obs} based on a model which will be discussed.

Previous Uptake Models

Prior to the mid 1980s very little experimental information was available to aid in the formulation of a model describing the uptake of gases by liquids. In a prevalent conceptualization of the process[22] it was assumed that since the impinging gas molecule does not have sufficient kinetic energy to displace solvent molecules, the molecule can enter the liquid only if a microscopic cavity of appropriate size forms at the surface near its location. Therefore, according to this model, the solvation rate, k_{sol}, would be proportional to the probability of cavity formation which in turn is proportional to $\exp(-\Delta_{cav} G/RT)$ where $\Delta_{cav} G = \Delta_{cav} H - T\Delta_{cav} S$ is the molar Gibbs energy of cavity formation in the liquid. The desorption rate k_{desorb} would be inversely proportional to intrinsic residence time, τ, of the gas molecule on the liquid surface. That is $\tau^{-1} = k_{desorp} = K \exp(\Delta_{surf} G/RT)$ where K is a pre-exponential factor and $\Delta_{sur} G$ is the surface Gibbs energy of the species. Therefore in this formulation of the process, $\alpha/(1-\alpha) = k_{sol}/k_{desorb} = \exp(-G^{\neq}_{obs})/RT = (-\Delta_{cav} G - \Delta_{sur} G/RT)$. This yields, $\Delta H_{obs} = \Delta_{cav} H - \Delta_{sur} H$ and $\Delta S_{obs} = \Delta_{cav} S - \Delta_{sur} S$. As will be shown, predictions based on this model are not in accord with the observed results.

We note that the enthalpy of cavity formation $\Delta_{cav} H$ is a positive number proportional to the size of the cavity required to accommodate the target molecule. The surface enthalpy, $\Delta_{sur} H$ is negative and is expected to be proportional to the solvation energy ($\Delta_{sol} H$) of the species, probably between 1/4 and 1/2 the magnitude of that parameter. While the conventional model does not preclude a negative value for ΔH_{obs}, neither does it suggest that this parameter is in all cases negative as is observed in the experiments.

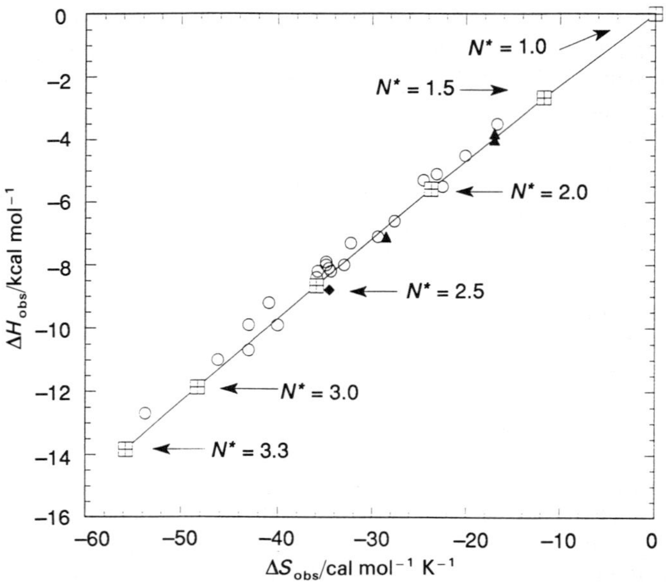

Fig. 4 Experimental and calculated values of ΔS_{obs} and ΔH_{obs}. ◆, HCl; ▲, halogens; ○, other experimental data; ⊞, calculations.

The conventional model does clearly predict that for molecules of comparable size, those which bind more strongly to water will have higher negative values for ΔH_{obs}. This is not the case. Hydrogen peroxide is more strongly bound to water than methanol (compare $\Delta_{sol} H$ values) yet the magnitude of its ΔH_{obs} is significantly lower. In fact, as stated earlier, the magitudes of ΔH_{obs} are inversely proportional to the expected hydrogen-bonding ability of the species.

The prediction of the conventional model regarding the dependence of ΔH_{obs} on the size of the molecule to be accommodated, is less clear. The parameter $\Delta_{cav} H$ increases directly with the size of the molecule. On the other hand while $\Delta_{sur} H$, which is expected to be proportional to $\Delta_{sol} H$, does tend to increase with species size. However, the increase is less than linear. In the sequence of the alkanols shown in Table 1, the molar volume of butanol is 2.3 times larger than that of methanol while the solvation enthalpy is larger by a factor of only 1.6. Therefore, one might expect a trend toward more positive values (smaller negative values) of ΔH_{obs} with increasing size of the species. No such trend is observed. Finally, the conventional model does not lead explicitly to the monotonic relationship between ΔH_{obs} and ΔS_{obs} shown in Fig. 4.

The following model based on nucleation theory provides an explanation for the observed uptake results.

New Model for the Gas Uptake Process

Both experimental[23,24] and modelling studies[25–28] show that the surface of water is a sharp but finite transition region several molecular diameters in thickness within which the density decreases from liquid to gas phase. The new model uses the concept that this narrow region is a dense gas-like state within which nucleation is continually occurring, forming clusters of varying size. The equilibrium density of clusters comprised of N molecules is proportional to $\exp(-\Delta G_N/RT)$ where ΔG_N is the molar Gibbs energy for the formation of clusters with N molecules. Nucleation theory shows that ΔG_N first increases with cluster size, reaches a maximum and then decreases. Therefore, the driving force is such that clusters smaller than a critical size (N^*) evaporate, whereas clusters larger than the critical size serve as condensation centres, grow in size and merge with the nearby bulk liquid. We suggest that gas uptake proceeds *via* the growth of critical clusters. The incoming gas molecules participate in this nucleation process. If a given molecule becomes part of a critical-sized cluster then it will be incorporated into the bulk liquid *via* condensation growth.

In this model, the ease with which a molecule can enter water depends on its ability to act as a centre for nucleation of water at the interface. The critical cluster consists of a specific number of molecules, N^*, which is the sum of the target molecule plus the additional number of water molecules required to form the critical cluster leading to cluster growth and subsequent uptake. This number, N^*, required to form a critical cluster, depends on the structure of the specific molecule undergoing the process of uptake. Thus, for example, ethylene glycol with two OH functional groups is a more effective nucleating agent than a simple alcohol with only one OH. The critical cluster size N^* for ethylene glycol is therefore smaller than for a simple alcohol. Consequently, a critical cluster is more readily formed around the former than the latter. The number of water molecules that have to collaborate with the incoming molecule to form a critical cluster is ($N^* - 1$). Thus a value of $N^* = 1$ implies that the molecule itself is in effect a critical cluster leading to growth by condensation. The parameters ΔH_{obs} and ΔS_{obs} are simply related to the enthalpy and entropy of forming a cluster of size N^*. A quantitative formulation of this model using simple clustering theory is described by Davidovits *et al.*[16]

In terms of transition-state theory, k_{sol} and k_{desorb} in eqn. (14) can be expressed as

$$k_{sol} = \frac{kT}{h}\exp\left(\frac{-\Delta G_{N^*}}{RT}\right) \quad \text{and} \quad k_{desorb} = \frac{kT}{h}\sum_{N=1}^{N^*-1}\exp\left(\frac{-\Delta G_N}{RT}\right) \tag{14}$$

Here k and h are the Boltzmann and Planck constants, respectively, and the term N^* is the number of molecules in the critical-sized cluster which grows to merge with the bulk liquid. For simplicity we will approximate the summation in eqn. (14) by the largest term, that is $N = 1$ which yields $k_{sol}/k_{desorb} = \exp(\Delta G_{N^*} - \Delta G_1)/RT$. From this relationship one obtains $\Delta G^{\neq}_{obs} = \Delta G_{N^*} - \Delta G_1$. A modified formulation for the Gibbs energy of a liquid cluster as discussed in ref. 16 yields expressions for ΔH_{obs} and ΔS_{obs}. In that study, the temperature dependence of the surface tension was not taken into account. Here, we expand the treatment by using the temperature-dependent expression for the surface tension β as $\beta = E^s - TS^s$. For water $E^s = 2.7 \times 10^{-6}$ cal cm^{-2} and $S^s = 3.3 \times 10^{-9}$ cal cm^{-2} K^{-1}. (Ref. 29, p. 57)

$$\Delta H_{obs} = (N^* - 1)\Delta_c H + (N^{*2/3} - 1)\left[4\pi E^s N_A\left(\frac{3V_m}{4\pi N_A}\right)^{2/3}\right] \tag{15}$$

and

$$\Delta S_{obs} = (N^* - 1)[\Delta_c S + R\ln(p/MR'T)] + (N^{*2/3} - 1)\left[4\pi S^s N_A\left(\frac{3V_m}{4\pi N_A}\right)^{2/3}\right] \tag{16}$$

Here N^* is the number size of the critical cluster and V_m is the molar volume of the condensing liquid (in this case water). The parameters $\Delta_c H$ and $\Delta_c S$ are the enthalpy and entropy of condensation obtained from the Gibbs energy of condensation for water ($\Delta_c G = \Delta_c H - T\Delta_c S$). As before, here $\Delta_c G$ is not the usually tabulated standard Gibbs energy of condensation. Rather, it is the Gibbs energy of transferring a mole of gas into liquid without volume change.[20,21] In this case equilibrium vapour pressure is given by $p_0 = MR'T\exp(\Delta_c G/R'T)$. Here M is the molarity of the liquid. (For water M = 55.5 mol dm^{-3}). The parameter, p, is the water vapour pressure in the surface region where condensation clustering is occurring. It is interpreted as the water vapour pressure (or equivalently the water density) in the surface region where the incoming species is stopped by collisions and participates in the nucleation process. It is assumed that p for all trace species is the same. A value of p valid for all species can be obtained from any of the measured ΔH_{obs} and ΔS_{obs} pairs. The value for this pressure obtained by averaging the entries in Table 1 is 0.33 atm.

Substituting numerical values as given in ref. 16 for the parameters in eqn. (15) and (16), yields

$$\Delta H_{obs} = -10(N^* - 1) + 7.53(N^{*2/3} - 1) \text{ kcal dm}^3 \text{ mol}^{-1} \tag{17}$$

$$\Delta S_{obs} = -29.5(N^* - 1) + 9.21(N^{*2/3} - 1) \text{ cal dm}^3 \text{ mol}^{-1} \tag{18}$$

It is evident that the formulations for ΔH_{obs} and ΔS_{obs} in eqn. (15) and (16) are in accord with the measurements. The values of the parameters are always negative as is observed. The magnitudes of the parameters do not depend on the size of the gas molecule. Rather, they are a function of the number of water molecules that have to be added to the target molecule in order to form a critical-sized cluster. Finally, the monotonic relationship between ΔH_{obs} and ΔS_{obs} is explicit in the formulation. We also note that in this formulation the uncomplexed gas molecule at the interface is taken as the reference state. Therefore $\Delta G^{\neq}_{obs}$ can be regarded as the Gibbs energy of the transition state between the interfacial trace molecule and the aqueous-phase solvated trace molecule.

Calculated values of ΔH_{obs} and ΔS_{obs} for selected values of N^* are shown in Table 2. These calculated values together with the measurements are plotted in Fig. 4. As can be seen the fit is remarkably good.

While individual critical clusters must contain an integral number of molecules, we interpret non-integer experimental N^* values as representing the average number of molecules in critical clusters across the interfacial region. Critical clusters formed in the region closer to the gas phase where the interfacial density is lower, will have larger N^* values than those formed in the region closer to the bulk liquid.

It should be noted that the formulas for ΔH_{obs} and ΔS_{obs} shown in eqn. (15) and (16) are approximations based on expressing k_{desorb} by the first term in the summation, that is the term with $N = 1$. This approximation is physically meaningful only if N^* is 2 or greater. In a case where $N^* = 1$, that is, when the incoming gas molecule can itself act as a critical cluster, k_{desorb} is in fact zero [see eqn. (14)] and α is unity in accord with eqn. (13). A more accurate but also more complex formulation takes this limit into account and is discussed in ref. 4.

Our treatment of nucleation raises several questions. The nucleation theory used here is highly simplified and it is doubtful that the bulk liquid-phase values for $\Delta_c S$, $\Delta_c H$ and β correctly represent the properties of the small clusters found in the vapour/liquid interface. The very concept of surface tension in such small clusters is certainly problematic. The justification for this simplified approach and other related issues are discussed in ref. 16 and citations therein.

Table 2 Calculated values for ΔH_{obs} and ΔS_{obs} obtained *via* eqn. (17) and (18)

N^*	ΔH_{obs}/kcal mol^{-1}	ΔS_{obs}/cal mol^{-1} K^{-1}
1	0	0
1.5	−2.7	−12
2	−5.6	−24
2.5	−8.7	−36
3	−11.9	−48
3.3	−13.8	−56

Kinetic Nature of the Aqueous Vapour/Liquid Interface

Over the past several years both experiments and theory have successfully begun to probe the interface between liquid water and its vapour. Surface-selective non-linear spectroscopy experiments have shown that *ca.* 25% of the hydrogen bonds at the surface of water are broken[30,31] and, as stated earlier, experiments and theoretical calculations indicate[23–28] that the transition between liquid and vapour phase occupies a region of several molecule diameters in thickness. So far, however, information about the kinetics of the interfacial region is not available. The clustering model may shed some light on this issue.

The mass accommodation coefficient for gas-phase water molecules on liquid water has so far not been measured. However, it is reasonable to assume that the clustering mechanism applies also to water itself. The values for ΔH_{obs} and ΔS_{obs} are likely to be in the region between those for H_2O_2 and the alkanols with a corresponding average critical cluster size N^* of *ca.* 2.3. Using the clustering model and detail-balance considerations, we can obtain a picture about the nature of the water molecules at the interface and about the mechanism of water evaporation.

In our earlier calculations of the clustering rate we assumed that the molecules in the interfacial transition region between liquid and gas behaved as a gas-phase system. In fact, the molecular density in the interfacial region is high and therefore the interfacial molecules are, on the average, likely to be in some way interconnected and bound. These interconnections may assume various configurations and aggregations. In this context a cluster can be thought of as a density fluctuation or a local tightening of the bond configurations.

In our initial discussion of the uptake process we assumed that in the first step a gas molecule which strikes the surface is adsorbed and becomes part of the surface species n_s. [See reaction (I).] From this perspective, the species n_s can be thought of as water molecules most weakly interconnected within the interface. Equilibrium water vapour consists primarily of monomers. Therefore, detailed-balance considerations lead to the assumption that water molecules leave the interface (evaporate) primarily as monomers *via* the breakage of interfacial bonds associated with the species n_s.

According to our model, a molecule (including a water molecule) moves from the interface into the bulk liquid *via* the merging of a critical cluster with the bulk liquid. Therefore, it seems reasonable to assume that the condensation–evaporation process involves primarily only two of the various interfacial configurations of water molecules; the most weakly interconnected species, n_s, and the critical clusters, n_s^*. In accord with detailed balance, evaporation must begin with the spontaneous emergence of a critical cluster from the bulk. Likewise since, according to the model, the critical clusters are formed by the aggregation of surface species, n_s, the species n_s must be formed by the dissociation of critical clusters. In other words a barrier exists to the direct formation of species n_s from the bulk liquid. A postulated Gibbs energy diagram for the relevant species in the region between liquid and vapour is shown in Fig. 5. The term $(\Delta_s G^* - \Delta_s G)$ is associated with the critical cluster energy which earlier we equated to $\Delta G^{\neq}_{obs}$. In this conceptualization, the surface tension is understood to be the average Gibbs energy of all the water molecules in their various configurations in the interfacial region. Likewise, the laser spectroscopy experiments which yield information about surface OH probe the average properties of the interfacial species.

We note that in the formulation of the gas-uptake process it has been assumed that every gas molecule that strikes the surface thermally accommodates and sticks, *i.e.* becomes a surface-adsorbed species n_s with unit probability. Since the barrier between n_s and the liquid (*i.e.* $\Delta_s G^* - \Delta_s G = \Delta G^{\ddagger}_{obs}$) is always positive, as expected the mass accommodation coefficient α is always less than 1 [see eqn. (13)].

In recent experiments Robinson *et al.*[32] measured the uptake of gas-phase HCl by

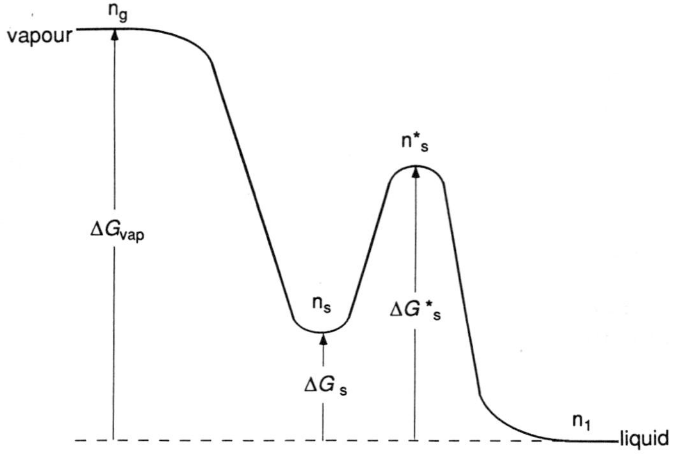

Fig. 5 Postulated Gibbs energy diagram for the liquid/vapour interface

aqueous sulfuric acid as a function of sulfuric acid concentration and temperature. Experiments were performed over a temperature range from 230 to 290 K and for 40, 50 and 60 wt.% sulfuric acid solutions. (This corresponds to respectively 0.11, 16 and 0.22 mol fraction of sulfuric acid.) Within the experimental accuracy (*ca.* 20%), the mass accommodation coefficient was found to be independent of the sulfuric acid concentration. The values for ΔH_{obs} and ΔS_{obs} are 8.8 kcal mol^{-1} and 34.6 cal mol^{-1} K^{-1}, respectively. These values are shown in Fig. 4 (◆). As can be seen, these results are also in accord with the clustering model for $N^* = 2.5$. This implies that even at the very low water vapour pressures over the cold sulfuric acid solutions, the interfacial region consists of water molecules and the kinetics of the surface is determined by this composition.

Uptake of Reactive Species

As was mentioned, the uptake of several species under reactive uptake conditions has been studied with the droplet-train technique. In some cases the measured uptake is in accord with bulk-phase species reaction while in other cases it is not. The Henry's law coefficient for several of the reactive species studied is less than 3 mol dm^{-3} atm^{-1}. With such small values of H, γ_{sol} is less than 10^{-4} on the ms timescale. Therefore the uptake due to γ_{sol} is not detectable with the droplet-train apparatus and can be neglected in the representation of the uptake process. If the uptake is governed by bulk-phase species reaction then upon substitution, and with γ_{sol} neglected, eqn. 9 becomes,

$$\frac{1}{\gamma_{meas}} = \frac{1}{\alpha} + \frac{1}{\gamma_{diff}} + \frac{c}{4RTH(kD_l)^{1/2}} \tag{19}$$

Here k is the pseudo-first-order rate constant. For an ionic solution $k = k_2(a_{[X^-]})$, where $a_{[X^-]}$ is the activity‡ of the liquid-phase ionic species X^-.

The uptake of O_3 by droplets containing I^- (from a KI solution) is an example of a system where the results are in accord with bulk-phase species reaction.[33] Since the Henry's law coefficient for O_3 is small, 0.02 mol dm^{-3} atm^{-1}, the uptake is predominantly due to the O_3–I^- reaction. In these studies, the I^- concentration ranged up to 3.0 mol dm^{-3} and the relationship between $1/\gamma_{meas}$ and $1/a_{I^-}^{1/2}$ remained linear over the

‡ In a highly concentrated ionic solution, activity $a_{[X^-]}$ rather than concentration $[X^-]$ is the appropriate measure of reactivity. The relationship between activity and concentration of the ionic solution is $a_{[X^-]} = \gamma[X^-]$, where γ is the activity coefficient. Activity coefficients are available for most common ions.

full range of I^- concentrations. The slope of the line gave a value for the second-order rate coefficient for the $O_3 + I^-$ reaction, $k_2 = 4 \times 10^9$ dm^3 mol^{-1} s^{-1} at 277 K. Within the experimental accuracy, this is in agreement with the value of 2×10^9 dm^3 mol^{-1} s^{-1} at 298 K given by Hoigne *et al.*[34]

Reactive uptake of SO_2 as a function of H_2O_2 concentration[7] at low pH and the uptake of formaldehyde as a function of OH^- and SO_3^{2-} concentrations[11] were also in accord with bulk-phase species reactions. In these cases the second-order reaction rate constants obtained from the uptake measurements were again within a factor of two of the values measured by more direct experimental techniques. Considering that the uptake is a function of the square root of the rate constant, a factor of two agreement is in the range of expected uncertainty.

The first observation[6] of a reactive uptake not in accord with bulk-phase species reaction was the SO_2 uptake by water at high and low pH. At high pH the uptake of SO_2 is governed by the reaction of SO_2 with H_2O to form H^+ and SO_3^-. The bulk water rate for this reaction has been measured by Eigen *et al.*[35] to be 3.4×10^6 s^{-1}. With this value of k and α assumed to be unity, the maximum calculated uptake coefficient (γ_{meas}) is 0.03. At high pH the measured value of the coefficient was 0.1. This implies that in the interfacial uptake process the SO_2–H_2O reaction rate is at least an order of magnitude higher than measured in the bulk phase. Likewise, at low pH (<2) the uptake was significantly greater than predicted. These observations together with measurements of uptake as a function of time, led to the conclusion that at the gas/liquid interface the SO_2–H_2O reaction is facile, forming an HSO_3^-–H^+ surface complex which is in equilibrium with the gas-phase SO_2. The SO_2 molecule enters the bulk liquid *via* this complex.[6] Recently, using second-harmonic generation and static surface tension measurements, Donaldson *et al.* presented direct experimental evidence for the existence of such a surface-bound partially hydrated SO_2 species.[36]

While the uptake of formaldehyde is in accord with the liquid-phase solubility and reaction kinetics of that species, the uptake of acetaldehyde at short gas–droplet contact times is significantly greater than predicted by the calculations based on bulk-phase parameters.[11] The time-dependent studies of the uptake suggest that here, as in the case of SO_2 at low pH, the greater uptake is due to a chemisorbed surface complex at the gas/water interface.

The reactive uptake of gas-phase Cl_2 and Br_2 by aqueous NaBr and NaI solutions are the most recent and detailed studies showing deviations from bulk-phase kinetics.[33] The gaseous halogen uptake was studied as a function of halide ion concentration (2.5×10^{-4} to 0.5 mol dm^{-3}), temperature (263–293 K) and gas–liquid interaction time (2–15 ms). Here again, the solubility of the halogen molecules X_2 (X = Cl or Br) is low, and therefore, the measured uptake is primarily due to the aqueous reaction of the species with the halide ion Y^- (Y = Br or I) *via* $X_2 + Y^- \rightarrow XY + Y^-$. Since the products are relatively insoluble halogen or interhalogen molecules, they partially evaporate back into the gas phase and are observed mass spectrometrically.

In Fig. 6(*a*) we plot the uptake coefficient γ_{meas} for Cl_2 at 293 K as a function of Br^- activity. The broken line in Fig. 6(*a*) is the calculated uptake coefficient obtained *via* eqn. (19) using a recently obtained value for the $Cl_2 + Br^-$ bulk aqueous-phase reaction rate coefficient of $k_2 = 7.7 \times 10^9$ dm^3 mol^{-1} s^{-1} at 298 K.[37] This rate coefficient is very close to the diffusion-limited rate calculated in terms of the water viscosity coefficient η ($k_2 = 8RT/3\eta$), which is 1×10^{10} dm^3 mol^{-1} s^{-1} at 298 K. Since the broken-line plot was calculated with $\alpha = 1$, it represents the maximum uptake attributable to bulk liquid-phase reaction. An unlikely large reaction rate coefficient of *ca.* 10^{11} dm^3 mol^{-1} s^{-1} would be needed to match the experimental results *via* eqn. (19).

Another aspect of this discrepancy becomes apparent when the experimental results are plotted in the form $1/\gamma_{meas}$ *vs.* $1/a^{1/2}_{[Br^-]}$ as shown in Fig. 6(*b*). The broken line in Fig. 6(*b*) corresponds to the broken line in Fig. 6(*a*). The relationship between the experimen-

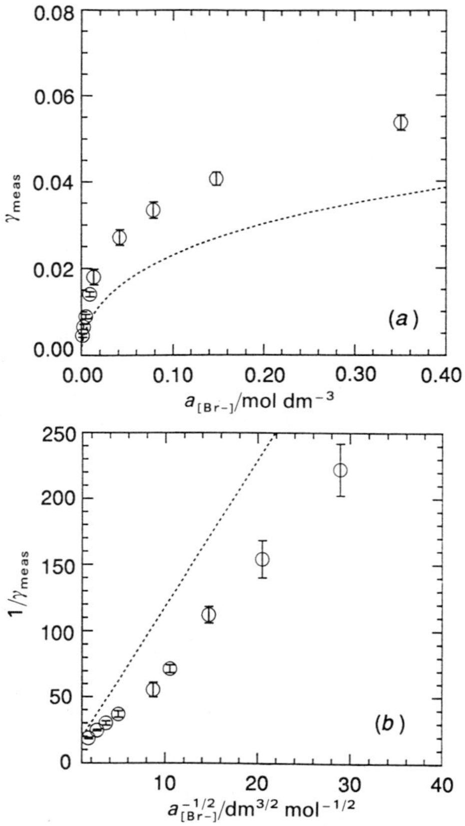

Fig. 6 (*a*) Uptake coefficient γ_{meas} for Cl_2 as a function of Br^- activity at 293 K. (*b*) $1/\gamma_{meas}$ *vs.* $a_{[Br^-]}^{-1/2}$ for Cl_2 at 293 K. The ion concentrations at the left and right extreme points of the plot are 0.5 and 1.25×10^{-3} mol dm^{-3}. The broken lines are the maximum calculated uptake coefficients resulting from bulk-phase reaction of Cl_2 with Br^-. (From ref. 33.)

tal $1/\gamma_{meas}$ and $1/a_{[Br^-]}^{1/2}$ is linear at low Br^- concentrations with the slope corresponding to k_2 in the expected magnitude range. However, at higher Br^- concentrations, the measured data clearly depart from this line. Similar results were obtained for the Cl_2–Br^- uptake at all the temperatures studied and also for the Cl_2–I^- and Br_2–I^- uptake. As is evident, the results of the X_2–Y^- uptake studies are not consistent with a model based on a solely bulk-phase reactive uptake described by eqn. (19).

Modelling the Cl_2 and Br_2 Surface Uptake

The Cl_2 and Br_2 uptake results can be explained by assuming that an additional channel participates in the uptake process. This channel involves a direct reaction of the halogen molecule (X_2) with the halide ion (Y^-) at the interface. That is for Cl_2–Br^- as an example

$$Cl_2(g) + Br^- \rightarrow (Cl_2Br)^-(\text{interface}) \rightarrow Cl^-(aq) + \text{products} \qquad (II)$$

For several reasons, the rates of such interfacial reactions may be expected to be more rapid than the rates of bulk-phase reactions. First, facile reactions that are diffusion controlled in the bulk may be more rapid at the interface where transport of the gas-phase molecule (X_2) to the aqueous reagent (Y^-) at the interface is not limited by liquid-phase diffusion. Secondly, the activity of the solvated molecules at the interface may be higher than in the bulk. Specifically the concentration of the halogen species [X_2(aq)] at the interface may be larger than given by Henry's law equilibrium. Such an enhanced concentration of a polar solute in water has been predicted and observed.[38] For butanol the concentration in the interface is at least 10 times higher than the equilibrium concentration in the bulk. While surface concentration studies for the less soluble non-polar species are not available, one might suggest from entropic arguments that in these cases surface enhancement is likely as well. Finally, a new reaction mechanism may be active at the interface, since in this region reaction hindrance due to steric effects is likely to be reduced and solvent reorientation necessary to achieve reaction may proceed more readily.

There are several possible ways of representing the effect of interfacial reactions on gas uptake. A feature of the surface uptake is its linear dependence on ion activity rather than the square-root dependence exhibited by the bulk reaction dependent uptake. Here we will represent the surface-enhanced uptake by the uptake coefficient, γ_s, which is simply connected as the resistor γ_s^{-1} across the terminals 1 and 2 in Fig. 1. The experimental results lead us to assume γ_s of the form

$$\gamma_s = p_s \frac{C_s a_{Y^-}}{1 + C_s a_{Y^-}} \tag{20}$$

Here C_s is a term proportional to the surface reaction rate. The denominator in the expression for γ_s takes into account possible saturation of the surface at high ion concentrations. This formulation is similar to one obtained from the Eley–Rideal type surface mechanism.[39] The halogen uptake studies present sufficient information to allow a semi-quantitative modelling of the surface-enhanced uptake process *via* this representation.

The introduction of the probability parameter, p_s with possible values between 0 and 1, into eqn. (20) is mandated by experimental observations. Without p_s, eqn. (20) implies that γ_s approaches unity at high ion concentrations. If that were the case, γ_{meas} would approach γ_{diff}. However, at high ion concentration the asymptotic value for γ_{meas} is less than γ_{diff}. The probability parameter p_s takes this into account.

The introduction of the p_s seems also reasonable. Chemical reaction at the interface between the ion and the halogen molecule is not likely to occur unless the halogen molecule is hydrated. We identify the parameter p_s as the probability of such a hydration. It is likely to be different from the mass accommodation coefficient α. The parameter, α, is the probability of hydration by pure water, while p_s is the hydration probability for the gas molecule which strikes the surface within reaction distance of the ion. Hydration in the latter case may be aided by the reacting halide ion at the interface, probably *via* a charge-induced dipole moment. In the current surface uptake model, we assume that p_s has a temperature dependence of the form

$$p_s = \exp(-\Delta_{su} G^{\neq}/RT) \tag{21}$$

As in the case of α, here also we associate $\Delta_{su} G^{\neq}$ ($\Delta_{su} G^{\neq} = \Delta_{su} H - T\Delta_{su} S$) with the Gibbs energy of the transition state toward hydration.

The measured quantity is γ_{meas}, which is obtained as a function of ion concentration and temperature. The unknown quantities are the mass accommodation coefficient, α, the surface parameters C_s and p_s, used to define γ_s, and the bulk liquid-phase reaction rate coefficient, k_2. (k_2 has been independently measured only for the Cl_2–Br^- reaction.) These parameters are extracted from the measured results using a non-linear least-square fit method. p_s was obtained by a global fit of the temperature-dependent data. Details of the modelling are described elsewhere.[33]

The two principal parameters that characterize the surface reaction are C_s and p_s. The parameter C_s can be expressed in terms of the ion–halogen surface-reaction cross-section, σ_s, which in turn can be written in terms of the geometric ion–halogen cross-section σ_{Y^-} as $\sigma_s = S\sigma_{Y^-}$. The geometric cross-section σ_{Y^-} can be calculated from the physical radii of the molecule (X_2) and the ion (Y^-). For the Br_2–I^-, Cl_2–I^- and Cl_2–Br^- reactions, the cross-sections are 3.2, 2.9 and 2.5×10^{-15} cm^2, respectively. The surface-reaction cross-section is highest for the Br_2–I^- reaction with the σ_s/σ_{Y^-} ratio, $S = 7.7$. For the Cl_2–I^- and Cl_2–Br^- reactions S is 3.3 and 0.33, respectively.

p_s is expressed with reference to eqn. (21) in the terms of $\Delta_{su}H$ and $\Delta_{su}S$. The relationships between $\Delta_{su}H$ and $\Delta_{su}S$ obtained for both Cl_2 and Br_2 are in accord with the cluster model of uptake and are included in Fig. 4. The critical cluster size for Br_2 is *ca.* 2.3 and for Cl_2 it is 1.8. As mentioned earlier, p_s is considered to be the probability of a halogen molecule to be hydrated at the interface owing to the reacting halide ion. The hydration in this case is essentially a cluster formation process about an ion-polarized halogen molecule. This view is consistent with the fact that the magnitude of the measured $\Delta_{su}H$ and $\Delta_{su}S$ values is characteristic of hydrophilic polar species rather than non-polar species such as an unperturbed halogen.

Finally, the question arises: why do certain species exhibit enhanced surface reactivity while others do not? We have examined several parameters that may provide clues to this difference. Systems with enhanced surface reactivity seem to possess additional reaction channels or possibilities of forming stable ionic complexes. For example, acetaldehyde which exhibits enhanced surface reactivity can undergo aldol condensation *via* a reaction with OH^- to form the enolate ion, whereas formaldehyde cannot. Likewise SO_2 reacts with H_2O to form the HSO_3^-–H^+ complex. A halogen molecule can form a stable complex with a halide ion whereas ozone cannot. In addition, while the X_2–Y^- reaction entails an electron transfer, the principal step in the O_3–I^- reaction is an oxygen atom transfer ($O_3 + I^- \rightarrow OI^- + O_2$). These observations provide interesting hints, however, a clear picture has so far not emerged.

Funding for this work was provided by the National Science Foundation Grant No. ATM-90-13073, the US Environmental Protection Agency Grant No. R-815469-01-0, and the department of Energy Grant No. DE-FG02-91ER61208.

References

1 See for example, J. Lelieveld and P. J. Crutzen, *Nature (London)*, 1990, **343**, 6225.
2 See for example, S. Solomon, *Rev. Geophys.*, 1988, **26**, 131.
3 See for example, R. J. Charlson, J. Langner and H. Rodhe, *Nature (London)*, 1990, **348**, 22.
4 C. E. Kolb, D. R. Worsnop, M. S. Zahniser, P. Davidovits, D. R. Hanson, L. R. Ravishankara, M.-T. Leu, L. R. Williams, M. J. Molina and M. R. Tolbert, in *Current Problems and Progress in Atmospheric Chemistry*, ed. J. R. Barker, *Adv. Phys. Chem.*, 1994, in the press.
5 J. A. Gardner, L. R. Watson, Y. G. Adewuyi, P. Davidovits, M. S. Zahniser, D. R. Worsnop and C. E. Kolb, *J. Geophys. Res.*, 1987, **92**, 10 887.
6 D. R. Worsnop, M. S. Zahniser, C. E. Kolb, J. A. Gardner, L. R. Watson, J. T. Jaynes, J. M. Van Doren and P. Davidovits, *J. Phys. Chem.*, 1989, **93**, 1159.
7 J. T. Jayne, J. A. Gardner, P. Davidovits, D. R. Worsnop, M. S. Zahniser and C. E. Kolb, *J. Geophys. Res.*, 1990, **95**, 20 559.
8 J. T. Jayne, P. Davidovits, D. R. Worsnop, M. S. Zahniser and C. E. Kolb, *J. Phys. Chem.*, 1990, **94**, 6041.
9 J. M. Van Doren, L. R. Watson, P. Davidovits, D. R. Worsnop, M. S. Zahniser and C. E. Kolb, *J. Phys. Chem.*, 1990, **94**, 3265.
10 J. T. Jayne, S. X. Duan, P. Davidovits, D. R. Worsnop, M. S. Zahniser and C. E. Kolb, *J. Phys. Chem.*, 1991, **95**, 6329.
11 J. T. Jayne, S. X. Duan, P. Davidovits, D. R. Worsnop, M. S. Zahniser and C. E. Kolb, *J. Phys. Chem.*, 1992, **96**, 5452.
12 W. J. De Bruyn, S. X. Duan, X. Q. Shi, P. Davidovits, D. R. Worsnop, M. S. Zahniser and C. E. Kolb, *Geophys. Res. Lett.*, 1992, **19**, 1939.
13 S. X. Duan, J. T. Jayne, P. Davidovits, D. R. Worsnop, M. S. Zahniser and C. E. Kolb, *J. Phys. Chem.*, 1993, **97**, 2284.

14 J. Hu, J. A. Shorter, P. Davidovits, D. R. Worsnop, M. S. Zahniser and C. E. Kolb, *J. Phys. Chem.*, 1993, **97**, 11037.
15 W. J. De Bruyn, J. A. Shorter, P. Davidovits, D. R. Worsnop, M. S. Zahniser and C. E. Kolb, *J. Geophys. Res.*, 1994, **99**, 16927.
16 P. Davidovits, J. T. Jayne, S. X. Duan, D. R. Worsnop, M. S. Zahniser and C. E. Kolb, *J. Phys. Chem.*, 1991, **95**, 6337.
17 P. V. Danckwerts, *Trans. Faraday Soc.*, 1950, **46**, 300.
18 S. E. Schwartz, in *Chemistry of Multiphase Atmospheric Systems*; NATO ASI series; ed. W. Jaeschke, Springer-Verlag, Berlin, 1986, p. 415.
19 H. Motz and H. Wise, *J. Chem. Phys.*, 1960, **32**, 1893.
20 A. Ben-Naim and Y. Marcus, *J. Chem. Phys.*, 1984, **15**, 2016.
21 A. Ben-Naim, *Solvation Thermodynamics*, Plenum Press, New York, 1988.
22 G. L. Pollack, *Science*, 1991, **251**, 1323.
23 A. Braslau, M. Deutsch, P. S. Pershan, A. H. Weiss, J. Als-Nielsen and J. Bohr, *Phys. Rev. Lett.*, 1985, **54**, 114.
24 A. Braslau, P. S. Pershan, G. Swislow, B. M. Ocko and J. Als-Nielsen, *Phys. Rev. A*, 1988, **38**, 2457.
25 M. Matsumoto and Y. Kataoka, *J. Chem. Phys.*, 1988, **88**, 3233.
26 M. A. Wilson, A. Pohorille and L. R. Pratt, *J. Phys. Chem.*, 1987, **91**, 4873.
27 R. M. Townsend and S. A. Rice, *J. Phys. Chem.*, 1991, **94**, 2207.
28 N. A. M. Besseling and J. Lyklema, *J. Phys. Chem.*, 1994, **98**, 11 610.
29 A. W. Adamson, *Physical Chemistry of Surfaces*, Wiley, New York, 1990.
30 K. B. Eisenthal, *Acc. Chem. Res.*, 1993, **26**, 636.
31 Q. Du, R. Supertine, E. Freysz and Y. R. Shen, *Phys. Rev. Lett.*, 1993, **70**, 2313.
32 G. Robinson, D. R. Worsnop, M. S. Zahniser, C. E. Kolb and P. Davidovits, in preparation.
33 J. Hu, Q. Shi, P. Davidovits, D. R. Worsnop, M. S. Zahniser and C. E. Kolb, *J. Phys. Chem.*, 1995, **99**, 8768.
34 J. Hoigne, H. Bader, W. R. Haag and J. Staehelin, *Water Res.*, 1985, **19**, (**8**) 993.
35 M. Eigen, K. Kustin and G. Z. Maass, *Phys. Chem.* (*Munich*), 1961, **30**, 130.
36 D. J. Donaldson, J. A. Guest and M. C. Goh, *J. Phys. Chem.*, submitted.
37 T. X. Wang, M. D. Kelley, J. N. Cooper and D. W. Margerum, *Inorg. Chem.*, 1994, **33**, 5872.
38 F. MacRitchie, *Chemistry at Interfaces*, Academic Press, New York, 1990, p. 34.
39 P. W. Atkins, *Physical Chemistry*, W. F. Freeman, New York, 3rd edn., 1986, p. 782.

Paper 5/00194C; *Received* 4*th January*, 1995

Faraday Discuss., 1995, **100**, 83–100

O_3 and OH Sensitivity to Heterogeneous Sinks of HO_x and CH_3O_2 on Aerosol Particles

Yulii M. Gershenzon, Vera M. Grigorieva, Andrey V. Ivanov and Rodion G. Remorov
Semenov Institute of Chemical Physics, Russian Academy of Sciences, 4 Kosygin Street, Moscow 117977, Russian Federation

A new modification of matrix-isolation EPR has been developed for the measurement of uptake coefficients of polyatomic radicals. The technique is combined either with a cylinder reactor having a movable source of radicals or a coaxial reactor having a movable rod and a fixed source of radicals. The walls of the cylinder reactor or the rod surface are covered with the aerosol material. A simple way to treat the experimental data on the basis of the kinetic resistance additivity formula is proved. The results of HO_2 and CH_3O_2 uptake coefficient measurements are given and the possible implications for atmospheric chemistry are discussed. The measured value of the HO_2 diffusion coefficient was closer to $D_{H_2O_2}$ than to D_{O_2}. The HO_2 uptake coefficient on the concentrated 80 and 96% w/w H_2SO_4 was >0.2 at 243 K. For NaCl, KCl, NH_4NO_3 and $(NH_4)_2SO_4$, the γ_{HO_2} values were in the range $(1–2) \times 10^{-2}$ and $\gamma_{CH_3O_2} = 0.3 \times 10^{-2}$ at 295 K for NaCl. No changes in γ_{HO_2} were found at $[HO_2] = 4 \times 10^9–3 \times 10^{11}$ cm^{-3}. The results are discussed from the viewpoint of stratospheric ozone in volcanic clouds and tropospheric hydroxyl abundance in urban regions.

1. Introduction

Heterogeneous chemistry is important for understanding and describing many atmospheric problems. Since the discovery of the Antarctic spring 'ozone hole' in 1985,[1] it has played a central role in studies of ozone and other minor constituents of the atmosphere. Heterogeneous reactions of reservoir gases containing chlorine and oxides of nitrogen are now recognized to have a strong impact on the ozone in polar and in middle-latitude regions.[2]

Among the tropospheric problems where heterogeneous reactions are expected to be significant we mention the following: The measured OH concentration in the urban region at noon[3] and the NO_3 abundance at night[4] are lower than their values calculated using the pure molecular model. Recently[5] the accordance of measured OH concentration with the model calculation has been achieved without introduction of heterogeneous reactions. On the other hand, the night-time HONO abundance exceeds the result of modelling and a heterogeneous source of HONO should probably be included in the model.[6] The NO_3 night-time deficit was explained by the fast heterogeneous reaction of N_2O_5 with moist aerosols on the basis of laboratory studies.[7] The same reaction with different types of stratospheric aerosol particles plays a central role in the problem of the ozone-layer depletion. Heterogeneous reactions of HCl, HOCl and $ClONO_2$ have been shown to be an effective source of photochemically unstable chlorine species which result in ozone-layer depletion.

Heterogeneous reactions of reservoir gases containing bromine have also become the subject of laboratory studies.[8] Reactions of sea-salt aerosols (NaCl, KCl, NaBr, KBr) with reservoir gases could be a source of active chlorine and bromine.[9] Thus, one could find that the heterogeneous reactions of reservoir gases with different atmospheric particles play a big role in the modern atmospheric chemistry.

Less attention has been paid to the heterogeneous reactions of free atmospheric radicals because their lifetime in gas phase is shorter than the lifetimes of the reservoir gases. Nevertheless, some laboratory studies have been performed to understand the possible role of the heterogeneous sink and transformation of atmospheric radicals in different situations.

The limiting stage of the chlorine cycle of ozone destruction in the polar stratosphere is thought to be the reaction

$$ClO + ClO \xrightarrow{M} Cl_2O_2$$

Leu[10] has observed that the ClO uptake probability, γ_{ClO}, on ice is high enough to compete with the gaseous association reaction under the conditions that prevail in the Antarctic stratosphere. The γ-value measured later[11] is at least two orders of magnitude smaller. Also, the rate of ClO uptake by the polar cloud particles should be slower than the rate of homogeneous recombination. Probably additional experiments with a lower level of ClO radicals are necessary to draw firm conclusions.

Both primary and secondary aerosols were found to be very active towards the OH uptake. The OH uptake coefficients were found to be close to 0.1 for tropospheric materials[12,13] and close to unity for stratospheric aerosols, ice and H_2SO_4.[13,14]

Several measurements of HO_2 uptake by atmospheric materials have been published.[12,14–18] A preliminary conclusion[19] announced that $\gamma_{HO_2} \approx 10^{-2}$ for some tropospheric aerosols. Recent measurements for liquid water, 28% w/w sulfuric acid in water[14] and for moist NH_4NO_3 show that the uptake coefficients can be even higher than 10^{-2}. There are no data for the decay of CH_3O_2 or other RO_2 radicals on aerosol materials.

The transfer of HO_x into the condensed phase could lead to formation of hydrogen peroxide[20] and could directly impact on $[HO_x]$ and the other minor constituents.[19]

The possible influence of OH and HO_2 heterogeneous sinks on the tropospheric gas composition was first mentioned in ref. 21 and 22. Having assumed that $\gamma_{OH} = \gamma_{HO_2} = 1$, Isaksen and Crutzen[22] concluded that the photochemical model is very sensitive to HO_x heterogeneous loss processes. Their calculations were, however, only of a preliminary manner because of a lack of adequate laboratory measurements.

In 1987 Elanskii *et al.*[23] published the first 'chemical' explanation for ozone depletion after a volcanic eruption based on the results of OH uptake coefficient measurements[13] for ice at 253 K and for concentrated sulfuric acid at room temperature.

The additional heterogenous OH sink, which could be at least of the same intensity as the homogeneous sink in gas-phase reactions, increases the NO_x concentration. According to ref. 23, it should intensify the NO_x cycle of ozone destruction.

The sensitivity of minor tropospheric constituents to the introduction of heterogeneous HO_x losses was analysed[19] for polluted regions using $\gamma_{OH} = 1$ or $\gamma_{HO_2} = 10^{-2}$. It was found in this qualitative sensitivity analysis that the tropospheric hydroxyl abundance can be essentially reduced in very polluted regions.

The possible influence of RO_2 removal on the gas composition in the marine boundary layer was mentioned in ref. 24.

The following conclusions on the heterogeneous radical chemistry can be made after this brief introduction: (1) Heterogeneous reactions of HO_x are expected to be important in the stratosphere and the troposphere, but the available data are insufficient for quantitative consideration. The HO_2 uptake data are more important than the OH uptake data because $[HO_2] \gg [OH]$, but unfortunately they are particularly poor. (2) There are no data on heterogeneous reactions of RO_2. (3) The mechanisms of heterogeneous sinks of radicals should be established for adequate application of the laboratory results to atmospheric reality. (4) There are no experimental data on the diffusion coefficients of the radicals which are important both for laboratory measurements of γ and for radical chemistry flux calculations into cloud-water droplets.

The aim of this work is to develop a universal technique for the study of heterogeneous decay of HO_x and RO_2 and to obtain new data for the HO_2 diffusion coefficient and uptake probability on H_2SO_4/H_2O, primary (NaCl, KCl) and secondary [NH_4NO_3, $(NH_4)_2SO_4$] aerosol materials and for CH_3O_2 on NaCl. A qualitative discussion of the possible role of the processes studied is available.

2. Modification of Matrix Isolation for Studying the Uptake of Polyatomic Radical

Earlier[13,15,17] we used two types of magnetic resonance technique for the investigation of HO_x uptake. The OH radicals were directly detected in the gas phase by EPR[13] with a sensitivity of *ca.* 10^{10} cm^{-3}. IR laser magnetic resonance (IRLMR) based on the CO_2 laser was used for studying the heterogeneous decay of HO_2.[17] The combined EPR–IRLMR set-up was also applied for investigating heterogeneous HO_2 losses.[15] This set-up permits active atoms and diatomic radical impurities to be observed in the course of studying HO_2 reactions by LMR.

Below we shall briefly describe a new magnetic resonance installation for studying the heterogeneous loss of polyatomic radicals. This technique is based on the matrix isolation of radicals trapped at liquid-nitrogen temperature and observation by EPR.

Usually in laboratory experiments, the carrier gas and the radicals move directly to the cold surface. The cold 'finger' is placed in the centre of the EPR cavity. This configuration restricts the carrier-gas flux to *ca.* 10^{-5} mol s^{-1}.[25] Any larger gas flux would heat the 'finger'. Therefore the EPR spectrum becomes unstable and the application of the matrix-isolation technique is practically impossible under fast flow conditions.

Our modification of radical trapping differs from earlier ones. The carrier gas bypasses the cold 'finger' which is shifted 5–7 mm from the axis of the flow. This allows the He carrier gas flux to be increased to 10^{-3} mol s^{-1} without any visible change in the radical trapping conditions.

Fig. 1 and 2 show two different types of reactors combined with diffusion modification of the matrix isolation. The typical gas flow velocity in 2.1 cm diameter reactors is *ca.* 2000 cm s^{-1} at a pressure of 1–3 Torr under usual fast flow conditions. A resolution time of a few ms can be achieved by changing the radical–surface interaction length.

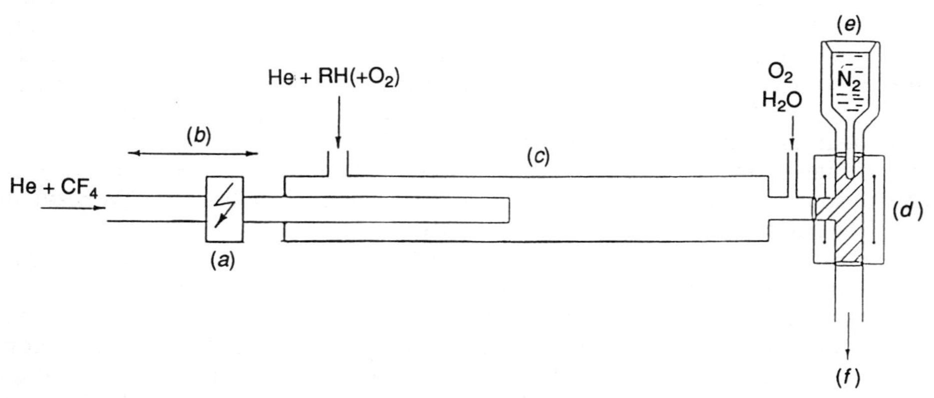

Fig. 1 Modification of matrix isolation combined with a movable source of manyatomic radicals: (*a*) microwave discharge cavity; (*b*) movable source of F atoms; (*c*) reactor; (*d*) EPR cavity; (*e*) Dewar flask with 'finger'; (*f*) to pump

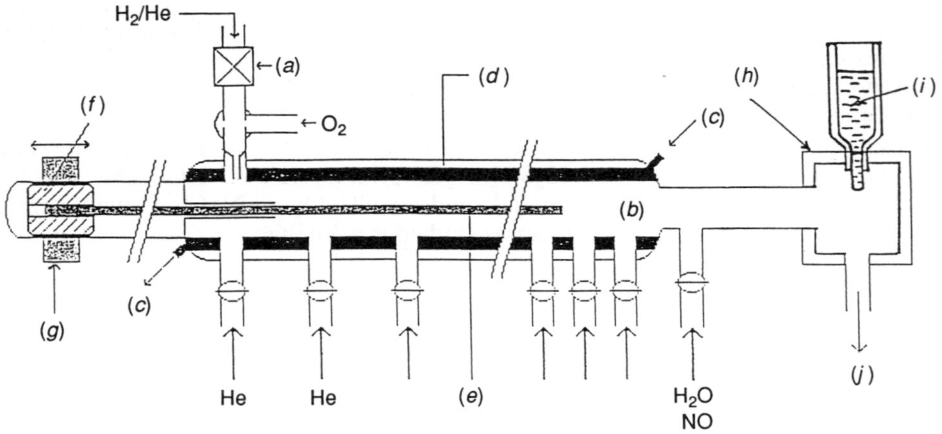

Fig. 2 Diffusion modification of matrix isolation combined with universal coaxial reactor: (*a*) microwave discharge cavity; (*b*) reactor; (*c*) liquid ethanol from the cryostat; (*d*) outer jacket pumped up to 10^{-4} Torr; (*e*) movable rod; (*f*) steel rod holder; (*g*) external magnet to move the rod; (*h*) EPR cavity; (*i*) Dewar flask with 'finger'; (*j*) to pump

The high sensitivity of the matrix-isolation technique is not lost when the diffusion radical delivery to the cold 'finger' occurs. However, this technique provides an opportunity to work under fast flow conditions and thus to measure high rates of chemical reactions of polyatomic radicals.

A cylinder reactor with a movable source of polyatomic radicals combined with the new modification of the matrix-isolation technique is shown in Fig. 1. This reactor was used for CH_3O_2 wall-decay studies for quartz covered with NaCl. The source of CH_3O_2 radicals is also shown in Fig. 1. The CH_3 radicals are formed by the reaction of F atoms with methane. The CH_3 radicals are directly transformed into the CH_3O_2 in the fast flow reactor in the course of the reactions

$$F + CH_4 \rightarrow CH_3 + HF$$

$$CH_3 + O_2 \rightarrow CH_3O_2$$

Some H_2O is added at the entrance of the EPR cavity for better matrix isolation of CH_3O_2. The flow and cavity axes are perpendicular, unlike the routine configuration where they are parallel.

To study the CH_3O_2 losses in reactions with surfaces, the wall of the cylinder was recovered with an aerosol material. The source of HO_2 shown in Fig. 2 can be also adjusted for this reactor and CH_3 chemistry can be studied if O_2 is added together with H_2O through the inlet nearest to the EPR.

The design shown in Fig. 2 is more universal. The reactor has eight sidearm inlets where the source of radicals can be adjusted. A high-pressure source of HO_2 is shown in Fig. 2. The HO_2 radicals are produced in the $H + O_2 + M$ reaction. The H atoms originate from a microwave discharge of He containing a very small amount of H_2. A thin capillary separates the reactor from the high-pressure HO_2 source.

The sidearm inlets are used to measure radical decay on the external wall of the coaxial reactor. To do this, the source of the radicals is placed at the inlet furthest from the cavity. An additional flux of He carrier gas is passed through the different sidearm inlets. The additional flux method[26] changes the time of contact of radicals with the cylinder wall and in this manner the uptake coefficient for the wall material is determined.

The coaxial reactor shown in Fig. 2 has to have a passive external wall so that the diffusion to the wall does not limit the wall-decay process, therefore the movable rod technique is the most simple for treatment.[27] The central rod of the coaxial reactor can be moved with the aid of an external magnet as shown in Fig. 2. The interaction time of the gaseous radicals with the rod is changed by this movement.

Two methods were used in the HO_2 heterogeneous reaction studies. The first is matrix isolation with diffusion transport of radicals and EPR in the solid state. This technique was applied in experiments at low HO_2 concentrations of $(3–5) \times 10^9$ cm^{-3}. In the second method, the fast titration reaction was used at $[HO_2] = (1–3) \times 10^{11}$ cm^{-3}. In this case, NO was added instead of H_2O at the sidearm inlet nearest to the EPR cavity to produce OH radicals by the reaction:

$$HO_2 + NO \rightarrow OH + NO_2$$

The hydroxyl radicals were detected by EPR in the gas phase. The reactor has a double jacket: the inner jacket is used to set the necessary temperature inside the reactor and the outer jacket is under vacuum (10^{-4} Torr) for thermal isolation. The temperature was maintained by means of an ethanol flux pumped by a cryostat through the inner jacket.

3. Additivity of Kinetic Resistances under Fast Flow Conditions

The mathematical treatment of the experimental data we have used in all cases is based on the rule of additivity of kinetic resistances.[28] We have shown[27] that the accuracy of this rule is even better under fast flow conditions than for an immovable static gas reactor. The theory of heterogeneous reaction in the coaxial reactor is described elsewhere.[27] Here we describe the features that are necessary for the experimental data treatment. The additivity formula for kinetic resistances[28] is

$$\frac{1}{k} = \frac{1}{k^{\mathrm{k}}} + \frac{1}{k^{\mathrm{d}}} \tag{1}$$

where k is the effective rate constant of heterogeneous decay, and k^{k} and k^{d} are its kinetic and diffusion limits. k (s^{-1}) is measured in the fast flow experiments. The radial diffusion rate constant, k^{d}, should be calculated for each reactor configuration and flow velocity.

3.1. Cylinder Reactor

According to ref. 29

$$k^{\mathrm{d}} = K^{\mathrm{d}}(Pe)D/R^2 \tag{2}$$

where D and R are the diffusion coefficient and radius, $Pe = 2v_{\mathrm{av}}R/D$ is the Peclet number, v_{av} is the average laminar flow velocity in the cylinder. K^{d} is the dimensionless rate constant.

The dependence of k^{d} on the Peclet number was studied in ref. 29 and 30 and under fast flow conditions ($Pe \rightarrow \infty$) when the axial diffusion can be neglected

$$K^{\mathrm{d}}(Pe \rightarrow \infty) = 3.66 \tag{3}$$

The kinetic regime is typical when the wall loss is slower than the radial diffusion rate. For the cylinder reactor

$$k^{\mathrm{k}} = \varepsilon c/2R \tag{4}$$

where c and R are the thermal velocity of molecular motion and the radius, respectively, $\varepsilon = 2\gamma/(2 - \gamma)$ and γ is the probability of radical uptake.

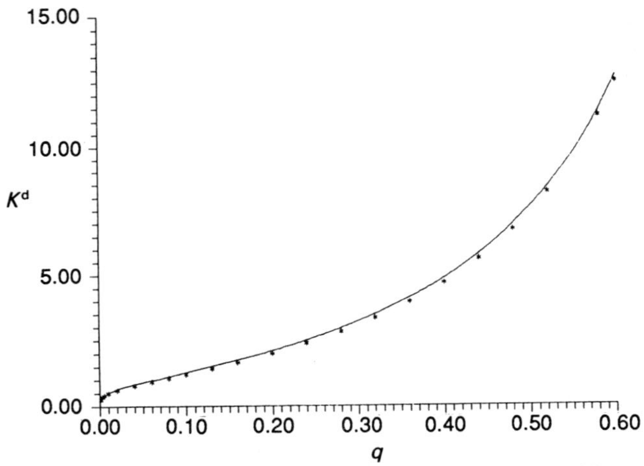

Fig. 3 Dimensionless radical diffusion rate constant for coaxial reactor with passive cylinder wall, $q = R_0/R$. Line and points are K^d under fast flow and static conditions, respectively.

The kinetic resistance additivity formula for fast flow conditions in a cylinder reactor is

$$\frac{1}{k} = \frac{2R}{c}\frac{1}{\varepsilon} + \frac{R^2}{3.66D_0}p \tag{5}$$

where D_0 is the diffusion coefficient at 1 Torr, p (Torr) is the total pressure of carrier gas. Note that this formula is approximate. Detailed analysis of the kinetic resistance additivity formula for solution of the reversed problem of ε determination will be published later.[30] The comparison between this formula and the exact solution is given in Fig. 4 in dimensionless form (see below).

3.2. Coaxial Reactor

A coaxial reactor with an active central rod and passive external wall was widely used in this study. For slow radical decay on the external wall the total rate constant of the heterogeneous reaction is the sum of the rate constant for a cylindrical reactor in the kinetic regime and the rate constant for the rod placed along the axis of a passive cylinder.[27] The external cylinder wall is 'absolutely' passive when

$$\left(\frac{\partial n}{\partial r}\right)\bigg|_{r=R} = 0 \tag{6}$$

where n is the radical concentration and r is the radius.

The problem of radial diffusion from the passive external wall to the active rod was solved by Grigorieva *et al.*[31]

The boundary condition for an active rod of radius R_0

$$n|_{r=R_0} = 0 \tag{7}$$

gives the maximum rate constant of the radical uptake with the rod under diffusion-limited conditions. It can be given in the form

$$k^d = K^d(q)D/R^2 \tag{8}$$

where $q = R_0/R$. The $K^d(q)$ dimensionless rate constant of radial diffusion is shown in Fig. 3. The solid line in Fig. 3 is the $K^d(q)$ dependence for fast flow conditions ($Pe \to \infty$)

and the points are the $K^d(q)$ values for static conditions ($Pe = 0$). As seen from Fig. 3, K^d is practically independent of the flow velocity in the coaxial reactor in contrast to the cylinder reactor.[29,30] This peculiarity is explained in ref. 27. One can see from Fig. 3 that the radical diffusion rate constant for the coaxial reactor becomes less than $K^d = 3.66$ for a cylinder reactor at $q < 0.34$. The present experiments were carried out at $q = 0.125$ and $K^d(q) = 1.45$. Under these conditions, K^d is almost three times smaller than for the cylinder reactor.

In the kinetic regime the heterogeneous rate constant for the rod is

$$k^k = \frac{\varepsilon c}{2R} \frac{q}{1 - q^2} \tag{9}$$

The kinetic resistance additivity formula for the movable rod is

$$\frac{1}{k} = \frac{2R}{c} \frac{1}{\varepsilon} \frac{1 - q^2}{q} + \frac{R^2}{K^d(q) D_0} p \tag{10}$$

where k is the rate constant measured by changing the time of the radical–surface interaction when the rod length in the coaxial reactor is varied.

In Fig. 4 the kinetic resistance additivity formula is compared with the exact solution of the reverse problem to determine ε (or γ) from the measured k value. The comparison is given in dimensionless form for both coaxial and cylinder reactors. Fig. 4(*a*) shows the approximate formula (1) in the dimensionless form

$$k^k/k^d = (k/k^d)/(1 - k/k^d) \tag{11}$$

In Fig. 4 curves (*b*) and (*c*) are the exact solutions for both coaxial ($q = 0.1$) and cylinder reactors under fast flow conditions.

If the k value is measured and k^d is known, k^k can easily be calculated from eqn. (1) or determinated exactly from the curves (*b*) and (*c*) in Fig.4. γ can then be calculated from eqn. (4) and (9) for the cylinder and coaxial reactors, respectively.

The high accuracy of the approximate eqn. (1) is seen from Fig. 4. The accuracy for the coaxial reactor is better than 3% even at $k/k^d = 0.9$. For the cylinder reactor, the accuracy of eqn. (1) is worse; however, it is better than our knowledge of the diffusion coefficient and experimental errors in k even in the region $k/k^d = 0.8$–0.85. Thus the

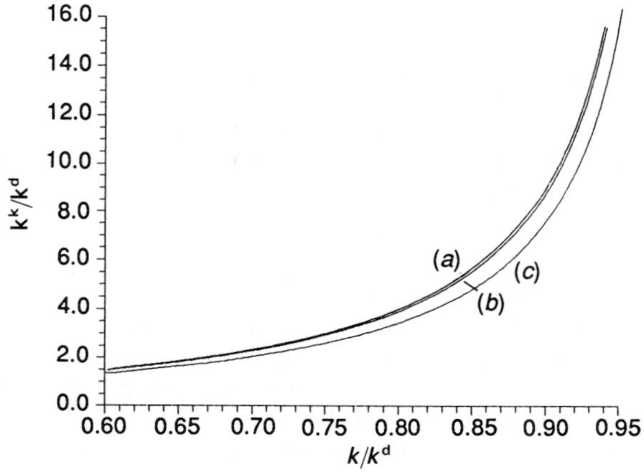

Fig. 4 Solution of the problem of k^d determination in dimensionless form. (*a*) Kinetic resistance additivity rule; (*b*) exact solution for coaxial reactor with $q = 0.1$; (*c*) exact solution for cylinder reactor. (*b*) and (*c*) are for fast flow conditions, $Pe \to \infty$.

kinetic resistance additivity formula can be recommended for wide application under fast flow conditions.

It is evident from Fig. 4 that the accuracy of γ measurements falls when k^k/k^d increases. To obtain an accurate γ value, k^k/k^d should be <3–5.

One can test using the above formulae that k^k/k^d is smaller for the coaxial reactor than for the cylinder, *e.g.* $Z = (k^k/k^d)_1/(k^k/k^d)_2 = 0.286$ at $q = 0.1$. Here the subscripts 1 and 2 refer to the cylinder and coaxial reactors, respectively. $Z = 0.286$ means that k^k/k^d decreases by a factor of 3.5 when we transfer from a cylinder reactor to a coaxial one with $q = 0.1$. Fig. 4 shows that it is practically impossible to determine γ when $k^k/k^d > 15$–20 for the cylinder reactor. However, if the coaxial reactor with $q = 0.1$ is used, k^k/k^d falls by a factor of 3.5 and γ can be determined with greater accuracy.

4. Experimental

4.1. Active Atoms and Radical Impurities

The identification of the gaseous radical(s) responsible for the EPR spectrum is difficult using the matrix-isolation technique. For example, H atoms can be transformed to HO_2 on the cold 'finger' if small oxygen impurities are present in the system. Another problem is to differentiate between HO_2 and other RO_2 species from the solid-state EPR spectrum. HO_2 and CH_3O_2 react with H, O and OH very fast and so their impurities should be reduced both from the point of view of radical identification and background RO_2 reaction. To reduce the active species impurities the He carrier gas and gases passing through the discharge had water and O_2 removed from them very carefully. The gases used were helium 99.995%, hydrogen 99.98% and oxygen 99.95%. He and H_2 additionally had oxygen removed by passing them through copper shavings at 670 K and water removal was performed with 4A molecular sieves at 77 K and 1 atm. O_2 was also removed from water at low pressure by trapping it at liquid-nitrogen temperature. After the gases had been cleaned, the H, O and OH radicals were not detected by EPR. Their impurity levels were estimated at $<10^{11}$, 3×10^{10} and 10^{10} cm^{-3}, respectively.

CF_4 was purified by repeated trapping at liquid-nitrogen temperature and low pressure. The atomic oxygen impurities from the He–CF_4 discharge were $<3 \times 10^{10}$ cm^{-3} when CF_4 was reduced so that the F-atom concentration was $<3 \times 10^{11}$ cm^{-3}. In all cases the gas-phase rates of HO_2 or CH_3O_2 reactions with active impurities were <0.5 s^{-1}. The heterogeneous rate constant was at least an order of magnitude higher.

4.2. Sensitivity of the Diffusion Modification of Matrix Isolation and EPR

To improve the efficiency of matrix isolation, water vapour was added at the entrance of the EPR cavity as shown in Fig. 1 and 2. This water, together with the radicals, was trapped by the cold surface. Complete trapping took place at $[H_2O] > (1$–$2) \times 10^{13}$ cm^{-3}. The HO_2 concentration in these experiments was *ca.* 3×10^{11} cm^{-3}. The carrier-gas flux and the pressure in the reactor were 6.8×10^{-4} mol s^{-1} and 1.8 Torr, respectively. The radical accumulation rate remained constant in the time interval up to 3×10^3 s at $[H_2O] < 10^{14}$ cm^{-3}. The $[H_2O]$ used for matrix isolation was usually $(0.3$–$1) \times 10^{14}$ cm^{-3}. The addition of H_2O improved the sensitivity *ca.* five-fold.

The gas-phase radical calibration and sensitivity estimation was performed twice. First, the NO_2 radicals were trapped and the NO_2 diffusion flux to the cold 'finger' was measured by comparing the NO_2 EPR signal with the known number of spins of the point-calibrating signal of Cu^{2+} placed inside the Dewar flask with liquid nitrogen. The diffusion flux was compared with the total NO_2 flux through the reactor measured in the usual manner. Under the experimental conditions ($p = 1.8$ Torr, $v = 18.6$ m s^{-1},

$d = 2.1$ cm) the portion of trapped radicals, α_{NO_2}, was *ca.* 15–17%. The second estimation of the fraction of stabilized radicals was carried out for HO_2. The H-atom flux was measured by gas-phase EPR when O_2 was replaced with He. It was compared with the HO_2 flux to the cold 'finger' determined by the point-spin calibrated signal of Cu^{2+}. Assuming [H] = [HO_2], the coefficient of HO_2 radical stabilization, α_{HO_2}, was *ca.* 15–17%. Thus, $\alpha_{NO_2} \approx \alpha_{HO_2} = 0.16 \pm 0.01$ and it did not depend on the concentrations of NO_2 and HO_2 in the range from 3×10^{11} to 2×10^{12} cm^{-3}. The sensitivities to NO_2 and HO_2 were roughly estimated as 10^8 cm^{-3} by means of these calibrations.

4.3. HO_2 Decay on the External Wall of the Coaxial Reactor

To be sure that the radical decay on the cylinder wall had no impact on the rod measurements, we had to measure the uptake coefficient for the reactor wall and the diffusion coefficient under experimental conditions. The external cylinder wall was covered with halocarbon wax to reduce the radical uptake on the wall. The wall covering was made by rinsing the wall with the halocarbon wax suspension in methyl ethyl ketone. After this treatment, the wall was dried at 210 °C for 30 min. The measurements of γ_w were carried out by the method of additional carrier-gas flow.[26] The change in the residence time inside the reactor was determined from[26]

$$\Delta t = \Delta l S/(W + W_{ad})(W_{ad}/W) \tag{12}$$

where Δl (in cm) is the change in the position of the additional flow, W and W_{ad} are the volume fluxes of the main and additional part of the He and S is the cross-section of the cylinder reactor.

The effective rate constant, k_w, was determined from the first-order wall decay and γ_w was calculated using eqn. (4) for the kinetic regime of the wall reaction in the cylinder. $\gamma_w = (7 \pm 2) \times 10^{-4}$ was measured at 295 K and the negative energy activation of γ_w was found to be $E_w = -2.9$ kcal mol^{-1} in the range 245–295 K. γ_w increased almost three-fold when the temperature decreased from 295 to 245 K. A further temperature decrease had a dramatic effect on γ_w. At $T < 240$ K γ increased sharply and any application of the rod technique became impossible. We do not know the reason for the γ_w break at 240 K, but all our attempts to decrease γ_w at low temperature were unsuccessful. Probably the method of surface covering should be changed. The lowest temperature in the experiments was 243 K.

4.4. Experimental Test of Kinetic Resistance Additivity for Fast Flow in the Coaxial Reactor

The applicability of the kinetic resistance additivity rule was tested in experiments with a steel rod. [HO_2] was *ca.* 3×10^{11} cm^{-3} and the NO titration method with OH detection was used in this experiment. The carrier-gas composition was He : O_2 = 5 : 1. The reversed rate constant of HO_2 decay on the steel rod is shown in Fig. 5 as a function of gas pressure. D_0 was determined from the slope of Fig. 5 using the relationship $k^d = K^d D_0/(pR^2)$ at $K^d(q = 0.125) = 1.45$. γ was determined from the value of $1/k$ at $p \to 0$. The results are:

$$D_0(HO_2\text{–He} : O_2 = 5:1) = 275^{+65}_{-40}\ \text{cm}^2\ \text{Torr s}^{-1}$$

$$\gamma_{HO_2}^{steel} = 0.3^{+0.5}_{-0.2}$$

The linear dependence of $1/k$ on p confirms the applicability of the kinetic resistance additivity rule. The diffusion coefficient of HO_2 was first experimentally determined here.

Next we calculated the Sherwood number

$$Sh = \gamma c R/4D$$

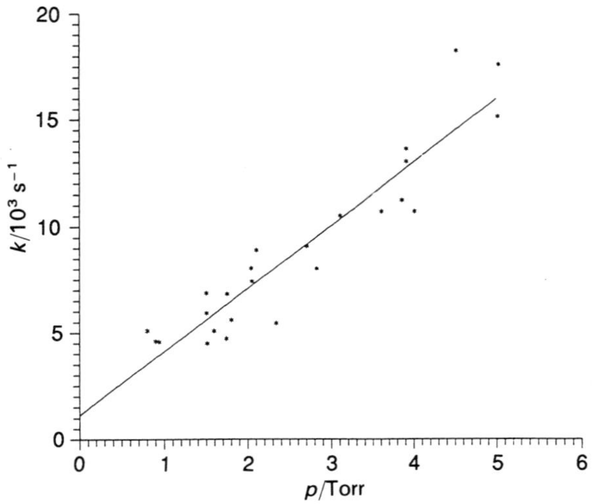

Fig. 5 Heterogeneous HO_2 loss constant on steel rod at different pressures

for the external wall of the coaxial reactor. For halocarbon wax at 2 Torr and $He:O_2 = 5:1$, $Sh = 0.1$ at 295 K and $Sh = 0.4$ at 243 K. The simplified theory of the coaxial reactor with a passive external wall has an accuracy $> 5\%$ if $Sh < 1$,[27] so all the experimental data obtained with the movable rod were treated as described in Section 3.2.

5. Results and Discussion

5.1. Diffusion Coefficient of HO_2 Radicals

The binary diffusion coefficients can be estimated[32]

$$D_{ij} = \frac{0.002628\,T^{1.5}}{(2\mu_{ij})^{0.5}\sigma_{ij}^2\langle\Omega_{ij}^{(1,\,1)*}\rangle}\ \text{atm cm}^2\ \text{s}^{-1} \tag{13}$$

where μ_{ij} is the reduced mass (u), T is temperature (K), σ_{ij} is the combined collision parameter (Å) and $\langle\Omega_{ij}^{(1,\,1)*}\rangle$ is the reduced collision integral. Unfortunately, the collision parameters are unknown for HO_2; however, we can use as an approximation the parameters for O_2 or H_2O_2. The collision parameters $\sigma_{O_2\text{–}He} = 3.096$ Å, $\sigma_{O_2\text{–}O_2} = 3.407$ Å, $(\varepsilon/k)_{O_2\text{–}He} = 27.94$ K, $(\varepsilon/k)_{O_2\text{–}O_2} = 121.1$ K[33] were used to calculate the binary diffusion coefficients $D_{O_2\text{–}He}$ and $D_{O_2\text{–}O_2}$. The diffusion coefficient in the $He:O_2 = 5:1$ mixture was calculated from the binary diffusion coefficients for O_2 in He and O_2 in O_2 and by using Blanc's formula[34]

$$\frac{1}{D_m} = \frac{y_{He}}{D_{O_2\text{–}He}} + \frac{y_{O_2}}{D_{O_2\text{–}O_2}} \tag{14}$$

where y_{He} and y_{O_2} are the mole fractions. We obtained $D_{O_2\text{–}He:O=5:1} = 375$ cm^2 Torr s^{-1}. This is *ca.* 35% larger than the experimental value.

There are no experimental data for binary diffusion coefficients $D_{H_2O_2\text{–}He}$ and $D_{H_2O_2\text{–}O_2}$ at room temperature and the binary collisional parameters are unknown. We have calculated them using the individual parameters σ and ε taken from ref. 35 [Table 1(*a*)]. The binary parameters were calculated from the individual parameters by using

Table 1 (*a*) Individual collisional parameters[35]

	σ/Å	(ε/k)/K
H_2O_2	4.196	298.3
He	2.551	10.22
O_2	3.467	106.7
N_2	3.798	71.40
H_2O	2.641	809.1

(*b*) Binary collisional parameters[a]

	σ/Å	(ε/k)/K
H_2O_2–He	3.374	54.38
H_2O_2–O_2	3.832	175.7
H_2O_2–H_2O	3.374	483.8
H_2O_2–N_2	3.997	143.7

[a] The binary collisional parameters were calculated from the individual parameters[35] using the combination rules.[33]

the combination rules:[33]

$$\sigma_{ij} = (\sigma_i + \sigma_j)/2; \quad \varepsilon_{ij} = (\varepsilon_i \varepsilon_j)^{1/2}$$

The results obtained are listed in Table 1(*b*). Table 2 lists the binary diffusion coefficients calculated for H_2O_2 and O_2 in He and O_2 at 295 K using eqn. (13). Using eqn. (14), we obtained at 295 K $D_{H_2O_2\text{–He}:O_2=5:1} = 272$ cm^2 Torr s^{-1}. This value is extremely close to the measured one.

To check the procedure of transfer from the individual parameters to the binary parameters, we compared the existing experimental data for H_2O_2–O_2 and H_2O_2–air at 333 K[36] and for H_2O_2–H_2O at 443 K[37] with the calculated values. This comparison ($D^{exp}_{H_2O_2–O_2}/D^{calc}_{H_2O_2–O_2} = 0.98$, $D^{exp}_{H_2O_2\text{–air}}/D_{H_2O_2\text{–air}} = 1.07$, $D^{exp}_{H_2O_2–H_2O}/D_{H_2O_2–H_2O} = 0.95$) yields good agreement between the experimental data and our simplified calculations for different collisional partners. From this comparison, we conclude that the binary diffusion coefficients $D_{H_2O_2\text{–He}}$ and $D_{H_2O_2–O_2}$ calculated in this manner are a good approximation of the HO_2 diffusion coefficients in He and O_2, respectively.

Now we compare the binary diffusion coefficients given in Table 2: $D^{calc}_{O_2–O_2}/D^{calc}_{H_2O_2–O_2} = 1.47$, $D^{calc}_{O_2\text{–He}}/D^{calc}_{H_2O_2\text{–He}} = 1.32$, $D^{calc}_{O_2\text{–He}:O_2=5:1}/D^{calc}_{H_2O_2\text{–He}:O_2=5:1} = 1.38$. The difference is large and the point is: Which approximation is better for HO_2, that from H_2O_2 or that from O_2? We think that $D_{HO_2–M}$ should be closer to $D_{H_2O_2–M}$ than to $D_{O_2–M}$. There are two reasons for this: (1) $\sigma_{HO_2–M}$ should be closer to $\sigma_{H_2O_2–M}$ than to $\sigma_{O_2–M}$ because the electron density and mass distributions are closer in H_2O_2 and HO_2 than in O_2 and HO_2. (2) If the dipole–dipole or dipole–quadrupole interaction has an influence on the HO_2 diffusion coefficient then the H_2O_2 approximation should, of

Table 2 Binary diffusion coefficients at 295 K

	D_{ij}/cm^2 Torr s^{-1}
H_2O_2–He	400.4
H_2O_2–O_2	105
O_2–He	529
O_2–O_2	154

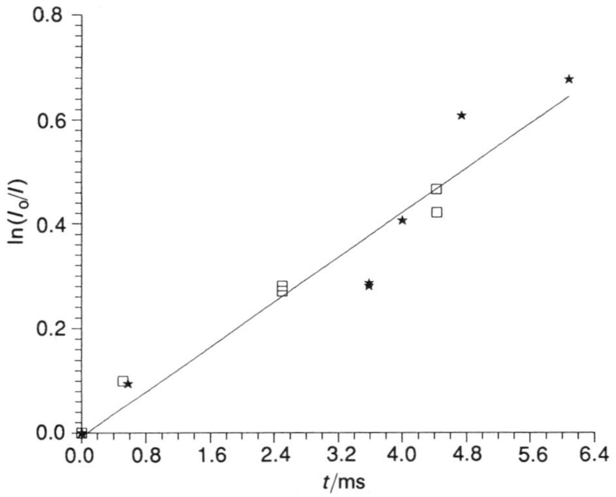

Fig. 6 Kinetics of HO_2 uptake by H_2SO_4–H_2O: ★, 96% w/w; □, 80% w/w

course, be better than the O_2 approximation because the dipole moments for HO_2 (2.1 D†) and H_2O_2 (2.26 D)[14] are close. If we take $D_{HO_2-M} = 1.1 \times D_{H_2O_2-M}$, we obtain $D_{HO_2-He:O_2=5:1} = 300$ cm^2 s^{-1}, in good agreement with the measured value.

We recommend the use of $D_{HO_2-M} = 1.1 \times D_{H_2O_2-M}$, although we understand that the error in the measurement of $D_{HO_2-He:O_2=5:1}$ is large and more accurate studies of the diffusion coefficient are necessary.

5.2. HO_2 Uptake by H_2SO_4–H_2O

Measurements of HO_2 uptake coefficients on liquid surfaces of sulfuric acid for two H_2SO_4–H_2O compositions, 96% and 80% w/w, were performed at 243 K by using the set-up shown in Fig. 2. The experimental conditions were: $p = 2.37$ Torr, He : O_2 = 6.8 : 1, $[HO_2] = 3 \times 10^{11}$ cm^{-3}. The rod surfaces were prepared by placing the rod into sulfuric acid for 10–20 min. Then it was moved the necessary distance inside the reactor with the aid of an external magnet. To support the H_2SO_4–H_2O composition on the rod surface, water vapour was added only for the 80% w/w H_2SO_4–H_2O composition [$p_{H_2O}(80\%) = 6 \times 10^{-4}$ Torr at 243 K[38]].

The first-order kinetic curves of the HO_2 decay on the rod surfaces covered with H_2SO_4–H_2O are shown in Fig. 6. All of them were obtained by changing the rod position. As is seen from Fig. 6, both compositions show a similar dependence of relative intensity on time. The average heterogeneous constant was found to be 105 ± 12 s^{-1}.

To obtain the uptake coefficients, we used following expression for k:

$$k^d = K^d(0.125)D/R^2 = 1.45D/R^2$$

The values of diffusion coefficients $D_{HO_2-He:O_2=6.8:1} = 1.1 \times D_{H_2O_2-He:O_2=6.8:1} =$ 96.8 cm^2 s^{-1} at 2.37 Torr and $D_{HO_2-He:O_2=6.8:1} = D_{O_2-He:O_2=6.8:1} = 122.4$ cm^2 s^{-1} at 2.37 Torr give the following results for γ: $\gamma_1 > 0.2$ and $\gamma_2 > 0.1$, for the first and the second cases, respectively.

It is not necessary to take into account the influence of H_2O on D in our measurements. The determination of γ is described in Section 3.2. For the first case: $k^d = 127.3$ s^{-1}, $k/k^d = 0.83$, $k^k/k^d = 4.5$ and $\gamma > 0.2$. There is only one measurement of γ_{HO_2} on 28% w/w H_2SO_4 at 249 K. The value $\gamma > 0.05$ was recommended.[14] The mixture $p_{H_2O} = 0.55$ Torr

† 1 D ≈ 3.335 64 × 10^{-30} C m.

and $p_{He} = 1$ Torr was used as carried gas in these experiments. The diffusion coefficients were taken as $D_{HO_2-H_2O} = D_{H_2O_2-H_2O}$ and $D_{HO_2-He} = D_{O_2-He}$. If we follow our recommendation that $D_{HO_2-He} = 1.1 \times D_{H_2O_2-He}$, we obtain $\gamma_{HO_2} > 0.1$ from the data given in Table 1 of ref. 14. The diffusion coefficient for the species HO_2–H_2O should be close to $D_{H_2O_2-H_2O}$, as reported in ref. 14.

Our results for concentrated 96 and 80% w/w H_2SO_4 and the results of ref. 14 for dilute 28% w/w sulfuric acid are similar. In both cases, the uptake coefficients are close to unity. The same situation is found for OH uptake by H_2SO_4–H_2O. For concentrated 96% H_2SO_4 $\gamma_{OH} = 1$,[13] and for 28% w/w H_2SO_4–H_2O $\gamma_{OH} > 0.08$.[14] Thus, both types of HO_x radical are active towards concentrated and dilute H_2SO_4–H_2O. Nevertheless, additional experiments for 50–80% w/w H_2SO_4 at lower temperatures are necessary.

5.3. HO_2 Uptake Coefficients for Salts

The uptake coefficients were determined for NaCl, KCl, NH_4NO_3 and $(NH_4)_2SO_4$ at 295 K. The set-up shown in Fig. 2 was used in these experiments under the following conditions: $p = 2.3$ Torr, He : O_2 = 2.72 : 1 and $v_{av} = 1056$ cm s^{-1}. For most measurements, the initial HO_2 concentrations were *ca.* 10^{11} cm^{-3} and in these experiments the NO titration technique was used. For NaCl, γ was also measured at $[HO_2] = 4 \times 10^9$ cm^{-3} by using the matrix-isolation technique.

The procedure of quartz rod covering was the same for all salts. A saturated solution of salt in water was used for surface preparation on the quartz rod. Before the rod was covered, its surface was purified with HF and washed with distilled water. Then the rod was dipped into the salt solution and dried for 20–30 min in air. The procedure was repeated several times until the surface became uniformly white.

The first-order kinetic curves were obtained by moving the rod inside the reactor. The measured heterogeneous rate constant for the rod and the γ values obtained in the experiments are listed in Table 3. The procedure of the experimental data treatment was carried out by using eqn. (8) and (9) at $K^d(q) = 1.45$, $D_{HO_2-He:O_2=2.72:1} = 1.1 \times D_{H_2O_2-He:O_2=2.72:1} = 109$ cm^2 s^{-1} and the kinetic resistance additivity rule. The radial diffusion rate constant $k^d = 143$ s^{-1} is 5–7 times higher than the measured value (k = 20–30 s^{-1}) and the correction for radical diffusion is not as large as for sulfuric acid (see Table 3).

The measured γ values for the salts are similar. For NaCl γ is independent of $[HO_2]$ in the range 4×10^9 to 10^{11} cm^{-3}. Similar values were determined for NaCl ($\gamma = 0.75 \times 10^{-2}$) in ref. 17 and for $(NH_4)_2SO_4$ ($\gamma = 1.3 \times 10^{-2}$) in ref. 12 for dry salts. For moist $NaNO_3$ and NH_4HSO_4 containing Cu^{2+} ions, the mass accommodation coefficient, α, which is larger than γ, was determined in ref. 39 to be >0.2. A comparison between γ_{HO_2} and γ_{OH} is given in Table 4. $\gamma_{HO_2} > \gamma_{OH}$ for the salts studied.

Table 3 Measurements of heterogeneous HO_2 losses on salts

salt	k/s^{-1}	k^k/s^{-1}	$10^2\gamma$
KCl	29.9	37.8	1.8
NaCl	25.5	31.0	1.5
NaCl[a]	28.0	34.8	1.65
NH_4NO_3	24.2	29.1	1.4
$(NH_4)_2SO_4$	22.7	26.9	1.1

[a] $[HO_2]_0 \approx 4 \times 10^9$ cm^{-3}, $p = 1.5$ Torr, $v_{av} = 800$ cm s^{-1}, He : O_2 = 2.72 : 1, matrix isolation and EPR technique.

Table 4 HO_x uptake coefficients for dry salts at 295 K

salt	$10^2\gamma_{HO_2}$	$10^2\gamma_{OH}$	ref.
KCl	1.8		this work
NaCl	1.6		this work
	0.75		17
		0.33	40
NH_4NO_3	1.4		this work
		1.0	12
		0.38	40
$(NH_4)_2SO_4$	1.1		this work
	1.3		12
		0.85	12

5.4. CH_3O_2 Heterogeneous Decay on Quartz and NaCl at 295 K

A movable source of CH_3O_2 shown in Fig. 1 was used to follow the kinetics for quartz and quartz covered with NaCl surfaces. The salt surface was prepared by washing the cylinder reactor wall with a saturated solution of the salt in water, after which the surface was dried for 30–50 min in air. The covering procedure was repeated several times until the surface became uniformly white. Before it was covered, the cylinder reactor wall was cleaned with $K_2Cr_2O_7$–H_2SO_4 for 20 min and then carefully washed with a large amount of distilled water.

Experiments were carried out at 295 K, 1.8–2.1 Torr, $[CH_3O_2] < (1–2) \times 10^{11}$ cm^{-3}. The second-order CH_3O_2 reaction as well as reactions with atom impurities were slow in comparison with the wall uptake. The CH_3O_2 uptake coefficient on quartz was measured twice by using both reactors.

Fig. 7 shows an example of the first-order kinetics of the CH_3O_2 heterogeneous losses on the quartz surface measured by using the additional carrier-gas flow method. The HO_2 wall decay kinetics for quartz studied in the same manner are also shown in

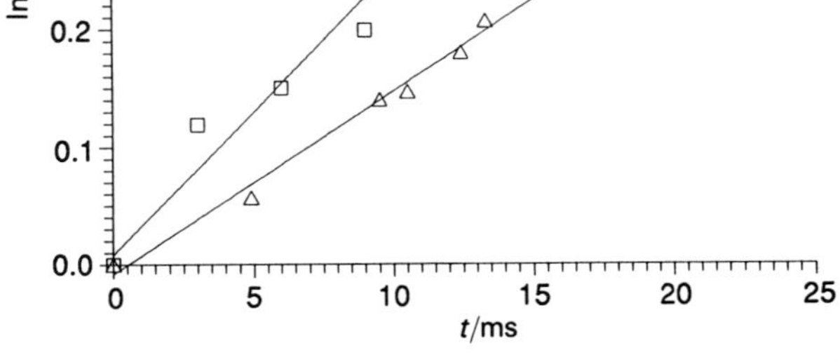

Fig. 7 Kinetics of CH_3O_2 (△) and HO_2 (□) decay on quartz measured by the additional flow method

Table 5 Uptake coefficients of CH_3O_2, HO_2 and CH_3 on quartz and NaCl at 296 K

surface	radical	$10^3\gamma$	method
quartz	CH_3O_2	0.9 ± 0.1	Fig. 2, matrix isolation, add. flow
	HO_2	1.12 ± 0.13	Fig. 2, matrix isolation, add. flow
	CH_3	0.50 ± 0.05	Fig. 1, matrix isolation
NaCl	CH_3O_2	4 ± 1	Fig. 1, matrix isolation
	HO_2	16 ± 3	Fig. 2, NO titration
	CH_3	<0.1	Fig. 1, matrix isolation

Fig. 7. The results of the CH_3O_2 uptake measurements for quartz and quartz covered with NaCl are listed in Table 5 and are compared with the uptake coefficients of CH_3 and HO_2 measured in this study. For NaCl $\gamma_{HO_2} > \gamma_{CH_3O_2} > \gamma_{CH_3}$.

6. Possible Atmospheric Implications

6.1. HO_2 Diffusion in Air

The highest rate of HO_x transformation on or inside the aerosol particles is restricted by mass transfer in air, which depends on the competition between mass transport by diffusion and surface accommodation processes. The effective first-order rate constant of HO_x mass transfer to the aerosol particles is given by the kinetics resistance additivity formula[41]

$$\frac{1}{k_{mt}} = \frac{1}{k_{mt}^{d}} + \frac{1}{k_{mt}^{k}} \tag{15a}$$

with

$$k_{mt}^{d} = \frac{3D}{a^2} V = 4\pi a D N \tag{15b}$$

$$k_{mt}^{k} = \frac{3\alpha c}{4a} V = \frac{\alpha c}{4} A; \qquad A = 4\pi a^2 N \tag{15c}$$

In eqn. (15*b*) and (15*c*) a and α are the radius of the aerosol particle and the mass accommodation coefficient, respectively. N, A and V are the number density, specific surface ($cm^2\ cm^{-3}$), and specific volume ($cm^3\ cm^{-3}$) of aerosols.

For the reaction of HO_x on/in an aerosol to be limited by mass transport in air, the following relationship must be fulfilled

$$\frac{k_{mt}^{k}}{k_{mt}^{d}} = \frac{3}{4}\frac{\alpha a}{\lambda} \gg 1; \qquad \alpha \gg \lambda/a \tag{16}$$

where λ is the free pathlength. For a typical cloud droplet of 10 μm diameter, diffusion limitation occurs at $\alpha > 10^{-2}$. The calculations of Chameides and Davis[20] and Schwartz[42] have shown that the rate of HO_2 uptake by cloud droplets is not strongly affected by the mass accommodation because usually $\alpha > \gamma > 10^{-2}$. In this case, a knowledge of the HO_x diffusion coefficient in air is particularly important for calculation of the radical's flux into the droplet. $D_{HO_2\text{–}air}$ is intermediate in value between $D_{H_2O_2\text{–}air}$ and $D_{O_2\text{–}air}$.

The first measurements of the HO_2 diffusion coefficient favour the use of $D_{H_2O_2\text{–}air}$ as the reference for calculation of $D_{HO_2\text{–}air}$. The simple formulae of the binary collisional parameters calculation for HO_2–M seems to be a good approximation for the calculation of $D_{H_2O\text{–air}}$. Assuming $D_{HO_2\text{–air}} = 1.1 \times D_{H_2O_2\text{–air}}$, and using the individual parameters from ref. 35, we obtained at 273 K for dry air $D_{HO_2\text{–air}} = 0.145$ atm cm^2 s^{-1}. This is *ca.* 30% lower than $D_{O_2\text{–air}}$. Additional studies of $D_{HO_2\text{–air}}$ and $D_{OH\text{–air}}$ are necessary.

6.2. Possible Impact of HO_x Heterogeneous Sinks on the Gas Composition in the Stratosphere

The mechanism of HO_x transformation in sulfuric acid droplets was discussed in ref. 14 and 43. The main reactions for OH and HO_2 are assumed to be:

$$OH + HSO_4^- \rightarrow H_2O + SO_4^-$$

$$HO_2 + HO_2 \rightarrow H_2O_2 + O_2$$

The uptake of OH by H_2SO_4–H_2O is a real first-order sink of OH, but HO_2 forms H_2O_2, which is desorbed in the gas phase and gives back H_2. For the stratospheric aerosols of 0.1 μm average radius,[44] the rate constant for heterogeneous loss is not limited by gas diffusion even for α (or γ) = 1 and is determined by eqn. (15*c*). At 220 K $k_{mt}(OH) = 5.1 \times 10^4 \varepsilon A$ and $k_{mt}(HO_2) = 3.7 \times 10^4 \varepsilon A$. Under background and postvolcanic (powerful eruption) conditions, $A_{bg} = 6 \times 10^{-9}$ cm^{-1} and $A_{pv} = 2 \times 10^{-7}$ cm^{-1} are typical.[43]

We have no data on γ_{OH} and γ_{HO_2} at stratospheric temperatures, but the preliminary measurements made at the Lille University[45] show a negative temperature dependence on the HO_2 uptake coefficient on the concentrated sulfuric acid surface. Taking into account the large values of γ_{OH} and γ_{HO_2} at $T > 240$ K, we could accept that $\gamma_{OH} \approx \gamma_{HO_2} = 1$ and $\varepsilon = 2$ at 205–215 K. Thus, we could have the highest rate constant of heterogeneous transfer onto droplets: $k_{mt}^{bg}(OH) = 6 \times 10^{-4}$ s^{-1}, $k_{mt}^{pv}(OH) = 2 \times 10^{-2}$ s^{-1} and $k_{mt}^{bg}(HO_2) = 4 \times 10^{-4}$ s^{-1}, $k_{mt}^{pv}(HO_2) = 1.4 \times 10^{-2}$ s^{-1}. Although the HO_x losses are very efficient, they can compete with the gas-phase losses only after large volcanic eruptions. On the other hand, even though the HO_2 sink onto the aerosol surface is fast it cannot change the gaseous H_2 if the only reaction in the droplet is second-order transformation of HO_2 to H_2O_2. The analysis of the heterogeneous reactions in sulfuric acid aerosols in ref. 43 showed that the assumption that H_2O_2 is formed in the particles has no significant effect on the gas composition. Only 7% change in OH abundance was found after large volcanic eruptions, which is too small to give noticeable ozone destruction due to HO_x fluxes into the aerosols.

Only OH losses were taken into account in ref. 23. It was announced that this process affects ozone fundamentally, but the role of this process was probably overestimated in ref. 23.

The HO_2 heterogeneous sink could affect the stratospheric gas composition more effectively than the OH sink because $[HO_2] > [OH]$, but to observe this effect, we should find HO_2 reactions with dissolved trace gases or with other species, *e.g.* with transition-metal ions.[39]

6.3. HO_x and CH_3O_2 Heterogeneous Sinks in the Troposphere

Earlier it was considered[19,46,47] that $\gamma_{OH} > \gamma_{HO_2}$ for tropospheric aerosol materials. As can be seen from Table 4, the situation with some salts is the opposite, *i.e.* $\gamma_{OH} < \gamma_{HO_2}$. The mechanism of interaction of HO_x with the dry surface of salts is unknown. If we assume that for both radicals the products of surface recombination are stable molecules (H_2O or $H_2 + O_2$), the role of the HO_2 sink should be recognized as more important because $[HO_2] \gg [OH]$ in the troposphere.

We have simulated the changes in the tropospheric gas composition under polluted urban conditions. For dry aerosols of radius 0.15 μm,[47] $k_{mt} = k_{mt}^{k} = (\gamma c A)/4$. The specific surface $A = 4 \times 10^{-6}$ cm^{-1} was measured at Jülich.[48] We calculated [OH] for molecular and aerosol models in the frame of 1D modelling.[46] For the aerosol model, we introduced $A = 4 \times 10^{-6}$ cm^{-1},[48] and $\gamma_{HO_2} = 10^{-2}$, as measured in this work for secondary aerosols, NH_4NO_3 and $(NH_4)_2SO_4$. The simulation gave *ca.* 40% reduction in [OH]. Other minor constituents were also changed in a reasonable manner.

There is only one measurement for CH_3O_2 given in Table 5. The uptake coefficient of CH_3O_2 on NaCl is *ca.* five times smaller than that for HO_2. If $\gamma_{CH_3O_2}$ is about the same for other aerosol particles, the heterogeneous uptake of CH_3O_2 cannot be very important in determining the tropospheric hydroxyl abundance.

Additional measurements are necessary to understand the possible role of heterogeneous CH_3O_2 reactions in urban regions and the marine boundary layer.

The conclusion of this section is that the tropospheric hydroxyl concentration should be sensitive to HO_2 heterogeneous sink.

7. Conclusion

(1) The diffusion coefficient of HO_2 is closer to the diffusion coefficient of H_2O_2 than to the that of O_2.

(2) The uptake coefficients of HO_2 and OH on sulfuric acid surfaces are close to unity. The impact of OH and HO_2 heterogeneous sinks on the stratospheric gas composition depends on the mechanisms of their reaction in the liquid state. If the second-order HO_2 reaction producing H_2O_2 is the most important and there is no linear reaction of HO_2 inside the droplets, the impact of heterogeneous OH and HO_2 reactions on the stratosphere may not be very great despite the very fast radical accommodation.

(3) According to our measurements, the HO_2 heterogeneous sink is the most important radical sink for the tropospheric hydroxyl abundance.

Further studies of the mechanisms of OH, HO_2, CH_3O_2 reactions on/in dry and moist salts are desirable.

D. Ehhalt, C. Howard, A. Ravishankara and S Solomon are acknowledged for helpful discussions. We thank S.Il'in for preparing the set-up and S. Umanskii for useful advice.

References

1 J. C. Farman, B. G. Gardiner and J. D. Shanklin, *Nature (London)*, 1985, **315**, 207.
2 J. M. Rodriguez, M. K. W. Ko and N. D. Sze, *Nature (London)*, 1991, **352**, 134.
3 D. Perner, U. Platt, M. Trainer, G. Hubler, J. W. Drummond, W. Ju. Kermann, J. Rudolf, B. Schubert, A. Volz, D. H. Ehhalt, K. J. Rumpel and G. Helas, *J. Atmos. Chem.*, 1987, **5**, 185.
4 W. L. Chameides, *J. Geophys. Res.*, 1986, **91**, 5331.
5 D. Poppe, J. Zimmerman, R. Bauer *et al.*, *J. Geophys. Res.*, 1994, **99**, (D8), 16 663.
6 A. Sjodin and F. Martin, *Atmos. Environ.*, 1985, **19**, 985.
7 M. Mozurkewich and J. G. Calvert, *J. Geophys. Res.*, 1988, **93**, 15 889.
8 D. R. Hanson and A. R. Ravishankara, *J. Phys. Chem.*, 1992, **96**, 9441.
9 R. S. Timonen, L. T. Chu, M-T. Leu and L. F. Keyser, *J. Phys. Chem.*, 1994, **98**, 9509.
10 M-T. Leu, *Geophys. Res. Lett.*, 1988, **15**, 851.
11 R.D. Kenner, I. C. Plumb and K. R. Rayn, *Geophys. Res. Lett.*, 1993, **20**, 193.
12 D. D. Jech, P. G. Easley and B. B. Kreiger, in *Heterogeneous Atmosphere Chemistry*, ed. D. C. Shryer, AGU, Washington, D.C., 1982.
13 Yu. M. Gershenzon, A. V. Ivanov, S. I. Kucheryavii and V. B. Rozenshtein, *Kinet. Catal.*, 1986, **27**, 923.
14 D. R. Hanson, J. B. Burkholder, C. J. Howard and A. R. Ravishankara, *J. Phys. Chem.*, 1992, **96**, 4979.
15 V. B. Rozenshtein, Yu. M. Gershenzon, S. D. Il'in, O. P. Kishkovich and R. T. Malkhasyan, *Kinet. Catal.*, 1985, **26**, 460.
16 E. V. Antzupov, G. I. Ksandopulo and Yu. M. Gershenzon, *Khim. Fiz.*, 1987, **6**, 1268.
17 E. V. Antzupov, *Khim. Fiz.*, 1988, **7**, 1082.
18 M. Mozurkewich, P. H. McMurry, A. Gupta and J. G. Calvert, *J. Geophys. Res.*, 1987, **92**, 4163.

19 Yu. M. Gershenzon, S. G. Zvenigorodskii, S. I. Kucheryavii, G. G. Politenkova, A. I. Poroikova, V. B. Rozenshtein and S. P. Smyshlyaev, *Khim. Fiz.*, 1987, **7**, 81.
20 W. L. Chameides and D. D. Davis, *J. Geophys. Res.*, 1982, **87**, 4867.
21 R. D. Cadle, P. J. Crutzen and D. H. Ehhalt, *J. Geophys. Res.*, 1975, **80**, 3381.
22 I. S. A. Isaksen and P. J. Crutzen, *Geophys. Norvegica*, 1977, **31**, 1.
23 N. F. Elanskii, S. G. Zvenigorodskii and S. P. Smyshlyaev, *Dokl. Acad. Nauk SSSR*, 1987, **294**, 1077.
24 N. M. Donahue, and R. G. Prinn, *J. Geophys. Res.*, 1990, **95**, 18, 387.
25 A. M. Bas and H. P. Broida, *Formation and Trapping of Free Radicals*, Academic Press, NY, 1960.
26 Yu. M. Gershenzon, S. D. Il'in, O. P. Kishkovich, R. T. Malkhasyan and V. B. Rozenshtein, *Kinet. Catal.*, 1982, **23**, 443.
27 V. M. Grigorieva, R. G. Remorov and Yu. M. Gershenzon, *Kinet. Catal.*, submitted; R. G. Remorov, PhD Thesis, Institute of Chemical Physics, Moscow, 1994.
28 D. A. Frank-Kamenetskii, *Diffusion and Heat Exchange in Chemical Kinetics*, Princeton University Press, Princeton, NJ, 1955.
29 Yu. M. Gershenzon and V. B. Rozenshtein, *Teor. Exp. Khim.*, 1974, **10**, 627.
30 A. Yu. Zasypkin, V. M. Grigorieva and Yu. M. Gershenzon, *Kinet. Catal.*, accepted.
31 V. M. Grigorieva, R. G. Remorov and Yu. M. Gershenzon, *Kinet. Catal.*, submitted; R. G. Remorov, PhD Thesis, Institute of Chemical Physics, Moscow, 1994.
32 J. O. Hirschfelder, C. F. Curtiss and R. B. Bird, *Molecular Theory of Gases and Liquids*, John Wiley, NY, 1954.
33 J. Bzowski, J. Kestin, E. A. Mason and F. J. Uribe, *J. Phys. Chem. Ref. Data*, 1990, **19**, 1179.
34 A. Blanc, *J. Phys.*, 1908, **7**, 825.
35 NASA Technical Report R-132, 1962.
36 R. L. McMurtie and F. G. Keyes, *J. Am. Chem. Soc.*, 1948, **70**, 3755.
37 S. Weissman, in *Proc. 4th Symp. on Thermophysical Properties*, ed. J. R. Moszynski, NY, 1968, p. 360.
38 *Handbook of Chemistry and Physics*, ed. R. C. Weast, CRC Press, Boca Raton, FL, 60th edn., 1979.
39 M. Mozurkevewich, P. H. McMurry, A. Gupta and J. G. Calvert, *J. Geophys. Res.*, 1987, **92**, 4163.
40 A. V. Ivanov, Yu. M. Gershenzon, P. Devolder and J-P. Sawerysyn, in book of *Abstracts of the Reunion Annuelle du Groupe de Cinetique et Photochimie en Phase Gazeuse*, Paris, June, 1994.
41 S. E. Schwartz, *Atmos. Environ.*, 1988, **22**, 2491.
42 S. E. Schwartz, *J. Geophys. Res.*, 1984, **89**, 11, 589.
43 D. R. Hanson, A. R. Ravishankara and S. Solomon, *J. Geophys. Res.*, 1994, **99**, 3615.
44 D. R. Hofmann and S. Solomon, *J. Geophys. Res.*, 1989, **94**, 5029.
45 A. V. Ivanov, Yu. M. Gershenzon and J-P. Sawerysyn, to be published.
46 Yu. M. Gershenzon, S. G. Zvenigorodskii and V. B. Rozenshtein, *Russ. Chem. Rev.*, 1990, **59**, 928.
47 Yu. M. Gershenzon and A. P. Purmal, *Russ. Chem. Rev.*, 1990, **59**, 1007.
48 U. Platt, M. Rateike, W. Junkerman, J. Rudolph and D. H. Ehhalt, *J. Geophys. Res.*, 1988, **93**, 5159.

Paper 5/00208G; *Received* 12*th January*, 1995

Faraday Discuss., 1995, **100**, 101–119

GENERAL DISCUSSION

Dr. Cox opened the discussion of Prof. Crutzen's lecture: Has the possibility that catalytic destruction by halogen atoms could be responsible for the very low ozone concentrations in the upper troposphere over the Pacific Ocean been eliminated? Ozone destruction by HO_x cycles (gas phase or liquid phase) is ultimately driven by ozone photolysis and will fall off in efficiency as the ozone concentration itself becomes very low, whilst halogen atom processes can completely remove ozone.

Prof. Crutzen replied: I have not considered this possibility. It cannot be ruled out, but it is impossible to pursue the issue until measurements become available.

Mr. Scheer and Dr. George communicated: It was mentioned that N_2O_5 is converted into NO_3^- by reaction with sea-spray aerosol and is therefore removed from the atmosphere. We have shown[1,2] that the reaction of N_2O_5 with NaCl solution (representing sea-spray) delivers nearly 100% $ClNO_2$, which is released to the gas phase again and forms Cl atoms by photolysis in the morning and additionally releases NO_2 again, so the loss of NO_x on sea-spray is less effective. Additionally, we observed a strong dependence of the uptake rates of $ClNO_2$ on the Br^- content of the solutions (in the concentration range of sea water) and observed[3] $BrNO_2$ and other Br-containing gaseous species such as BrCl or Br_2 as gas-phase products; NO_2^- in the liquid phase may escape as HONO as the solution is acidified. These additional reactions may contribute to the release of two halogen atoms and probably HONO from the reaction of one N_2O_5 on sea-spray.

We wonder how important the impact of such recently identified reactions could be in activating halogenated radicals in polluted marine-air masses. Especially, we would like to know if you have modelling results that take into account such catalytic conversion of N_2O_5 to Br_2, BrCl and HONO and how they contribute to morning chemistry.

1 C. Zetzsch and W. Behnke, *Ber. Bunsen-Ges. Phys. Chem.*, 1992, **96**, 488.
2 W. Behnke, V. Scheer and C. Zetsch, *J. Aerosol Sci.*, 1993, **S24**, 115.
3 W. Behnke, V. Scheer and C. Zetsch, *J. Aerosol Sci.*, 1994, **S25**, 277.

Prof. Crutzen communicated in reply: In a paper[1] coauthored with Dr. Rolf Sander, and based on his Ph.D. thesis, we did indeed consider the production of $ClNO_2$ in the reaction of N_2O_5 with Cl^- as one of the alternatives and agree with this remark. It is good to know that the channel leading to $ClNO_2$ is the only significant one and we will consider this in our future work. The finding of a possible reaction $ClNO_2 + Br^- \rightarrow BrCl + NO_2^-$ with the possibility of escape of HONO to the gas phase is a very interesting and potentially important one. We have not yet included this reaction in our model, but will do so as soon as possible, as kinetic data become available.

1 R. Sander and P. J. Crutzen, *J. Geophys. Res.*, 1995, submitted.

Prof. Kerr opened the discussion of Dr. Atkinson's paper: I have two questions: First, in your analysis of the products of the photooxidation of *n*-pentane, did you find any evidence for the formation of a hydroxypentyl nitrate? This would be predicted to be formed by the reaction of the isomerized radical, first with O_2, then with NO, *i.e.* a second channel of your reaction (11). I would point out that in our recent analogous study of *n*-hexane (your ref. 6), we detected hydroxyhexyl nitrate using absorption tubes followed by GC–MS analysis, but we were unable to quantify this compound owing to its strong tendency to be adsorbed on the walls of our Teflon bag reactor.

Secondly, your detection of the C_5-hydroxycarbonyl by the MS–MS technique is good evidence for the isomerization of the 2-pentoxyl radical but were you able to quantify this product? If not, is it the case that your derived rate constant ratios *e.g.* k_{isomer}/k_{O_2} are based solely on the analyses of the pentanones from the O_2 reactions taken together with the measured consumption of *n*-pentane?

Dr. Atkinson replied: We obtained no evidence for the presence of 1-nitratopentan-4-ol or 4-nitratopentan-1-ol from our GC or atmospheric pressure ionization (API) MS analyses. However, because the API MS–MS instrument is insensitive to alkyl nitrates in the positive-ion detection mode (although not to carbonyl nitrates) under the instrumental conditions we employed and we had no authentic standards of the hydroxypentyl nitrates, this does not rule out the formation of these products. It may be noted that Carter[1] has concluded from computer modelling of the NO_x–air photooxidations of *n*-heptane and *n*-octane that the formation of hydroxynitrates from the reactions of δ-hydroxyalkyl peroxyl radicals with NO [the second channel to reaction (11) in our paper] is minor.

As noted in our paper, the API MS analyses of the carbonyl and hydroxycarbonyl products were semi-quantitative, and are at best good to a factor of 2–3. You are correct in noting that the rate constant ratios k_{decomp}/k_{O_2} and $(k_{isomer} + k_{decomp})/k_{O_2}$ are determined solely from the formation yields of pentan-2-one and pentan-3-one, obtained from GC-FID analyses of pentanones and *n*-pentane, at the two O_2 concentrations used. The formation yields of pentan-2-one and pentan-3-one and pentan-2-yl and pentan-3-yl nitrate obtained, extrapolate to $[O_2]^{-1} = 0$, together with the calculated fraction of the overall OH radical reaction with *n*-pentane forming the pentan-1-yl radical, account for all the pentanyl radicals formed. Although the products arising from the decomposition and isomerization rections were not quantified or, in some cases, identified, the formation of products subsequent to the alkoxyl radical decomposition or isomerization reactions (such as the hydroxynitrates mentioned above) should be of no consequence in the derivation of the rate constant ratios k_{decomp}/k_{O_2} and $(k_{isomer} + k_{decomp})/k_{O_2}$.

1 W. P. L. Carter, unpublished data (1985), cited in W. P. L. Carter and R. Atkinson, *J. Atmos. Chem.*, 1985, **3**, 377.

Prof. Zellner commented: From measurements of integrated hydrocarbon oxidation mechanisms using time-resolved laser-based techniques,[1] we have recently obtained evidence that the rate coefficients for reactions of *n*-alkoxyl radicals with O_2 at 298 K are of the order of 8×10^{-15} cm^3 s^{-1}, essentially independent of the size of the oxyl radical (C_2–C_8).[2]

1 A. Hoffmann, V. Mörs and R. Zellner, *Ber. Bunsen-Ges. Phys. Chem.*, 1992, **96**, 437.
2 R. Zellner, *Proc. EUROTRAC Symp.*, 1992, ed. P. M. Borrell, P. Borrell, T. Cvitas and W. Seiler, Academic Press, The Hague, 1993.

Dr. Atkinson responded: As discussed in our paper, based on the literature rate constants for the reactions of O_2 with the methoxyl, ethoxyl and prop-2-oxyl radicals[1] we estimated the rate constants for the reactions of O_2 with the but-1-oxyl, but-2-oxyl, pent-2-oxyl, pent-3-oxyl, hex-2-oxyl and hex-3-oxyl radicals as being in the range (8–10) $\times 10^{-15}$ cm^3 molecule^{-1} s^{-1} at room temperature.[2] These estimated rate constants, used to derive the rate constants for the alkoxyl radical decomposition and isomerization reactions, are then consistent with those measured by you and your group for the alkoxyl radicals originating from *n*-butane, *n*-pentane and *n*-hexane of (7.3–8) $\times 10^{-15}$ cm^3 molecule^{-1} s^{-1}.[3] The alkoxyl radical reaction rates given in Table 3 of our paper are therefore not changed to any significant extent.

1 R. Atkinson, D. L. Baulch, R. A. Cox, R. F. Hampson Jr., J. A. Kerr and J. Troe, *J. Phys. Chem. Ref. Data*, 1992, **21**, 1125.
2 R. Atkinson, *J. Phys. Chem. Ref. Data*, 1994, **Monograph 2**, 1.
3 *EUROTRAC Annual Report*, 1993, Part 8, (LACTOZ) Table 12 (p. 60), EUROTRAC, International Scientific Secretariat, Garmisch-Partenkirchen, Germany, July 1994.

Dr. Cox said: You present some very elegant measurements of the products of isomerization of longer chain alkoxyl radicals. You draw attention to the mismatch between the observed differences in reactivity towards internal abstraction from $-CH_2-$ and $-CH_3$ groups, and that calculated (presumably by thermochemical kinetic arguments) some time ago. This would require estimation of activation energies for the isomerization, which raises the question of the temperature dependence of the relative rate of isomerization compared with reaction with O_2. Can anything be said at this stage about the relative importance of the isomerization *vs.* reaction with O_2, for this class of alkoxyl radical, as the temperature is varied over the range encountered in the troposphere?

Dr. Atkinson answered: The original rate constant estimates of Baldwin *et al.*[1] for alkoxyl radical isomerization *via* a six-member transition state were $2.4 \times 10^{11} \exp(-3875/T)$ s^{-1} for H-atom abstraction from a $-CH_3$ group and $1.6 \times 10^{11} \exp(-2315/T)$ s^{-1} for H-atom abstraction from $-CH_2-$ group (5×10^5 and 7×10^7 s^{-1}, respectively, at 298 K). The room-temperature isomerization rate constants given in Table 3 of our paper can be combined with the pre-exponential factors estimated by Baldwin *et al.*[1] to yield rate constants of $2.4 \times 10^{11} \exp(-4100/T)$ s^{-1} (2.5×10^5 s^{-1} at 298 K) for H-atom abstraction from a $-CH_3$ group and $1.6 \times 10^{11} \exp(-3250/T)$ s^{-1} (3×10^6 s^{-1} at 298 K) for H-atom abstraction from a $-CH_2-$ group (note that these rate constants may apply only to isomerizations of alkoxyl radicals formed from alkanes[2]). Using the rate constants recommended[3] for the reactions of O_2 with the ethoxyl ($6.0 \times 10^{-14} \exp(-550/T)$ cm^3 molecule^{-1} s^{-1}) and prop-2-oxyl ($1.5 \times 10^{-14} \exp(-200/T)$ cm^3 molecule^{-1} s^{-1}) radicals as being representative of those for primary and secondary alkoxyl radicals, respectively, and taking the variation of the O_2 concentration with altitude into account, the rates of reaction of alkoxyl radicals with O_2 are calculated to be (4–5) $\times 10^4$ s^{-1} at 298 K and 760 Torr total pressure of air (0 km) and (4–5) $\times 10^3$ s^{-1} at 210 K and 90 Torr total pressure of air (15 km; the tropopause). These rates of reaction with O_2 can be compared with those for isomerization; for H-atom abstraction from a $-CH_3$ group, *ca.* 2.5×10^5 s^{-1} at 0 km and *ca.* 8×10^2 s^{-1} at 15 km, and for H-atom abstraction from a $-CH_2-$ group, *ca.* 3×10^6 s^{-1} at 0 km and *ca.* 3×10^4 s^{-1} at 15 km. Clearly, these calculations indicate that alkoxyl radical isomerization proceeding *via* H-atom abstraction from either a $-CH_3$ or $-CH_2-$ group dominates over reaction with O_2 at *ca.* 298 K and atmospheric pressure; however, at the tropopause the O_2 reaction will dominate over isomerization *via* H-atom abstraction from $-CH_3$ groups, but not over isomerization *via* H-atom abstraction from $-CH_2-$ groups. Therefore, for the but-1-oxyl, pent-2-oxyl, hex-3-oxyl and hept-4-oxyl radicals the dominant reaction pathway will change from isomerization in the lower troposphere to reaction with O_2 in the upper troposphere, while for the pent-1-oxyl, hex-1-oxyl and hex-2-oxyl, hept-1-oxyl, hept-2-oxyl and hept-3-oxyl and all of the alkoxyl radicals formed from the $\geqslant C_8$ *n*-alkanes, the dominant reaction pathway throughout the troposphere is expected to be isomerization. Note also that, while decomposition of the alkoxyl radicals formed from the *n*-alkanes is slightly less important than reaction with O_2 in the lower troposphere (Table 3 of our paper), decomposition is expected to be of negligible importance at higher altitudes.

1 A. C. Baldwin, J. R. Barker, D. M. Golden and D. G. Hendry, *J. Phys. Chem.*, 1977, **81**, 2483.
2 R. Atkinson and S. M. Aschmann, *Int. J. Chem. Kinet.*, 1995, **27**, 261.
3 R. Atkinson, D. L. Baulch, R. A. Cox, R. F. Hampson Jr., J. A. Kerr and J. Troe, *J. Phys. Chem. Ref. Data*, 1992, **21**, 1125.

Dr. Eberhard and Prof. Ravishankara communicated: No results from the GC-MS analysis of the samples collected on the Tenax TA adsorbent are reported. Could there be any evidence for the formation of the δ-hydroxyalkyl nitrate from the reaction of the isomerized pentan-2-oxyl radical with oxygen and NO? Could the δ-hydroxy-carbonyl compound formed from the pentan-2-oxyl radical be detected in this analysis? If not, do the authors have any idea why they could detect the 1-hydroxypentan-4-one compound by APCI only?

When presenting the expected reactions of the pentan-2-oxyl radical the possible decomposition reaction to form a methyl radical and butanal is omitted. Is this reaction assumed to be negligible based on estimated[1] rate constants.

When discussing the yields of pentan-2-yl and pentan-3-yl radicals from the reaction of OH radicals with pentane, it is implicitly assumed that the rate constants of their reaction with oxygen to form the corresponding peroxyl radicals are the same.

1 W. P. L. Carter and R. Atkinson, *J. Atmos. Chem.*, 1985, **3**, 377.

Dr. Atkinson communicated in reply: As noted in our paper (GC-FID Analyses section in the Results), GC–MS analyses confirmed the formation of pentan-2-one, pentan-3-one, pentan-2-yl nitrate and pentan-3-yl nitrate. As noted in the response to the comment of Prof. Kerr, our GC and API MS analyses showed no evidence for the formation of δ-hydroxynitrates, though the formation of these products is not ruled out. Subsequent to the data reported in our paper, we have obtained an authentic sample of 5-hydroxypentan-2-one and have shown that while this compound can be readily detected by API MS (and gives an API MS–MS spectrum in agreement with the API MS–MS spectrum of the 103 u ion peak in the irradiated CH_3ONO–NO–*n*-pentane–air mixtures), we cannot detect 5-hydroxypentan-2-one by gas chromatography (by collection from a Teflon chamber onto Tenax solid adsorbent with subsequent thermal desorption and GC-FID or by liquid injection in dichloromethane solution onto a GC column with flame ionization detection).

Decomposition of the pent-2-oxyl radical to $\dot{C}H_3 + CH_3CH_2CH_2CHO$ is *ca.* 1 kcal mol^{-1} more endothermic than decomposition to $CH_3CHO + CH_3CH_2\dot{C}H_2$, and decomposition to $CH_3CHO + CH_3CH_2\dot{C}H_2$ has been shown to dominate (by a factor of 80 ± 30 at 413 K.)[1]

Under the conditions employed in our study, the pentanyl radicals react only with O_2 [reaction (7) in our paper] and are converted immediately (within < 1 μs)[2] to the corresponding pentanyl peroxyl radicals. Providing that the pentanyl radicals quantitatively converted to the corresponding peroxyl radicals, the rate constants for reaction with O_2 are immaterial and the reactions of importance are then reaction (8) in our paper (RO_2 + NO), the reactions of the alkoxyl radicals (decomposition, isomerization and reaction with O_2) and the subsequent reactions of the alkyl nitrates and of the products arising from the reactions of the alkoxyl radicals.

1 S. Dóbé, T. Bérces and F. Márta, *Int. J. Chem. Kinet.*, 1986, **18**, 329.
2 R. Atkinson, *J. Phys. Chem. Ref. Data*, 1994, **Monograph 2**, 1.

Dr. Horie communicated: Although not discussed in the paper, β-hydroxyalkoxyl radicals formed subsequent to the addition of OH radicals to the alkenes belong to an important and general class of alkoxyl radicals, for which very few experimental data are available about the isomerization.[1] Understandably, the relative importance of the isomerization reaction would increase as the radical becomes larger. It is interesting, therefore, to know to what extent the conclusions in the paper about the pentanoxyl radical isomerization apply to large ($> C_4$) β-hydroxyalkoxyl radical isomerizations.

1 R. Atkinson and W. P. L. Carter, *J. Atmos Chem.*, 1991, **13**, 195.

Dr. Atkinson communicated in reply: We have studied the products formed from the gas-phase reactions of the OH radical with the alk-1-enes but-1-ene through oct-1-ene using gas chromatography and Fourier-transform IR absorption spectroscopy[1,2] and API MS–MS.[3] The molar yields of the carbonyls, RCHO + HCHO, formed by decomposition of the β-hydroxyalkoxyl radicals, $RCH(OH)CH_2\dot{O}$ and $RCH(\dot{O})CH_2OH$, formed after OH radical addition to the alkenes $RCH{=}CH_2$, decrease with increasing carbon number from *ca.* 0.94 for but-1-ene to 0.21–0.39 for oct-1-ene.[1,2] API MS and MS–MS analyses show the formation of dihydroxycarbonyls (in addition to the aldehydes RCHO) from these alk-1-enes.[3] For the pent-1-ene and oct-1-ene reactions, the dihydroxycarbonyl/aldehyde ratios are independent of O_2 concentration,[3] and the dihydroxycarbonyls are interpreted as being formed subsequent to isomerization of the intermediate β-hydroxyalkoxyl radicals. The importance of the isomerization pathway increases with increasing carbon number from but-1-ene to oct-1-ene.[3] API MS has therefore allowed *in situ* detection of products arising from the isomerization reactions of the alkoxyl and hydroxyalkoxyl radicals formed from the OH radical-initiated reactions of alkanes and alkenes, respectively.

1 R. Atkinson, E. C. Tuazon and W. P. L. Carter, *Int. J. Chem. Kinet.*, 1985, **17**, 725.
2 R. Atkinson, E. C. Tuazon and S. M. Aschmann, *Environ. Sci. Technol.*, 1995, **29**, 1674.
3 E. S. C. Kwok, R. Atkinson and J. Arey, *Environ. Sci. Technol.*, submitted.

Dr. Doughty initiated the discussion of Prof. Ravishankara's paper: The pre-exponential factor used for the unimolecular decomposition of CH_3SO_x in the paper is 5×10^{13} s^{-1}. This implies a $\Delta S^{\ddagger}$ of 0 J K^{-1} mol^{-1}. Considering that the CH_3SO_x would be loose, such a low $\Delta S^{\ddagger}$ would not be expected. By analogy with reactions involving simple bond fission to yield methyl + fragment, we would expect a pre-exponential *A* factor of *ca.* 10^{16} s^{-1}.[1] This is a factor of 200 greater than the estimate used for CH_3SO_x decomposition in the DMS oxidation model. What would the effects be on the prediction of the DMS oxidation model if an *A* factor of 10^{16} s^{-1} were used for the unimolecular decomposition of CH_3SO_x?

Are the predictions of the model insensitive to the rate of unimolecular decomposition of CH_3SO_x?

The argument presented in the paper relies to an extent on expected branching ratios between DMS oxidation channels as an explanation for observed [MSA]/[SO_4^{2-}] ratios. How would an increase of a factor of 200 in the unimolecular decomposition rate of CH_3SO_x, as a consequence of using a pre-exponential *A* factor of 10^{16} s^{-1} for these reactions, affect the expected branching ratios?

1 S. W. Benson, *Thermochemical Kinetics*, Wiley, New York, 1976.

Prof. Ravishankara replied: I will answer all three questions together. Dr. Doughty's point that the *A* factor we chose for the decomposition of CH_3SO_x is much smaller than that expected for a decomposition such as this is well taken. It should be remembered that neither the *A* factor nor the activation energy for this reaction is known. Our approach was to assume an endothermicity as a good estimation of the activation energy and then choose an *A* factor to make the process competitive with other atmospheric loss processes of the radical. The prediction of the model in terms of the 'absolute' yield of MSA relative to non-sea salt (nss) SO_4^{2-} will be very sensitive to the *A* factor, but the temperature dependence of the ratio will not. If we use an *A* factor of 1×10^{16} s^{-1} with an endothermicity of 22 kcal mol^{-1} the ratio of MSA/nss SO_4^{2-} will be reduced by 10%. It should be remembered, however, that if the activation energy is 24 kcal mol^{-1}, we will have a completely different off-set for the factor of 200 change in the *A* factor. To calculate absolute branching ratios as functions of temperature and composition in the atmosphere will require a much greater understanding of the fundamental elementary reactions involved in the oxidation mechanism.

Your argument is much more important with respect to the decomposition of CH_3SO_2. However, since we do not know the rate of conversion of CH_3SO_2 to CH_3SO_3, playing with the *A* factor is of little consequence. Two points are worth noting: (*a*) If the activation energy for a process gets very large, then the importance of that process in the atmosphere will be negligible even though the branching would be highly temperature dependent. (*b*) We suggest that the major branching is not in the step that involves CH_3SO_3 decomposition.

Dr. Herrmann asked: Could you comment on the possiblity of phase partitioning of not only the intermediates such as $CH_3S(O)_xOONO_2$ but also the smaller peroxyl radical species shown in Fig. 2 of your contribution? Earlier modelling studies[1] indicated that phase transfer of not only stable species but also peroxyl radical species like HO_2 (*cf.* Fig. 2 in the above reference) and CH_3O_2 (*cf.* Fig. 4 in the above reference) may significantly alter the understanding of the odd hydrogen cycle as well as the methane oxidation when cloud chemistry is considered. Could something similar be of importance in the oxidation of DMS?

1 D. J. Jacob, *J. Geophys. Res.*, 1986, **91**, 9807.

Prof. Ravishankara replied: It is possible that the peroxyl species such as $CH_3S(O)_xOO$, are soluble in the water droplets of the marine boundary layer. Hence, scavenging by droplets could represent a removal mechanism of sulfur in the atmosphere and influence the observed MSA to SO_4^{2-} ratio in the atmosphere. It should be remembered, of course, that free radical gas-phase chemistry and photochemistry are quite rapid and, hence, reactive uptake will have to compete with this fast chemistry. Considering such processes for the peroxyl radicals is, therefore, secondary when the uptake of stable species have not yet been elucidated. We would like to add that measuring MSA in the gas phase will help evaluate the role of heterogeneous chemistry in DMS oxidation.

Prof. Elias said: You postulate that adduct formation plays a significant role in the atmospheric oxidation of DMS. From the chemical point of view, DMS is a 'soft' Lewis base and the OH radical a 'hard' Lewis acid. What is the chemical nature of the adduct DMS · OH you postulate?

Prof. Ravishankara said in reply: Maybe we should define what we mean by an adduct. The adducts we are talking about are bound by 10–15 kcal mol^{-1}. They are not 'hydrogen' bonds, but they may not be the strong adducts you may be referring to.

The interaction between the OH radical and DMS could possibly be described in terms of a Lewis acid–base interaction. However, an *ab initio* structure calculation on the OH + DMS reaction by McKee[1] suggests that this is not the case. McKee looked at DMS adducts with OH, Cl and Br. The Br and Cl adducts appear to fit better with the classic Lewis acid–base description. The theoretical description of the bonding in the OH · DMS adduct requires a higher level of theory than a simple donor–acceptor interaction. All we know is that the OH is attached to the sulfur in DMS, and the formation of HO_2 in the OH · DMS + O_2 reaction suggests a bond between the O and the S. The *ab initio* electronic structure calculations bear this out. It will be very useful to measure the IR spectrum of this adduct.

1 M. L. McKee, *J. Phys. Chem.*, 1993, **97**, 10971.

Dr. Hynes said: Given the difficulties involved in this experiment, the data presented by Barone *et al.* are in excellent agreement with our previously presented results on the

$[^2H_6]$DMS equilibration.[1–3] Although the derived van't Hoff expression and thermochemistry differ, the differences in the extrapolated equilibrium constant between 300 and 250 K are small. The agreement between the reported rates for the reaction of the adduct with O_2 is also excellent. Our 1986 rate[4] for this reaction was based on deriving elementary rates from a single observation of the deviation in linearity of pseudo-first-order rate *vs.* $[^2H_6]$DMS. The equilibrium constant is obtained by modelling the 'roll-off' and is extremely sensitive to the abstraction rate chosen. Re-analysis of these data[3] shows that they are consistent with the van't Hoff expressions presented by us and by Barone *et al.* The value for the adduct + O_2 rate, derived from the 1986 data on the 'O_2 enhancement' to the observed rate, is then found to be consistent with the current value. While the agreement on HO_2 and inferred DMSO yields is excellent, direct observation of DMSO under realistic atmospheric conditions is imperative. We have measured the OH + DMSO rate coefficient[3,5] and obtain a value of 1×10^{-10} cm^3 $molecule^{-1}$ s^{-1}. Consequently, if DMSO is produced in significant yield, its main sink is reaction with OH.

Of more concern is the idea that there is a clear and well defined inverse temperature dependence in the [MSA] : [nss-sulfate] ratios observed in field measurements and that a mechanism based largely on estimates can explain this. The interpretation of such field measurements is controversial. While there are measurements[6] at a variety of latitudes which support an inverse *T* dependence, seasonal measurements[7] at high latitude show an opposite dependence. Similarly, the data from ice cores show a distinct inverse relationship in the data from Vostoc[8] but the opposite behaviour is evident in data obtained in Greenland.[9] There is, typically, a distinct difference in the aerosol size distribution of nss-sulfate and MSA.[10] Hence the interpretation of observed ratios in terms of a purely chemical mechanism may not be justified. Meteorological influences may also be important.

Note that there has only been one direct study of the observed rate of the reaction of OH with DMS under atmospheric conditions.[4] The only clean competitive rate study (performed only at room temperature), gave a significantly larger 'O_2 enhancement'.[11] Modelling studies of the diurnal variation in DMS concentrations observed in the MAGE campaign found that use of the faster rate, obtained in the competitive rate study, gives much better agreement with the observed amplitude of the diurnal variation.[12] Note that if this discrepancy between the direct and competitive studies is real, being due for instance to a channel which regenerates the OH radical, then our 1986 study would have underestimated the temperature dependence of the branching ratio. In the short term, the clearest resolution of this problem is likely to come from 'clean' competitive rate studies of the temperature dependence of the observed rate coefficient and the [SO2] : [gas phase MSA] ratio.

1 A. J. Hynes, R. E. Stickel, A. J. Pounds, Z. Zhao, T. McKay, J. D. Bradshaw and P. H. Wine, in *Proccedings of the International Symposium on Dimethylsulphide. Oceans, Atmosphere and Climate*, ed. G. Restelli and G. Angeletti. Kluwer Academic Publishers, Dordrecht, 1993.

2 A. J. Hynes, in *Spectroscopy in Environmental Science*, ed. R. J. H. Clark and R. E. Hester. Wiley, Chichester, 1995.

3 A. J. Hynes, R. B. Stoker, A. J. Pounds, T. McKay, J. D. Bradshaw, J. M. Nicovich and P. H. Wine, *J. Phys. Chem.*, submitted.

4 A. J. Hynes, P. H. Wine and D. H. Semmes, *J. Phys. Chem.*, 1986, **90**, 4148.

5 A. J. Hynes and P. H. Wine, *J. Atmos. Chem.*, submitted.

6 T. S. Bates, J. A. Calhoun and P. K. Quinn, *J. Geophys. Res.*, 1992, **97**, 9859.

7 D. L. Savoie, J. M. Prospero, R. J. Larsen and E. S. Saltzman, *J. Atmos. Chem.*, 1992 **14**, 181.

8 M. Legrand, C. Feniet-Saigne, E. S. Saltzman, C. Germain, N. I. Barkov and V. N. Petrov, *Nature (London)*, 1991, **350**, 144.

9 M. E. Hansen and E. S. Saltzman, *Geophys. Res. Lett.*, 1993, **20**, 1163.

10 E. S. Saltzman, D. L. Savoie, R. G. Zika and J. M. Proepero, *J. Geophys. Res.*, 1983, **88**, 10897.

11 I. Barnes, V. Bastian and K. H. Becker, *Int. J. Chem. Kinet.*, 1988, **20**, 415.

12 S. A. Yvon, E. S. Saltzman, D. J. Cooper, T. S. Bates and A. M. Thompson, *J. Geophys. Res.*, submitted.

Prof. Ravishankara responded: I agree with some of the comments made by Dr. Hynes and disagree with others. It may be worth reiterating the goal of our paper. Because the atmospheric oxidation of DMS involves many adducts and because a unimolecular reaction is likely to be more temperature dependent than a bimolecular reaction, the ratio of MSA to nss from the gas-phase oxidation of DMS will vary with temperature.

Our studies agree with those of the Georgia Tech group. They should, because the experiments are similar! Note that a third-law analysis, which we used, is inherently a better way to analyse the equilibrium constant data. Some of the differences in the values reported by us and Hynes and Wine[1] are due to this factor. We agree that direct observation of DMSO under realistic atmospheric conditions is imperative. We hasten to add that there are many other such 'imperatives' in DMS chemistry!

Dr. Hynes points to the disparities in field measurements. It is clear, however, that the MSA to nss ratio varies in the atmosphere. The trend in this ratio with temperature is not always in agreement. Remembering that MSA measured in the atmosphere is in particles, the whole question of whether there is a variation in the MSA to nss ratio with temperature may be ill posed! (It is only very recently that MSA has been measured in the gas phase.) Interpreting the ice-core data, in the absence of suitable tracer coordinate systems to normalize for transport and lifetimes, is a very tricky business. This is especially true for the Greenland ice-core because of the meteorology. In the absence of coherent data, maybe we should restate the aim of our paper: it is to see if the changes in the atmospheric DMS oxidation mechanism could lead to changes in the gas-phase ratios of MSA to SO_2.

The point about differences in the O_2 enhancement observed by Barnes *et al.*[2] and the Georgia Tech group[3] is unlikely to change our qualitative understanding about the variation of the oxidation products with temperature. We are very reluctant to emphasize matching the observations in the field with different rate data when the field studies do not measure all the needed parameters. We do not share Dr. Hynes' view that a 'clean' competitive rate study will resolve the OH regeneration issue, because such 'clean' studies are nearly impossible to carry out. Our recent work using $^{18}O_2$, ^{16}OH, and deuteriated DMS suggests that there is no regeneration of OH in this reaction.

1 A. J. Hynes, T. Pounds, T. McKay, J. D. Bradshaw and P. H. Wine, *Detailed Mechanistic Studies of the OH-Initiated Oxidation of Biogenic Sulfur Compounds under Atmospheric Conditions*, 12th International symposium on gas kinetics, Reading, UK, 1992.
2 I. Barnes, V. Bastian and K. H. Becker, *Int. J. Chem. Kinet.*, 1988 **20**, 415.
3 A. J. Hynes, P. H. Wine and D. H. Semmes, *J. Phys. Chem.*, 1986, **90**, 4148.

Dr. Plane commented (**Miss Allan, Miss Smith, Dr. Coe, Miss Burgess, Miss Carpenter, Dr. Clemitshaw** and **Prof. Penkett** in part communicated): Dr. Ravishankara states in his paper that more observations of NO_3 in the marine boundary layer are required. We have, in fact, carried out a series of field campaigns at the Weybourne Atmospheric Observatory on the North Norfolk coast to study the interaction between NO_3 and DMS. NO_3 was measured by differential optical absorption spectroscopy[1] over a 5.5 km path-length and other species by point sampling instruments at the observatory. Fig. 1 illustrates the time-resolved profiles of NO_3, NO_2, O_3 and peroxyl radicals during a night in late autumn. By referring to the solar zenith angle plotted in the figure, it will be seen that NO_3 appears above the detection limit (2 ppt) at sunset and disappears at dawn, as expected because of its rapid photolysis during daytime. Note that the NO_3 concentration during the night is more closely correlated with O_3 than with NO_2 (this point is discussed in the later paper by Prof. Platt). A rather striking observation is the coincident appearance and disappearance of peroxyl radicals with NO_3. Indeed, at this time of year more peroxyl radicals were observed at night than during the day. The peroxyl radicals were monitored by the chemical amplifier technique, which measures

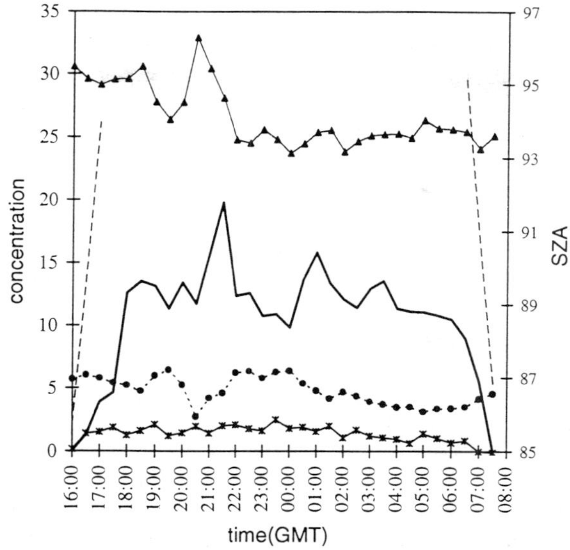

Fig. 1 Concentration profiles at the Weybourne Atmospheric Observatory (Norfolk, England) during the night 31 October–1 November, 1994: (——), NO_3 in ppt; (–×–), peroxyl radicals in ppt; (– ● –), NO_2 in ppb; (–▲–), O_3 in ppb; (– – –), solar zenith angle in degrees

HO_2 and CH_3O_2 with close to 100% efficiency. However, we believe that more complex peroxyl radicals, such as $CH_3SCH_2O_2$ produced from the reaction between NO_3 and DMS, are probably not measured efficiently by radical amplification, and this explains the low levels of peroxyl radicals observed (Fig. 1).

Fig. 2 is a plot of the measured lifetime[2] of NO_3, compared with the removal lifetime due to either reaction with DMS, reactions with unsaturated hydrocarbons such as butenes and pentenes, or the heterogeneous loss of N_2O_5 on aerosols.[2] This shows that the reactions with DMS and hyrocarbons do not account for the short lifetime of NO_3. The heterogeneous removal rate was calculated by assuming an accommodation coeffi-

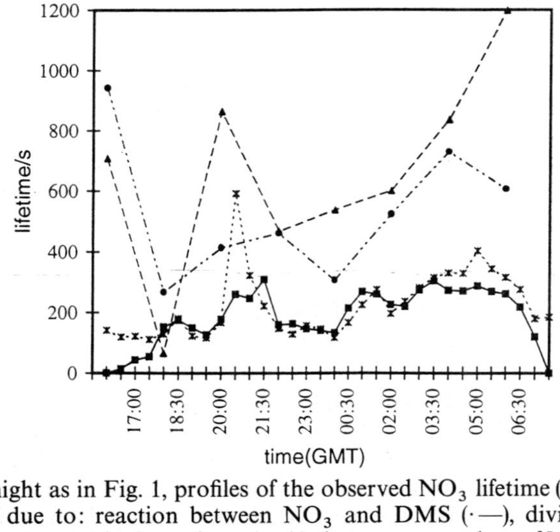

Fig. 2 For the same night as in Fig. 1, profiles of the observed NO_3 lifetime (–■–), compared with the removal lifetimes due to: reaction between NO_3 and DMS (·—), divided by 10; reactions between NO_3 and unsaturated hydrocarbons (—▲); heterogeneous loss of N_2O_5 on aerosols (∗)

cient for N_2O_5 of 0.05, and estimating the relative aerosol surface area over the course of the night from the observed attenuation of the DOAS signal due to Mie scattering over the wavelength range 500–800 nm.[3] Inspection of Fig. 2 shows a rather convincing correlation over much of the night between the measured NO_3 lifetime and that due to heterogeneous loss of N_2O_5. The absolute removal rate shown in the figure would require the aerosol surface area to have ranged from 34 to 120 μm^2 cm^{-3}, in sensible accord with aerosol measurements in the marine boundary layer.[4]

During spring and early summer the situation changes. There are much higher levels of DMS (>100 ppt) associated with phytoplankton blooms along this part of the Norfolk coast, and the reaction between DMS and NO_3 then becomes the dominant removal process of NO_3.

1 J. M. C. Plane and N. Smith, in *Spectroscopy in Environmental Sciences*, ed. R. E. Hester and R. J. H. Clark, Wiley, London 1995, pp. 223–262.
2 N. Smith, J. M. C. Plane, C-F. Nien and P. A. Solomon, *Atmos. Environ.*, in the press.
3 M. A. Box and S. Lo, *J. Appl. Met.*, 1976, **15**, 1068.
4 W. L. Chameides and A. W. Stelson, *J. Geophys. Res.*, 1992, **97**, 20565.

Prof. Ravishankara responded: It is wonderful to see the NO_3 measurements of you and your co-workers. I am sure that you would agree that more such measurements are needed, especially in the remote marine boundary layer. Simultaneous measurements of DMS would also be useful in fully resolving the initiation chemistry and the observed diurnal variation of DMS.

Dr. Rattigan communicated: The authors suggest the reason for the variation in [MSA]/[SO_4^{2-}] with temperature lies predominantly in the competition between thermal decomposition of adducts such as CH_3SO_2 and their bimolecular reaction with molecular species. However, it is also possible that a competitive pathway between the thermal decomposition of CH_3SCH_2O and its reaction with O_2 exists and that this is, in part, responsible for the variation in the ratio [MSA]/[SO_4^{2-}] with temperature.

Prof. Ravishankara communicated in response: It is indeed possible that two competing processes determine the atmospheric fate of the CH_3SCH_2O:

$$CH_3SCH_2O(+M) \rightarrow CH_3S + (CH_2O(+M) \quad (1)$$

$$CH_3SCH_2O + O_2 \rightarrow HO_2 + CH_3SCHO \quad (2)$$

We did not consider such a competition for the following reasons. Recently, McKee[1] calculated that reaction (1) is exothermic by 6.6 kcal mol^{-1}. Our direct observation of the CH_3S radical from the OH + DMS reaction in O_2 and NO support this finding and we can place a lower limit on the rate coefficient of reaction (1) of $>1 \times 10^5$ s^{-1} at 298 K. Zhao[2] reports a similar limit of $>3 \times 10^4$ s^{-1}, from his observation of the CH_2O product. Indirect evidence for reaction (2) has been reported by Butkovskaya and LeBras.[3] Our experiments monitoring OH in the presence of NO are consistent with a value of 8×10^{-15} cm^3 $molecule^{-1}$ s^{-1} for reaction (2). However, this interpretation of our data is not unique and the rate coefficient could be much lower. We cannot rule out the occurrence of reaction (2) in the atmosphere from our data. However, if we assume an A factor of reaction (1) of 5×10^{13} s^{-1}, a barrier of *ca.* 12 kcal mol^{-1} is needed to have equal rates between reactions (1) and (2). If the A factor is 1×10^{16} s^{-1}, then the barrier for reaction (1) has to be even higher for reaction (2) to compete. Such a barrier seems unrealistically high for an exothermic bond fission process.

Moreover, if the competition between reactions (1) and (2) is responsible for the branching between MSA and SO_2, the two channels must lead to different stable end products. By analogy with hydrocarbon oxidation chemistry, the atmospheric fate of the

CH_3SCHO is likely to be photolysis and reaction with OH. It appears likely that both these processes lead to CH_3S and, hence, they are identical with respect to stable product formation. Only if the photolysis led to cleavage of both C—S bonds, or led to OCS formation, and if the activation energy for reaction (1) was *ca.* 15 kcal mol^{-1} would the proposal by Rattigan be important. Experimental and theoretical work to check this possibility would be useful.

1 M. L. McKee, *J. Phys. Chem.*, 1993, **97**, 10971.
2 Z. Zhao, Doctoral Dissertation, Georgia Institute of Technology, 1995.
3 N. I. Butkovskaya and G. LeBras, *J. Phys. Chem.*, 1994, **98**, 2582.

Dr. Heal and **Mr. Nicholson** communicated: We have also been conducting modelling studies, similar to those you present on the formation of methane sulfonic acid (MSA) and sulfuric acid in the gas-phase oxidation of DMS. Our reaction scheme is based on that recently proposed by Koga and Tanaka[1] and simulations were run using the FACSIMILE software from AEA Harwell. Our results confirm a number of observations reported in the above paper but differ in others.

It is clear that the atmospheric fate of the CH_3SO_2 radical is absolutely pivotal in the determination of both the ratio of MSA to SO_4^{2-} and of the temperature dependence of this ratio. Although unimolecular decomposition of the higher oxidation state CH_3SO_3 radical can also lead to SO_4^{2-} formation, the overwhelming majority of CH_3SO_3 proceeds to form MSA. Therefore, the MSA to SO_4^{2-} ratios presented in the model of Barone *et al.* are almost entirely determined by the branching between CH_3SO_2 oxidation to CH_3SO_3 and its unimolecular decomposition to CH_3 and SO_2. However, aside from the trends in temperature dependence of branching ratios, our model results suggest that at a fixed temperature (of 298 K) the absolute branching of oxidation to decomposition at CH_3SO_2 is too small in comparison with field observations of the MSA to SO_4^{2-} ratio. This contrasts with the absolute ratios presented in the Discussion which, in most cases, exceed unity. Our results indicate that either reaction of other oxygenated (and possibly nitrated) sulfur radicals [*e.g.* CH_3SO, CH_3SOO, $CH_3S(O)OO$, $CH_3S(O)OONO_2$] must lead directly to CH_3SO_3 and/or MSA formation and so bypass CH_3SO_2 (this may include heterogeneous conversion of such radical species to MSA), or that significant amounts of MSA are indeed produced following the addition part of the reaction of OH to DMS in the initial oxidation step (a route not included in the Boulder model). If the latter is the case any decrease in the rate of formation of CH_3SO_2 from the initial abstraction route will serve to increase the ratio of MSA produced by the initial adduct formation route. We have included in our model the formation of the CH_3SOO peroxyl radical from the reaction of CH_3S with O_2 as experimentally observed by Turnipseed *et al.*[2] At lower temperatures the proportion of sulfur present as peroxy increases significantly, which lowers the flux through to CH_3SO_2 and increases the MSA to SO_4^{2-} ratio. The effect is slightly offset, however, when the possibility of isomerization from CH_3SOO to CH_3SO_2 is also included, since this clearly re-establishes a route to CH_3SO_2. The energetics of the isomerization reaction are uncertain but it presumably passes through a three-membered oxythirane ring. We used a rate of 25 s^{-1} at 298 K.[3] Note again, though, that the rate of isomerization from CH_3SOO to CH_3SO_2 will decrease as the temperature is lowered, which will act again to increase the MSA to SO_4^{2-} ratio.

In conclusion, we reiterate the absolute need for quantitative measurement of the unimolecular formation and decomposition reactions of CH_3SO_2 in particular, and as a function of temperature.

1 S. Koga and H. Tanaka, *J. Atmos. Chem.*, 1993, **17**, 201.
2 A. A. Turnipseed, S. B. Barone and A. R. Ravishankara, *J. Phys. Chem.*, 1992, **96**, 7502.
3 A. A. Turnipseed, S. B. Barone and A. R. Ravishankara, *J. Phys. Chem.*, 1993, **97**, 5926.

Prof. Ravishankara communicated in reply: It is very nice to hear about your work. We agree with all the comments. Indeed, constraining the DMS oxidation model by incorporating bounds on the absolute magnitudes of the stable end-products is a potentially powerful way of gaining insight. We would like to caution that the uncertainty in the amount of SO_2 produced from DMS oxidation is a major hindrance to calculating absolute yields. We suggest that your calculations could be evidence that heterogeneous reactions of $CH_3S(O)_xOO$ and $CH_3S(O)_xOONO_2$ to form MSA could indeed play a significant role in the overall product branching.

In the modelling presented here, we deliberately left out the possibility of the OH·DMS adduct leading to MSA because so little is known about the subsequent chemistry and this would be a good way to suppress the SO_2 yield from DMS oxidation. However, it is likely that its oxidation will yield CH_3SO_x species in the atmosphere. The chemistry of DMSO and the other products of OH·DMS reactions in the atmosphere need to be better defined to include this process in the models.

In regard to the influence of the O_2 addition to the CH_3S radical, our laboratory measurements of the reactivity of the CH_3SOO radical suggest that $CH_3S + O_3$ will dominate the overall loss of the CH_3S species.[1] The O_2-addition channel will slow the flux through CH_3S and make other slower competitive reactions more important (such as the irreversible reaction of O_2 with CH_3S and CH_3SOO, or thermal isomerization of CH_3SOO).

1 A. A. Turnipseed, S. B. Barone and A. R. Ravishankara, *J. Phys. Chem.*, 1994, **97**, 5926.

Prof. Ashfold and **Dr. Wilson** communicated: Barone *et al.* report a value of $\Delta_f H^0$ (CH_3SOO, 298 K) derived from their measured temperature dependence of the equilibrium constant for the reaction

$$CH_3S + O_2 + M \rightarrow CH_3SOO + M$$

[which yields $\Delta_r H^0$(298 K)] and the recommended literature value for $\Delta_f H^0(CH_3S$, 298 K) = 29.8 ± 0.4 kcal mol^{-1}).[1] We wish to report the results of recent experiments in our group which serve to confirm the correctness of this latter value and thus the reliability of the calculation of the enthalpy of formation of the CH_3SOO adduct.

For the past few years we have been developing and exploiting the high-resolution H-atom photofragment translational spectroscopy technique[2] as a means of studying the photodissociation dynamics of hydride molecules, including methanethiol, CH_3SH. Near-UV photolysis of this molecule yields $H + CH_3S(\tilde{X})$ fragments as the major decomposition products. Fig. 3 shows a spectrum of the internal energy distribution within the $CH_3S(\tilde{X})$ fragments resulting from 274.0 nm photolysis of a jet-cooled sample of CH_3SH, derived from time-of-flight (TOF) measurements of the partner H atoms.[3] This clearly shows that most of the $CH_3S(\tilde{X})$ fragments are formed in their zero-point vibrational level, with a spin–orbit branching ratio $^2E_{3/2} : {}^2E_{1/2} \approx 2:1$. Band contour simulations suggest that the $CH_3S(\tilde{X})$ fragments are formed with a near-Boltzmann rotational state population distribution described by a rotational temperature $T_{rot} \approx 100$ K, whilst energy conservation arguments yield a value for $D_0(H{-}SCH_3) = 30250 \pm 100$ cm^{-1} (86.7 ± 0.3 kcal mol^{-1}). $D_0(H{-}SCH_3)$ is related to the room-temperature dissociation enthalpy $\Delta H_{298}(H{-}SCH_3)$ by the integrated heat capacity,[1] from whence we deduce $\Delta H_{298}(H{-}SCH_3) = 88.0 \pm 0.3$ kcal mol^{-1}. Given $\Delta_f H^0(CH_3SH$, 298 K) $= -5.5 \pm 0.1$ kcal mol^{-1} and $\Delta_f H^0$(H, 298 K) = 52.10 kcal mol^{-1} [1] we derive $\Delta_f H^0(CH_3S$, 298 K) = 30.4 ± 0.4 kcal mol^{-1}, in very reasonable accord with the current literature recommendation.

The H-atom TOF spectra obtained at shorter photolysis wavelengths, around 216.3 nm, show clear evidence for the secondary photodissociation of the primary $CH_3S(\tilde{X})$ fragments, yielding H atoms in association with H_2CS molecules in their excited $\tilde{A}\,^1A_2$

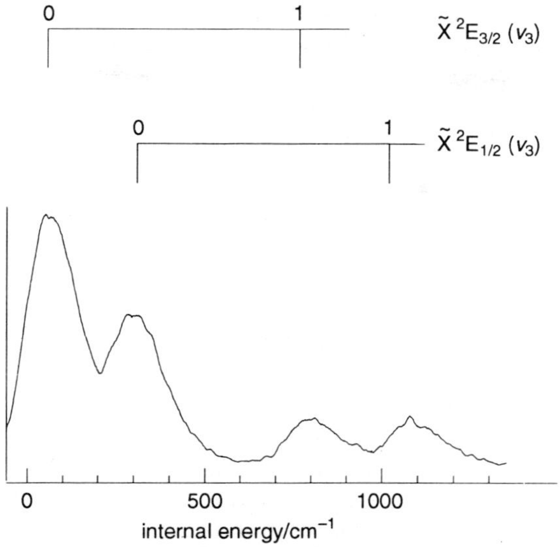

Fig. 3 Internal energy spectrum of $CH_3S(\tilde{X}\,^2E)$ fragments resulting from linearly polarized photolysis (ε_{phot} perpendicular to the TOF axis) of jet-cooled methanethiol molecules at 274.0 nm. Superimposed above the spectrum are combs indicating the short (spin–orbit split) progression in υ_3, the C—S stretching vibration.

electronic state. Analysis of these features yields a refined value for $D_0(H—CH_2S) = 17\,150 \pm 200$ cm^{-1} (48.9 ± 0.5 kcal mol^{-1}).

The S—H bond strength in methanethiol may be compared with that in hydrogen sulfide and that of the SH radical [$D_0(H—SH) = 31\,480 \pm 40$ cm^{-1}, $D_0(H—S) = 29\,300 \pm 100$ cm^{-1}], both of which have also been refined recently using the high-resolution H-atom photofragment translational spectroscopy technique.[4]

1 J. Berkowitz, G. B. Ellison and D. Gutman, *J. Phys. Chem.*, 1994, **98**, 2744 and references therein.
2 M. N. R. Ashfold, I. R. Lambert, D. H. Mordaunt, G. P. Morley and C. M. Western, *J. Phys. Chem.*, 1992, **96**, 2938 and references therein.
3 S. H. S. Wilson, M. N. R. Ashfold and R. N. Dixon, *J. Chem. Phys.*, 1994, **101**, 7538.
4 G. P. Morley, I. R. Lambert, D. H. Mordaunt, S. H. S. Wilson, M. N. R. Ashfold, R. N. Dixon and C. M. Western, *J. Chem. Soc., Faraday Trans.*, 1993, **89**, 3865.

Dr. Hancock opened the discussion of Dr. Wallington's paper: Several degradation products of HFCs produce 'stable' species such as CF_2O and FNO in quite exothermic reactions. Some of these products are vibrationally excited in non-local thermodynamic equilibrium; we have recently observed considerable excitation in these products formed in the exothermic reactions of $CF_3 + NO_2$ and $CF_3O + NO$ by time-resolved Fourier-transform IR emission. Are these species de-excited too rapidly to be of importance in potential chemical reactions or photochemical processes in the atmosphere?

Dr. Wallington replied: Ozone is present in the stratosphere in a mixing ratio of 1–10 ppm. Hence, any vibrationally excited HFC degradation products will suffer 10^5–10^6 collisions with 'inert' diluents (N_2 and O_2) before colliding with an O_3 molecule. While the rate of vibrational relaxation is dependent on the identities of the excited and third-body species, based upon the available data 10^2–10^4 is a reasonable estimate for the number of collisions necessary for vibrational relaxation.[1] Hence, vibrational relaxation is expected to dominate over chemical reactions.

It is worth noting that the atmospheric chemistry of CF_3 radicals is dominated by reaction with O_2.[2] The reaction of CF_3 radicals with NO_2 cited by you is of no atmospheric importance.

1 J. R. Barker and B. M. Toselli, *Int. Rev. Phys. Chem.*, 1993, **12**, 305.
2 E. W. Kaiser, T. J. Wallington and M. D. Hurley, *Int. J. Chem. Kinet.*, 1995, **27**, 205.

Prof. Ravishankara commented: The efficiency of a stratospheric catalytic ozone destruction cycle is dictated by two factors: (1) the length of the catalytic chains and (2) the ability to keep the critical ingredient(s) in the active form. Therefore, rates (not just rate constants) of the chain-carrying cycles, the chain-termination reactions and activation steps are all important. For example, the rate coefficients for the reactions of O_3 with CF_3O and CF_3O_2 radicals are not too different frrom those with OH and HO_2, which contribute most to O_3 loss in the lower stratosphere. However, while OH and HO_2 are abundant because they are made from the very abundant O_3 and water, the concentrations of CF_3O would be limited by the small strength of the sources, as in the case of chlorine. What makes chlorine so efficient in destroying ozone is that (a) the chain lengths are large and (b) the inactive forms are activated quite rapidly. In the case of CF_3O or fluorine, once the sink reaction occurs to make CF_2O and HF, respectively, they are not converted back into the active form. I suggest that this simple way of thinking about the efficiencies is very useful and can tell you how efficient a chemical will be in destroying stratospheric ozone without elaborate modelling studies. Such studies are, of course, needed to quantify the extent of ozone depletion. In this regard, it should be remembered that sophisticated 2D models are being routinely used to make such assessments.

I would like to point out that CFCs contain fluorine and, hence, the chemistry of fluorine has been discussed for a very long time. The strength of the HF bond and, hence, the inability of atomspheric processes [with the exception of $O(^1D) + HF$] to re-release F are the major factors in making fluorine inefficient in destroying O_3.

Dr. Wallington responded: You are correct in your comment that the relative stability of reservoirs (*e.g.* HCl and HF) is an important factor in determining the efficiency of ozone destruction. This point was actually noted in the paper.

The chemistry of fluorine has indeed been discussed for a very long time. However, the chemistry of radicals containing both carbon and fluorine, *e.g.* CF_3O_x and FCO_x has only recently been considered. We have shown that all these fluorinated radicals readily form HF which is an effective sink and that there are no reservoirs that can convert fluorine back into an active form.

Prof. Gershenzon added: I agree with the comment of Prof. Ravishankara. The length of the fluorine chain of ozone destruction is indeed short owing to the fast termination in F-atom reactions with H_2O and CH_4 and because HF, unlike HCl, does not reproduce F atoms. The pure $FO–FO_2$ cycle proposed in ref. 1

$$FO_2 + O_3 \rightarrow FO + 2O_2 \quad (1)$$

$$FO + O_3 \rightarrow FO_2 + O_2 \quad (2)$$

cannot be realized because K_2 is at least five orders of magnitude smaller than the rate constant of the reaction

$$FO + NO \rightarrow F + NO_2 \quad (3)$$

which, in reality, is a termination step for the above cycle.

1 Yu. R. Bedzhanyan, E. M. Markin and Yu. M. Gershenzon, *Kinet. Catal.*, 1992, **33**, 594.

Dr. Wallington responded: You are correct in stating that the termination steps are sufficiently rapid that the FO–FO_2 cycle will not be effective.

Prof. Okumura opened the discussion of Prof. Davidovits paper: You have suggested that a new mechanism may be operable at the gas/liquid interface for the reaction of halide ions with adsorbates. This mechanism would be a surface reaction, in contrast to a bulk reaction involving dissolved solutes. Recent theoretical and experimental investigations into microscopic aspects of halide ion solvation through the study of clusters may shed some light on the hypothesized surface mechanism by raising intriguing possibilities concerning the proposed surface reactivity of halide ions.

Molecular dynamics calculations by several groups[1–3] have found that when halide ions are solvated by clusters of water molecules, the halide ions are preferentially solvated on the exterior of the cluster rather than completely surrounded by H_2O. These effects arise from a combination of the relatively low heats of hydration of anions coupled with sizeable negative entropy effects. These results have also been confirmed by *ab initio* calculations for smaller clusters.[4] The exterior solvation state is increasingly favoured as one goes down the periodic table, with iodide ions exhibiting the strongest tendency for remaining outside a water cluster.

The experimental evidence for these effects is still forthcoming. Chesnovsky and co-workers[5] have found that photoelectron spectra of $X^-(H_2O)_n$ are consistent with exterior solvation of X^-, but extracting structural information from electron binding energies has proven difficult. We have recently obtained vibrational spectra of small, size-selected chloride ion hydrate clusters $Cl^-(H_2O)_n$.[6] These spectra unambiguously support the prediction that the chloride ion is asymmetrically solvated.

The results from the cluster studies suggest that halide ions at the gas/liquid interface may exist in a state of partial solvation. Such a picture is consistent with the cluster model of the interface that you proposed. In the gas phase, reactions such as $Cl_2 + Br^- \rightarrow Cl^- + ClBr$ are rapid and highly exothermic; thus, partially solvated ions at the surface are likely to act as highly reactive sites, although one expects that the concentration of such species will be less than the bulk halide concentration. Although such a model of the microscopic dynamics is highly speculative, it suggests new lines of inquiry into mechanisms of ionic surface reactions.

1 P. Lybrand and P. A. Kollman, *J. Chem. Phys.*, 1985, **83**, 2923.
2 L. Perera and M. Berkowitz, *J. Chem. Phys.*, 1991, **95**, 1954; 1992, **96**, 8288; 1994, **100**, 3085.
3 W. L. Jorgensen and D. L. Severance, *J. Chem. Phys.*, 1993, **99**, 4233.
4 J. E. Combariza, N. R. Kestner and J. Jortner, *J. Chem. Phys.*, 1994, **100**, 2851.
5 G. Markovich, R. Giniger, M. Levin and O. Chesnovsky, *J. Chem. Phys.*, 1994, **101**, 9344.
6 J. H. Choi, K. Kuwata, Y. Cao and M. Okumura, submitted.

Prof. Davidovits replied: You point out that the exterior solvation state is more favoured as one goes down the periodic table. This is consistent with our measurements which show that the surface reactivity of I^- is an order of magnitude higher than that of Br^-.

Dr. Rieley said: You draw attention to the fact that certain species exhibit enhanced surface reactivity (and are, therefore, taken up with higher efficiency) while others do not. Further, you conclude that this enhanced surface reactivity seems to result when the possibility exists for the formation of a stable ionic complex.

We wish to draw out some similarities from our work on ice films which indicates that the ice/gas interface seems to behave in an analogous fashion.[1] We find that the surface reactivity of ice films to incident gaseous species depends very much on whether or not the film has been pre-prepared with a potential ionic reagent, and on the nature of the pre-prepared ionic surface. We have recently performed some measurements of the sticking coefficients of statospherically relevant species on ice films at low temperature (*ca.* 100 K). While the temperature at which these measurements were conducted is somewhat removed from those in the stratosphere (*ca.* 200 K) we expect that the chemistry which we elucidate will be relevant at these more elevated temperatures.

In Table 1 we present the results for sticking coefficients of HCl and N_2O_5 on pure ice and on an ice film first prepared with each reagent in turn, before being exposed to the second reagent. It is clear that gaseous HCl sticks with high efficiency to form a hydrated, or 'ionic', form at the surface (a subject to which we will turn later in this discussion), whereas gaseous N_2O_5 does not. On the other hand, gaseous N_2O_5 reacts efficiently with a surface prepared with hydrated hydrogen chloride while gaseous HCl does not appear to react with a surface prepared with an overlayer of the ionic form of dinitrogen pentoxide.

These observations are consistent with the findings of other workers which highlight the importance not only of the role of ionic species in such heterogeneous processes but also of the availability of water of hydration. When we prepare an ionic overlayer of $NO_2^+ \cdot NO_3^-$ which, unlike the hydrated forms of dinitrogen pentoxide, does not expose any free water at the surface, gaseous HCl is unable to hydrate and react on this surface. Conversely, we believe that HCl becomes highly hydrated at the surface of our ice, being associated with at least four water molecules, which would present a reactive, yet still relatively water-rich surface, for potential reaction. In fact, this concurs with Prof. Okumura's comment that recent experiments and calculations suggest the Cl^- associated with four water molecules exists as a stable surface species.

There are clearly useful analogies to be drawn between reactive uptake on liquid droplets and on low-temperature ice films, both in the chemical mechanisms which occur and in the kinetic models appropriate for the description of these uptake processes. We expect these similarities to become even more important in the understanding of uptake processes on stratospheric aerosols.

1 H. Rieley, H. D. Aslin and S. Haq, to be published.

Prof. Davidovits replied: I think that your observations are interesting and important. I agree that under certain circumstances the surface of ice is not much different from the surface of liquid water, except that perhaps the surface mobility is reduced. Therefore, processes that have been observed to occur on liquid surfaces are likely to occur on ice surfaces. In an analysis of results for $ClONO_2$ and N_2O_5 uptake into H_2SO_4 solutions, both from our own laboratory and those of others, we have concluded that hydration of ionic intermediates (*e.g.* NO_2^+/NO_3^-) controls the rate-limiting kinetic step in the hydrolysis of the $ClONO_2$ and N_2O_5 species in an aqueous environment, much as discussed in your comment.[1]

1 G. N. Robinson, D. R. Worsnop, M. S. Zahniser, C. E. Kolb and P. Davidovits, in preparation.

Table 1 Sticking coefficients for the reactive uptake of HCl and N_2O_5 on an ice film

reaction[a]	sticking coefficient
$HCl(g) + (n + 1)H_2O(s) \rightarrow [(H_2O)_n \cdot H_3O]^+ \cdot Cl^-$	0.95
$N_2O_5(g) + H_2O(s) \rightarrow NO_2^+ \cdot NO_3^-(a) + H_2O(s)$	$\leqslant 0.1$
$N_2O_5(g) + [(H_2O)_n \cdot H_3O]^+ \cdot Cl^- \rightarrow ClNO_2(g) + HNO_3(s) + (n + 1)H_2O(s)$	~ 0.5
$HCl(g) + NO_2^+ \cdot NO_3^-(a) + H_2O(s) \rightarrow$ no reaction	0

[a] We are uncertain about the precise value of *n* but, on the basis of FTIR data, believe it likely that $n \geqslant 3$.

Dr. Clary addressed Dr. Rieley: You referred to 'low temperature'. It is known from the work of *e.g.* Delzeit *et al.*[1] that the covalent or ionic nature of HCl on ice surfaces is very sensitive to temperature. What temperatures of ice were appropriate to your measurements and were the sticking coefficients you measured sensitive to temperature?

1 L. Delzeit, B. Rowland and J. P. Devlin, *J. Phys. Chem.*, 1993, **97**, 10312.

Dr. Rieley responded: The temperature range over which we have studied the sticking coefficients of HCl and HBr on ice films is 80 to 130 K. Other measurements were conducted from our lower temperature limit of 80 K up to the desorption temperature of the ice films under the high-vacuum conditions we employ of *ca.* 170 K. On no occasion were we able to observe any evidence for the existence of covalent HCl or HBr as witnessed by Delzeit *et al.* However, we note that the surface temperatures in this work were generally somewhat lower (15–90 K) and the HCl partial pressures somewhat higher (*ca.* 10^{-3} Torr) than in our work.

Dr. Borrell addressed Prof. Davidovits: Your work teaches us a lot about the details of absorption from the gas into a liquid. Is it, however, possible to generalize the numerical results for use in cloud models so that one can say, for example, that one class of compounds is diffusion-controlled while for another the individual accommodation coefficient is required? Such generalizations are crucial in practical modelling.

Prof. Davidovits replied: Whether the uptake of a gas-phase is controlled by gas-phase diffusion or the mass accommodation coefficient depends on the size of the droplet and on the ambient pressure of the atmosphere, the relative importance of the two processes can be most simply evaluated in terms of the uptake coefficients discussed in our paper [see eqn. (6)–(8)]. Simple calculations show that, in the troposphere, the mass accommodation coefficient (α) governs the uptake for particles smaller than 10 μm if α is smaller than 10^{-2}. For smaller aerosols and at lower atmospheric pressures such as found in the stratosphere for example, the effect of the mass accommodation coefficient becomes increasingly more important.

Dr. Peter asked: The new uptake model presented in your paper describes the incorporation of incoming molecules at a liquid surface as a nucleation process. The term 'nucleation' usually refers to the start of some kind of phase transition and involves new species appearing, but in the case of gas uptake the species are all present. I would have no problem if, instead of the formation of a critical cluster, the establishment of a critical number of molecular bonds between the incoming molecule and the liquid was meant. Please explain.

Prof. Davidovits explained: Dr. Peter's comment certainly brings attention to the difficulty of using conventional terms to describe events at the gas/liquid interface. We are groping for an appropriate terminology. The use of the term nucleation is perhaps inappropriate; we only use an analogous kinetic formalism. However, the critical cluster size at the interface is inherent in the model. Still, I believe that equating the concept of a critical surface cluster with a critical number of molecular bonds between the incoming molecule and the liquid molecules at the surface is equivalent and, under certain conditions, perhaps more useful.

Prof. Vaida communicated: Recently, Donaldson *et al.*[1] have obtained experimental evidence for a surface-bound SO_2 species in aqueous solution. A combination of surface second-harmonic generation[2] (SHG) and static surface tension[3] measurements were used to probe the surface properties of the air/water interface in aqueous solutions of S^{IV}

oxides of varying concentration. Both techniques indicated that a surface layer of a neutral S^{IV} species is formed as the aqueous concentration increases. A single Langmuir adsorption isotherm can be fitted to both sets of results, as shown in Fig. 4.

The resonance condition for efficient SHG, the value of the surface tension at higher S^{IV} concentrations and the unfavourable thermodynamics associated with small, soluble anions existing preferentially at the surface all suggest that the adsorbed layer is composed of SO_2. An upper limit to the magnitude of the Gibbs energy of adsorption from solution can be estimated from these results: $\Delta_{ads} G^0$ (from solution) ≈ -14 kJ mol^{-1} at room temperature. In combination with the physical Henry's Law constant for SO_2 in water,[4] an upper limit for the magnitude of $\Delta_{ads} G^0$ (from gas) is *ca.* -23 kJ mol^{-1} at room temperature.

1 D. J. Donaldson, J. A. Guest and M. C. Goh, *J. Phys. Chem.*, 1995, **99**, 9313.
2 Y. R. Shen, *Annu. Rev. Phys. Chem.*, 1989, **40**, 327; K. B. Eisenthal, *Annu. Rev. Phys. Chem.*, 1992, **43**, 627.
3 A. W. Adamson, *Physical Chemistry of Surfaces*, Wiley, New York, 1982.
4 E. Wilhelm, R. Battino and R. J. Wilcox, *Chem. Rev.*, 1977, **77**, 219.

Prof. Platt opened the discussion of Prof. Gershenzon's paper: The ratio of HO_2/OH is a strong function of the $NO_x(=NO + NO_2)$ concentration, in particular, if NO_x is in the range of 1–10 ppb, HO_2/OH should approach unity at the higher end of this NO_x concentration range (a value of $HO_2/OH \approx 100$ only occurs at NO_x levels far below 1 ppb). Thus the effect of heterogeneous loss of $HO_x(=OH + HO_2)$ should be more pronounced at low NO_x levels. The Jülich OH-data you are referring to were made at NO_x levels in the above range; therefore, the deviation between measurement and calculation (not including heterogeneous losses of HO_x) should be larger at low NO_x, which is not seen.

Prof. Gershenzon replied: Your consideration supposes that the aerosol surface loading does not correlate with NO_x abundance. I am not sure that the absence of such a correlation is well established today.

Dr. Horie said: In your measurements of the uptake coefficients for salts using the coaxial reactor, the coefficients have been determined assuming the geometrical surface area of the rod for all cases. However, from the method of preparation of the salt sur-

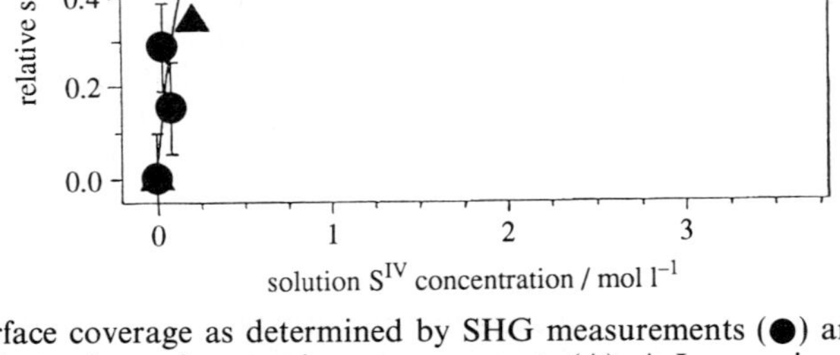

Fig. 4 Relative surface coverage as determined by SHG measurements (●) and from the surface excess, calculated from the surface tension measurements (▲). A Langmuir adsorption isotherm consistent with both sets of data is shown as the solid line.

faces, it can be considered probable that the actual or effective surface area of the salts available for HO_2 radicals may be significantly larger than that of a plain, uncoated quartz rod.

How would this increase in the surface area affect your results on the uptake coefficient of HO_2 on the salt surfaces?

Also, how would the reduction of the uptake coefficient of HO_2 on the salt surfaces affect your conclusions about the possible atmospheric importance of the heterogeneous loss of the HO_2 radicals?

Prof. Gershenzon replied: We have no exact data on the real active surface area of the salt-covered rod. However, our preliminary observation made by a microscope shows that it is close to the geometrical area. The difference is not more than 20–25%. We believe that it does not essentially change the measured value of γ_{HO_2}. We are also sure that the rate of the HO_2 sink on sea-salt aerosols can be calculated with good accuracy using the results of our measurements.

Faraday Discuss., 1995, **100**, 121–127

Mechanistic Study of the Heterogeneous Conversion of NO_2 into HONO and N_2O on Acid Surfaces

Peter Wiesen,* Jörg Kleffmann, Ralf Kurtenbach and Karl H. Becker
Physikalische Chemie/Fachbereich 9, Bergische Universität-Gesamthochschule Wuppertal, D-42097 Wuppertal, Germany

The heterogeneous conversion of NO_2 into HONO and N_2O on acid surfaces has been investigated in an 11 l Pyrex glass reactor under simulated atmospheric conditions and in a 64 l quartz glass reactor at reduced pressure. NO_2, N_2O and HONO were monitored with time by long-path tunable IR diode laser absorption spectroscopy using single vibrational rotational lines at 1641.0, 1254.47 and 1254.46 cm^{-1}, respectively.

The experimental results show that NO_2 is heterogeneously converted on acid surfaces into HONO and then into N_2O. HONO formation in the reaction system has been found to occur by the heterogeneous reaction of NO_2 with adsorbed water on the surface and does not require NO. The experimental findings indicate that the nitrosyl ion probably plays an important role in converting HONO into N_2O and HNO_3. A mechanism is proposed which can explain the experimental findings.

During the last decade the great importance of heterogeneous processes, involving water droplets in clouds, fog, sulfuric acid and other particulates, for the chemical transformation of atmospheric trace gases has been demonstrated. For the removal of atmospheric nitrogen compounds heterogeneous reactions leading to the final end-product HNO_3 are of special interest. However, these reactions may also lead, at least in part, to the formation of chemically active nitrogen species such as nitrous acid (HONO).

Nitrous acid is of particular importance in atmospheric chemistry, because the photolysis of HONO significantly enhances photooxidation processes early in the morning owing to the rapid production of OH radicals:

$$HONO + h\nu \rightarrow OH + NO \qquad (1)$$

The mechanisms by which HONO is formed in the atmosphere are not yet understood. Two gas-phase reactions have been suggested:

$$NO + NO_2 + H_2O \rightarrow 2HONO \qquad (2)$$

$$2NO_2 + H_2O \rightarrow HONO + HNO_3 \qquad (3)$$

However, the above reactions are much too slow to explain the observed night-time HONO production rates.[1] Evidence of heterogeneous HONO production by the surface reaction of NO_x with water vapour was found in smog chamber studies.[2–5] In addition, heterogeneous production of HONO has also been proposed in order to explain the high HONO concentrations which have been measured in the atmosphere.[1,6,7] The influence of HONO on atmospheric oxidation processes by the accelerated production of OH radicals has also been recently inferred from field measurements.[8]

From laboratory experiments it has been concluded that HONO formation does not require the presence of NO[9,10] as suggested in reaction (3), whereas results from field measurements tend to favour reaction (2) as the source of HONO.[7] However, very recently Harrison and Kitto[11] concluded from their field experiments that HONO is formed at the surface only by heterogeneous reactions of NO_2. In addition, indirect evidence that HONO formation does not require NO has also been obtained from field measurements where either high ozone concentrations were present, which automatically infers low NO concentrations, or low NO concentrations were directly measured. The knowledge on HONO formation in the atmosphere has recently been reviewed by Calvert *et al.*[12] These authors proposed the reaction of N_2O_3, formed by NO + NO_2 + M, with moist aerosols or other surfaces as the most important reaction for night-time HONO formation.

Among the variety of atmospheric trace gases nitrous oxide (N_2O) is of particular importance because it is directly involved in global warming as well as in the destruction of ozone in the stratosphere. Nitrous oxide has an atmospheric lifetime of *ca.* 150 years and a global warming potential of 206.[13] The transport of tropospheric N_2O to the stratosphere and its subsequent reaction with singlet oxygen atoms provide the major source of stratospheric NO.

N_2O is produced by industrial and biological processes. The atmospheric N_2O concentration has been observed to be increasing at a rate of 0.2–0.3% per year. This is probably caused by the use of fertilisers, combustion of fossil fuels, biomass burning and changes in land use in the tropics. Although a number of N_2O sources have been quantified during the last few years *ca.* 30–50% of the emission rates in the N_2O budget are still unknown.[14]

In the present study the heterogeneous conversion of NO_2 on different surfaces was investigated by tunable IR diode laser absorption spectroscopy (TDLAS). By studying, initially, the production process for N_2O it was discovered that HONO occurs as an intermediate product. Accordingly, both HONO and N_2O were studied in the system. The observed N_2O formation suggests a possible new atmospheric source of nitrous oxide and the proposed chemical mechanism also gives an explanation for heterogeneous HONO formation.

Experimental

The heterogeneous conversion of NO_2 into HONO and N_2O was studied in both an 11 l Pyrex glass reactor [surface/volume (s/v) ratio *ca.* 22 m^{-1}] under simulated atmospheric conditions and in a 64 l quartz glass reactor (White cell) (s/v ratio *ca.* 21 m^{-1}) at reduced pressure, see Fig. 1. For some experiments the 11 l Pyrex glass reactor was connected to a temperature-controlled bubbler. The quartz glass reactor was equipped with a White mirror system (base length 2.0 m) which allows *in situ* concentration measurements of the different species by long-path tunable diode laser absorption spectroscopy. The Pyrex glass reactor was connected to a 5 l path-length Herriott cell. Spectroscopic data were collected using either a total path length of 80 m in the quartz glass reactor or 43 m in the Herriott cell. In order to exclude the influence of pressure broadening of the absorption lines the total pressure in the quartz glass reactor and the Herriott cell was always kept between 6 and 7 Torr.

During the experiments in the Pyrex glass reactor a mixture of NO_2 in synthetic air or nitrogen was pumped continuously through different acid solutions in the bubbler. In some experiments acid aerosols were generated in the Pyrex glass reactor, either by injecting water into a mixture of gaseous SO_3 in synthetic air or by mixing NH_3 and HCl. Gas mixtures were sampled at regular intervals ($\geqslant$10 min) from the Pyrex glass reactor through a Teflon pipe with a Teflon needle valve into the Herriott cell where the spectra were recorded.

The absolute concentrations of NO_2 and HONO were measured by TDLAS by using the line strengths of single rotational vibrational lines of NO_2 and HONO. The

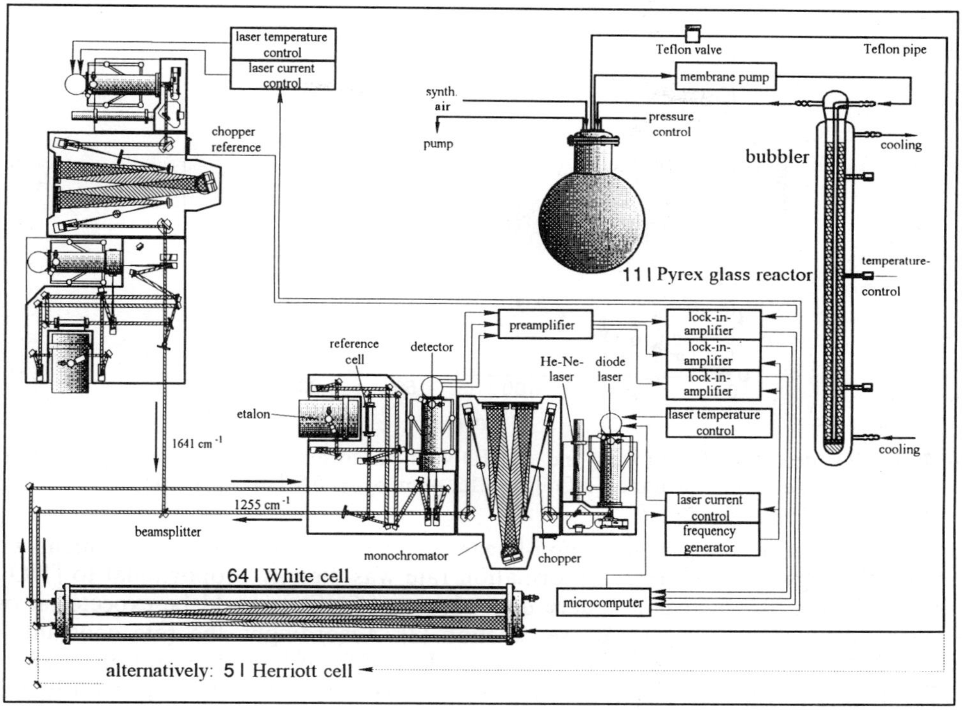

Fig. 1 Schematic diagram of the experimental set-up

absolute concentration of N_2O was measured relative to either the absorption signal of N_2O in ambient air or to gas mixtures containing known concentrations of N_2O. The detection limits achieved in the experiments were 10, 30 and 300 ppb for N_2O, HONO and NO_2, respectively. The details of the diode laser spectrometer are described elsewhere.[15]

Results and Discussion

Experiments in the Quartz Glass Reactor

In a first set of experiments the heterogeneous conversion of NO_2 was studied in the quartz glass reactor (White cell) in the dark under reduced total pressure. Concentration–time profiles of NO_2, HONO and N_2O were determined for initial NO_2 concentrations which were varied from 1.5 to 26 ppm in dry synthetic air. Fig. 2 shows a typical concentration–time profile for NO_2, HONO and N_2O. In all experiments the formation of HONO was observed to reach a maximum concentration followed by a slow decay whereas N_2O was formed continuously. The concentration–time profiles of HONO and N_2O shown in Fig. 2 indicate a consecutive reaction process:

$$NO_2 \rightarrow HONO \rightarrow N_2O \tag{4}$$

The HONO behaviour with time can be fitted by the difference of two exponentials:

$$[HONO]_t = \frac{k_a[NO_2]_{t=0}}{k_b - k_a}\,[\exp(-k_a t) - \exp(-k_b t)] \tag{I}$$

where t represents the reaction time and k_a and k_b the rate constants for HONO formation and destruction, respectively. This equation was fitted to the HONO concentrations

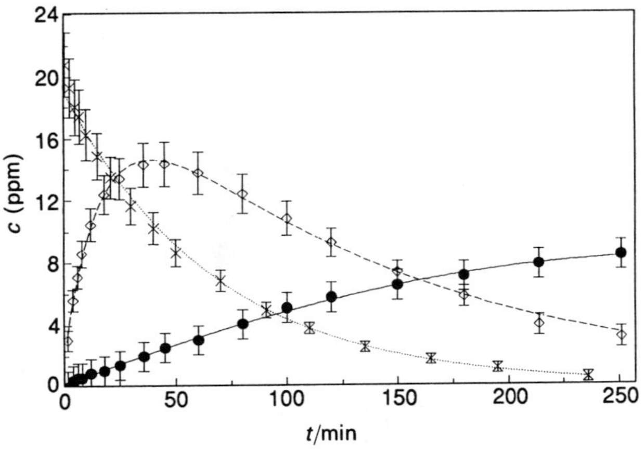

Fig. 2 ×, NO_2 decay, ●, N_2O (×20) and ◇, HONO (×5) formation in the quartz glass reactor (White cell) in the dark; $[NO_2]_0 = 21.2$ ppm, $p_{tot} = 6.5$ Torr synthetic air

for each set of conditions using a non-linear least-squares method. In agreement with previous findings[2,3,6,7] the HONO formation rate was directly proportional to $[NO_2]$, as shown in Fig. 3, whereas a square dependence of $[NO_2]$ would be expected if one assumes the thermolecular gas-phase reaction (3). Accordingly, the rate constant for HONO formation, k_a, was found to be independent of the NO_2 concentration, as shown in Table 1. This indicates that the HONO formation occurs heterogeneously with adsorbed water on the reactor surface.

Experiments in the Presence of Acid Aerosols

In order to show whether acidity of the surface of the quartz glass reactor was required for the chemical transformation of NO_2, acid aerosols were generated in the 11 l Pyrex glass reactor which was connected to the White cell for montoring the different species. Using sulfuric acid and NH_4Cl aerosols behaviour similar to that in the quartz glass reactor was observed. Starting with an initial NO_2 concentration of 13 ppm a final N_2O

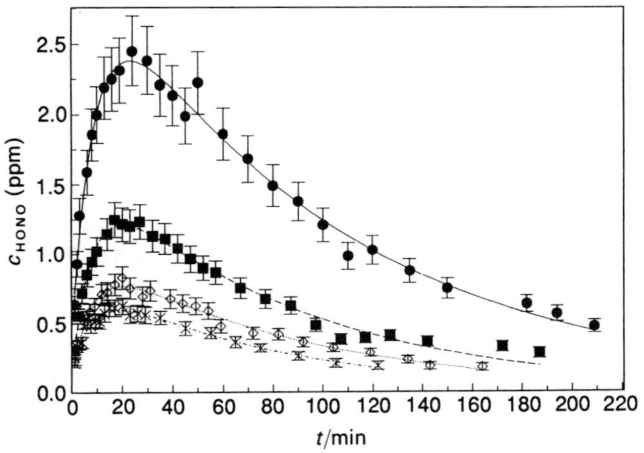

Fig. 3 Dependence of the concentration–time profiles of HONO on the initial NO_2 concentration: ×, 1.6; ◇, 3.3; ■, 6.6 and ●, 13.2 ppm; measured in the quartz glass reactor (White cell); $p_{tot} = 6.5$ Torr synthetic air

Table 1 Dependence of the formation (k_a) and destruction (k_b) rate constants of HONO on the initial NO_2 concentration

$[NO_2]$/ppm	k_a/min^{-1}	k_b/min^{-1}
13.2	0.10 ± 0.02	0.010 ± 0.004
6.6	0.14 ± 0.02	0.011 ± 0.004
3.3	0.12 ± 0.03	0.012 ± 0.003
1.6	0.12 ± 0.03	0.013 ± 0.003
mean:	0.12 ± 0.02	0.012 ± 0.003

concentration of *ca.* 1.3 ppm was obtained after a reaction time of 1 day. This behaviour can be described quantitatively by the formal disproportionation reaction:

$$8NO_2 + 3H_2O \rightarrow N_2O + 6HNO_3 \qquad (5)$$

from which a maximum N_2O concentration of 1.6 ppm would be expected. However, in the experiments with sulfuric acid aerosols no HONO was observed.

Experiments in the Pyrex Glass Reactor connected to a Bubbler

Since it was found to be very difficult to control the surface activity of the different reactors and also the aerosols, the experimental set-up was modified. Instead of the White cell, a long-path absorption Herriott cell was used for monitoring the different species. In addition, the gas mixtures in the 11 l Pyrex glass reactor were pumped continuously through a bubbler containing different solutions using well defined parameters such as concentration, temperature, volume, flow velocity and bubble diameter. It was observed that N_2O formation took place in acid solutions such as H_2SO_4, HCl and H_3PO_4 whereas for water or NaCl solutions no N_2O formation was observed.

In addition, the dependence of the HONO formation rate on the NO_2 concentration was studied in the presence of sulfuric acid in the bubbler. The NO_2 behaviour with time could be described by a single-exponential decay. Again, the HONO formation rate was found to be directly proportional to $[NO_2]$ as was already observed in the experiments in the White cell.

In another set of experiments the dependence of the maximum HONO concentration and the N_2O formation rate on the concentration of sulfuric acid was studied. HONO was observed in the gas phase only for H_2SO_4 concentrations <55%. The N_2O formation rate was found to have a maximum at *ca.* 55% H_2SO_4, as shown in Fig. 4.

Possible Reaction Mechanism

Based on the experimental observations described above the following reaction mechanism is proposed:

$$8NO_2 + 4H_2O \underset{}{\overset{\text{(surface)}}{\rightleftharpoons}} 4HONO + 4HNO_3 \qquad (6)$$

$$2HONO + 2H_2SO_4 \rightleftharpoons 2H_2O + 2NO^+HSO_4^- \qquad (7)$$

$$2NO^+HSO_4^- + 2HONO + 2H_2O \longrightarrow (HNO)_2 + 2HNO_3 + 2H_2SO_4 \qquad (8)$$

$$(HNO)_2 \longrightarrow N_2O + H_2O \qquad (9)$$

$$8NO_2 + 3H_2O \longrightarrow N_2O + 6HNO_3 \qquad \text{(net reaction, 10)}$$

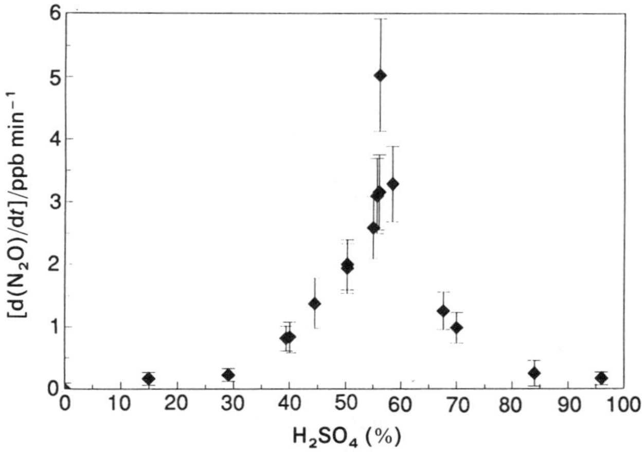

Fig. 4 Dependence of the N_2O formation rate on the concentration of H_2SO_4 measured in the modified experimental set-up (Pyrex glass reactor, bubbler, Herriott cell); $[NO_2]_0 = 15$ ppm, $p_{tot} = 740$ Torr, $V_{H_2SO_4} = 500$ ml

In a first step, reaction (6), NO_2 is converted on the surface into HONO and HNO_3. In the reaction system the NO_2 decay constant was found to be independent of the NO_2 concentration and the volume of the sulfuric acid solution but was found to be directly proportional to the surface area. Accordingly, the observed NO_2 decay should mainly be caused by an uptake of NO_2 in the sulfuric acid solution.

It should be pointed out that HNO_3 was not detected in any of the experiments in the gas phase which is probably caused by its high solubility in sulfuric acid. However, HONO was observed in the gas phase with maximum concentration which was found to be proportional to the inverse volume of the sulfuric acid solution.

In a second step, reaction (7), dissolved HONO reacts with sulfuric acid forming $NO^+HSO_4^-$. It is proposed that the nitrosyl cation (NO^+), which is a strong oxidiser, oxidises HONO to form HNO_3. By this reaction step NO^+ is reduced to $(HNO)_2$ which can easily decompose into N_2O and water, reaction (9).[16]

For H_2SO_4 concentrations $>55\%$ HONO is converted almost completely into $NO^+HSO_4^-$. Accordingly, no HONO can be observed in the gas phase and reaction (8) slows down because no HONO is available. As a consequence, also, less $(HNO)_2$ can be formed resulting in less formation of N_2O. For low H_2SO_4 concentrations the N_2O yield is also small. This is probably caused by the low formation yields of $NO^+HSO_4^-$, reaction (7), since it is known that $NO^+HSO_4^-$ is not stable in an excess of water.[17]

It should be pointed out that the proposed reaction mechanism can also explain the experimentally observed formation of N_2O in the presence of other acid solutions such as HCl and H_3PO_4.

Conclusion

The experimental results from the present study show that NO_2 is heterogeneously converted into HONO and then into N_2O. This conversion seems to take place only on acid surfaces. The kinetic behaviour of both HONO and N_2O indicate that N_2O is formed *via* HONO as an intermediate.

The findings from the present study imply that HONO formation occurs by the reaction of NO_2 with water and does not require NO. This is in agreement with previous laboratory studies and the most recent findings from field measurements, from which the involvement of NO in the surface formation of HONO was excluded. The results from the present study do not confirm the formation of N_2O_3 as an intermediate as recently suggested by Calvert *et al.*[12]

A reaction mechanism explaining the experimental results is proposed. The nitrosyl ion probably plays the major role in converting HONO into N_2O and HNO_3.

Heterogeneous N_2O formation which has been observed for the first time in the present study could represent a new atmospheric source of this greenhouse gas and could help to explain, at least in part, the existing discrepancy between known sources and sinks in the global N_2O budget. Since NO_x emissions from combustion sources are often associated with the emission of sulfate and soot particles it is possible that surface reactions reduce the concentration of nitrogen oxides and lead to N_2O formation downwind in the plume.

In addition, the proposed mechanism can explain the significant HONO formation before sunrise in foggy urban air in the presence of large abundances of NO_2. Further experiments are presently underway in order to validate the proposed reaction mechanism.

Financial support by the 'Deutsche Forschungsgemeinschaft (DFG)' within the DFG-Priority Programme 'Basics of the Impact of Air and Space Transportation on the Atmosphere', by the 'Bundesministerium für Forschung und Technologie (BMFT)' and by the 'Gesamtverband des Deutschen Steinkohlebergbaus' is gratefully acknowledged.

References

1 W. Junkermann and T. Ibusuki, *Atmos. Environ. A*, 1992, **26**, 3099.
2 F. Sakamaki, S. Hatakeyama and H. Akimoto *Int. J. Chem. Kinet.*, 1983, **15**, 1013.
3 J. N. Pitts, E. Sanhueza, R. Atkinson, W. P. L. Carter, A. M. Winer, G. W. Harris and C. N. Plum, *Int. J. Chem. Kinet.*, 1984, **16**, 919.
4 H. Akimoto, H. Takagi and F. Sakamaki, *Int. J. Chem. Kinet.*, 1987, **19**, 539.
5 J. P. Killus and G. Z. Whitten, *Int. J. Chem. Kinet.*, 1990, **22**, 547.
6 G. Lammel, D. Perner and P. Warneck, in *Physico-chemical Behaviour of Atmospheric Pollutants*, ed. G. Restelli and G. Angeletti, Kluwer, Dordrecht, 1988, p. 469.
7 J. Notholt, J. Hjorth and F. Raes, *Atmos. Environ. A*, 1992, **26**, 211.
8 F. Febo, personal communication.
9 R. Svensson, E. Ljungström and O. Lindqvist, *Atmos. Environ.*, 1987, **21**, 1529.
10 M. E. Jenkin, R. A. Cox and D. J. Williams, *Atmos. Environ.*, 1988, **22**, 487.
11 R. M. Harrison and A. N. Kitto, *Atmos. Environ.*, 1994, **28**, 1089.
12 J. G. Calvert, G. Yarwood and A. M. Dunker, *Res. Chem. Intermed.*, 1994, **20**, 463.
13 R. T. Watson, H. Rodhe, H. Oeschger and U. Sieganthaler, in *Climate Change: The IPCC Scientific Assessment*, ed. J. T. Houghton, G. J. Jenkins and J. J. Ephraums, Cambridge University Press, Cambridge, 1991, p. 1.
14 M. A. K. Khalil and R. A. Rasmussen, *J. Geophys. Res.*, 1992, **97**, 14 651.
15 K. H. Becker, J. Kleffmann, R. Kurtenbach and P. Wiesen, in *Physico-Chemical Behaviour of Atmospheric Pollutants*, ed. G. Restelli and G. Angeletti, Report EUR 15609/1 EN, ECSC-EC-EAEC, ISBN 92-826-7922-5, Luxembourg, 1994, vol. 1, p. 300.
16 N. N. Greenwood and A. Earnshaw, *Chemistry of Elements*, Pergamon, Oxford, 1984.
17 G. H. Coleman, G. A. Lillis and G. E. Goheen, in *Inorganic Syntheses*, ed. H. S. Booth, McGraw-Hill, New York, 1939, vol. 1, p. 55.

Paper 5/00199D; *Received 6th January*, 1995

Faraday Discuss., 1995, **100**, 129–153

Laboratory Studies of Atmospheric Aqueous-phase Free-radical Chemistry: Kinetic and Spectroscopic Studies of Reactions of NO_3 and SO_4^- Radicals with Aromatic Compounds

H. Herrmann, M. Exner,† H-W. Jacobi, G. Raabe, A. Reese and R. Zellner
Institut für Physikalische und Theoretische Chemie, Universität GH Essen, Universitätsstr. 5, D-45117 Essen, Germany

Recent results from kinetic and spectroscopic laboratory studies of reactions of free radicals such as NO_3 and SO_4^- with aromatic compounds in aqueous solution at $T = 298$ K are presented. Three different experimental approaches, (i) laser photolysis–visible long-path laser absorption, (ii) laser photolysis–UV long-path laser absorption and (iii) time-resolved broadband diode-array spectroscopy, have been used to investigate the kinetics of reactions of NO_3 and SO_4^- with aromatics, as well as the formation of oxidation intermediates.

For the reactions of NO_3 with benzene (1), anisole (2), *p*-xylene (3), *p*-cresol (4), toluene (5), mesitylene (6) and 1,4-dimethoxybenzene (7), rate coefficients of $k_1 = (4.0 \pm 0.6) \times 10^8$ l mol^{-1} s^{-1}, $k_2 = (1.0 \pm 0.4) \times 10^9$ l mol^{-1} s^{-1}, $k_3 = (1.6 \pm 0.1) \times 10^9$ l mol^{-1} s^{-1}, $k_4 = (8.4 \pm 2.3) \times 10^8$ l mol^{-1} s^{-1}, $k_5 = (1.2 \pm 0.3) \times 10^9$ l mol^{-1} s^{-1}, $k_6 = (1.3 \pm 0.3) \times 10^9$ l mol^{-1} s^{-1} and $k_7 = (1.0 \pm 0.3) \times 10^9$ l mol^{-1} s^{-1} were obtained. For the corresponding reactions of SO_4^- with benzene (8), toluene (9), *p*-xylene (10), *p*-cresol (11) and mesitylene (12), rate coefficients of $k_8 = (6.4 \pm 2.5) \times 10^8$ l mol^{-1} s^{-1}, $k_9 = (1.3 \pm 0.6) \times 10^9$ l mol^{-1} s^{-1}, $k_{10} = (2.7 \pm 0.9) \times 10^9$ l mol^{-1} s^{-1}, $k_{11} = (2.8 \pm 0.8) \times 10^9$ l mol^{-1} s^{-1} and $k_{12} = (1.3 \pm 0.4) \times 10^9$ l mol^{-1} s^{-1} were determined. Reactivity correlations for both radical species with the aromatic compounds are presented and potential applications of such correlations are discussed. In addition, the effect of ionic strength in the reactions of NO_3 with benzene and toluene has been investigated. Transient intermediates have been spectroscopically identified in the reactions of the sulfate radical anion in aerated as well as in oxygen-free solutions. The nature of these intermediates is discussed in view of the existing literature. Finally, possible impacts of free-radical reactions with aromatic compounds on the current understanding of chemical conversion processes within tropospheric multiphase systems are considered.

Tropospheric chemical conversions can be described not only by homogeneous gas-phase reactions but also by a variety of heterogeneous processes. The latter include processes within the so-called 'tropospheric aqueous phase', *i.e.* in the aqueous droplets of clouds, fog and rain, and also in the aqueous layer of aqueous aerosol. Moreover, heterogeneous tropospheric chemistry is also concerned with interactions between gaseous species and solid particles, but this is not the subject of the present investigation. In this study, aqueous-phase reactions between free radicals such as the nitrate

† Present address: Fa. Coloplast, Mannheim, Germany.

radical (NO_3) and the sulfate radical anion (SO_4^-) with a number of aromatic compounds are investigated. Such reactions may represent important reaction pathways in the tropospheric oxidation of aromatics, if these compounds show some solubility in water. Soluble aromatic compounds have been identified in field investigations of rain- and cloud-water[1–3] and it is possible that such compounds are readily oxidized in a variety of aqueous-phase reactions. To our knowledge, the present paper is the first to report systematic, laboratory-based, laser kinetic and spectroscopic investigations of the reactions between free radicals and aromatic compounds in aqueous solutions, with an emphasis on the consequences of such conversion processes on our present understanding of tropospheric chemistry.

Experimental

Laser Photolysis–Long-path Laser Absorption Study of Nitrate Radical (NO_3) Kinetics in Aqueous Solution

The kinetics of the reaction of NO_3 with aromatics were studied using a laser photolysis–long-path laser absorption (LP–LPLA) apparatus. Details of the experimental arrangement have been fully described before[3] and only the main characteristics are summarized here.

An excimer laser (Lambda Physik LPX 100) operated at 351 nm (active medium XeF) with typical pulse energies of 50 mJ was used to photolyse peroxodisulfate anions ($[S_2O_8^{2-}] = 3 \times 10^{-2}$ mol l^{-1}):

$$S_2O_8^- + h\nu(\lambda = 351 \text{ nm}) \rightarrow 2SO_4^- \quad (13)$$

NO_3 is then produced *via* the consecutive reaction of sulfate radical anions with nitrate ions:

$$SO_4^- + NO_3^- \rightarrow NO_3 + SO_4^{2-} \quad (14)$$

Because reaction (14) is relatively slow [$k_{16} = (5.0 \pm 0.5) \times 10^4$ l mol^{-1} s^{-1}, $I \rightarrow 0$[3]] nitrate ions were added in large excess, *i.e.* $[NO_3^-] = 0.1$ mol l^{-1}. The reaction cell used in the present work consisted of a glass cylinder with an inner diameter of 50 mm and an inner length of 60 mm. The front windows of the cell are plane plates made of fused silica (Spindler und Hoyer). A small pump is mounted behind the cell to enable the system to be operated as a flow system.

The absorption spectrum of NO_3 in aqueous solution shows a maximum at 635 nm.[4] A helium–neon laser (Spindler und Hoyer 35-2, power 2 mW) operated at 632.8 nm was applied as the analysing light source. The laser light was passed into a combination of two dielectrically coated mirrors in White configuration[5] and multiply folded through the reaction cell generating a total optical pathlength of $d = 192$ cm. Because the photolysis and analysis laser beams were passed through the cell coaxially, they were separated by a lens ($f = 200$ mm), a prism and a small aperture which allowed only the HeNe laser beam to pass. In order to block any of the rest of the excimer laser radiation a cut-off filter (Schott OG 550) was placed between the aperture and the detector. A series of apertures was mounted in front of the silicon photodiode (Hamamatsu S 1337-33) to shield the diode from ambient light. The electrical output of the photodiode was intensified using a fast amplifier (Burr-Brown OPA 102) and fed to a digital storage oscilloscope (Gould 4035), connected to an IBM-AT compatible computer. The computer is capable of triggering the excimer laser and thus enabled the results from any desired number of individual experiments to be averaged. In the present study the results of, typically, eight experiments were averaged.

LP–LPLA Study of Radical-anion (SO_4^-) Kinetics in Aqueous Solution

The same principles described above, also apply here. A schematic representation of the set-up is shown in Fig. 1. SO_4^- is again generated by excimer laser (Lambda Physik EMG 110 iCC) mediated photodissociation of peroxodisulfate ($[S_2O_8^{2-}] = 3 \times 10^{-2}$ mol l^{-1}) at $\lambda = 351$ nm, *cf.* reactions (13) and (14). The reaction cell was made of Teflon, with a rectangular geometry ($V = 112.5$ cm^3) and equipped with four Suprasil optical windows. Solutions were stored in a thermostatted glass reservoir before the experiments and were made to flow through the cell from the bottom to the top.

A two-colour HeCd laser (Omnichrome Mod. 2056 15/35) was applied in this experiment as the analysing light source. This laser simultaneously emits blue light at $\lambda = 442$ nm ($P = 35$ mW) and in the UV at $\lambda = 325$ nm ($P = 15$ mW). The analysing light was passed through the reaction cell perpendicular to the excimer laser radiation. In the present study, aluminium mirrors (Laser Optik, Hannover) were applied to reflect the 442 nm radiation 16 times through the reaction cell leading to an overall pathlength of $d = 96$ cm. For the study of other radical anions such as Cl_2^-, White mirrors equipped with dielectrical coatings (Laser Optik) were applied, resulting in the same optical pathlength at $\lambda = 325$ nm.

In either case the analysing light was fed onto a fast photodiode–amplifier combination. The photodiode was shielded from the excimer radiation at 351 nm by means of appropriate blocking filters. The output of the amplifier was fed to a digital storage oscilloscope (Gould 4050), which served as the signal recorder. The oscilloscope was connected to an IBM-AT-compatible computer which controlled the whole experiment.

Laser Photolysis–Broadband Diode-array Absorption Study of Transient Absorptions

A schematic representation of the apparatus used for the spectroscopic identification of transients formed in reactions of SO_4^- with aromatic compounds is shown in Fig. 2 This experiment has been optimized over the last couple of years and descriptions of former versions may be found elsewhere.[6] Again, excimer laser photodissociation of peroxodisulfate anions [reaction (13)] is applied; however, the photolysis wavelength

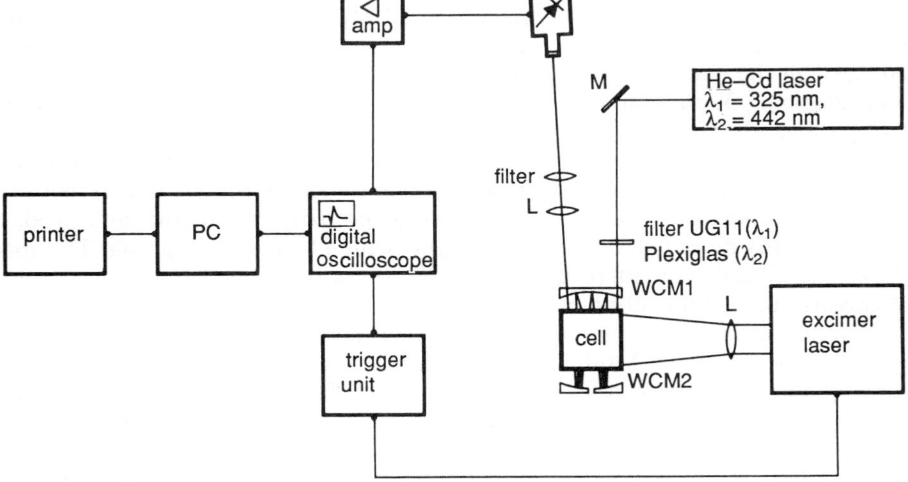

Fig. 1 Laser photolysis–long-path two-colour laser absorption (LP–LPLA) apparatus for the study of radical-anion (SO_4^-) kinetics in aqueous solution

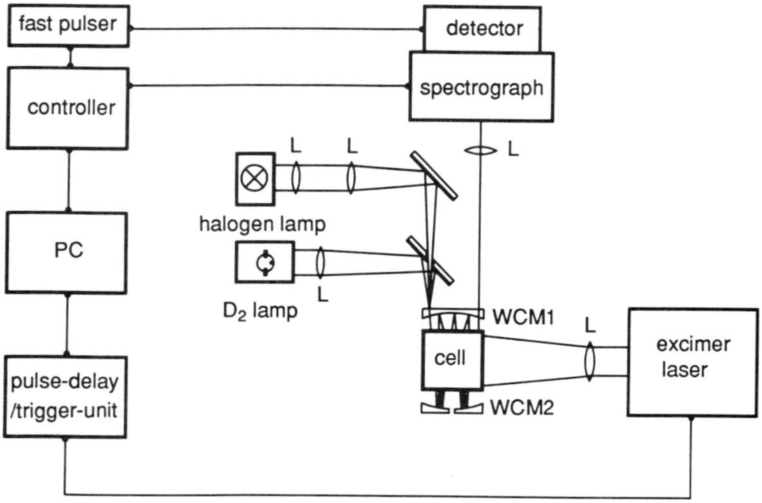

Fig. 2 Laser photolysis–broadband diode-array absorption apparatus for the study of transient spectroscopy of aromatic compounds in aqueous solution

chosen was $\lambda = 248$ nm, because the excimer laser employed (Lambda Physik COMPEX 200) was optimized for operation at 193 and 248 nm. At $\lambda = 248$ nm, typical pulse energies as high as 600 mJ are available from this laser. The geometry of the excimer light beam was adjusted to fit to the reaction cell geometry by a combination of a bi-convex quartz lens ($f = 50$ mm) with a plano-convex quartz cylinder lens ($f = 100$ mm).

The reaction cell is formed by four quartz glass plates of 4 mm thickness and has inner dimensions of 30 mm × 40 mm × 20 mm ($V_{cell} = 24$ ml). It is designed as a flow-through system. Solutions were stored in a thermostatted tank ($V = 5$ l), connected to the reaction cell with thermostatted glass pipes. Flow rates were measured using an FM 2000 flowmeter (Stellar Instruments) and were usually adjusted to 400 ml min^{-1} in the present experiments.

A 150 W water-cooled high-power deuterium lamp (Hamamatsu L 1314) and a 400 W quartz tungsten halogen lamp (Osram Xenophot HCX 64663) were used as analysing light sources. The output of both lamps was combined using a hole-mirror and then folded through the reaction cell eight times by means of two aluminium-coated concave mirrors (diameter, 30 min) in White configuration.[5] This optical configuration results in a total pathlength of $d = 32$ in the wavelength region $215 \leqslant \lambda/nm \leqslant 750$. The light emerging from the output port of the White cell was focussed onto the entrance slit of a 0.3 m spectrograph (Acton Research Mod. SP 275) by an appropiate quartz lens ($f = 50$ mm). A gated diode-array detector (700 diodes, Spectroscopy Instruments Mod. IRY-690S-B-PAR) was mounted at the focal output plane of the spectrograph. The spectrograph was equipped with three gratings (600, 300 and 150 grooves mm^{-1}), allowing simultaneous optical absorption measurements at wavelength intervals of about 108, 216 and 432 nm, respectively. The diode-array detector was connected to a controller (Spectroscopy Instruments Mod. Si 180) which again was linked to an IBM-compatible microcomputer by means of a fast parallel data bus. The multi-channel plate inside the gateable diode-array detector head is driven by a fast pulser (Spectroscopy Instruments FG 100) which allows generation of gates for irradiation of the detector diode array in the range between 5 ns and 3.5 μs. With the optical configuration of the experiment as described earlier, however, the shortest usable gate-time was 100 ns, because at shorter exposure times the detected light intensity becomes too small to generate usable spectra.

Table 1 Decadic molar absorption coefficients for aromatic compounds in aqueous solution

aromatic compound	ε/l mol^{-1} cm^{-1} (solvent, wavelength/nm)	ref.	ε/mol^{-1} cm^{-1},[a] (wavelength/nm)	ref.
benzene	204 (H_2O, 254)	7	204 (254)	7
	177 (H_2O, 254)	10		
	212 (hexane, 254)	8		
anisole	1480 (H_2O, 269)	7	1480 (269)	7
	2200 (octane, 271)	10		
p-xylene	535 (heptane, 266)	10	430 (267)	this work
p-cresol	1950 (methanol, 278)	9	1650 (277)	this work
toluene	229 (H_2O, 262)	8	229 (262)	8
	260 (hexane, 262)	10		
	300 (ethanol, 260)	7		
mesitylene	219 (ethanol, 265)	10	143 (286)	this work
1,4-dimethoxybenzene	3420 (cyclohexane, 289)	this work	2442 (286)	this work

[a] Selected for H_2O as solvent.

The timing of the experiment was controlled by a delay-pulse generator (SRS DG 535) which was triggered by another pulse generator (HP 3325 A) at frequencies between 1 and 10 Hz. In the present investigations spectra were taken at a rate of 10 Hz and usually several thousand individual spectra were averaged (see Results section for the exact numbers).

The software of the detector control allows the delay-pulse generator to be driven in a mode which automatically sets the 'gate'-position to different times after the excimer laser pulse. All time-resolved measurements for kinetic studies were performed in this mode of operation.

Other Experimental Procedures

In all experiments the number of incident excimer laser photons was measured with an appropiate energy monitor head (Radiant Dyes Mod. PEM 50 M). All aqueous solutions were freshly prepared. $K_2S_2O_8$, KNO_3, KOH and the aromatic compounds were otained from Fluka and were of analytical grade (p.a.). All chemicals were used as received. The water used was distilled twice using a quartz distilling apparatus and subsequently treated by means of a Millipore ('Milli-Q') water purification system.

Concentrations of the aromatic compounds used in the present investigation were determined directly for each experiment by UV–VIS absorption after dissolving a given amount of aromatic reactant in water. The concentrations were calculated by use of the molar decadic absorption coefficients listed in Table 1. As can be seen from Table 1, for some compounds the available literature data show considerable scatter and differences between different solvents. Furthermore, for a number of reactants, literature absorption coefficients in the aqueous phase are not available. In this case, the absorption coefficients were determined in separate measurements within the present work. The results of these determinations are also listed in Table 1.

Results

Rate Constants for Reactions of the Nitrate Radical (NO_3) with Aromatic Compounds

The first-order rate coefficients measured for the decay of NO_3 as a function of aromatic reactant concentrations are summarized in Fig. 3. Fig. 3(*a*) shows the data for the reac-

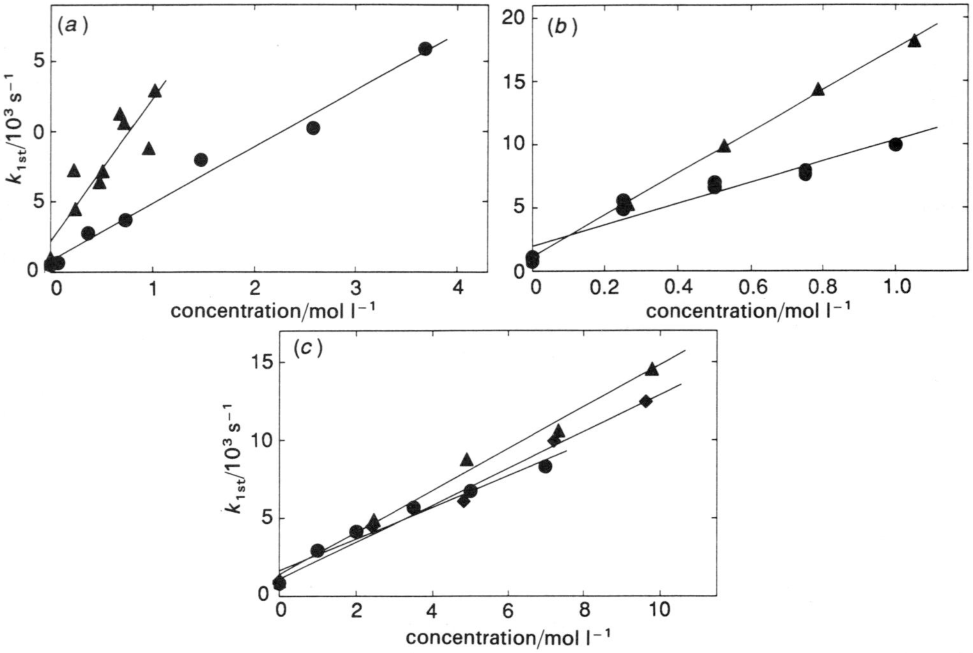

Fig. 3 (*a*) Pseudo-first-order rate coefficients as a function of reactant concentration for the reactions of NO_3 with benzene (●) and anisole (▲). (*b*) Pseudo-first-order rate coefficients as a function of reactant concentration for the reactions of NO_3 with *p*-xylene (▲) and anisole (●). (*c*) Pseudo-first-order rate coefficients as a function of reactant concentration for the reactions of NO_3 with toluene (◆), mesitylene (▲) and 1,4-dimethoxybenzene (●).

tions with benzene and anisole, *viz.*

$$NO_3 + C_6H_6 \rightarrow \text{products} \tag{1}$$

$$NO_3 + C_6H_5OCH_3 \rightarrow \text{products} \tag{2}$$

From the slopes of the straight lines, second-order rate coefficients of $k_1 = (4.0 \pm 0.6) \times 10^8$ l mol^{-1} s^{-1} and $k_2 = (1.0 \pm 0.4) \times 10^9$ l mol^{-1} s^{-1} can be derived at $T = 298$ K, pH 4 and an ionic strength of $I = 0.19$ mol l^{-1}. The corresponding rate constants for the reactions of the nitrate radical with *p*-xylene and *p*-cresol [*cf.* Fig. 3(*b*)], *i.e.*

$$NO_3 + 1,4\text{-}C_6H_4(CH_3)_2 \rightarrow \text{products} \tag{3}$$

$$NO_3 + 1,4\text{-}CH_3C_6H_4OH \rightarrow \text{products} \tag{4}$$

were determined as $k_3 = (1.6 \pm 0.1) \times 10^9$ l mol^{-1} s^{-1} and $k_4 = (8.4 \pm 2.3) \times 10^8$ l mol^{-1} s^{-1}. Finally, in Fig. 3(*c*), the first-order rate coefficients determined for the reactions of NO_3 with toluene, mesitylene and 1,4-dimethoxybenzene, *viz.*

$$NO_3 + CH_3C_6H_5 \rightarrow \text{products} \tag{5}$$

$$NO_3 + 1,3,5\text{-}(CH_3)_3C_6H_3 \rightarrow \text{products} \tag{6}$$

$$NO_3 + 1,4\text{-}(CH_3O)_2C_6H_4 \rightarrow \text{products} \tag{7}$$

are presented. From the slopes of the straight lines, second-order rate coefficients of $k_5 = (1.2 \pm 0.3) \times 10^9$ l mol^{-1} s^{-1}, $k_6 = (1.3 \pm 0.3) \times 10^9$ l mol^{-1} s^{-1} and $k_7 = (1.0 \pm 0.3) \times 10^9$ l mol^{-1} s^{-1} were obtained. All errors given above are statistical ones (precision only) for a confidence interval of 0.95.

A large number of reactions of NO_3 radicals with aromatic compounds have been investigated in previous studies. However, most of the measurements have been performed in organic solvents such as acetonitrile, *cf.* the review by Wayne *et al.*[11] Moreover, for the reagents studied in the present work there is only one single determination of k_2 (NO_3 + anisole → products) available in the literature for direct comparison. Neta and Huie[12] found a value of $k_2 = 3.2 \times 10^9$ l mol^{-1} s^{-1}, which is a factor of three larger than the one found in the present investigation. However, these data were obtained in pulse radiolysis studies in the presence of 6 mol l^{-1} HNO_3. It has been shown earlier that high ionic strengths may lead to higher reaction rate coefficients not only in ion–ion reactions but also in ion–dipole or dipole–dipole reactions.[3,13] Therefore, it is suggested that at least part of the difference between the two available determinations of k_2 may be attributed to a secondary ionic strength effect. This effect has been studied in the present work for the reactions of NO_3 with benzene and toluene and will be discussed later. No literature data for the reactions of NO_3 with the other reagents investigated in the present work are available.

Rate Constants for Reactions of the Sulfate Radical Anion (SO_4^-) with Aromatic Compounds

To compare the reactivity of the nitrate radical in reactions with aromatics with that of the sulfate radical anion, the reactions of SO_4^- with benzene, toluene, *p*-xylene, *p*-cresol and mesitylene, *viz.*

$$SO_4^- + C_6H_6 \rightarrow \text{products} \quad (8)$$

$$SO_4^- + CH_3C_6H_5 \rightarrow \text{products} \quad (9)$$

$$SO_4^- + 1,4\text{-}C_6H_4(CH_3)_2 \rightarrow \text{products} \quad (10)$$

$$SO_4^- + 1,4\text{-}CH_3C_6H_4OH \rightarrow \text{products} \quad (11)$$

$$SO_4^- + 1,3,5\text{-}(CH_3)_3C_6H_3 \rightarrow \text{products} \quad (12)$$

were also investigated.

For these reactions the first-order rate coefficients obtained are plotted as a function of reactant concentration in Fig. 4. Fig. 4(*a*) shows the data for the reactions with benzene [reaction (8)] and toluene [reaction (9)], while data for the reactions with *p*-xylene [reaction (10)], *p*-cresol [reaction (11)] and mesitylene [reaction (12)] are summarized in Fig. 4(*b*). As can be seen from Fig. 4, some of the straight lines fitted to the data points lead to negative ordinate intercepts, and a linear dependence of the pseudo-first-order rate coefficient for the decay of the sulfate radical was found only above a

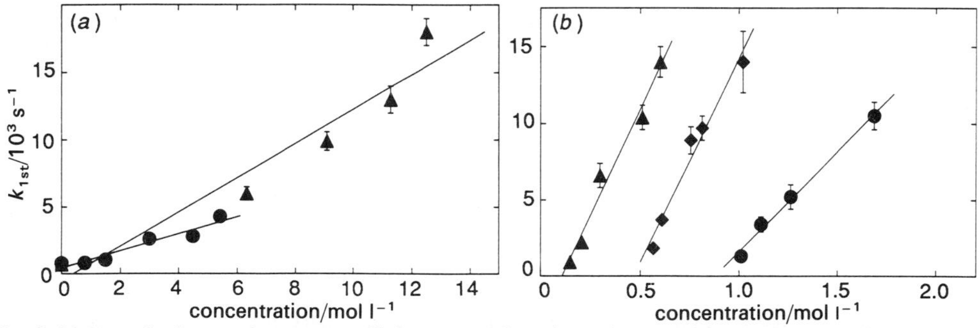

Fig. 4 (*a*) Pseudo-first-order rate coefficients as a function of reactant concentration for the reactions of SO_4^- with benzene (●) and toluene (▲). (*b*) Pseudo-first-order rate coefficients as a function of reactant concentration for the reactions of SO_4^- with *p*-cresol (▲), *p*-xylene (◆) and mesitylene (●).

certain limiting concentration. At low concentrations, the measured k_{1st} values showed some non-linear concentration dependences and these data points were not used for the derivation of second-order rate coefficients. A possible explanation for this behaviour is that at low concentrations of aromatics (in the μmol l^{-1} range) there are loss processes, *i.e.* adsorption at the walls of the cell, which significantly change the reactant concentrations. At higher concentrations, however, the loss of reactant becomes negligible and a linear dependence of sulfate radical loss rate constant on reagent concentration is observed.

From the plots shown in Fig. 4, rate coefficients of $k_8 = (6.4 \pm 2.5) \times 10^8$ l mol^{-1} s^{-1}, $k_9 = (1.3 \pm 0.6) \times 10^9$ l mol^{-1} s^{-1}, $k_{10} = (2.7 \pm 0.9) \times 10^9$ l mol^{-1} s^{-1}, $k_{11} = (2.8 \pm 0.8) \times 10^9$ l mol^{-1} s^{-1}, and $k_{12} = (1.3 \pm 0.4) \times 10^9$ l mol^{-1} s^{-1}, were determined at $T = 298$ K, pH 4 and $I = 0.09$ mol l^{-1}. There are only two previous determinations of rate coefficients available from literature for direct comparison. Both former investigations dealt with the reaction of SO_4^- with benzene, *i.e.* reaction (8). Roebke *et al.* determined a value of $k_8 = 8 \times 10^8$ l mol^{-1} s^{-1} at 298 K and pH 7.[14] Whereas this value is consistent with the result from the present study within the error limits, Neta *et al.* determined a value of $k_8 = 3 \times 10^9$ l mol^{-1} s^{-1} at 298 K, pH 7 and $I = 0.03$ mol l^{-1}.[15] This result is considerably higher than our value and that of Roebke *et al.*. In their study, Neta *et al.* studied reaction (8) in the presence of 0.1 mol l^{-1} *tert*-butyl alcohol (Bu^tOH). It may hence be speculated that their results were influenced by this high radical scavenger concentration. Bu^tOH is known to be a very effective scavenger for OH [$k(Bu^tOH + OH) = 6 \times 10^8$ l mol^{-1} s^{-1}].[16] Although the rate constant for its reaction with the sulfate radical anion is smaller by several orders of magnitude [$k(Bu^tOH + SO_4^-) = 8.4 \times 10^5$ l mol^{-1} s^{-1}],[17] the observed decay rates of SO_4^- have probably still been accelerated by the addition of Bu^tOH in high concentrations. Under the experimental conditions of the study of Neta *et al.*, a first-order loss rate for the sulfate radical anion of $k_{1st} = 8.4 \times 10^4$ s^{-1} results from the reaction with Bu^tOH.

Ionic Strength Effects

The influence of ionic strength on reaction (1) was studied at $T = 298$ K in the range $0.19 \leqslant I$/mol $l^{-1} \leqslant 4$. The second-order rate coefficients obtained are listed in Table 2 and are presented graphically in Fig. 5. As can be seen from the data, higher ionic strength lead to significantly higher reaction rates, although there is no linear dependence. From Fig. 5 the following values for the limiting rate constants for zero ionic strength [$k_1(I \rightarrow 0)$] and for high ionic strength [$k_1(I \rightarrow \infty)$] are derived: $k_1(I \rightarrow 0) = 5.2 \times 10^8$ l mol^{-1} s^{-1} and $k_1(I \rightarrow \infty) = 1.4 \times 10^9$ l mol^{-1} s^{-1} *i.e.* an increase in ionic strength results in an increases in the reaction rate coefficient by a factor of 2.7. Similar effects have been observed in other reactions of the nitrate radical with anions[3,13] and also with neutral species,[18] which may be attributed to an increase in the activity coefficients of the reactants, in accordance with the theoretical prediction of the 'secondary

Table 2 Second-order rate constants for the reaction of NO_3 with benzene as a function of ionic strength, $T = 298$ K, pH 4

I/mol l^{-1}	k/l mol^{-1} s^{-1}	log(k/l mol^{-1} s^{-1})
0.19	4.0×10^8	8.60
1.00	8.5×10^8	8.93
2.00	1.4×10^9	9.15
3.00	1.2×10^9	9.08
4.00	1.5×10^9	9.18

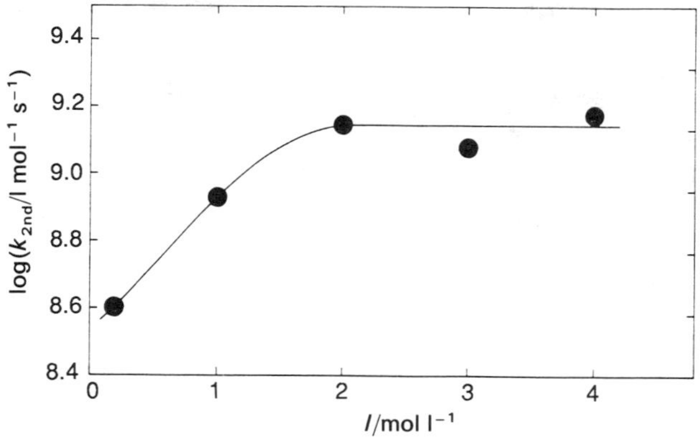

Fig. 5 Second-order rate coefficients for the reaction of NO_3 with benzene as a function of ionic strength

kinetic salt effect' of Debye and McAulay.[19] The present results represent the first study of the effect of ionic strength on the rate of reaction (1). No literature data are currently available for comparison.

The influence of ionic strength on reaction (5) was studied at $T = 298$ K in the range $0.19 \leqslant I/\text{mol l}^{-1} \leqslant 4$. The second-order rate coefficients obtained are summarized in Table 3. As can be seen from the data, higher ionic strengths lead to significantly higher reaction rates, just as in the case of the reaction with benzene. The data of Table 3 are represented graphically in Fig. 6. From Fig. 6 the following values for the limiting rate constants for zero ionic strength $[k_5(I \rightarrow 0)]$ and for high ionic strength $[k_5(I \rightarrow \infty)]$ are obtained: $k_1(I \rightarrow 0) = 1.1 \times 10^9$ l mol^{-1} s^{-1} and $k_5(I \rightarrow \infty) = 1.7 \times 10^9$ l mol^{-1} s^{-1}.

Therefore, the rate constant for the reaction of nitrate radicals with toluene $[k_5(I \rightarrow \infty)/k_5(I \rightarrow 0) = 1.5]$ is much less influenced by the secondary ionic strength effect than the reaction of NO_3 with benzene $[k_1(I \rightarrow \infty)/k_1(I \rightarrow 0) = 2.7]$. The present results represent the first study of the effect of ionic strength on the rate of reaction (5). No literature data are currently available for comparison.

Spectroscopy of Transients formed in the Reaction of SO_4^- with Aromatics

Transient Absorption Spectra observed following the Reaction of SO_4^- *with Benzene*

In addition to the kinetic studies the transient absorption spectra of reaction (8) have been investigated. Fig. 7 shows the transient absorption obtained after 248 nm photolysis of peroxodisulfate ($[K_2S_2O_8] = 10^{-4}$ mol l^{-1}) in the presence of 10^{-6} mol l^{-1} benzene in an oxygen-free solution for $\Delta t = 60$ μs. The solid curve shown in Fig. 7 represents the mathematically smoothed average of 15 000 individual experiments. The

Table 3 Second-order rate constants for the reaction of NO_3 with toluene as a function of ionic strength, $T = 298$ K, pH 4

I/mol l^{-1}	k/l mol^{-1} s^{-1}	log(k/l mol^{-1} s^{-1})
0.19	1.2×10^9	9.08
0.60	1.7×10^9	9.23
1.00	1.8×10^9	9.26
2.00	1.6×10^9	9.20
3.00	1.7×10^9	9.23

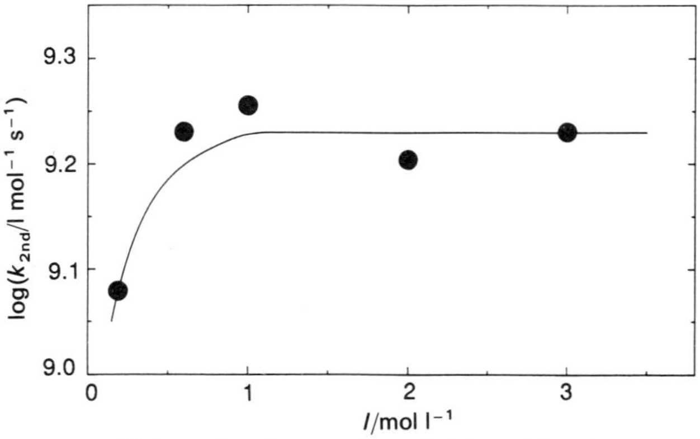

Fig. 6 Second-order rate coefficients for the reaction of NO_3 with toluene as a function of ionic strength

spectrum was measured at a delay of $\Delta t = 60$ μs because at that time the absorption of the sulfate radical was no longer detected. This was tested in separate experiments where the absorption around 450 nm, which corresponds to the sulfate radical anion, was measured. In other control experiments, benzene solutions were photolysed directly but no significant absorptions were detected in the wavelength region $245 \leqslant \lambda/\text{nm} \leqslant 350$. Fig. 7 also contains the spectrum of the hydroxycyclohexadienyl (HCHD) radical (**1**) which was obtained by Mantaka *et al.*[20] in a study of reaction of OH with benzene in aqueous solutions.[15]

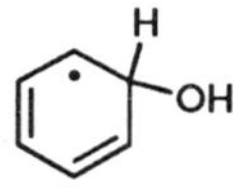

(**1**)

Hydroxycyclohexadienyl radical (HCHD)

The literature spectrum from ref. 15 was normalized to the one observed in reaction (8). The maximum absolute absorption detected in this experiment was $A = 1 - I/I_0 =$

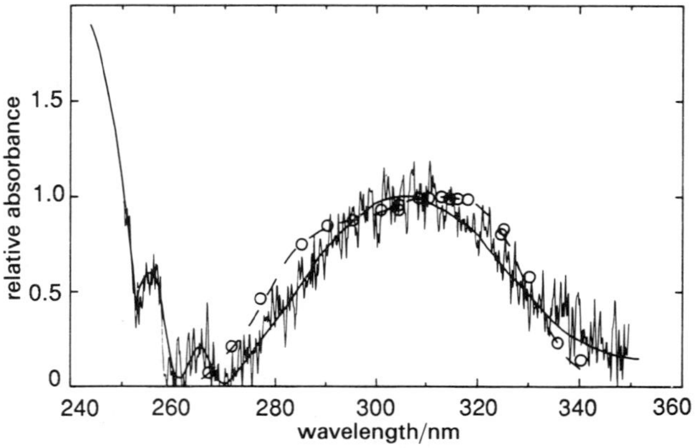

Fig. 7 Transient absorption spectrum obtained at $\Delta t = 60$ μs following the reaction of SO_4^- with benzene. (—O—O—) absorption of the HCHD radical[20]

0.04 at $\lambda = 310$ nm. A comparison reveals that both spectra have essentially the same shape at wavelengths between 270 and 340 nm. However, the spectrum obtained by Mantaka *et al.* seems to consist of the sum of two separate bands. This feature is not found in the present investigation. Moreover, the present study indicates two smaller absorption maxima at 265 and 255 nm and also a strong absorption at wavelengths below 250 nm. These features represent new spectral information. Unfortunately, the spectrum of HCHD for wavelengths smaller than 270 nm is currently not available. In time-resolved experiments the half-life of the transient species detected was determined as $\tau \approx 1$ ms. The observed kinetics were identical for all absorption bands detected. A reaction scheme which may rationalize the formation and disappearance of HCHD radicals subsequent to reaction (8) will be discussed later.

In addition to the spectrum taken at short delay times, spectra were also recorded at much longer times after the laser photolysis, *i.e.* for $\Delta t = 4$ ms. The results are shown in Fig. 8(*a*) and (*b*); both are difference spectra because the photolysis of aqueous benzene solutions leads to significant transient absorption at this delay time. In order to record the background spectrum, *i.e.* that obtained in the photolysis of pure benzene solutions, the benzene concentrations were adjusted in such a way that the same number of photons were absorbed by benzene as in the experiments with peroxodisulfate–benzene solutions. For each spectrum 3000 single experiments were averaged. The spectrum shown in Fig. 8(*a*) (top) was obtained as the difference spectrum between the background (4.7×10^{-7} mol l^{-1}) and the spectrum measured following the photolysis of a mixture containing 10^{-4} mol l^{-1} $K_2S_2O_8$ and 5×10^{-7} mol l^{-1} benzene under oxygen-free conditions.

The spectrum shown in Fig. 8(*b*) (bottom) was measured at the same concentrations of peroxodisulfate and benzene, but the solutions were saturated with oxygen. Both data sets shown in Fig. 8 represent new results for the transient absorption spectra observed after the reaction of sulfate radical anions with benzene. Note that in both cases the absolute absorption detected are much higher compared with the spectrum shown in Fig. 7. As can be seen from Fig. 8, a peak absorption of $A(234\text{ nm}) = 0.36$ in oxygen-free solutions and of $A(238\text{ nm}) = 0.21$ in oxygen-saturated solutions is observed. Hence, both maximum absorption peaks are located at nearly the same wavelength around $\lambda_1 = (236 \pm 2)$ nm. In oxygen-free solutions, smaller absorption band maxima are found at $\lambda_2 = 265$ nm ($A_2 = 0.28$), $\lambda_3 = 340$ nm ($A_3 = 0.10$) and $\lambda_4 = 500$ nm ($A_4 = 0.04$). A λ_2 maximum is also found in the spectrum of Fig. 8(*b*), but is less intense ($\lambda_2 = 270$ nm, $A_2 = 0.12$). The same is also true for the λ_3 maximum ($\lambda_3 = 346$ nm, $A_3 = 0.04$). The most pronounced difference between the spectra, however, concerns the combined λ_4/λ_5 maximum in Fig. 8(*b*) ($\lambda_4 = 500$ nm, $A_4 = 0.04$ and $\lambda_5 = 540$ nm, $A_5 = 0.06$). In contrast

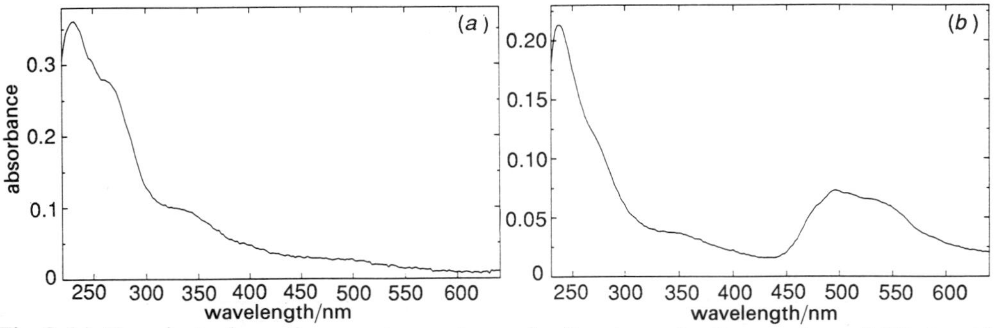

Fig. 8 (*a*) Transient absorption spectrum observed after 4 ms in the reaction of SO_4^- with benzene in oxygen-free solutions. (*b*) Transient absorption spectrum observed after 4 ms in the reaction of SO_4^- with benzene in oxygen-saturated solutions.

to the relative strengths of the other transitions compared with the λ_1 transition, this absorption band amounts to about 30% of the λ_1 absorption in oxygen-saturated solutions, but only to about 10% of the λ_1 absorption in oxygen-free solutions. A similar effect has been observed recently in a study concerned with reactions of SO_4^- with some substituted benzenes.[21] However, there are no comparable literature data available for spectroscopic identification of the corresponding transient species following the reaction of sulfate radical anions with benzene. Note that the observed absorptions shown in Fig. 8 did not change over a timescale of up to 60 ms.

Transient Absorption Spectra observed following the Reaction of SO_4^- *with Toluene*

The transient absorption spectrum detected following reaction (9) is shown in Fig. 9. This spectrum was obtained following the 248 nm photolysis of peroxodisulfate ($[K_2S_2O_8] = 10^{-4}$ mol l^{-1}) in the presence of 1.5×10^{-6} mol l^{-1} toluene in oxygen-free solution for $\Delta t = 30$ μs after the laser pulse. The solid curve shown in Fig. 9 represents the mathematically smoothed average of 15 000 individual experiments. As in the case of the reaction of SO_4^- with benzene (Fig. 7), the spectrum shown in Fig. 9 was recorded at delay times for which absorptions caused by the sulfate radical could no longer be detected ($\Delta t = 30$ μs). The actual time delay, shortened by a factor of two compared with the benzene system, corresponds to the ratio of the rate constants for reactions (9) and (8), *i.e.* $k_9 = (1.3 \pm 0.6) \times 10^9$ l mol^{-1} s^{-1} and $k_8 = (6.4 \pm 2.5) \times 10^8$ l mol^{-1} s^{-1}, *vide supra.* In control experiments toluene solutions were photolysed directly, but no significant absorptions were detected. The maximum absorption found in this experiment is located at $\lambda = 308$ nm with an absolute absorption of $A(308 \text{ nm}) = 0.04$. Very recently, a transient absorption spectrum following the pulse radiolysis of mixtures of 10^{-3} mol l^{-1} toluene, 1.5×10^{-2} mol l^{-1} $K_2S_2O_8$ and 0.2 mol l^{-1} Bu^{t}OH was observed by Merga *et al.*[21] The absorption detected in this study exhibited a maximum at 315 nm. From a comparison with earlier work on the reaction of OH with toluene,[22] Merga *et al.*[21] ascribed the observed spectrum to the OH adduct, *i.e.*

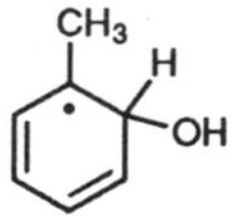

(2)

o-methylhydroxycyclohexadiencyl radical (*o*-MeHCHD)

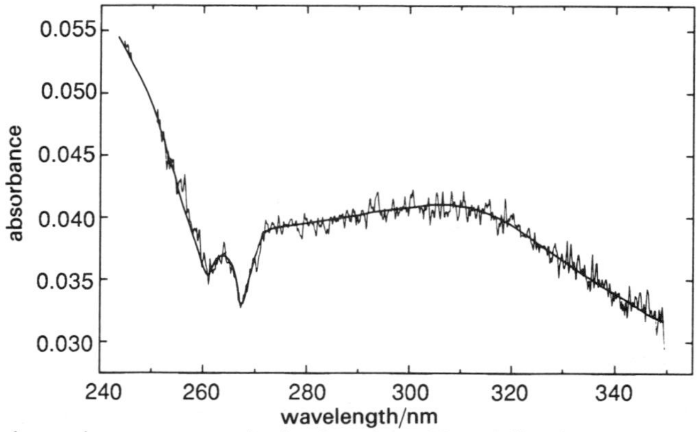

Fig. 9 Transient absorption spectrum obtained at $\Delta t = 30$ μs following the reaction of SO_4^- with toluene

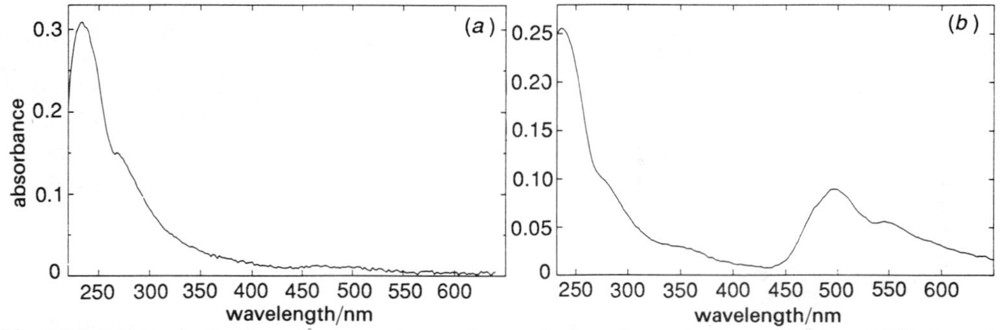

Fig. 10 (*a*) Transient absorption spectrum observed after 4 ms in the reaction of SO_4^- with toluene in oxygen-free solutions. (*b*) Transient absorption spectrum observed after 4 ms in the reaction of SO_4^- with toluene in oxygen-saturated solutions.

This would indicate that the reaction of sulfate radical anions with toluene leads to an HCHD derivative in direct analogy with the reaction of SO_4^- with benzene. However, the spectrum observed in the present study is different from that observed by Merga *et al.* Although the location of the maximum absorption peak is almost identical, the spectrum observed in the present work is much broader and of a different shape. Furthermore, the band observed at wavelengths below 260 nm was not detected in the previous study.

As in the case of benzene, transient absorption spectra were also recorded in the SO_4^- + toluene reaction as difference spectra for $\Delta t = 4$ ms. Concentrations of $[K_2S_2O_8] = 10^{-4}$ mol l^{-1} in the presence of 5×10^{-6} mol l^{-1} toluene (4.4×10^{-6} mol l^{-1} toluene for the reference spectra) were applied. The spectra resulting from 2500 single experiments are summarized in Fig. 10, where Fig. 10(*a*) and Fig. 10(*b*) show the spectra obtained in oxygen-free and oxygen-saturated solutions, respectively. The spectrum obtained in oxygen-free solutions is of a remarkably simple structure. Apparently, it consists of only two absorption bands at $\lambda_1 = 234$ nm ($A_1 = 0.31$) and a weaker band at $\lambda_2 = 270$ nm ($A_2 = 0.15$), and therefore is of a considerably simpler structure than that obtained in the corresponding experiment with benzene [*cf.* Fig. 8(*a*)]. However, when experiments are performed in oxygen-saturated solution, the resulting spectrum shows the same five transitions as in the case of the benzene reaction, *i.e.* $\lambda_1 = 234$ nm ($A_1 = 0.26$), $\lambda_2 = 270$ nm ($A_2 = 0.11$), $\lambda_3 = 350$ nm ($A_3 = 0.03$), $\lambda_4 = 498$ nm ($A_4 = 0.09$) and $\lambda_5 = 550$ nm ($A_5 = 0.06$). The only difference is that the transition at 550 nm is not as strong as in the benzene case. Moreover, the temporal behaviour is the same as observed in the benzene system, *i.e.* no changes in the observed spectra occurred within 60 ms. Hence, it may well be possible that the observed spectra are due to 'stable' reaction products. Finally, note that it was not possible to investigate directly the nature of the transients formed in the reaction of aromatics with NO_3, because of the strong absorption of the corresponding precursors $\{(NH_4)_2[Ce(NO_3)_6]$ or $K_2S_2O_8/KNO_3$ (0.1 mol l^{-1})$\}$.

Discussion

Comparison of the Kinetics of Some Reactions of the Sulfate Radical Anion and the Nitrate Radical with Aromatic and Aliphatic Organic Compounds

To test if a correlation of the reaction rate constants, as found for aliphatic substrates,[18] is also found for the reactivity of NO_3 and SO_4^- towards aromatic compounds, a double-logarithmic plot of the corresponding rate coefficients [listed in Table 4(*a*)] is

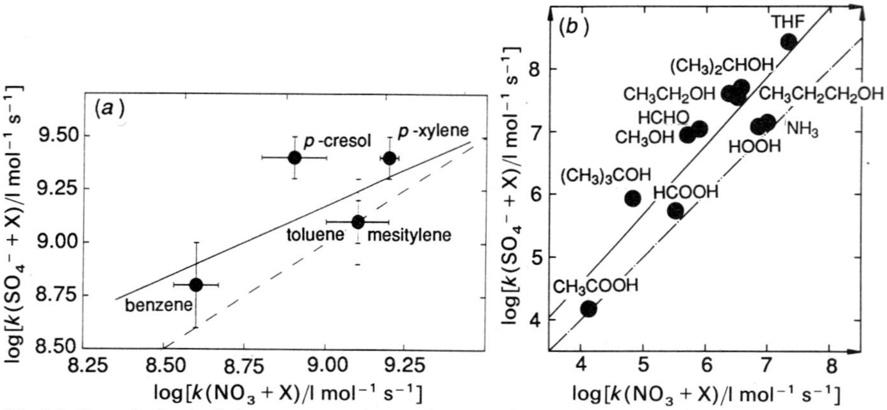

Fig. 11 (*a*) Correlation of the rate constants for reactions of SO_4^- and NO_3 with aromatic substrates. (*b*) Correlation of the rate constants for H-abstraction reactions of SO_4^- and NO_3 with aliphatic substrates.

shown in Fig. 11(*a*) for the five subtrates, the reaction rates of which have been determined in the present study for both NO_3 and SO_4^-. From this plot it is concluded that generally the sulfate radical anion reactions are somewhat faster than the corresponding nitrate radical reactions. However, for very fast reactions the rate constants for SO_4^- and NO_3 with the same aromatic substrate become more or less identical and the reactions have to be looked upon as being controlled by diffusion. Since the data set currently available for the reactivity of NO_3 and SO_4^- towards aromatic substrates is rather limited, it is not possible to provide a reliable quantitative relationship between the two sets of rate constants. It is also possible that the radicals do not react by the same mechanism, in constrast to their identical reactivity with aliphatic compounds, *vide infra.*

Results from recent studies in our laboratory on the kinetics of nitrate radical reactions with a number of aliphatic organic compounds are summarized in Table 4(*b*). This table also contains the rate coefficients for the reactions of the sulfate radical anion with the same substrates. In addition to the tabular summary, the kinetic parameters are presented graphically in Fig. 11(*b*), where in a double-logarithmic plot the rate constants for the reactions of the sulfate radical anion are plotted against the corrresponding quantities for the nitrate radical.

As can be seen, both sets of rate constants are strongly correlated. The correlation may be described by the expression:

$$\log[k(SO_4^- + HX)] = (0.18 \pm 0.03) + (1.1 \pm 0.3)\log[k(NO_3 + HX)]$$

with $N = 10$ data points and a resulting correlation coefficient of $r = 0.91$. The dotted line in F•g. 11(*b*) represents a 1 : 1 reactivity correlation. As can be seen, the fit according to the above equation results in a line nearly parallel to the 1 : 1 reactivity line. Both radical species are expected to react with aliphatic compounds by H-atom abstraction.[23,24] Therefore, it may be concluded that, in this type of reaction, the sulfate radical anion is a somewhat stronger oxidizing agent than the nitrate radical although the dependence on the nature of the hydrocarbon and the strength of the C—H bond is similar in both cases. Generally, rate constants for hydrogen abstractions are larger by about one order of magnitude for the sulfate radical anion compared with the nitrate radical.[17,24] From the good correlation of the rate constants, as apparent from Fig. 11(*b*) [*cf.* Table 4(*b*)], it is also confirmed that both SO_4^- and NO_3 react with organic aliphatic subtrates by the same mechanism, *i.e.* hydrogen-atom abstraction.

Table 4 Comparison of kinetic data for the reactions of NO_3 and SO_4^-

(*a*) With aromatic organic compounds[a]

reactant	k_{NO_3+X}/l mol^{-1} s^{-1}	$k_{SO_4^-+X}$/l mol^{-1} s^{-1}
benzene	$(4.0 \pm 0.6) \times 10^8$	$(6.4 \pm 2.5) \times 10^8$
toluene	$(1.2 \pm 0.3) \times 10^9$	$(1.3 \pm 0.6) \times 10^9$
p-xylene	$(1.6 \pm 0.1) \times 10^9$	$(2.7 \pm 0.9) \times 10^9$
p-cresol	$(8.4 \pm 2.3) \times 10^8$	$(2.8 \pm 0.8) \times 10^9$
mesitylene	$(1.3 \pm 0.3) \times 10^9$	$(1.3 \pm 0.4) \times 10^9$

(*b*) With aliphatic organic componds

species	$k(SO_4^-)$/l mol^{-1} s^{-1}	ref.	$k(NO_3)$/l mol^{-1} s^{-1}	ref.
methanol	8.8×10^6	26	5.1×10^5	23
ethanol	4.0×10^7	17[b]	2.4×10^6	23
propan-1-ol	3.5×10^7	17[b]	3.2×10^6	23
propan-2-ol	5.0×10^7	17[b]	3.7×10^6	23
tert-butyl alcohol	8.4×10^5	27	6.6×10^4	this work
formic acid	5.4×10^5	26	3.3×10^5	25
acetic acid	1.5×10^4	26	1.3×10^4	25
hydrogen peroxide	1.2×10^7	17	7.0×10^6	23
formaldehyde	1.1×10^7	26	7.8×10^5	23
THF	2.8×10^8	28	2.1×10^7	23

[a] Data from present investigation. [b] Average of literature data from ref. 17.

Ionic Strength Effects in the Reactions of NO_3 with Benzene and Toluene

Both of the reactions of NO_3 with benzene and toluene involve neutral species and hence should not be affected by a 'primary kinetic salt effect'. However, according to Debye and McAulay[19] not only for ion–ion reactions but also for ion–neutral and neutral–neutral reactions the experimentally observed rate coefficient depends on the ionic strength. Usually this effect, which is referred to as the 'secondary ionic strength effect', is small for $I < 0.1$ mol l^{-1}. Hence, it may usually be negligible in many environments in which aquatic chemistry occurs, such as natural surface waters. However, the effect might explain differences in experimental results obtained in kinetic studies by different investigators on the same reaction studied at drastically different ionic strength. Such differences were observed in the rate coefficient for the reaction of nitrate radicals with chloride anions and could be ascribed to the secondary ionic strength effect.[3,24]

Debye and McAulay[19] have derived an expression for the activity coefficient, f_i, of a neutral species in solution where the activity coefficients of the neutral species are influenced by ions not participating in the reaction:

$$\log f_i = \alpha\left(\frac{e^2}{\varepsilon_0 akT}\right)\frac{1}{2}\sum c_j z_j^2 = \beta' I$$

where α is the experimental constant, e the elementary charge, ε_0 the relative permittivity of water, k the Boltzmann constant, T the temperature, a the radius of the ion, c_j the concentration of ions, z_j the ionic charge. The rate of reaction of a neutral species X with a neutral species Y can be described with a rate constant k', independent of the ionic strength:

$$\frac{d[X]}{dt} = k' a_x a_y = k' f_x [X] f_y [Y] = k[X][Y]$$

This equation shows that the measured rate constant k is correlated to k' by:

$$k = k'f_x f_Y$$
$$\log k = \log k' + \log f_x + \log f_Y$$
$$= a + (\beta_x + \beta_Y)I$$
$$= a + \beta I$$

Hence, from the linear increase of log k with I, a 'salting coefficient β' for a specific reaction is obtained. From the slopes of the straight lines interpolated through the data points for the smallest ionic strengths for reactions of NO_3 with benzene (1) and toluene (2), salting coefficients of $\beta_1 = 0.13$ l mol^{-1} and $\beta_5 = 0.22$ l mol^{-1} are obtained. These results have to be looked upon as preliminary since the error in these data is still large (*ca.* 50%) and further measurements are necessary to characterize the salting coefficients with greater confidence. However, the results obtained for β still indicate that the acceleration of the reaction with increasing ionic strength is much more pronounced for toluene than for benzene. The salting coefficients derived may be compared with data available from the literature.[29] It may then be seen that the coefficients derived in the present work are essentially smaller than those found in reactions of NO_3 with anions like chloride (Cl^-) or formate (HCO_2^-), but they are of a similar magnitude to those found for the reaction of NO_3 with acetaldehyde (CH_3CHO), for which a value of $\beta = (0.19 \pm 0.08)$ l mol^{-1} has been determined.[29] It may hence be concluded that the stronger interactions (ion–dipole) in the case of reactions of NO_3 with anions are more strongly influenced by the presence of additional electrolyte than the weaker van der Waals-type interactions which are expected to occur when NO_3 and the neutral reactant approach each other.

Higher ionic strength can increase the reaction rate constants of important reactions. For certain reactions in the tropospheric aqueous phase the influence of ionic strength on the reaction rate can be even more important than the effect of temperature.[29] Up to now, it has not been possible to estimate *a priori* the order of magnitude of the influence of ionic strength. Information on the secondary ionic strength effect for a specific reaction has yet to be obtained by individual experiments. The results in Fig. 5 and 6 showing the influence of ionic strength on the rate coefficients of the reactions of NO_3 with benzene (Fig. 5) and toluene (Fig. 6) constitute the first systematic investigation of this effect for these reactions. As can be seen from Fig. 5, the limiting rate constant for the reaction of NO_3 with benzene [reaction (1)] is reached at about $I = 2$ mol l^{-1}, while in the case of the toluene reaction it is already reached at about $I = 0.5$ mol l^{-1}. This feature has also been observed in the study of the I dependence of the reaction rate constants for the reactions of NO_3 with chloride and formate.[29]

From Fig. 5 and 6 it has also been deduced that the ratios of the high ionic strength limiting rate constants [$k(I \rightarrow \infty)$] to the limiting values for low ionic strengths [$k(I \rightarrow 0)$] are different by about a factor of two for reactions (1) and (5). This can be explained by considering the absolute values of the rate coefficients determined for these reactions. While at low ionic strength, the benzene reaction rate coefficient is $k_1(I \rightarrow 0) = 5.2 \times 10^8$ l mol^{-1} s^{-1}, the corresponding value for the reaction with toluene is already twice as high, *i.e.* $k_5(I \rightarrow 0) = 1.1 \times 10^9$ l mol^{-1} s^{-1}.

The rate coefficient for a diffusion-controlled reaction between the nitrate radical (NO_3) and an aromatic reactant (AR) may be calculated from the Smoluchowski equation:[30]

$$k_{DIFF}/\mathrm{l\ mol^{-1}\ s^{-1}} = 4 \times 10^3 \pi N_A (D_{NO_3} + D_{AR})(r_{NO_3} + r_{AR})$$

where D_{NO_3} and D_{AR} represent the diffusion coefficients (in m^2 s^{-1}), r_{NO_3} and r_{AR} are the reaction radii (in m) of the reaction partners and N_A is the Avogadro number. For an estimate of $D_{NO_3} + D_{AR} \approx 3 \times 10^{-9}$ m^2 s^{-1} and $r_{NO_3} + r_{AR} \approx 7 \times 10^{-10}$ m (derived from data provided by Elliot *et al.*[31] a diffusion rate coefficient of $k_{DIFF} = 1.6 \times 10^{10}$ l mol^{-1} s^{-1} would result for pure water at $T = 298$ K. However, this value should be reduced significantly at ionic strengths of several mol l^{-1} (adjusted by the addition of $NaClO_4$) owing to the changing viscosity of the medium and hence smaller diffusion coefficients. It is known that the viscosity of sodium salt solutions increases by about 20% at concentrations of 1 mol l^{-1}.[32] It is therefore suggested that the relative acceleration of reaction (5) is smaller than that of reaction (1) because a nearly identical diffusion limit for the reaction is reached for these aromatic molecules at about 1.5×10^9 l mol^{-1} s^{-1}. A detailed treatment of these effects, which also includes the secondary kinetic salt effects observed in the reactions with anions studied so far in our laboratory, is in preparation.[33]

Proposed Reaction Scheme for Reactions of Free Radicals X, (*e.g.* X = NO_3, SO_4^-) with Aromatic Compounds and the Nature of the Detected Transient Absorptions

Fig. 12 summarizes possible pathways for the reactions of free radicals, X (X = NO_3, SO_4^-), with aromatic compounds in aqueous solution. The scheme presented is a compilation of reaction pathways postulated in the course of several studies available from literature, *e.g.* ref. 20–22, 34, 35. As can be seen, the HCHD radical (1) may be formed by addition of X followed by reaction with water and the elimination of HX, *i.e.* HNO_3 or HSO_4^-. It is also possible that HCHD is formed as a result of a direct electron transfer from the aromatic ring to the oxidant X. This reaction type results in a cation which subsequently reacts with water and eliminates HX. Note that in a recent study by Baciocchi *et al.* it has been shown that this mechanism dominates in the reactions of NO_3 with aromatics in acetonitrile.[36] Electron transfer was assigned to the reactions as the predominant mechanism because a typical Rehm–Weller reactivity pattern,[37] *i.e.* an acceleration of electron-transfer rates with increasing Gibbs energy change in the reaction, was observed. In contrast to this earlier study, only very little variation of the rate coefficients for the reaction of NO_3 with the same aromatics studied by Baciocchi *et al.* was found with water as solvent in the present study. All rate coefficients determined in the present study for the reactions of NO_3 with aromatics are in the range $0.4 \leqslant k/10^9$ l mol^{-1} s^{-1} $\leqslant 1.6$, where the lower and the upper limit are given by the reaction rate coefficients for benzene and *p*-xylene, respectively. This means that the reactions of the nitrate radical with aromatic compounds in aqueous solution are nearly diffusion controlled. In the study by Baciocchi *et al.* with acetonitrile as solvent, on the other hand, rate coefficients varied over five orders of magnitude depending strongly on the Gibbs energy of reaction ($\Delta_r G^0$).

Unfortunately, it was not possible to investigate directly the nature of the transients formed in the reaction of aromatics with NO_3, *vide supra*, and hence transient absorption spectra could be recorded only for the sulfate radical reactions. From a comparison of the spectrum obtained following the reaction of SO_4^- with benzene [reaction (8)] we conclude that HCHD is formed in a fast primary reaction step which is completed after about 60 µs. HCHD then decays with a half-life of about 1 ms. As can be seen from Fig. 12, three reaction pathways may occur: (i) formation of cresols and HO_2 in the presence of oxygen, (ii) formation of non-aromatic cyclodienylperoxy intermediates and (iii) formation of aromatic radical cations after protonation of HCHD and the elimination of water.

In the case of toluene, the formation of a transient with the absorption spectrum shown in Fig. 9 is completed in about 30 µs after laser irradiation. However, the observed spectrum is not in agreement with the only spectrum available in the literature.[21] The reason for this discrepancy is not obvious. It may be speculated that in the toluene case in addition to formation of *o*-MeHCHD by either addition/elimination or

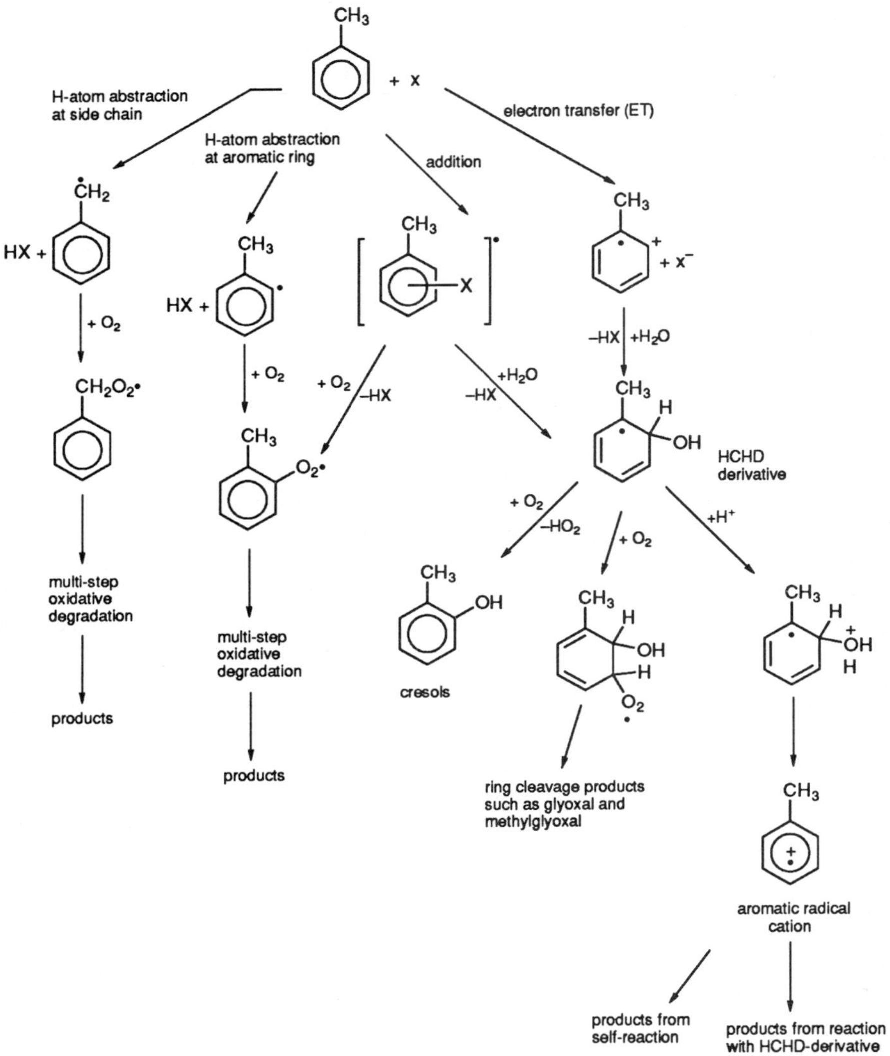

Fig. 12 Possible reaction mechanisms in reactions of free radicals X (*e.g.* X = SO_4^-, NO_3) in aqueous solution

electron transfer, H-atom abstraction from the side chain also occurs, which in oxygenated solution leads to the formation of a benzylperoxy radical. The ratio of these two pathways may be different owing to the totally different experimental approaches applied in the present study and that of Merga *et al.*[21] The pulse radiolysis technique applied by these authors may have led to direct fragmentation of toluene and the formation of products different from those observed in the present study. However, it is not possible to differentiate at the present time between the spectral contributions of *o*-MeHCHD and other unknown species to the experimentally observed transient spectra in either the present study or that of Merga *et al.* It may also be possible that other unknown products contribute to the observed spectra following H-abstraction from the side chain. This is supported by the findings of several authors on the kinetics of NO_3 with aromatics in acetonitrile.[36,38,39]

In the spectra of both HCHD (Fig. 7) and *o*-MeHCHD (Fig. 9) it was possible to detect another absorption band with its maximum below 250 nm in both systems. Apparently, this absorption is much stronger than the absorptions observed around 310 nm. The spectra indicate a broader absorption in the case of the toluene system compared to benzene. In both spectra an intermediate region between the two strongest absorption bands is also found in the wavelength range 250–270 nm for benzene and 260–270 nm for toluene. In the HCHD spectrum absorption bands are found at $\lambda_2 = 258$ nm and at $\lambda_3 = 263$ nm. For *o*-MeHCHD, only one smaller absorption band is found at $\lambda_3 = 263$ nm, *i.e.* at the same wavelength as the λ_3 band in the benzene system. To date, there is no information available on the absorption spectra of HCHD or *o*-MeHCHD in aqueous solution below $\lambda = 270$ nm. Comparable information from gas-phase studies is also not available.

As mentioned above, all the spectra observed 4 ms after laser photolysis, *i.e.* the spectra shown in Fig. 8(*a*) and (*b*) [obtained after the reaction of SO_4^- with benzene, reaction (8)] and those shown in Fig. 10(*a*) and (*b*) [obtained after the reaction of SO_4^- with toluene, reaction (9)], remain unchanged on a timescale of 60 ms. In an additional set of experiments, the absorption spectra were recorded by means of a diode-array absorption experiment (Fig. 2) over much longer timescales, *i.e.* 1000 spectra were taken every 5 min after laser irradiation. In a control experiment it was found that the observed behaviour does not change when the analysing light was blocked in between the measurements at 5 min intervals. Hence, the possibility that the analysing light caused photochemical reactions can be ruled out. From these findings it is concluded that the spectra shown in Fig. 8(*a*) and (*b*) and Figs 10(*a*) and (*b*) are due to stable reaction products which are formed following the reactions of SO_4^- with benzene and toluene in either oxygen-free [Fig. 8(*a*) and 10(*a*)] or oxygen-saturated [Figs. 8(*b*) and 10(*b*)] solutions.

Despite numerous studies on free-radical reactions with aromatics in aqueous solution, only relatively little information on the absorption spectra of stable products has been published, which in some cases has led to the identification of products and thus reaction pathways, see *e.g.* ref. 20 and 21. Most of the existing literature on reactions of free radicals with aromatics in solution, however, deals with reactions in organic solvents (see Wayne *et al.*[11] and references therein) or describes pulse radiolytic investigations in the aqueous phase (*e.g.* ref. 20 and 21). The latter experiments are often performed at concentration levels of the initially formed free radicals (like OH or SO_4^-)[20,21] of the order of 10^{-5} mol l^{-1} and even higher. In contrast to these experiments, in the transient absorption detection experiments performed in the present study, initial concentrations of the sulfate radical anion of *ca.* 5×10^{-7} mol l^{-1} were generated. The high concentrations of primary radicals in pulse-radiolytic experiments, on the one hand, leads to a simplified detection of transients, since optical pathlengths of a few cm are sufficient for detection. On the other hand, high concentrations of primarily generated radicals also lead to the formation of high concentrations of reaction products *via* complicated reaction schemes such as shown in Fig. 12. If transients like HCHD and its derivatives are formed in high concentrations, radical–radical recombinations will occur and possibly even govern the temporal decays of the observed transients. This was the case in the study by Mantaka *et al.*[20] where dimerization of HCHD was observed as the most important fate of this radical. The non-radical intermediate dimer was subsequently decomposed under acid catalysis to form either biphenyl *via* one reaction channel or phenol and benzene *via* another. This study has been performed in oxygen-free solutions and hence can be compared with the spectrum shown in Fig. 8(*a*). In contrast to the study by Mantaka *et al.*[20] the absorption of biphenyl ($\lambda_{max} = 247$ nm, $\varepsilon_{max} = 1.7 \times 10^4$ l mol^{-1} cm^{-1},[20] where ε_{max} is the molar absorption coefficient at λ_{max}) is not observed in the present study. Hence, we conclude that dimerization of HCHD is not an important fate of this radical under the experimental conditions of the present study. However, we suggest that the λ_2 absorption band observed as a 'shoulder' in both Fig. 8(*a*) ($\lambda_2 = 265$ nm) and Fig. 8(*b*) ($\lambda_2 = 270$ nm) is attributed to phenol, which is known to show an absorption peak at $\lambda = 270$ nm with $\varepsilon_{270} = 1.5 \times 10^3$ l mol^{-1} cm^{-1}.[10] This indicates that HCHD undergoes a rearrangement in which the ring is retained to form the more stable phenol and HO_2. The formation of HO_2 might have important consequences for tropospheric aqueous-phase chemistry, as will be discussed later. If ring-retention is also an important fate of *o*-MeHCHD, which is formed in the course of the reaction of SO_4^- with toluene [reaction (9)], *o*-cresol would be expected to be the product. *o*-Cresol shows a broad absorption at around 270 nm with $\varepsilon = 1820$ l mol^{-1} cm^{-1}.[11] We therefore conclude that in both reactions HCHD and *o*-MeHCHD undergo rearrangements to form the more stable ring-retaining products phenol and *o*-cresol, respectively.

From the observed absorptions of HCHD [A_{max}(310 nm) = 0.04] and of *o*-MeHCHD [A_{max}(308 nm) = 0.04] and absorption coefficients available from literature of ε_{313} (HCHD) = 4700 l mol^{-1} cm^{-1},[20] and $\varepsilon_{317.5}$ (*o*-MeHCHD) = 5500 l mol^{-1} cm^{-1},[22] $[HCHD]_{max} = 1.2 \times 10^{-7}$ mol l^{-1} and $[o\text{-MeHCHD}]_{max} = 10^{-7}$ mol l^{-1} may be calculated. These are consistent with an initial concentration of the sulfate radical anion of $[SO_4^-]_0 = 5 \times 10^{-7}$ mol l^{-1}. The remaining sulfate radicals probably recombine [$k(SO_4^- + SO_4^-) = 1.6 \times 10^8$ l mol^{-1} s^{-1},[40]]. From the absorptions observed in oxygen-saturated solutions at $\lambda_2 = 270$ nm of $A_2 = 0.12$ (benzene) and $A_2 = 0.11$ (toluene), concentrations of the products phenol and *o*-cresol can also be estimated, giving $[C_6H_5OH] = 1.2 \times 10^{-6}$ mol l^{-1} and $[o\text{-}C_6H_4CH_3OH] = 8.7 \times 10^{-7}$ mol l^{-1}. These concentrations exceed the maximum concentrations of HCHD and *o*-MeHCHD as calculated earlier. Therefore, it is concluded that only part of the absorptions detected at the λ_2 bands at 270 nm can be ascribed to the absorption of phenol and *o*-cresol, respecitvely. The branching ratio for the pathway leading to phenol/*o*-cresol and HO_2 cannot be quantified only by the observed absorptions at $\lambda_2 = 270$ nm because this absorption is not solely caused by two of the possible products, *i.e.* phenol or *o*-cresol.

Note that the λ_1 absorption at $\lambda_1 = (236 \pm 2)$ nm is observed in all four spectra observed at 4 ms after irradiation, *i.e.* in both oxygen-free and oxygen-saturated solutions. This absorption cannot easily be ascribed to phenol or *o*-cresol because these species show short-wavelength absorption maxima at $\lambda = 211$ nm (phenol[10]) and $\lambda = 214$ nm (*o*-cresol[9]). At around 236 nm both species even show absorption minima totally different from the absorption spectra observed in the present study.

Finally, both the spectra obtained in oxygen-saturated solution [Fig. 8(*b*) and 10(*b*)] show another interesting feature: absorption at $\lambda_4 \approx (500 \pm 2)$ nm and $\lambda_5 \approx (545 \pm 5)$ nm. In both spectra the λ_4 band is the more intense one and the λ_5 band is significantly smaller for the toluene system compared with that of the benzene reaction. These bands have also been observed in control experiments, to remain unchanged after long reaction times (60 min). Therefore, it is concluded that they also have to be assigned to stable reaction products.

To summarize, the observed spectra are assigned to stable reaction products. They are not inconsistent with the formation of phenol/*o*-cresol and HO_2, but it is not possible at present to determine the branching ratio for this reaction pathway. The nature of the absorptions at about 236, 500 and 540 nm requires further investigation, which will have to employ not only UV–VIS spectroscopy but also analytical instrumental analysis such as HPLC.

Comparison of Reactivity of NO_3 in the Gas Phase and in Aqueous Solution

Kinetic data for the reactions of NO_3 with aliphatic and aromatic organic compounds in the gas phase are summarized in Tables 5(*a*) and (*b*). These rate constants are present-

Table 5 Gas-phase rate coefficients used for Fig. 13

(*a*) aliphatic organic compounds[11]

reactant	k_{NO_3}/l mol^{-1} s^{-1}	k_{NO_3}/cm^3 molecule^{-1} s^{-1}
methanol	1.3×10^5	2.1×10^{-16}
ethanol	$\leqslant 5.4 \times 10^5$	$\leqslant 9 \times 10^{-16}$
propan-2-ol	$\leqslant 1.4 \times 10^5$	$\leqslant 2.3 \times 10^{-16}$
acetone	$\leqslant 5.1 \times 10^3$	$\leqslant 8.5 \times 10^{-18}$
formaldehyde	4.1×10^5	6.8×10^{-16}
NH_3	$\leqslant 3.6 \times 10^5$	$\leqslant 6 \times 10^{-16}$
H_2O_2	$\leqslant 1.2 \times 10^6$	$\leqslant 2 \times 10^{-15}$
THF	2.94×10^6	4.88×10^{-15}
CH_2Cl_2	3.0×10^3	5×10^{-18}
$CHCl_3$	1.8×10^4	3×10^{-17}

(*b*) aromatic organic compounds

reactant	k_{NO_3}/l mol^{-1} s^{-1}	k_{NO_3}/cm^3 molecule^{-1} s^{-1}	ref.
benzene	$<6.6 \times 10^3$	$<1.1 \times 10^{-17}$	41
toluene	$(4.1 \pm 1.7) \times 10^4$	$(6.8 \pm 2.8) \times 10^{-17}$	43
anisole	$(1.3 \pm 0.2) \times 10^5$	$(2.1 \pm 0.4) \times 10^{-16}$	43
mesitylene	$(1.4 \pm 0.7) \times 10^5$	$(2.3 \pm 1.1) \times 10^{-16}$	42
p-xylene	$(2.7 \pm 0.5) \times 10^5$	$(4.50 \pm 0.77) \times 10^{-16}$	43

ed with the corresponding aqueous-phase rate coefficients in a double-logarithmic plot in Fig. 13. As can be seen, the rate coefficients for H-atom abstraction reactions in the different media are very well correlated. The broken line shown in Fig. 13 represents a 1 : 1 reactivity correlation of NO_3(aq) and NO_3(g). From a fit of the kinetic data for the reactions with aliphatic compounds the regression line

$$\log(k_{NO_3(g)}/\text{l mol}^{-1}\text{ s}^{-1}) = (0.78 \pm 1.21) + (0.74 \pm 0.20)\log(k_{NO_3(aq)}/\text{l mol}^{-1}\text{ s}^{-1})$$

with $N = 10$ and $r = 0.94$ was derived. This result is essentially the same as a similar fit recently published from our laboratory.[23] From the strong correlation of rate coefficients in both media and the good correlation of the rate coefficients with bond dissociation energies (E_{bd}) of the organic substrates in aqueous solution[23] and in the gas phase[11] it is concluded that the reaction of nitrate radicals with saturated aliphatic organic compounds proceeds by H-atom abstraction in both media. The absolute values of the H abstraction rate coefficients are somewhat larger in the aqueous phase than in the gas phase but the difference is, in most cases, smaller than one order of magnitude.

As can be seen from Fig. 13, the rate constants for the reactions of NO_3 with aromatic compounds investigated in the present study follow a totally different pattern when compared with the corresponding gas-phase rate constants. In contrast to the behaviour of aliphatic compounds, the absolute reaction rate constants are considerably larger in the aqueous phase than in the gas phase. Differences range from about two orders of magnitude (naphthalene[45]) to about five orders of magnitude in the case of benzene. From this it is concluded that H-abstraction is not the dominant mechanism for reactions of NO_3 with aromatics in either the aqueous or the gas phase. NO_3 in the gas phase is generally believed to react with aromatic substrates *via* an addition mechanism.[11] From the drastically increased rate coefficients it may be concluded that the dominant mechanism in aqueous solution may be electron transfer from the aromatic ring. This would also be consistent with the finding that electron transfer has been

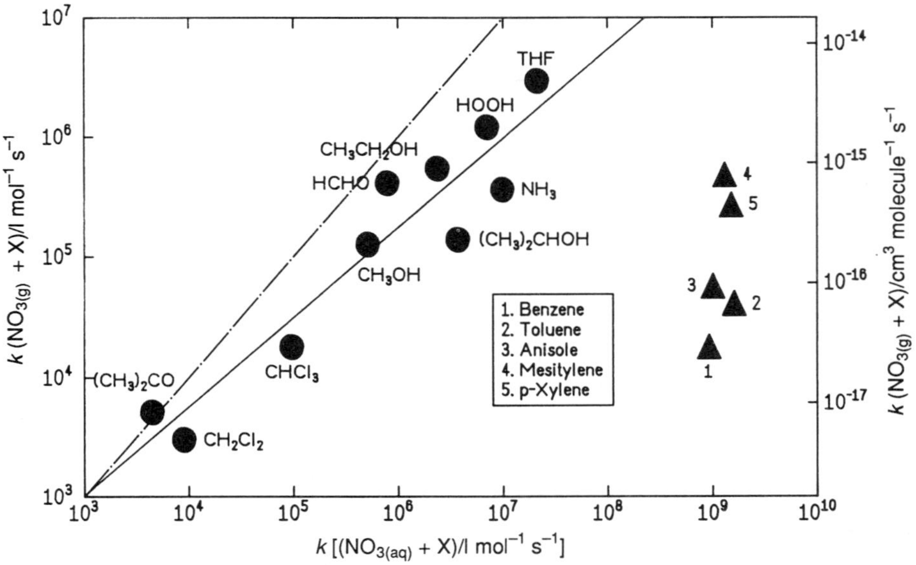

Fig. 13 Correlations between NO_3 reaction rate constants towards aliphatic and aromatic compounds in the gas and the aqueous phase

identified as a dominant mechanism in acetonitrile solutions. However, the influence of different solvents should be studied further.

The equations derived above may be applied to the prediction of nitrate radical reaction rate constants with organic substrates when kinetic parameters for the reaction of NO_3 with the same substrate in the other phase (gas phase or aqueous phase) are available from literature.

Atmospheric Implications, Conclusions and Outlook

Aromatic organic compounds have been studied intensively in order to understand their gas-phase atmospheric chemistry. They are thought to contribute significantly to the formation of photo-oxidants, especially in urban areas.[46,47] The present study is the first part of an extensive study to elucidate if aqueous-phase reactions involving aromatics also influence the free-radical concentration levels within the troposheric aqueous phase significantly. As has been stated earlier, aromatic compounds have been identified at very significant concentration levels within cloud- and rain-water,[1,2,48] so that it may well be possible that aromatic compounds in the tropospheric aqueous phase are important reaction partners for the free radicals thought to drive the numerous interconversion reactions in this environment, see Zellner and Herrmann[24] for a recent overview. Despite some controversies in the literature, the formation of HO_2 and ring-retaining products like cresols has to be looked upon as an important pathway resulting from the interaction of free radicals with aromatics in the gas phase. The resulting (substituted) cresols may subsequently undergo phase transfer into the aqueous environment and react with free radicals such as OH, NO_3, SO_4^-. According to this picture, the gas-phase oxidation of aromatics (*e.g.* benzene, toluene) by free radicals leads to the formation of more water-soluble ring-retaining products (phenols, cresols) which to a certain extent might be transferred into the tropospheric aqueous phase and undergo further degradation reactions in this environment. High concentrations of glyoxal, methylglyoxal and related compounds have often been found in cloud water.[48,49] In polluted areas these species often exist in the form of their condensation products with sulfur(IV).[50] These species are usually looked upon as tracers for aromatic compounds,[51] *i.e.* they indicate the former presence of aromatic compounds. It may be speculated that some of these dialdehydes originate from the degradation of aromatic compounds within the aqueous phase, *i.e.* a partitioning of aromatics between both phases may take place with possibly different degradation pathways in either phase, see Fig. 12. Note that large concentrations of oxalic acid are usually found in cloudwater samples from air masses that originate from heavily polluted areas such as the LA basin.[52] Oxalic acid is the oxidation product of glyoxal in aqueous solution and previous studies from our laboratory indicate that the conversion of glyoxal to oxalic acid by the nitrate radical is moderately fast [$k = (1.1 \pm 0.1) \times 10^6$ l mol^{-1} s^{-1}].[53]

The present study focussed on a number of aromatic model compounds (benzene, toluene, *p*-xylene, *p*-cresol, mesitylene). Kinetic investigations of the reactions of NO_3 and SO_4^- with these compounds have shown that aromatics react with both radical species at diffusion-controlled rates in aqueous solution. Reactions of NO_3, a radical which is thought to be present in both the tropospheric aqueous and the gas phases,[24,54] have been shown, in general, to proceed faster (by several orders of magnitude) in the aqueous phase. Hence, the conclusion may be drawn that the concentration levels of NO_3 in the tropospheric aqueous phase may be significantly influenced not only by reactions with inorganic species like chloride, bromide, sulfur(IV) or aliphatic organic compounds, but also by its reactions with aromatics. On the other hand, the presence of aromatics in the tropospheric aqueous phase may have considerable impact on the interconversions of radicals and radical anions. Because in the course of the reactions of free radicals X (X = OH, NO_3, SO_4^-, *e.g.*) with aromatics, HO_2 is thought to be formed either by the 'prompt' reaction of the HCHD derivative with molecular oxygen or by a 'delayed' reaction from an intermediate peroxy radical (see Fig. 12 and ref. 55), the reaction of X with aromatics results in an interconversion of the free radicals X into HO_2/O_2^-. This again may have consequences for the processing of ozone within the tropospheric aqueous phase[24,56,57] which is thought to be controlled by the reaction with superoxide radical anions, *viz.*

$$O_2^- + O_3 + H^+ \rightarrow OH + 2O_2 \tag{15}$$

This reaction is known to be very fast ($k_{15} = 1.5 \times 10^9$ l mol^{-1} s^{-1},[58]). Reaction (15) has two consequences: it leads to the destruction of ozone within tropospheric droplets and it produces OH radicals from HO_2/O_2^- which originates from the oxidation of the aromatic compounds. OH again can undergo a wide variety of reactions within the tropospheric aqueous environment and it may also be possible that a chain reaction occurs, because OH could undergo further reactions with aromatics. Such a reaction chain would result in the simultaneous consumption of ozone and aromatics within the tropospheric aqueous phase.

Because the highest gas-phase concentrations of aromatics are usually found in urban environments, the above conclusions might apply not only to the aqueous droplets of clouds, fog and rain, but also to the water layer surrounding solid aerosol nuclei in so-called 'urban aerosols'.[59] In this particular environment, ionic strengths of between 8 and 16 mol l^{-1} have been identified in field studies.[60] Therefore, in the present work, the effect of high ionic strength on the rate coefficients of the reactions of NO_3 with benzene and toluene in aqueous solution has been examined for the first time and a substantial secondary kinetic salt effect has been found. Hence, possible conversions of aromatics might be especially effective within aqueous urban aerosols owing to the synergistic effect of enhanced reaction rates and high concentrations of the aromatic substrate.

The reactions between aromatic compounds and the sulfate radical anion studied in the present work might also lead to a different understanding of the coversion of sulfur(IV) in the tropospheric aqueous phase. Because the sulfate radical is known to be a major chain-carrier in the free-radical-driven S^{IV} oxidation, the importance of the latter pathway, in comparison with other oxidation routes ($S^{IV} + H_2O_2$, O_3), may be reduced. However, the radical interconversions discussed earlier should be taken into account in future reaction schemes. The inclusion of the complicated chemistry following the reactions of SO_4^- with aromatics will not necessarily result in a smaller S^{IV} oxidation efficiency, because not only are sulfate radical anions converted to sulfate(VI), but also other radicals like HO_2/O_2^- and OH emerge in the course of the reaction when the coupling with ozone reactions is considered. In particular, OH might cause further S^{IV} oxidation and/or oxidative degradation of aromatics, as outlined above.

To summarize, we conclude from the present study that the reactions of free radicals with aromatic compounds within the tropospheric aqueous phase should be taken into consideration in an effort to understand this particular part of atmospheric 'heterogeneous' chemistry better. Reactions have been shown to proceed very rapidly. A strong secondary ionic strength effect was found in the reactions of NO_3 with both benzene and toluene. The reactions of the sulfate radical anion with aromatics generally proceed even faster than those of the nitrate radical and this might have consequences for our understanding of the aqueous-phase conversion of S^{IV}. The occurrence of dialdehydes and oxalic acid in cloudwater samples might be due to the degradation of aromatics in the gas phase coupled with subsequent phase transfer of the degradation products. However, phase transfer of substituted soluble aromatics followed by their highly efficient degradation by free radicals within the aqueous phase is also a possibility.

Future laboratory studies should focus on the enlargement of the kinetic database not only for model compounds but also for those species often identified in field measurements (*e.g.* nitrophenols). Furthermore, time-resolved spectroscopic studies are necessary to develop further the reaction mechansims for the interactions of free radicals with aromatic compounds in solution under conditions where self-reaction of the transients formed does not govern the observed chemistry. Future collaborative projects combining laboratory studies, field measurements and modelling efforts might enhance our understanding of the interactions and phase partitioning of free radicals and aromatics in the multiphase atmospheric system.

The present study was performed within the EUROTRAC sub-project HALIPP. The authors gratefully acknowledge support by the Bundesministerium für Forschung und Technologie (BMFT) under contracts 07EU7801 and 07EU7801A, the Commission of the European Communities (Project RINOXA, contract EV5V-CT93-0317) and the Fonds der Chemischen Industrie (FCI).

References

1 A. Geißler and H. F. Schöler, *Vom. Wasser*, 1993, **80**, 357.
2 A. Geißler and H. F. Schöler, *Water Res.*, 1994, **28**, 2047.
3 M. Exner, H. Herrmann and R. Zellner, *Ber. Bunsenges. Phys. Chem.*, 1992, **96**, 470.
4 H. Herrmann, M. Exner and R. Zellner, *Ber. Bunsenges. Phys. Chem.*, 1991, **95**, 598.
5 J. U. White, *J. Opt. Soc. Am.*, 1942, **42**, 285.
6 H. Herrmann, A. Reese and R. Zellner, *Air Pollution Research Report 45: Chemical Mechanisms Describing Tropospheric Processes*, ed. J. Peeters, CEC, Brussels, 1993, p.253.
7 M. Hesse, H. Meier and B. Zeeh, *Spektroskopische Methoden in der Organischen Chemie*, Thieme Verlag, Stuttgart, 3rd edn., 1987.
8 N. S. Bayliss and L. Hulme, *Aust. J. Chem.*, 1053, **6**, 257.
9 W. W. Simmons, *The Sadler Handbook of Ultraviolet Spectra*, Sadler Res. Laboratories, Philadelphia, 1979.
10 *UV Atlas Organischer Verbindungen*, Verlag Chemie/Butterworths, Weinheim/London, 1966.
11 R. P. Wayne, I. Barnes, P. Biggs, J. P. Burrows, C. E. Canosa-Mas, J. Hjorth, G. LeBras, G. K. Moortgat, D. Perner, G. Poulet, G. Restelli and H. Sidebottom, *Atmos. Environ. A*, 1991, **25**, 1.

12 P. Neta and R. E. Huie, *J. Phys. Chem.*, 1986, **90**, 4644.
13 K. J. Laidler, *Reaktionskinetik II: Reaktionen in Lösung*, BI Hochschultaschenbücher Band 291, B. I. Wissenschaftsverlag, Mannheim, 1973.
14 W. Roebke, M. Renz and A. Henglein, *Int. J. Radiat. Phys. Chem.*, 1969, **1**, 39.
15 P. Neta, V. Madhavan, H. Zemel and R. W. Fessenden, *J. Am. Chem. Soc.*,1977, **99**, 163.
16 G. V. Buxton, C. L. Greenstock, W. P. Helman and A. B. Ross, *J. Phys. Chem. Ref. Data.*, 1988, **17**, 513.
17 P. Neta, R. E. Huie and A. B. Ross, *J. Phys. Chem. Ref. Data.*, 1990, **19**, 413.
18 G. Raabe, Dissertation, Essen, 1995, in preparation.
19 P. Debye and J. McAulay, *Phys. Z.*, 1925, **26**, 22.
20 A. Mantaka, D. G. Marketos and G. Stein, *J. Phys. Chem.*, 1971, **75**, 3886.
21 G. Merga, C. T. Aravindakumar, B. S. M. Rao, H. Mohan and J. P. Mittai, *J. Chem. Soc., Faraday Trans.*, 1994, **90**, 597.
22 K. Seheseted, H. Corfitzen, H. C. Christensen and E. J. Hart, *J. Phys. Chem.*, 1975, **79**, 310.
23 H. Herrmann, M. Exner and R. Zellner, *Geochim. Cosmochim. Acta*, 1994, **58**, 3229.
24 R. Zellner and H. Herrmann, in *Advances in Spectroscopy*, ed. R. J. H. Clark and R. E. Hester, vol. 24, 1995, in the press.
25 M. Exner, H. Herrmann and R. Zellner, *J. Atmos. Chem.*, 1994, **18**, 359.
26 P. H. Wine, Y. Tang, R. P. Thorn and J. R. Wells, *J. Geophys. Res. D*, 1989, **94**, 1085.
27 C. L. Clifton and R. E. Huie, *Int. J. Chem. Kinet.*, 1989, **21**, 677.
28 R. E. Huie, C. L. Clifton and S. A. Kafafi, *J. Phys. Chem.*, 1991, **95**, 9336.
29 H. Herrmann, M. Exner, H-W. Jacobi and R. Zellner, in *Proc. EUROTRAC Symp. 94*, ed. P. M. Borrell, P. Borell, T. Cvitas and W. Seiler, SPB Academic Publishing, The Hague, 1994, p. 1056.
30 A. M. North, *The Collison Theory of Chemical Reactions in Liquids*, Methuen, London, 1964.
31 A. J. Elliot, D. R. McCracken, G. V. Buxton and N. D. Wood, *J. Chem. Soc., Faraday Trans.*, 1990, **86**, 1539.
32 J. D'Ans and E. Lax, *Taschenbuch für Chemiker und Physiker*, ed. M. D. Lechner, Band 1, Springer Verlag, Berlin, 4th edn., 1992, p.680 ff.
33 H. Herrmann, H-W. Jacobi and R. Zellner, in preparation.
34 P. O'Neill, S. Steenken and D. Schulte-Frohlinde, *J. Phys. Chem.*, 1975, **79**, 2773.
35 S. Steenken, C. J. Warren and B. C. Gilbert, *J. Chem. Soc., Perkin Trans. 2*, 1990, 335.
36 E. Baciocchi, T. Del Giacco, S. M. Murgia and G. V. Sebastiani, *J. Chem. Soc., Chem. Commun.*, 1987, 1246.
37 D. Rehm and A. Weller, *Ber. Bunsenges. Phys. Chem.*, 1969, **73**, 834.
38 T. Del Giacco, E. Baciocchi and S. Steenken, *J. Phys. Chem.*, 1993, **97**, 5451.
39 O. Ito, S. Akiho and M. Iino, *J. Org. Chem.*, 1989, **54**, 2436.
40 H. Herrmann, A. Reese and R. Zellner, ref. 29, p. 1017.
41 R. Atkinson, C. N. Plum, W. P. L. Carter, A. M. Winer and J. N. Pitts Jr., *J. Phys. Chem.*, 1984, **88**, 1210.
42 R. Atkinson, C. N. Plum, W. P. L. Carter, A. M. Winer and J. N. Pitts Jr., *Int. Jl Chem. Kinet.*, 1984, **16**, 887.
43 R. Atkinson, S. M. Aschmann and J. N. Pitts Jr., *J. Phys. Chem.*, 1988, **92**, 3554.
44 W. P. I. Carter, A. M. Winer and J. N. Pitts Jr.; *Env. Sci. Technol.*, 1981, **15**, 819.
45 G. Raabe, H. Herrmann and R. Zellner, in preparation.
46 R. G. Derwent and M. E. Jenkin, *Atmos. Environ. A*, 1991, **25**, 1661.
47 K-H. Becker, ref. 29, p. 67.
48 K. Kawamura and I. R. Kaplan, in *Organic Compounds in Rainwater*, ed. L. D. Hansen and D. J. Eatough, CRC Press, Boston, 1991, p. 233.
49 J. W. Munger, C. Tiller and M. R. Hoffmann, *Science*, 1986, **231**, 247.
50 T. M. Olson and M. R. Hoffmann, *Atmos. Environ. A*, 1989, **23**, 985.
51 T. E. Graedel, *Chemical Compounds in the Atmosphere*, Academic, New York, 1978.
52 D. L. Sedlak and J. Hoignè, *Atmos. Environ. A*, 1993, **27**, 2173.
53 M. Exner, H. Herrmann, G. Raabe and R. Zellner, ref. 29, p. 149.
54 W. L. Chameides, *J. Geophys. Res.*, 1986, **91**, 5331.
55 C. Zetsch, ref. 29, p. 115.
56 J. Lelieveld and P. J. Crutzen, *Nature (London)*, 1990, **343**, 227.
57 J. Lelieveld and P. J. Crutzen, *J. Atmos. Chem.*, 1991, **12**, 229.
58 K. Sehested, J. Holcman and E. J. Hart, *J. Phys. Chem.*, 1983, **87**, 1951.
59 G. P. Ayers, in *Atmospheric Chemistry*, ed. E. D. Goldberg, Springer, New York, 1982.
60 A. W. Stelson and J. H. Seinfeld, *Environ. Sci. Technol.*, 1981, **15**, 671.

Paper 5/00286K; *Received* 13th *January*, 1995

Faraday Discuss., 1995, **100**, 155–174

Measurements of Peroxides in the Atmosphere and their Relevance to the Understanding of Global Tropospheric Chemistry

S. A. Penkett, B. J. Bandy and C. E. Reeves
School of Environmental Sciences, University of East Anglia, Norwich, UK NR4 7TJ

D. McKenna† and P. Hignett
DRA, Meteorological Research Flight, Bldg Y46, Farnborough, Hants, UK GU14 6TD

Peroxides are produced as termination products in atmospheric chain reactions involving peroxy radicals, both organic and inorganic. They are the principal sink for radicals produced in the troposphere from the photolysis of ozone in the presence of water vapour and as such are excellent indicators of the extent of free radical chemistry taking place at any given location. Their measurement is relatively simple and data on the concentration of peroxides in the atmosphere with respect to time and space can be collected easily and extensively.

New data on peroxide measurements collected at different parts of the atmosphere, principally by the Meteorological Office C-130 Hercules aircraft are presented. They indicate that the extent of hydroxyl radical chemistry during the summer is controlled mostly by the water vapour content of the atmosphere. Both negative and positive correlations are observed between ozone and peroxide concentrations in vertical profiles over the North Atlantic Ocean and the equatorial Pacific. The negative correlations demonstrate that the ozone concentration throughout the troposphere is determined mostly by *in situ* photochemistry. This is borne out by the close correlation between calculated and measured concentrations of peroxides in vertical profiles. Positive correlations over the North Atlantic allow us to make an estimate of the amount of ozone, present there in the summer, which is formed from tropospheric as opposed to stratospheric chemistry.

Hydrogen peroxide is present ubiquitously in nature because it is the main termination product of chain reactions involving hydroxyl radicals. In the atmosphere, peroxide is formed by a photochemical process linking together oxygen in the form of ozone, which has its origin primarily in the stratosphere, and water vapour with a source at the ocean surface at the base of the troposphere. Several other trace gases are also involved in the overall chain reaction which has the capacity either to destroy or produce ozone in the troposphere.

An in-depth study of the distribution of atmospheric peroxides in space and in time can reveal many factors associated with the extent of free radical chemistry and its controlling features, and with the state of ozone in the troposphere, including particularly the balance between photochemical production and loss. These are currently questions of considerable interest in many research programmes on the chemistry of the

† Present address: Institute 1: Stratospharische Chemie, Forschungszentrum Jülich GmbH, D-52425 Jülich, Germany.

troposphere and between them hold the key to quantifying the 'oxidative capacity' of the atmosphere.

Measurement of gas-phase peroxides in the atmosphere has been possible for a number of years with instrumentation developed by Lazrus *et al.* at NCAR.[1] This instrument was initially developed to study the efficiency of the production of sulfuric acid (the major component of acid rain) from oxidation of sulfur dioxide by hydrogen peroxide in cloud droplets.[2,3] The combination of measurement of a critical parameter with inexpensive and effective instrumentation, which is a permanent fixture in an aircraft deployed on many missions throughout the troposphere, has enabled the collection of a large database on peroxides.

Various earlier aircraft studies have been made over continental areas mostly in connection with the role of peroxides in the oxidation of sulfur dioxide. These include those by Kleinman,[4] Heikes *et al.*[5] and Daum *et al.*[6] More recently, the focus of the study of peroxides has been more concerned with their role as indicators of basic tropospheric photochemistry.[7,8] A few studies have also been made in the marine atmosphere but mostly at ground-based sites or from ships[9–11], again in connection with studies of background troposheric chemistry. Studies made at Cape Grim in Tasmania[10] have provided strong evidence for the dominant role played by photochemistry in controlling ozone concentrations in the lower troposphere. This work is quoted extensively in the present paper.

We report here many co-measurements in the marine atmosphere of peroxides, ozone and water vapour concentrations collected by the C-130 Hercules aircraft operated by the Meteorological Office Research Flight at altitudes between the surface and 9 km over a wide range of latitudes from 20°S, through the equatorial region, to 70°N. The measurements also incorporate the annual cycle of peroxides at mid- to high-latitudes in the northern hemisphere. The measurements, both in the extent of their coverage and in the combination of gases measured, represent a unique data set.

The objectives of this measurement programme as formulated in the paper title are ambitious and will not reach fulfilment in this single manuscript. However, the data patterns obtained so far do reveal some extremely interesting phenomena capable of being interpreted in a simple empirical manner: they clearly show the controlling influence of water vapour on tropospheric free radical chemistry and, in doing so, provide a photochemical reaction rather than a transport interpretation of the layered structure of ozone observed previously over the remote ocean.[12] They also show the perturbation of ozone from the natural state in the background atmosphere at mid-latitudes of the northern hemisphere. In future papers the experimental data will be compared with models of tropospheric chemistry to test their capacity to portray present reality, from which it will be possible to check their credibility to predict future atmospheric composition.

Experimental

The experimentally determined altitude profiles of peroxide and ozone together with a wide range of meteorological parameters, including particularly dew point, were recorded during flights on board the Meteorological Research Flight, Hercules C-130 aircraft, operated by the UK Meteorological Office based at Farnborough. The Hercules C-130 has a maximum altitude of 10 000 m and a minimum of 15 m over the oceans; it has a range of 5500 km operating at a typical cruise speed of 150 m s^{-1}. Measurements were made at constant air speeds of typically 100 m s^{-1}. Profile descents or ascents were performed at rates of 5 m s^{-1} slowing to 2.5 m s^{-1} below 1500 m in descent or in ascent dependent on aircraft performance. The aircraft has an extensive fit of instrumentation including positional, altitudinal and dew-point sensors.

Aircraft velocity, heading and ground-speed components were measured using a Honeywell H423 inertial navigation unit with latitude, longitude and altitude being obtained from a Navstar XR5 global positioning system backed by a Litton LTN-211 Omega navigation system. The dew-point measurements were made with a General Eastern 1011B thermoelectric hygrometer, with a range of -60 to $+50\,°C$ and a resolution of 0.03 °C. This instrument works by direct measurement of the temperature of a Peltier cooled mirror which is dependent on the surface coating. The time constant of the dew-point analyser is variable, dependent on temperature and humidity. Typically the instrument performs best in the mid-free troposphere.

The ozone monitor used throughout the measurement campaigns was a Bendix 8002, chemiluminescence detector operated by the Meteorological Research Flight (MRF). This instrument records an ozone value every second which is averaged up to produce 10 s values. The air flow rate through the analyser is maintained at $1\ l\ s^{-1}$. The inlet (see below) runs directly from the main air-sampling pipe located forward of the engines which results in a time lag on measurements of *ca.* 15 s. During the investigations made in OCTA (see below) this instrument was calibrated, using a Dasibi calibrator, before and after each flight, whilst for the measurements taken in the southern hemisphere during the TOGA COARE (see below) intensive calibration was performed at the beginning and the conclusion of the study.

Peroxide measurements were made using a fluorescence monitor described by Lazrus *et al.*[1] and built at the University of East Anglia specifically for aircraft use. The peroxide inlet system consists of $\frac{1}{4}$ in Teflon tube which runs 2 m into the main aircraft gas-sampling pipe. The main aircraft sampling inlet is a stainless-steel 5 cm id tube which during flight has a throughput of air with a linear velocity in excess of $100\ m\ s^{-1}$. Peroxides in air from the inlet system are stripped by a stripping coil and fed into a chemical mixing system in which *p*-hydroxyphenylacetic acid dimerises in a reaction catalysed by the enzyme peroxidase. The presence of the dimer is detected by fluorescence and the fluorescence signal is proportional to the concentration of peroxide. The peroxide analyser is operated with a sampling frequency of 1 s. Air flow rate is maintained constant throughout each flight, usually at $1.5\ l\ min^{-1}$. There is a time lag on the measurements of 190 s which is due to the time taken for an air sample to be collected and processed chemically before measurement by the in-built fluorometer. The analyser detection limit is 40 ± 20 pptv at a signal/noise ratio of 2 : 1. For more detail on the performance of this instrument, including a discussion of errors, refer to Ayers *et al.*[13] The calibration of this instrument was performed using a liquid standard for a span check before each flight with multi-point determinations performed at the beginning of each intensive campaign. The air-inlet system was checked for peroxide loss down the Teflon pipe by injecting a gas-phase sample immediately into the analyser and then also down the length of the Teflon tubing connecting the instrument to the main aircraft sampling tube. It was not possible to perform checks on this tube but with a diameter of 5 cm and a large flow rate, losses are considered to be most unlikely. Note also that the aircraft flew in the air space between clouds to minimise losses due to the presence of droplets in the main sampling tube.

Results

The results were collected in the course of the OCTA (oxidising capacity of the tropospheric atmosphere) programme of atmospheric chemistry field measurements made over the North Atlantic Ocean throughout 1993 and 1994, and in the TOGA COARE experiment carried out over the West Pacific Ocean in January 1993. The TOGA COARE experiment provided data in the southern hemisphere tropics and in equatorial regions when the C-130 was based in Townsville, Australia and in the Solomon Islands. The locations of many of the flights are shown approximately in the world map in Fig. 1.

The OCTA programme caused the C-130 aircraft to cross the Atlantic Ocean in the summer of 1993 from the UK to Halifax, Nova Scotia and to recross to West Africa *via* the Azores before returning to the UK. Data were collected both north and south of the polar front in the winter and spring of 1994. Table 1 gives a summary of the flights made between January 1993 and April 1994 with location of vertical profiles, altitude range and other comments. Overall, 32 separate vertical profiles were obtained with continuous measurements of peroxides, ozone and water vapour content. Continuous measurements were made in horizontal flight for which data are also available, but the size of this database is too large for convenient presentation in most cases.

The data set shown contains measurements made in locations almost devoid of pollutants capable of disturbing the natural situation (A227, A229, A232 and A233), measurements made over the Atlantic Ocean which are subject to pollution by emission from Europe and North America (A269, A270, A272 and A273) and measurements north and south of the polar front in winter (A305, A307, A315 and A322). Many interesting features of basic tropospheric chemistry are revealed by this data set which is best considered as a set of vertical profiles covering a large range of concentrations of peroxide, water vapour and photon flux with both positive and negative ozone–peroxide correlations (see Table 1).

The peroxide instrument used in these experiments measures total peroxides, *i.e.* organic plus hydrogen peroxide. Attempts to use a similar instrument to differentiate between hydrogen and organic peroxides at Cape Grim in Australia proved to be only partially successful[13] and although some data are available on speciation from the aircraft experiments, it is not considered sufficiently reliable for presentation. The term peroxide in this paper therefore refers to total peroxide. It may well be an underestimate of this quantity, because of the lower solubility of organic peroxides to hydrogen peroxide, which reduces the efficiency of collection from air of the former species in the stripper, which is part of the inlet to the measurement instrument. Measurements in the free troposphere over North America[8] suggest that on average the organic peroxide contributes 31% to the total peroxide concentration. These measurements on many occasions,

Fig. 1 Global map showing position of Hercules flight

Table 1 Flight data showing date, location, latitude, longitude and altitude

flight no.	latitude	longitude	altitude/km	O_3/H_2O_2*	date	experiment	location[b]
A224	3°S[a]	157°E	0–5	+ve	Jan 1993	TOGA COARE	EQ
A226	19°S	150°E	0–6	?	Jan 1993	TOGA COARE	ST
A227	2°S	156°E	0–6	−ve	Jan 1993	TOGA COARE	EQ
A229	3°S	156°E	0–6	−ve	Jan 1993	TOGA COARE	EQ
A232	4°S	157°E	0–5	+ve	Jan 1993	TOGA COARE	EQ
A233	4°S	154°E	0–5	+ve	Jan 1993	TOGA COARE	EQ
A234	19°S	150°E	0–5	−ve	Feb 1993	TOGA COARE	ST
A235	19°S	151°E	0–6	−ve	Feb 1993	TOGA COARE	ST
A266 (1)	52°N	4°W	0–7	+ve	Aug 1993	OCTA	NA
A266 (2)	52°N	4°W	0–7	+ve/−ve	Aug 1993	OCTA	NA
A267 (1)	45°N	57°W	0–9	−	Aug 1993	OCTA	WA
A268 (1)	45°N	61°W	0–5	+ve	Aug 1993	OCTA	WA
A268 (2)	43°N	42°W	0–7	−	Aug 1993		
A269 (1)	47°N	55°W	0–7	+ve/−ve	Aug 1993	OCTA	WA
A269 (2)	44°N	60°W	0–7	−ve/+ve	Aug 1993	OCTA	WA
A270 (1)	38°N	30°W	0–7	−	Aug 1993	OCTA	WA
A270 (2)	38°N	30°W	0–7	(+ve)	Aug 1993	OCTA	WA
A270 (3)	37°N	27°W	0–8	−	Aug 1993	OCTA	WA
A271 (1)	30°N	15°W	0–7	+ve/−ve	Aug 1993	OCTA	EA
A272 (1)	32°N	13°W	0–6	−(−ve)	Aug 1993	OCTA	EA
A272 (2)	39°N	11.5°W	0–6	+ve(−ve)	Aug 1993	OCTA	EA
A273 (1)	42°N	12°W	0–7	(+ve)	Sept 1993	OCTA	EA
A273 (2)	49°N	10°W	0–7	+ve/−ve	Sept 1993	OCTA	EA
A305	45°N	11°W	0–9	(−)	Jan 1994	OCTA	EA/SP
A307	60°N	5°E	0–9	none	Jan 1994	OCTA	EA/NP
A315	40°N	12°W	0–9	+ve	Mar 1994	OCTA	EA/SP
A316	60°N	7°W	0–9	none	Mar 1994	OCTA	EA/NP
A322	68°N	0	0–9	+ve	Apr 1994	OCTA	EA/NP
A323 (1)	58°N	3°W	0–9	(?)	Apr 1994	OCTA	EA
A323 (2)	58°N	3°W	0–10	+ve to 6 km	Apr 1994	OCTA	EA
A325	40°N	17°W	0–8	(+ve)	Apr 1994	OCTA	MA
A326	36°N	25°W	0–7	+ve	Apr 1994	OCTA	MA

[a] This column indicates the presence of positive and/or negative correlations between peroxide and ozone in the profile data. [b] EQ = Equatorial; ST = Southern Tropics; NA = North Atlantic; EA = Eastern Atlantic; WA = Western Atlantic; MA = Mid-Atlantic; NP = North of Polar Front; SP = South of Polar Front.

particularly in the free troposphere, can therefore be considered to represent mostly hydrogen peroxide.

Relationships between Peroxide and Water Vapour Concentrations

Measurements of the peroxide and water vapour content of the atmosphere were made in January 1993 as a component of the TOGA COARE experiment. The location of the study area is shown in the dashed-line box in Fig. 2, being north and east of the Bismark Archipelago between 1°N and 4°S and between 152 and 154°E. This area is a region of general atmospheric convergence associated with the Walker circulation. Persistent easterly winds are interrupted by outbreaks of westerlies with enhanced convective activity. Flights A227 and A229 took place during a period of easterly winds.

Profiles of peroxide and water vapour for flight A227 and A229 are shown in Fig. 3(*a*) and (*b*). In both cases there is a high degree of correlation between the two compounds above 1500 m. The water vapour profile, in particular, indicates the presence of a complex layered structure in the free troposphere between 1500 and 6000 m. This period coincided with a pronounced minimum in the degree of large-scale convective

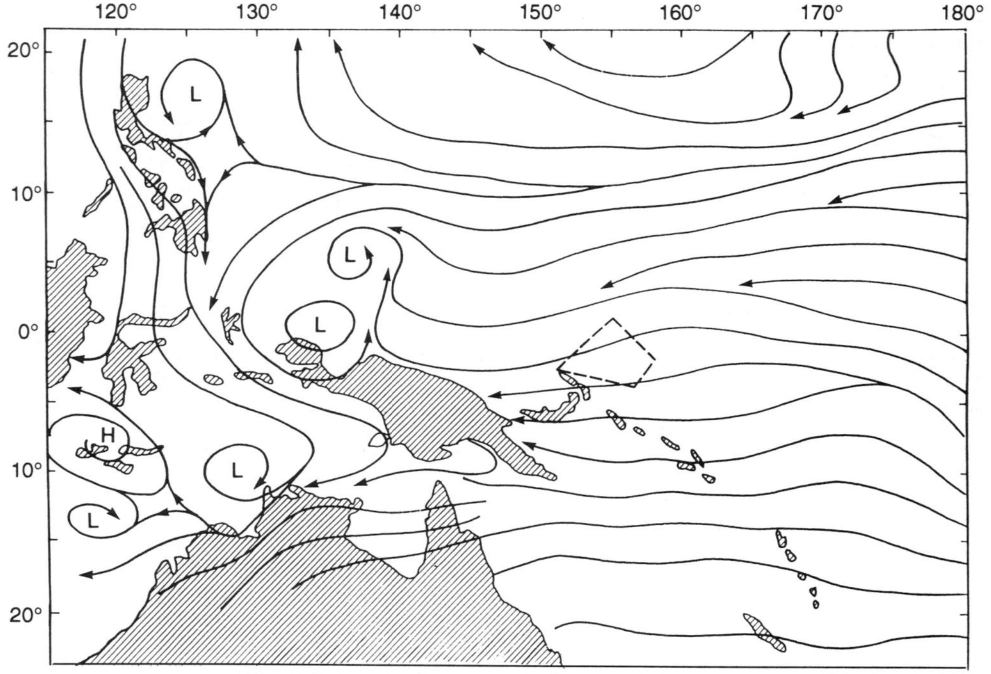

Fig. 2 Synoptic diagram referring to Flight A227

activity in this area; hence over the previous few days, the air above the boundary layer had not been subject to any significant vertical mixing resulting from deep convection. It is likely, therefore, that the free tropospheric air was steadily subsiding, warming and stabilising. Vertical transport of water vapour was restricted to the boundary layer with a steady drying of the free tropospheric air between A227 and A229. In the boundary layer the positive correlation between water vapour and peroxide disappears; the water

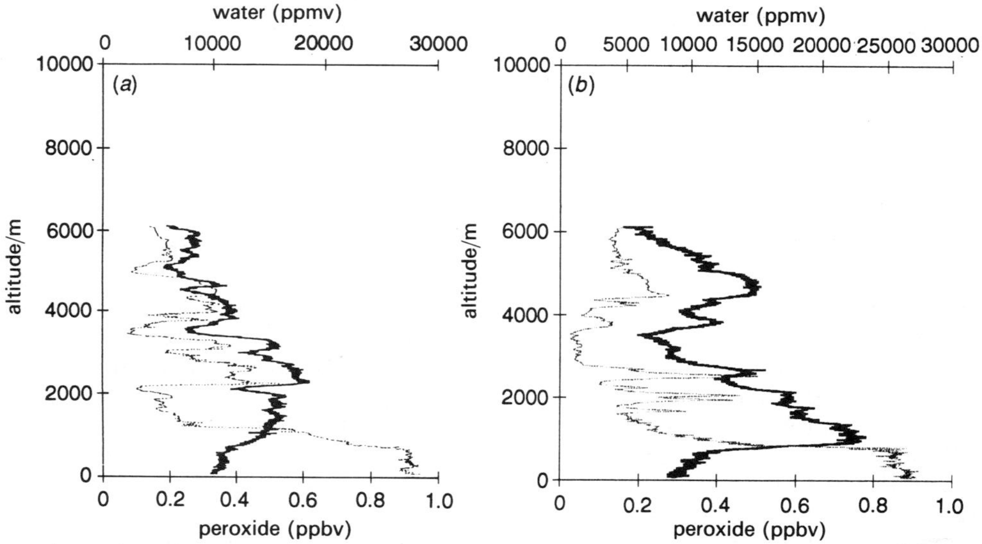

Fig. 3 Vertical profiles of (——) peroxide and (· · ·) water vapour for flights (*a*) A227 and (*b*) A229

vapour concentration increases towards the ocean surface, which is a source, and the peroxide concentration declines owing to deposition. The rate of peroxide loss at the surface is determined by its deposition velocity and the rate of vertical exchange of air in the boundary layer.

For flights A227 and A229 plots of water vapour *vs.* peroxide are shown in Fig. 4(*a*) and (*b*). For Flight A229 the average slope of a line drawn asymptotically to the lowest water vapour concentration at any given peroxide level is *ca.* 1 ppbv peroxide for 10 000 ppmv H_2O. Peroxide concentrations to the left of the maximum value, at any water content, are almost certainly caused by loss by physical removal processes such as dry deposition in the case of the upper lobe in the data set which can extend to 30 000 ppmv (3%) H_2O in the boundary, and possibly by rain-out and photolysis in the case of the bulk of the measurements. Rain-out also effects water vapour but the distribution of peroxide is such that, within cloud, most of it is present in the liquid phase, whereas most water is always present in the vapour phase. Peroxide loss in rain will therefore be more pronounced than water loss, even though this represents the major removal process for water from the atmosphere. Low peroxide concentrations have also been

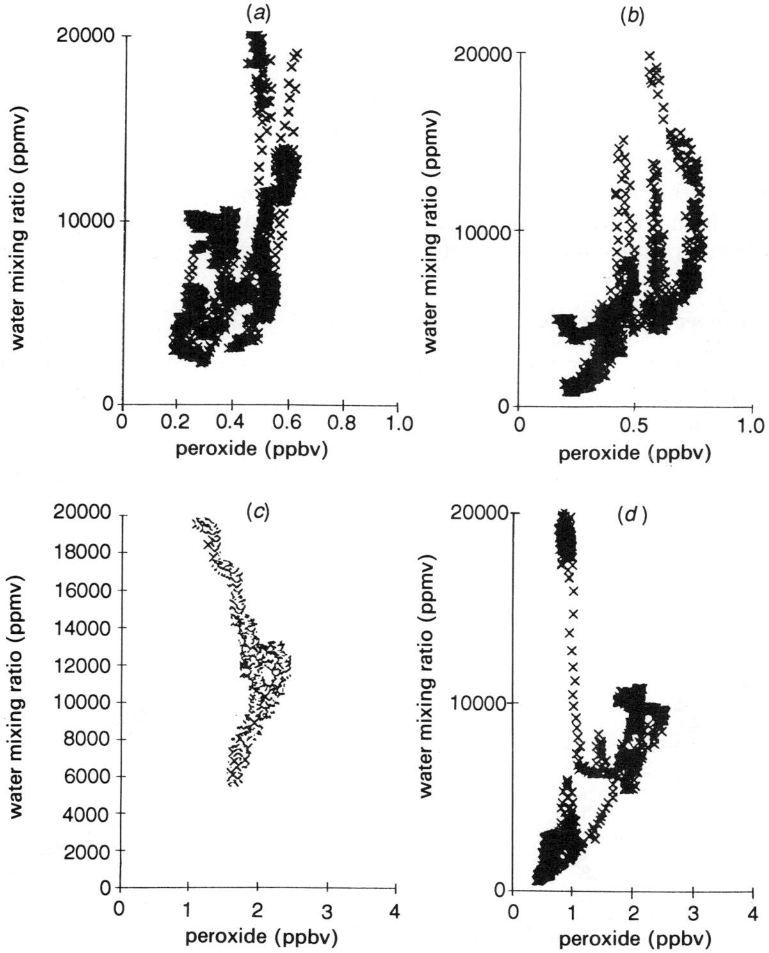

Fig. 4 Correlation plots of peroxide and water vapour for flights (*a*) A227, (*b*) A229, (*c*) A269 and (*d*) A272

recorded in a set of independent measurements made in the same location[14] and it is speculated that they are partly due to efficient rain-out over a long period. There is no particular reason to suspect the measurements since the transfer efficiency of the inlet lines carrying peroxide-laden air to the instrument on the aircraft were checked on the ground in Townsville (see Experimental).

The location of the OCTA flights made over the North Atlantic in the summer of 1993 is shown in Fig. 1. Fig. 4(*c*) and (*d*) show the water vapour *vs.* peroxide plots for vertical profiles carried out in flights A269 and A272, respectively. These flights were made with a wide difference in longitude between 10 and 60°W, at similar latitudes (*ca.* 35–40°N) (see Table 1). An interesting feature of these Atlantic plots is that the abundance of peroxide is roughly twice that observed in equatorial regions for a given water vapour concentration, *i.e.* 2 ppbv peroxide at 10 000 ppmv (1%) water vapour *vs. ca.* 0.8 ppbv peroxide at 10 000 ppmv (1%) water vapour. The slopes of the asymptotes to the lowest water vapour concentrations in the Atlantic plots are therefore very similar to those found over the North American continent[7] in August and September (2.6 ppbv H_2O_2 at 1% H_2O), although on one occasion the slope was in the same range as the Pacific data (1.36 ppbv H_2O_2 at 1% H_2O) when the free troposphere water vapour content was high, *ca.* 5000 ppmv, and similar to the Pacific values (see Fig. 3). In other respects the plots for the North Atlantic Ocean are the same as those for the Pacific Ocean showing a high degree of correlation, with an upper anticorrelation lobe due to losses of peroxide in the boundary layer. These plots suggest that the maximum amount of peroxide formation, and therefore HO_x free radical chemistry, in the free troposphere is limited by the water vapour content. Losses of peroxide in the boundary layer limit its usefulness as a true indicator of the free radical chemistry occurring there, although the large extent of the latter is indicated from calculated peroxide concentration (see Discussion), which in the absence of major physical sinks is strongly correlated with measured concentrations above the boundary layer.

Relationship between Peroxide and Ozone Concentrations

The layered structures over the equatorial Pacific reveal evidence for the simultaneous production of peroxide in the course of ozone destruction by photochemistry in the background clean troposphere that was first observed at Cape Grim.[10] Fig. 5(*a*) and (*b*) for flights A227 and A229 show a remarkable anticorrelation between ozone and perox-

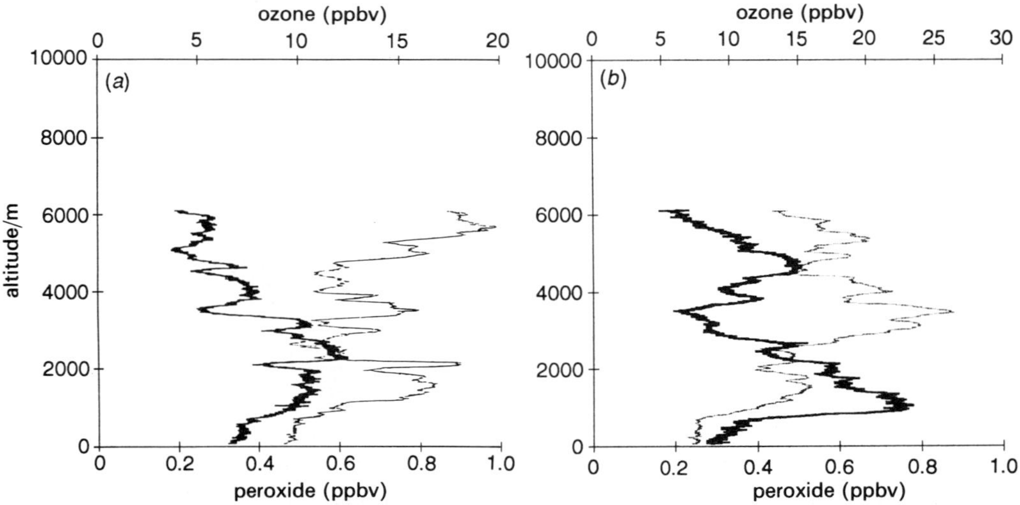

Fig. 5 Vertical profiles of (——) peroxide and (——) ozone for flights (*a*) A227 and (*b*) A229

ide above 1500 m in vertical profiles of these two trace gases. Similar layered structures have also been observed in other data records over the American continent[7] and by ourselves in previous Atlantic flights but they have not been seen to the same extent nor commented on previously in the literature, even though they are of such fundamental importance to the verification of basic tropospheric photochemical theory. Fig. 6(*a*) and (*b*) show plots of ozone *vs.* peroxide for flight A227 and A229. The anticorrelation is clearly seen in most of the data, but is much clearer in Fig. 6(*b*) for Flight A229. The positive correlation lobe at the bottom of Fig. 6(*b*) is where peroxide and ozone are both being removed close to the top of the boundary layer.

From Fig. 6 the quantitative relationship between molecules of peroxide gained and molecules of ozone lost is *ca.* 1 : 20 in the equatorial Pacific data (see Flight A229). This compares with a summertime average diurnal variation value of 1 molecule of peroxide gained during the day for 4 molecules of ozone lost, and of a ratio in amplitude of the seasonal cycle for peroxide to ozone of 1 : 20 at Cape Grim in 1991–1992.[10] Differences in the observed ratios of peroxide production to ozone loss, which represents transfer of odd oxygen from one molecular form to another, are influenced by the relative lifetimes of the two molecules. Since ozone has a lifetime *ca.* 10 times that of peroxide, the ratio will change with time from the low instantaneous value of the diurnal observation at Cape Grim, towards the averaged seasonal values; these latter are much more sensitive to differential loss processes for the two molecules. The ratio is also a function of the chain length of the free radical destruction of ozone by HO_2 and OH radicals, which is subject to the concentration of nitrogen oxides.

In contrast to the ozone removal observed in flights A227 and A229, flights A232 and A233 over the same area of the Pacific Ocean showed a positive correlation between ozone and peroxide in the vertical profiles of the two gases [Fig. 7(*a*) and (*b*)]. These positive ozone–peroxide correlations occurred in the absence of any correlation between peroxide and water vapour, which declined monotonically from the surface to the top of the flight (5 km) with no structure on each occasion. The ratio of molecules of peroxide produced to molecules of ozone produced [Fig. 8(*a*) and (*b*)] for the two flights is between 1 : 12 and 1 : 16, which compares with values of 1 : 10 measured in real time at Cape Grim in air reaching Tasmania from the Australian continent.[13] It differs, however, from ratios of peroxide and ozone gain observed in flights over the USA which show values in the range of 1 : 60 when the NO concentration varied between 0.4 and 1.6 ppbv.[7] This no doubt reflects the increased efficiency of reaction of peroxy radicals

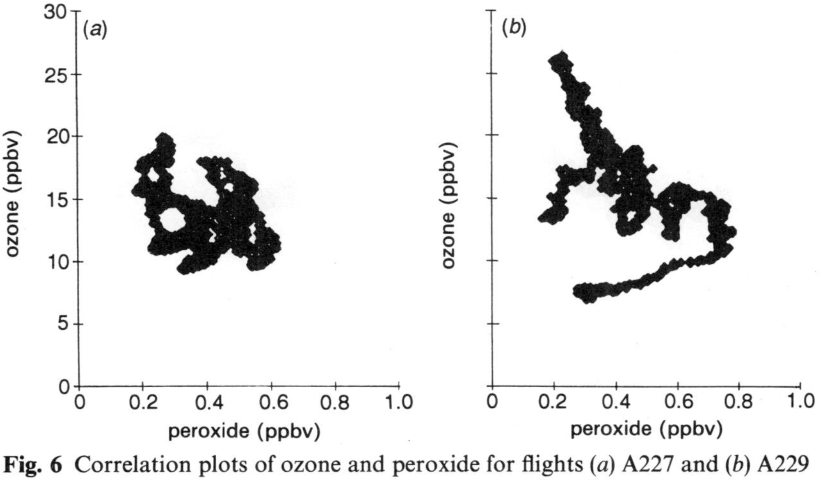

Fig. 6 Correlation plots of ozone and peroxide for flights (*a*) A227 and (*b*) A229

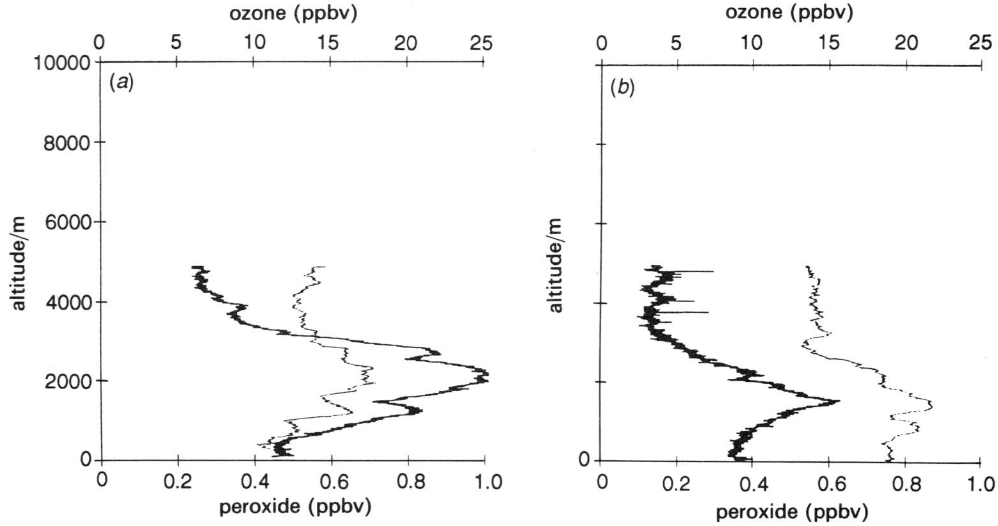

Fig. 7 Vertical profiles of (——) ozone and (——) peroxide for flights (*a*) A272 and (*b*) A233

with NO rather than with themselves to form peroxides in the more polluted continental air.

A major difference between the meteorological conditions for the flights showing positive ozone–peroxide correlation [Fig. 7(*a*) and (*b*)] and those showing a negative correlation [Fig. 5(*a*) and (*b*)], was the wind direction (Fig. 2). In the latter case it was from the direction of the largest ocean in the world and would contain concentrations of nitrogen oxides of less than 5 pptv,[15] whilst in the former case the air had its origin over Indonesia and had recently passed over the large Island of New Guinea (Fig. 9), both of which are significant sources of common pollutants from biomass burning capable of producing peroxide and ozone simultaneously on photolysis.

The flights over the Atlantic Ocean in the summer of 1993 (Fig. 1 and Table 1) show evidence of both positive and negative ozone–peroxide correlations, with the former predominating below 500 m. Fig. 10 shows the correlation found between ozone and

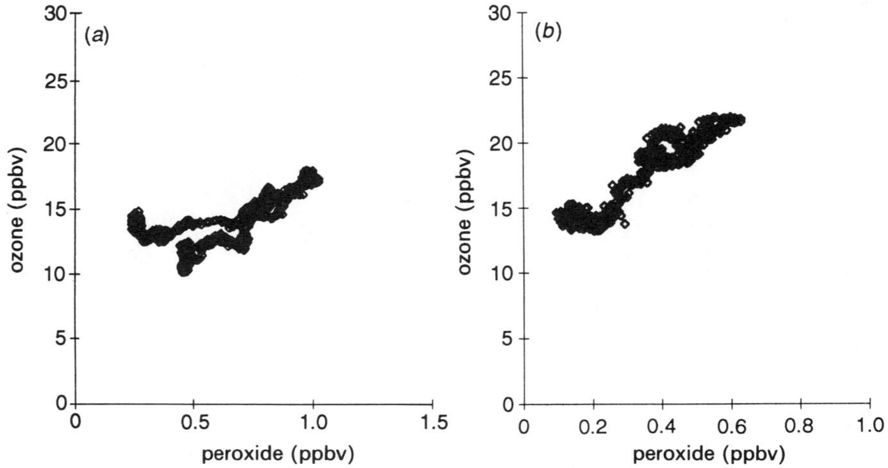

Fig. 8 Correlation plots for ozone and peroxide for flights (*a*) A232 and (*b*) A233

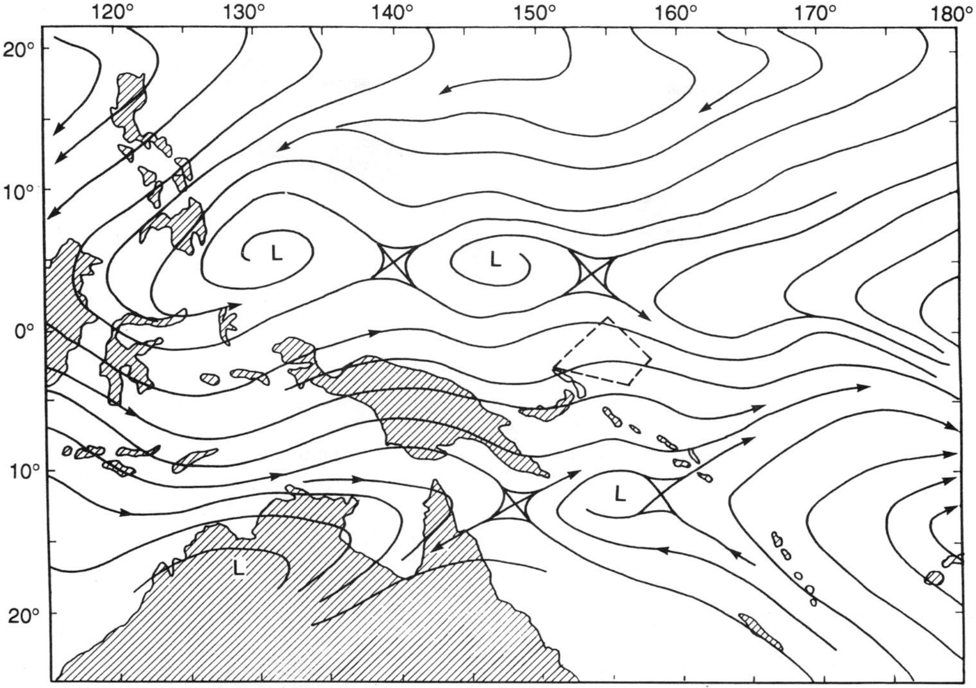

Fig. 9 Synoptic diagram referring to Flight A233

peroxide in the western Atlantic [A269, Fig. 10(*a*)] and in the eastern Atlantic [A273, Fig. 10(*b*) and A272, Fig. 10(*c*)]. Flight A269 is positive and negative, A272 is positive as is A273 with slopes indicating the production of *ca.* 12–16 molecules of ozone produced for 1 molecule of peroxide, in agreement with both the equatorial and the southern mid-latitude Tasmanian data. The value of this ratio will be critically dependent on the NO_x concentrations since it reflects the tendency of peroxy radicals such as HO_2 to react with NO to form NO_2, which creates ozone, or with themselves to form H_2O_2. Other flights made in August, A266, A271 and components of A269, A271, A272 and A273 show correlations that were either neutral or slightly negative. Only on Flight A270 were anticorrelations seen which all resemble the equatorial data. The origin of the air on this occasion was from latitudes south of 30°N, *i.e.* somewhat similar to that seen on the equatorial flights.

Flight A272, which took place *ca.* 100 miles west of the Portuguese coast, provided some very interesting data which have been interpreted as a case of low-level (<3000 m) transport of polluted air from Europe and higher-level (>4000 m) transport of air from North America. Both of these phenomena can be seen in the profile shown in Fig. 11(*a*). The lower plume is obvious, with a high degree of correlation between peroxide and ozone; the upper one is less obvious for peroxide but is clear for ozone, and for other data on hydrocarbons, organic nitrates, carbon monoxide and particle count which have been used to identify the presence of polluted air at these high altitudes. In line with this, the peroxide concentration at 6 km is *ca.* 0.8 ppbv which is a concentration that is too high for stratospheric air.

Seasonality of Atmospheric Peroxide Measurements and the Influence of the Polar Front

The concentrations of peroxide in the atmosphere show a strong seasonal dependence at mid- to high-latitudes in the northern hemisphere, particularly north of the polar front.

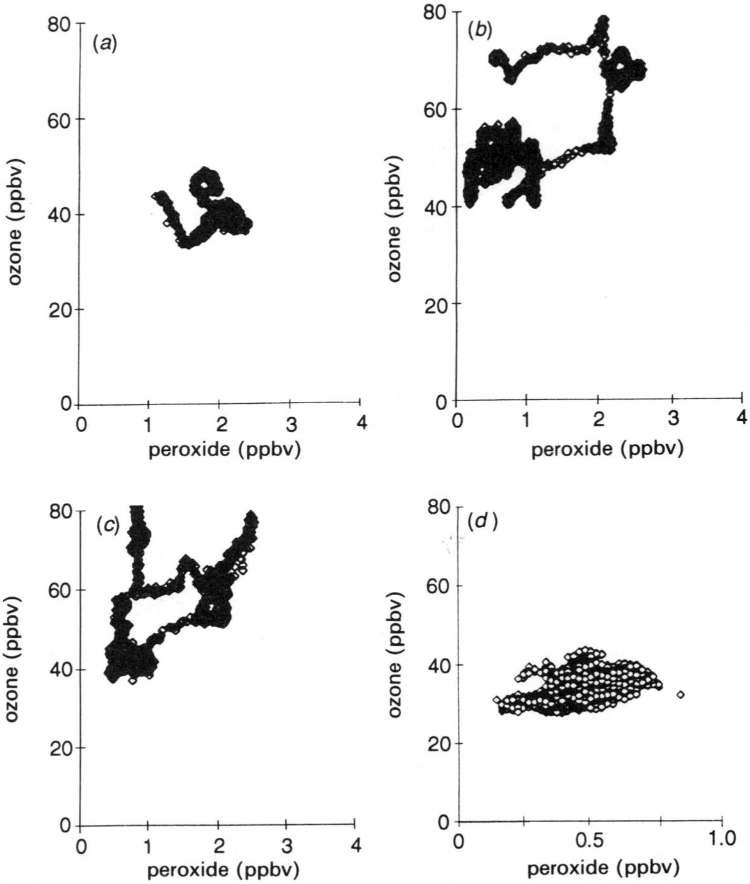

Fig. 10 Correlation plots for ozone and peroxide for flights (*a*) A209, (*b*) A272, (*c*) A273 and (*d*) A315

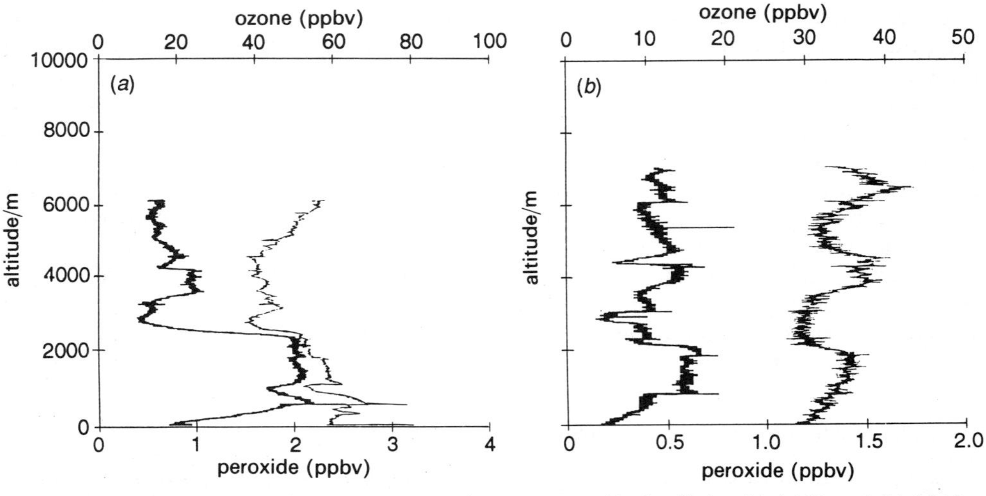

Fig. 11 Vertical profiles of (——) ozone and (——) peroxide for flights (*a*) A272 and (*b*) A315

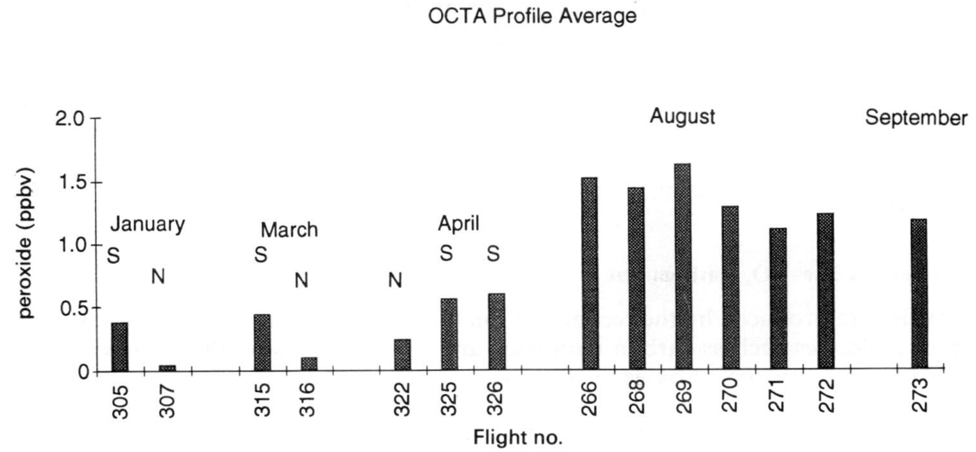

Fig. 12 Average peroxide concentration recorded on various flights over the North Atlantic between January and September (N refers to north of front, S refers to south of front)

In Fig. 12 average peroxide concentrations are shown from January through to September and they vary from less than 0.1 ppbv to over 1 ppbv. In the January flight north of the polar front (A307) there is no correlation at all between water vapour and peroxide, and the peroxide concentration does not exceed 0.1 ppbv. There is no correlation in these quantities for the March flight (A316) but the peroxide has increased to 0.2 ppbv average; a weak correlation is observed in April (A323) such that, at a water vapour concentration of 10 000 ppmv, the peroxide concentration is *ca.* 0.5 ppbv. This compares with summer values over the North Atlantic of 2 ppbv peroxide at 10 000 ppmv water vapour and in the equatorial Pacific of *ca.* 1 ppbv peroxide at 10 000 ppmv water vapour. It is clearly seen, therefore, that the nature of the photon flux into the troposphere also plays a significant role in peroxide production and that this is the limiting factor in winter at high latitudes. It is also clear that the polar front acts as an efficient barrier for transport of air from lower latitudes to high latitudes since the average peroxide concentrations shown in Fig. 12 north of the polar front are substantially lower than those south of the front in the winter months.

The degree of correlation of peroxide and ozone profiles on flight A315 south of the polar front, slightly north and west of the Spanish coast bordering the Bay of Biscay, is shown in Fig. 10(*d*). The two substances are highly correlated with a peroxide to ozone production ratio of 1 : 16 in air with its origin from the direction of the mid Atlantic Ocean. The correlation persists up to and above 7000 m [Fig. 11*b*] except for the region below 1000 m where boundary layer effects predominate, and in areas where sudden peroxide loss has occurred possibly owing to removal by cloud chemistry. Significant peroxide concentrations (*ca.* 0.4 ppbv) were present even at the top of the profile (9000 m) coincident with high concentrations of hydrocarbons and carbon monoxide (C_2H_6 = 1500 pptv, C_3H_8 = 170 pptv, C_2H_2 = 120 pptv, CO = 110 ppbv) that were typical of the troposphere rather than the stratosphere. North of the front on flight A315 there is also a positive correlation between peroxide and ozone, with a ratio closer to 1 : 50, *i.e.* similar to the ratios experienced in air over the continental USA.[7]

Discussion

These results, collected in an extensive set of aircraft flights, covering a wide range of latitude and longitude, have some clear messages on factors controlling the production of free radicals in the atmosphere, and on the contrast between the ozone budget over the North Atlantic and that over the unpolluted equatorial Pacific. The discussion here will focus on the correlations between peroxide and water vapour and peroxide and ozone, and it will show that many observations of ozone over remote oceans are the result of ozone loss by *in situ* photochemistry, rather than by transport of high ozone and low water vapour from the stratosphere. It will also show that, in the free troposphere in summer, there is a high degree of correlation between measured and calculated peroxide concentrations using very simple photochemical theory.

Peroxides in a Low-NO_x Environment

Peroxides are produced by the recombination of peroxy radicals produced in the oxidation of molecules such as carbon monoxide and hydrocarbons by the hydroxyl radical. In the case of hydrogen peroxide (H_2O_2)

$$HO_2 + HO_2 \rightarrow H_2O_2 + O_2 \tag{1}$$

with most of the HO_2 coming from CO oxidation.

$$CO + OH \rightarrow CO_2 + H \tag{2}$$

followed by $H + O_2 \rightarrow HO_2$. A very simple expression for the $[H_2O_2]$ concentration at time t can be derived from these two processes: $[H_2O_2] = \frac{1}{2}k_2[CO][OH]t$, suggesting that the peroxide concentration reflects the integrated extent of hydroxyl radical chemistry that has occurred within an element of the atmosphere. In reality the expression is more complex and includes terms for methane oxidation by OH, for oxidation of non-methane hydrocarbons by OH and for loss of peroxide by various processes. However, these do not alter the link between peroxide production and the extent of oxidation by hydroxyl chemistry.

The hydroxyl radical is mostly formed from reaction of $O(^1D)$ atoms with water vapour.

$$O(^1D) + H_2O \rightarrow 2OH \tag{3}$$

and in our CO-dominated atmosphere is removed by reaction with CO, so that

$$[OH] = \frac{2k_3[O(^1D)][H_2O]}{k_2[CO]}$$

$O(^1D)$ is formed by photolysis of ozone with light of wavelength <310 nm

$$O_3 \xrightarrow[\lambda<310\ \text{nm}]{h\nu} O(^1D) + O_2 \tag{4}$$

and is mainly deactivated to $O(^3P)$ by reaction with atmospheric molecules (M).

$$O(^1D) + M \rightarrow O(^3P) + M \tag{5}$$

so that the expression for $[H_2O_2]$ becomes

$$[H_2O_2] = \frac{k_3 j_4[O_3][H_2O]t}{k_5[M]} \tag{I}$$

The factors controlling the $[H_2O_2]$ concentration in the sunlit troposphere are therefore seen to be the ozone concentration $[O_3]$; the photon flux between 295 and 310 nm, which is a component of j_4; the atmospheric pressure [M], and the water vapour concentration $[H_2O]$. The variable which will have the largest influence is clearly water vapour since this changes by a factor of *ca.* 100 between the ground and the tropopause. This is borne out in the plots in Fig. 3 and 4 which suggest that the concentration of atmospheric peroxides between the boundary layer and the tropopause is dominated in the summer by the upwards diffusion of water vapour. Eqn. (I) suggests, however, that other factors are involved and this is demonstrated in the overall set of results on ozone and on photon flux as related to the annual cycle of peroxide.

Using eqn. (I) it is possible to calculate a peroxide concentration for comparison with the measured concentration of ozone and water vapour in any profile. Fig. 13 shows such a comparison of measured and calculated concentrations for flight A227 in which the product of the ozone and water vapour concentration in parts per billion have been divided by a factor of 2×10^8 [$\equiv K$ in eqn. (II)] to produce a calculated peroxide concentration. The overlap between measured and calculated data is remarkable at all altitudes between 2000 and 6000 m in a situation where peroxide production is dominated by ozone loss in the unpolluted marine atmosphere.

The same factor K (2×10^8) derived from the unpolluted Pacific data is then used on the product of ozone and water vapour concentration to produce the comparisons shown in Fig. 14(*a*) and (*b*) for flights A267 and A269 and in Fig. 15(*a*) and (*b*) for flights A272 and A273, respectively, over the Atlantic Ocean in summer. This time the calculated and measured data track closely above 1000 m but with differences in magnitude. At higher altitudes the measured concentrations are greater than the calculated values

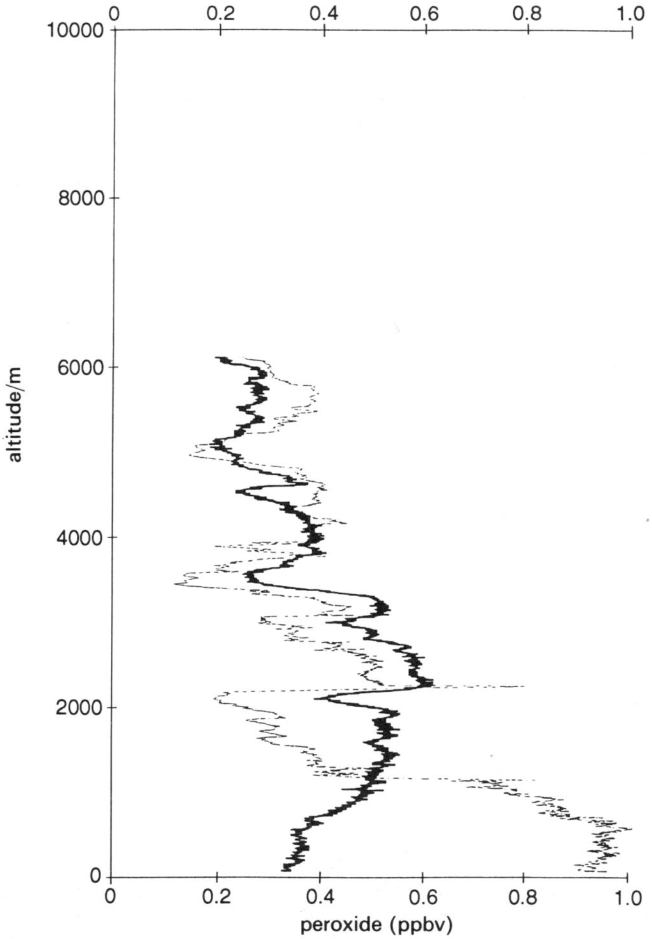

Fig. 13 Measured (——) and calculated (– – –) plots of peroxide *vs.* altitude for flight A227

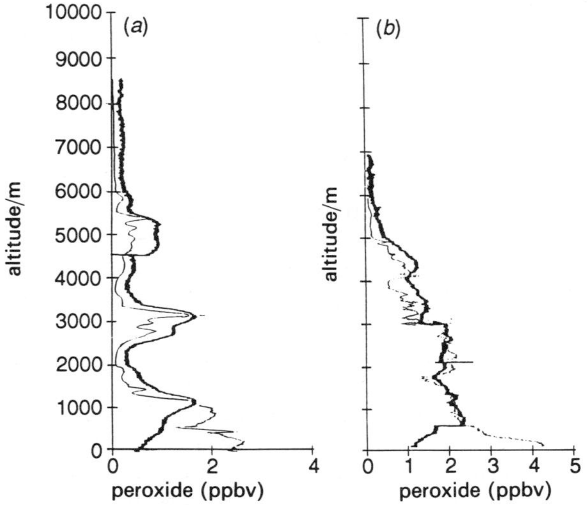

Fig. 14 Measured (——) and calculated (– – –) plots of peroxide *vs.* altitude for flights (*a*) A267 and (*b*) A269

and at lower altitudes the calculated values usually exceed the measured values. There are various possible explanations for the differences which lie beyond this simple analysis, however; for the case where the calculated values exceed the measured values it is likely that peroxide production is suppressed by reaction of the peroxy radicals with nitric oxide to produce ozone *via* NO_2 photolysis. Also some of the difference will be caused by physical removal of the soluble peroxide. Irrespective of the differences in magnitude the diagrams do illustrate well the potential for extensive peroxy radical production and hence ozone production at lower altitudes in the troposphere, which then falls away with increasing altitude. The profile for flight A272 shows extra calcu-

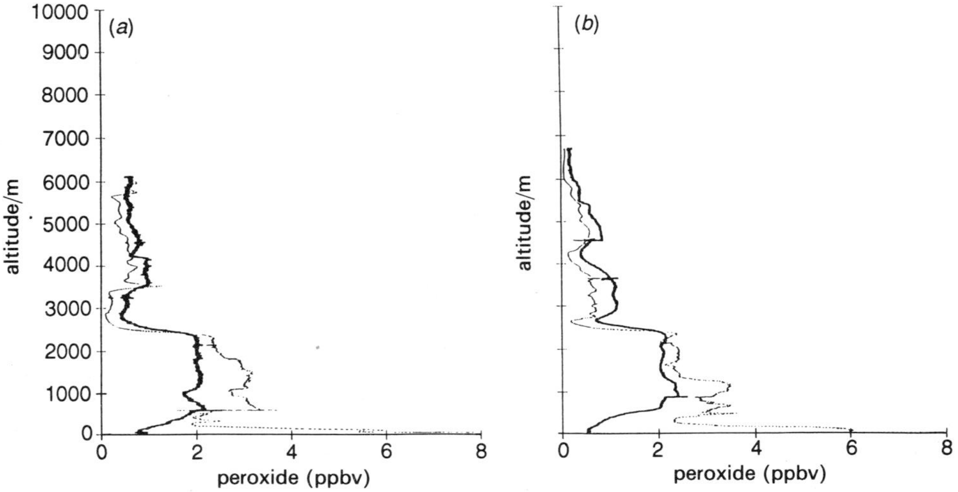

Fig. 15 Measured (——) and calculated (– – –) plots of peroxide *vs.* altitude for flights (*a*) A272 and (*b*) A273

lated peroxide over measured peroxide both at low and high altitudes, which correlates well with the co-presence of additional ozone and other tracers of polluted air in that profile.

The peroxide concentrations are calculated using the following expression:

$$[H_2O_2] = K[O_3][H_2O] \tag{II}$$

$K = 2 \times 10^8$ ppbv^{-1} when all concentrations are in ppbv. By analogy with expression (I) and

$$K = \frac{k_3 j_4}{k_5[M]} t$$

substituting values of 1×10^{-5} s^{-1} for j_4, 10^9 ppbv for [M] and 8.46 for the ratio of k_3/k_5 then t is calculated to be *ca.* 16 h, corresponding to 2 days' reaction in the atmosphere. The value of t will increase when loss terms are taken into account and of course it is strongly dependent on the value chosen for j_4. Its value, however, is in line with peroxide production rates of 200 pptv (% H_2O)$^{-1}$ h^{-1} calculated by Tremmel *et al.*[7] It should be a reflection of the time for the peroxide profiles to develop in a sunlit atmosphere containing ozone and water vapour during the summer. The close correlation between measured and derived peroxide concentration in these vertical profiles above the boundary layer suggests that the peroxide concentration does indeed reflect the extent of free radical chemistry which has taken place; it also emphasises the extent of this chemistry which is occurring in the lower troposphere where ozone can either be destroyed or created efficiently. This is particularly true in the boundary layer where measured and calculated peroxide concentrations diverge but calculated peroxide concentrations are very large (up to 6 ppbv). The calculated concentrations are not realised due to the presence of large physical loss processes in the boundary layer, but the potential for ozone formation, if sufficient NO_x is present, is shown to be enormous (see Fig. 14 and 15).

Another feature of the observations, which is of considerable importance, is that they allow a different interpretation to be placed on the complex layering structures of ozone, and its anticorrelation with water vapour, first observed by Routhier and Davis.[12] They interpreted the phenomenon in terms of transport of high ozone and low water vapour from the stratosphere. A more likely explanation, in view of the close correlation between calculated and measured peroxide, is photochemical destruction of ozone in the troposphere stimulated by high water vapour concentration and leading to peroxide formation.

Reaction (4) combined with reaction (3) and also reaction (6) constitutes a loss of ozone due to photochemistry. Measurements made at Cape Grim show an anticorrelation between ozone and peroxide in time, both annually and diurnally, due to photolysis.[10] These aircraft measurements made in clean air over the Pacific Ocean clearly demonstrate the anticorrelation in space rather than in time caused by photochemistry, which is more efficient at high water vapour concentrations.

Low ozone concentrations have been recorded previously in oceanic equatorial regions[16–19] and this, as discussed above, is probably caused by efficient photolytic removal of ozone. The low peroxide measurements recorded simultaneously could imply a low level of integrated hydroxyl chemistry which, in turn, could imply that removal of trace gases over the remote oceans in the tropical regions is not as efficient as elsewhere. At present there are too few simultaneous measurements of ozone and peroxide in these regions to draw definite conclusions. Surface measurements by Thompson *et al.*[11] in the course of the SAGA3 cruise in a similar area of the Pacific revealed average total peroxide concentration in the region of 1 ppbv at the surface, with minimum values down to 600 pptv. These are not dissimilar to our own data.

Peroxide–Ozone Relationships at Significant NO_x Concentrations

The demonstration of this anticorrelation between peroxide and ozone is very important in showing that the budget of ozone in the troposphere is largely dominated by *in situ* photochemistry, rather than simply by injection from the stratosphere followed by loss at the Earth's surface. The large amount of destruction is calculated to be slightly exceeded by production, especially in areas where the concentration of NO exceeds *ca.* 10 pptv depending on the simultaneous ozone concentration.[20] This is because, in addition to formation of peroxides [reaction (1)] and ozone destruction [reaction (6)], HO_2 can also react with NO to form NO_2, [reaction (7)]

$$HO_2 + O_3 \rightarrow OH + 2O_2 \tag{6}$$

$$HO_2 + NO \rightarrow NO_2 + OH \tag{7}$$

Reaction (7) followed by reactions (8) and (9) leads to ozone production in the troposphere.

$$NO_2 \xrightarrow[\lambda < 4000 \text{ nm}]{h\nu} NO + O(^3P) \tag{8}$$

$$O(^3P) + O_2 + M \rightarrow O_3 + M \tag{9}$$

A recent model calculation[21] in connection with the 1995 UNEP Ozone Assessment produced the ozone budget shown in Table 2. The net production of ozone (343 Tg year^{-1}) is the difference between the very large terms in excess of 4000 Tg year^{-1} for production and loss; one of the main challenges for tropospheric chemistry today is to try to verify these figures and to assess the extent of perturbation of the natural ozone budget by emissions of nitrogen oxides from anthropogenic sources as observed over the remote Pacific and over the North Atlantic. This will entail the deployment of aircraft able to measure peroxy radicals and NO simultaneously to determine the rate at which ozone is being produced in reaction (7), as well as the rate at which it is being destroyed by reactions (3), (4) and (6).

Table 2 Globally integrated production and loss terms for tropospheric ozone in the current atmosphere, as calculated in one 2D model (Derwent, 1994) (in Tg)

term	strength
$HO_2 + NO$	3117
$CH_3O_2 + NO$	1006
$RO_2 + NO$	462
total production	4585
$O(^1D) + H_2O$	1704
$O_3 + OH$	410
$O_3 + HO_2$	1719
$O_3 + NMHC$	178
$O_3 + NO$ (net loss)	177
$O(^3P)$ (net loss)	54
total loss	4242
net chemical production	343

The measurements reported here also clearly indicate that ozone production is occurring in the troposphere remote from source regions. The co-production of peroxide and ozone through reactions (1) and (7)–(9) is observed over the Pacific in air contaminated with land-based pollutants from either industrial or biomass burning activities (see Fig. 8). Unfortunately, no measurements which could act as tracers of polluted air were made in these equatorial flights, but the conclusion is inescapable, that ozone is being produced by a perturbation of the natural photochemistry which otherwise would destroy it. The same conclusion applies to ozone production observed at Cape Grim in air contaminated with emissions of pollutants from Australia.[10] It is also known that ozone concentrations are increasing in Asia due to anthropogenic activities, with positive trends in concentrations being observed with time.[22]

Over the North Atlantic the peroxide–ozone relationships are mostly positive suggesting that continental emissions from both Europe and North America are perturbing the natural ozone budget and causing larger ozone concentrations in the background atmosphere of significant parts of the northern hemisphere. Even when the relationship is not positive it is rarely negative indicating, at least, that ozone destruction is occurring less efficiently than it would in the unperturbed situation. This constitutes evidence from a study of *in situ* chemical activity for the ozone increases in tropospheric air in the northern hemisphere that have been observed in many measurement programmes over the last 20 years[23] and from comparisons of present-day ozone records with historical records.[24]

The slopes of the peroxide *vs.* ozone plots shown in Fig. 8 and 10 cover the range 1:12 to 1:16 in greatly different parts of the atmosphere and they compare with ratios of 1:14 at Cape Grim in aged continental air.[13] Using 1:14 as an average ratio and combining it with the peroxide concentration data in Fig. 12 for summer flights, then up to 20 ppbv of ozone could be present over the remote Atlantic from tropospheric photochemistry. This constitutes *ca.* 40% of the ozone observed in the OCTA Atlantic flights up to altitudes of 6 km and agrees quantitatively with conclusions reached from a recent study of the ozone and CO budgets of air leaving North America.[25] This study calculated the number of moles of ozone transported into Atlantic air based upon a measured average ratio of ozone to CO concentration in the air of 0.3 and the amount of CO emitted into the atmosphere in North America east of the Mississippi River in the summer months. The calculated ozone was then compared with the amount of ozone transferred from the stratosphere to the troposphere over the area of the North Atlantic during the summer. Both these independent studies would suggest that the ozone concentration in background tropospheric air of the northern hemisphere in summer is now roughly twice what it was in the last century, and therefore corroborate the studies comparing present day and historical ozone records.[24,26] The situation in other seasons of the year is more difficult to interpret from measurements of peroxide since the lifetimes of ozone and peroxide diverge greatly.

Conclusions

An extensive set of data has been collected, over a wide latitude range, on the variation in concentration of peroxides in the maritime atmosphere bewtween the surface and the tropopause.

When combined with co-measurements of water vapour and ozone it is possible to determine the principal factors limiting the extent of hydroxyl radical chemistry throughout the troposphere, based on a simple photochemical theory that shows that peroxides act as integrators of photochemical activity.

Peroxide and ozone show negative and positive correlations which can be used to quantify both ozone loss and ozone gain due respectively to natural and anthropogenically perturbed tropospheric chemistry. Over the equatorial Pacific the negative correlations occur in layers, which also show negative correlations between water vapour and ozone. We suggest that this phenomenon is mostly a result of tropospheric photochemistry rather than transport of air in the stratosphere. Over the North Atlantic in summer, estimates from the positive correlations of the two oxidants suggest that *ca.* 40% of the ozone, in the atmosphere up to 6 km, is formed by *in situ* photochemistry.

We gratefully acknowledge the expert flying capabilities of the RAF aircrew in the successful outcome of these experiments. We also acknowledge the measurements of ozone and other trace gases by the Met. Office personnel (Joss Kent and Colin Hord). Funds to carry out this research were provided by the European Union, the Department of the Environment and the Natural Environment Research Council. This paper is a contribution to the EUROTRAC TOR Project.

References

1 A. L. Lazrus, G. L. Kok, J. A. Lind, S. N. Gitlin, B. G. Heikes and R. E. Shetter, *Anal. Chem.*, 1985, **58**, 594.
2 S. A. Penkett, B. M. R. Jones, K. A. Brice and A. E. J. Eggleton, *Atmos. Environ.*, 1979, **13**, 123.
3 A. S. Chandler, T. W. Choularton, G. J. Dollard, M. J. Gay, T. A. Hill, A. Jones, B. M. R. Jones, A. P. Morse, S. A. Penkett and B. J. Tyler, *Atmos. Environ.*, 1988, **22**, 683.
4 L. L. Kleinman, *J. Geophys. Res.*, 1986, **91**, D10, 10 889.
5 B. G. Heikes, G. L. Kok, J. G. Walega and A. L. Lazrus, *J. Geophys. Res.*, 1987, **92**, D1, 915.
6 P. Daum, L. L. Kleinman, A. J. Hills, A. L. Lazrus, A. C. D. Leslie, K. Busness and J. Boatman, *J. Geophys. Res.*, 1990, **95**, D7, 9857.
7 H. G. Tremmel, W. Junkermann and F. Slemr, *J. Geophys. Res.*, 1993, **98**, D1, 1083.
8 H. G. Tremmel, W. Junkermann and F. Slemr, *J. Geophys. Res.*, 1994, **99**, D3, 5295.
9 B. G. Heikes, *J. Geophys. Res.*, 1992, **97**, D16, 18,001.
10 G. P. Ayers, S. A. Penkett, R. W. Gillett, B. Bandy, I. E. Galbally, C. M. Meyer, M. Elsworth, S. T. Bentley and B. W. Forgan, *Nature (London)*, 1992, **360**, 446.
11 A. M. Thompson, J. E. Johnson, A. L. Torres, T. S. Bates, K. C. Kelly, E. Atlas, J. P. Greenberg, N. M. Donahue, S. A. Yon, E. S. Saltzman, B. G. Heikes, B. W. Mosher, A. H. Shashkov and V. I. Yelorov, *J. Geophys. Res.*, 1993, **98**, D9, 16 955.
12 F. Routhier and D. D. Davis, *J. Geophys. Res.*, 1980, **85**, C12, 7307.
13 G. P. Ayers, S. A. Penkett, R. W. Gillett, B. J. Bandy, I. E. Galbally, C. P. Meyer, C. M. Elsworth, S. T. Bentley and B. W. Forgan, *J. Atmos. Chem.*, submitted.
14 B. Heikes, personal communication, 1994.
15 A. L. Torres and A. M. Thompson, *J. Geophys. Res.*, 1994, **98**, D9, 16 949.
16 F. Routhier, R. Dennett, D. D. Davis, A. Wartburg, P. Haagenson and A. C. Delany, *J. Geophys. Res.*, 1980, **85**, C12, 7307.
17 S. R. Piotrowicz, H. F. Bezdek, G. R. Harvey, M. Springer-Young and K. J. Hanson, *J. Geophys. Res.*, 1991, **96**, D10, 679.
18 J. E. Johnson, R. H. Gammon, J. Larsen, T. S. Bates, S. J. Oltmans and J. C. Farmer, *J. Geophys. Res.*, 1990, **95**, D8, 11 847.
19 S. C. Liu, M. McFarland, D. Kley, O. Zafiriou and B. Huebert, *J. Geophys. Res.*, 1983, **88**, C2, 1360.
20 P. J. Crutzen and P. H. Zimmerman, *Tellus*, 1991, **43AB**, 136.
21 R. G. Derwent, *Philos. Trans. R. Soc. London, A*, 1995, **351**, 205.
22 H. Akimoto, H. Nakane and Y. Matsumoto, in *The Chemistry of the Atmosphere: Its Impact on Global Change*, ed. J. G. Calvert, Blackwell Scientific Publications, Oxford, 1994, p. 261.
23 D. Kley, H. Geiss and V. A. Mohnen, ref. 22, p. 245.
24 A. Marenco, H. Gouget, P. Nédélec, J-P. Pagés and F. Karcher, *J. Geophys. Res.*, 1994, **99**, D8, 16 617.
25 D. D. Parrish, J. S. Holloway, M. Trainer, P. C. Murphy, G. L. Forbes and F. C. Fehsenfeld, *Science*, 1993, **259**, 1436.
26 A. Volz and D. Kley, *Nature (London)*, 1988, **332**, 240.

Paper 5/00232J; Received 12th January, 1995

Faraday Discuss., 1995, **100**, 175–198

Observation and Role of the Free Radicals NO_3, ClO, BrO and IO in the Troposphere

Ulrich Platt and Christof Janssen
Institute für Umweltphysik, Univ. Heidelberg, INF 366, D-69120 Heidelberg, Germany

An attempt is made to obtain estimates of the average tropospheric abundance of nitrate radicals (NO_3) and reactive halogen species (halogen oxide radicals and halogen atoms) on a global scale and thus to quantify their possible impact on the oxidation potential of the troposphere. For that purpose we combine new measurements with data that recently became available. It is found that the integrated formation of the radicals is comparable to that of OH radicals in certain situations. However, it is argued that the effects of these radicals on the oxidation capacity of the troposphere are not necessarily additive. For instance, BrO could be an NO_3-scavenger.

Free radicals are known to be the driving force for most chemical processes in the atmosphere. Since the pioneering work of Weinstock[1] and Levy[2], photochemically generated HO_x (OH and HO_2) radicals have been thought to play a key role in tropospheric chemistry. Although direct, unequivocal identification of these species in the atmosphere could (to date) only be performed in a few cases (*e.g.* ref. 3), there is an enormous amount of indirect evidence for their presence. The importance of HO_x for atmospheric chemistry can be assumed to be proven beyond reasonable doubt. Nevertheless, there has always been a possibility of other reactive species playing a part, beginning with the (historical) idea of oxygen atoms $O(^3P)$ or excited oxygen molecules $O_2(^1\Delta)$. In this context, recent investigations indicate a possibly important influence of several other radicals. In particular the nitrate radical (NO_3) and the halogen oxide radicals BrO, IO and ClO (which are in a stationary state with the corresponding halogen atoms) can make a considerable contribution to the oxidising capacity of the troposphere. For instance reaction with NO_3 or BrO can be an important sink of dimethyl sulfide (DMS) in the marine environment. Also, night-time reactions of nitrate radicals with organic species can be a source of peroxy radicals (like HO_2 or CH_3O_2) and even OH radicals.[4,5] In contrast to the HO_x chemistry, it is likely that heterogeneous reactions play an important role in the atmospheric cycles of the above radicals. On the other hand, the effect of several oxidising species need not be additive; in fact it is possible that the presence of halogen oxides can reduce the OH concentrations, owing to an increase in the photostationary-state NO_2/NO ratio.[6] Also halogen oxide radicals may be involved in the recently discovered phenomenon of sudden loss of boundary layer ozone in the Arctic.[7] In addition, reactive bromine species (Br, BrO), and possibly iodine species, could be effective NO_3 scavengers under certain conditions.

Obviously, knowledge of the concentration of these species in the atmosphere is essential for understanding atmospheric chemistry. Unfortunately the low concentration of free radicals makes measurements particularly difficult; few measurements have been reported to date.[3] However, the method of deriving the global (or hemispheric) average radical concentration from the known budget of certain trace gases (like CO, ^{14}CO,[8,9]

CH_3Cl[10]), that proved very successful for OH radicals appears to be difficult to apply for NO_3 radicals and halogen oxides.

Here an attempt will be made to quantify the possible impact of the above radicals on a global scale and to identify local environments, where chemical processes in the atmosphere are influenced or even dominated by those radicals. First we describe the tropospheric chemistry of the respective species and try to derive representative ambient concentration levels from new measurements as well as data available from the literature. Then the consequences of the presence of those radicals for the oxidation capacity of the troposphere will be discussed. Finally, the effects of the radicals will be compared with the influence of OH reactions.

Tropospheric Chemistry of Nitrate Radicals, NO_3

The prerequisite for nitrate radical production is the simultaneous presence of nitrogen dioxide and ozone in the same air mass. The reaction scheme of oxidised nitrogen species and reactive halogen species is given in Fig. 1. Nitrate radicals are formed by the reaction of NO_2 with ozone

$$NO_2 + O_3 \rightarrow NO_3 + O_2 \tag{1}$$

which constitutes the only significant source of NO_3 in the troposphere. Therefore the rate of NO_3 production in a given airmass can be calculated simply from the concentrations of NO_2 and O_3:

$$\frac{d[NO_3]}{dt} = [NO_2][O_3]k_1 = P_{NO_3} \tag{I}$$

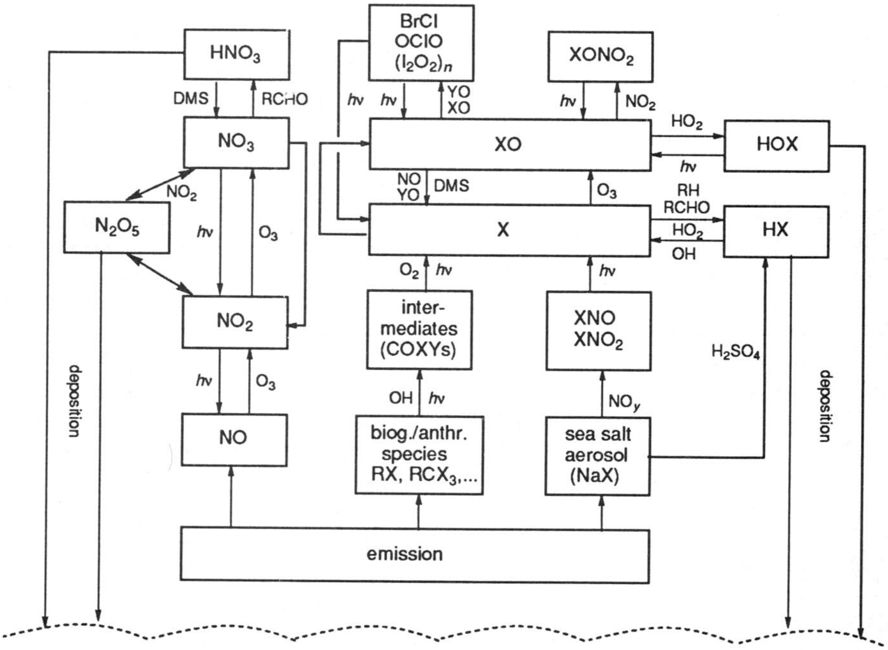

Fig. 1 Simplified scheme of NO_x and halogen reactions in the troposphere

Chemical Cycles of NO_3

During daytime, NO_3 radicals are effectively destroyed by photolysis:

$$NO_3 + h\nu \rightarrow NO + O_2 \tag{2a}$$

$$\rightarrow NO_2 + O \tag{2b}$$

(photolytic lifetime *ca.* 5 s) and rapid reaction with NO ($k_3 = 2.6 \times 10^{-11}$ cm^3 s^{-1} at 293 K):

$$NO_3 + NO \rightarrow NO_2 + NO_2 \tag{3}$$

At night $J_2 = 0$ and NO levels are usually low; therefore the NO_3 lifetime, and thus concentration, are likely to become much larger. Loss of NO_3 due to gas-phase reaction with nitrogen dioxide or self reaction is slow:

$$NO_3 + NO_2 \rightarrow NO + NO_2 + O_2 \tag{4}$$

$$NO_3 + NO_3 \rightarrow NO_2 + NO_2 + O_2 \tag{5}$$

Likewise, the unimolecular decay of the NO_3 radical postulated by Johnston *et al.*[11] upon reinterpretation of older measurements is probably of minor importance in the atmosphere:

$$NO_3 \xrightarrow{M} NO + O_2 \tag{6}$$

The rate of decay (in particular at elevated ambient temperature) corresponds to much shorter NO_3 lifetimes than are actually observed in the atmosphere.[12]

Gas-phase reaction of NO_3 with water vapour is endothermic, however, NO_3 is quite water-soluble; in solution it might react with anions, for instance:[13]

$$(NO_3)_{aq} + (Cl^-)_{aq} \rightarrow (NO_3{}^-)_{aq} + (Cl)_{aq} \tag{7}$$

The main destruction mechanisms are then reaction with organic species or heterogeneous loss (*i.e.* reaction at the ground, or at the surface of aerosol or cloud particles) of either NO_3 or N_2O_5 (see below). Reaction with alkanes and aldehydes proceeds *via* H-abstraction:[14]

$$NO_3 + RH \rightarrow R + HNO_3 \tag{8}$$

The alkyl radicals produced in reaction (8) are transformed to peroxy radicals (RO_2) by reaction with O_2; RCO radicals originating from NO_3–aldehyde reactions react to give $HO_2 + CO$, in the case of R = H, or to RCO_3-type acyl peroxy radicals for R = higher alkyl groups. Reaction with alkenes (including isoprene and terpenes) proceeds *via* NO_3 addition to a C=C double bond. Further reaction steps are O_2 addition followed by reaction either with other peroxy radicals or with NO_2. Reasonably stable end-products are probably organic nitrates and peroxy radicals.[14] Reaction with aromatic compounds is also important; while reaction with benzene is quite slow (rate constant of the order of 10^{-17} cm^3 molecule^{-1} s^{-1}), hydroxy-substituted benzenes react very fast with rate constants of the order of 10^{-12} cm^3 molecule^{-1} s^{-1} (furan, phenol) to 10^{-11} cm^3 molecule^{-1} s^{-1} (cresols).

Finally NO_3 radicals react rapidly with DMS (rate constant, 10^{-12} cm^3 s^{-1}); this can be significant for the budgets of both DMS and NO_3, in particular in the marine atmosphere. The exact reaction mechanism is still subject to investigation.[15] Recent investigations show that the primary step is probably H-abstraction, thus peroxy radicals should be a dominant product.[16]

Since most of the above reactions of NO_3 with organic species appear to lead to the formation of peroxy radicals, an interesting consequence of the night-time reaction of

NO_3 with organic species is a considerable contribution to the HO_x-budget.[4] In air masses with elevated levels of NO_x (which is a prerequisite for formation of NO_3) during daytime a likely fate of RO_2 radicals is reaction with NO:

$$RO_2 + NO \rightarrow RO + NO_2 \tag{9}$$

However, the presence of NO is practically incompatible with the existence of NO_3 in an airmass during night-time. Therefore NO_3-produced peroxy radicals are likely to undergo ractions other than reaction (9), probably RO_2 self-reaction or reaction with NO_3:

$$RO_2 + NO_3 \rightarrow RO + NO_2 + O_2 \tag{10}$$

In this sense reaction (10) can play a role in the RO_2 chemistry analogous to the role of reaction (9) during daytime.

NO_3–N_2O_5 Equilibrium

An important property of NO_3 is its equilibrium with N_2O_5:

$$NO_2 + NO_3 \xleftrightarrow{M} N_2O_5 \tag{11}$$

The equilibrium concentration of N_2O_5 is given by:

$$[N_2O_5] = \frac{[NO_2][NO_3]}{K} \tag{II}$$

Under 'moderately polluted' conditions (NO_2 mixing ratio of *ca.* 1 ppb) typical for rural areas in industrialised countries NO_3 and N_2O_5 concentrations are roughly equal at 293 K. While homogeneous reactions of N_2O_5 with water vapour and other atmospheric trace gases are believed to be extremely slow, heterogeneous reaction with water contained in aerosol or cloud particles will convert N_2O_5 to nitric acid:

$$N_2O_5 + H_2O \xrightarrow{\text{het}} HNO_3 \tag{12}$$

Accommodation coefficients for N_2O_5 and NO_3 on water (droplet) surfaces are in the range of 0.04 to 0.1[17,18] (the higher values applying to water containing H_2SO_4), and >0.0025,[19] respectively.

NO_3 Lifetime

From measured concentrations of O_3, NO_2 and NO_3 the atmospheric lifetime of NO_3 (τ_{NO_3}) as limited by the combination of any first-order loss process:

$$NO_3 + Z \rightarrow \text{products} \tag{13}$$

can be calculated. Assuming stationary-state conditions with respect to NO_3, τ_{NO_3} becomes

$$\tau_{NO_3} = \frac{[NO_3]}{P_{NO_3}} = \frac{[NO_3]}{[NO_2][O_3]k_1} \tag{III}$$

Since k_1 is known from laboratory measurements and the concentrations of O_3, NO_2 and NO_3 can be simultaneously measured [for instance by the differential optical absorption spectroscopy (DOAS) technique], τ_{NO_3} can readily be calculated for various atmospheric conditions. It has to be emphasized that the steady-state assumption is actually an approximation, which, however, is a very good one since usually the actual change in NO_3 is much smaller than the production and loss rates. Note that the NO_3

lifetime can as well be limited by any irreversible loss of N_2O_5 (for instance $k_{12} > 0$), because both species are in a rapidly established thermodynamic equilibrium (for reaction 11 and its reverse reaction, the time constant to reach the equilibrium is of the order of 1 min at ambient temperatures). Under conditions where both NO_3 and N_2O_5 are lost, the observed NO_3 lifetime, τ'_{NO_3}, is given by:

$$\tau'_{NO_3} = \left(\frac{1}{\tau_{NO_3}} + \frac{[NO_2]}{K}\frac{1}{\tau_{N_2O_5}}\right)^{-1} \quad \text{(IV)}$$

Of course, in the case of negligible loss of N_2O_5 ($\tau_{N_2O_5} \rightarrow \infty$) $\tau'_{NO_3} = \tau_{NO_3}$. For conditions where loss of NO_3 molecules can be neglected ($\tau_{NO_3} \rightarrow \infty$), eqn. (IV) reduces to:

$$\tau'_{NO_3} = \frac{K}{[NO_2]} \cdot \tau_{N_2O_5} \quad \text{(V)}$$

It is interesting to note that the simple stationary-state assumption predicts an observed NO_3 lifetime inversely proportional to the NO_2 concentration in cases where not NO_3 but N_2O_5 is actually removed from the gas phase. Since formation of NO_3 is proportional to $[NO_2]$ [reaction (1)] under those conditions, $[NO_3]$ should become independent of $[NO_2]$.

Since the NO_3 source strength is relatively well known, the central problem in attempting to estimate the tropospheric NO_3 level lies in the correct quantification of the atmospheric NO_3-loss mechanisms (and thus the NO_3 lifetime). These can be summarized as follows:

(1) Reaction with organic species. (2) Reaction with night-time NO; this loss process is likely to be of great importance close to sources of pollution (where P_{NO_3} is also large). (3) Heterogeneous loss of N_2O_5; this process is quite effective in air masses containing elevated NO_x concentrations, but is probably unimportant at background NO_x levels. (4) Heterogeneous loss of NO_3; there is very little information on this process available. (5) Thermal decay of NO_3; there is evidence showing this process to be of little importance.[12] (6) Other loss mechanisms, for instance reaction with halogen oxides (see later).

Tropospheric Chemistry of Reactive Halogen Species (RHS)

It is common knowledge that halogen species, in particular halogen oxides, play a pivotal role in stratospheric ozone chemistry.[20] In comparison, the possible influence of RHS on tropospheric chemistry has received little attention. Here an attempt is made to summarize presently available information on sources, chemical cycles, and concentration levels of RHS in the troposphere. While this includes halogen atoms (X,Y), halogen monoxide radicals (XO), halogen hydroxides (HOX), nitrosyl and nitryl halides (XNO, XNO_2) and halogen nitrates ($XONO_2$) we will focus on XO and X atoms here.

Tropospheric Sources of RHS

Unlike in the stratosphere, where the dominating source of reactive halogen species is photochemical degradation of fully halogenated compounds (*e.g.* CF_2Cl_2 *etc.*), in the troposphere RHS originate from less stable precursors:

(*a*) Degradation of partially halogenated organic compounds (*e.g.* chloroform) originating from a variety of natural and artificial sources.[21–23] Natural sources include the oceans, which emit a wide variety of halocarbons such as the methyl halides, CH_3Cl, CH_3Br and CH_3I, [estimated source strength (100–400) $\times$ 10^9 g CH_3Br year^{-1},[24–26] $\approx 4 \times 10^9$ g CH_3I year^{-1},[24]] as well as some polyhalogenated species such as bromoform, $CHBr_3$, or CH_2Br_2.[26] Methyl halides are also produced from the burning of biomass, mainly in tropical regions [(10–50) $\times$ 10^9 g CH_3Br year^{-1},[27]]. Finally, industrial processes lead to the release of halogen-containing species (including fluorinated compounds), which can be degraded in the troposphere. Once in the atmosphere these molecules are broken down by photochemical processes to release atomic halogens.

(*b*) Liberation of halogen compounds from sea salt aerosol. Sea salt contains (by weight) 55.7% Cl, 0.19% Br and 0.00002% I and, given the abundance of sea salt particles in the marine atmosphere, these could be a significant source of gaseous halogen species. The magnitude of this source has not been accurately established but there is substantial evidence for its existence. For example, the depletion of the halogen content (Cl and Br) of marine aerosol relative to bulk seawater (*i.e.* X/Na ratios) gives some indication that the net flux of Cl and Br is to the gaseous phase. As discussed in detail by Cicerone[21] in marine aerosol the Cl/Na and Br/Na ratios are frequently depleted relative to seawater. On the other hand I/Na is always much greater (typically by a factor of 1000 to 10 000[28]). The large iodine enrichment indicates transfer of gas-phase inorganic I to the aerosol. In particular, reactions of the nitrogen oxides NO_2 or N_2O_5 with NaX contained in sea salt aerosol can lead to the formation of XNO or XNO_2, respectively:

$$N_2O_5 + (NaX)_{aerosol} \rightarrow XNO_2 + NaNO_3 \qquad (14)$$

$$NO_2 + (NaX)_{aerosol} \rightarrow XNO + \text{products} \qquad (15)$$

These reactions have been demonstrated in laboratory studies for X = Cl and Br.[29–34]

(*c*) Conversion of hydrogen halides, HX, to halogen atoms: There are substantial direct sources of HF and HCl, as well as of Cl_2, from industrial processes (*e.g.* ref. 22). Probably more important is production of HX by halogen-atom attack on saturated hydrocarbons or aldehydes. It is also likely that liberation from sea salt aerosol occurs by the action of strong acids, such as H_2SO_4 and HNO_3 originating from man made or natural sources.[34]

$$H_2SO_4 + 2(NaX)_{aerosol} \rightarrow 2HX + Na_2SO_4 \qquad (16)$$

Heterogeneous reactions may convert reservoir species such as HX or $XONO_2$ to halogen atoms or X_2 molecules and thus recycle reactive halogen species, as suggested by Fan and Jacob[35] and Mozurkewich.[36] Homogeneous reactions are less efficient in converting HX to X; only reaction with OH [see reaction (23) below] and possibly, in the case of X = I, with NO_3, can provide this pathway.

$$NO_3 + HI \rightarrow HNO_3 + I \qquad (17)$$

Finally, reaction (22) (see below) will convert the halogen atoms to XO radicals.

Chemical Cycles of RHS

The halogen species believed to be of importance in the troposphere and their chemical reaction pathways are summarized in Fig. 1. Many of the tropospheric RHS sources produce either halogen atoms or molecules readily converted to X atoms, *e.g.* molecular halogen (X_2) or XNO_n species (see above):

$$X_2 + h\nu \rightarrow X + X \qquad (18)$$

or:

$$XNO_2 + h\nu \rightarrow X + NO_2 \qquad (19)$$

Consequently, the tropospheric fate of halogen atoms has to be analysed. In particular, light X atoms are highly reactive towards hydrocarbons, leading to the formation of hydrogen halides through hydrogen abstraction (addition occurs in reactions with alkenes):

$$X + RH \rightarrow HX + R \qquad (20)$$

Table 1 Rate constants, k, and reaction frequencies, R, for reaction of halogen atoms (X) with hydrocarbons (RH) and O_3, and the relative importance of $X + O_3$ *vs.* $X + RH$

X	RH (O_3)	k_{X+RH}[a] (k_{X+O_3})/cm^3 molecule^{-1} s^{-1}	typ [RH] (typ [O_3])/cm^{-3}	$R_{RH} = k_{X+RH}$ [RH] ($R_{O_3} = k_{X+O_3}$ [O_3])/s^{-1}	$\sum R_{RH}$/s^{-1}	$\frac{R_{O_3}}{R_{O_3} + \sum R_{RH}}$
F	H_2O	1.4×10^{-11}	2.0×10^{17}	2.8×10^{6}		
	CH_4	8.0×10^{-11}	4.0×10^{13}	3.2×10^{3}		
	H_2	2.6×10^{-11}	1.2×10^{13}	312	2.8×10^{6}	
	(O_3	1.3×10^{-11}	7.2×10^{11}	9.4)		3.3×10^{-6}
Cl	CH_4	1.0×10^{-13}	4.0×10^{13}	4.0		
	C_2H_6	5.9×10^{-11}	3.0×10^{10}	1.8		
	C_3H_8	1.4×10^{-10}	1.0×10^{10}	1.4		
	CH_2O	7.3×10^{-11}	1.0×10^{10}	0.73		
	C_2H_2[b]	1.0×10^{-10}	1.0×10^{10}	1.0		
	C_2H_4[b]	3.0×10^{-10}	5.0×10^{9}	1.5	10.4	
	(O_3	1.2×10^{-11}	7.2×10^{11}	8.6)		0.45
Br	CH_2O	1.0×10^{-12}	1.0×10^{10}	0.01		
	C_2H_2	1.5×10^{-13}	1.0×10^{10}	1.5×10^{-3}		
	HO_2	2.0×10^{-12}	1.0×10^{8}	2.0×10^{-4}	0.012	
	(O_3	1.2×10^{-12}	7.2×10^{11}	0.86)		0.988
I	HO_2	3.8×10^{-13}	1.0×10^{8}	3.8×10^{-5}	3.8×10^{-5}	
	(O_3	1.0×10^{-12}	7.20×10^{11}	0.72)		0.99994

[a] Rate constants from Atkinson *et al.*,[70] evaluated for 298 K. [b] Reaction leads to Cl addition.

Table 2 Self- and cross-reactions of halogen oxides

XO	ClO	BrO	IO
ClO	Cl_2O_2, unstable in the troposphere	reaction (29), three product channels: *a*: $Br + Cl + O_2$ *b*: $BrCl + O_2$ *c*: $OClO + Br$	unknown
BrO	—	$Br + Br + O_2$	unknown
IO	—	—	I_2O_2

For example:

$$X + CH_4 \rightarrow HX + CH_3 \tag{21}$$

F and Cl atoms react readily in this way, while Br atoms can only abstract hydrogen atoms from HO_2 or aldehydes (see Table 1); I atoms are even less reactive. The alternative to reaction (20) is oxidation of the halogen atom (X = Cl, Br, I) by ozone:

$$X + O_3 \rightarrow XO + O_2 \tag{22}$$

Because of the decreasing reactivity of the heavier halogens towards H-containing compounds to form HX, the fraction of halogen atoms being converted to XO by reaction with atmospheric ozone [reaction (22)] is virtually zero for F, *ca.* 50% for Cl, *ca.* 99% for Br and virtually 100% for iodine (see Table 1). The fate of the resulting HX is important when assessing stationary-state RHS concentration. Since HX is highly water-soluble it can be irreversibly removed from the atmosphere by wet or dry deposition. Alternatively, heterogeneous reactions can recycle RHS from HX. The only important gas-phase removal mechanism of HX is reaction with OH,

$$HX + OH \rightarrow X + H_2O \tag{23}$$

which will return the halogen atom to the RHS reservoir. Reaction (23) is endothermic for X = F, and F to HF conversion is extremely rapid in the troposphere, so essentially any reactive F species will be quickly and irreversibly converted to HF. Other reactions of F atoms are therefore not considered here. Lower limits to the branching ratios of XO/HX formation can be estimated from the reactions of halogen atoms with hydrocarbons (R_{RH}) and ozone (R_{O_3}), respectively (see Table 1).

The resulting halogen oxide radicals can undergo a number of reactions including oxidation of organic species (like DMS), photolysis, or recombination with HO_2,

$$XO + HO_2 \rightarrow HOX + O_2 \tag{24}$$

or with another XO. Kinetics and products of XO self-reactions and XO–YO reactions have been investigated for a number of combinations as summarized in Table 2. Another important loss mechanism of halogen oxide radicals is reaction with NO (rate constants given in cm^3 $molecule^{-1}$ s^{-1}):

$$XO + NO \rightarrow X + NO_2 \qquad k_{25} = 1.7 \times 10^{-11} \ (X = Cl) \tag{25}$$

$$k_{25} = 2.1 \times 10^{-11} \ (X = Br, I)$$

In addition, photolysis of XO can be of importance for X = I, Br and, to a minor extent, Cl:

$$XO + h\nu \rightarrow X + O \tag{26}$$

where $J_{26} \approx 3 \times 10^{-5}$ s^{-1}, 4×10^{-2} s^{-1} and 0.2 s^{-1} for X = Cl, Br and I, respectively. Ultimately, the XO/X ratio is determined by reactions (22), (25) and (during daytime)

(26). The self-reactions of XO may also play a role (see Table 2), as in the case of BrO:

$$BrO + BrO \rightarrow 2Br + O_2 \quad (27)$$

Here k_{27} ($=3.2 \times 10^{-12}$ cm^3 molecule^{-1} s^{-1}) is the overall rate constant of the self-reaction of BrO including the channel giving Br_2 (*ca.* 20% at 250 K), which is assumed to photolyse rapidly into Br atoms (18). Iodine oxide appears to produce a stable dimer

$$IO + IO \xrightarrow{M} I_2O_2 \quad (28)$$

and possibly higher oxides of the type $(I_2O_2)_n$.[37] In contrast to the case for bromine, the self-reaction of ClO does not lead to the direct production of Cl atoms. A dimer is formed which is unstable at tropospheric temperatures [$\tau(Cl_2O_2)$ is of the order of seconds], and quantitatively decomposes to ClO + ClO, thus rendering ClO self-reaction unimportant in the troposphere.

However, the BrO–ClO reaction can be important (rate constants are given in cm^3 molecule^{-1} s^{-1}):

$$ClO + BrO \rightarrow Br + Cl + O_2; \quad k_{29a} = 7.0 \times 10^{-12} \quad (29a)$$

$$\rightarrow BrCl + O_2; \quad k_{29b} = 1.2 \times 10^{-12} \quad (29b)$$

$$\rightarrow OClO + Br; \quad k_{29c} = 9.0 \times 10^{-12} \quad (29c)$$

Taking into account the reactions (22), (25), (12) and XO + XO and XO + YO reactions, where applicable, the stationary state [XO]/[X] ratio can be estimated:

$$\frac{[XO]}{[X]} = \frac{k_{22}[O_3]}{k_{25}[NO] + k^*[XO] + k'_{29}[YO] + J_{26}} \quad (VI)$$

where k^* denotes the rate constant for XO self-reaction, while k'_{29} denotes that of XO–YO reactions leading to the formation of X atoms (for instance in the case of X = Br during daytime $k'_{29} = k_{29a} + k_{29b} + k_{29c}$). At night for X = Cl, reaction (29*c*) actually serves as an irreversible sink of ClO, such that there will be no stationary state between Cl and ClO, given a certain abundance of BrO. However, in the case of night-time Br, reaction (29) cannot be expected to be an efficient removal process, assuming equal amounts of BrO and ClO at the beginning of the night, since ClO will be removed much more quickly *via* reaction (29*c*) than BrO is *via* reaction (29*b*). Therefore a BrO/Br steady state will be established. Note that the conversion of XO to X by reaction (29) will set an upper limit to the [XO]/[X] ratio at the very low NO concentrations to be expected in the background troposphere. Table 3 summarizes some estimates of XO/X ratios to be expected for different levels of NO calculated according to eqn. (VI). The night-time ratios were calculated ignoring the formation of other X-containing species such as BrCl or OClO, thus in particular [ClO]/[Cl] could be larger.

Table 3 Ratios of XO/X for various NO concentrations (and 40 ppb O_3)

	night ($J_{26} = 0$)		day	
ratio	[NO] = 0	[NO] = 10 ppt	[NO] = 10 ppt	[NO] = 1 ppb
ClO/Cl	>3600[a]	>1600[a]	1000	30
BrO/Br	130[a,b]	80[a,b]	20	2
IO/I	—	≤190	≤5	1

[a] Assuming 17 ppt BrO (5×10^8 molecule cm^{-3}) as measured in Alert by Hausmann and Platt.[71] [b] Assuming 17 ppt ClO (5×10^8 molecule cm^{-3}).

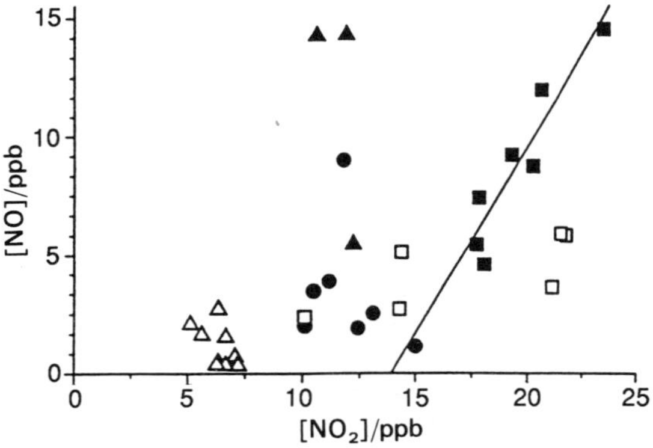

Fig. 2 Correlation between NO near the local source B3 (UMEG[40]) and NO_2 measured by DOAS. Only data corresponding to NO_3 values well above the detection limit are shown (26–27 February, squares; 28 February–1 March, circles; 4–5 March, triangles). Extremely low NO values on the last night coinciding with low NO_2 values are marked by open symbols. The first part of the night 26–27 February (filled squares) reveals the influence of local sources: NO_2 above a 'background' level of 16 ppb is highly correlated with NO, indicating that a portion of the measured NO_2 is due to local emission.

In airmasses containing elevated NO_x levels (more than a few tens of ppt) storage of XO as $XONO_2$ can become important:

$$XO + NO_2 \overset{M}{\longleftrightarrow} XONO_2 \tag{30}$$

At tropospheric temperatures the thermal lifetime of $XONO_2$ species is of the order of minutes, thus photolysis of $XONO_2$ should be of minor importance and the equilibrium $XONO_2$ concentration will normally be reached. The $XONO_2$/XO ratio in thermal equilibrium is *ca.* unity at background NO_x levels.

Experimental Results

A series of observations made at Heidelberg provides evidence for the limitation of the NO_3 lifetime by the reaction of NO_3 with NO [reaction (3)]. Measurements of NO_3 and its precursors (O_3 and NO_2) using the DOAS technique[38,39] were performed over a period of more than five weeks during winter 1994. The lightpath with a total length of 7250 m allowed the detection of 3 ppt NO_3. It was situated northwest of the Heidelberg city centre (49.4°N, 8.7°E) extending mainly over grasslands and cultivated fields. However there was a major county road (B3) running more or less parallel to the lightpath at an average distance of *ca.* 500 m and intersecting the lightpath at the northernmost end. During rush hours in the early mornings and evenings this road is expected to be the major local pollution source. Additional measurements of NO, O_3 and meteorological parameters were made next to the road (B3) near the southern end of the lightpath by UMEG.[40] Half-hour averages of NO were taken by chemiluminescence. These measurements allowed us to evaluate the influence of NO_x emissions from the B3 by correlating the NO values to the NO_2 measurements at the lightpath (Fig. 2). Spectra for 26 out of 36 nights of observation were evaluated. During three of these NO_3 exceeded the detection limit. The corresponding τ'_{NO_3} ranged from below 50 s during the nights (where no NO_3 could be detected) up to 300 s when NO_3 exceeded the detection limit. These values are in the lower range of observed values (see Table 4). By looking at the nights when NO_3 was detected (26/27 February, 28 February/1 March and 4/5 March, 1994) it can be concluded that scavenging of NO_3 by NO originating

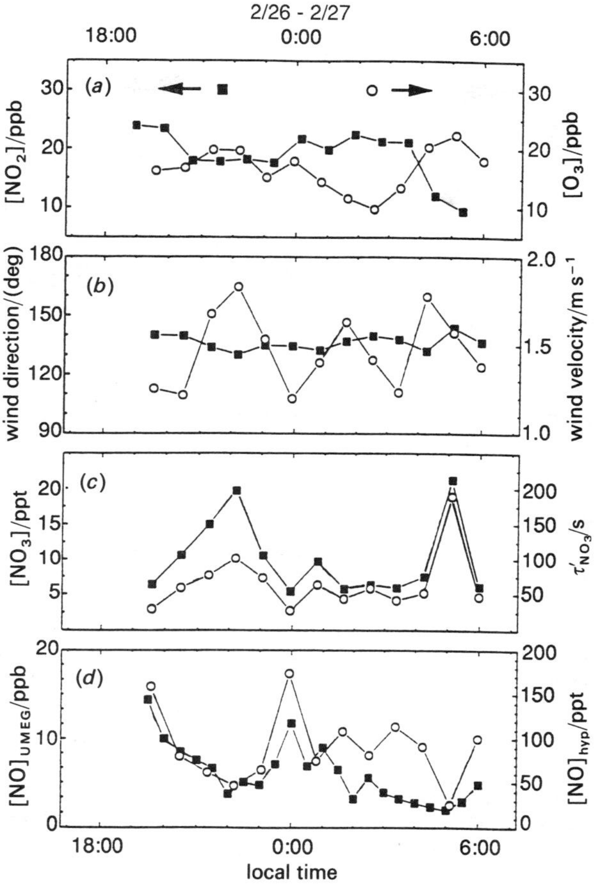

Fig. 3 Time series of NO_3 and related data during the night 26–27 February, 1994. Solid squares refer to the axes on the left, open circles refer to the axes on the right. (*a*) DOAS measurements of NO_2 and O_3. (*b*) Wind speed and direction taken by UMEG[40] reflecting the steady meteorological situation. (*c*) NO_3 and τ_{NO_3} (evaluated according to eqn. 3). (*d*) [NO] (UMEG) and $[NO]_{hyp}$ required to explain the observed τ_{NO_3} solely by reaction with NO [reaction (3)]. During the first half of the night [NO] and $[NO]_{hyp}$ track each other remarkably well, but with $[NO]/[NO]_{hyp} \approx$ 60–100. This agreement disappears during the second half of the night, probably reflecting the influence of NO sources other than the B3.

from nearby sources is the likely cause for those short lifetimes. Fig. 3 shows the time series of NO_3 and related data during the night 26/27 February, where the meteorological situation indicated a steady NO emission from the B3, thus probably making reaction (3) the dominant sink of NO_3. Indeed this assumption leads to remarkable agreement between the NO variations observed at the B3 and the hypothetical NO concentration, $[NO]_{hyp}$, required to explain the observed NO_3 lifetimes solely by losses *via* reaction (3), at least during the first half of the night. From Fig. 3 a ratio between the NO observed at the source (B3) and $[NO]_{hyp}$ of about 60–100 can be derived. This reduction in [NO] is, of course, mostly due to oxidation of NO by O_3 [reaction (35)]. In fact, from the good correlation between NO_2 and $(\tau'_{NO_3})^{-1} \propto [NO]_{hyp}$ [line (*c*) in Fig. 4] a similar reduction factor in NO can be deduced.

In addition, the average NO concentration on the lighpath as calculated from simple chemistry (taking an average travel time of an airmass from the B3 to the lighpath of 470 s) is in the range of $[NO]_{hyp}$. The night beginning 28 February shows a similar

Table 4 Observations of NO_3 in the atmosphere

site of measurement	pollution level of airmass	typical $[NO_2]$ /ppb	typical $[O_3]$ /ppb	typical[a] $[NO_3]^b$ /ppt	max: $[NO_3]^b$ /ppt	typical τ_{NO_3} /s	reference
Hawaii	none[c]	0.03	30	0.2	—	690	42
Schauinsland[d]	none[c]	$\leqslant 0.05$	40	7	7	$\geqslant 6100$	72
Schauinsland[d]	slight	0.8	50	20	45	870	72
Schauinsland[d]	slight	0.7	80	8	9	250	73
Loop Head, Ireland (westerly wind)	slight	0.15	30	1.5	<3	580	74
Death Valley USA	slight	<0.3	30	30	40	>5800	75
Colorado Mtns., USA	slight–mod.	2	40	20	70	440	42, 43
Deuselbach, Germany	slight–mod.	5	60	200	250	1700	76
Edwards AFB, USA	slight–mod.	1.5	30	70	90	2700	75
Jülich, Germany	mod.	15	30	20	70	80	76
Key Biscane, Florida, USA	slight–mod.	20	20	10	20	40	77, 78
Pointe de Penmarc'h Brittany, France	slight–mod.	1	40	5	18	220	79, 80, 81
Rügen, Germany	slight–mod.	3	30	9	90	170	41, 44
Phelan, Mojave USA	mod.	2	30	10	50	290	75
Whitewater, USA	mod.	2	30	100	160	2900	75
Edison, Calif., USA	mod.	30	>30	30	80	<60	77, 78
Los Angeles, USA	strongly polluted	60	60	200	300[e]	10	82, 83, 84

[a] Second half of the night. [b] Values corrected for σ_{NO_3} (662 nm) = 2.0×10^{-17}. [c] Above the inversion. [d] Black Forest, Germany. [e] Two hours after sunset.

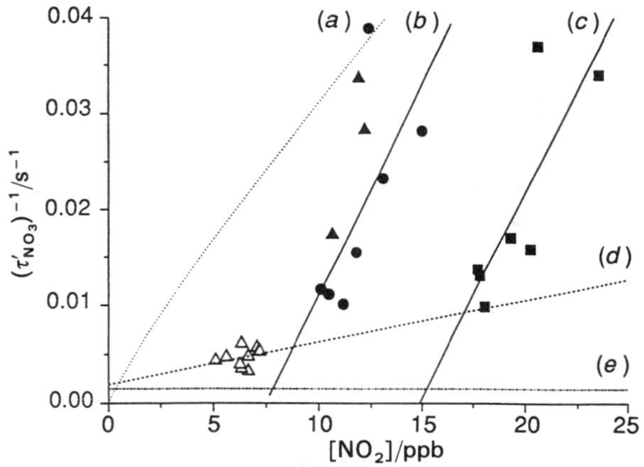

Fig. 4 Inverse lifetime, $(\tau'_{NO_3})^{-1}$, *vs.* NO_2. Measurements in Heidelberg are represented by filled symbols (26–27 February, squares; 28 February–1 March, circles; 4–5 March, triangles), open symbols indicate relatively unpolluted situations. Straight lines (*c*), (*b*) and (*d*) refer to the best linear-square fits to the respective data. As expected from the correlation of NO with NO_2 during the night 26–27 February the inverse lifetime is linearly correlated with NO_2 $\{(\tau'_{NO_3})^{-1} = (0.033 + 4.3 \times 10^{-3} \times [NO_2]/\text{ppb})\ \text{s}^{-1}\}$ indicating direct NO_3 removal by NO. Similarly the night 28 February–1 March gives $(\tau'_{NO_3})^{-1} = (0.063 + 4.4 \times 10^{-3} \times [NO_2]/\text{ppb})\ \text{s}^{-1}$. Line (*d*): best fit to the slightly polluted situation $\{(\tau'_{NO_3})^{-1} = (2 \times 10^{-3} + 4 \times 10^{-4} \times [NO_2]/\text{ppb})\ \text{s}^{-1}\}$. Owing to large scatter the parameters are not very well determined, a constant removal rate of *ca.* $4.7 \times 10^{-3}\ \text{s}^{-1}$ would also be consistent with the data. For comparison, data from other measurements are included. (*a*) Fit to observed lifetimes in Rügen,[42] $(\tau'_{NO_3})^{-1} = 0.004\ \text{s}^{-1} \times ([NO_2]/\text{ppb})^{0.89}$ (*e*) Rate of removal derived by Noxon[42,43] from data taken in Hawaii and at the Colorado mountains $(\tau'_{NO_3})^{-1} = 1.5 \times 10^{-3}\ \text{s}^{-1}$.

$(\tau'_{NO_3})^{-1}$ to NO_2 relation [line (*b*) in Fig. 4], also indicating NO_3 loss being dominated by reaction with NO.

The relatively low pollution level during most of the night 4/5 March allows a rough estimate of τ'_{NO_3} due to additional sink mechanisms and can be used as an upper limit for possible loss of N_2O_5 or NO_3 due to other (possibly heterogeneous) mechanisms. The data are compatible with a first order NO_3 loss rate of $4.7 \times 10^{-3}\ \text{s}^{-1}$ as well as with a combined loss of NO_3 and N_2O_5 at rates of $2 \times 10^{-3}\ \text{s}^{-1}$ and $2 \times 10^{-4}\ \text{s}^{-1}$, respectively [Fig. 4, line (*d*)].

A fit to NO_3 lifetimes observed during a one year continuous observation program of NO_3, NO_2 and O_3 in Rügen[41] is also shown in Fig. 4 [line (*a*)]:

$$\tau_{NO_3}{}^{-1} = 4.0 \times 10^{-3}([NO_2]/\text{ppb})^{0.89}\ \text{s}^{-1} \qquad \text{(VII)}$$

In contrast to the measurements close to source of NO in Heidelberg the near-inverse proportionality between τ'_{NO_3} and NO_2 is here probably due to loss of N_2O_5 as pointed out earlier [eqn. (5)]. For comparison, the rate of NO_3 removal derived by Noxon[42,43] from data taken in Hawaii and in the Colorado mountains is $(\tau'_{NO_3})^{-1} = 1.5 \times 10^{-3}\ \text{s}^{-1}$, [line (*e*)].

Tropospheric Concentration of Nitrate Radicals

To date, explorative measurements of the NO_3 concentrations in a variety of atmospheric environments ranging from slightly to heavily polluted have been made. Table 4 gives a characterisation of airmass type and NO_3 peak concentrations encountered.

Note that the range of NO_3 concentrations observed at the various measurement sites is much smaller than the variation in the rate of NO_3 production. On the other

Table 5 Indirect estimates of X-atom and XO concentration (X = Cl, Br) derived from relative rates of hydrocarbon degradation in the atmosphere or measurement of other X-containing species

ref.	site	[Cl][a]/ molecule cm^{-3}	d[Cl]/d*t*/ molecule cm^{-3} s^{-1}	[ClO][a]/ molecule cm^{-3}	[Br][a]/ molecule cm^{-3}	[BrO][a]/ molecule cm^{-3}
85[b]	Point Arena, CA, 39°N	1.5×10^4	1.5×10^5	3×10^7	—	—
50	Point Arena	$< \times 10^3$	$< \times 10^4$	$<2 \times 10^6$	—	—
51	Alert and north > 82°N	$(1–8) \times 10^4$	$(1–8) \times 10^5$	$(2–16) \times 10^7$	$(1–6) \times 10^7$	$(1–6) \times 10^9$
52	Ny-Alesund, Spitsbergen, 79°N	$(3–6) \times 10^4$	$(3–6) \times 10^5$	$(6–12) \times 10^7$	$(3–10) \times 10^7$	$(3–10) \times 10^9$
48	Virginia Key, Florida, 26°N	$(1–10) \times 10^4$	$(1–10) \times 10^5$	$(2–20) \times 10^7$	—	—

Data are 24 h averages. [a] Estimated. [b] Interpreting data from Parrish *et al.*[86]

hand, the present data set of NO_3 radical measurements, largely consisting of measurements over periods of a few days to weeks, is by no means representative. This is true for both the seasonal variation and the spatial distribution of NO_3 radicals.

The actually observed NO_3 concentrations are, therefore, largely influenced by the loss mechanisms discussed above. Any attempt to derive meaningful global average tropospheric NO_3 concentrations appears to be a hopeless task. However the observation that the NO_3 lifetime appears to vary inversely with the rate of NO_3 production close to sources of NO indicates that polluted areas with large NO_3 production (constituting only a small fraction of the earth's surface) should not contribute out of proportion to the global NO_3 concentration.

An estimate of the NO_3 concentration in the remote troposphere could, therefore, be a meaningful approximation of the global average NO_3 concentration. It can be derived from available data by considering the following assumptions:

1. Assuming a constant lifetime of NO_3. For instance, using the value reported by Noxon[42,43] of about 700 s, combined with a background NO_x (night-time NO_2) level of 30 ppt would result in a background NO_3 concentration of 0.4 ppt. However, NO_3 lifetimes up to an order of magnitude higher were observed (see Table 4), thus a range of 0.5 to 5 ppt NO_3 is more likely.

2. Assuming an NO_3 lifetime inversely proportional to the NO_2 concentration (see Fig. 4). The relationship between $\tau_{NO_3}{}^{-1}$ and NO_2 derived from the extensive data set taken at Rügen,[41,44] eqn. (VII) can be simplified to $\tau_{NO_3}{}^{-1} \propto [NO_2]$. Combination with eqn. (I) yields:

$$[NO_3] \approx 250 \text{ s} \times [O_3] \times k_1 \text{ ppb} \qquad \text{(VIII)}$$

At a typical concentration of 40 ppb of O_3 the above relationship predicts an NO_3 concentration of 5 ppt coinciding with the upper limit of the range obtained from the constant lifetime consideration above.

Tropospheric Concentration of RHS

Overall there is not much information on the abundance of RHS in the troposphere. The available data come from (1) Indirect evidence for the presence of halogen atoms, mostly from observation of hydrocarbon ratios in airmasses of known chemical age. Also, some indirect evidence for the presence of reactive Cl species comes from observation of tropospheric HCl. (2) Direct observation of XO species in the troposphere.

Indirect Evidence for XO

Some estimates of Cl, Br, ClO and BrO concentrations are summarized in Table 5 (assuming [XO]/[X] ratios of 1000 and 100 for typical background conditions, for X = Cl and Br, respectively, see Table 3) based on the following considerations.

Chlorine

Since HCl is the major product of reactions of chlorine atoms with atmospheric trace gases [see for instance reaction (20)], its detection in the atmosphere provides indirect evidence for the presence of Cl atoms. Over the Northwest Atlantic HCl levels between 0.1 and 0.5 ppb were found below 3 km altitude and from 0.05 to 0.1 ppb at 3–7 km.[45] Also from ground based IR total column measurements aimed at the detection of stratospheric HCl at European mid-latitudes[46] and at Spitsbergen,[47] tropospheric total column values of the order of $(5–7) \times 10^{14}$ cm^{-2} can be derived, which correspond to an average mixing ratio of 30 to 45 ppt, or around 100 ppt assuming 3 km scale height. Cicerone[21] quoted values *ca.* 10 times higher, also derived from total column measurements. Pszenny *et al.*[48] reported *in situ* data ranging from 50 to 250 ppt HCl in the marine boundary layer near Florida. During the same measurements up to 0.25 ppb Cl in the form of Cl_2, HOCl or ClO (which could not be differentiated between by their technique) were also found. Further indirect evidence for the formation and reaction of Cl atoms in the troposphere comes from recent observations of time-dependence and relative surface concentrations of C_2–C_6 hydrocarbons (Table 5). Frequently the relative concentrations of, for instance, butane/ethane *vs.* propane/ethane are more consistent with Cl atoms rather than OH reactions accounting for the removal of certain hydrocarbons in coastal air. While there is criticism of this method (see for instance ref. 49 and 50), the results nevertheless provide indications for the presence of other oxidants in the troposphere. From volatile organic compound (VOC) measurements in the Arctic (>80°N) springtime, Jobson *et al.*[51] concluded that Cl atoms are likely to play a significant role in VOC removal. Similar conclusions were reached by Solberg *et al.*[52] based on a large set of observations made during 1989–1993 in Ny-Alesund (Spitsbergen).

Bromine

There is indirect evidence for the presence of BrO_x species gained during campaigns observing recently discovered episodic destruction of boundary-layer ozone in the Arctic.[7,53–55] During these episodes, high concentrations of 'filterable bromine' (f-Br, Br that can be collected on cellulose filters) are observed, the active component of which might be BrO radicals. Further evidence for the participation of Br atoms in atmospheric chemistry, at least in the Arctic, comes from observations of the behaviour of acetylene relative to other non-methane hydrocarbons,[51,52] see Table 5. The deviation of acetylene from the trend in degradation of the light alkanes was consistent with its removal by the reaction of Br with C_2H_2, at average [Br] of the order of 10^7 molecule cm^{-3}. Such levels of [Br] are consistent with the calculations of McConnell *et al.*[56] using a model in which HBr and brominated organic compounds undergo heterogeneous reactions on ice particles releasing gaseous Br_2.

Iodine

The tropospheric iodine cycle appears to be composed of a major oceanic source, mainly as CH_3I, which is converted to gaseous inorganic iodine by photochemistry (see tropospheric chemistry of RHS section). The gaseous iodine transfers to the particulate or aerosol phase, and is subsequently removed by deposition. The comparatively rapid photo-oxidation of CH_3I in the troposphere results in low ambient concentrations (typical background level,[57,24] *ca.* 2 ppt). The photolysis of CH_3I is known to generate I atoms, thereby providing evidence for the participation of IO radicals in the mechanisms forming the inorganic iodine reservoirs.[21,37,58,59] On the basis of available laboratory kinetics data, HOI has recently been calculated to be the major component, with significant contributions from $IONO_2$ under elevated NO_x conditions.[60]

Direct Observation of XO

Although all halogen oxide radicals (XO) have strong banded absorption spectra in the near-UV only very few attempts have been made to detect halogen oxides in the troposphere. From Long-Path DOAS (LP-DOAS) measurements of IO at the French Atlantic coast in 1989, upper limits of 10^7 molecular cm^{-3} were reported.[79]

Recently, direct measurements of bromine oxide and ozone by LP-DOAS performed at Alert (82.3°N, 62.2°W), located in the Canadian Arctic at the shore of the Arctic ocean, in April 1992 were reported.[61,71] Bromine oxide levels range from the detection limit of *ca.* 4 ppt up to a maximum of *ca.* 17 ppt.

Table 6 Estimates of atmospheric halogen oxide radical concentrations

species	indirect data/ppt	direct observation/ppt
FO	$<10^{-7}$	—
ClO	1–10	—
BrO	30–300	<4–17
IO	≈0.05	⩽0.5

In summary, XO concentrations for X = Cl, Br, I of the order of ppt can be derived from the available indirect and direct data. FO concentrations can be estimated to be <1 molecule cm^{-3} assuming a source strength similar to the heavier X-atoms (Table 5) together with the low FO/HF branching ratio (Table 1). Table 6 gives a synopsis of the resulting estimates. Note, however, that the data presented in Tables 5 and 6 are derived from observations in areas with presumably high levels of halogens, thus global average concentrations are likely to be much smaller (see below). For instance an upper limit can be placed on the global average Cl concentration (and thus on the average ClO levels) by assuming that in the global CH_4 level an additional loss of ⩽20% due to reaction (21) could go unnoticed. This consideration yields $[Cl]_{glob.av.} \leqslant 8 \times 10^3$ molecule cm^{-3} (or $[ClO]_{glob.av.} \leqslant 8 \times 10^6$ molecule cm^{-3}), which is about one order of magnitude lower than most estimates in Table 5.

Interaction between NO_3 and RHS

There are interesting effects of the hypothetical simultaneous presence of NO_3 and RHS in the same airmass. Since the NO_3 can only reach appreciable concentrations at night, the importance of reactions between NO_3 and RHS species depends critically on the conversion of X or XO into reservoir species in the absence of sunlight. At low NO_x levels the currently known reactions would not necessarily remove XO species. For instance, reaction between halogen atoms or halogen oxide radicals and NO_3 may influence the chemical cycle of either species due to the reactions:

$$NO_3 + X \rightarrow NO_2 + XO \tag{31}$$

$$NO_3 + XO \rightarrow X + NO_2 + O_2 \tag{32a}$$

$$\rightarrow OXO + NO_2 \tag{32b}$$

In the case of X = Cl, channel (32*b*) could account for about 30%, thus possibly creating a tropospheric night-time OClO source. Table 7 summarizes the rate constants for the

Table 7 Rate constants for reactions of halogen atoms and halogen oxide radicals with NO_3 and NO_3 lifetime caused by the presence of 'typical' concentrations of XO_x species (see Tables 6, 5) in the troposphere

XO_x species	typical concentration /molecule cm^{-3}	rate constant $k_{XO_x+NO_3}$ /cm^3 molecule^{-1} s^{-1}	$\tau_{NO_3}(XO_x)$/s
Cl	10^3–10^5	2.6×10^{-11}	3.8×10^7–3.8×10^5
ClO	10^6–10^8	4.0×10^{-13}	2.5×10^6–2.5×10^4
Br	10^7–10^8	1.6×10^{-11}	6.3×10^3–630
BrO	10^9–10^{10}	1.0×10^{-12}	1000–100
I	—	—	—
IO	10^7	—	—

Table 8 Upper limits for the concentration of Br atoms and BrO radicals derived from observed NO_3 lifetimes

site[a]	τ_{NO_3}/s	$[Br]_{max}$ /molecule cm^{-3}	$[BrO]_{max}$ /molecule cm^{-3}
Schauinsland	1.6×10^4	1×10^7	1.6×10^8
Deuselbach	1.7×10^3	4×10^7	6×10^8
Hawaii	10^3	9×10^7	10^9
Mojave	1.5×10^3	2×10^8	3.4×10^9
Brittany	220	3×10^8	4.5×10^9
Rügen	170	4×10^8	6×10^9

[a] Designation refers to Table 4.

above reactions as well as the NO_3 lifetimes that would result if typical (daytime) XO or X concentrations were also present at night (see Tables 6 and 5).

In particular BrO could well be limiting for the NO_3 lifetime. In fact BrO could be the 'mysterious' NO_3 scavenger postulated by several authors.[42,43,75] On the other hand, field observations of NO_3 lifetimes (see above) may be used to place upper limits on the BrO or Br concentration in tropospheric airmasses. Table 8 summarizes some upper limits of BrO_x derived in this way. The data are at the lower end of the ranges given in Table 5.

Consequences of the Presence of RHS and NO_3 in the Troposphere

There are numerous possible consequences of the presence of these radicals in the troposphere, most of them are related to changes in the oxidation capacity of the atmosphere. In the following we list a few examples.

Tropospheric Ozone Destruction

Episodic destruction of boundary-layer ozone has been observed at various arctic sites (Alert, Canada;[7,53–55] Barrows, Alaska;[62] Ny-Alesund, Spitsbergen[52]). During these episodes, ozone levels drop from the normal *ca.* 40 ppb to unmeasurable (<2 ppb) levels. While the reasons for this phenomenon are far from being understood, it is being hypothesized by several authors that BrO_x[7,54,63] or $BrO_x + ClO_x$[64] catalysed reaction cycles [involving the reaction sequences reactions (27), (22) or (29), (22), respectively] may be responsible for the observed ozone destruction. Recently the presence of BrO radicals was verified experimentally.[71] In addition, evidence is accumulating that significant amounts of chlorine atoms are present in the arctic troposphere during times when ozone depletion events occur.[51,52]

Thus the total loss of ozone rate limited by reactions (27) and (29) is given by:

$$-\frac{d}{dt}[O_3] = 2(k'_{29}[BrO][ClO] + k_{27}[BrO]^2) \qquad (IX)$$

Fig. 5 illustrates the efficiency of the BrO and BrO + ClO catalysed ozone destruction mechanisms. It is interesting to note, that the O_3 destruction is zero order in $[O_3]$, as long as there are more than a few ppb of O_3.[71] Therefore a combined BrO + ClO mechanism could remove the ozone within *ca.* two days, while a hypothetical first order loss of O_3 with $\tau_{O_3} \approx$ two days would require about one week before $[O_3]/[M] < 2$ ppb would be reached.

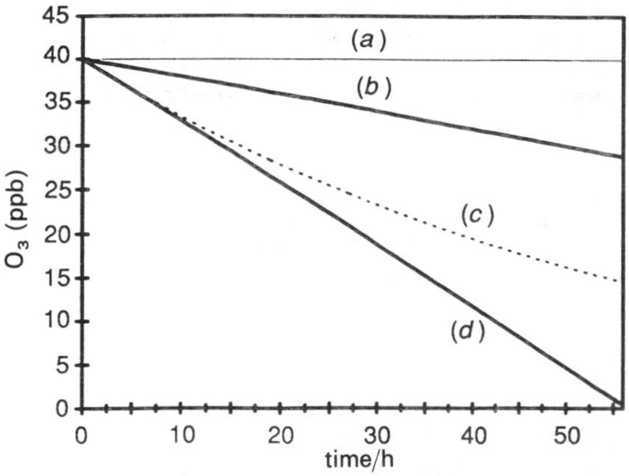

Fig. 5 Calculated rates of ozone destruction in several scenarios: (*a*) 17 ppt ClO, no BrO; (*b*) no ClO, 17 ppt BrO; (*d*) 17 ppt ClO and 17 ppt BrO; (*c*) other process causing a first order loss of O_3 corresponding to $\tau = 56$ h.

Shift of the NO_2/NO Photostationary State

The presence of halogen atoms in the troposphere will change the Leighton ratio (L) as determined by the reactions

$$NO_2 + h\nu \longrightarrow NO + O \tag{33}$$

$$O + O_2 \xrightarrow{M} O_3 \tag{34}$$

$$O_3 + NO \longrightarrow NO_2 \tag{35}$$

$$L = \left(\frac{NO_2}{NO}\right)_{\text{Leighton}} = \frac{[O_3]k_{35}}{J_{33}} \tag{X}$$

It has been noted by several authors that L' calculated from observed values of NO, NO_2, O_3 and J_{33} is always larger than L given by eqn. (X). This observation is usually attributed to the reaction of NO with peroxy radicals (RO_2) like HO_2 or CH_3O_2 leading to the modified Leighton ratio:

$$L' = \frac{[O_3]k_{35} + \sum_i [R_iO_2]k_{R_i}}{J_{33}} \tag{XI}$$

where $[R_iO_2]$ denotes the concentration of R_iO_2 radicals and k_{R_i} the reaction constants R_iO_2 + NO. As pointed out by Harder *et al.*[6] the enhancement of L' by peroxyl radicals can also be due to the reaction of halogen oxides with NO [reaction (25)]. Thus the concentrations of XO have to be included in the sum over $[R_iO_2] \times k_{R_i}$ in eqn. (XI). Comparing the rate constants of (25): 1.7×10^{-11} cm^3 molecule^{-1} s^{-1} (X = Cl), 2.1×10^{-11} cm^3 molecule^{-1} s^{-1} (X = Br, I) to $k_{HO_2+NO} = 0.8 \times 10^{-11}$ cm^3 molecule^{-1} s^{-1}, it becomes clear that XO radicals will be two to three times more effective in modifying L than RO_2. In particular, like RO_2 they will have a noticeable effect on L even at ppt-mixing ratios. An example of the required RO_2 and/or BrO concentrations was deduced from measurements in Brittany during 1992;[6,65] the BrO concentrations required to explain the (*ca.* +36%) deviation of L' from L (by the reaction of BrO alone) are of the order of 30 ppt.

An important consequence of this increase in L due to XO is the reduction of the OH concentration. An estimate of the absolute magnitude of this effect can be derived from photochemical model calculations of the sensitivity of the OH concentration against changes of J_{33}[66] resulting in $S = (\Delta[OH]/[OH])/(\Delta[J_{33}]/[J_{33}]) \approx 0.5$ at $NO_x \approx 0.3$–2 ppb. Since $L' \propto (J_{33})^{-1}$ changes in L' will influence the OH with the same sensitivity as does J_{33}. Assuming that half of the deviation between L' and L reported by Harder *et al.*[6] is due to BrO (the other half being caused by RO_2), then 15 ppt BrO radicals would cause a shift in L' of +18% resulting in a $\Delta[OH]/[OH] \approx -S \times 18\%$ or a *ca.* 9% reduction in the OH concentration.

Degradation of Organic Species

Obviously the presence of halogen atoms, even at low concentrations, can have a significant influence on the oxidation capacity of the troposphere, since the rate constants, in particular of Cl atoms towards hydrocarbons (see Table 1) are much larger than the corresponding OH + HC reactions. Halogen oxide radicals can also play a role in the degradation of volatile organic compounds like DMS; k_{XO+DMS} is anomalously high for X = Br, as noted by Toumi.[67]

Conclusions

In summary, there is a growing body of evidence that halogen species and nitrate radicals are present in the troposphere at concentration levels sufficient to influence chemical reaction cycles, in particular those related to the oxidation capacity of the boundary layer air. On the other hand, to date there is no direct proof for the involvement of halogen compounds in tropospheric chemistry. Areas where RHS are likely to be relevant include polluted coastlines, where anthropogenic acids and NO_x can liberate RHS from sea spray. Likewise, plumes from biomass burning extending over the oceans may produce elevated levels of RHS. There are not enough data to quantify the additional oxidation capacity due to RHS and NO_3, let alone the changes in the total oxidation capacity due to indirect effects of those species. Nevertheless, it is tempting to compare the removal rate of certain species by OH reaction to that by the radicals discussed here.

There are two ways to get a rough idea of the oxidation potential of a particular radical: (1) Determining the average (daily) production. This number will give an upper limit on the total amount of species (for instance in ppb day^{-1}) that can be degraded. (2) Determining the average concentration. Multiplying this number by the concentration of species potentially degraded by the radical and the respective rate constant will give an estimate of the rate of removal. In Table 9 average radical concentrations estimated according to approach (2) are summarized

Following the first approach we obtain:

OH

The OH production integrated over a clear day is about 1.2×10^{11} molecule cm^{-3} or *ca.* 5 ppb day^{-1} [from photolysis of 40 ppb O_3, assuming 10% of the $O(^1D)$ to be converted[3] to OH]. While in the case of HO_x recycling of radicals can be important, this will only enhance the daily integrated OH production at NO_x concentrations around 1 ppb (then enhancement will be about six-fold). At NO_x levels much lower (for instance in clean air, $NO_x \approx 30$ ppt) or much higher than 1 ppb the enhancement of the integrated OH production is small (<20% in the case of $NO_x < 30$ ppt and >100 ppb, respectively[68]). On the other hand the above value is a clear day peak value, which has to be scaled by $[OH]_{av}/[OH]_{peak}$ or about 1/6 leading to 800 ppt OH/day (see Table 10).

Table 9 Estimated lifetimes (in h) of selected species against degradation by the radicals discussed in the text (evaluated at 283 K)

species	OH (av.)	OH (peak)	NO_3 (av.)	NO_3 (peak)	ClO (av.)	ClO (peak)	Cl (av.)	Cl (peak)	BrO (av.)	Br (av)
conc./molecule cm^{-3}	6×10^5	1×10^7	1.5×10^8	1×10^{10}	10^6	10^8	10^3	10^5	10^8	5×10^6
DMS	110^a	6.3^a	1.7^a	0.03^a	$3 \times 10^{4\,c}$	280^c	870^f	8.7^f	10^c	0.43^g
n-butane	190^a	11^a	$2.8 \times 10^{4\,a}$	430^a	—	—	1400^h	14^h	—	—
C_2H_4	56^b	3.4^b	$1 \times 10^{4\,h}$	160^h	—	—	2800^b	28^b	—	—
isoprene	4.6^h	0.3^h	1.9^h	0.03^h	—	—	$<1100^d$	$<11^d$	—	0.9^e
toluene	77^h	4.6^h	$3.1 \times 10^{4\,h}$	460^h	—	—	—	—	—	—
CH_4	$7 \times 10^{4\,a}$	$4 \times 10^{3\,a}$	$>5 \times 10^{4\,h}$	$>700^h$	$>7 \times 10^{7\,b}$	$>7 \times 10^{5\,b}$	$3 \times 10^{6\,a}$	$3 \times 10^{4\,a}$	—	—

[a] Coefficients from ref. 70. [b] Constants from ref. 87. [c] Coefficients from ref. 88. [d] Based on $k(Cl + \text{Isoprene}) \geqslant 2.5 \times 10^{-10}$ cm^3 $molecule^{-1}$ s^{-1} from ref. 89. [e] Based on $k(Br + \text{isoprene}) = 6 \times 10^{-11}$ cm^3 $molecule^{-1}$ s^{-1} from ref. 89. [f] Coefficients from ref. 90. [g] Constants from ref. 91. [h] Constants from ref. 92. [i] Coefficient from ref. 93.

Table 10 Estimated radical production, ppt per day (24 h average)

species	OH	Cl, ClO	Br, BrO	I, IO	NO_3
average	800	4	40	—	30
maximum	5000	400	4000	—	—

Halogen Atoms

Data for d[Cl]/d*t* (and thus d[ClO]/d*t*) can be taken from Table 5; the corresponding values for Br are probably ten times as large, while estimates for iodine species are difficult. Since those data are 24 h averages, integration is straighforward, leading to 4–400 ppt Cl day^{-1} and up to 4 ppb Br day^{-1}.

NO_3

Assuming an NO_x to NO_3 conversion time of *ca.* one night combined with a background NO_x level of 30 ppt leads to an estimated integral production of 30 ppt NO_3 day^{-1}. It appears that this value is a lower limit in the background atmosphere with much higher levels occurring in airmasses affected by pollution. However, in areas close to NO_x sources NO_3 loss due to reaction with NO, and thus returning NO_3 to the NO_x reservoir, can be significant. The enhanced heterogeneous conversion of NO_3 to HNO_3 *via* N_2O_5 can be regarded as NO_3-radical initiated oxidation of NO_x (analogous to the $OH + NO_2$ recombination).

Considering the information in Tables 9 and 10, it becomes clear that the halogen oxide radicals can provide a considerable fraction of the oxidation capacity provided by the HO_x system. Although the available information on the tropospheric concentration of RHS pertains mostly to the boundary layer it should be kept in mind that the short-lived trace species are degraded there. While the rate of NO_3 formation appears to be small, when compared with that of OH on a global scale, this radical can be quite efficient under conditions of low insolation, *i.e.* at night and in mid- and northern latitudes during wintertime. Also, the NO_x–NO_3 conversion *via* N_2O_5 initiated by NO_3 proceeds at twice the rate of NO_3 production. Additionally it should be stressed that the oxidation of DMS over the free ocean could be dominated by the NO_3 pathway. At typical DMS concentrations of 20–60 ppt above the free ocean,[69] 20 to 30 ppt DMS will react with NO_3 during the night of a 24 h period, if no other significant NO_3 scavengers are assumed. The lower value corresponds to a total DMS consumption, the higher to a total consumption of NO_3, since the NO_3–DMS reaction is very fast. In comparison the integral amount of DMS oxidized by OH during daytime will be 2–6 ppt, thus making oxidation by NO_3 5–10 times as effective as oxidation by the OH radical.

The considerations presented here clearly show that more research is required and warranted to reduce the uncertainties in the concentration levels of those free radicals referred to in this study.

We are especially grateful to the UMEG[40] for providing observational data from Heidelberg. Helpful discussion and the provision of kinetic data prior to publication by G. Le Bras is gratefully acknowledged.

References

1 B. Weinstock, *Science*, 1969, **166**, 244.
2 H. Levy II, *Science*, 1971, **173**, 141.
3 D. H. Ehhalt, H-P. Dorn and D. Poppe, *Proc. Royal Soc. Edinburgh*, 1991, **97B**, 17.

4 U. Platt, G. LeBras, G. Poulet, J. P. Burrows and G. Moortgat, *Nature* (*London*), 1990, **348**, 147.
5 G. Le Bras, C. Gölz and U. Platt, *Proc. Internat. Symp. Dimethyl Sulphide: Oceans, Atmosphere and Climate*, Belgirate, Italy, ed. G. Restelli and G. Angeletti, Kluwer Academic Publishers, 1993, 251.
6 H. Harder, A. Volz-Thomas and U. Platt, *Geophys. Res. Lett.*, in preparation.
7 L. A. Barrie, J. W. Bottenheim, R. C. Schnell, P. J. Crutzen and R. A. Rasmussen, *Nature* (*London*), 1988, **334**, 138.
8 A. Volz, D. H. Ehhalt and R. G. Derwent, *J. Geophys. Res.*, 1981, **86**, 5163.
9 C. A. M. Brenninkmeijer, M. R. Manning, D. C. Lowe, G. A. Wallace, R. J. Sparks and A. Volz-Thomas, *Nature* (*London*), 1992, **356**, 50.
10 R. Prinn, D. Cunnold, R. Rasmussen, P. Simmonds, F. Alyea, A. Crawford, P. Fraser and R. Rosen, *Science*, 1987, **238**, 945.
11 H. S. Johnston, C. A. Cantrell and J. G. Calvert, *J. Geophys. Res.*, 1986, **91**, 5159.
12 A. G. Russell, G. R. Cass and J. H. Seinfeld, *Environ. Sci. Technol*, 1986, **20**, 1167.
13 W. L. Chameides, *J. Geophys. Res.*, 1986, **91**, 5331.
14 R. Atkinson, *Atmos. Environ.*, 1990, **24A**, 1.
15 N. R. Jensen, J. Hjorth, C. Lohse, H. Skov and G. Restelli, *Int. J. Chem. Kinet.*, 1992, **24**, 839.
16 N. I. Butkovskaya and G. Le Bras, *J. Phys. Chem.*, 1994, **98**, 2582.
17 J. M. Van Doren, L. R. Watson, P. Davidovits, D. R. Worsnop, M. S. Zahniser and Ch. E. Kolb, *J. Phys. Chem.*, 1990, **94**, 3265.
18 M. Mozurkewich and J. G. Calvert, *J. Geophys. Res.*, 1988, **93**, 15 889.
19 K. Thomas, A. Volz-Thomas and D. Kley, Dissertation, K. Thomas, Univ. Wuppertal D468, 1993, 2755.
20 S. Solomon, *Rev. Geophys.*, 1988, **26**, 131.
21 R. J. Cicerone, *Rev. Geophys. and Space Phys.*, 1981, **19**, 123.
22 H. B. Singh and J. F. Kasting, *J. Atmos. Chem.*, 1988, **7**, 261.
23 W. T. Sturges, G. F. Cota and P. T. Buckley, *Nature* (*London*), 1992, **358**, 660.
24 C. Schall and K. G. Heumann, *Fresenius J. Anal. Chem.*, 1993, **346**, 717.
25 M. A. K. Khalil, R. A. Rasmussen and A. Gundwardena, *J. Geophys. Res.*, 1993, **98**, 2887.
26 T. Oertel, PhD Thesis, Universität Bremen, 1992.
27 S. Manö and M. O. Andreae, *Science*, 1994, **263**, 1255.
28 W. T. Sturges and L. A. Barrie, *Atmos. Environ.*, 1988, **22**, 1179.
29 B. J. Finlayson-Pitts and S. N. Johnson, *Atmos. Environ.*, 1988, **22**, 1107.
30 B. J. Finlayson-Pitts, M. J. Ezell and J. N. Pitts, *Nature* (*London*), 1989, **337**, 241.
31 C. Zetzsch and W. Behnke, *NATO-ASI Series, Subseries I 'Global Environmental Change'*, ed. H. Niki and K. H. Becker, Springer-Verlag, 1993, vol. 7, pp. 291–306.
32 W. Behnke, H-U. Krüger, V. Scheer and C. Zetzsch, *J. Aerosol Sci.*, 1991, **22**, S609.
33 W. Behnke, H-U. Krüger, V. Scheer and C. Zetzsch, *J. Aerosol Sci.*, 1992, **23**, S933.
34 W. Behnke, V. Scheer and C. Zetzsch, *J. Aerosol Sci.*, 1993, **24**, S115.
35 S-M. Fan and D. J. Jacob, *Nature* (*London*), 1992, **359**, 522.
36 M. Mozurkewich, *J. Geophys. Res.*, 1994, in the press.
37 M. E. Jenkin, R. A. Cox and D. E. Candland, *J. Atmos. Chem.*, 1985, **2**, 359.
38 U. Platt and D. Perner, *Measurements of atmospheric trace gases by long path differential UV/visible absorption spectroscopy: Optical and Laser Remote Sensing*, ed. D. K. Killinger and A. Mooradian, Springer Ser. Optical Sci., 1983, **39**, 95–105.
39 U. Platt, in *Air Monitoring by Spectroscopic Techniques*, ed. M. W. Sigrist, Chemical Analysis Series, Wiley, 1994, vol. 127.
40 UMEG, Data from *Gesellschaft für Umweltmessung und Umwelterhebung mbH*, Karlsruhe, 1994.
41 F. Heintz, J. Stutz, U. Platt, H. Flentje and R. Dubois, *Proc. EUROTRAC Symp. on Transport and Transformation of Pollutants in the Troposphere*, Garmisch-Partenkirchen, 1994, 395–398.
42 J. F. Noxon, R. B. Norton and E. Marovich, *Geophys. Res. Lett.*, 1980, **7**, 125.
43 J. F. Noxon, *J. Geophys. Res.*, 1983, **88**, 11 017.
44 F. Heintz, U. Platt, H. Flentje and R. Dubois, presentation at TOR meeting, Oslo, October 1993.
45 B. Vierkorn-Rudolph, K. Bächmann, B. Schwarz and F. X. Meixner, *J. Atmos. Chem.*, 1984, **2**, 47.
46 R. Zander, G. Roland, L. Delbouille, A. Sauval, C. B. Farmer and R. H. Norton, *J. Atmos. Chem.*, 1987, **5**, 395.
47 J. Notholt and O. Schrems, *Geophys. Res. Lett.*, 1994, **21**, 1355.
48 A. A. P. Pszenny, W. C. Keene, D. J. Jacob, S. Fan, J. R. Maben, M. P. Zetwo, M. Springeryoung and J. N. Galloway, *Geophys. Res. Lett.*, 1993, **20**, 699.
49 D. D. Parrish, C. J. Hahn, E. J. Williams, R. B. Norton, F. C. Fehsenfeld, H. B. Singh, J. D. Shetter, B. W. Gandrud and B. A. Ridley, *J. Geophys. Res.*, 1993, **98**, 14 995.
50 S. A. McKeen, M. Trainer, E. Y. Hsie, R. K. Tallamraju and S. C. Liu, *J. Geophys. Res.*, 1990, **95**, 7493.
51 B. T. Jobson, H. Niki, Y. Yokouchi, J. Bottenheim, F. Hopper and R. Leaitch, *J. Geophys. Res.*, 1994, in the press.
52 S. Solberg, O. Hermansen, E. Joranger, N. Schmidtbauer, F. Stordal and O. Hov, NILU Report OR 27/94, 1994, ISBN 82–425–0575–6.

53 L. A. Barrie, S. M. Li, D. M. Toom, S. Landsberger and W. Sturges, *J. Geophys. Res.*, 1994, in the press.
54 J. W. Bottenheim, L. W. Barrie, E. Atlas, L. E. Heidt, H. Niki, R. A. Rasmussen and P. B. Shepson, *J. Geophys. Res.*, 1990, **95**, 18 555.
55 R. E. Mickle, J. W. Bottenheim, R. W. Leaitch and W. Evans, *Atmos. Environ.*, 1989, **23**, 2443.
56 J. C. McConnell, G. S. Henderson, L. Barrie, J. Bottenheim, H. Niki, C. H. Langford and E. M. J. Templeton, *Nature* (*London*), 1992, **355**, 150.
57 R. A. Rasmussen, M. A. K. Khalil, R. Gunawardena and S. D. Hoyt, *J. Geophys. Res.*, 1982, **87**, 3086.
58 W. L. Chameides and D. D. Davis, *J. Geophys. Res.*, 1980, **85**, 7383.
59 R. B. Chatfield and P. J. Crutzen, *J. Geophys. Res.*, 1990, **95**, 22 319.
60 M. E. Jenkin, Report AEA Harwell, AEA-EE-0405, Nov. 1992.
61 U. Platt and M. Hausmann, *Res. Chem. Intermed.*, 1994, **20**, 557.
62 W. T. Sturges, R. C. Schnell, G. S. Dutton, S. R. Garcia and J. A. Lind, *Geophys. Res. Lett.*, 1993, **20**, 201.
63 B. J. Finlayson-Pitts, F. E. Livingston and H. N. Berko, *Nature* (*London*), 1990, **343**, 622.
64 G. Le Bras and U. Platt, *Geophys. Res. Lett.*, 1995, in the press.
65 U. Platt, *LACTOZ/HALIPP Workshop*, Leipzig, 1995, in the press.
66 A. Hofzumahaus, U. Platt and D. Perner, Poster at Bunsentagung, 1986.
67 R. Toumi, *Geophys. Res. Lett.*, 1994, **21**, 117.
68 P. D. Lighfoot, R. A. Cox, J. N. Crowley, M. Destriau, G. D. Hayman, M. E. Jenkin, G. K. Moortgat and F. Zabel, *Atmos. Environ.*, 1992, **26**, 1805.
69 R. Staubes and H. W. Georgii, in *Dimethyl Sulfide: Oceans, Atmosphere and Climate, Proc. Int. Symp.*,ed. G. Restelli and G. Angeletti, Kluwer Academic Publishers, Dordrecht, 1993, pp. 95–102.
70 R. Atkinson, D. L. Baulch, R. A. Cox, R. F. Hampson, J. A. Kerr and J. Troe, *J. Phys. Chem. Ref. Data*, 1992, **21**, 1125.
71 M. Hausmann and U. Platt, *J. Geophys. Res.*, 1994, **99**, No. D12.
72 U. Platt, D. Perner and S. Semke, in *Ozone in the Atmosphere, Proc. of the Quadrennial Ozone Symposium 1988*, ed. R. D. Bojkov and P. Fabian, Deepak Publishing, Hampton, 1989, pp. 512–515.
73 D. Mihelcic, D. Klemp, P. Müsgen, H. W. Pätz and A. Volz-Thomas, *J. Atmos. Chem.*, 1993, **16**, 313.
74 U. Platt and D. Perner, *J. Geophys. Res.*, 1980, **85**, 7453; J. F. Noxon, *Geophys. Res. Lett.*, 1981, **8**, 1223.
75 U. Platt, A. M. Winer, H. W. Biermann, R. Atkinson and J. N. Pitts, *Environ., Sci. Technol.*, 1984, **18**, 365–369.
76 U. Platt, D. Perner, J. Schröder, C. Kessler and A. Toennissen, *J. Geophys. Res.*, 1981, **86**, 11 965.
77 J. M. C. Plane and C. F. Nien, *Study of Night-time* NO_3 *Chemistry by Differential Optical Absorption Spectroscopy, SPIE Proc.*, ed. H. I. Schiff, 1991, vol. 1433.
78 J. M. C. Plane and C. F. Nien, *Rev. Sci. Instr.*, 1992, **63**, 1867.
79 T. Brauers, H-P. Dorn and U. Platt, in *Physico-Chemical behaviour of Atmospheric pollutants, Proceedings of the 5th European Symp.*, Varese, Italia, ed. G. Restelli and G. Angeletti, Kluwer Academic, Dordrecht, 1990, pp. 237–242.
80 T. Brauers, H-P, Dorn, R. Neuroth and U. Platt, in *Optical Methods in Atmospheric Chemistry*, ed. H. I. Schiff and U. Platt, 1993, vol. 1715 Proc. SPIE, 292–302.
81 C. Gölz, C. Janssen and U. Platt, *Atmos. Environ.*, submitted.
82 U. Platt, D. Perner, G. W. Harris, A. M. Winer and J. N. Pitts, *Geophys. Res. Lett.*, 1980, **7**, 89.
83 J. N. Pitts, Jr., H. W. Biermann, R. Atkinson and A. M. Winer, *Geophys. Res. Lett.*, 1984, **11**, 557.
84 H. W. Biermann, E. C. Tuazon, A. M. Winer, T. J. Wallington and J. N. Pitts, *Atmos. Environ.*, 1988, **22**, 1545.
85 B. J. Finlayson-Pitts, *J. Geophys. Res.*, 1993, **98**, 14 991.
86 D. D. Parrish, C. J. Hahn, E. J. Williams, R. B. Norton, F. C. Fehsenfeld, H. B. Singh, J. D. Shetter, B. W. Gandrud and B. A. Ridley, *J. Geophys. Res.*, 1992, **97**, 15 883.
87 W. B. De More, S. P. Sander, D. M. Golden, R. F. Hampson, M. J. Kurylo, C. J. Howard, A. R. Ravishankara, C. E. Kolb and M. J. Molina, *Chemical Kinetics and Photochemical Data for Use in Stratospheric Modeling*, JPL Publication 92–20, 1992.
88 I. Barnes, B. Bonsang, T. Brauers, P. Carlier, R. A. Cox, H-P. Dorn, M. E. Jenkin, G. Le Bras and U. Platt, in *Air Pollution Series*, ed. G. Le Bras, European Communities, Brussels, 1991.
89 Y. Bedzkanyan, G. Poulet and G. Le Bras 1995, personal communication.
90 O. J. Neilsen, H. W. Sidebottom, L. Nelson, O. Rattigan, J. J. Treacy and D. J. O'Farrel, (1990).
91 R. E. Stickel, J. M. Nicovich, S. Wang, Z. Zhao and P. H. Wine, *J. Phys. Chem.*, 1992, **96**, 9875.
92 R. P. Wayne, I. Barnes, P. Biggs, J. P. Burrows, C. E. Canosa-Mas, J. Hjorth, G. Le Bras, G. K. Moortgat, D. Perner, G. Poulet and H. Sidebottom, *Atmos. Environ., Part A*, 1991, **25**, 1.
93 R. Atkinson, S. M. Aschmann and J. N. Pitts, *J. Phys. Chem.*, 1988, **92**, 3454.

Paper 5/00547G; *Received* 30*th January*, 1995

Faraday Discuss., 1995, **100**, 199–213

GENERAL DISCUSSION

Prof. Davidovits opened the discussion of Dr. Wiesen's paper: You suggest that, at the surface, two NO_2 molecules react heterogeneously with one water molecule to form an HONO molecule and HNO_3 [see eqn. (6)]. It is difficult to visualize how such a concerted reaction can readily occur. I would like to suggest that surface complexes may play a role in the surface formation of HONO. Previous studies have indicated (see our paper) that surface complexes of gas-phase species can form at the gas/liquid interface. A complex of SO_2 at a water surface may take the form shown:

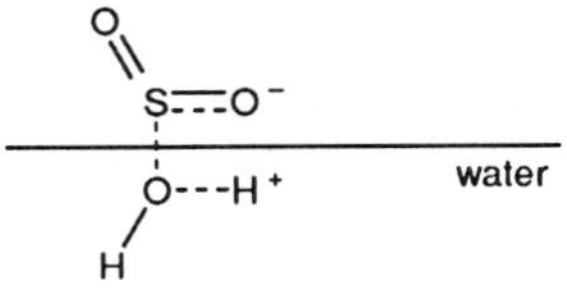

Recently such a complex was detected using surface second harmonic generation techniques.

The molecule NO_2 may likewise form a surface complex. For example, in a highly acidic solution a complex with H^+ could take the form shown. (Other complexes could be envisioned.)

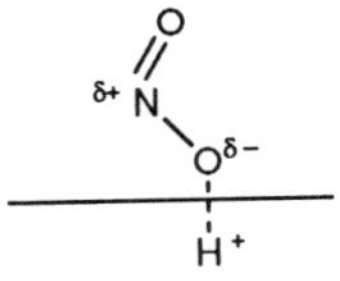

Such a complex might then react with water to form HONO. That is $ONOH^+ + H_2O \rightarrow HONO + OH + H^+$. We have found that surface reactions are more facile than the corresponding reactions in the bulk phase. This is not surprising since orientational hindrances are less and solvent reorientation is easier to accomplish at the interface.

Dr. Wiesen replied: From our experimental findings it is very difficult to make any assumptions concerning the possible formation of a surface complex. However, the reaction which you suggest should lead to an HONO yield which would be a factor of two higher than one would expect from our mechanism. Consequently, the N_2O yield should be higher than observed in our experiments.

In addition, recent experiments carried out in our laboratory in which NO_3^- was detected by ion chromatography in a sulfuric acid solution showed that the NO_3^- yield was in agreement with the proposed net reaction (10):

$$8NO_2 + 3H_2O \rightarrow N_2O + 6HNO_3$$

Dr. van Albada said: The proposed intermediate $(HNO)_2$ formed in reaction (8) is known to be a very strong oxidant, which gives the half-reaction $(NOH)_2 + 2H^+ + 2e \rightarrow N_2 + 2H_2O$. If, in parallel with reaction (9), a reduction of $(NOH)_2$ occurs (by *e.g.* NO_2 or HONO), then part of the expected N_2O would be replaced by N_2; the ratio HNO_3/N_2O could be >6, as seems to be found (Fig. 2 of your paper).

Dr. Wiesen responded: The reaction you propose should indeed have a direct influence on the N_2O concentration in our system leading to $HNO_3/N_2O > 6$. However, in the experiments which were carried out in the bubbler, the N_2O yield was found to be in agreement with the proposed mechanism.

It is correct that the N_2O yield in Fig. 2 is smaller than one would expect from the mechanism. However, it should be pointed out that this particular experiment was carried out in the quartz glass reactor (White cell) for which the characterisation of the surface is very difficult and, therefore, quantitative evidence is more speculative.

Dr. Cox commented: Concerning the reaction of NO_2 with water to form nitrous acid, I would like to draw attention to kinetic measurements made by us. We found that the reaction involved two molecules of NO_2 for the production of one molecule of HONO, consistent with the reaction:

$$2NO_2 + H_2O \rightarrow HONO + HONO_2$$

$HONO_2$ remained absorbed on the surface in what we believe to be a heterogeneous reaction. The formation of HONO was, however, first order in $[NO_2]$ with a slight negative temperature coefficient of the rate constant. A mechanism involving the formation of a surface complex of NO_2 was proposed, which is consistent with the idea put forward by Paul Davidovits. The rates measured by us were too slow by at least an order of magnitude to explain HONO formation as observed in the atmospheric boundary layer.

The observation of N_2O formation from the $NO_2 + H_2O$ reaction on acid surfaces is valuable new information. The reaction rate under atmospheric conditions is probably slow since there is no evidence for a large *in situ* source of N_2O in the atmosphere.

1 M. E. Jenkin, R. A. Cox and D. Williams, *Atmos. Environ.*, 1988, **22**, 487.

Dr. Wiesen responded: Our experimental results confirm those of Jenkin *et al.* However, up to now we have carried out experiments only at room temperature. We are planning to study the temperature dependence of the reaction system in the range 320–230 K.

Concerning the comment on an *in situ* source for N_2O in the atmosphere, I would like to emphasise that the existence of atmospheric sources and sinks of N_2O has been discussed in the literature for a long time. Recently, Prasad[1] evaluated 10 different laboratory experiments in order to clarify the present uncertainties in the global N_2O budget.

It should be pointed out that Khalil and Rasmussen[2] who studied the N_2O emission from coal-fired power plants, reported a conversion of nitrogen oxides into N_2O in the plumes farther downwind. The magnitude of the effect could not however, be quantitatively evaluated with their data.

At present it is very difficult to quantify the possible contribution of N_2O formation *via* HONO to the global N_2O budget. However, it can be concluded that a conversion of a few percent of the emitted NO_2 into N_2O *via* the proposed mechanism would result in an important contribution to the global N_2O budget.

1 S. S. Prasad, *J. Geophys. Res.*, 1994, **99**, 5285.
2 M. A. K. Khalil and R. A. Rasmussen, *J. Geophys. Res.*, 1992, **97**, 14645.

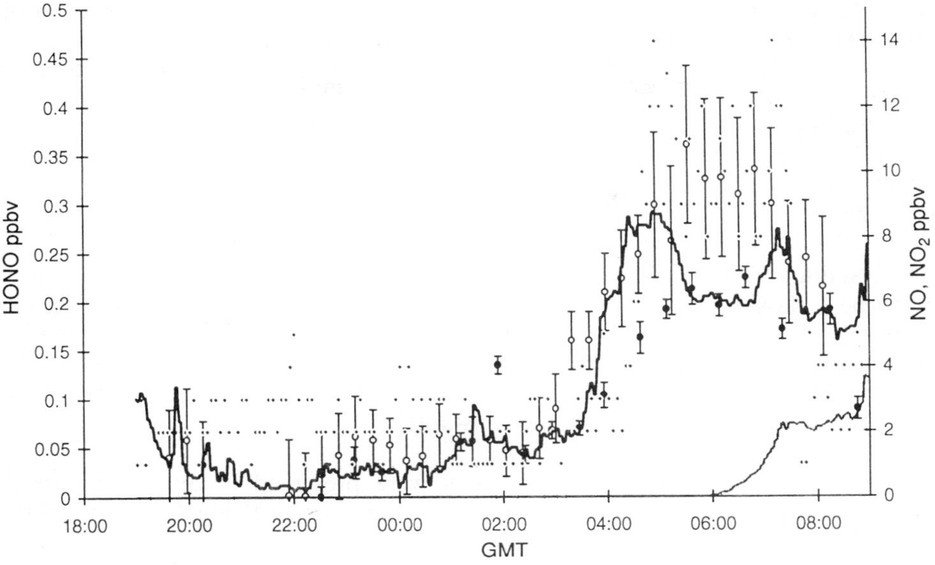

Fig. 1 Time profiles of HONO concentration (left-hand axis) measured by the University of Heidelberg (●), the University of Birmingham (◆) and the University of East Anglia (○). Also shown are the NO_2 (——) and NO (——) concentrations (right-hand axis). The measurements were made at the Weybourne Atmospheric Observatory, Norfolk on September 25/26, 1994.

Dr. Plane said (**Dr. Coe** in part communicated): The conclusion in your paper that HONO can be formed by heterogeneous reactions in the absence of NO is supported by the results of a recent field campaign at the Weybourne Atmospheric Observatory in North Norfolk, England. An intercomparison of several differential optical absorption spectrometers was organised under the auspices of the TOPAS subproject of EUROTRAC. In Fig. 1 we present measurements of HONO made by two DOAS instruments over a pathlength of 5.5 km. The instruments were from the University of Heidelberg[1] (Mr. J. Stutz and Prof. U. Platt) and the University of East Anglia[2] (Ms. B. Allan and Ms. N. Smith). In addition, HONO was measured by a point sampling instrument from the University of Birmingham[3] (Dr. J. D. Peake and Prof. R. M. Harrison). Fig. 1 shows remarkably good agreement between the instruments, even at HONO concentrations below 100 ppt.

Also illustrated in the figure are the NO_2 and NO concentrations (Dr. K. Clemitshaw and Prof. S. A. Penkett, University of East Anglia). Between 00:00 and 06:00 hours it will be seen that the increase in HONO was closely correlated with an increase in NO_2, while the NO concentration remained below the detection limit (60 ppt). After sunrise at 06:00 hours, when NO first appeared, the HONO concentration began to decrease, presumably because of photolysis. The night-time production of HONO in the absence of NO has also recently been reported in a suburban air mass.[3]

1 U. Platt, in *Air monitoring by Spectroscopic Techniques*, ed. M. W. Sigrist, Wiley, London, 1994.
2 J. M. C. Plane and N. Smith, in *Spectroscopy in Environmental Sciences*, ed. R. E. Hester and R. J. H. Clark, Wiley, London, 1995, pp. 223–262.
3 R. M. Harrison and A. M. N. Kitto, *Atmos. Environ.*, 1994, **28**, 1089.

Dr. Wiesen replied: That is an important additional example showing that HONO formation in the atmosphere probably occurs *via* the reaction of NO_2 with water vapour and does not require NO.

Prof. Pilling said: You have much more kinetic information in your kinetic profiles for NO_2, HONO and N_2O than you have used. Have you examined

(i) The decay of NO_2. Does the first-order rate constant agree with that for the growth of HONO (k_a)?

(ii) The ratio $[HONO](t)/[NO_2](t = 0)$ which should provide information on the yield of HONO from NO_2 and hence should indicate whether other decay channels for NO_2 are important.

(iii) The growth of N_2O. Is it comparable with the decay of HONO?

Prof. Smith added: The kinetic curves shown in Fig. 2 of your paper appear, at first sight, to be a classic case of a consecutive reaction. However, on closer examination, it appears that the rise in the intermediate HONO concentration is more rapid than the fall in the concentration of the primary reagent, NO_2. I wonder if you could comment on whether this impression is correct and, if so, does the proposed mechanism accommodate this difference in first-order rate constants?

Dr. Wiesen replied: The reactions which were presented in our mechanism (or, better, reaction sequence) are not elementary reactions. Accordingly, it is very questionable whether kinetic information from the profiles for NO_2, HONO and N_2O can be extracted from the data in a simple manner. The possibility that adsorption and desorption phenomena have a significant influence on the overall behaviour of the system cannot be excluded.

In addition, in order to draw quantitative conclusions, *e.g.* the HONO yield, it is of paramount importance to know the concentration of dissolved HONO in sulfuric acid. This requires, however, knowledge of Henry's law constant for $HONO/H_2SO_4$ *e.g.* as a function of temperature and sulfuric acid concentration, which are, unfortunately, not known at present.

Dr. Rieley addressed Dr. Wiesen: I note two interesting observations made in your paper. First, that NO_2 is taken up by water through reaction, in the absence of NO to form HONO; secondly, that the occurrence of this reaction does not appear to depend on the presence of acid while the subsequent conversion to N_2O, which relies on HONO, is sensitive to acid concentration.

On the first point, I am surprised that the uptake and subsequent reaction of NO_2 on pure water seems to be so high in the study presented for two reasons. The first is that, as mentioned already by Dr. Plane we are aware of a study of the uptake of NO_2 on NO on stratospheric aerosols by Saastad *et al.*, recently presented in the literature, which concluded that the amount of uptake is negligible under atmospheric conditions.[1] The second reason for some surprise on our part stems from some of our own work regarding the adsorption and photochemistry of oxides of nitrogen on ice films at low temperature.[2] We find that NO_2, when incident on a cold ice film (< 140 K) forms a surface layer of the dimer N_2O_4. If this layer is subjected either to heat or to light, in the wavelength region 300–400 nm, the layer simply desorbs from the surface, *i.e.* there is no reaction of NO_2 with low-temperature ice.

On the second point, we have had considerable experience in the preparation and photochemical study of pure HONO[3,4] and were reminded of two classic preparations of nitrous acid. These are as follows:

$$NO_2 + NO + H_2O \rightarrow 2HONO \qquad (1)^5$$

$$H_2SO_4 + 2NaNO_2 \rightarrow 2HONO + Na_2SO_4 \qquad (2)^{6,7}$$

The first of these is a wall-catalysed reaction in which the equilibrium mixture at room temperature unavoidably contains significant amounts of NO_2 and NO, while the second is the preparation of choice if NO_2 is to be excluded. We note further that the first reaction is considerably accelerated at dry-ice temperatures.[5] Taken together these reactions highlight two aspects of the formation of HONO which are relevant to your paper: the formation of HONO on the water-coated reactor wall will be extremely sensitive to the presence of NO (a common impurity in commercial samples of NO_2) and the presence of acid would appear to be beneficial to the hydrolysis of NO_2. If NO were present, this should be detectable as N_2O_3 which forms from the equilibrium of NO_2 with NO in the presence of a third body or wall.[8,9] There is an apparent inconsistency in the argument concerning the role of acid in the formation of HONO and thence N_2O. The formation of N_2O apparently depends on the concentration of sulfuric acid present (Fig. 4 of your paper) but the formation of HONO, the species required for N_2O formation according to the proposed mechanism [eqn. (9)], does not!

Following on from these comments, I have three questions:

(1) What is the role of the acid in the formation of HONO and N_2O?

(2) Given our observations of the null reaction of NO_2 with low-temperature water ice, over what temperature range were your experiments carried out and how do you, therefore, rationalise your observations with ours and those of Saastad *et al.*?

(3) Have you any evidence for the formation of N_2O_3 in your experiments which would suggest the presence of an NO impurity and a contributing role for reaction (1) above in the production of HONO?

1 O. W. Saastad, T. Ellerman and C. J. Nielsen, *Geophys. Res. Lett.*, 1993, **20**, 1191.
2 H. Rieley, D. P. McMurray, H. D. Aslin and S. Haq, to be published.
3 R. N. Dixon and H. Rieley, *J. Chem. Phys.*, 1988, **91**, 2308.
4 R. N. Dixon and H. Rieley, *Chem. Phys.*, 1989, **137**, 307.
5 R. Varma and R. F. Curl, *J. Phys. Chem.*, 1976, **80**, 402.
6 R. A. Cox and R. G. Derwent, *J. Photochem.*, 1976/77, **6**, 23.
7 R. D. Kenner, F. Rohrer and F. Stuhl, *J. Phys. Chem.*, 1986, **90**, 2635.
8 A. J. Vosper, *J. Chem. Soc., Dalton Trans.*, 1976, 135.
9 I. W. M. Smith and G. Yarwood, *Chem. Phys. Lett.*, 1986, **130**, 24.

Dr. Wiesen responded: Of course, sulfuric acid has an influence on both N_2O and HONO formation. The dependence of the N_2O formation rate on the H_2SO_4 concentration is shown in Fig. 4 of our paper and has been described in the text. The influence of the H_2SO_4 concentration on the gas-phase HONO concentration is shown in Fig. 2 below.

It was observed that the HONO concentration increases with increasing sulfuric acid concentration, reaching a maximum concentration at *ca.* 45% H_2SO_4. This behaviour is probably caused by expulsion of the weaker acid HONO from the solution by the stronger acid H_2SO_4. Above 50% H_2SO_4, the gas-phase HONO concentration starts to decrease which can be explained by reaction (7). It should be pointed out that Seel and Winkler[1] who investigated various HONO/H_2SO_4 solutions by UV absorption spectroscopy reported equal concentrations of NO^+ and HONO in 55% H_2SO_4 which is in excellent agreement with the dependences of HONO and N_2O on the sulfuric acid concentration observed in the present study.

Since our experiments were all carried out at room temperature they cannot be compared directly with those of Dr. Rieley who studied the adsorption of NO_2 on ice films at 140 K. However, we have estimated an uptake coefficient from our experiments in the quartz glass reactor which were carried out at reduced total pressure. Our estimated value of 1×10^{-7} is much smaller than the lower limit of 5×10^{-6} reported by Saastad *et al.*[2] which means that the disappearance of NO_2 is indeed very slow.

Concerning your question about the possible formation of N_2O_3 in our system we would like to emphasise that we checked that the NO content in our NO_2 cylinder was $<1\%$ by FTIR spectroscopy. Accordingly, the maximum HONO yield resulting from

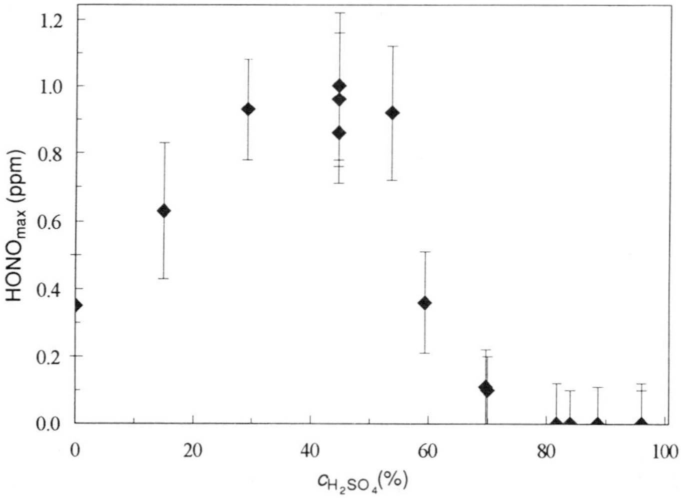

Fig. 2 Dependence of the maximum observed HONO concentration on the H_2SO_4 concentration measured in the modified experimental set-up (Pyrex glass reactor, bubbler, Herriott cell); $[NO_2]_0 = 25$ ppm, $p_{tot} = 740$ Torr, $V_{H_2SO_4} = 200$ ml

the N_2O_3–H_2O pathway which was recently suggested by Calvert *et al.*[3] is expected to be <2%.

1 F. Seel and R. Winkler, *Z. Phys. Chem.*, 1960, **25**, 217.
2 O. W. Saastad, T. Ellerman and C. J. Nielsen, *Geophys. Res. Lett.*, 1993, **20**, 1191.
3 J. G. Calvert, G. Yarwood and A. M. Dunker, *Res. Chem. Intermed*, 1994, **20**, 463.

Dr. Tuckett communicated: As an aside to your paper, at least two groups (and, I suspect, others) working in the gas phase have observed that it is almost impossible not to get heterogeneous conversion of NO_2 and any hydrogen-containing precursor to HONO on almost any surface, even if that is not the desired product. In 1989, working in the laboratory of Prof. Curt Wittig at USC, Los Angeles, we attempted to observe the OH rovibronic product state distribution from the reaction of hot H atoms with NO_2 in a standard two-laser pump and probe experiment:

$$H + NO_2 \rightarrow OH(v', J') + NO(v', J'); \qquad \Delta H_0^\circ = -123.6 \text{ kJ mol}^{-1}$$

The interest was to see how the product state distributions of this radical–radical reaction changed from experiments under thermal conditions.[1] The source of hot H atoms was 248 nm dissociation of one of HI, HBr or H_2S, generating H atoms with kinetic energies of 1.24, 1.95 and 1.10 eV, respectively. However, in all cases, immaterial of how we flowed the HX-precursor and the NO_2 into the interaction region and immaterial of the type of material we used for the inlet system (*e.g.* metal, glass, Teflon), all we could ever observe was 'prompt' unimolecular photodissociation of HONO at 248 nm producing $OH(v = 0)$. Never were we able to observe OH produced from the bimolecular reaction, which should grow in as the time delay between the two lasers increases. Similar problems were encountered by McKendrick *et al.*[2] in their study of $O(^3P) + HBr \rightarrow OH(v', J') + Br$, where the source of O atoms was 355 nm photolysis of NO_2. However, in their case, because $OH(v' = 0)$ is a very minor product of the bimolecular reaction, they were able to live with the heterogeneous production of HONO in their gas inlet system, however inconvenient its presence was. In our case, where $OH(v' = 0)$ is the most populated vibrational level of the bimolecular process,[1] it was a killer!

1 A. M. L. Irvine, I. W. M. Smith, R. P. Tuckett and X. F. Yang, *J. Chem. Phys.*, 1990, **93**, 3177; 3187.
2 K. G. McKendrick, D. J. Rakestraw and R. N. Zare, *Faraday Discuss. Chem. Soc.*, 1987, **84**, 39.

Dr. McCoustra communicated: Prof. Smith commented that the kinetic data in Fig. 2 of your paper provide an almost ideal example of a consecutive reaction system. I would like to suggest that your Fig. 4 is an ideal example of a volcano plot. Such behaviour is commonly seen in studies of heterogeneous catalysis and arises from the balance of adsorption and desorption processes involving both reactants and products.[1]

1 G. C. Bond, *Heterogeneous Catalysis: Principles and Applications*, Oxford University Press, Oxford, 1987), ch. 6.

Ms. Titcombe opened the discussion of Dr. Herrmann's paper: Your paper emphasizes the importance of aqueous-phase reactions of radicals with aromatic compounds. In the atmosphere most aromatics are emitted in boundary-layer regions where aerosol loadings are relatively high. The significance of droplet heterogeneous processes for the tropospheric degradation of aromatics will depend directly on the kinetics of gas to liquid mass transfer for each species and for the products of their gas-phase oxidation (which will usually be more polar oxygenated compounds).

We have made measurements of mass accommodation coefficients, α, for a series of substituted aromatic trace gases onto aqueous droplets. We are using the droplet train technique described by Prof. Davidovits in his paper. A novel feature of our experiment is that changes in trace-gas concentration as a function of exposed liquid surface area, gas pressure and gas–liquid contact time are detected by laser-induced fluorescence.

At present we have measured α for aniline (0.013 ± 0.005) and phenol (0.019 ± 0.005) onto aqueous droplets at 283 K. We have also investigated the accommodation of toluene but the value of α is at or below our current detection limit of 10^{-3}.

We would like to ask Dr. Herrmann which aromatic species he considers to have the most important aqueous-phase chemistry, and what ranges in values of α are necessary for these liquid-phase processes to be of significance in polluted air?

Dr. Herrmann replied: The work of your group represents an elegant approach to the understanding of tropospheric conversions of aromatics and their oxidation products better. It is complementary to the study performed by our laboratory in that it characterizes mass accommodation coefficients for aromatics and their oxidation products.

We consider those aromatic compounds to be most important which have been identified in field measurements of cloudwater, rainwater and tropospheric aerosols.[1] Phenols from automobile emissions followed by benzoic acid from the same source and ketones from fuel combustion are identified as the most abundant organics in samples from Los Angeles rainwater. Other authors[2,3] report relatively high concentrations of various substituted nitrophenols in urban-area rainwater samples in Germany.

It is difficult to identify the aromatic compounds which have the most important aqueous-phase chemistry because systematic investigations (including product studies) of the interactions of soluble aromatics with small radicals in water are only just beginning. In general, the effective uptake of any aromatic compound into the tropospheric aqueous phase will change our current understanding of tropospheric aqueous-phase free-radical chemistry. This is because the reactivity of aromatics with free radicals in water is extremely high and may result in increased formation of HO_2/O_2^- which subsequently may react with dissolved ozone leading to the regeneration of OH. It is possible that chain reactions take place which consume aromatics as well as ozone in the atmospheric aqueous phase.

To our understanding, mass accommodation coefficients above 0.01 indicate that phase transfer is not the rate-limiting step in the sequence (i) gas-phase diffusion, (ii) uptake, (iii) aqueous-phase diffusion, (iv) aqueous phase chemical reaction. From the data presented for aniline and phenol by the Leeds group it is concluded that phase transfer is not rate limiting for these compounds.

1 K. Kawamura and I. R. Kaplan, in *Organic Compounds in Rainwater*, ed. L. D. Hansen and D. J. Eatough, CRC Press, Boston, 1991, p. 233.
2 A. Geißler and H. F. Schöler, *Vom Wasser*, 1993, **80**, 357.
3 A. Geißler and H. F. Schöler, *Wat. Res.*, 1994, **28**, 2047.

Dr. Buxton said: In the context of methods for generating free radicals in aqueous solution, photochemistry and radiation chemistry are complementary. One should not, therfore, be unduly dismissive of one method over the other.

You suggest that ionising radiation may have led to direct fragmentation of toluene and the formation of products different from those observed in your photolytic experiments. I want to point out that absorption of energy from a beam of high-energy electrons, such as is used in pulse radiolysis experiments, is determined by the electron density of the material which is *ca.* 0.5 for elements of low atomic weight. Thus, for an aqueous solution containing 10^{-3} mol dm^{-3} toluene, the fraction of energy absorbed by the solute is less than 1 part in 55000, so direct fragmentation of the toluene will be immeasurably small. Because of this property, the radiolysis of water provides an excellent method for generating free radicals in dilute aqueous solution.

I do not understand why you attribute the negative intercepts in Fig. 4 to possible loss of the aromatic solutes through their adsorption on the walls of the cell when it clearly does not apply to the data in Fig. 3. Since $SO_4^{\bullet -}$ is known to react with aromatic molecules in aqueous solution to form radical cations (ref. 34) which absorb at wavelengths greater than 400 nm, is it possible that when you measure the decay of absorbance due to $SO_4^{\bullet -}$ at 442 nm the signal is distorted by absorbance due to the product radical cation? It is interesting to note that the negative intercepts in Fig. 4 are larger for those molecules carrying substituents that would increase the stability of the radical cations through electronic effects.

Dr. Herrmann responded: The negative intercepts in the case of sulfate radical measurements are attributed to adsorption of the reactants on the wall of the reactor. No such effect has been observed in studies of reactions of the NO_3 radical. However, in the sulfate radical anion measurements another apparatus (shown in Fig. 1 of the paper) was applied which utilizes other surfaces in contact with the solution (Teflon, quartz). It may well be possible that adsorption of reactants takes place in the experimental set-up applied for the sulfate radical anion studies and that this effect is negligible in the case of the nitrate radical experiments.

With regards to the possibility of distortion of the measured time-resolved absorption signals by the absorption of radical anions in the case of benzene and toluene, the HCHD-derivatives were identified spectroscopically. In these cases no significant absorption of the HCHD-derivatives above 360 nm was monitored. It may be possible that in the first reaction step when the aromatic substrates react with the sulfate radical anion, a radical cation is formed. However, we were not able to detect any measurable absorption above 360 nm at short timescales in any of the aromatics' reactions with SO_4^-. Therefore, we conclude that under the experimental conditions (especially the low concentrations of transients) of the present study, the initially formed radical cation is rapidly (within 5 μs of the laser flash) converted to the HCHD derivative, which does not absorb above 360 nm and hence does not disturb the detection of SO_4^- at 442 nm.

Dr. Atkinson commented: I believe that the correlation between the gas- and aqueous-phase NO_3 radical reaction rate constants for aromatic compounds is significantly less good than shown (although this does not affect the conclusions of this paper concerning the mechanism of the aqueous-phase NO_3 radical reactions with aromatic compounds). The rate constants given in Table 5 and plotted in Fig. 13 for the gas-phase reactions of the NO_3 radical with *p*-cresol and naphthalene are incorrect. The rate constant for reaction of the NO_3 radical with *p*-cresol is 1.1×10^{-11} cm^3 $molecule^{-1}$ s^{-1}.[1,2] The NO_3 radical reaction with naphthalene is complex,[3,4] with $k_b \gg (k_c[NO_2] + k_d[O_2])$ for NO_2 concentrations $\leqslant 1 \times 10^{15}$ molecule cm^{-3} [3] and $k_d/k_c < 4 \times 10^{-7}$ at room temperature.[4] This results in a measured rate constant of $k_{obs} = k_a k_c[NO_2]/k_b$ for NO_2 concentrations $>2 \times 10^{12}$ molecule cm^{-3},[4] which depends on the NO_2 concentration. An analogous reaction scheme is expected to apply for the NO_3 radical reaction with benzene.[3]

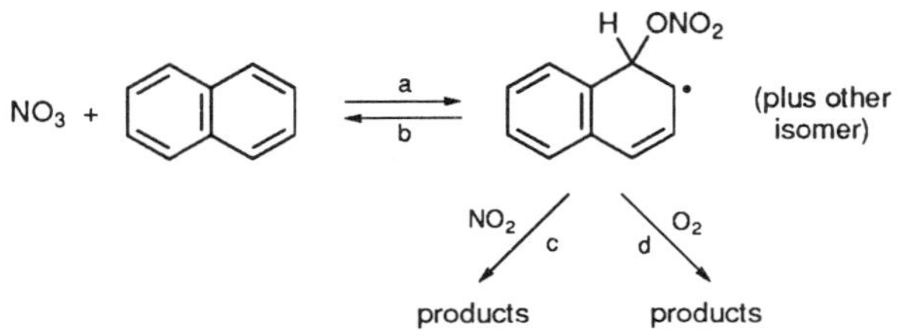

Rate constants for the gas-phase reactions of the NO_3 radical with toluene, $[^2H_8]$-toluene and $[^2H_3]$-toluene ($C_6H_5CD_3$) show that the rate-determining step is H (or D) atom abstraction from the C—H (or C—D) bonds of the methyl substituent group,[2,3] and similar observations have been made for *p*-xylene.[5,6]

Phenol and the cresols are much more reactive towards the NO_3 radical than expected from correlations with H-atom abstraction by OH radicals,[3] and it has been postulated that initial addition of the NO_3 radical to the ring occurs.[1–3] This may also be the case for aromatic compounds not containing unsaturated substituent groups,[3] but with the initial NO_3 radical addition not being the rate-determining step, except possibly for phenol and the cresols.[3]

1 R. Atkinson, S. M. Aschmann and J. Arey, *Environ. Sci. Technol.*, 1992, **26**, 1397.
2 R. Atkinson, *J. Phys., Chem. Ref. Data*, 1994, **Monograph 2**, 1.
3 R. Atkinson, *J. Phys., Chem. Ref. Data*, 1991, **20**, 459.
4 R. Atkinson, E. C. Tuazon, I. Bridier and J. Arey, *Int. J. Chem. Kinet.*, 1994, **26**, 605.
5 B. Rindone, F. Cariati, G. Restelli and J. Hjorth, *Fresenius J. Anal. Chem.*, 1991, **339**, 673.
6 G. Chiodini, B. Rindone, F. Cariati, S. Polesello, G. Restelli and J. Hjorth, *Environ. Sci. Technol.*, 1993, **27**, 1659.

Dr. Herrmann replied: In the original version of the manuscript we used a rate coefficient for the gas-phase reaction of NO_3 with *p*-cresol of $(6.4 \pm 2.5) \times 10^{-15}$ cm^3 $molecule^{-1}$ s^{-1}.[1] This rate coefficient has now been shown to be much too low and a value of $= 1.1 \times 10^{-11}$ cm^3 $molecule^{-1}$ s^{-1} [2] seems to be presently accepted. We are grateful to Dr. Atkinson for pointing out this mistake in our paper to us.

Table 5*b* of the paper has been corrected accordingly and a new version of Fig. 13 comprising the gas-phase kinetic data for the nitrate radical reactions with benzene, toluene, anisole, mesitylene and *p*-xylene only has been prepared. A 'straight-line fit' linking the reactivity of NO_3 towards aromatics in the gas phase with that in aqueous solution has been removed from the paper.

1 R. Atkinson, C. N. Plum, W. P. L. Carter, A. M. Winer and J. N. Pitts Jr., *Int. J. Chem. Kinet*, 1984, **16**, 887.
2 R. Atkinson, S. M. Aschmann and J. Arey, *Environ. Sci. Technol.*, 1992, **26**, 1379.

Dr. Hudson communicated: I should like to draw attention to the value of EPR studies in liquid-phase free-radical chemistry. The reaction of SO_4^- with a variety of organic molecules including aromatics was reported 25 years ago.[1] The advantage of the EPR technique is that it can provide an unambiguous identification of free-radical intermediates. Kinetic studies are also possible. Preliminary studies at Sussex indicate that it is possible to obtain and analyse spectra from quite complicated radicals involved in the oxidation of monoterpenes.

1 R. O. C. Norman, P. M. Storey and P. R. West, *J. Chem. Soc. B*, 1970, 1087.

Dr. Herrmann communicated in reply: We fully agree with your comments. EPR studies are very valuable in studies of complex oxidation mechanisms in solution. They provide structural information on intermediates and, with the option of time-resolved measurements, the possibility of gaining kinetic information.

Miss Smith said (**Dr. Coe** and **Dr. Plane** in part communicated): We have carried out a study of NO_3 chemistry at Izaña de Tenerife in the Canary Islands (28°18′N and 16°29′W) during May, 1994. The meteorological station at Izaña lies on a mountain platform at an altitude of 2376 m above sea-level and 400 km west of Africa. Air reaching the observatory is usually representative of the free troposphere, as the station rests above an inversion at around 2000 m which acts as a cap for the pollution generated below.[1] Levels of NO_x are a little higher than would be expected for true background air, with typically up to 1 ppb during the day and maybe 100 ppt at night-time. The site is also characterised by very low levels of reactive hydrocarbons.

During the campaign, the concentration of NO_3 was measured by the DOAS technique which is described in more detail elsewhere.[2] Results from the evening of the 1–2 May are presented. Fig. 3 shows the 30 min average values of NO_3 along with the O_3 concentration measured at the meteorological station. The NO_3 can be seen to increase after sunset which occurred between 18:30 and 19:00 hours and then rapidly decrease at sunrise between 06:00 and 06:30 hours. It can be seen that higher concentrations of O_3 appear to be associated with higher NO_3. This relationship has often been noted by researchers at the Weybourne Atmospheric Observatory, North Norfolk in the UK

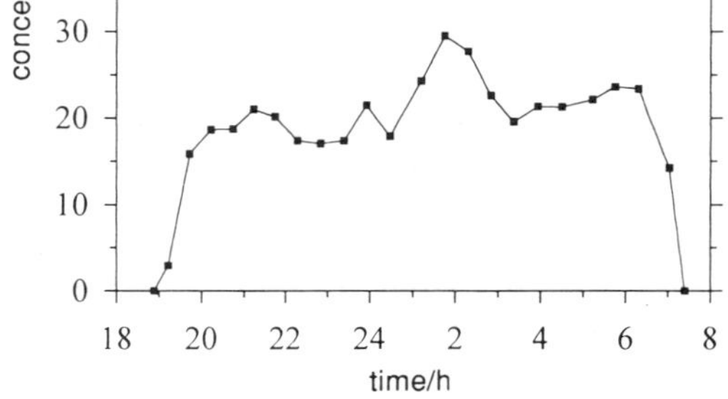

Fig. 3 Relationship between NO_3 and O_3 on May 1–2, 1994 in the free troposphere at Izaña de Tenerife

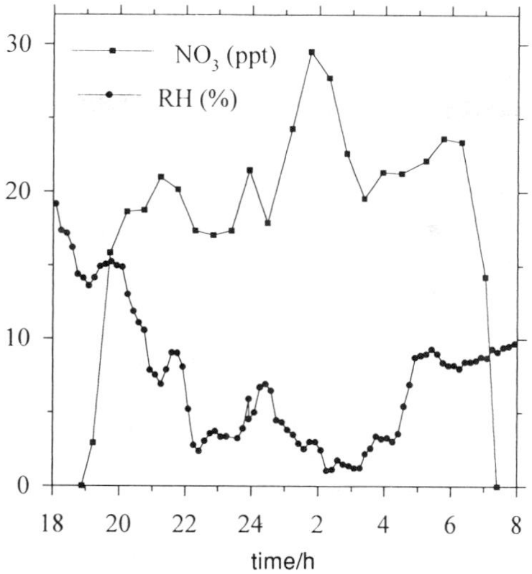

Fig. 4 Inverse relationship between NO_3 and relative humidity (RH) on May 1–2, 1994 in the free troposphere at Izaña de Tenerife

where the instrument is usually based. Fig. 4 shows clearly the negative correlation between NO_3 and relative humidity (RH) that has been noted previously[3] and is mentioned in Prof. Platt's paper. The relative humidities experienced during this campaign were much lower than those listed in Table 4 of Prof. Platt's paper, with NO_3 being measured at RHs below 10% for the first time. NO_3 would be expected to have rather a long lifetime under these conditions, since heterogeneous loss should be unimportant. For instance, on this evening $[O_3] = 65$ ppb, $[NO_2] = 50$ ppt, $[NO_3] = 20$ ppt, the rate coefficient for the reaction between NO_2 and O_3 to produce NO_3 is *ca.* 2.3×10^{-17} cm^3 $molecule^{-1}$ s^{-1} and hence $\tau(NO_3)$ is of the order of 4 h. This lifetime is rather long, but the humidity for much of this night was less than 5% (Fig. 2).

Plausible loss mechanisms responsible for this long lifetime can be investigated. The concentration of NO was not above the detection limit of 20 ppt. Although only a few hydrocarbons were measured, some indication of the low concentrations is given by the concentration of propene being 3 ppt with isoprene below the detection limit of 3 ppt. Significant removal of NO_3 by either NO or hydrocarbons can probably be discounted. The homogeneous gas-phase reaction between N_2O_5 and H_2O vapour is another possibility, with an upper limit of 2×10^{-21} cm^3 $molecule^{-1}$ s^{-1} at 298 K.[4] The conditions experienced at Izaña give us an opportunity to estimate this rate coefficient in the atmosphere, by assuming that it is the only important loss route. By using the observed lifetime of NO_3 of 4 h, $[NO_2] = 50$ ppt and an equilibrium constant between NO_3 and N_2O_5 of 1.81×10^{-10} cm^3 $molecule^{-1}$ at 285 K,[5] the first-order loss rate of N_2O_5 is *ca.* 4×10^{-4} s^{-1}. By assuming RH is 10% at a temperature of 12 °C, the water vapour concentration is estimated to be 4.4×10^{16} molecule cm^{-3}, and hence the rate coefficient for the reaction between N_2O_5 and water vapour is of the order of 9×10^{-21} cm^3 $molecule^{-1}$ s^{-1}. This is in good accord with the laboratory study.[4] The rest of the data set will be used to investigate this further.

These measurements appear to be the first direct observations of NO_3 in the free troposphere. The absence of scavenging mechanisms in the free troposphere indicate that the chemistry of NO_3 in this region should not be neglected.

1 R. Schmitt, B. Schreiber and I. Levin, *J. Atmos. Chem.*, 1988, **7**, 335.
2 J. M. C. Plane and N. Smith, in *Spectroscopy in Environmental Sciences*, ed. R. E. Hester and R. J. H. Clark, Wiley, London, 1995.

3 U. F. Platt, A. M. Winer, H. W. Biermann, R. Atkinson and J. N. Pitts Jr., *Environ. Sci. Technol.*, 1984, **18**, 365.
4 R. Atkinson, D. L. Baulch, R. A. Cox, R. F. Hampson Jr., J. A. Kerr and J. Troe, *J. Phys., Chem. Ref. Data*, 1989, **18**, 881.
5 W. B. Demore, S. P. Sander, D. M. Golden, R. F. Hampson, M. J. Kurylo, C. J. Howard, A. R. Ravishankara, C. E. Kolb and M. J. Molina, *Chemical Kinetics and Photochemical Data for Use in Stratospheric Modelling*: Evaluation Number 10, Jet Propulsion Laboratory, California, 1992.

Prof. Penkett said: In connection with the measurements of high concentrations of the nitrate radical in the free troposphere at an observatory in the Canary Islands described by Miss Smith, I would refer to some of my own published data[1] on the hydrocarbon composition of air over the Atlantic Ocean which suggested a role for nitrate radical removal of some species. In particular the concentration of 10 extensive iso-isomers of common alkanes, such as pentane, hexane and heptane, was lower than would be expected with respect to that of the normal isomers. The selective increased removal of isopentane over *n*-pentane from Atlantic air masses with a common continental source would require *ca.* 40 pptv of nitrate radicals and a reaction time of four months. The measurements made by John Plane's group provide direct evidence for this possibility and suggest that nitrate radical chemistry may indeed be having an impact on trace gas composition over substantial parts of the northern hemisphere.

1 S. A. Penkett, N. J. Blake, P. Lightman, A. R. W. Marsh, P. Anwyl and G. Butcher, *J. Geophys. Res.*, 1993, **98D**, 2865.

Dr. Fish asked: Would you like to comment on the discrepancy between your estimate, that 40% of tropospheric ozone is formed by *in situ* photochemistry and Professor Crutzen's estimate, that *ca.* 90% of tropospheric ozone is formed *in situ*?

Prof. Penkett replied: The estimate for a tropospheric origin for 40% of the ozone observed over the Atlantic in our experiment is extremely conservative. It was derived simply from an arithmetical relationship of 20 : 1 observed between ozone and peroxides in the southern hemisphere when they were formed simultaneously. In other experiments over North America referred to in our text, the observed relationship was nearer 60 : 1. This would allow a much larger proportion of the Atlantic ozone to be of tropospheric origin. Also, it has to be remembered that if ozone production is observed, ozone destruction is prevented. Overall I would guess that Prof. Crutzen's estimate that 90% of the ozone in the troposphere is formed by *in situ* processes is a little too high; at the same time it is not grossly in error.

Prof. Crutzen commented: In a recent study,[1] Rolf Sander and I have called attention to the possibility of Cl- and Br-activation when air polluted with NO_x is transported over the marine boundary layer where it may interact with the NaCl and NaBr in seasalt. We found that after a period of about one week, following photochemical destruction of alkenes, high levels of gas-phase, activated Br can form, leading to rapid destruction of O_3 *via* the catalytic cycle $(Br + O_3 \rightarrow BrO + O_2)$ + $(BrO + HO_2 \rightarrow HOBr + O_2)$ + $(HOBr + h\nu \rightarrow OH + Br)$ + $(OH + CO + O_2 \rightarrow HO_2 + CO_2) = O_3 + CO \rightarrow CO_2 + O_2$.

Such reactions may contribute to the polar sunrise ozone losses in the Arctic, as discussed by Prof. Platt. However, we do not rule out that they may take place elsewhere under meteorologically stable conditions in the outfloor of polluted air over the oceans, *e.g.* in some subtropical high-pressure areas.

1 R. Sander and P. J. Crutzen, *J. Geophys. Res.*, 1995, submitted.

Dr. Slemr said: I would like to make a comment in response to Prof. Crutzen's remark that positive correlation of peroxide and ozone concentrations observed by Prof.

Penkett in moderately polluted air masses is in apparent disagreement with the fact that peroxide formation is hampered in the presence of NO_x, a necessary ingredient for ozone formation. However, the ozone formation rate is proportional to the product of the nitric oxide and peroxyl radical concentrations and because of the latter the ozone formation will always be accompanied by peroxide formation by recombination of peroxyl radicals. In fact during our flights over eastern USA, we could directly observe[1] simultaneous production of ozone and peroxides which were consistent with each other.

From the difference between the afternoon and morning mixing ratios averaged over the free tropospheric part of the flight altitudes we could calculate average H_2O_2, ROOH and O_3 formation rates of 22, 5.6 and 650 pptv h^{-1}, respectively. An average concentration of HO_2 radicals of 2.1×10^8 molecule cm^{-3} can be derived from the H_2O_2 formation rate with the rate constant of 3×10^{-12} cm^3 $molecule^{-1}$ s^{-1} for recombination of two HO_2 radicals at 280 K, 800 mbar and 0.1% water content. From the average HO_2 concentration and the ROOH formation rate, an average concentration of 0.2×10^8 molecule cm^{-3} can be derived for RO_2 radicals assuming an effective rate constant of 8.3×10^{-12} cm^3 $molecule^{-1}$ s^{-1} for recombination of RO_2 with HO_2 radicals. From the ozone formation rate and the measured average NO mixing ratio of 90 pptv, the sum of peroxyl radical concentrations of 2.4×10^8 molecule cm^{-3} could be independently derived using an effective rate constant of 8.3×10^{-12} cm^3 $molecule^{-1}$ s^{-1} for the reaction of peroxyl radicals with NO. This was in excellent agreement with the sum of peroxyl radical concentrations 2.3×10^8 molecule cm^{-3} derived from the peroxide formation rates.

1 Tremmel *et al.*, *J. Geophys. Res.*, 1994, **99**, 5295.

Prof. Penkett responded: My earlier comments to Paul Crutzen were in the context of ozone loss in the absence of nitrogen oxides ($NO + NO_2$). Franz Slemr's observations of co-production of ozone and peroxide in slightly polluted air are very much in line with our own observations in time at Cape Grim, and in space over the equatorial Pacific. His calculated peroxyl radical concentrations are also largely in agreement with our recent direct measurements in air at Cape Grim on Tasmania. Overall, I believe we are beginning to place together a good understanding of tropospheric photochemistry confirmed by experiment.

Prof. Smith said: I should like to give Stuart Penkett the opportunity to expand on the difference between measured and calculated peroxide concentrations in the lowest kilometre of the troposphere. Is this due entirely to diffusion-controlled physical removal or is there the possibility of some chemical removal which is not being allowed for in the models?

Prof. Penkett replied: The equation to calculate the peroxide concentrations, used in our paper, contains only a chemical production term. In the boundary layer powerful physical loss mechanisms operate and this explains the divergence between the calculated and measured peroxide concentrations. Over continental areas where concentrations of nitrogen oxides are significant, the divergence would also be caused by production of ozone from the peroxyl radical precursors. The high peroxide concentrations which result from the calculations do illustrate the potential for large ozone production if sufficient NO_x is also present in the lower troposphere.

Prof. Platt commented: In a vertical profile reported by Tremmel *et al.*, both NO and water vapour were high at ground level, and fell off to low levels at certain altitudes. The fall-off of water vapour occurred at a somewhat higher altitude. Hydrogen peroxide levels showed a marked maximum in the (thus formed) layer with high water vapour and low NO. This observation constitutes another 'textbook case' of atmospheric chemistry at work: in the lowermost layer (with high NO) HO_2 levels are suppressed owing to reaction with NO, thus HO_2 recombination (and therefore H_2O_2 formation) are low. In the uppermost layer (with low NO and water vapour) HO_2 levels are also low because of lack of water vapour. In the intermediate layer (high water, low NO) HO_2 levels, and thus H_2O_2 production, reach their peak.

1 H. G. Tremmel, W. Junkermann, F. Slemr and U. Platt, *J. Geophys. Res.*, 1988, **98**, 1083.

Prof. Vaida asked: The measurements of peroxides presented in your paper can be used to monitor radicals produced in the troposphere from the photolysis of ozone in the presence of water vapour. We studied the photolysis of the ozone–water weakly bound complex and found it to be a possible source of tropospheric OH. This photochemical reaction occurs at lower energy than that needed for $O(^1D)$ formation from unclustered ozone. Consequently the [OH] from the ozone–water complex will have a diurnal profile different from that due to O_3–$O(^1D)$, and this profile could be observed most easily at sunrise and sunset. Our estimates of the contribution of the photochemistry of this complex suggest that field techniques must be sensitive to *ca.* 5×10^4 OH molecules cm^{-3}. In this context, can tropospheric measurements of peroxides and of OH be made at sunrise and sunset with fast-enough time response and high-enough sensitivity so that they could measure OH from ozone–water photolysis?

1 G. J. Frost and V. Vaida, *J. Geophys. Res.*, in the press.
2 V. Vaida, G. J. Frost, L. A. Brown, R. Naaman and Y. Hurwitz, *Ber. Bunsen-Ges. Phys. Chem.*, 1994, **99**, 1.

Prof. Penkett responded: I think it is unlikely that peroxide measurements at sunrise and sunset would be sufficiently deterministic to establish whether an additional source for the hydroxyl radical exists in the troposphere. The small amount of peroxide produced by minor sources could easily be confused with fluctuations in a background value associated with transport *etc.* It may be possible to demonstrate the production of additional peroxyl radicals however in very low light conditions using instrumentation currently under development.

Dr. Borrell communicated to Prof. Penkett: Do the most impressive negative and positive correlations between ozone and hydrogen peroxide, found on the various flights, allow you to determine the 'crossover point' between ozone formation and destruction at low NO_x concentrations in the free troposphere?

Prof. Penkett communicated in response: The experimental data are indeed remarkable and we were fortunate to collect them so easily in such quantities. In that respect I must compliment my co-workers from UEA and the Meteorological Office. The crossover point he mentions refers to the point at which ozone begins to accumulate rather than being destroyed by atmospheric photochemistry, and it is critically dependent on the amount of nitric oxide on the air at levels around 10 pptv. Unfortunately, we did not have instruments available for making the necessary measurements at the low levels encountered on the aircraft flights. We did observe instances of both ozone production and loss on some flights, however, and in future experiments with more sensitive instrumentation for NO_x measurements we are hopeful that we can provide him with an answer.

Dr. Roscoe addressed Prof. Platt: First, I would like to take up your comment at the start of your talk, about the paucity of evidence that OH is indeed the principal oxidizer in the troposphere. This is, of course, related to the paucity of measurements of OH in the troposphere. To my knowledge, there are only five operational instruments for measuring tropospheric OH which have published data that the community accept as reliable; one in Germany, two PHAGE in USA, and two in USA that have recently published a detailed intercomparison.[1] It is to their great credit that in their paper Eisele's *in situ* instrument and Mount's long-path UV absorption instrument gave results which agreed within their error bars; nevertheless, the world needs more instruments for measurement of tropospheric OH, and particularly it needs cheaper and more portable ones.

With others, I have proposed a development of long-path UV absorption measurement, with a Ti–sapphire laser instead of an excimer laser, that might lead to better portability and lower cost. Either this or other comparable ideas, of which there are several, should be carried out with all expediency, so that may look forward to frequent and reliable measurements of tropospheric OH by many workers at many locations.

Secondly, your illustration of how the altitude distribution of NO can depend critically upon the altitude distribution of other gases, even within the boundary layer, leads me to mention a facility of the UK Met Office which is available for boundary layer research. The Met Research Unit's tethered balloon facility routinely measures 3D turbulence and fluxes of heat and humidity from up to 12 probes suspended from their balloon at Cardington. Measurements of ozone, NO_2 and aerosol will soon be added, and they also go on field campaigns.

1 F. L. Eisele, G. H. Mount, F. C. Fehsenfeld, J. Harder, E. Marovich, D. D. Parrish, J. Roberts, M. Trainer and D. Tanner, *J. Geophys. Res.*, 1994, **99**, 18605.

Prof. McKenna said: There are at least three quite distinct mechanisms whereby the H_2O_2 may become suppressed in the lowest layer of the atmosphere. (i) Close to a source of recent NO_x emissions the high concentrations of NO react with HO_2 to produce ozone and shift the balance between OH and HO_2 in favour of OH (this is the only purely photochemical mechanism, as pointed out by Platt). (ii) In polluted air-masses, which contain high concentrations of SO_2, aqueous-phase reactions between SO_2 and aqueous O_2^+ from H_2O_2 result in an effective reaction between SO_2 and H_2O_2. This reaction is very important because unlike many other sulfur oxidation reactions the oxidation process is not limited by low pH. (iii) Finally, when air is exposed to open water the high solubility of H_2O_2 means that it is rapidly lost to the open water.

The observations reported by Penkett *et al.* which exhibit this pronounced reduction in peroxide were all made at a great distance from source regions where both mechanisms (i) and (ii) are unlikely to be effective. Confirmation of this supposition [at least with respect to mechanism (i)] comes from the coincident observations of NO and NO_y made by scientists from KFA Jūlich (ICG-2), which demonstrated that NO concentrations were low throughout the boundary layer. These observations suggest that the oceans are a major sink of peroxide. Therefore, loss to the oceans after conversion to H_2O_2 must be a major sink of OH radicals, which will be regulated primarily by the rate at which peroxide is exposed to the oceans.

Faraday Discuss., 1995, **100**, 215–227

Product Channels in the Near-UV Photodissociation of Ozone

Stephen M. Ball, Gus Hancock and Fiona Winterbottom
Oxford Centre for Applied Kinetics, Physical and Theoretical Chemistry Laboratory, Oxford University, South Parks Road, Oxford, UK OX1 3QZ

The relative quantum yields for the formation of $O_2(a\,^1\Delta_g)$ from the photolysis of ozone have been measured between 270 and 329 nm at room temperature, and between 300 and 322 nm at 227 K, near the fall-off region for the formation of spin-allowed singlet products $O_2(a\,^1\Delta_g)$ and $O(^1D_2)$. The molecular fragment was detected by resonance enhanced multiphoton ionisation at 331.5 nm. The measurements were put on an absolute scale by comparison with previous measurements in the short-wavelength region. The results at room temperature are in excellent agreement with the recommended quantum yields for $O(^1D_2)$ production at wavelengths up to 310 nm, but at longer wavelengths exhibit a pronounced tail of 10–20% out to at least 329 nm. Measurements at 227 K are identical to those at room temperature between 300 and 309 nm, and do not show a shift in the fall-off curve to shorter wavelengths as has been reported in the literature for $O(^1D_2)$. For wavelengths between 309 and 319 nm the yield of $O_2(a\,^1\Delta_g)$ is smaller than that at room temperature and this, together with the results of measurements at fixed wavelengths as a function of temperature, confirms that the photolysis of internally excited ozone provides a major source of $O_2(a\,^1\Delta_g)$ at wavelengths just beyond 310 nm. For wavelengths $\geqslant$320 nm the quantum yield is found to be approximately constant at the two temperatures, and suggests that spin-forbidden dissociation of ozone is taking place as the dominant process in this long-wavelength region. The results are compared with recent modelling calculations for the formation of the $O(^1D_2)$ product in the fall-off region which take into account the spin-allowed dissociation of internally excited ozone molecules. Good agreement is found at wavelengths up to 320 nm, particularly at room temperature, and suggests that the formation of singlet products extends noticeably beyond the fall-off region. The implications of this for stratospheric and atmospheric modelling are briefly discussed.

The UV photodissociation of ozone is of crucial importance in determining the oxidising capacity of both the stratosphere and the polluted troposphere, and the evaluation of the quantum yields of the various product channels has been an area of considerable activity for over twenty years.[1] Within the strong Hartley band between 220 and 310 nm the quantum yield for dissociation of ozone is close to unity, and is believed to take place through two spin-allowed photodissociation channels:

$$O_3 + h\nu \rightarrow O(^1D_2) + O_2(a\,^1\Delta_g) \quad (1)$$

$$O_3 + h\nu \rightarrow O(^3P_{2,1,0}) + O_2(X\,^3\Sigma_g^-) \quad (2)$$

Process (1) forming two electronically excited species dominates within most of the Hartley band. Photofragment spectroscopy experiments have determined the quantum yields, Φ, to be $\Phi(1) = 0.9$, $\Phi(2) = 0.1$ at wavelengths of 274,[2] and 266 nm,[3] coherent anti-Stokes Raman scattering (CARS) spectroscopy has been used to infer an average value of $\Phi(1) = 0.89 \pm 0.03$ over the wavelength range 266–311 nm,[4] and time-resolved resonance fluorescence experiments on the $O(^3P)$ product have shown values of the relative $O(^1D_2) : O(^3P_{2,1,0})$ ratio [equal to the relative quantum yields of processes (1) and (2) if only spin-allowed photodissociation takes place] close to 0.9 : 0.1 at 248,[5] 266,[6] and 308 nm.[7] In addition there have been a large number of studies of the relative quantum yields of the $O(^1D_2)$ product as a function of wavelength, both in the Hartley bands and at longer wavelengths (for a recent compilation of the results see ref. 8). The main feature of these measurements is that the quantum yield of $O(^1D_2)$ remains close to unity between 220 and 305 nm, and drops at longer wavelengths; as the thermodynamic threshold for process (1) is 310 nm for ozone molecules with zero internal energy, this result appears to vindicate the dominance of the spin-allowed processes in the photodissociation mechanism.

However, there is still considerable uncertainty in the quantum yields of the products at wavelengths above 310 nm. Some sets of results[9] which form the basis for the JPL–NASA recommended quantum-yield curve[10] show the quantum yield for the $O(^1D_2)$ product falling essentially to zero at 320 nm, whereas two sets of independent measurements show a persistent 10–20% quantum yield in this wavelength region.[11,12] Although the absorption coefficient of ozone is far lower at wavelengths above 310 nm than at its peak in the Hartley band at 255 nm,[13] the consequences of $O(^1D_2)$ formation in this region are of great importance for the chemistry of the troposphere and lower stratosphere. For example, the wavelength dependences of the absorption cross-section of ozone, the solar actinic flux and the $O(^1D_2)$ quantum yield combine to give a window close to 310 nm where tropospheric formation of $O(^1D_2)$ can take place, and its calculated production rate can vary by up to a factor of two, depending upon whether the NASA-recommended quantum-yield curve, or that including the long-wavelength tail, is incorporated in the model.[14] This production step is the rate-determining process for the formation of the major tropospheric scavenger, the OH radical, by reaction of $O(^1D_2)$ with water.

Although it has been known for some years that the absorption spectrum of ozone is dependent upon temperature, particularly in the structured Huggins bands between 310 and 360 nm,[13,15–17] there have been few investigations of the temperature dependence of the various photodissociation quantum yields. Measurements have been confined to the $O(^1D_2)$ product, and have reported that, as the temperature is reduced, the fall-off near 310 nm is shifted to shorter wavelengths[18–20] and, at a fixed wavelength of 313 nm, the $O(^1D_2)$ quantum yield decreases smoothly.[20–22] There is no evidence for a temperature-dependent quantum yield at wavelengths before the fall-off region.[23] These results imply that internally excited ozone molecules, with absorption features shifted to the red of those of the ground-state species, are playing an important role in the photodissociation process. Calculations of the absorption spectra of the vibrationally excited states have been made from the observed temperature dependence of the cross-sections[24,25] and from the known properties of the potential-energy surfaces of the ground and electronically excited states.[26,27] All three states with one quantum of vibrational energy have absorptions which are red-shifted with respect to the ground state and, in addition, the (100) level exhibits a bimodal structure similar to that extracted from experimental measurements.[15] Recently, a calculation of the quantum yield of $O(^1D_2)$ as a function of both wavelength and temperature in the fall-off region has been evaluated from the absorption measurements and from the measured $O(^1D_2)$ quantum yields in the (uncontroversial) wavelength region below 311 nm.[25]

We have recently reported our preliminary measurements on the wavelength dependence of the excited molecular fragment, $O_2(a\,^1\Delta_g)$, in the photolysis of ozone.[28] The results show that at wavelengths between 270 and 310 nm (the latter wavelength being in the middle of the fall-off region) the quantum yield matches that for the $O(^1D_2)$ product as recommended by NASA,[10] but that at longer wavelengths there is a persistent tail of 10–20% out to at least 329 nm. Here we describe these results in more detail and include measurements of the quantum yield of the $O_2(a\,^1\Delta_g)$ product as a function of temperature. These show that at wavelengths above 309 nm the yield at lower temperatures falls more rapidly than at room temperature, showing the importance of internal-energy-mediated dissociation of ozone just beyond the fall-off region. However, at longer wavelengths a persistent temperature-independent quantum yield of *ca.* 10% is found, and suggests the participation of a spin-forbidden dissociation channel. We discuss the importance of these results for the retrieval of ozone concentrations from observations of the $O_2(a\,^1\Delta_g)$ atmospheric band emission at 1.27 μm[29] [where current methods use the NASA-recommended tailless $O(^1D_2)$ quantum-yield curve] and their implications for the quantum yield of $O(^1D_2)$ at wavelengths longer than 310 nm. We also outline the strategy currently being used in our laboratory for unravelling the four possible combinations of atomic [$O(^1D_2)$ and $O(^3P_{2,1,0})$] and molecular [$O_2(a\,^1\Delta_g)$ and $O_2(X\,^3\Sigma_g^-)$] oxygen fragments formed in spin-allowed and spin-forbidden photodissociation of ozone.

Experimental

The experimental strategy used was a laser photolysis–laser probe method. The photolysis radiation was generated by frequency doubling the output of a pulsed tunable dye laser (Lambda Physik 2002) which was pumped at 308 nm by an excimer laser (Questek 2000). The $O_2(a\,^1\Delta_g)$ fragment was detected by (2 + 1) resonance-enhanced multiphoton ionisation (REMPI) at wavelengths near 331.5 nm, again produced from a frequency doubled pulsed dye laser (Lambda Physik 2002) and excimer laser (Lambda Physik 101MSC) combination. $O_2(a\,^1\Delta_g)$ has no known laser-induced fluorescence transitions: the REMPI scheme produces O_2^+ ions from the intermediate $3s\sigma_g\,^1\Pi_g(v = 1)$ level, produced in a coherent two-photon absorption step from the ground vibrational level of the $(a\,^1\Delta_g)$ state.[30] The laser beams counterpropagated through the pyrex photolysis cell and were focused to the same point between two ion collection plates (separation 1 cm), biased at 120 V. The cell could be cooled to temperatures of *ca.* 220 K, and evacuated double-window sections were used to prevent frosting at low temperatures. The temperature measured at several points around the photolysis cell by T-type thermocouples was stable to ± 2 K during a run. The pulsed REMPI current and the intensities of the two lasers were simultaneously recorded.

Ozone was prepared from molecular oxygen with a commercial ozoniser, passed through a trap at 196 K to remove oxides of nitrogen and collected on silica gel at 196 K. Once sufficient ozone had been produced, residual oxygen was pumped away, and the ozone desorbed into a Pyrex storage bulb. Samples were analysed regularly for ozone concentrations by UV absorption at 255 nm and normally contained *ca.* 50% molecular oxygen present from the gradual decomposition. Total pressures in the reaction vessel ranged from 1–30 Torr, with stable ozone partial pressures during each experiment.

Results

Spectral Identification and Data Treatment

Johnson *et al.*[30] have identified the carrier of a REMPI spectrum near 331 nm as $O_2(a\,^1\Delta_g)$ in its ground vibrational state by means of time-of-flight spectroscopy on the molecular ion. In the present experiments, in which total ion current is measured as a function of wavelength, we have observed very similar spectra under three sets of conditions, namely: from a discharge in molecular oxygen, from the nascent products of ozone photolysis (carried out with low or zero delay between photolysis and probe lasers) and from the internally relaxed photofragments, generally carried out with a time delay of 3.5 μs between laser pulses for total pressures ranging from 1–30 Torr. In this wavelength region the nascent spectrum was somewhat broader than that taken after collisional relaxation, and this illustrates that for the present measurements of the $O_2(a\,^1\Delta_g)$ quantum yields it is important to establish that the signals measured at a single wavelength of 331.5 nm truly represent the total amount of the species formed in the photolysis. Measurements were taken at a combination of pressure and time delay between laser pulses such that rotational relaxation would be complete, but that little electronic relaxation of the fortuitously metastable electronic state would have taken place. If, however, vibrationally excited $O_2(a\,^1\Delta_g)$ is formed in a fashion which is wavelength dependent and vibrational relaxation is incomplete, then monitoring the $v = 0$ population will introduce errors. In the wavelength region 270–293 nm the fractional population of $v \geqslant 1$ is small (*ca.* 30%) but approximately constant,[4] and thus the measurements of the $v = 0$ level should be representative of the true quantum yield. The rate constant for quenching of $O_2(a\,^1\Delta_g)\,v \geqslant 1$ by ozone is $4.7 \times 10^{-12}\ cm^3\ molecule^{-1}\ s^{-1}$,[31] and measurements carried out between 270 and 293 nm under conditions such that the calculated amount of vibrational quenching of $O_2(a\,^1\Delta_g)$ was changed between 25 and 65% showed no change in the essentially constant quantum yield. However, if $O_2(a\,^1\Delta_g)$ is formed in a spin-forbidden process, for example, at wavelengths above the thermodynamic threshold for process (1), 310 nm, then substantial internal excitation might be expected, and the effects of vibrational quenching could lead to the measured signals being underestimates of the true quantum yield. No such effect was seen in experiments carried out under differing collisional conditions. Fig. 1 shows an example of such a check, in which the $O_2(a\,^1\Delta_g)$ REMPI signal is shown as a function of time between the laser pulses (with increasing time corresponding to increasing numbers of collisions). The similarity of the decay curves for three wavelengths spanning the fall-off region, where such effects might be most pronounced, appears to rule out this possibility and the shape of the decay curve is entirely consistent with transport of $O_2(a\,^1\Delta_g)$ molecules out of the region of overlap of the laser beams being the dominant loss process. We should, however, point out that if vibrationally excited $O_2(a\,^1\Delta_g)$ formed at wavelengths

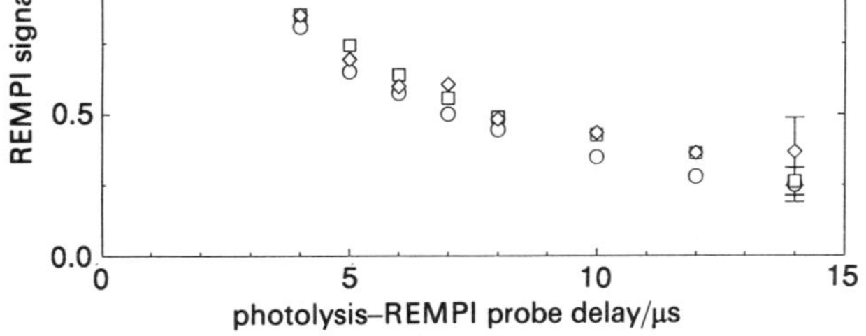

Fig. 1 Time dependences of the 331.5 nm REMPI signal from $O_2(a\,^1\Delta_g)$ formed from the photolysis of ozone at three different wavelengths ○, 303; □, 308 and ◇, 313 nm spanning the 310 nm threshold for formation of excited singlet products in process (1). $O_3 = 0.54$ Torr, $O_2 = 0.44$ Torr. Representative error bars are shown on the early- and late-time data. The decay of the $O_2(a\,^1\Delta_g)$ signal is caused by physical removal of the species from the overlap region of the laser beams and its invariance with photolysis wavelength is used to argue against the formation of vibrationally excited $O_2(a\,^1\Delta_g)$ influencing the measured quantum yields (see text).

above 310 nm can be removed in a state-specific fashion which does not yield molecules in $v = 0$, then our estimates of the quantum yields at these longer wavelengths could be too low. We consider other possible interferences from collisional processes below.

As well as the 331.5 nm REMPI band, we also observe strong, structured bands corresponding to ionisation *via* the $3s\sigma_g\,^1\Pi_g(v = 2)$ level at wavelengths near 321 nm.[30] Both these and the 331.5 nm bands also appear (at reduced intensity) when a single laser experiment is carried out, *i.e.* the same laser being used for photodissociation and for REMPI. Under these conditions the spectra correspond to the nascent distributions of internal energy and thus cannot be used for the quantum-yield measurements. A marked feature of the nascent single laser spectra was a series of sharp lines near 328 nm, absent from the relaxed spectra and which, although as yet unassigned, appears to come from $O_2(a\,^1\Delta_g)$ with considerable internal excitation. The possible consequences of this are discussed below.

Experiments were carried out under conditions such that the photolysis process was normally far from saturation. This was checked both by calculation, making use of the measured maximum photon flux at the centre of the laser beam and the ozone cross-section, and from experimental observation that the REMPI signal varied linearly with photolysis intensity. The one exception to this was for data taken between 270 and 280 nm, the region of highest ozone absorption cross-section, where a correction of up to 25% was needed to account for the measured non-linearity. An additional experiment carried out in this wavelength region at lower laser intensities (where saturation was negligible) showed exactly the same wavelength dependence of the signals, but with larger error bars. The ion signal was found to depend upon the intensity of the REMPI laser to the power 2.5, consistent with a (2 + 1) REMPI process in which one of the steps (almost certainly that of ionisation of the Rydberg intermediate state) is partially saturated. REMPI signals were corrected appropriately for variations of the intensities of both lasers, and divided by the ozone absorption cross-section[13] to produce a value proportional to the quantum yield of $O_2(a\,^1\Delta_g)$.

Room-temperature Data

Fig. 2 shows the variation with wavelength of the relative quantum yield of $O_2(a\,^1\Delta_g)$ from the photolysis of ozone at 298 K.[28] Four different runs are shown, taken with four different dye solutions in the photolysis laser, each dye producing a range of wavelengths which overlapped. Within each run the ozone partial pressure and ozone plus oxygen total pressure were kept constant: the separate runs were normalised to each other over the ranges of overlap of the dye wavelengths. Each data point was the average of some 400 laser shots. Our data do not measure the absolute quantum yield and thus were put on an absolute scale by normalisation in the range 290–300 nm (where the data from several runs overlap) to an absolute value of $\Phi = 0.95$, that recommended by NASA for the $O(^1D_2)$ quantum yield,[10] and very similar to the measurements of Φ for channel (1) reported in the literature.[2–4] In Fig. 2 the solid line represents the NASA $O(^1D_2)$ quantum yield recommendation. It can be seen that this is in extremely good agreement with the present data (and provides good evidence for the domination of the spin-allowed process forming the two singlet products) up to *ca.* 311 nm, but at longer wavelengths a persistent $O_2(a\,^1\Delta_g)$ yield of 10–20% is observed. It should be noted that the ozone absorption coefficient varies by over a factor of 2000 between the low- and high-wavelength data points shown in Fig. 2.

Collisional Processes

As our detection method necessarily relies on collisional processes, we need to consider their possible interference with $O_2(a\,^1\Delta_g)$ formation. The data of Fig. 2 show measurements at a variety of O_3/O_2 pressures, chosen to maximise the signal-to-noise ratios in

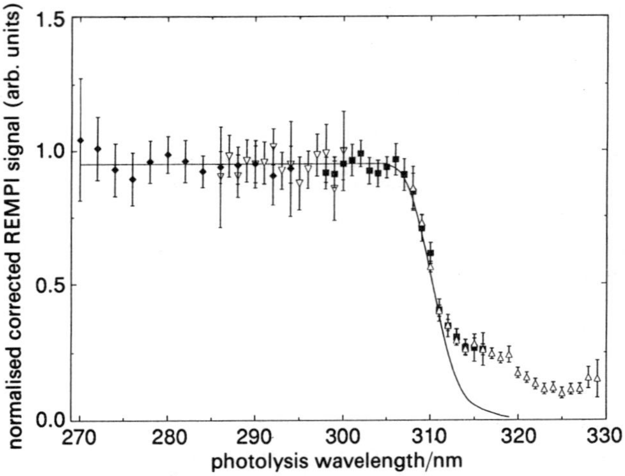

Fig. 2 REMPI signal of $O_2(a\,^1\Delta_g)$ measured at 331.5 nm as a function of the wavelength of the photolysis of ozone. The solid line is the quantum yield of $O(^1D_2)$ recommended by NASA for use in stratospheric modelling.[10] The data for $O_2(a\,^1\Delta_g)$ have been normalised to the $O(^1D_2)$ quantum yield over the range 290–300 nm. The symbols refer to the wavelength range spanned by various laser dyes in separate experimental runs (◆, Coumarin 153, $O_3 = 0.6$ Torr and $O_2 = 1.5$ Torr; ▽, Rhodamine 6G, $O_3 = 2.5$ Torr, $O_2 = 1.4$ Torr; ■, Sulforhodamine B, $O_3 = 2.8$ Torr, $O_2 = 1.8$ Torr; △, Sulforhodamine 101, $O_3 = 12.7$ Torr, $O_2 = 17.3$ Torr).

the various wavelength regions. Total pressures for the data shown in Fig. 2 ranged from 2.1–30 Torr. Experiments were also carried out at a constant pressure of 5 Torr of an equimolar O_3/O_2 mixture and showed the same variation with wavelength (although with higher error bars at the high- and low-wavelength limits). No clear variation of Φ with pressure was thus apparent, and this serves to eliminate a number of collisional processes. First, production of $O_2(a\,^1\Delta_g)$ by reactions involving the co-product in the spin-allowed process (1), $O(^1D_2)$, need to be considered. The sequence involving the $b\,^1\Sigma_g^+$ state of O_2:

$$O(^1D_2) + O_2(X\,^3\Sigma_g^-) \rightarrow O(^3P) + O_2(b\,^1\Sigma_g^+) \quad (3)$$

$$O_2(b\,^1\Sigma_g^+) + O_2(X\,^3\Sigma_g^-) \rightarrow O_2(a\,^1\Delta_g) + O_2(X\,^3\Sigma_g^-) \quad (4)$$

$$O_2(b\,^1\Sigma_g^+) + O_3 \rightarrow O_2(a\,^1\Delta_g) + O_3 \quad (5)$$

does not appear to compete with the fast quenching process

$$O(^1D_2) + O_3 \rightarrow \text{products} \quad (6)$$

the evidence coming from both literature values of the rate constants and quantum yields for the processes,[10,32] and the results of our experimental measurements carried out at different O_3/O_2 pressures and delay times. Possible formation of an energy-rich ozone intermediate in the steps involving an internally excited molecular oxygen species, O_2^*

$$O_3 + h\nu \rightarrow O_2^* + O \quad (7)$$

$$O_2^* + O_3 \rightarrow O_2 + O_3^* \quad (8)$$

followed by photolysis of O_3^* at 331.5 nm to produce $O_2(a\,^1\Delta_g)$ and detection within the same 'probe' laser pulse needs to be considered. Here O_3^* represents a molecule with vibrational/rotational energy; it cannot be $O_2(a\,^1\Delta_g)$ in $v = 0$ as the electronic quenching-rate constant for process (8) is too slow.[10] Experiments were carried out to measure the relative quantum yields as a function of delay time between photolysis and probe lasers at photolysis wavelengths of 303, 308 and 313 nm as shown in Fig. 1. At 308 nm the quantum yield for singlet formation is close to unity, and $O_2(a\,^1\Delta_g)$ is formed with little excitation energy,[4] insufficient to transfer to ozone to allow the latter to be photolysed in an efficient spin-allowed channel to give $O_2(a\,^1\Delta_g)$. At longer wavelengths, dissociation of ozone to form $O_2^*(X\,^3\Sigma_g^-)$ could provide this mechansim: it seems to be unimportant as no variation of relative quantum yield with delay time could be seen. All experiments in which the gas composition, pressures and delay times were varied yielded results which are consistent with the measured $O_2(a\,^1\Delta_g)$ signals representing the true primary dissociative quantum yields.

Measurements near 320 nm

The absorption spectrum of ozone is considerably structured, particularly in the Huggins bands at wavelengths above 310 nm. Fig. 3 shows part of the absorption spectrum near 320 nm, together with $O_2(a\,^1\Delta_g)$ quantum-yield measurements taken at 298 K and at higher resolution than those of Fig. 2 over the structured feature. The structure has been attributed[33] to transitions to vibrationally excited levels of the upper electronically excited state, now assigned as having $^1B_2$ symmetry in both the Huggins and Hartley bands,[34,35] and features visible in Fig. 3 include transitions from the (0, 0, 0) ground state to the (5, 0, 2) and (6, 0, 1) levels in the excited state at 319.4 and 320.0 nm, respectively.[33] The quantum-yield data show no obvious positive correlation with these marked features, with the general trend the same as that seen in Fig. 2, namely of decreasing values as the wavelength increases. Formation of $O_2(a\,^1\Delta_g)$ in a fashion dependent on the upper-state vibronic level does not appear to be dominant in this wavelength region.

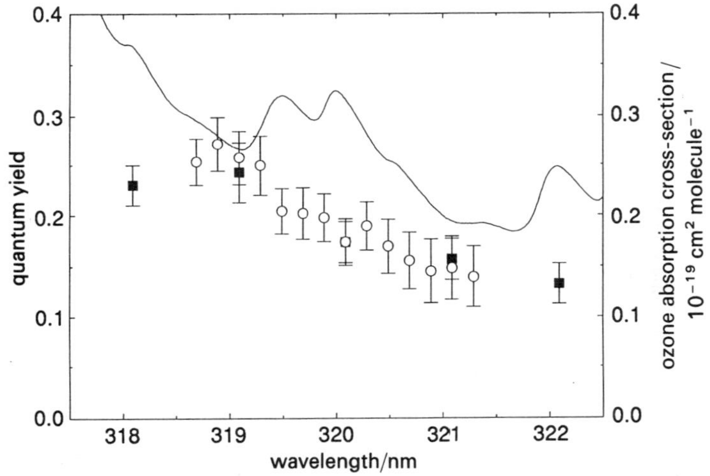

Fig. 3 Quantum yield of $O_2(a\,^1\Delta_g)$ from the photolysis of room-temperature ozone as a function of wavelength over the structured absorption feature near 320 nm. The solid line shows the ozone absorption cross-section with the scale on the right-hand axis, taken from the data of Brion *et al.*[17] ■, Quantum yield measurements at 1 nm intervals from Fig. 2 and ○, an additional set of measurements at 0.2 nm intervals. Within the error limits of the experiment, the quantum yield shows a monotonic decline between 319 and 321 nm and has no obvious correlation with either the structured absorption features[33] or the changing contributions from absorption by vibrationally excited molecules as calculated from the temperature dependence of the absorption cross-section.[25]

Low-temperature Data

Similar measurements to those described above were carried out over a smaller wavelength range, 300–322 nm, for a sample of ozone cooled to 227 K. There were minor differences in the detailed structures of the 331.5 nm band observed with the typical pressure (5 Torr) and delay time between the laser pulses (3.5 μs) at 227 and 298 K because of the differing rotational-state populations. Data were converted to relative quantum yields in the same way as described for the room-temperature results, except that ozone absorption cross-sections at 226 K[13] were taken from the literature. The data are shown on Fig. 4, together with the room-temperature data of Fig. 2 over the same wavelength range. We normalise the low-temperature data between 303 and 307 nm to the value recommended by NASA for the quantum yield of $O(^1D_2)$ at 305 nm and 227 K, 0.94.[10] This implies that at both temperatures the spin-allowed dissociation process (1) dominates at wavelengths below the fall-off region. The low-temperature data show the same, approximately constant, quantum yield between 300 and 307 nm and the same initial fall-off behaviour up to 309 nm as the room-temperature measurements. No marked shift in the fall-off to shorter wavelengths at the lower temperature is observed, in contrast to that reported for the $O(^1D_2)$ fragment.[19,20] However, a marked temperature effect is seen between 309 and 318 nm, the quantum yields at the lower temperature being up to a factor of two lower than those at 298 K. It appears that photodissociation of internally excited ozone contributes markedly to the quantum yield in this region, but at longer wavelengths, $\geqslant$320 nm, temperature has a much smaller effect, with quantum yields being equal, to within the experimental error. The solid and dashed lines on Fig. 4 are the results of model calculations of the quantum yield of $O(^1D_2)$ by Michelsen *et al.*[25] and will be discussed below.

Several sets of measurements of the REMPI signals were carried out at fixed wavelengths of photodissociation as a function of temperature. Fig. 5 shows results at 305, 312 and 313 nm, demonstrating that temperature has no effect on the signal at 305 nm

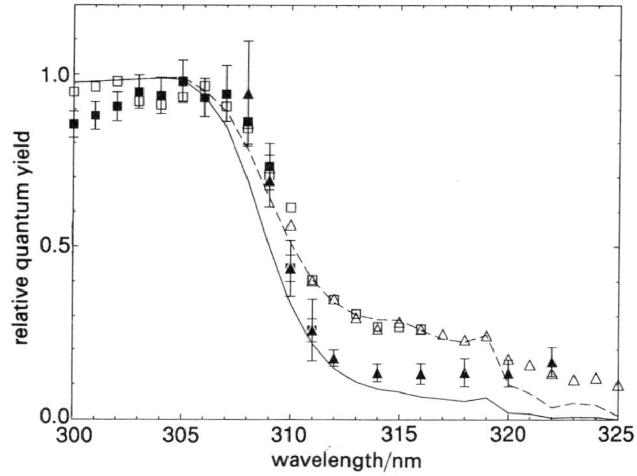

Fig. 4 The filled points with error bars are the relative quantum yields of $O_2(a\,^1\Delta_g)$ as a function of photolysis wavelength of ozone at 227 K. Also shown, as open points but without error bars are the $O_2(a\,^1\Delta_g)$ yields at room temperature from Fig. 2. The symbols refer to the wavelength range spanned by the two laser dyes in separate experimental runs (■ and □, Sulforhodamine B, ▲ and △, Sulforhodamine 101). The average yield near 305 nm at each temperature has been set equal to the NASA recommendation for the yield of $O(^1D_2)$ (see text).[10] The solid and broken lines are the predicted yields of $O(^1D_2)$ at 227 and 298 K, respectively, calculated from the model of Michelsen *et al.*[25]

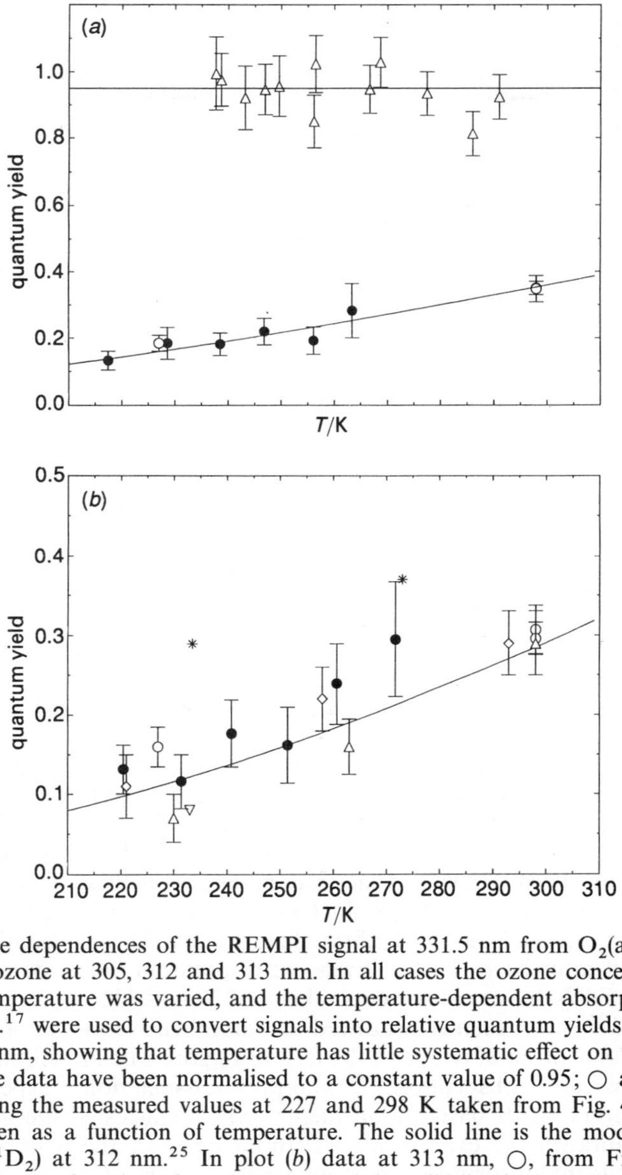

Fig. 5 Temperature dependences of the REMPI signal at 331.5 nm from $O_2(a\,^1\Delta_g)$ formed from the photolysis of ozone at 305, 312 and 313 nm. In all cases the ozone concentration remained constant as the temperature was varied, and the temperature-dependent absorption cross-section data of Brion *et al.*[17] were used to convert signals into relative quantum yields. In plot (*a*), △ are data taken at 305 nm, showing that temperature has little systematic effect on the signal between 240 and 292 K, the data have been normalised to a constant value of 0.95; ○ and ●, data taken at 312 nm, ○, being the measured values at 227 and 298 K taken from Fig. 4 and ●, separate measurements taken as a function of temperature. The solid line is the modelled temperature dependence of $O(^1D_2)$ at 312 nm.[25] In plot (*b*) data at 313 nm, ○, from Fig. 4; ●, separate measurements taken as a function of temperature, and the solid line is the modelled $O(^1D_2)$ dependence.[25] Data are also shown from the $O(^1D_2)$ measurements of Lin and DeMore[18] (▽); Moortgat *et al.*[19,20] (△); Kuis *et al.*[21] (◇) and Kajimoto and Cvetanovic[22] (*).

[where the quantum yield for $O_2(a\,^1\Delta_g)$ is assumed to be close to unity], but at the wavelengths beyond the fall-off limit, an increase in temperature causes a marked increase in signal. The data have been put on an absolute scale by comparison with those at 227 K shown in Fig. 4 and, as can be seen from the curves at 312 and 313 nm, the temperature dependences for the two types of experimental run (one at constant temperature with wavelength varying, the other at constant wavelength and number density with temperature varying) are in good agreement. This observation is important, as REMPI data taken under the latter condition with temperature varying need to be interpreted with caution. For them to be converted directly into relative quantum yields needs the assumption that the spectroscopy of the thermalised quantum-state distribution is such that at 331.5 nm the signal represents the same fraction of the total molecules at each temperature, and that the efficiency of ion production in the REMPI cell is independent of temperature. Within the precision of the measurements, the first point appears to be valid, as estimated from the spectra at the two extremes of temperature. For the second point we have no information on the temperature dependence of the way in which the initially formed $O_2{}^+$ ions are amplified in the dc field to form the detected current pulse. The agreement between the two types of results in Fig. 5 appears to suggest that any such effect is small. Modelled temperature dependences at 312 and 313 nm are shown,[25] together with a number of results on the temperature dependences of the quantum yields of $O(^1D_2)$ at 313 nm.[18–22] These are discussed below.

Discussion

The main conclusions from this study can be summarised as follows: (1) At room temperature the quantum yield for $O_2(a\,^1\Delta_g)$ formation in the photolysis of ozone follows closely that recommended by NASA for the formation of $O(^1D_2)$ between 270 and 311 nm, namely a constant value up to *ca.* 306 nm followed by a marked drop near the thermodynamic threshold for process (1) of 310 nm. The observations suggest that the spin-allowed step (1) dominates the dissociation process in the Hartley bands. However, at wavelengths above 311 nm our measurements are consistently higher than the NASA recommendation and show a persistent 10–20% quantum yield out to the longest wavelength quantitatively measured, 329 nm. This long-wavelength tail is very similar to that observed in two sets of measurements of the $O(^1D_2)$ quantum yield which conflict with the NASA recommendation.[11,12]
(2) At 227 K the quantum yields show the same behaviour between 300 and 309 nm as at 298 K and do not exhibit the marked shift in the fall-off to lower wavelengths at low temperatures as reported for the quantum yield of $O(^1D_2)$.[18–20] For wavelengths beyond 309 nm, and up to at least 318 nm, temperature has a marked effect on the $O_2(a\,^1\Delta_g)$ quantum yields, showing that internal-energy-mediated photolysis of ozone plays a major role in forming this product in this wavelength region. Studies at temperatures between 227 and 298 K confirm these trends. At wavelengths $\geqslant$320 nm the temperature effect appears to be far lower, with quantum yields being approximately equal at 227 and 298 K for measurements taken at 320 and 322 nm.

Michelsen *et al.*[25] have recently produced a comprehensive model of the $O(^1D_2)$ yield from the photodissociation of ozone in the fall-off region. The model builds upon earlier IR–UV double-resonance absorption[24,36] and photodissociation[37] experiments, and calculations of quantum yields based upon them[26] which suggested the importance of the dissociation of vibrationally excited species. Input data for the new model[25] were the temperature-dependent absorption cross-sections of ozone, together with the $O(^1D_2)$ quantum yield data of Brock and Watson at 298 K[11] and Lin and DeMore at 233 K,[18] but only at wavelengths up to 311 nm to avoid the controversy concerning the long-wavelength tail: the model was used to predict quantum yields at longer wavelengths. As in the previous analysis of Adler-Golden *et al.*[24] a pronounced tail was predicted at room temperature, in close agreement with two sets of experimental data[11,12] but not with the NASA recommendation.[10] Fig. 4 shows that the model is in excellent agreement with our findings at 298 K for the $O_2(a\,^1\Delta_g)$ fragment, at least up to 320 nm, and again suggests that process (1) dominates under these conditions. Fig. 4 also shows the modelled $O(^1D_2)$ yield at 227 K. The predicted wavelength shift in the threshold in the model is a direct consequence of the input data including the low-temperature quantum-yield measurements of Lin and DeMore[18] who observe such a shift. For the $O_2(a\,^1\Delta_g)$ fragment we observe no such effect. Near 312 nm the predicted lowering of the quantum yield reproduces our experimental findings extremely well, and this is further illustrated by the excellent agreement between the 312 nm temperature-dependence results illustrated in Fig. 5 and those predicted by the model.

At 313 nm, where several sets of $O(^1D_2)$ experiments have been reported, our measured values are in good agreement with those of Kuis *et al.*,[21] show somewhat higher values at lower temperatures than the data of Moortgat and co-workers[19,20] and Lin and Demore,[18] and are slightly above the modelled values of Michelsen *et al.*,[25] as shown on Fig. 5 (the higher quantum yield values of Kajimoto and Cvetanovic[22] also shown in the figure have been called into question[19]). At longer wavelengths the disagreement between model and experiment is more pronounced. Fig. 4 shows that for the low-temperature data at wavelengths above 314 nm, and for all the data above 320 nm, the experimental quantum yields are greater than the predicted values, and there appears to be little temperature dependence at the longest wavelengths studied, 320 and 322 nm. We suggest that although internal-energy-mediated photolysis of ozone provides the major source of $O_2(a\,^1\Delta_g)$ in the region 310–318 nm, there exists another source for the species which is dominant at higher wavelengths. The obvious candidate is the spin-forbidden process

$$O_3 + h\nu \rightarrow O(^3P_{2,1,0}) + O_2(a\,^1\Delta_g) \qquad (9)$$

which has a thermodynamic threshold corresponding to absorption of a photon at wavelengths $\leqslant$611 nm, and which has been suggested as taking place at 334 nm in experiments carried out some twenty years ago.[38] Our observations of REMPI signal in single-laser experiments (*i.e.* where the same laser is used both to photolyse ozone and to ionise the fragments) at 321, 328 and 331.5 nm for both room temperature and cooled ozone support this proposal, although accurate quantum-yield measurements were not possible under these single-laser experimental conditions.

The existence of spin-forbidden dissociation has some grounding in theoretical calculations of the potential-energy surfaces of ozone. Several repulsive triplet surfaces correlating with $O(^3P_{2,1,0}) + O_2(a\,^1\Delta_g)$, $O(^3P_{2,1,0}) + O_2(b\,^1\Sigma_g^+)$ and $O(^1D_2) + O_2(X\,^3\Sigma_g^-)$ fragments have been found to cross the $\tilde{D}\,^1A'$ surface, which correlates in C_s symmetry with the $^1B_2$ state populated in absorption in the Hartley and Huggins bands.[35] Mixing of these triplet states with the $\tilde{D}$ state has been advanced as a possible explanation for the perturbation of lines in the Huggins band.[35] Direct spin-forbidden absorption or curve-crossing processes, to access one or more of these triplet surfaces, could produce the majority of the yields of $O_2(a\,^1\Delta_g)$ observed for wavelengths above 320 nm and cannot be ruled out as being a minor pathway for ozone dissociation at shorter wavelengths.

One more point concerning the model calculations deserves discussion. Michelsen *et al.*[25] found that the best fit to the data of temperature-dependent absorption spectra and quantum yields required the assumption that ozone molecules in the (001) or (100) stretching vibrational levels were able to use the whole of this energy in shifting their energetic thresholds for process (1) to longer wavelengths, but that for molecules in the (010) bending level, only 30% of the vibrational energy was so utilised. If this is a true vibrational-state effect, and not caused by uncertainties in the experimental data used in the fitting procedures, then it implies that the states reached by good Franck–Condon overlap from ozone molecules with one quantum of bending vibration in the ground state partition their available energy into photofragments in a way which is different from the behaviour of other levels. No clear effect of this kind, however, is seen in the $O_2(a\,^1\Delta_g)$ quantum yield data shown in Fig. 3 and taken over a region where the vibrational-state-dependent absorption coefficients recovered from the model calculations are markedly dependent upon wavelength.[25]

We now consider the implications of the present results for atmospheric chemistry. From a practical viewpoint, there will be some (but probably small) changes in ozone profiles recovered from airglow measurements and arising from the tail in the quantum yield for $O_2(a\,^1\Delta_g)$ production above 310 nm. Such changes will be most marked in the lower stratosphere and at high solar zenith angles. The effects, however, will not be as large as the factor of *ca.* two calculated[39] some years ago if spin-forbidden $O_2(a\,^1\Delta_g)$ dominated the ozone dissociation process at wavelengths above 310 nm. The more important consequence concerns the effect on the quantum yield of $O(^1D_2)$. Since the present results indicate the importance of the dissociation of vibrationally excited ozone, and since there is no reason to doubt that this will take place dominantly by a spin-allowed process, then our results support the studies[11,12] which show the existence of a long-wavelength tail in the $O(^1D_2)$ quantum yield above 310 nm. Recently reported simultaneous measurements of solar actinic UV radiation and $O(^1D_2)$ formation rates support these conclusions.[14] We expect the $O(^1D_2)$ tail to be temperature dependent; it will affect the calculated production rates of $O(^1D_2)$ and hence OH in the atmosphere by a marked amount over those resulting from the NASA-recommended quantum-yield curve.

The most obvious need for further confirmation of these conclusions is for a comprehensive study of the wavelength and temperature dependence of the $O(^1D_2)$ quantum yield to be carried out. We have detected this species formed from ozone photolysis in a highly sensitive fashion by REMPI at 205.47 and 203.81 nm *via* the $^1P_1$ and $^1F_3$ states, respectively,[40] and are currently using this method to measure the $O(^1D_2)$ quantum yield directly. In addition we have recently constructed a time-of-flight (TOF) mass spectrometer and observed molecular oxygen ions formed from REMPI of $O_2(a\,^1\Delta_g)$ produced by photolysis of ozone. The instrument has been constructed so that the kinetic energy release in the single-photon dissociation step can be measured from the TOF profiles.[41] This will enable the relative contributions of processes (1) and (9) to be determined as a function of wavelength in the fall-off region. Although no spin-forbidden formation of $O(^1D_2)$ (with ground-state molecular oxygen as co-product) has been directly observed in the photolysis of ozone, analogy with the present results indicates that such a process cannot be ruled out, and in principle this TOF method can be used to estimate its importance. Ground-state oxygen atoms can be easily detected by REMPI near 226 nm and ion imaging of the resultant O^+ has been used recently to determine the dissociation dynamics of ozone photolysis at this wavelength.[42] In principle, ion-detection methods can be used for all oxygen fragments of ozone photolysis (although no scheme has yet been devised for the highly vibrationally excited levels of the $X\,^3\Sigma_g^-$ ground state formed in the photolysis) and thus it may be expected that REMPI methods will enable considerable progress to be made in unravelling the complexities of the different channels expected near the fall-off region.

This work was supported by the Atmospheric Chemistry Initiative of the SERC and NERC, and by the European Commission Environmental Programme. We are grateful to Mat Heal, Ian Murphy and Steve Rayner for their help in the early stages of this project.

References

1 R. P. Wayne, in *The Handbook of Environmental Chemistry*, ed. O. Hutzinger, Springer-Verlag, Berlin, 1989, vol. 2, Part E, p. 1.
2 C. E. Fairchild and E. K. C. Lee, *Chem. Phys. Lett.*, 1978, **60**, 36.
3 R. K. Sparks, L. R. Carlson, K. Shobatake, M. L. Kowalczyk and Y. T. Lee, *J. Chem, Phys.*, 1980, **72**, 1401.
4 J. J. Valentini, D. P. Gerrity, D. L. Phillips, J-C. Nieh and K. D. Tabor, *J. Chem, Phys.*, 1987, **86**, 6745.
5 S. T. Amimoto, A. P. Force, J. R. Wiesenfeld and R. H. Young, *J. Chem, Phys.*, 1980, **73**, 1244.
6 J. C. Brock and R. T. Watson, *Chem. Phys. Lett.*, 1980, **71**, 371.
7 G. D. Greenblatt and J. R. Wiesenfeld, *J. Chem, Phys.*, 1983, **78**, 4924.
8 I. A. Cooper, P. J. Neil and J. R. Wiesenfeld, *J. Geophys. Res.*, 1993, **98**, 12 795.

9 G. K. Moortgat and P. Warneck, *Z. Naturforsch., A*, 1975, **30**, 835; I. Arnold, F. J. Comes and G. K. Moortgat, *Chem, Phys.*, 1977, **24**, 211.
10 W. B. DeMore, S. P. Sander, D. M. Golden, R. F. Hampson, M. J. Kurylo, C. J. Howard, A. R. Ravishankara, C. E. Kolb and M. J. Molina, *Chemical Kinetics and Photochemical Data for Use in Stratospheric Modeling*, Evaluation Number 10, JPL Publication 92-20, 1992.
11 J. C. Brock and R. T. Watson, *Chem, Phys.*, 1980, **46**, 477.
12 M. Troiler and J. R. Wiesenfeld, *J. Geophys. Res. D*, 1988, **93**, 7119.
13 L. T. Molina and M. J. Molina, *J. Geophys. Res. D*, 1986, **91**, 14 501.
14 M. Müller, A. Kraus and A. Hofzumahaus, *Geophys. Res. Lett.*, 1995, **22**, 679.
15 J. W. Simons, R. J. Paur, H. A. Webster III and E. J. Bair, *J. Chem, Phys.*, 1973, **59**, 1203.
16 D. E. Freeman, K. Yoshino, J. R. Esmond and W. H. Parkinson, *Planet. Space Sci.*, 1984, **32**, 239; M. Cacciani, A. Di Sarra, G. Fiocco and A. Amoruso, *J. Geophys. Res. D*, 1989, **94**, 8485.
17 J. Brion, B. Coquart, D. Daumont, A. Jenouvrier, J. Malicet and M. F. Merienne, in *The Proceedings of EUROTRAC Symposium' 92.* ed. P. M. Borrell, P. Borrell, T. Cvitaš and W. Seiler, SPB Academic, The Hague, 1993, p.423.
18 C-L. Lin and W. B. DeMore, *J. Photochem.*, 1973, **2**, 161.
19 G. K. Moortgat, E. Kudsus and P. Warneck *J. Chem. Soc., Faraday Trans. 2*, 1977, **73**, 1216.
20 G. K. Moortgat and E. Kudsus, *Geophys. Res. Lett.*, 1978, **5**, 191.
21 S. Kuis, R. Simonaitis and J. Heiklen, *J. Geophys. Res.*, 1975, **80**, 1328.
22 O. Kajimoto and R. J. Cvetanovic, *Chem. Phys. Lett.*, 1976, **37**, 533.
23 J. I. Steinfeld, S. M. Adler-Golden and J. W. Gallagher, *J. Phys. Chem. Ref. Data*, 1987, **16**, 911.
24 S. M. Adler-Golden, E. L. Schweitzer and J. I. Steinfeld, *J. Chem, Phys.*, 1982, **76**, 2201.
25 H. A. Michelsen, R. J. Salawitch, P. O. Wennberg and J. G. Anderson, *Geophys. Res. Lett.*, 1994, **21**, 2227.
26 S. M. Adler-Golden, *J. Quant. Spectrosc. Radiat. Transfer*, 1983, **30**, 175.
27 M. G. Sheppard and R. B. Walker, *J. Chem, Phys.*, 1983, **78**, 7191; I. C. McDade and E. J. Llewellyn, *J. Photochem.*, 1986, **32**, 133.
28 S. M. Ball, G. Hancock, I. J. Murphy and S. P. Rayner, *Geophys. Res. Lett.*, 1993, **20**, 2063.
29 D. W. Tarasick and W. F. J. Evans, *Adv. Space Res.*, 1993, **13**, 145.
30 R. D. Johnson III, G. R. Long and J. W. Hudgens, *J. Chem, Phys.*, 1987, **87**, 1977.
31 O. Klais, A. H. Laufer and M. J. Kurylo, *J. Chem, Phys.*, 1980, **73**, 2696.
32 L. C. Lee and T. G. Slanger, *J. Chem, Phys.*, 1978, **69**, 4053; T. G. Slanger and G. Black, *J. Chem, Phys.*, 1979, **70**, 3434; M. B. Knickelbein, K. L. Marsh, O. E. Ulrich and G. E. Busch, *J. Chem, Phys.*, 1987, **87**, 2392.
33 D. H. Katayama, *J. Chem, Phys.*, 1979, **71**, 815.
34 A. Sinha, D. Imre, J. H. Goble Jr. and J. L. Kinsey, *J. Chem, Phys.*, 1986, **84**, 6108; F. Le Quéré and C. Leforestier, *Chem. Phys. Lett.*, 1992, **189**, 537.
35 A. Banichevich, S. D. Peyerimhoff and F. Grein, *Chem, Phys.*, 1993, **178**, 155.
36 I. C. McDade and W. D. McGrath, *Chem. Phys. Lett.*, 1980, **73**, 413.
37 P. F. Zittle and D. D. Little, *J. Chem, Phys.*, 1980, **72**, 5900.
38 I. T. N. Jones and R. P. Wayne, *J. Chem, Phys.*, 1969, **51**, 3617; I. T. N. Jones and R. P. Wayne, *Proc. R. Soc. London, A.*, 1970, **319**, 273; E. Castellano and H. J. Schumacher, *Chem. Phys. Lett.*, 1972, **13**, 625.
39 P. J. Crutzen, I. T. N. Jones and R. P. Wayne, *J. Geophys. Res.*, 1971, **76**, 1490.
40 S. T. Pratt, P. M. Dehmer and J. L. Dehmer, *Phys. Rev. A*, 1991, **43**, 282; 1991, **43**, 4702.
41 M. Mons and I. Dimicoli *J. Chem, Phys.*, 1989, **90**, 4037.
42 R. L. Miller, A. G. Suits, P. L. Houston, R. Toumi, J. A. Mack and A. M. Wodtke, *Science*, 1994, **265**, 1831.

Paper 5/00191I; *Received 5th January*, 1995

Faraday Discuss., 1995, **100**, 229–251

State-to-state Rate Constants for Relaxation of Highly Vibrationally Excited O_2 and Implications for its Atmospheric Fate

C. A. Rogaski,† J. A. Mack and A. M. Wodtke‡
Department of Chemistry, University of California at Santa Barbara, Santa Barbara, CA 93106, USA

Collisional relaxation of highly vibrationally excited $O_2(X\,^3\Sigma_g^-, v)$ has been investigated for vibrational states from $v = 18$ to 27, using O_2 and N_2 as quenching partners. These excited states of O_2 are now known to be produced in stratospheric ozone photolysis and this study gives information on their eventual atmospheric fate. State-to-state relaxation rate constants were obtained using the stimulated emission pumping method in the pump–dump–probe geometry. For self-relaxation, O_2 ($v = 19$) was found to decay in single vibrational-quantum steps and the rate constants were found to agree well with theoretical predictions which rely on an *ab initio* potential-energy surface. For vibrational levels above $v = 25$, a dark channel is observed, where molecules prepared in a state v do not appear in vibrational states $v - 1$ or $v - 2$. The most likely explanation to this dark channel is the onset of the chemical reaction

$$O_2(X\,^3\Sigma_g^-, v \geqslant 26) + O_2(X\,^3\Sigma_g^-) \rightarrow O_3 + O(^3P) \quad (1)$$

The temperature-dependent data presented in this paper allow the derivation of activation energies for the dark channel when the reactants are prepared in vibrational states just below the known activation barrier of reaction (1). For these states, a fraction of the molecules 'trickle down' into lower vibrational states, allowing their kinetics to be probed. The derived activation energies are quantitatively consistent with the assignment of the dark channel to reaction (1). The first vibrational state which is above the activation barrier to reaction (1), $v = 28$,, is not observed to 'trickle down', suggesting that it only follows the dark channel, *i.e.* reaction (1). The measurements diverge from the theory successful at $v = 19$ just at the energetic onset of reaction (1). Relaxation rate constants for N_2 quench gas decrease with O_2 vibrational quantum number and theory is not successful in reproducing these results. It appears that a more accurate O_2–N_2 potential-energy surface is needed.

One of the fundamental assumptions used in atmospheric modelling is local thermodynamic equilibrium (LTE).[1] When LTE conditions exist, the internal quantum-state distribution of all atmospheric constituents can be described by a Maxwell–Boltzmann distribution characterized by a local temperature. Considering that molecules in excited states can react with vastly different rate constants than molecules in quantum states normally populated in a Maxwell–Boltzmann distribution at atmospheric temperatures

† Current address: SRI International, Molecular Physics Department, Menlo Park, CA 94025, USA.
‡ NSF Presidential Young Investigator; Camille and Henry Dreyfus Teacher-Scholar; Alfred P. Sloan Research Fellow.

(*ca.* 200–300 K),[2–11] LTE represents a massive simplification. Without LTE, each quantum state of an atmospheric constituent should be treated as a unique atmospheric species with its own chemical and photochemical rate constants. Similarly, if one were to attempt a concentration retrieval from infrared emission data collected from a satellite, it would be necessary to know accurately the steady-state distribution of emitting vibrational levels before the observed radiance could be interpreted.

Mechanisms for production of non-LTE energy distributions of atmospheric constituents are becoming better understood.[12–15] For example, production of ground electronic state $O_2(X\,^3\Sigma_g^-)$ in vibrational states as high as $v = 27$ has been reported recently.[12] If such high vibrational states of O_2 exhibit chemical or photochemical behaviour not normally attributed to LTE O_2, no amount of modelling within the LTE assumption would accurately reproduce actual atmospheric events. Indeed, trying to use an LTE model to fit an inherently non-LTE set of data might yield misleading results.

The observation of highly vibrationally excited O_2, hereafter referred to as O_2(vib), takes on added significance since atmospheric methods for modelling stratospheric O_3 concentrations have consistently fallen short of measured values.[16] At 35–50 km, almost all models underestimate the O_3 concentrations by a factor of 1.2–1.5,[17] and above 50 km, estimates fall short by as much as a factor of 1.5–2.0.[18] This 'ozone deficit'[19] has been a source of concern on the part of atmospheric chemists as it implies an unknown mechanism for production of stratospheric O_3. In other words, non-LTE production of ozone, *e.g. via* O_2(vib), could explain the systematic deficit of the LTE models.

There have been two suggestions how O_2(vib) could produce O_3. In the absence of collisional relaxation, photolysis of O_2(vib) is much more efficient than that of vibrationless O_2, since 'hot' O_2 can be photolysed by light at longer wavelengths, which is more prevalent in the middle atmosphere.[15,20] Consequently, the presence of O_2(vib) could, in principle, raise the 'normal' LTE O_2 photolysis rate. Early attempts to incorporate this so-called 'Slanger mechanism' into predictions for stratospheric ozone production and infrared emission were promising,[21] based on assumptions regarding the collisional relaxation of O_2(vib). However, when experimental measurements of the collisional relaxation rates were carried out (some of which are presented here), it became clear that collisional relaxation was much more rapid than solar photodissociation and this idea had to be abandoned.[15,22]

The second suggestion for the formation of O_3 from O_2(vib) implicated its reaction with O_2 at vibrational levels above $v = 25$.[22]

$$O_2(X\,^3\Sigma_g^-, v \geqslant 26) + O_2(X\,^3\Sigma_g^-) \rightarrow O_3 + O(^3P) \quad (1)$$

which takes part in the following cycle.

$$O_3 + h\nu \rightarrow O_2(^3\Sigma_g^-, v \geqslant 26) + O(^3P) \quad (2)$$

$$O_2(^3\Sigma_g^-, v \geqslant 26) + O_2(^3\Sigma_g^-, v = 0) \rightarrow O_3 + O(^3P) \quad (1)$$

$$2 \times [O(^3P) + O_2(^3\Sigma_g^-, v = 0) + M \rightarrow 2O_3 + M] \quad (3)$$

$$\text{net:} \quad 3O_2 + h\nu \rightarrow 2O_3$$

This mechanism is also clarified by the reaction cycle diagram given in Fig. 1. In previous work the stimulated emission pumping method was used to produce and study individual vibrational states of O_2 from $v = 18$ to 27. A marked threshold for enhanced collisional relaxation was found coincident with the threshold of reaction (1). However, direct observation of chemical reaction products was impossible, and proof of the existence of this chemical reaction remained elusive.

The present work presents the results of a further series of stimulated emission pumping experiments designed to probe the dynamics O_2(vib)–O_2 and O_2(vib)–N_2. These experiments make direct measurements of the total and state-to-state relaxation rate constants at temperatures of 295 and 465 K, for O_2(vib)–O_2, and 295 K for

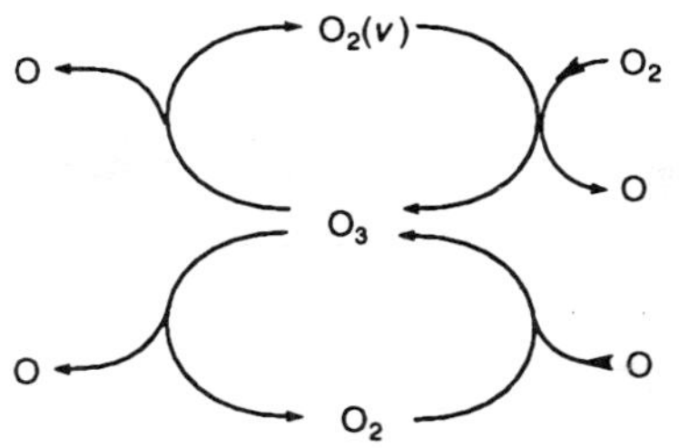

Fig. 1 Schematic to describe the chemical cycle given as reactions (1)–(3): The upper ring of this cycle represent reactions (1)–(3), while the lower ring represents the photochemical equilibrium of O_2 with O_3. This diagram shows how, despite the fact that the mechanism destroys O_3 in the initial step, the overall mechanism produces O_3 in the return leg of the upper ring by converting stable O_2 into one O atom.

O_2(vib)–N_2 collisional relaxation. For self-relaxation, good agreement with theoretical work is found at lower vibrational quantum numbers, but large deviations are found above the threshold of reaction (1). Clear evidence of a 'dark' channel is found whose most likely explanation is the onset of reaction (1). Poor agreement with theoretical work is found for the O_2–N_2 collisional system.

Experimental

The stimulated emission pumping method was used to prepare $O_2(X\,^3\Sigma_g^-,\ v = 18\text{–}27)$. The energy-level diagram is shown schematically in Fig. 2. The experimental set-up is shown in Fig. 3. The approach is similar to previous experiments.[23,24] Ground-state O_2

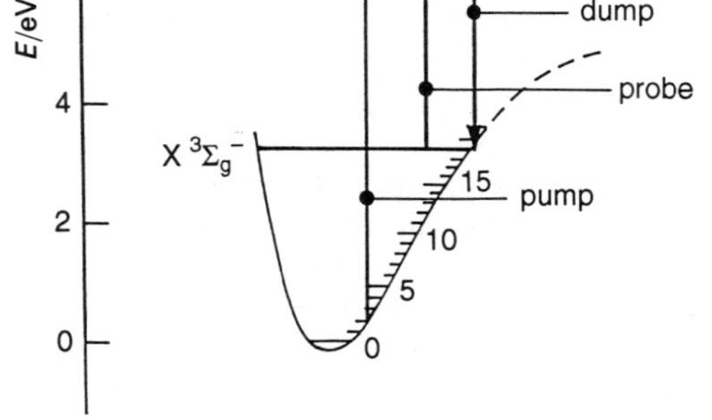

Fig. 2 Electronic States of O_2 and the experimental pump–dump–probe scheme showing the stimulated emission pumping scheme for measuring the collisional properties of O_2(vib). The pump laser induces population transfer from the $X\,^3\Sigma_g^-$ to the $B\,^3\Sigma_u^-$ excited state. The dump laser stimulates emission back to high vibrational levels of the $X\,^3\Sigma_g^-$ state. A third probe laser observes the time evolution of the prepared state. The experimental approach is quite sensitive, as evidenced by the good signal to noise ratios found despite the facts that ns pulsed lasers are used and that all the vibrational levels of the $B\,^3\Sigma_u^-$ state predissociate on the ps timescale.

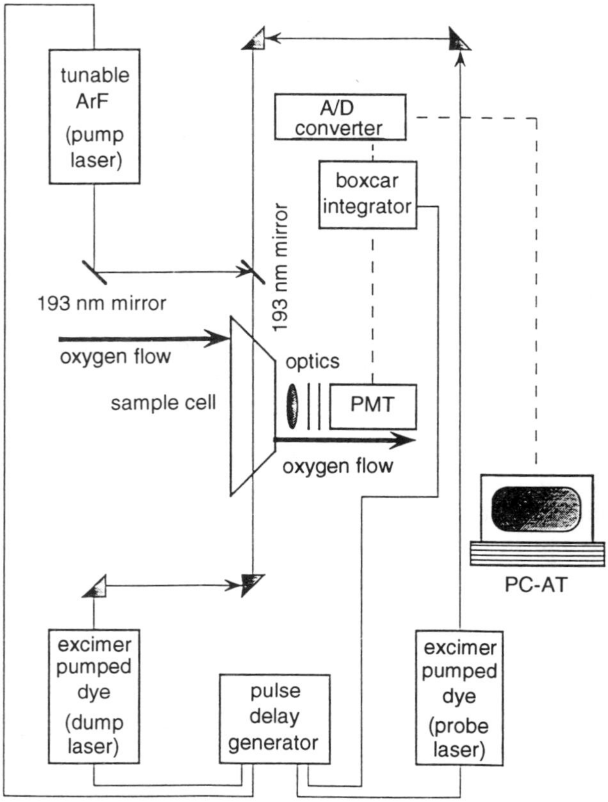

Fig. 3 Schematic of the experimental set-up, for measuing the vibrational relaxation rate constants with stimulated emission pumping the output of a tunable ArF laser and two excimer pumped dye lasers were overlapped in a low-pressure, temperature-controlled flow cell. At 90° to the overlapped laser beams, a PMT assembly monitors fluorescence induced by the probe laser. Collection optics and optical filters were used to minimize interference from scattered laser light. A boxcar integrator was used to monitor the probe fluorescence signal measured by the PMT from the evolving population of vibrationally excited molecules.

molecules ($X\,^3\Sigma_g^-$, $v = 0$, 1, or 2) passing through a temperature and pressure controlled flow cell are excited to a given rovibronic level of the predissociative $B\,^3\Sigma_u^-$ state by a pulsed, tunable argon fluoride (TArF) pump laser. (Lambda Physik, EMG 150 TMSC: *ca.* 15 ns pulse width, *ca.* 1.0 cm^{-1} spectral resolution, *ca.* 60 mJ per pulse at 10 Hz). A 5 mm diameter circular aperture was used to pick out 2–3 mJ of the TArF light for use in these experiments. Once excited, the molecules are stimulated back down to a specific rovibronic level of the gound electronic state by means of an excimer-pumped dye laser (Lambda Physik, LPX205i pumping a Lambda Physik FL3002 dye laser: *ca.* 20 ns pulse width, *ca.* 0.2 cm^{-1} spectral resolution and 3–15 mJ per pulse at 10 Hz) spatially and temporally overlapped with the pump laser beam. Finally, at a given time delay after the pump–dump preparation event a second tunable dye laser probes the population created in the final state by laser-induced fluorescence (LIF). Probe LIF was monitored by a phototube assembly (Hamamatsu R212) equipped with collection optics to enhance signal and optical filters to reduce scattered light from the lasers. The photomultiplier tube (PMT) assembly is set at 90° to the overlapped pump–dump and probe beams (see Fig. 3). The output of the PMT assembly is sent to a boxcar integrator gated on the fluorescence signal resulting from the probe pulse. The output from the integrator is then sent to an A/D converter sampled by an IBM-compatible computer where it is collected and stored for analysis. By varying the time delay between the dump and probe lasers, the time-dependent occupation of the prepared or a collisionally relaxed vibrational state is measured.

Results

Preparation of O_2(vib) by Stimulated Emission Pumping

The first step in the preparation of highly vibrationally excited oxygen molecules is excitation through the well known Schumman–Runge system of O_2 to the intermediate B $^3\Sigma_u^-$ state. The B $^3\Sigma_u^-$ state has a much distended O—O bond length providing large off-diagonal elements in the B–X Franck–Condon overlap matrix. Unlike all previous stimulated emission pumping experiments the intermediate electronic state used in these experiments is strongly predissociative, with state-specific predissocation lifetimes dependent on v', J' and N', varying between *ca.* 1 and 50 ps.[25] Normally, one can experimentally determine the two laser frequencies needed to pump the two-photon stimulated emission pumping transition using one-photon LIF followed by fluorescence dip spectroscopy. The short lifetime of the B state prevents fluorescence dip measurements. Consequently other means had to be found to find the stimulated emission transitions.

Fig. 4 shows the O_2 B–X LIF spectrum coincident within the 1 nm tuning range of the TArF laser at 295 and 465 K. At 295 K, the broad features in the spectrum are due to rovibronic transitions in the B(4)–X(0) vibronic band. The large width is due to the homogeneous broadening associated with the *ca.* 1 ps predissociation lifetime of this vibronic level. At 465 K, the broad features of the B(4)–X(0) transitions are masked by the sharper and stronger rovibronic transitions of the B(10)–X(2) and B(7)–X(1) bands.

Upon excitation, spontaneous emission from the B state back down to a manifold of X-state vibrational levels may occur before predissociation, albeit with low probability. The amount of population created in a particular X-state vibrational level by this means

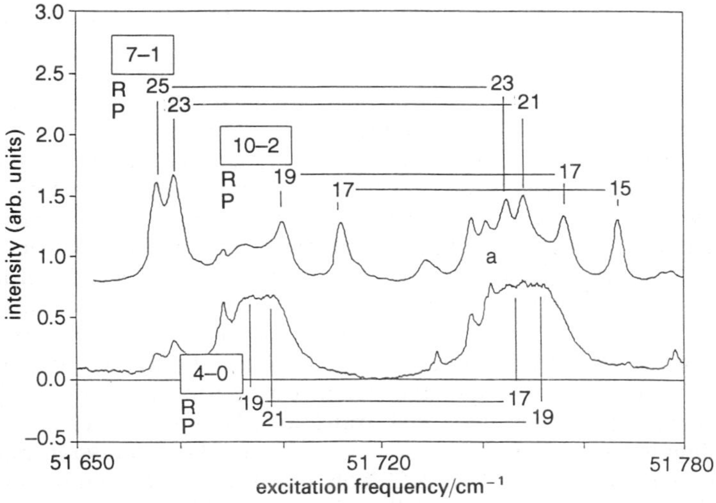

Fig. 4 Schumann–Runge excitation spectra of O_2 at 295 K (lower trace) and 465 K (upper trace), the wavelength of the tunable ArF pump laser is scanned while observing the gated fluorescence from a cell containing 100 Torr of O_2. Hot band transitions are much more prominent in the elevated temperature spectrum and can be readily assigned from previous measurements.

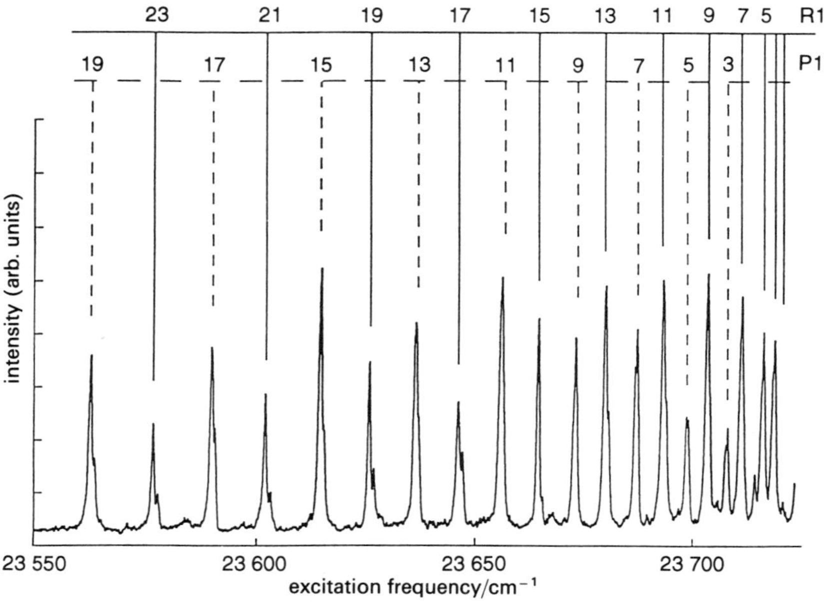

Fig. 5 Franck–Condon pumping of O_2 showing the $0 \leftarrow 19$ vibronic band of the Schumann–Runge system. The population in the $v'' = 19$ state has been induced by the small amount of spontaneous emission from the $B\,^3\Sigma_u^-$ state to levels higher than or equal to the $v'' = 19$ which were prepared in the pump step described above. The presence of this state is detected using the excimer pumped probe dye laser. The spectral features can easily be assigned to P and R branches. The fine structure best observed in the high-*J* transitions is due to the well known spin–spin and spin–rotation interactions of this triplet molecule.

is determined by the Franck–Condon factor of the final level with the pump-laser-prepared state and the rate of the excited-state predissociation and can be substantial. Fig. 5 shows the B(0)–X(19) LIF spectrum obtained using spontaneous emission to populate X(19). A full complement of rovibronic transitions is observed, indicating that rotational equilibration with the room-temperature background gas is already complete. Vibrational equilibration is obviously still underway. Equilibration of the rotational degrees of freedom, which is much more rapid than vibrational equilibration, is estimated to be complete after only *ca.* 5 ns under our conditions.

If the probe laser is now tuned to the absorption maximum of one of the features in such a spectrum as that of Fig. 5, we may look for enhancement of this two-laser signal as a function of the frequency of the dump laser. Fig. 6 shows the spectrum obtained from this experiment. As the wavelength of the dump laser is scanned over the allowed electronic transitions from the selected rovibronic level of the B state, a stong enhancement of the detected population in the final vibrational level is observed. Typically, a 100-fold enhancement in the population of a single highly vibrationally excited level of O_2 can be obtained using this technique. The appearance of only two spectral features in this experiment is clear evidence that the signal is a multiple resonance, corresponding to the P and R lines of the parallel $\Sigma \leftarrow \Sigma$ transition. The frequencies of the transitions allows unambiguous spectral identification of each stimulated emission pumping transition using known molecular constants.[26] Note that there is no rotational equilibration within the B state during the course of the experiment. This a result of the short predissociation lifetime of the B state. Despite the fact that the experiment is carried out with 20 ns long laser pulses, the average time that a molecule spends in the B state is only a few ps, much shorter than the average time required for a collision to take place.

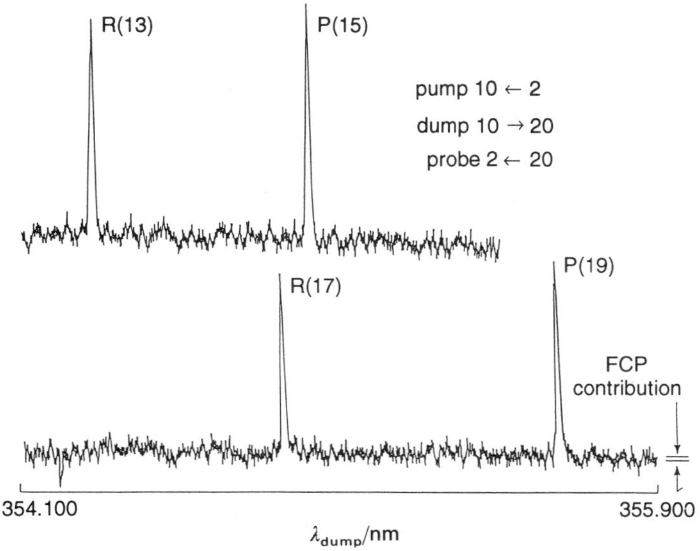

Fig. 6 Stimulated emission pumping of O_2 showing that the probe signal of Fig. 4 is dramatically enhanced at specific dump laser frequencies. The two transitions are P- and R-branch emission features allowed by optical selection rules from the pump-laser-prepared quantum state. The dependence of the stimulated emission pumping transition frequencies on the choice of the pump transition is shown in the two traces. The contribution of Franck–Condon pumping to the total emission signal is indicated.

Time-dependence Measurements of Collisional Relaxation

By varying the delay time between the probe pulse and the pump–dump preparation event, the decay of the population of the initially prepared state can be obtained. Fig. 7 shows an exponential decay for molecules in $v = 20$ obtained at 465 K and 170 Torr by monitoring the magnitude of the stimulated emission pumping enhanced population in the prepared state as a function of delay time. Fitting these decays yields the pressure-dependent lifetime of the prepared vibrational state from which the rate constant for vibrational relaxation may be derived. Fig. 8 shows a Stern–Volmer-like plot of the data obtained from $v = 20$ at 465 K. The slope of this line yields the total collisional deactivation rate constant for $v = 20$ which includes contributions from V–V and V–T relaxation. The assumption that only self-relaxation with ground-state O_2 molecules is being probed by this technique is supported by the fact that the plot is quite linear, and that the y intercept of the plot goes through zero to the limit of the experimental uncertainties. If impurities were present that contributed to the relaxation process, this would not be the case.

For $v = 19$–26 at both 295 and 465 K, the vibrational relaxation rate constants are observed to increase with vibrational quantum number, v, and are tabulated in Tables 1 and 2 along with results for $v = 27, 28$. Fig. 9 and 10 show the rate constants graphically along with a comparison to the theory of Billing and Kolesnick discussed below.

For every state below $v = 27$ the relaxation can be observed directly with the scheme described above. For $v \geqslant 27$, however, the rate of relaxation at all pressures is too fast to be observed directly. LIF detection of $v = 27$ is also less sensitive than the other experiments presented here. Consequently, a different method was chosen for the measurement of the $v = 27, 28$ rate constants.

Molecules that have been prepared in highly excited vibrational levels will relax through a cascading mechanism of single or multi-quantum steps and eventually reach

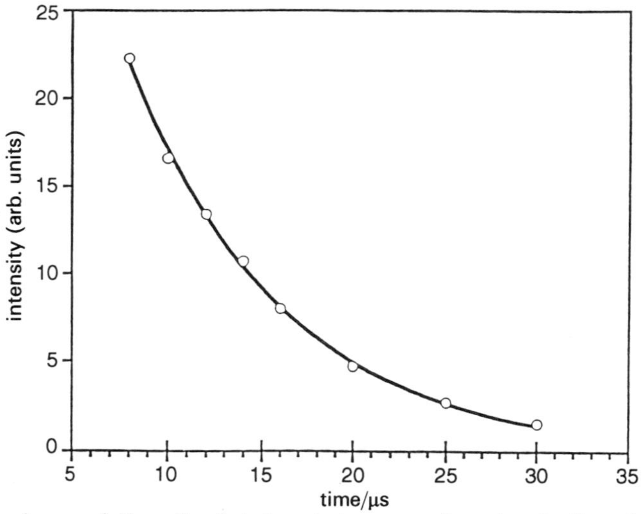

Fig. 7 Time dependence of the stimulated emission pumping signal: the stimulated emission pumping enhanced population of $v'' = 20$ is plotted *vs.* delay time between the dump and probe lasers. The data are shown as circles. The curve is the best fit to the exponential decay from which a pressure-dependent lifetime can be derived. This experiment was carried out at 465 K and 170 Torr.

$v = 0$ through a combination of V–T and V–V processes. If the probe laser is tuned to observe the population in a level just below the stimulated emission pumping prepared state, the time dependence of the stimulated emission pumping enhanced portion of the signal can be used to extract the vibrational relaxation rate constant of the higher state provided the rate constant for the lower level is already known.

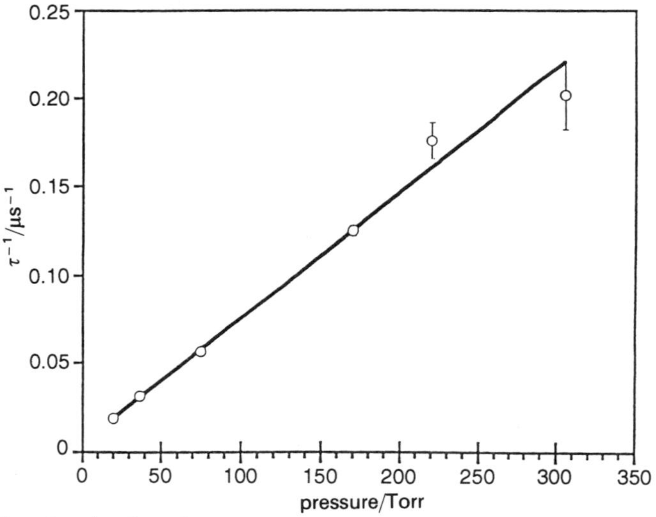

Fig. 8 Derived vibrational relaxation rate constant: a Stern–Volmer analysis for $v'' = 20$. The experimental temperature is 465 K. The inverse of the derived pressure-dependent collisional lifetime for $v'' = 20$ (prepared by stimulated emission pumping) is plotted *vs.* pressure. The slope of the line of best fit gives the relaxation rate constant.

Table 1 Relaxation rate constants ($cm^3\ s^{-1}$) for individual vibrational levels of O_2 ($X\ ^3\Sigma_g^-$) colliding with O_2 ($v'' = 0$) at 295 K

v''	$k(v'')$ (exp.)	$k(v'')$ (ref. 29)
19	$(4.7 \pm 0.3) \times 10^{-15}$	$(3.6 \pm 1.1) \times 10^{-15}$
20	$(3.2 \pm 0.3) \times 10^{-15}$	$(3.7 \pm 1.1) \times 10^{-15}$
21	$(5.8 \pm 1.2) \times 10^{-15}$	$(4.1 \pm 1.2) \times 10^{-15}$
22	$(5.4 \pm 0.8) \times 10^{-15}$	$(4.8 \pm 1.4) \times 10^{-15}$
23	$(1.2 \pm 0.4) \times 10^{-14}$	$(5.7 \pm 1.7) \times 10^{-15}$
24	$(0.84 \pm 0.04) \times 10^{-14}$	$(7.1 \pm 2.1) \times 10^{-15}$
25	$(1.8 \pm 0.05) \times 10^{-14}$	$(9.1 \pm 2.7) \times 10^{-15}$
26	$(4.7 \pm 0.2) \times 10^{-14}$	$(13 \pm 3.9) \times 10^{-15}$
27	$(3.0 \pm 1) \times 10^{-13}$	$(17 \pm 5.1) \times 10^{-15}$

Table 2 Relaxation rate constants ($cm^3\ s^{-1}$) for individual vibrational levels of O_2 ($X\ ^3\Sigma_g^-$) colliding with O_2 ($v'' = 0$) at 465 K

v''	$k(v'')$ (exp.)	$k(v'')$ (ref. 29)
19	$(2.3 \pm 0.1) \times 10^{-14}$	$(1.8 \pm 0.5) \times 10^{-14}$
20	$(3.1 \pm 0.08) \times 10^{-14}$	$(3.1 \pm 0.9) \times 10^{-14}$
21	$(2.2 \pm 0.9) \times 10^{-14}$	$(3.5 \pm 1.1) \times 10^{-14}$
22	$(3.7 \pm 0.3) \times 10^{-14}$	$(4.3 \pm 1.3) \times 10^{-14}$
23	$(4.1 \pm 0.6) \times 10^{-14}$	$(5.1 \pm 1.5) \times 10^{-14}$
24	$(6.9 \pm 0.5) \times 10^{-14}$	$(7.4 \pm 2.2) \times 10^{-14}$
25	$(11.7 \pm 0.2) \times 10^{-14}$	$(8.4 \pm 2.4) \times 10^{-14}$
26	$(16.4 \pm 2) \times 10^{-14}$	$(13.4 \pm 4.0) \times 10^{-14}$
27	$>83 \times 10^{-14}$	$(17.4 \pm 5.2) \times 10^{-14}$
28	$>1.2 \times 10^{-11}$	$(22.0 \pm 7.0) \times 10^{-14}$

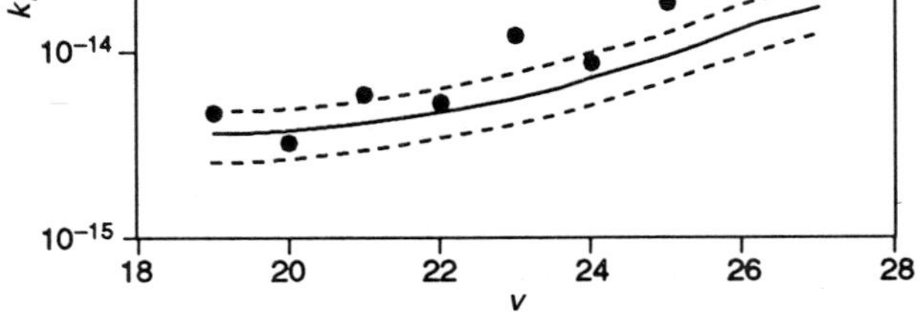

Fig. 9 Vibrational state specific relaxation rate constants for O_2(vib) relaxing by collisions with O_2 (295 K). ●, this study and (——) extrapolations from the theoretical calculation of Billing[38]. Estimated errors in the calculation are shown by dashed lines. The deviation from theory increases as one approaches energies near the threshold to bimolecular chemical reaction.

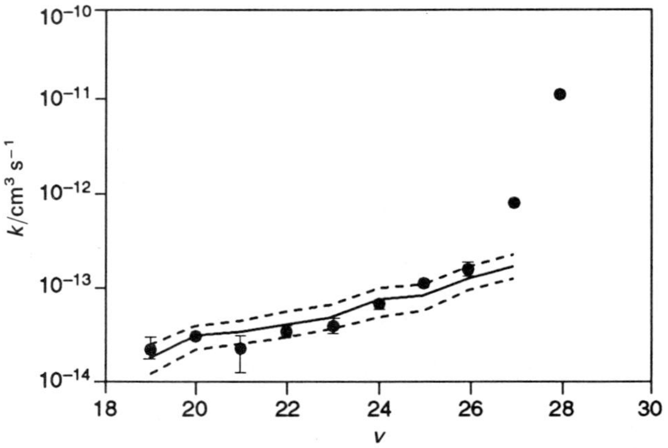

Fig. 10 Vibrational state specific relaxation rate constants for O_2(vib) relaxing by collisions with O_2 (465 K). Symbols: this study (●, experimental and ■, modelled results) and (——) extrapolations from the theoretical calculation of Billing[38]. Estimated errors in the calculation are shown by dashed lines. The deviation from theory increases as one approaches energies near the threshold to bimolecular chemical reaction.

To understand this quantitatively, we must consider a simple kinetic model for the relaxation. The first step is to solve the set of coupled differential equations that govern the cascading relaxation, the solutions of which yield the time dependence of the vibrational states' populations: $N(v'', t)$, $N(v'' - 1, t)$, $N(v'' - 2, t)$ after initial preparation in $|v''>$, with population, $N(v'', 0)$. It is assumed that the amount of v'' population that gains a vibrational quantum by collisional excitation is negligibly small. The resulting equations describe an unconstrained collisional relaxation model that allows collisional deactivation to proceed by single or double quantum steps. A schematic diagram for these processes is given in Fig. 11. The master equation for this type of system has been

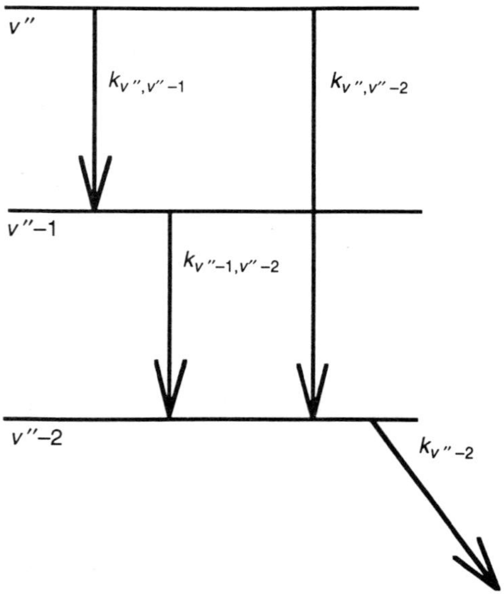

Fig. 11 Schematic to show the meaning of the individual rate constants used in the modelling of the experimentally observed populations using eqn. (5)–(10)

solved[27] and the results are presented here:

$$N(v'', t) = N(v'', 0)\exp(-k_{v''} pt) \tag{4}$$

$$N(v'' - 1, t) = N(v'', 0)\left(\frac{k_{v'', v''-1}}{k_{v''} - k_{v''-1}}\right)[\exp(-k_{v''-1}pt) - \exp(-k_{v''}pt)] \tag{5}$$

$$N(v'' - 2, t) = N(v'', 0)[\alpha \exp(-k_{v''-2} pt) - \beta \exp(-k_{v''-1}pt) + \gamma \exp(-k_{v''} pt)] \tag{6}$$

where

$$\alpha = \frac{k_{v'', v''-1}k_{v''-1, v''-2}}{(k_{v''} - k_{v''-2})(k_{v''-1} - k_{v''-2})} + \frac{k_{v'', v''-2}}{(k_{v''} - k_{v''-2})} \tag{7}$$

$$\beta = \frac{k_{v'', v''-1}k_{v''-1, v''-2}}{(k_{v''} - k_{v''-1})(k_{v''-1} - k_{v''-2})} \tag{8}$$

$$\gamma = \frac{k_{v'', v''-1}k_{v''-1, v''-2}}{(k_{v''} - k_{v''-1})(k_{v''} - k_{v''-2})} - \frac{k_{v'', v''-2}}{(k_{v''} - k_{v''-2})} \tag{9}$$

In this notation, the total vibrational deactivation rate constants are given as $k_{v''-n}$, where $n = 0, 1, 2$, and the state-to-state vibrational deactivation rate constants are given as $k_{v''-n, v''-q}$, where $n = 0, 1, 2$ and $q = 1$ for a single quantum rate constant and $q = 2$ for a two-quantum rate constant.

The previous equations can easily be adapted to describe a single quantum cascading mechanism by setting the state-to-state $k_{v''-n, v''-n-1}$ ($n = 0, 1$) rate constants equivalent to $k_{v''-n}$, and by setting all $k_{v'', v''-2}$ rate constants to zero. The total vibrational depopulation rate constants, $k_{v''-n}$, for $n = 0, 1, 2\ldots$, were previously measured and held fixed at their measured values. The state-to-state vibrational rate constants were obtained by simultaneously fitting the observed populations to the above sets of equations. The results of the fitting procedure are described for each set of levels measured.

In order to fit the observed time-dependent signals to a kinetic model, the measured signal intensities must first be converted to relative vibrational populations. This takes into account the different transition strengths for probing different vibrational states of O_2. The Schumann–Runge transition line strength is given by the following expression:[28]

$$S_{i, v', J'; j, v'', J''} \equiv \sum_{M', M''} |\langle i, v', J', M' | \mu | j, v'', J'', M'' \rangle|^2 \tag{10}$$

which can be easily factored into the more recognizable expression:

$$S_{i, v', J'; j, v'', J''} = q_{v,', v''} |\mu_e^{i, j}|^2 S_{J', J''} \tag{11}$$

In this expression $q_{v', v''}$ is the Franck–Condon factor, $|\langle v' | v'' \rangle|^2$, and $S_{J', J''}$ is the Hönl–London rotational line strength factor. This expression was derived assuming isotropic excitation and is applicable to our experimental arrangement.

Each set of vibrational populations was obtained by probing through the same upper rovibronic state and at the same laser pulse energy. This was done in order to decouple the complicated upper-state predissociation dynamics from the observed relative populations. Additionally, by choosing the same upper rovibronic state for the probe step, which cancels the Hönl–London rotational line-stength factors, and the assumption that the change in the transition dipole between probing vibrational state v'' and $v'' - 1, 2$ is negligible and can be ignored, simplifies the relative normalization of the measured vibrational populations to each other as only depending on the different Franck–Condon factors for the different probe transitions used.

We may now return to the derivation of the total removal rate constants for vibrational levels above $v = 26$. Fig. 12 shows the experimental results for preparation of

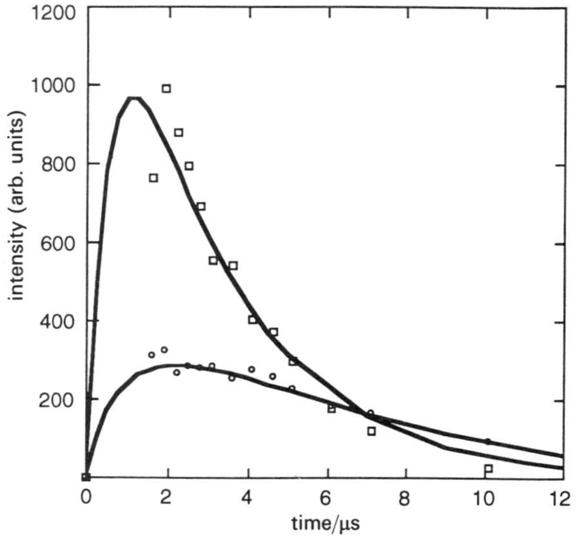

Fig. 12 Time dependence of ○, $v'' = 25$ and □, $v'' = 26$ after preparation of $v'' = 27$ at 465 K: The solid lines are the result of a model calculation that assumes multiple quantum relaxation. In this case, the only adjustable parameters are the rate constant for $v'' = 27$ and the state-to-state rate constants. The rate constants for $v'' = 26$ and $v'' = 25$ are held at their measured values. The analysis of this data is described in more detail in the text.

$v'' = 27$, followed by observation of $v = 25, 26$ ($T = 465$ K, $P = 93$ Torr) along with the modelled time dependence of $v = 25, 26$. The direct measurement of $v'' = 27$ at $T = 465$ K was not feasible because the collisional depopulation of this level was too fast to be observed experimentally. It was readily apparent by examining the observed vibrational populations for $v'' = 26, 25$ that a single quantum vibrational energy transfer model does not adequately describe the data. A fit of the observed vibrational populations was undertaken using the multiquantum vibrational transfer model in which the vibrational depopulation constants k_{26} and k_{25} were held fixed at their previously measured values. The parameters that were allowed to vary were k_{27}, $k_{27,26}$, $k_{27,25}$ and $k_{26,25}$.

The value of $N(27, 0)$ has to be fixed in order for the state-to-state collisional rate constants, $k_{27,26}$ and $k_{27,25}$, to be independently determined by the fitting procedure. If this is not done, these rate constants are all highly correlated and lose their physical meaning. The value of $N(27, 0)$ was estimated by comparison to measurements of the exponential decay of $v'' = 26$ at the same pressure and laser pulse energies, using the same spectroscopic transitions to determine the initial population prepared in $v'' = 26$, $N(26, 0)$. The value for $N(27, 0)$, is then calculated from $N(26, 0)$ by taking into account the different Franck–Condon factors of the differing transitions used in the dump step for preparation of the two states.

Identical experiments were also carried out at 295 K. As before, the direct vibrational depopulation of $v'' = 27$ was too fast to be observed and the vibrational population that appears in $v'' = 26, 25$ after preparation of $v'' = 27$, under normal conditions, was too small to be observed. However, operating at the largest possible dump and probe laser pulse energies and having the PMT sensitivity as high as possible, initial population prepared in $v'' = 27$ could be observed appearing in $v'' = 26$. The signal level in comparison to the experimental noise, though, was poor allowing only the determination of k_{27} and $k_{27,26}$. The experimental data and fit are given in Fig. 13.

An additional set of state-to-state experiments was carried out at 295 K, where $v'' = 26$ was initially prepared. The experimental results and fit are shown in Fig. 14.

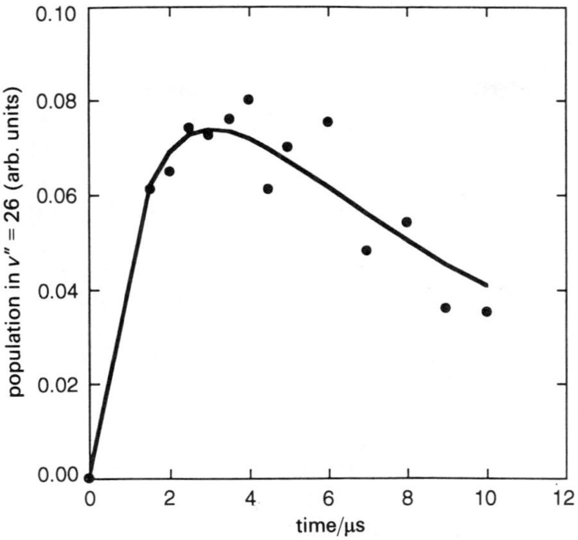

Fig. 13 Normalized observed time dependence of the initially prepared vibrational state, $v'' = 27$, and the subsequent appearance and decay of $v'' = 26$ at 295 K. The solid line is the result of modelling the observed population to eqn. (5)–(10). Fixing the initial population in $v'' = 27$, $N(v'' = 27, t = 0)$, allows kinetic rate constants to be extracted. These constants are reported in Table 3.

Similar experiments were carried out in an effort to determine the relaxation rate constant for $v = 28$. It should be possible to produce significant quantities of $v = 28$ in comparison to $v = 27$ since the Franck–Condon factors for preparation are a factor of two more favourable. Despite this, essentially no population is observed to 'trickle

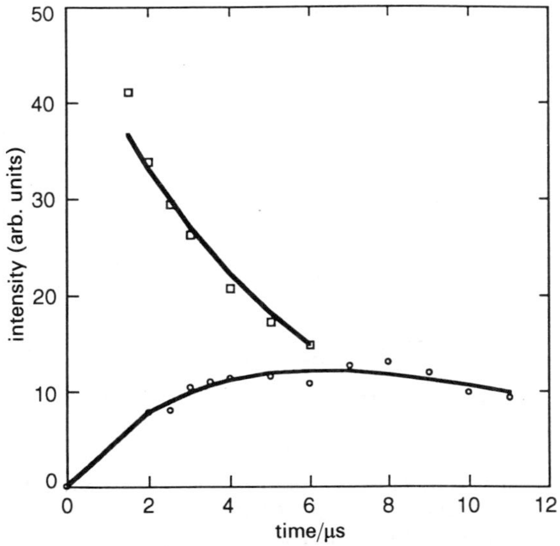

Fig. 14 Normalized observed time dependence of the initially prepared vibrational state, $v'' = 26$, (□) and the subsequent appearance and decay of $v'' = 25$ (○) at a temperature of 295 K. The solid lines are the result of modelling the observed populations to eqn. (5)–(10). The constants obtained from the modelling are reported in Table 3. It is readily apparent that not all of the population prepared in $v'' = 26$ relaxes into $v'' = 25$.

down' into lower vibrational levels when we prepare $v = 28$ with stimulated emission pumping. Fig. 15 shows a comparison between the two experiments. The dump laser is tuned over the resonances for preparation of $v = 28$ and 27. The transfer of population to the prepared state is detected by probing the longer lived $v = 26$ vibrational state with the probe laser set at a 2–3 μs delay. While it is clear that some fraction of the stimulated emission pumping prepared $v = 27$ population 'trickles down' to $v = 26$, there is no evidence that $v = 28$ 'trickles down' at all. This gives powerful evidence of a dark channel. By comparing the intensities of the two signals in Fig. 15 we can estimate a lower limit to the $k_{dark}(28)$. After correcting for the relative preparation efficiencies of $v = 27$ and 28 and modelling the vibrational relaxation as was done in the analysis above, we find that in order to reduce the absolute signal from $v = 28$ enough so that it will be less than the observed noise levels requires a $k_{dark} > 1.2 \times 10^{-11}$ cm^3 s^{-1} or *ca.* 2% of the gas-kinetic cross-section at these temperatures. The actual rate could be higher.

Table 3 summarizes the results of the state-to-state measurements for $v = 27$ and 26. The fate of $v'' = 27$ can be seen for both experimental temperatures. At 465 K the molecules that relax do so predominantly through $\Delta v = -1$ transitions, although $\Delta v = -2$ is clearly present as well, occurring with *ca.* 20% of the probability of $\Delta v = -1$. Attempts to observe $\Delta v = -3$ relaxation were unsuccessful. It is reasonable to assume that $\Delta v < -2$ collisions are insignificant, although we have not been able to show this explicitly. $\Delta v < -3$ is, in principle, possible although considered very unlikely to be more important than $\Delta v < -2$. Note that only 41% of the population initially prepared in $v'' = 27$ undergoes vibrational relaxation through $v'' = 26$ and 25. The majority has disappeared by another mechanism. Note that $k_{dark, v''}$ for this 'dark channel' are also presented in Table 3.

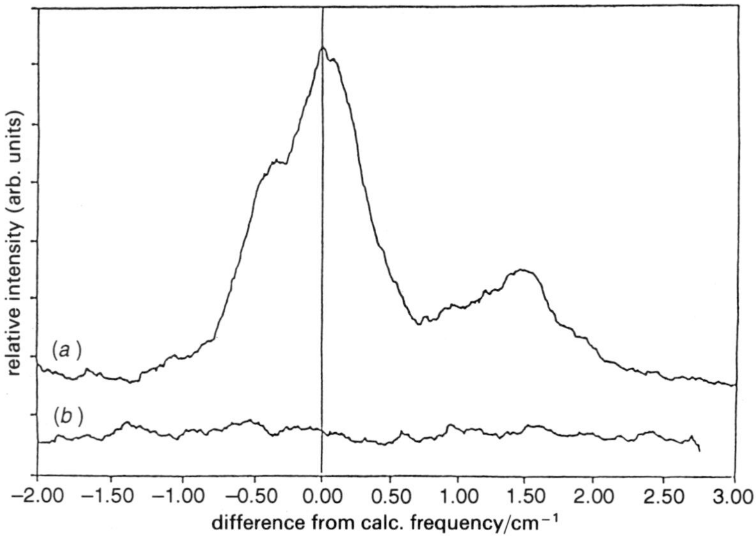

Fig. 15 'Trickle down' spectroscopy of $v'' = 27$ and $v'' = 28$ at 460 K. The probe laser is tuned to detect the presence of $v'' = 26$, which was observed through the 2–26 band 2 μs after stimulated emission pumping preparation. The dump laser is scanned over the transition producing $v'' = 27$ (*a*), which is easily detected after $v'' = 27$ 'trickles down' to $v'' = 26$. The identical experiment for production of $v'' = 28$ (*b*) yields almost no 'trickle-down' signal in $v'' = 26$. The $v'' = 27$ data is at least a factor of 10 larger than that of $v'' = 28$, providing strong evidence that the $v'' = 28$ population is being removed through a different channel. Reaction of the vibrationally excited O_2(vib) with ground-state O_2 to form O and O_3 is the proposed mechanism.

Table 3 Results of modelling $v'' = 27, 26, 25$ vibrational populations

constants[a]	465 K	295 K
k_{27}	$6.3\ (5)^b \times 10^{-13}$	$3\ (0.3) \times 10^{-13}$
$k_{27,26}$	$2.2\ (5) \times 10^{-13}$	$4\ (3.1) \times 10^{-15}$
$k_{27,25}$	$4\ (3.1) \times 10^{-14}$	NA[c]
$k_{\text{dark},27}$	3.7×10^{-13}	3×10^{-13}
k_{26}	$16\ (2) \times 10^{-14}$	$4.7\ (2) \times 10^{-14}$
$k_{26,25}$	$3\ (3.1) \times 10^{-14}$	$1.3\ (4) \times 10^{-14}$
$k_{\text{dark},26}$	1.3×10^{-13}	3.4×10^{-14}

[a] Units: $cm^3\ s^{-1}$. [b] Denotes 1σ statistical error in the last digit. [c] This number was not experimentally available.

The competition between two qualitatively different relaxation mechanisms is also evident in the temperature dependence. At 295 K one finds that the 'dark' channel is taken by 99% of the initially prepared $v'' = 27$. The dark channel is obviously less temperature dependent than is the vibrational relaxation mechanism. The rate constant for the dark channel, $k_{\text{dark},v''}$, at two different temperatures can be used to derive an activation energy for the dark channel and result in $E_{\text{a, dark, 27}} = 0.3 \pm 1$ kcal mol^{-1} and $E_{\text{a, dark, 26}} = 2 \pm 1$ kcal mol^{-1}. The significance of this will be discussed below.

Relaxation of O_2 $v'' = 19$, $T = 295$ K

Because of the very surprising result that a dark channel exists for $v = 26$ and 27, we decided to carry out similar state-to-state measurements for a set of lower vibrational states which are expected to decay only by vibrational relaxation. Theoretical predictions of the self-relaxation of highly vibrationally excited O_2 suggest that $v'' = 19$ relaxes by single quantum V–T energy transfer.[29] Our experiment initially prepared $v'' = 19$ of the ground state of molecular oxygen. The population in $v'' = 19$ and $v'' = 18$ was then monitored as a function of probe–pump/dump delay time. The observed populations were normalized to each other as described earlier and then fit to the single quantum vibrational energy transfer model, *i.e.* the above mentioned multiquantum model was constrained to permit only single-quantum relaxation. The observed populations were described quite well by the single-quantum vibrational energy transfer model (Fig. 16). The only parameters allowed to vary in this fit are the initially prepared population in $v'' = 19$, $N(19, 0)$, and the vibrational depopulation rate constant, k_{18}, which had not been determined previously. The validity of the single-quantum model was then tested when the observed $v'' = 19, 18$ vibrational populations were fitted to the multiquantum vibrational energy transfer model. The state-to-state vibrational rate constant, $k_{19,18}$, returned by the multiquantum fitting procedure, was equal to the vibrational depopulation rate constant, k_{19}, and the calculated vibrational populations were indistinguishable from the single-quantum model calculations. The results of the fitting procedure are given in Table 4. In summary, O_2 ($v = 19$) relaxes dominantly by a $\Delta v = -1$ mechanism, consistent with the theory of Billing and Kolesnik.[29] These results offer a kind of control experiment which gives us further confidence in the state-to-state experiments presented above for $v'' = 27$ and 26.

O_2 ($v = 19$–26) N_2 Collisional Relaxation

By varying the delay time between the probe and pump/dump pulses, the decay of the initially prepared level can be followed for the O_2–N_2 collision system. Extraction of the

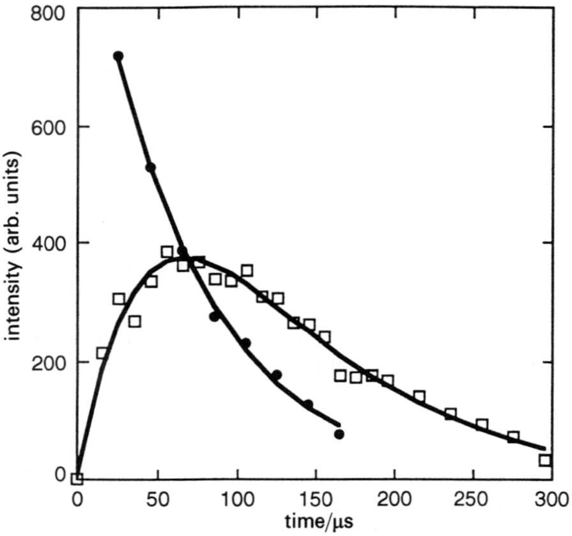

Fig. 16 Observed time dependence of the initially prepared vibrational population in $v'' = 19$ (●) and the appearance and decay of the vibrational population in $v'' = 18$ (□). The lines are the result of modelling the normalized populations to the kinetic model presented as eqn. (5)–(10). The results of the modelling indicate that all of the initial population prepared in $v'' = 19$ undergoes single quantum transfer to $v'' = 18$.

relevant rate constants, k_{N_2} and k_{O_2}, is not as straightforward as the O_2 self-relaxation system. The model used to analyse the O_2–N_2 decays is described as following.

Pseudo-first-order kinetics are assumed and the following rate equation is solved

$$\frac{d[O_2(vib)]}{dt} = -k_{O_2}[O_2] - k_{N_2}[N_2] \tag{12}$$

and we arrive at the result that the initial population of vibrationally excited O_2 decays exponentially with the lifetime, τ, defined as:

$$\frac{1}{\tau p_t} = k_{O_2} + (k_{N_2} - k_{O_2})\alpha \tag{13}$$

As can be seen from the expression for the collisional lifetime, a plot of $(\tau p_t)^{-1}$ *vs.* α, the mole fraction of N_2, will yield a straight line with $(k_{N_2} - k_{O_2})$ as the slope and k_{O_2} given as the intercept. The k_{O_2} obtained from the intercept is used as verification of the fitting procedure by comparison with the previous direct measurements of k_{O_2} and typically

Table 4 Results of modelling $v'' = 19$, 18 vibrational population as a function of time (T = 295 K)

constant[a]	$k_{v''}$ [b]	ref. 15[c]
k_{19}	$4.7\ (3) \times 10^{-15}$	1.6×10^{-15}
$k_{19,18}$	$4.7\ (5) \times 10^{-15}$	
k_{18}	$4.7\ (3.1) \times 10^{-15}$	2.5×10^{-15}

[a] Units: $cm^3\ s^{-1}$. [b] Denotes 1σ statistical error in the last digit. [c] Results obtained from ref. 15.

Table 5 Relaxation rate constants ($cm^3\ s^{-1}$) for individual vibrational levels of $O_2(X\ ^3\Sigma_g^-)$ colliding with N_2 ($v'' = 0$) at 295 K

v''	$k(v'')$ (exp.)	$k(v'')$ (ref. 30)
19	$(2.8 \pm 0.3) \times 10^{-14}$	$(8.8 \pm 1.8) \times 10^{-15}$
20	$(2.3 \pm 0.2) \times 10^{-14}$	$(9.2 \pm 1.8) \times 10^{-15}$
21	$(1.9 \pm 0.5) \times 10^{-14}$	$(9.8 \pm 2.0) \times 10^{-15}$
22	not measured	$(12 \pm 2.0) \times 10^{-15}$
23	$(0.81 \pm 0.5) \times 10^{-14}$	$(14 \pm 2.8) \times 10^{-15}$
24	$(0.67 \pm 0.1) \times 10^{-14}$	$(16 \pm 3.2) \times 10^{-15}$
25	$(0.68 \pm 0.3) \times 10^{-14}$	$(18 \pm 3.6) \times 10^{-15}$
26	$(0.31 \pm 0.05) \times 10^{-14}$	$(20 \pm 4.0) \times 10^{-15}$

agreed to within 10% of the values obtained directly. Fig. 17 is a plot of $(\tau p_t)^{-1}$ *vs.* α for $v = 19$, 21 and 24–26 of ground-state O_2.

The rate constants for collisional deactivation of O_2(vib) at $T = 295$ K decrease as the O_2 vibrational quantum number is increased. The results are shown in Fig. 18 and tabulated in Table 5 along with the results of the semiclassical trajectory calculations of Billing.[30] One can see that N_2, in comparison to O_2, is efficient in relaxing O_2 ($v \leqslant 24$), but is relatively inefficient for collisional relaxation of O_2 ($v > 24$). The implications of these results and the comparison with theory is described below.

Discussion

Comparison of the Results with Theoretical Predictions

The present data permit a direct comparison between experiment and theory. One of the most widely used approximate models is that of Rapp and Englander-Golden.[31] Evaluation of this model for the case of O_2(vib) is of particular importance since it was previously used to estimate the vibrational relaxation rates in the Slanger mechanism discussed above.[20,21] While originally designed only to calculate the probability of near-resonant energy transfer between excited- and ground-state molecules in the zeroth and first vibrational levels ($0 \rightarrow 1$, $1 \rightarrow 0$), this theory has been applied to collisions between ground and highly vibrationally excited species. How well does it perform? According to this theory, the probability of near-resonant V–V energy transfer for substances like N_2, CO and O_2 is approximated by the expression

$$P_{v,v-1} = (v + 1)(3.7 \times 10^{-6})T\ \mathrm{sech}^2(0.174\gamma/T^{1/2}) \tag{14}$$

where T is the temperature in K, and γ is the energy mismatch in cm^{-1} between the energy levels of the colliding partners. This probability expression is a very strong function of γ, rapidly dropping off as γ becomes large. The rate constant for collisional deactivation can be found by multiplying $P_{v,v-1}$ by the hard-sphere collision cross-section and the mean velocity of the molecules.

Perhaps not too surprisingly, since the theory is not expected to describe V–V relaxation when the energy gap, ΔE, exceeds *ca.* 350 cm^{-1},[32] the results of this approximate theory agree rather poorly with the measured values for the relaxation rate constants, underestimating the observed rates by one to two orders of magnitude for O_2(vib) relaxation with O_2.[20] In order to compensate partially for the inadequacies of this model, Toumi made use of a modified form of the Rapp, Englander-Golden model including expressions for two-quantum relaxation, which is shown in this work to play a significant role, as well as V–T energy transfer.[21] Despite these improvements, the model under predicts the observed relaxation rates by more than a factor of ten.

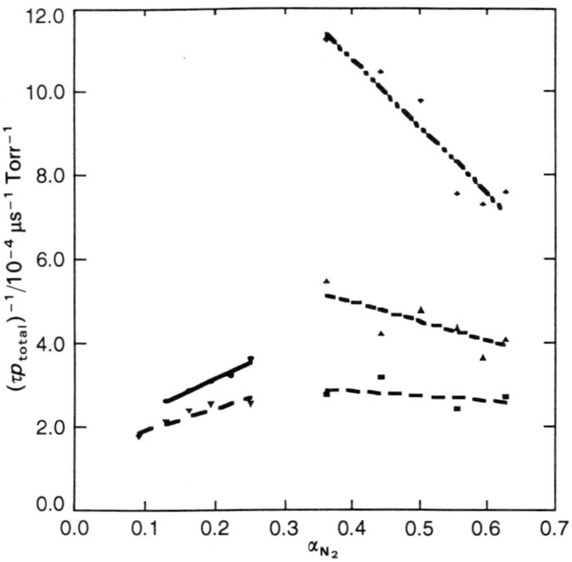

Fig. 17 $(\tau p_t)^{-1}$ *vs.* mole fraction of N_2 for v'' = ●, 19; ▼, 21; ■, 24; ▲, 25; and ◆, 26 of ground-state O_2 showing the experimental results for O_2 collisional relaxation by N_2. Note the change in sign of the slope as the O_2 vibrational quantum number is increased. Using eqn. (15), this shows quite clearly that for $v'' \geqslant 25$, O_2 vibrational relaxation occurs mainly due to self-relaxation.

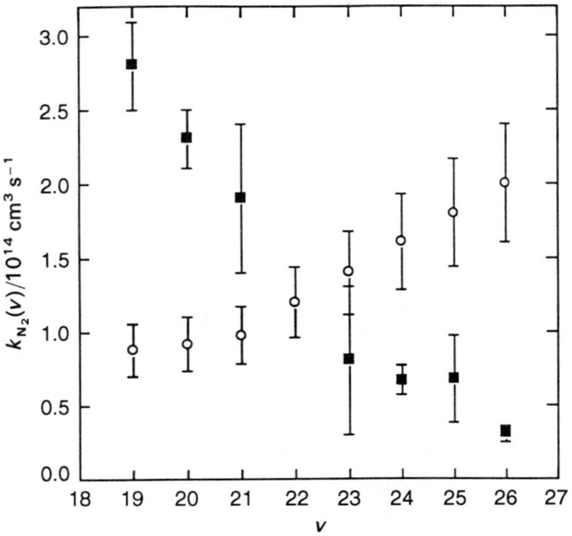

Fig. 18 Vibrational state specific relaxation rate constants for O_2(vib) relaxing by collisions with N_2 (295 K) showing the experimentally measured collisional deactivation rate constants (■) by nitrogen (N_2) and the theoretical calculations of Billing (ref. 30, ○). The error bars on the calculated values of Billing are given as 20%. It is readily clear that the calculated values do not correctly model the vibrational quantum number dependence for N_2 collisional relaxation of vibrationally excited O_2.

A more recent treatment used semiclassical trajectory calculations to probe the temperature dependence of the O_2 vibrational relaxation rates for both V–V and V–T processes.[29] These calculations make use of a new *ab initio* potential-energy surface[33] and take into account long-range dispersion forces and quadrupole–quadrupole interaction terms. This approach allows for single quantum V–V and both single and two-quantum V–T relaxation. Using similar methods, relaxation rate constants for CO(vib)–CO and CO(vib)–N_2 compared favourably with experiment.[34] For O_2(vib)–O_2 the comparison between experiment and theory is shown in Fig. 9 and 10. Fig. 9 shows the experiment (295 K) and the theory (300 K). Fig. 10 shows the experiment (465 K) and the theory (500 K). The imperfect matching of the temperatures in the comparison is not considered significant. The calculations were only carried out up to $v = 20$, while our experimental results were obtained up to $v = 27$. The comparison is therefore only strictly possible for $v = 19$ and 20, where one can see that agreement is excellent. It is interesting to note that the theory can easily be extrapolated allowing comparison with our experimental values. The extrapolation of Billing's calculation is shown as a solid line with his error estimate of $\pm 30\%$ shown as a pair of dashed lines above and below the extrapolated values. The calculated results are in excellent agreement with our measured rate constant for vibrational levels below $v = 26$. This agreement is encouraging and suggests that the theoretical treatment is valid up to as high as $v \approx 24$ for O_2 self-relaxation.

We note in passing that the same methods have been applied very recently in an attempt to gain theoretical understanding of the O_2(vib)–N_2 collisional system and this comparison is shown in Fig. 16. Although the correct magnitude results, the vibrational quantum number dependence is reversed from the experiment. It appears that the O_2–N_2 potential-energy surface used in this calculation was not sufficiently accurate to reproduce experiment.[30]

Evidence for Chemical Reaction of O_2(vib)

The most likely explanation for the observed threshold in the vibrational self-relaxation data presented in Fig. 9 and 10 is the onset of reaction (1). Let us consider the reasons for this conclusion.

First, the experimental results presented above give clear evidence for the presence of a 'dark' channel. The threshold is clearly the onset of a new chemical process and not the enhancement of V–V and V–T energy transfer. The data allow us to calculate the fraction of molecules that do not trickle down the vibrational ladder, $k_{dark}(v)/k_v$ (assuming $\Delta v < -2$ collisions are insignificant) and we observe that this fraction is temperature dependent. This can only be the case if two different processes with different temperature dependences, *i.e.* activation energies, are competing with one another.

Secondly, the activation energies for the dark process can be determined as a function of the vibrational quantum number. These were presented above and are $E_{a,\,dark,\,27} = 0.3 \pm 1$ kcal mol^{-1} and $E_{a,\,dark,\,26} = 2 \pm 1$ kcal mol^{-1}. These values are consistent with reaction (1), since one would certainly expect that as vibrational excitation brings one closer to the top of the activation barrier of a vibrationally promoted chemical reaction, the activation energy will become smaller and smaller, as observed. It is interesting to compare these experimentally derived activation energies to known thermodynamical quantities of reaction (1). One can easily calculate that O_2 ($v = 27$, $N = 0$) and O_2 ($v = 26$, $N = 0$) are 0.9 and 3.5 kcal mol^{-1} below the top of the activation barrier of the reaction, respectively. This compares favourably with the derived activation energies of the dark channel.

Thirdly, the first vibrational state that is above the chemical reaction's activation barrier is not observed to trickle down the vibrational ladder. As mentioned above, despite the fact that one should be able to prepare $v = 28$ more efficiently than one can prepare $v = 27$, we observe no $v = 28$ trickling to $v = 26$ while we do see $v = 27$ trickling to $v = 26$.

Fourthly, we see that the predictions of the theory of Billings and Kolesnick are remarkably successful up to the threshold of the reaction (1). Since this theory does not employ a potential-energy surface that allows for reaction, this deviation between experiment and theory is actually indirect evidence for the presence of reaction (1) in the experiment.

Further evidence for the chemical reaction comes when one considers the problems associated with the other two possible explanations: enhanced vibrational energy transfer and collisional deactivation to another electronic state. In a similar series of experiments on nitric oxide[32,33] a threshold for enhanced vibrational relaxation rates was also found for NO(vib) that was coincident with the lowest reaction threshold. This increase in the relaxation rate was explained in terms of the increased interaction between collision partners on portions of the potential surface near the transition state of the chemical reaction. Anharmonicities near the transition state of the potential surface were thought to lead to a qualitatively different collision event than assumed under simple V–V or V–T energy transfer theories which often rely on a more impulsive picture of the collision. In that work the phrase 'transient chemical bonding' was coined to describe these sorts of interactions and to emphasize the importance of the transition-state region of the intermolecular potential under this proposed mechanism. A similar explanation of the threshold observed in this work is also possible, at least in principle.

However, for NO, state-to-state vibrational relaxation rates for levels well above the energy of the reaction threshold could still be measured, indicating that no dark channel was present. Indeed, above and below the threshold signal levels were comparable, indicating that only a small fraction of the prepared molecules reacted away. For O_2 at 465 K, the jump in the relaxation rates between vibrationally excited molecules one quantum below, and one quantum above the barrier is proportionally much greater than for NO, indicating a stronger interaction near the barrier. For O_2 it may be the case that most of the vibrationally excited species above the barrier do react rather than undergo vibrational relaxation.

One might also argue that multiquantum relaxation (*i.e.* $\Delta v \ll -1$) would appear as a 'dark' channel. For a single quantum vibrational energy transfer mechanism

$$k_{v''} = k_{v'', v''-1} \tag{15}$$

On the other hand, for multiquantum relaxation eqn. (16) is required.

$$k_{v''} = \sum_{n=1}^{v''} k_{v'', v''-n} \tag{16}$$

As can be seen by examining eqn. (15) and (16), multiquantum transfer could lead to situations where $k_{v''} \neq k_{v'', v''-1}$ as is observed in this work. Let us consider the 465 K data for preparation of $v = 27$, where the data are most complete. Adding the two rate constants, $k_{27, 26}$ and $k_{27, 25}$, we find that we can only account for *ca.* 40% of k_{27}. If multiquantum vibrational energy transfer is to be consistent with the observations, the summation of the rate constants for larger quantum transfer, *i.e.* $\Delta v < -2$, must make up the remaining 60% of k_{27}. The experiments show, however, that the multiquantum rate constant $k_{27, 25}$ is almost an order of magnitude smaller than $k_{27, 26}$, the single-quantum rate constant. This suggests that $\Delta v < -2$ collisions are unimportant. The probability for $\Delta v < -2$ could become significant if there existed an energetic resonance that could facilitate the transfer of three quanta of vibrational energy. However, no such resonance is available in the O_2–O_2 system. We conclude that it is unlikely that large rate constants for $\Delta v < -2$ energy transfer processes are responsible for the observed difference between the total depopulation rate constant of $v'' = 26, 27$ and the state-to-state rate constants.

It is possible that the observed thresholds are due to near resonant vibrational-to-electronic (V–E) energy transfer:

$$O_2(X\,^3\Sigma_g^-, v'' \geqslant 26) + O_2(X\,^3\Sigma_g^-) \rightarrow O_2(c\,^1\Sigma_u^-, v'' = 0) + O_2(X\,^3\Sigma_g^-) \qquad (17)$$

This process has an energetic threshold at 32 664 cm^{-1}, 230 cm^{-1} below $O_2(X\,^3\Sigma_g^-, v'' = 26)$.[35] Experimentally, this mechanism cannot be dismissed or confirmed; however, before lending credence to this mechanism, there are considerations that have to be made. Theoretical and experimental work has given evidence for the approximate validity of the Franck–Condon principle in V–E and E–E energy transfer[36–38]. Applying the Franck–Condon principle to reaction (17) we calculate:

$$\langle c\,^1\Sigma_u^-, v' = 0 \,|\, X\,^3\Sigma_g^-, v'' = 26\rangle^2 = 3 \times 10^{-12}$$

This small Franck–Condon factor is believed to reduce greatly the probability that reaction (17) is playing a role. Furthermore, the good agreement between experiment and theory below $v'' = 22$ in O_2 self-relaxation is evidence that V–E energy transfer to a $^1\Delta_g$ and $b\,^1\Sigma_g^+$ is unimportant. These V–E processes have much better Franck–Condon factors than reaction (17). Theoretical work[39] has also shown that cross-sections for V–E and E–E scale with the square of the electronic transition dipole moments of the isolated molecules. The fact that $O_2(c\,^1\Sigma_u^-) \leftarrow O_2(X\,^3\Sigma_g^-)$ is an electric dipole forbidden transition further reduces the likelihood that reaction (17) is playing a role in determining the vibrational population distributions observed. There is also no reason to believe that the V–E energy transfer mechanism would coincidentally possess activation energies, $E_{a,\,dark,\,26}$ and $E_{a,\,dark,\,27}$, which are so clearly suggestive of reaction (1).

Unfortunately, this section cannot give unambiguous proof that the rate constants for reaction (1) have been obtained. Indeed, any future work which can prove or disprove this point would be a significant contribution. Despite this, given all of the information that is currently available, the 'reactive explanation' of the observed threshold is clearly the most likely explanation.

Conclusions

The collisional relaxation dynamics of O_2(vib) with O_2 ($v = 0$) have been studied using a stimulated emission pumping, pump–dump–probe technique. Vibrational state-specific vibrational relaxation rate constants have been directly measured at 295 and 465 K for $v = 19$–27, and additionally $v = 28$ at 465 K for O_2 self-relaxation. Vibrational state-specific vibrational relaxation rate constants have been directly measured at 295 K for N_2 relaxation of $O_2(X\,^3\Sigma_g^-, v'' = 19$–$26)$.

Using a mechanism put forth by Slanger *et al.*,[20] a recent atmospheric modelling calculation by Toumi[21] found a significant effect on stratospheric O_3 concentrations above 50 km when the presence of vibrationally excited O_2 was included. While he stressed that the model was very sensitive to the vibrational relaxation rate constants and the distribution of vibrational levels produced by O_3 photolysis, his results showed up to *ca.* 10% enhancement of O_3 at 60 km, resulting from the photodissociation of vibrationally excited O_2. This calculation made use of an equation for the vibrational state-specific rate constants based on a modified form of the approximate V–V energy transfer model of Rapp and Englander-Golden.[31] The output of the expression used in Toumi's calculation is in poor agreement with our measured values at the temperatures of our measurements, underestimating the observed rates by over an order of magnitude. Our measured values for the rate constants are in quite good agreement, however, with Billing and Kolenik's semiclassical trajectory calculation on a recent potential surface for all values below $v = 26$.[29] This leads us to suggest that Billing and Kolesnick's temperature-dependent rate constants should be used in this atmospheric modelling calculation rather than the approximate values. Unfortunately, substituting the larger values for the collisional relaxation rate constants from Billing and Kolesnick's calculation greatly reduces the importance of the Slanger mechanism.

There is considerable evidence that the deviations in the results of this study from theory at $v \geqslant 26$ are due to the reactive part of the bimolecular potential-energy surface on the interactions between the ground, and highly vibrationally excited molecules. A rapid (five-fold) increase in the rate of vibrational relaxation is observed at energies near the onset of the bimolecular chemical reaction (1). Clear evidence of a dark channel was found which is consistent with reaction (1). The fraction of molecules following the dark channel is: 60% for $v = 27$ at 465 K, 99% for $v = 27$ at 295 K, 80% for $v = 26$ at 465 K, 70% for $v = 26$ at 295 K and 100% for $v = 28$ at 465 K. Activation energies for this dark channel could be extracted from the data and are found to compare favourably with thermodynamic quantities of reaction (1). The first vibrational state which is above the activation barrier is not observed, indicating the dark channel's dominance in its disappearance.

This paper summarizes all of the work that has been carried out bearing on the importance of reaction (1). We hope this work will add fuel to the growing fire that non-LTE distributions of atmospheric constituents can play an important role in atmospheric chemistry.

This work was supported by a grant from the US National Science Foundation ATM-88922214. The Santa Barbara Laser Pool helped provide some of the equipment necessary for this work under grant CHE-8411302.

References

1 G. Brasseur and S. Solomon, *Aeronomy of the Middle Atmosphere*, D. Reidel, Boston, 1984.
2 F. F. Crim, M. C. Hsiao, J. L. Scott, A. Sinha and R. L. Vander Wal, *Philos. Trans. R. Soc. London Ser. A*, 1990, **332**, 259.
3 F. F. Crim, A. Sinha, M. C. Hsiao and J. D. Thoemke, *Proc. 24th Jerusalem Symp. in Quantum Chemistry and Biochemistry*, 1992.
4 M. C. Hsiao, A. Sinha and F. F. Crim, *J. Phys. Chem.*, 1991, **95**, 8263.
5 A. Sinha, *J. Phys. Chem.*, 1990, **94**, 4391.
6 A. Sinha, M. C. Hsiao and F. F. Crim, *J. Chem. Phys.*, 1990, **92**, 6333.
7 A. Sinha, J. D. Thoemke and F. F. Crim, *J. Chem. Phys.*, 1992, **96**, 372.
8 R. L. Vander Wal and F. F. Crim, *J. Phys. Chem.*, 1989, **93**, 5331.
9 R. L. Vander Wal, J. L. Scott and F. F. Crim, *J. Chem. Phys.*, 1991, **94**, 3548.
10 R. L. Vander Wal, J. L. Scott, F. F. Crim, K. Weide and R. J. Schinke, *Chem. Phys.*, 1991, **95**, 3548.
11 M. J. Bronikowski, W. R. Simpson, B. Girard and R. N. Zare, *J. Chem. Phys.*, 1991, **95**, 8467.
12 R. L. Miller, A. G. Suits, P. L. Houston, R. Tuomi, J. A. Mack and A. M. Wodtke, *Science*, 1994, **265**, 1831.
13 T. Kinugawa, T. Sato, T. Arikawa, Y. Matsumi and M. Kawasaki, *J. Chem. Phys.*, 1990, **93**, 3289.
14 M. J. Daniels and J. R. Wiesenfeld, *J. Chem. Phys.*, 1993, **98**, 321.
15 H. Park and T. G. Slanger, *J. Chem. Phys.*, 1994, **100**, 287.
16 L. Froidevaux, M. Allen and Y. L. Yung, *J. Geophys. Res.*, 1985, **90**, 12999; R. T. Clancy, D. W. Rusch, R. J. Thomas, M. Allen and R. S. Eckman, *J. Geophys. Res.*, 1987, **92**, 3067.
17 NASA Reference Publication 1292, Vol. II, *The Atmospheric Effects of Stratospheric Aircraft*: Report of the 1992 Models and Measurements Workshops. March 1993.
18 *Atmos. Ozone Rep.* No. 16, World Meteorological Organisation, Geneva, 1986.
19 J. Eluszkiewicz and M. Allen, *J. Geophys. Res.*, 1993, **98**, 1069.
20 T. G. Slanger, L. E. Jusinski, G. Black and G. E. Gadd, *Science*, 1988, **241**, 945.
21 R. Toumi, B. J. Kerridge and J. A. Pyle, *Nature (London)*, 1991, **351**, 217.
22 C. A. Rogaski, J. M. Price, J. A. Mack and A. M. Wodtke, *Geophys. Res. Lett.*, 1993, **20**, 2885.
23 X. Yang, E. H. Kim and A. M. Wodtke, *J. Chem. Phys.*, 1990, **93**, 4483; X. Yang, E. H. Kim and A. M. Wodtke, *J. Chem. Phys.*, 1992, **96**, 5123; J. M. Price, J. A. Mack, C. A. Rogaski and A. M. Wodtke, *Chem. Phys.*, 1993, **175**, 83.

24 X. Yang and A. M. Wodtke, *J. Chem. Phys.*, 1990, **92**, 116; X. Yang, D. McGuire and A. M. Wodtke, *J. Mol. Spectrosc*, 1992, **154**, 361.

25 S. S-L. Chiu, A. S-C. Cheung, M. Finch, M. J. Jamieson, K. Yoshino, A. Dalgarno and W. H. Parkinson, *J. Chem. Phys.*, 1992, **97**, 1787; P. C. Cosby, H. Park, R. A. Copeland and T. G. Slanger, *J. Chem. Phys.*, 1993, **98**, 5117.

26 D. M. Creek and R. W. Nichols, *Proc. R. Soc. London, Ser. A*, 1975, **341**, 517.

27 G. M. Jurisch and F. F. Crim, *J. Chem. Phys.*, 1981, **74**, 4455.

28 H. Lefebvre-Brion and R. W. Field, *Peturbations in the Spectra of Diatomic Molecules*, Academic Press, New York, 1986.

29 G. D. Billing and R. E. Kelesnick, *Chem. Phys. Lett.*, 1992, **200**, 382.

30 G. D. Billing, *Chem. Phys.*, 1994, **179**, 463.

31 D. Rapp and P. Englander-Golden, *J. Chem. Phys.*, 1964, **40**, 573; D. Rapp, *J. Chem. Phys.*, 1965, **43**, 316.

32 J. T. Yardley, *Introduction to Molecular Energy Transfer*, Academic Press, New York, 1980.

33 A. van der Avoird, *J. Chem. Phys.*, 1984, **81**, 1929; and personal communications.

34 M. Cacciatore and G. D. Billing, *Chem. Phys.*, 1981, **58**, 395; G. D. Billing, *Chem. Phys.*, 1980, **50**, 165.

35 K. P. Huber and G. Herzberg, *Molecular Spectra and Molecular Structure IV: Constants of Diatomic Molecules*, Von Nostrand Reinhold, New York, 1979.

36 I. Despersinska, J. A. Beswick and A. Tramer, *J. Chem. Phys.*, 1979, **71**, 2477.

37 D. H. Katayama, T. A. Miller and V. E. Bondeybey, *J. Chem. Phys.*, 1979, **71**, 1662.

38 V. E. Bondybey and T. A. Miller, *J. Chem. Phys.*, 1978, **69**, 3597.

39 R. G. Gordon and Y. N. Chiu, *J. Chem. Phys.*, 1971, **55**, 1469.

Paper 5/00173K; *Received* 10*th January*, 1995

Faraday Discuss., 1995, **100**, 253–267

Modelling the Photonucleation of Water Vapour by UV in the Presence of Oxygen and the Absence of Pollutants

William Byers Brown, Ian H. Hillier, Andrew J. Masters, Ian J. Palmer, David H. V. Dos Santos, Matthias Stein and Mark A. Vincent
Department of Chemistry, University of Manchester, Manchester, UK M13 9PL

The progress on modelling work based on the hypothesis that the photonucleation of water vapour in the presence of oxygen is due to the formation of charge-transfer complexes of molecular oxygen and water, $H_2O^+O_2^-$, is reported with argument in favour of the relevance of the phenomenon to aerosol formation in the atmosphere. The following results are described: (*a*) potential-energy curves for the complexes $H_2O \cdot O_2$ and $H_2O^+O_2^-$, and the calculated cross-section for the UV absorption $H_2O \cdot O_2 \rightarrow H_2O^+O_2^-$; (*b*) possible alternative photochemical products arising from water and oxygen using *ab initio* electronic structure methods; (*c*) molecular dynamics calculations on small clusters of water molecules containing $H_2O^+O_2^-$ complexes or H_2O_2 molecules; (*d*) an initial estimate of the rate of aerosol formation and of hydrogen peroxide production due to photonucleation in the absence of pollutants.

1. Introduction

The phenomenon of the photo-induced nucleation of pure water vapour in the presence of air or oxygen by UV radiation in the wavelength range 200–300 nm has been neglected by most atmospheric scientists despite its potential importance for aerosol formation. This is remarkable in view of the major pioneering investigation by Wilson which included the use of solar UV at sea level in one series of experiments,[1] and which was repeated with essentially the same results by Farley.[2] The phenomenon has been rediscovered and studied by other workers, in particular more recently by Hoppel and Dinger[3] in the troposphere, by Reiss *et al.*[4] in the laboratory and, quantitatively, by Katz and co-workers[5] using a thermal-diffusion cloud chamber. A typical photonucleation spectrum measured by the last-named group is shown as a series of circular data-points in Fig. 1. The left-hand axis gives the UV light intensity required to cause a constant nucleation rate of 2 drops cm^{-2} s^{-1} in Katz's apparatus[5] under the stated conditions of temperature and supersaturation.

There is a major difference of opinion on the role of oxygen in the photonucleation of water vapour which has been discussed by one of us.[6] In this paper we side with Wilson[1] and Farley[2] in accepting that it is essential to the phenomenon in the wavelength range cited above. Further evidence for this view is provided by comparing the photonucleation spectrum and the UV absorption spectrum of oxygenated liquid water.[7,8] The heavy line in Fig. 1 refers to the right-hand axis, and is $-\log A_{10}$, where A_{10} is the decadic absorbance, of oxygenated liquid water plotted against wavelength. The scales are arbitrary, but the almost featureless monotonic behaviour is the same in both cases, and lends strong support to the idea that the absorption seen in oxygenated

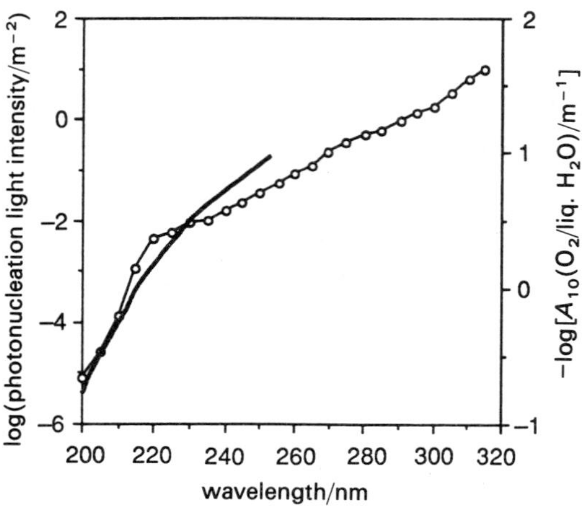

Fig. 1 Left-hand axis, ○, log (UV light intensity) required to cause a constant nucleation rate of 2 drops cm^{-2} s^{-1} as a function of wavelength for pure water vapour in a thermal diffusion cloud chamber at a supersaturation of 3.059 at 295.44 K. The data are reproduced with permission from Fig. 1 of ref. 5. Right-hand axis, heavy line, $-\log A_{10}$ of oxygenated liquid water at 297 K as measured by Heidt and co-workers.[8]

liquid water is the same as that involved in photonucleation. For comparison, Katz and co-workers have established that when photonucleation is caused by *o*-tolualdehyde and by CS_2, the photonucleation spectra follow the highly structured absorption spectra remarkably closely.[9,10]

The hypothesis has been proposed[6] that in both cases UV is absorbed by the collision (van der Waals) complex of water and oxygen, $H_2O \cdot O_2$, to form a charge-transfer complex $H_2O^+O_2^-$. In the vapour phase, in view of its large dipole moment, the complex will immediately attract water molecules to form clusters which it is suggested should be called 'Wilson clusters'. A preliminary quantum chemistry study of $H_2O \cdot O_2$ and $H_2O^+O_2^-$ confirmed the existence of both, with an excitation energy corresponding to *ca.* 200 nm and the latter with a dipole of *ca.* 6 D†,[11] thus strongly supporting the hypothesis. The subsequent chemistry in Wilson clusters is unknown, but will almost certainly lead finally to hydrogen peroxide, as originally suggested by Wilson.[1] The vapour pressure of the clusters will be lowered both before and after reaction, and this will allow coalescence with other Wilson clusters leading to condensation; alternatively the H_2O_2 may diffuse out and evaporate, or decompose photochemically. The observations of Wilson and of Katz and co-workers can thus be understood qualitatively.

In this paper we summarise the results of high-level quantum mechanical and molecular simulation methods applied to various aspects of the phenomenon. Detailed accounts of the calculations in the following sections will be published separately.

2. *Ab initio* Water–Oxygen Potential Surfaces

Here, we describe the results of some *ab initio* electronic structure calculations on the water–oxygen potential-energy surface which have been used in order to model the UV absorption spectrum. The aim was to try to establish if the charge-transfer absorption of

† 1 D (debye) ≈ $3.335\,64 \times 10^{-30}$ C m.

water–oxygen collision pairs exhibits the broad featureless behaviour observed in the experimental UV spectrum of oxygenated water in the range 200–250 nm, as shown in Fig. 1. The charge-transfer phenomenon results from the existence of an electronic state of the super-molecule dominated by an electronic configuration in which an electron originating in one fragment has been transferred to the orbital space of the other. Low-lying charge-transfer states are most likely for pairs of molecules consisting of an electron donor which possesses a low ionisation potential and an acceptor which possesses a high electron affinity.

The ground state of the water–oxygen super-molecule comprises water and oxygen in their $^1A_1$ and $^3\Sigma_g^-$ ground states, respectively. The charge-transfer absorption may be simply viewed as the excitation of an electron from the b_2 lone pair orbital of water to the π^* system of oxygen. Since there are two orthogonal π^* orbitals in oxygen, two possible charge-transfer states exist which will be degenerate over large regions of geometry space. This absorption process results in a change in the electronic structure from the neutral ground state $H_2O \cdot O_2$ collision pair to what is essentially a zwitter ion, $H_2O^+O_2^-$. In this state the super-molecule will possess a large dipole moment and the interaction of the fragments will have long-range Coulombic attraction.

A precise simulation of the water–oxygen UV spectrum would require the generation of the electronic potential-energy surface and solution of the nuclear Hamiltonian in the full nine linearly independent coordinates of the system. This is clearly not realistic, and in order to simulate the spectrum a simple one-dimensional model has been adopted which only considers motions along a single inter-fragment coordinate. In making this simplification we are assuming that the major influence on the general shape of the spectrum is from the inter-fragment motion and not from intramolecular motions of water and oxygen. Since the interaction between the two monomers on the ground-state surface is found to be extremely weak, and the electronic transition is to an excited state of the super-molecule, it would appear reasonable that the inter-fragment coordinate will have a major effect on the absorption spectrum.

Despite the apparently simple nature of the system a rigorous treatment of the electronic structure requires complex *ab initio* techniques. Since the triplet surfaces of ground and excited electronic states are to be studied, a multi-configuration reference wavefunction must be generated. A suitable method is the complete active space self-consistent field method (CASSCF)[12] and the implementation of this method in the GAUSSIAN92[13] and MOLPRO94[14] packages have been used in this study. The active space chosen comprised the σ/σ^* and π/π^* orbitals and electrons of oxygen and the b_2 lone pair of water. This therefore corresponds to ten electrons distributed in seven orbitals. The effects of electron correlation must be added by a higher-order method using the CASSCF reference wavefunction. In this study, electron correlation effects are important to allow inclusion of dispersion forces in the ground state van der Waals interaction and to improve the calculation of relative energetics. Calculations have therefore been performed using the internally contracted multi-reference configuration interaction (MR-CI)[15,16] method and multi-reference averaged coupled pair theory (MR-ACPF)[17] available in MOLPRO94.

The C_{2v} geometries of the van der Waals molecule studied previously[11] have been reoptimised at the MP2/6-311++G(2d,p) level relaxing the internal geometries of the molecules. Using this method it is also possible to calculate analytic second derivatives which show the ground-state potential surface to be extremely flat in the region of the C_{2v} minimum. In this paper we shall present the results for a single intermolecular coordinate corresponding to the structure shown in Fig. 2. The MR-CI/6-311++G(2d,p) energy[18] of this structure is -226.311991 H and the dipole moment is 2.16 D, calculated from the CASSCF reference energy of -225.796075 H and dipole moment of 2.20 D. Along this coordinate the ground state has $^3B_1$ symmetry and the charge-transfer states are of $^3A_1$ and $^3B_1$ symmetry. The potential energy surfaces of the $^3B_1$

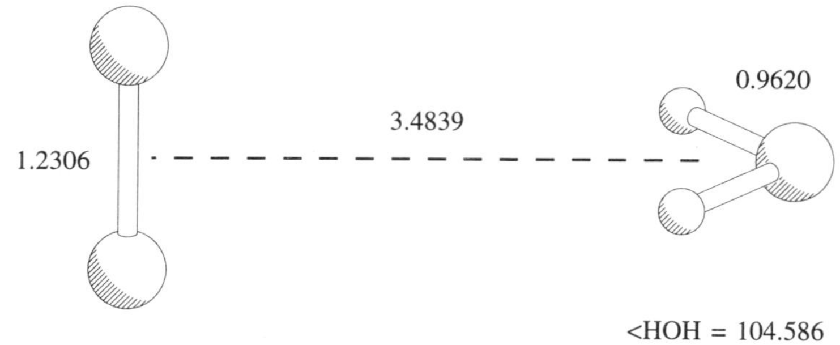

Fig. 2 MP2/6-311+ +G(2d,p) optimised geometry of the $^3B_1$ ground state C_{2v} van der Waals coordinate. All distances are in Å and angles in degrees.

ground state and $^3A_1$ excited charge-transfer state, presented later in this section, have been generated by fixing the internal geometries at the optimum ground-state values of Fig. 2 and varying the inter-fragment distance.

The CASSCF energy of the optimum $^3A_1$ excited-state structure shown in Fig. 3 is $-225.570\,502$ H and the dipole moment is calculated as 6.09 D. The MR-CI energy of this structure was calculated as $-226.094\,586$ H and the dipole moment at 6.04 D using *ca.* 3×10^5 contracted and 11.5×10^6 uncontracted configurations. This gives an energy separation from the ground-state minimum at the MR-CI level of 5.92 eV corresponding to an excitation wavelength of 208 nm. The MR-ACPF energies of the two minima are $-226.371\,306$ H and $-226.159\,960$ H, respectively, corresponding to an energy separation of 5.75 eV (214 nm). These results are consistent with the results obtained in the preliminary study on this system[11] and the experimentally observed absorption.[8]

The potential-energy curves generated along the inter-fragment coordinate for the $^3B_1$ ground state and $^3A_1$ charge transfer state are shown in Fig. 4(*a*) and (*b*), respectively. Two curves are shown for each state, one corresponding to the MR-CI energy and the other to the MR-ACPF energy. Although the MR-CI method is variational it has the problem that it is not size consistent. Corrections can be made by using methods such as a multi-reference version of the Davidson correction,[19] however, ACPF, which is approximately size consistent but non-variational, has found successful application in the study of inter-fragment coordinates.[20] The ACPF curves of Fig. 4(*a*) and (*b*) have therefore been used for the spectrum simulation and these curves have also been corrected for basis set superposition error (BSSE) using the counterpoise correction.[21] The

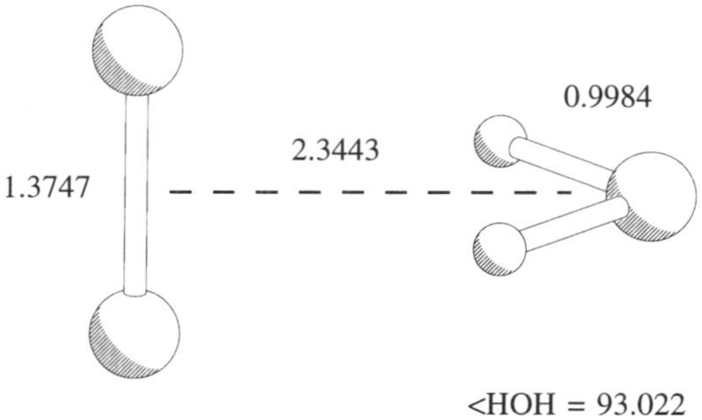

Fig. 3 CASSCF/6-311+ +G(2d,p) optimised C_{2v} geometry for the $^3A_1$ CT state. All distances are in Å and angles in degrees.

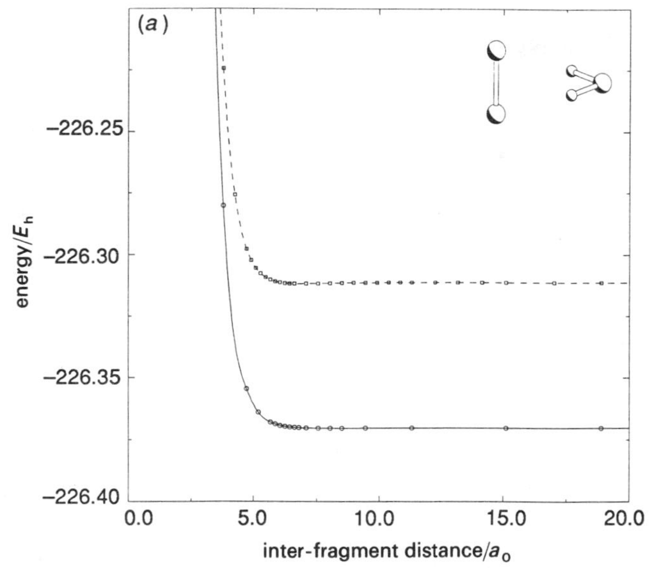

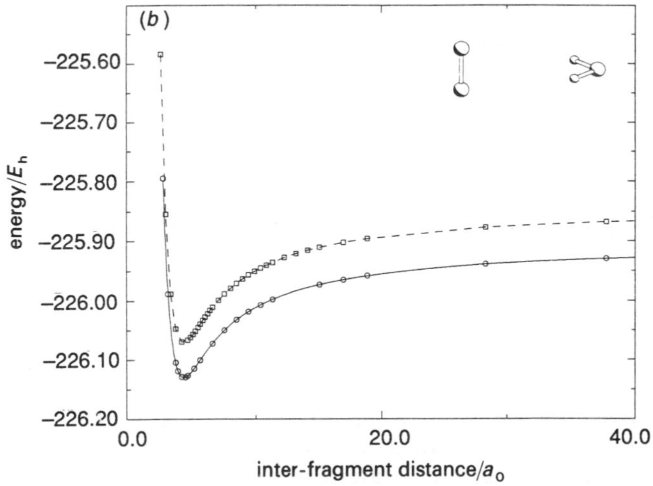

Fig. 4 Spline fits and *ab initio* data for (*a*) the $^3B_1$ ground-state energy and (*b*) the $^3A_1$ CT state *vs.* intermolecular coordinate calculated at the MRCI/6-311++(2d,p) [□, (– – –)] and MR-ACPF/6-311++G(2d,p) [○, (——)] level. The MR-ACPF results have been corrected with an estimate of the BSSE error calculated *via* the counterpoise method. The inset shows the orientation of the molecules along the intermolecular coordinate.

inclusion of the BSSE correction essentially eliminates the very shallow potential well of the MP2 calculations.

In order to calculate the absorption spectrum it is also necessary to determine the electronic transition dipole moment function between the two states. The results of calculations performed at the CASSCF and MR-CI levels using the active space described earlier, but with state-averaged orbitals,[22] are shown graphically in Fig. 5. A spline fit to

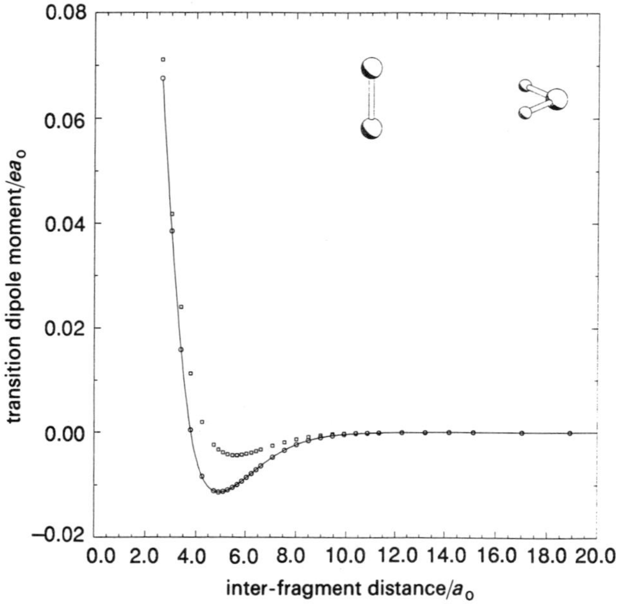

Fig. 5 Spline fits and *ab initio* data for the $^3A_1 \leftarrow {}^3B_1$ electronic transition dipole moment SA-CASSCF/6-311++G(2d,p) (□) and MR-CI/6-311++G(2d,p) [○, (——)] level. The inset shows the orientation of the molecules along the intermolecular coordinate.

the MR-CI data was used in the simulation of the UV spectrum which is discussed in the next section.

3. Model Absorption Spectrum Calculation

The *ab initio* calculations detailed in the previous section showed the ground-state potential surface to be extremely flat in the region of the C_{2v} minimum. The potential curve along the intermolecular coordinate was also found to be essentially unbound, which suggests the spectrum will be dominated by absorption of collision pairs. In this section we describe the formalism we have used to calculate a model absorption spectrum. The development of the model is essentially the same as the treatment of collision-induced spectra of molecules,[23] which are normally associated with roto-vibrational transitions on the electronic ground state. Collision-induced electronic spectra do not seem to have been so well studied, although molecular oxygen offers a classic example of a collision-induced electronic transition which exhibits a typical quadratic dependence on pressure.[23] The important distinction between the process under study here and most examples of collision-induced spectra, is that the excitation is of the super-molecule as a whole and not absorption activated in one of the fragments. This makes the treatment of the absorption spectrum slightly different.

The starting point is the familiar Lambert Law

$$I(\nu) = I_0(\nu)\exp[-A(\nu)l] \tag{3.1}$$

which relates the transmitted intensity, $I(\nu)$, to the incident intensity, $I_0(\nu)$, for a sample path length l, at frequency, ν, and defines the absorbance, $A(\nu)$, of the sample. Beer's Law states that the absorbance is proportional to the concentration of the absorbing species. However, the absorption due to the pair of weakly interacting H_2O and O_2 molecules will have contributions from bound van der Waals (vdW) and unbound collisional pairs

$H_2O + O_2$. The total absorbance is therefore given by

$$A(\nu) = \alpha(\nu)[\text{vdW}] + \beta(\nu)[H_2O][O_2] \tag{3.2}$$

where $\alpha(\nu)$ is the unary absorption coefficient and $\beta(\nu)$ is the binary absorption coefficient, and the square brackets indicate the molar concentrations of the species.

Expressions for the unary and binary absorption coefficients may be derived from Fermi's Golden Rule. For the unary molar coefficient at temperature, T, we find

$$\alpha(\nu) = \frac{4\pi^2 L}{3\hbar cn}\nu[1 - \exp(-h\nu/kT)]\sum_f \sum_i P_i \frac{|\langle f|\mu|i\rangle|^2}{4\pi\varepsilon_0}\delta(\nu_{fi} - \nu) \tag{3.3}$$

where $\nu_{fi} = (E_f - E_i)/h$ is the frequency of the excitation $f \leftarrow i$, n is the refractive index of the sample, $\langle f|\mu|i\rangle$ is the transition dipole moment, P_i is the probability of the system occupying state $|i\rangle$ with energy E_i, and the other symbols have their usual meanings. P_i is given by

$$P_i = \frac{\exp(-E_i/kT)}{Z_{\text{vdW}}(T)} \tag{3.4}$$

where $Z_{\text{vdW}}(T)$ is the dimer partition function, which involves a sum over all internal bound states of the vdW complex

$$Z_{\text{vdW}}(T) = \sum_i \exp(-E_i/kT) \tag{3.5}$$

In reality, the delta functions in eqn. (3.3) should be replaced by unit line-shape functions with finite width, the coefficients of which are the integrated frequency absorptions α_{fi} of the various transitions.

For transitions from the unbound continuum states of the collision pair to the upper bound states, the formula for the absorbance must be modified to account for the continuous spectrum of energies available as the initial state. The final expression for the binary molar absorption coefficient is given by

$$\beta(\nu) = \frac{4\pi^2 L^2 \Lambda^3}{3\hbar cn} h\nu[1 - \exp(-h\nu/kT]\sum_f \exp(-E/kT)\frac{|\langle f|\mu|E\rangle|^2}{4\pi\varepsilon_0} \tag{3.6}$$

where $E = E_f - h\nu$ is the relative kinetic energy of the absorbing collision pair $H_2O + O_2$, and Λ is their thermal de Broglie wavelength given by

$$\Lambda = \frac{h}{\sqrt{(2\pi\mu_{\text{pair}} kT)}} \tag{3.7}$$

with μ_{pair} the reduced mass. Note that it has been assumed that the continuum wavefunctions $|E\rangle$ are delta-normalised with respect to energy, so that the dimensions of β are $m^5\ mol^{-2}$, as implied by eqn. (3.2).

The relative contributions of the vdW and collision pairs to the spectrum are dependent on their concentrations. In thermal equilibrium these quantities may be determined from a knowledge of the equilibrium constant for the formation of the van der Waals dimer

$$K_\nu(T) = \frac{[\text{vdW}]}{[H_2O][O_2]} \tag{3.8}$$

In order to determine the transition dipole moment integrals which appear in the expressions for the unary and binary absorption coefficients it is necessary to calculate the bound and continuum nuclear wavefunctions, as well as the electronic transition dipole moment function. The bound-state nuclear wavefunctions have been generated on a grid using the Fourier grid Hamiltonian method of Balint-Kurti *et al.*[24] and the energy-normalised continuum wavefunctions have been generated within the semi-classical Langer approximation.[25] Both these methods allow use of the *ab initio* potential for the generation of the nuclear wavefunctions. The electronic transition dipole moment integrals were obtained from spline fits of the *ab initio* MR-CI data presented in the last section. The integration to obtain the transition dipole moment integrals for the calculation of the absorption coefficient was performed using numerical quadrature.

The ground-state potential along the intermolecular coordinate has only a shallow minimum which has been found to support only a single bound state. This suggests that the major contribution to the absorption spectrum will be from continuum to bound-state transitions. The calculated unary and binary absorption coefficients are plotted in Fig. 6 and 7. The plots are essentially smoothed histograms in which all transitions within small energy intervals (corresponding to wavelength intervals of *ca.* 1 nm) are collected and processed appropriately; the unary transitions are assumed to have small linewidths less than the interval width. The solid lines are the curves calculated from the MR-ACPF curves and the dashed lines correspond to a calculation in which the energy separation of the two curves has been reduced to be in line with the calculated separation of the optimum minima structures. The binary absorption coefficient demonstrates the general features of the experimentally observed spectrum, having a peak at the short wavelength side, falling off slowly over a range of 50 nm toward the longer wavelength end of the absorption. However, both the unary and binary absorptions are shifted to shorter wavelengths than experiment, and are numerically much smaller than the (unary) absorption coefficient for oxygenated liquid water.[8b] The major origin of these discrepancies is no doubt the crudity of the one-dimensional model we have used to calculate the spectrum, which excludes all contributions from non-zero angular momentum states. Also from Fig. 2 and 3 it is clear that there are differences in the internal geometries of the ground and charge-transfer state structures, particularly in the O—O bond length. This means that by constraining the internal geometry in the charge-transfer state we will not only increase the energy separation of the curves but also

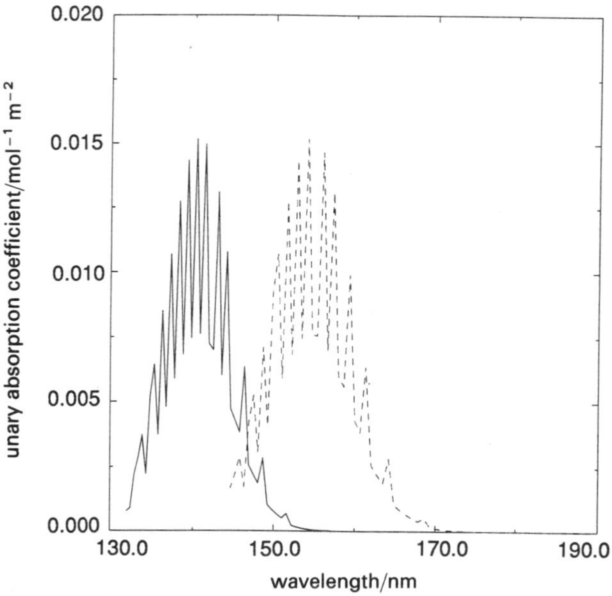

Fig. 6 Calculated unary absorption coefficient. (——) Corresponds to the result obtained from the MR-ACPF curves and (– – –) corresponds to the result with separation of the PE curves adjusted.

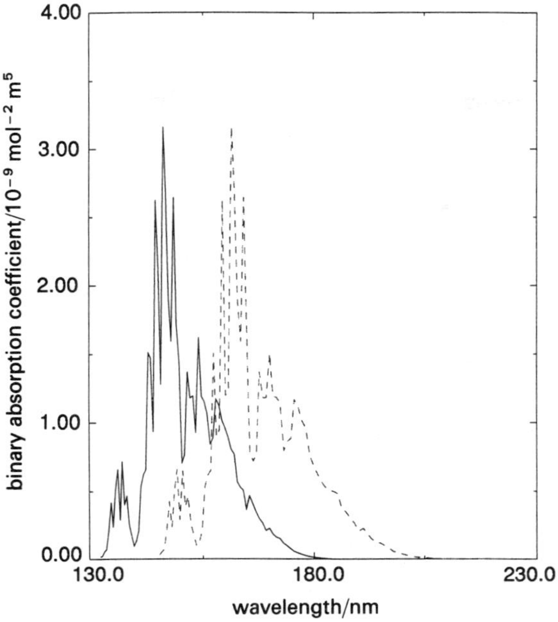

Fig. 7 Calculated binary absorption coefficient. (——) Corresponds to the result obtained from the MR-ACPF curves and (– – –) corresponds to the result with separation of the PE curves adjusted.

change their shape. Given these factors, it is not surprising that the energetics of this initial simulation of the spectrum are not exact. What is pleasing, however, is that the general behaviour toward longer wavelengths appears to agree with experiment and suggests that a charge-transfer band can explain the differences between the absorption spectra of pure water and oxygenated water and also offer a source of nucleators in the photonucleation of atmospheric water vapour.

4. Decomposition Products of the Charge-transfer Complex

The calculations described in Sections 2 and 3 have confirmed that formation of the charge-transfer complex ($H_2O^+O_2^-$) may indeed be associated with the observed absorption spectrum. However, in view of the possible short lifetime of this complex we have calculated the energies of possible products arising from its rearrangement, which themselves may act as nucleating agents. Using a high level of electron correlation [QCISD(T)][26] and a 6-311++G(2d,p) basis[18] we have examined all possible energy minima on the singlet and triplet potential-energy surface of H_2O_3.[27] We find a range of possible products close in energy to that of the charge-transfer complex. These include (hydrogen peroxide + O), [oxywater ($H_2O—O$) + O], and [1-hydrotrioxy (HO_3) + H]. Of particular note are both hydrogen peroxide, frequently implicated in atmospheric phenomena, and oxywater, which by virtue of its large dipole moment (4.4 D) is also a strong candidate as a nucleator.

5. Molecular Dynamics Calculations on Wilson Clusters

In order to gain some insights into the structural and energetic properties of Wilson clusters, we performed some molecular dynamics (MD) studies on water droplets containing either an $H_2O^+O_2^-$ charge-transfer (CT) complex or one or more hydrogen peroxide molecules. A necessary prerequisite for doing this was the choice or development of models for the various potentials of interaction involved.

Two models in common usage for water are the TIP4P model[28] and the SPC model.[29] Both of these take water to be a rigid molecule and the pair interaction between two molecules is of the form,

$$U = \sum_{a,\, b} \left(\frac{A_{ab}}{r_{ab}^{12}} - \frac{B_{ab}}{r_{ab}^{6}} + \frac{q_a q_b}{r_{ab}} \right) \tag{5.1}$$

where a and b refer to sites on the different molecules. A_{ab} and B_{ab} are the site–site Lennard-Jones parameters, q_a is the partial charge situated on site a, and r_{ab} is the distance between sites a and b. The models differ in the values of the Lennard-Jones parameters and in the number and positions of the partial charges. Both models have been parametrised to reproduce the main thermodynamic and structural properties of bulk water at *ca.* 1 atm and 300 K.

The main deficiency of these models lies in the poor prediction of the surface tension. To our knowledge, there exists no published water model capable of giving both good bulk and surface properties and the introduction of flexible and/or polarisable models has not, as yet, improved this situation. Given that an accurate description of a water cluster requires a good account of both the interior and surface regions, we must accept at the outset that this simulation work has its limitations. Rather, our strategy has been to carry out simulations involving both TIP4P and SPC water models, to see whether the general trends observed are sensitive to the exact water model used. If they are insensitive, then we are hopeful that our studies will provide some basic insights into the properties of real clusters.

There is insufficient space here to describe the *ab initio* SCF calculations for $H_2O^+O_2^-$ and H_2O_2, their interactions with TIP4P- and SPC-model H_2O molecules, and the fitting of the results to analytical functions of the form of eqn. (5.1) to obtain approximate pair potentials. This and the details of the MD methods used will be given elsewhere,[30] and we proceed directly to the results. In Fig. 8 the average potential energy per molecule for a mole of clusters is plotted against the total aggregation number (*i.e.* the sum total of all species in the droplet). We present results for pure water, for droplets with a single CT complex in them and for droplets containing a single H_2O_2 molecule. The figure shows data for SPC water, but the TIP4P data show the same general trends. For a given aggregation number, replacing a water molecule with a CT complex marginally decreases the energy whereas replacement with an H_2O_2 mol-

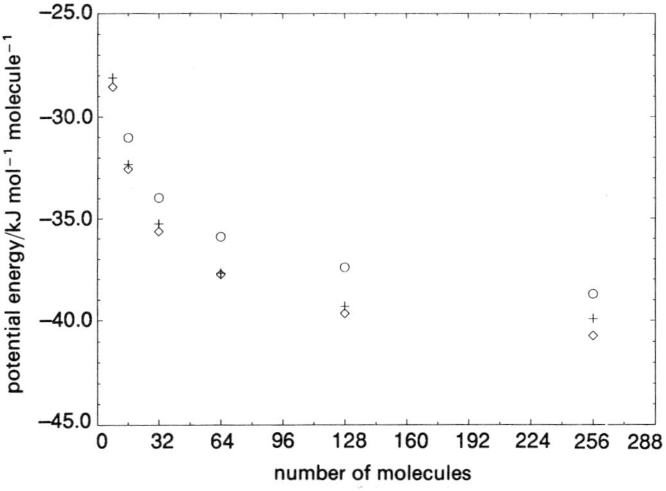

Fig. 8 Average potential energy per molecule *vs.* aggregation number for clusters of +, pure water, ◇, water and a CT complex and ○, water and a hydrogen peroxide molecule

ecule increases the potential energy. This result is especially surprising for the highly polar CT complex, which one would expect to be strongly attracted to a cluster of water molecules. It means that the water molecules can pack and orient themselves more favourably energetically with each other than with either a CT complex or an H_2O_2 molecule.

We further calculated the average position and orientation of the solute molecule within the cluster. For the CT complex, for all cluster sizes studied (up to 256 molecules) and for both SPC and TIP4P water, a striking result was that the complex spent all its time on the surface of the droplet. From its initial central position, it immediately proceeded to migrate to the surface and thereafter remained there for the rest of the simulation. The tendency was for the O—O bond to lie flat on the surface and for the dipole moment to point slightly out of the cluster. We found these observations to be independent of the starting configuration.

This was an unexpected result, so as a further check we prepared a cluster of one CT complex and seven water molecules. The initial configuration was first quenched to 0 K, yielding a structure with a central CT complex. We prepared a second structure by first equilibrating at 300 K and then quenching to 0 K, this time with the complex lying on the surface. A single CI calculation with a 6-31G basis set indicated that the second structure was lower in energy by 75.5 kJ mol^{-1}. Naturally these calculations are not accurate enough to be definitive, but they do lend considerable weight to the non-intuitive conclusion that the complex lies preferentially at the surface of the droplet.

We now turn to the hydrogen peroxide results, where replacement of a water molecule by an H_2O_2 molecule increases the average potential energy, both for SPC and TIP4P water. Unlike the CT complex, however, the position of the H_2O_2 molecule does depend on which water model is used. In the TIP4P case the molecule again lies on the surface; statistically independent initial configurations, all with a central H_2O_2 molecule, show this final equilibrated result. On the other hand, for SPC water, the H_2O_2 molecule remains, on average, in the interior of the droplet.

Simulations were also carried out on TIP4P water droplets containing more than one H_2O_2 molecule. In Fig. 9, we plot the average potential energy per molecule in the cluster against total aggregation number, for clusters containing between zero and four H_2O_2 molecules. Again the temperature was set to 300 K. The results indicate that

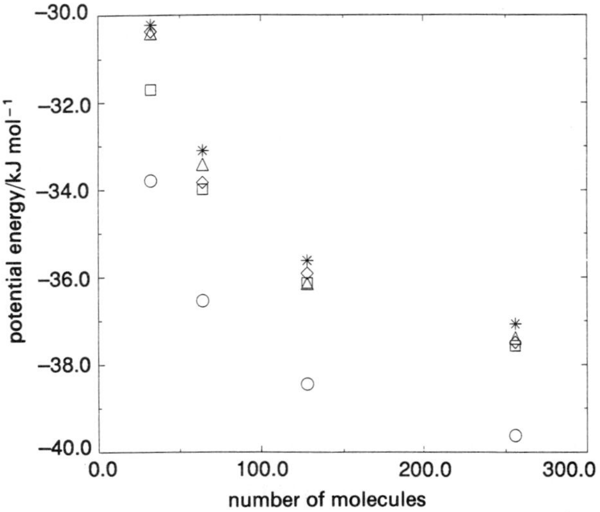

Fig. 9 As in Fig. 8 but for clusters containing ○, 0; □, 1; △, 2; ◇, 3 and ∗, 4 hydrogen peroxide molecules

replacing an H_2O molecule with an H_2O_2 molecule leads to an increase in energy, the energy increasing slightly the more H_2O_2 molecules are added. The simulations also indicate that the H_2O_2 molecules form a cluster at the surface of the drop (at least for the aggregation numbers up to 256 molecules and for up to four H_2O_2 molecules).

To us, these results are also surprising. Experimentally, bulk liquid hydrogen peroxide is a strongly associated liquid: at 298 K $\Delta_{vap}H$ is 51.7 kJ mol^{-1} as compared with the value of 44.0 kJ mol^{-1} for pure water. Furthermore, hydrogen peroxide and water are miscible in all proportions and $\Delta_{vap}H$ for the mixture increases monotonically with increasing mole fraction of hydrogen peroxide.[31] From this one might have expected that replacing an H_2O molecule with an H_2O_2 molecule would decrease rather than increase the energy of a cluster! Also the bulk miscibility would lead one to expect a much more homogeneous drop than that observed from simulation.

Concluding this section, our simulation results indicate that neither the CT complex nor H_2O_2 molecules appear to stabilise a water cluster energetically to any great extent. Many of the simulations also showed the solute to migrate to the surface. Although one may claim that these effects are simply artefacts of our chosen forms for the potentials of interaction, we believe that our results suggest that applying a homogeneous solution theory to Wilson Clusters might well be most inappropriate.

6. Rate of Formation of Wilson aerosols in the Atmosphere

In the previous sections we have been concerned to review the results of modelling work based on the hypothesis that the photonucleation of water vapour by UV in the presence of oxygen is due to the formation of CT complexes of molecular oxygen and water, $H_2O^+O_2^-$, which attract water molecules to form clusters stabilised against evaporation (Wilson clusters), which under sufficiently humid conditions go on to form stable aerosol droplets or to condense. In this section we attempt to argue for the possible relevance of the phenomenon to atmospheric physics and chemistry.

It is not difficult to make a crude estimate of the rate at which aerosols formed from Wilson clusters will be generated initially in the atmosphere from a given intensity of solar UV. The end result of the process will almost certainly be the presence of extra hydrogen peroxide in aerosol water. The importance of H_2O_2 for atmospheric chemistry is being increasingly recognised, and as its concentration is fairly easily measured experimentally, it is appropriate to choose its rate of production as a possible indicator of the importance of the photonucleation process under discussion.

The intensity of the solar UV in the troposphere in the range 200–300 nm is very low, owing to stratospheric absorption by O_2 and O_3. It is conceivable that, despite the low concentration of water, the process plays some role in the stratosphere. However, our knowledge of the Wilson photonucleation mechanism is too incomplete to attempt to model its role in the atmosphere as a whole. We therefore choose to estimate the rate of production of H_2O_2 in air saturated with water vapour at 0 °C exposed to one tenth of the total solar radiation between the wavelengths 200 and 250 nm.

The theoretical results of Section 3 on the UV absorption spectrum of mixtures of water and oxygen are not definitive enough to decide whether the absorption arising from the bound vdW complex is more or less important than that induced by collisions of the free H_2O and O_2 molecules. For simplicity we adopt the vdW route, and assume the following chemical and photochemical equations:

$$H_2O + O_2 \leftrightarrow \text{vdW (fast equilibrium)}, \tag{6.1}$$

$$\text{vdW} + h\nu \rightarrow \text{CTC} \rightarrow \text{vdW (de-excite, lifetime } \tau) \tag{6.2}$$

where CTC = $H_2O^+O_2^-$ and vdW = $H_2O \cdot O_2$. Let the intensity of UV radiation of frequency ν be I, the molar absorption coefficient of the vdW complex be $\alpha(\nu)$, and τ be the lifetime of the CT complex. Neglecting reaction and diffusion, the rate equation for CTC is

$$d[CTC]/dt = (\alpha I/Lh\nu)[vdW] - [CTC]/\tau \quad (6.3)$$

where L is the Avogadro constant, so that the steady state concentration is

$$[\overline{CTC}] = (\tau\alpha I/Lh\nu)[vdW] \quad (6.4)$$

The equilibrium concentration of vdW complexes is given by eqn. (3.8), namely

$$[vdW] = K_V[H_2O][O_2] \quad (6.5)$$

where the equilibrium constant, K_V, can be estimated by statistical mechanics from the calculated energy and geometry of the vdW complex. Assuming it is a 'loose' complex, so that the internal vibrations and rotations of H_2O and O_2 are unaffected by complexing, we get[6]

$$K_V = 5.3\sqrt{(T)}\exp(433/T)\ \mathrm{cm^3\ mol^{-1}} \quad (6.6)$$

We focus on the fate of the CT complexes (whether they react to yield H_2O_2 or not) as they form Wilson clusters, W_1, which go on to coalesce with other Wilson clusters, and which are assumed to contain the number of H_2O molecules necessary for vapour equilibrium.

$$CTC + H_2O \xrightarrow{k_1} W_1 \quad (6.7)$$

The rate of production of W_1 will be given by

$$d[W_1]/dt = k_1[CTC][H_2O] \quad (6.8)$$

If we assume that every collision leads to a Wilson cluster, then $k_1 = Z$, the collision number, and the rate can be written

$$d[W_1]/dt = Z(\tau\alpha I/Lh\nu)K_V[H_2O]^2[O_2] \quad (6.9)$$

We estimate collision diameters of 200 and 500 pm for H_2O and $O_2 \cdot H_2O$ in order to calculate Z. The lifetime, τ, of the CTC is unknown and has not yet been estimated thereoretically, but for a strong broad absorption at short wavelength it is likely to lie between 1 ns and 1 μs; let us make the conservative assumption that $\tau = 1$ ns. In saturated oxygenated liquid water nearly all the dissolved oxygen will be present as vdW complexes, so that α can be found from the results of Heidt and Johnson[8b] over the range 200–250 nm. We use an average value of $\alpha I/\nu$ for this wavelength range obtained by assuming the sun behaves as a black-body at a temperature 5785 K with solar constant 1368 W m^{-2}, and integrating numerically,

$$\frac{\alpha I}{\nu} = \frac{1}{C}\int_{200\ \mathrm{nm}}^{250\ \mathrm{nm}} \alpha(c/\lambda)I_\lambda(\lambda)\lambda\ d\lambda \quad (6.10)$$

where I_λ is the intensity distribution function for wavelength. This yields $\alpha I/Lh\nu = 2 \times 10^{-5}\ \mathrm{s^{-1}}$. By substituting into eqn. (6.9) we find

$$d[W_1]/dt = 5 \times 10^8\ \text{molecules cm}^{-3}\ \mathrm{s^{-1}} \quad (6.11)$$

If we suppose that 90% of the solar UV is already absorbed, then the initial rate of aerosol production becomes 5×10^7 molecules cm^{-3} s^{-1}. The concentration of H_2O_2, which will be produced at twice this rate, will therefore reach 1 ppb by volume in *ca.* 4 min, neglecting loss by diffusion. Typical mean gas-phase values observed in the atmosphere range from 0.1 to 1 ppb,[32] so it is not inconceivable that the Wilson mechanism could contribute to the total amount of hydrogen peroxide in the atmosphere.

7. Conclusions

The calculations described herein lead to a number of conclusions concerning the possible role of an excited CT complex ($H_2O^+O_2^-$) in the photonucleation of water vapour in the presence of oxygen.

(i) The calculated gas-phase absorption spectrum is in qualitative agreement with the experimental absorption spectrum of oxygenated liquid water, and with the photonucleation spectrum of Katz and co-workers.

(ii) Rearrangement of the complex may be possible to form other species such as hydrogen peroxide and oxy-water which can also act as nucleating agents.

(iii) Molecular simulations of water clusters containing hydrogen peroxide and the CT complex, using admittedly approximate potentials, show unexpected and intriguing features: the solutes accumulate on the surface, and do not appear to stabilise the clusters much energetically.

(iv) In spite of uncertainty in the magnitude of the necessary parameters, a simple kinetic scheme involving the CT complex leads to an estimate of the rate of aerosol formation due to photonucleation, and a rate of H_2O_2 production which could be a significant contribution to the observed concentration of H_2O_2 in the atmosphere.

If such a CT complex is indeed important, as the calculations reported here suggest, then other atmospheric species with appropriate ionisation potentials may be involved in place of water as a source of primary nucleators, as suggested by studies involving hydrogen sulfide.[33]

We thank Prof. J. N. L. Connor for helpful discussions and the NERC for financial support.

References

1 C. T. R. Wilson, *Proc. Camb. Phil. Soc.*, 1897, **9**, 392; *Phil. Trans. R. Soc. A*, 1899, **192**, 403.
2 F. J. M. Farley, *Proc. R. Soc. A*, 1951, **207**, 527.
3 W. A. Hoppel and J. E. Dinger, *J. Atmos. Sci.*, 1973, **30**, 331.
4 H. Reiss, D. C. Martin and R. H. Heist, *J. Colloid Interface Sci.*, 1977, **58**, 125.
5 F. C. Wen, T. McLaughlin and J. L. Katz, *Science*, 1978, **200**, 769.
6 W. Byers Brown, *Chem. Phys. Lett.*, 1995, **235**, 94.
7 T. I. Quickenden and J. A. Irvin, *J. Chem. Phys.*, 1980, **72**, 4416.
8 (*a*) L. J. Heidt and L. Ekstrom, *J. Am. Chem. Soc.*, 1957, **79**, 1260; (*b*) L. J. Heidt and A. M. Johnson, *J. Am. Chem. Soc.*, 1957, **79**, 5587.
9 J. L. Katz, F. C. Wen, T. McLaughlin, R. J. Reusch and R. Partch, *Science*, 1977, **196**, 1203.
10 F. C. Wen, T. M. McLaughlin and J. L. Katz, *Phys. Rev. A*, 1982, **26**, 2235.
11 W. Byers Brown, M. A. Vincent, K. Trollope and I. H. Hillier, *Chem. Phys. Lett.*, 1992, **192**, 213.
12 (*a*) H-J. Werner and P. J. Knowles, *J. Chem. Phys.*, 1985, **82**, 5053; (*b*) P. J. Knowles and H-J. Werner, *Chem. Phys. Lett.*, 1985, **115**, 259; (*c*) D. Hegarty and M. A. Robb, *Mol. Phys.*, 1979, **38**, 795; (*d*) R. H. E. Eade and M. A. Robb, *Chem. Phys. Lett.*, 1981, **83**, 362; (*e*) M. J. Frisch, I. N. Ragazos, M. A. Robb and H. B. Schlegel, *Chem. Phys. Lett.*, 1992, **189**, 524.
13 GAUSSIAN 92, Revision E.1, M. J. Frisch, G. W. Trucks, M. Head-Gordon, P. M. W. Gill, M. W. Wong, J. B. Foresman, B. G. Johnson, H. B. Schlegel, M. A. Robb, E. S. Replogle, R. Gomperts, J. L. Andres, K. Raghavachari, J. S. Binkley, C. Gonzalez, R. L. Martin, D. J. Fox, D. J. Defrees, J. Baker, J. J. P. Stewart and J. A. Pople, Gaussian, Inc., Pittsburgh, PA, 1992.
14 MOLPRO 94, a package of *ab initio* programs by H-J. Werner and P. J. Knowles, with contributions from J. Almlof, R. Amos, S. Elbert, W. Meyer, E. A. Reinsch, R. Pitzer and A. Stone.
15 (*a*) H-J. Werner and P. J. Knowles, *J. Chem. Phys.*, 1988, **89**, 5803; (*b*) P. J. Knowles, H-J. Werner, *Chem. Phys. Lett.*, 1988, **145**, 514.
16 P. J. Knowles and H-J. Werner, *Theor. Chim. Acta*, 1992, **84**, 95.
17 (*a*) R. J. Gadnitz and R. Ahlrichs, *Chem. Phys. Lett.*, 1988, **143**, 413; (*b*) H-J. Werner and P. J. Knowles, *Theor. Chim. Acta*, 1990, **78**, 175.
18 (*a*) R. Krishnan, J. S. Binkley, R. Seeger and J. A. Pople, *J. Chem. Phys.*, 1980, **72**, 650; (*b*) T. Clark, J. Chandrasekhar, G. W. Spitznagel and P. v. R. Schleyer, *J. Comput. Chem.*, 1983, **4**, 294.
19 E. R. Davidson and D. W. Silver, *Chem. Phys. Lett.*, 1977, **53**, 403.

20 See for example, E. Huang, Y-L. Hwang, P. J. Dagdigian and M. H. Alexander, *J. Chem. Phys.*, 1993, **98**, 8484.
21 S. F. Boys and F. Bernardi, *Mol. Phys.*, 1970, **19**, 553.
22 (*a*) R. N. Diffenderfer and D. R. Yarkony, *J. Phys. Chem.*, 1982, **86**, 5098; (*b*) B. H. Lengsfield, *J. Chem. Phys.*, 1982, **77**, 4073; (*c*) H-J. Werner and W. Meyer, *J. Chem. Phys.*, 1981, **74**, 5794.
23 *Collision-Induced Absorption in Gases*, ed. Lothar Frommhold, Cambridge University Press, Cambridge, 1993.
24 (*a*) G. G. Balint-Kurti, C. L. Ward and C. C. Marston, *Comput. Phys. Commun.*, 1991, **67**, 285; (*b*) C. C. Marston and G. G. Balint-Kurti, *J. Chem. Phys.*, 1989, **91**, 3571; (*c*) G. G. Balint-Kurti, R. N. Dixon and C. C. Marston, *Int. Rev. Phys. Chem.*, 1992, **11**, 317.
25 J. N. L. Connor, *J. Chem. Phys.*, 1981, **74**, 1047.
26 J. A. Pople, M. Head Gordon and K. Raghavachari, *J. Chem. Phys.*, 1987, **87**, 5968.
27 M. A. Vincent and I. H. Hillier, *J. Phys. Chem.*, 1995, **99**, 3109.
28 W. L. Jorgensen, J. Chandrasekhar, J. D. Madura, R. W. Impey and M. L. Klein, *J. Chem. Phys.*, 1983, **79**, 926.
29 H. J. C. Berendsen, J. R. Grigera and T. P. Straatsma, *J. Phys. Chem.*, 1987, **91**, 6269.
30 D. dos Santos, M. Stein and A. J. Masters, to be published.
31 P. A. Giguere, B. G. Morissette, A. W. Olmos and O. Knop, *Can. J. Chem.*, 1955, **33**, 804.
32 K. J. Olszyna, J. F. Meagher and E. M. Bailey, *Atmos. Environ.*, 1988, **22**, 1699.
33 H. Cardy, C. Chevaldonnet, R. Sainte-Cluque and E. Poquet, *Chem. Phys.*, 1985, **94**, 385.

Paper 5/00297D; *Received* 16*th January*, 1995

Faraday Discuss., 1995, **100**, 269–278

HCl Vapour Pressures and Reaction Probabilities for $ClONO_2$ + HCl on Liquid H_2SO_4–HNO_3–HCl–H_2O Solutions

M. J. Elrod, R. E. Koch, J. E. Kim and M. J. Molina
Department of Earth, Atmospheric and Planetary Sciences, and Department of Chemistry, Massachusetts Institute of Technology, Cambridge, MA 02139, USA

Henry's Law solubility constants for HCl have been measured for liquid H_2SO_4–HNO_3–HCl–H_2O solutions; the results are in good agreement with predictions from published semiempirical models. The $ClONO_2$ + HCl reaction on the surfaces of such solutions with compositions simulating those of stratospheric aerosols has been investigated; as the composition changes following the temperature drop characteristic of the high-latitude stratosphere the reaction probability γ increases rapidly. Furthermore, the γ values remain essentially unchanged when HNO_3 uptake is neglected; the controlling factor appears to be the solubility of HCl. These results corroborate our earlier suggestion that supercooled liquid sulfate aerosols promote chlorine activation at low temperatures as efficiently as solid polar stratospheric cloud particles.

The importance of heterogeneous reactions involving $ClONO_2$, HCl, N_2O_5 and H_2O on the surfaces of polar stratospheric cloud (PSC) particles has been well established in connection with the observation of large-scale ozone depletion during the Antarctic springtime.[1,2] Laboratory studies have shown that both solid and liquid particles can promote heterogeneous reactions such as the following, converting the relatively inert chlorine reservoir molecules ($ClONO_2$ and HCl) into species (Cl_2, HOCl) which are readily photolysed to yield chlorine radicals:

$$ClONO_2 + H_2O \rightarrow HOCl + HNO_3 \quad (1)$$

$$HOCl + HCl \rightarrow Cl_2 + H_2O \quad (2)$$

$$ClONO_2 + HCl \rightarrow Cl_2 + HNO_3 \quad (3)$$

In particular, heterogeneous reactions on sulfuric acid aerosols have recently been under intense study since these particles may play a role in mid-latitude ozone depletion.[3–6] Hanson *et al.*[7] have described a framework for applying the results of these experimental studies to the modelling of the actual micrometre sized liquid sulfuric acid aerosols found in the stratosphere. It was concluded in these studies that reactions (1)–(3) become important at sulfate aerosol compositions corresponding to low atmospheric temperatures ($T < 200$ K), in agreement with our earlier suggestion that stratospheric aerosol particles need not freeze in order for chlorine activation reactions to proceed efficiently at high latitudes.[8]

It is important to determine accurately the composition of the aerosols as a function of the temperature and partial pressures of their constituents in order to model the various possible heterogeneous reactions adequately. Several groups have investigated uptake of HNO_3[9] and HCl[4,10,11] by sulfuric acid solutions. Vapour pressures of these constituents dissolved in bulk solutions have also been measured.[12] In addition, some of this data has been used to construct semi-empirical thermodynamic models which are intended to predict sulfate aerosol composition.[13,14] Although it has been shown that HNO_3 and HCl uptake by stratospheric sulfate aerosols is negligible for equilibrium temperatures above 210 K, the solubility of these species increases rapidly with decreasing temperature. Field observations of total aerosol volume at high northern latitudes have commonly been interpreted as evidence for solid PSC formation, since the aerosol volume is observed to rise quickly over a temperature range near the nitric acid trihydrate (NAT) frost point.[15] However, these observations are more consistent with the existence of supercooled liquid sulfate aerosols with substantial nitric acid content than with the existence of solid PSC particles.[16]

Since the reaction probability for the $ClONO_2$ + HCl reaction depends very sensitively on the HCl concentration in the aerosol, it is particularly important to determine accurately the solubility of HCl in solutions with compositions of atmospheric relevance. The existing results for the HCl solubility in sulfuric acid solutions are in conflict; in particular, the Henry's Law coefficients determined earlier in our laboratory[12] deviate significantly from similarly extensive measurements of Hanson and Ravishankara[4] and from the thermodynamic model predictions of Luo *et al.*[13]

Here we describe further measurements designed to resolve the discrepancy in the solubility results for the ternary H_2SO_4–H_2O–HCl system, and report the first measurements of Henry's Law coefficients for HCl for the quaternary (H_2SO_4–HNO_3–H_2O–HCl) system. We also report measurements of $ClONO_2$ + HCl reaction probabilities on the surface of such ternary and quaternary supercooled solutions with compositions relevant to the stratosphere.

Experimental

The HCl vapour pressures and the reaction probabilities for the $ClONO_2$ + HCl reaction were measured using a vertical wetted-wall flow tube coupled to a differentially pumped molecular-beam-sampling quadrupole mass spectrometer. The vertical flow tube is similar to that employed by Utter *et al.*[17] for measurements of ozone uptake on aqueous surfaces.

The top of the vertically aligned flow tube (id 2.4 cm) was fitted with a liquid delivery system that contained inlets for the helium carrier gas, a movable injector and the sulfuric acid solution. The delivery system consisted of an annular cup which, when filled with liquid, overflows to create a falling cylindrical film of acid on the inner wall of the flow tube. This wetted-wall technique has been widely used in the field of chemical engineering, and Danckwerts[18] has described the fluid dynamics of such flow. In order to maintain cold, stable temperatures, it was necessary to shut off the acid solution flow during actual measurement of HCl vapour pressures. The temperature of the acid was assumed to be the temperature of the coolant (which was sampled at the inlet to the flow tube) when the HCl signal was observed to stabilize (of the order of a few minutes). The flow-tube wall was rinsed with fresh solution for each measurement in order to minimize the possible effects of dehydration: there is insignificant dehydration of such acid solutions on such short timescales, as shown by Hanson and Ravishankara.[19] The carrier gas was sampled at the end of the flow tube and directed to the detection system, while the slowly moving acid film was collected in a bulb maintained at temperatures below that of the flow tube in order to prevent backstreaming of the vapours.

Most measurements were made with flow tube pressures of *ca.* 1 Torr and carrier gas flow velocities of *ca.* 1000 cm s^{-1}. The saturation of the carrier gas with HCl vapour was checked by directing the helium flow through the moveable injector; the observed HCl signal did not change with injector position, thus confirming the equilibrium condition. In order to determine absolute values for the HCl vapour pressures, the mass spectrometer was calibrated using the extrapolated HCl–H_2O data of Fritz and Fuget.[20] The detection limit for HCl (monitored at mass 36) using these methods was *ca.* 10^{-7} Torr.

$ClONO_2$ partial pressures of *ca.* 10^{-6} Torr were used for the kinetics measurements and the possible reactive depletion of HCl was investigated for each solution. Because the electron-impact ionization scheme results in the production of NO_2^+ for both $ClONO_2$ and HNO_3, it was necessary to detect the reaction product, Cl_2, for the quaternary solutions (since the ambient HNO_3 vapour pressures were high enough to interfere with the $ClONO_2$ signals). Reaction probabilities were determined using standard cylindrical flow tube analysis techniques.[21]

Bulk solutions were prepared by diluting 96% H_2SO_4, 70% HNO_3 and 37% HCl with deionized water. The solutions were spot-checked for composition with density and standard acid–base titration measurements.

Results and Discusssion

HCl Vapour Pressures

Henry's law solubility constants for HCl were determined for two ternary solutions and for two quaternary solutions over a range of temperatures (*ca.* 205–235 K) of stratospheric relevance. A 50% H_2SO_4 ternary solution was chosen for study to allow direct comparison with previous measurements. The other solutions were chosen in order to test the thermodynamic model of Luo *et al.*[13] for compositions of atmospheric interest. Neglecting uptake of HNO_3, it may be inferred that at a stratospheric temperature of *ca.* 190 K at 50 mbar the aerosols will contain 43% H_2SO_4 by weight, assuming 5 ppm by volume H_2O. We therefore measured solubility constants for this composition, for which the uptake of HCl is almost surely large enough for the $ClONO_2$ + HCl reaction to be quite important. In order to spot-check the Luo *et al.* model further, we chose two quaternary solutions that correspond to aerosol compositions at equilibrium temperatures of 198 and 196 K at 100 mbar, with 10 ppb by volume HNO_3 and 2 ppb by volume HCl; these solutions contain significant levels of nitric acid (3.5% and 12.5%, respectively).

Table 1 contains the measured Henry's Law solubility constants (H^*) for the four solutions considered. In Fig. 1, the various results for the 50% H_2SO_4 solution are presented. In earlier work in our laboratory[12] the vapour pressure data was parametrized as a function of temperature, and it is shown in Fig. 1 as a dashed line. The results from the model of Luo *et al.*[13] are shown as a solid line, and each data point from Hanson and Ravishankara,[4] Williams and Golden,[11] and this work is shown individually. Our new results form a consensus with two of the previous measurements and with the model predictions, while the Zhang *et al.*[12] data lies well outside the range of the consensus findings. The reason for the discrepancy is not clear; it is most likely a consequence of HCl calibration errors in our earlier work. In Fig. 2, the H^* values for the 43% H_2SO_4 solution are plotted together with the model predictions of Luo *et al.*;[13] as can be seen in the figure, the agreement is very good.

In Fig. 3, our measured HCl Henry's Law constants for the two quaternary solutions are presented together with the predictions from the thermodynamic model; once again, for these atmospherically relevant compositions, the agreement is excellent. These measurements are not intended to be an exhaustive test of the accuracy of the model; nevertheless, as has been shown to be the case for the ternary solutions it appears that the model also predicts the HCl solubility for quaternary systems very accurately.

Reaction Probabilities for HCl + $ClONO_2$

The compositions of the bulk solutions used for the kinetics investigations simulated those of typical stratospheric aerosols. These compositions were determined by using the Luo *et al.*[13] model to search for mixtures whose vapour pressures correspond (within

Table 1 Experimental Henry's law solubility constants

composition	T/K	H^*/mol dm^{-3} atm^{-1}
50% H_2SO_4	225	4.0×10^5
9.2×10^3 mol dm^{-3} HCl	222	5.2×10^5
	219	7.2×10^5
	216	9.8×10^5
	214	1.2×10^6
	211	1.8×10^6
	208	2.3×10^6
43% H_2SO_4	226	2.5×10^6
3.0×10^{-2} mol dm^{-3} HCl	222	4.1×10^6
	218	6.8×10^6
	214	9.2×10^6
	208	1.8×10^7
48% H_2SO_4	233	8.8×10^4
3.5% HNO_3	231	1.1×10^5
3.9×10^{-3} mol dm^{-3} HCl	226	1.7×10^5
	222	2.7×10^5
	216	5.4×10^5
36.2% H_2SO_4	228	5.5×10^5
12.5% HNO_3	223	1.2×10^6
6.2×10^{-3} mol dm^{-3} HCl	218	1.7×10^6
	213	2.8×10^6
	208	4.7×10^6

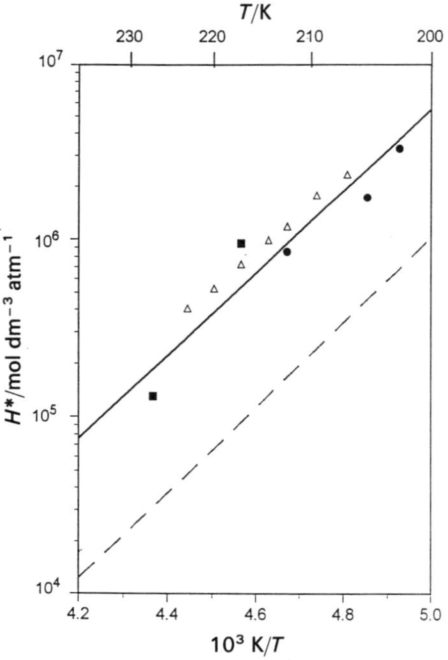

Fig. 1 Henry's Law constants as a function of temperature for a 50% H_2SO_4 solution. (— —) data from Zhang *et al.*,[12] and (——) from the semi-empirical model of Luo *et al.*[13] △, our data, ■, data from Williams and Golden[11] and ●, from Hanson and Ravishankara.[4]

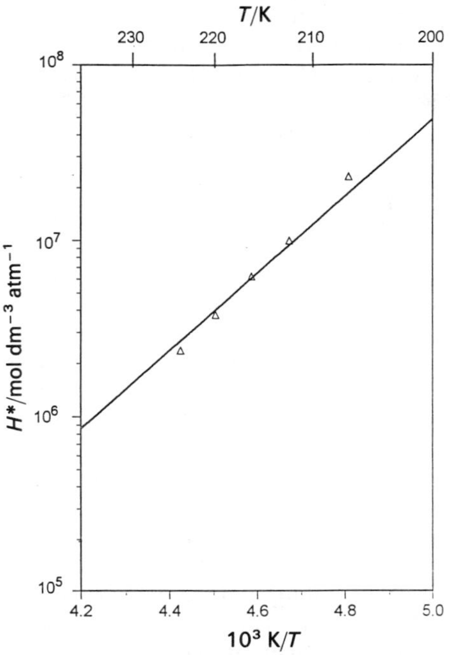

Fig. 2 Henry's Law constants as function of temperature for a 43% H_2SO_4 solution. The line represents estimates from the Luo *et al.* model[13] and △, represent our data.

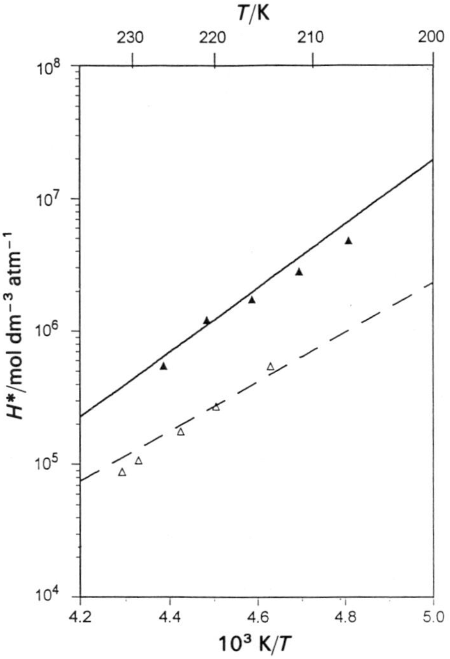

Fig. 3 Henry's law constants as function of temperature for the two quaternary solutions studied in this work. The lines represent estimates from the Luo *et al.* model[13] and △, our data. (——) and ▲, 36.2% H_2SO_4–12.5% HNO_3 solution, (— —) and △, 48.0% H_2SO_4–3.5% HNO_3 solution.

10%) to various atmospheric equilibrium temperatures T(eq), assuming 5 ppm by volume H_2O and 2 ppb by volume HCl at 100 mbar (*ca.* 16 km altitude), with no HNO_3 in Table 2 and 10 ppb by volume HNO_3 in Table 3. As the equilibrium temperature is lowered from 199 to 195 K, the weight fraction of HNO_3 in the solution increases from 2.2% to 28.3%, and the amount of HCl increases by more than a factor of 30. For aerosols at equilibrium with an atmosphere of 50 mbar total pressure and the same mixing ratios, similar compositional changes are expected at temperatures shifted by *ca.* 4 K below the 100 mbar values. We did not account for depletion of gas-phase HNO_3 or HCl in the calculation of the equilibrium compositions; it should not be significant for background aerosol sulfate loadings in the temperature range considered in this work.[16] For ease of discussion, we will hereafter refer to each solution by its equilibrium temperature and by whether it contains nitric acid [*i.e.*, 197 K (no HNO_3)].

In Fig. 4 we present loss and production curves for $ClONO_2$ and Cl_2, respectively, for the 198 K (no HNO_3) solution; the two curves yield similar first-order rate constants. The reaction probability values (γ) for the two nitric acid scenarios are listed in Tables 2 and 3, and are the result of an average of at least three separate measurements. Through the use of bulk solutions prepared with specific compositions, we have effectively decoupled the atmospheric equilibrium temperature from the laboratory temperature; the former has a much larger effect on the reaction probabilities. Because of the tendency of the supercooled solutions to freeze, it was necessary to measure the reaction probabilities at laboratory temperatures which are higher than the atmospheric equilibrium temperatures. As shown in Table 4, the variation in the reaction probability for the 197 K (no HNO_3) solution over the laboratory temperature range 203–233 K was quite small, of the order of the experimental uncertainty. The 195 K (HNO_3) solution lies well within the nitric acid trihydrate (NAT) stability region of the phase diagram,[8] and we were unable to cool the solution below 223 K without freezing. All other reaction probability measurements were carried out at 213 K and usually also at

Table 2 Reaction probabilities for the ternary H_2SO_4–H_2O–HCl system measured at 203 K, as a function of atmospheric equilibrium temperature

T(eq)/K	weight fraction H_2SO_4	weight fraction HCl	reaction probability
201	0.552	3.00×10^{-6}	0.010
199	0.520	1.60×10^{-5}	0.022
198	0.510	2.40×10^{-5}	0.038
197	0.490	6.00×10^{-5}	0.061
196	0.470	1.40×10^{-4}	0.083
195	0.450	3.44×10^{-4}	0.12
194	0.430	8.24×10^{-4}	0.30

Table 3 Reaction probabilities for the quaternary H_2SO_4–HNO_3–H_2O–HCl system measured at 203 K, as a function of atmospheric equilibrium temperature

T(eq)/K	weight fraction H_2SO_4	weight fraction HNO_3	weight fraction HCl	reaction probability
199	0.503	0.022	1.38×10^{-5}	0.021
198	0.480	0.035	2.70×10^{-5}	0.026
197	0.440	0.061	6.18×10^{-5}	0.058
196	0.362	0.126	1.76×10^{-4}	0.070
195	0.202	0.283	4.88×10^{-4}	0.110

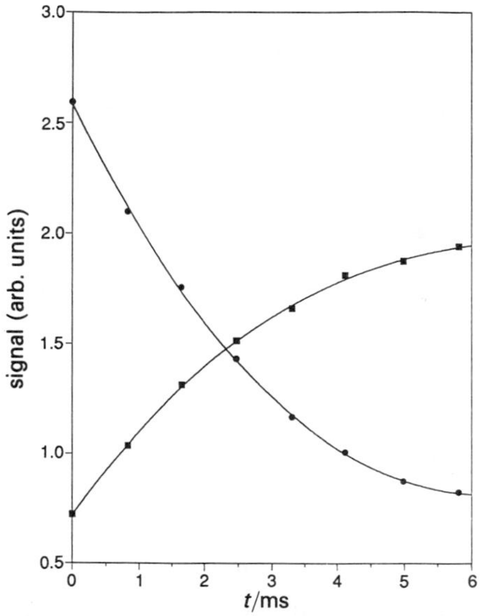

Fig. 4 ●, Loss ($ClONO_2$) ($k = 315\ s^{-1}$) and ■, production (Cl_2) ($k = 322\ s^{-1}$) curves for the $HCl + ClONO_2$ reaction on the 198 K (no HNO_3) solution measured at a laboratory temperature of 213 K

203 K, so that the extrapolation to the atmospheric temperature was not extreme. Hanson and Ravishankara[5] had already noted that the reaction probability is nearly independent of temperature for solutions of a given composition; they attributed this behaviour to a fortuitous compensation of effects.

In Fig. 5, the results from our reaction probability measurements are presented and compared with values estimated from the results reported by Hanson and Ravishankara[5] for ternary (H_2SO_4–H_2O–HCl) compositions which closely correspond to our 196, 198 and 201 K solutions. It may be seen that our reaction probabilities for solutions without HNO_3 are slightly lower than those of Hanson and Ravishankara.[5] Fig. 5 also shows that the incorporation of HNO_3 into the sulfate solutions lowers the reaction probability relative to solutions without HNO_3 only to a small extent, despite the presence of HNO_3 levels as high as 28.3%; the difference is of the order of the experimental uncertainty, although it appears to be statistically significant. Zhang *et al.*[6] also measured reaction probabilities without HNO_3 at a temperature of 198 K, for one solution containing *ca.* 4% HNO_3, and found no change in the $ClONO_2$ + HCl reaction probability.

In Fig. 6 the reaction probabilities from Tables 2 and 3 are instead plotted as a function of HCl content. The line in the figure is a least-squares fit; its slope indicates that γ is directly proportional to the square root of the HCl concentration. Hanson and Ravishankara[5] measured γ over a wide HCl partial pressure (p_{HCl}) range: as expected, at very low p_{HCl} the $ClONO_2$ decay rate is dominated by reaction (1), and hence is nearly independent of p_{HCl}. For intermediate and high p_{HCl} values γ was found to be proportional to $(p_{HCl})^{1/2}$ and to p_{HCl}, respectively; Hanson and Ravishankara[5] interpreted these results in terms of a bulk and a surface component to γ. The measurements of Zhang *et al.*[6] were carried out at relatively high HCl concentrations, and their data indicate close to linear proportionality between γ and p_{HCl}. Considering the validity of Henry's law for these conditions, the same functional forms should apply to [HCl]: our measurements were carried out at intermediate [HCl] values, yielding a square-root

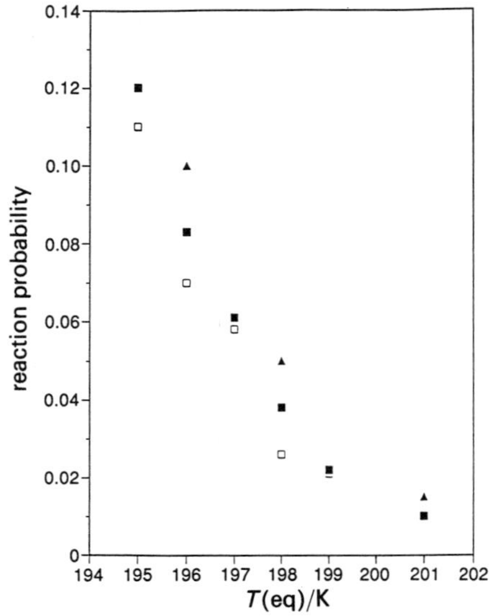

Fig. 5 Reaction probability measurements for HCl + $ClONO_2$ as a function of atmospheric temperature *T*(eq), assuming 5 ppm by volume H_2O and 2 ppb by volume HCl at 100 mbar. □, Our data, for 10 ppb by volume HNO_3; ■, our data, assuming no HNO_3; ▲, data estimated from the results of Hanson and Ravishankara[5] (no HNO_3).

proportionality, in agreement with the results obtained by Hanson and Ravishankara[5] for similar [HCl] values.

In an attempt to ascertain a dependence of γ on the H_2SO_4 content, we chose a fixed HCl concentration of 2.3×10^{-3} mol dm^{-3} [corresponding to the moderately efficient reaction probability of the 197 K (no HNO_3) solution] and varied the H_2SO_4 content from 40 to 60%. This data is listed in Table 4; it is clear that the reaction probability

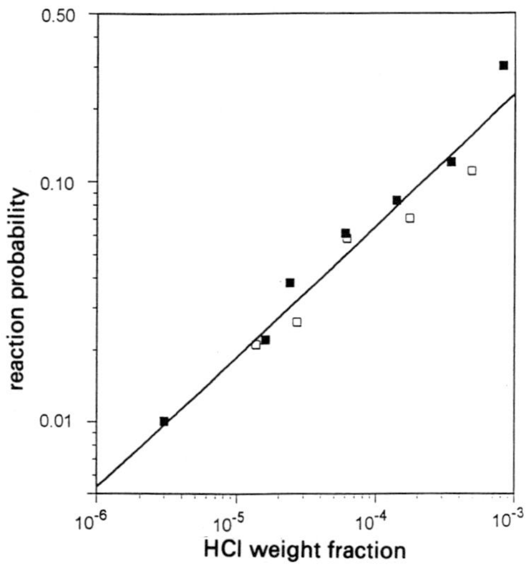

Fig. 6 Reaction probability measurements for HCl + $ClONO_2$ as a function of HCl weight fraction. □ and ■, experimental data for solutions with and without HNO_3, respectively. The line is a fit through the data; its slope is 0.545.

Table 4 Reaction probabilities for the H_2SO_4–H_2O–HCl system as a function of temperature and sulfuric acid content, measured at a fixed HCl concentration of 2.3×10^{-3} mol dm^{-3}

T(exp)/K	weight fraction H_2SO_4	reaction probability
203	0.49	0.060
213	0.49	0.061
223	0.49	0.050
233	0.49	0.049
213	0.40	0.057
213	0.45	0.046
213	0.51	0.061
213	0.55	0.059
213	0.60	0.054

does not change significantly over the stratospherically relevant H_2SO_4 content range given in the table. This result, in combination with the laboratory temperature-dependence studies, suggests that the atmospheric temperature (and thus the composition of the aerosols) affects the $ClONO_2$ + HCl reaction probability predominantly through its effect on the HCl solubility. Although the uptake of HNO_3 serves to increase the solubility of HCl, it may be seen (Tables 2 and 3) that reaction probabilities are already very high at the atmospheric temperatures where this effect becomes important.

Conclusions

The Henry's Law solubility constants for HCl reported in this work for the ternary H_2SO_4–H_2O–HCl and quaternary H_2SO_4–HNO_3–H_2O–HCl systems have been shown to agree with the thermodynamic model of Luo *et al.*[13] for selected atmospherically relevant compositions and temperatures. Furthermore, our new results for the ternary system are in agreement with the previous measurements of Hanson and Ravishankara[4] and Williams and Golden,[11] while our earlier results[12] deviate significantly from these consensus findings.

Our results show that the uptake of substantial levels of nitric acid (as much as 28%) does not appreciably affect the reaction probabilities for the $ClONO_2$ + HCl process on sulfuric acid surfaces. The results also show that the reaction probabilities are not sensitive to temperature for solutions with a given composition. The controlling parameter is the concentration of HCl in the solution, which for a given HCl partial pressure depends in turn on temperature; in the lower stratosphere this concentration increases rapidly below *ca.* 200 K.

Since it is now clear that the $ClONO_2$ + HCl reaction proceeds efficiently on supercooled liquid sulfuric acid surfaces with compositions relevant to the cold stratosphere, including the possibility of extensive HNO_3 uptake, it is not necessary to invoke the presence of solid PSC particles to explain the high chlorine activation rates associated with large scale ozone depletion. As suggested earlier, the key factor is the stratospheric temperature;[8] as it approaches the HNO_3–H_2O dew point, the composition of the liquid aerosols changes rapidly,[22] with the consequent increase in HCl solubility. If the aerosols are solid, it is the approach to the ice frost point that matters; chlorine activation occurs readily on the ice-like surfaces. We note, however, that a change in phase for the stratospheric aerosols might have other important consequences: the surface area available for reaction can change with freezing and, more importantly, the mechanism for efficient denitrification might involve nucleation of a small fraction of the liquid particles; the solid ones would then selectively grow and fall out of the stratosphere.

This work was supported by a NASA grant (NAG2-632) and by an NSF grant (ATM-9017150). We thank S. Clegg for sending us a computer program to calculate vapour pressures using the model described by Luo *et al.*[13]

References

1 S. Solomon, R. R. Garcia, F. S. Rowland and D. J. Wuebbles, *Nature* (*London*), 1986, **321**, 755.
2 M. J. Molina, in *The Chemistry of the Atmosphere: Its Impact on Global Change*, ed. J. G. Calvert, Blackwell Scientific, Boston, MA., 1994, p. 27.
3 M.A. Tolbert, M. J. Rossi and D. M. Golden, *Geophys. Res. Lett.*, 1988, **15**, 847.
4 D. R. Hanson and A. R. Ravishankara, *J. Chem. Phys.*, 1993, **97**, 12 309.
5 D. R. Hanson and A. R. Ravishankara, *J. Phys. Chem.*, 1994, **98**, 5728.
6 R. Zhang, M-T. Leu and L. F. Keyser, *J. Phys. Chem.*, 1994, **98**, 13 563.
7 D. R. Hanson, A. R. Ravishankara and S. Solomon, *J. Geophys. Res.*, 1994, **99**, 3615.
8 M. J. Molina, R. Zhang, P. J. Wooldridge, J. R. McMahon, J. E. Kim, H. Y. Chang and K. D. Beyer, *Science*, 1993, **261**, 1418.
9 C. M. Reihs, D. M. Golden and M. A. Tolbert, *J. Geophys. Res.*, 1990, **95**, 16 545; J. M. Van Doren, L. R. Watson, P. Davidovits, D. R. Worsnop, M. S. Zahniser and C. E. Kolb, *J. Phys. Chem.*, 1991, **95**, 1684.
10 L. R. Watson, J. M. Van Doren, P. Davidovits, D. R. Worsnop, M. S. Zahniser and C. E. Kolb, *J. Geophys. Res.*, 1990, **95**, 5631.
11 L. R. Williams and D. M. Golden, *Geophys. Res. Lett.*, 1993, **20**, 227.
12 R. Zhang, P. J. Wooldridge and M. J. Molina, *J. Phys. Chem.*, 1993, **97**, 8541.
13 B. Luo, K. S. Carslaw, T. Peter and S. L. Clegg, *Geophys. Res. Lett.*, 1995, **22**, 247.
14 A. Tabazadeh, R. P. Turco and M. Z. Jacobson, *J. Geophys. Res.*, 1994, **99**, 12 897.
15 S. R. Kawa, D. W. Fahey, K. K. Kelly, J. E. Dye, D. Baumgardner, B. W. Gandrud, M. Loewenstein, G. V. Ferry and K. R. Chan, *J. Geophys. Res.*, 1992, **97**, 7925.
16 K. S. Carslaw, B. P. Luo, S. L. Clegg, T. Peter, P. Brimblecombe and P. J. Crutzen, *Geophys. Res. Lett.*, 1994, **21**, 2479; K. Drdla, A. Tabazadeh, R. P. Turco, M. Z. Jacobson, J. E. Dye, C. Twohy and D. Baumgardner, *Geophys. Res. Lett.*, 1994, **21**, 2475.
17 R. G. Utter, J. B. Burkholder, C. J. Howard and A. R. Ravishankara, *J. Phys. Chem.*, 1992, **96**, 4973.
18 P. V. Danckwerts, *Gas–Liquid Reactions*, McGraw-Hill, New York, 1970.
19 D. R. Hanson and A. R. Ravishankara, *J. Geophys. Res.*, 1991, **96**, 17 307.
20 J. J. Fritz and C. R. Fuget, *Chem. Eng. Data Ser.*, 1956, **1**, 10.
21 R. L. Brown, *J. Res. Natl. Bur. Stand.*, 1978, **83**, 1; C. J. Howard, *J. Phys. Chem.*, 1979, **83**, 3.
22 K. D. Beyer, S. W. Seago, H. Y. Chang and M. J. Molina, *Geophys. Res. Lett.*, 1994, **21**, 871.

Paper 5/01024A; *Received* 20*th February*, 1995

Faraday Discuss., 1995, **100**, 279–294

GENERAL DISCUSSION

Mr. Ball opened the discussion of Dr. Hancock's paper: It should be noted that the two pathways to $O_2(a\,^1\Delta_g)$ production in the near-UV photolysis of ozone proposed by Dr. Hancock, namely the spin-allowed photodissociation

$$O_3 + h\nu \rightarrow O(^1D_2) + O_2(a\,^1\Delta_g) \quad (1)$$

and the spin-forbidden dissociation

$$O_3 + h\nu \rightarrow O(^3P_{2,1,0}) + O_2(a\,^1\Delta_g) \quad (9)$$

have very different thermodynamic wavelength thresholds. The thresholds are 310 and 611 nm,[1] respectively.

In a series of experiments[2] building on the work presented by Dr. Hancock, we have used a Wiley–McLaren time-of-flight (TOF) mass spectrometer to measure the kinetic energy release accompanying the photodissociation of ozone at a number of wavelengths either side of the 310 nm threshold for process (1). A strong correlation is known to exist between the transition dipole moment in the parent ozone molecule, $\boldsymbol{\mu}$, and the photofragments' velocities, $\boldsymbol{v}$: the anisotropy parameter $\beta = 1.18$,[3] *i.e.* it is a predominantly parallel dissociation. In our present experiment, ozone is photolysed with polarised radiation such that the electric vector ε_{photo} is directed along the TOF axis. Half of the $O_2(a\,^1\Delta_g)$ fragments are produced with their velocities directed towards the TOF tube and half with their velocities directed away from it. Shortly after photodissociation, the $O_2(a\,^1\Delta_g)$ fragments are ionised by the REMPI probe laser beam. Under the influence of the electric fields within the instrument, both sets of ionised fragments are directed down the TOF tube and onto the detector. Owing to the anisotropy of the dissociation, a time separation exists between the arrival of the two fragment sets, and this separation provides information on the kinetic energy release accompanying the dissociation. For photolysis far from threshold, the excess energy is large and the two fragment sets arrive well separated in time: as the threshold is approached, the separation decreases. A clear narrowing of the fragment ions' TOF profile can be seen when the photolysis wavelength is changed from 299 to 305 nm, Fig. 1.

However, for photolysis well beyond the 310 nm threshold for process (1) in the region where spin-forbidden dissociation is believed to be taking place, the separation of the peaks in the $O_2(a\,^1\Delta_g)$ ions' TOF profile becomes very wide indeed. The results at 328 and 331 nm show that the photolysis of ozone at these wavelengths produces $O_2(a\,^1\Delta_g)$ fragments with high kinetic energy. The photolysis must therefore be taking place far from threshold, and is clear evidence for the production of $O_2(a\,^1\Delta_g)$ *via* the spin-forbidden process (9).

1 R. Atkinson, D. L. Baulch, R. A. Cox, R. F. Hampson Jr., J. A. Kerr and J. Troe, *J. Phys., Chem. Ref. Data*, 1992, **21**, 1125.
2 S. M. Ball, G. Hancock, J. C. Pinot de Moira, C. M. Sadowski and F. Winterbottom, *Chem. Phys. Lett.*, in the press.
3 C. E. Fairchild, E. J. Stone and G. M. Lawrence, *J. Chem. Phys.*, 1978, **69**, 3632; A. G. Suits, R. L. Miller, L. S. Bontuyan and P. L. Houston, *J. Chem. Soc., Faraday Trans.*, 1993, **89**, 1443.

Prof. Ashfold said: Hancock and co-workers provide clear evidence for the formation of $O_2(a\,^1\Delta_g)$ fragments *via* a spin-forbidden process following photoexcitation of ozone within its structured Huggins band system at wavelengths longer than *ca.* 320 nm. I invite comments on the reliability of our current understanding of the spectroscopy of this band system. For example, there are previous literature assignments[1] of two vibronic

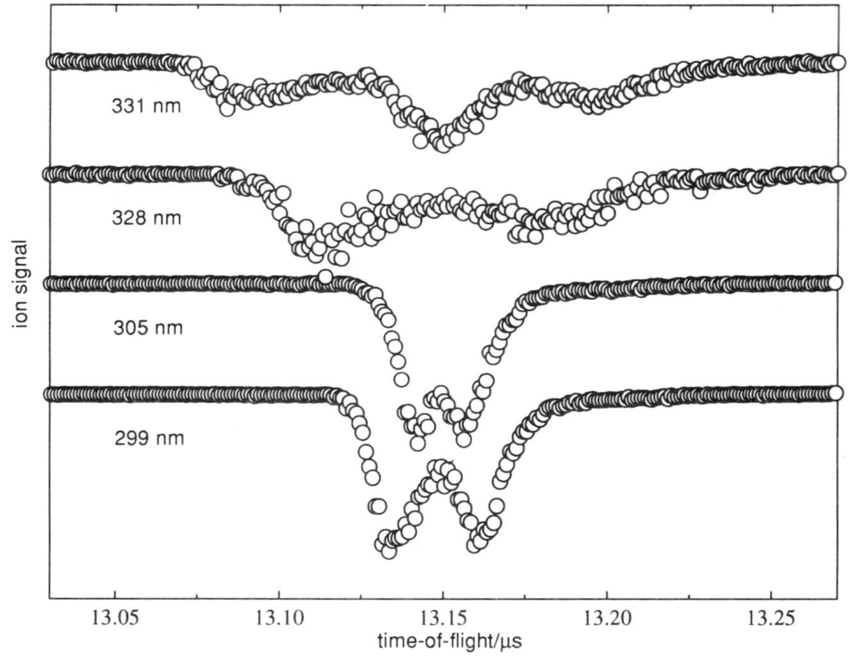

Fig. 1 $O_2(a\,^1\Delta_g)$ TOF profiles at various ozone photolysis wavelengths as indicated on the figure. The electric vectors of the polarised photolysis radiation and REMPI probe radiation lie along the TOF axis.

features appearing in absorption around 320 nm. These both originate from the zero-point level of the ground ($\tilde{X}\,^1A_1$) state but reportedly terminate on two different (v_1', v_2', v_3') excited-state vibrational levels, namely (5, 0, 2) and (6, 0, 1). These differ by one quantum of the asymmetric stretching vibration, v_3, and, since this vibration transforms as b_2 in the C_{2v} point group of the molecule, these two bands should display different rotational band contours. Rotational congestion, compounded by predissociation broadening, means that it is not possible to test this hypothesis with the room-temperature spectrum, but jet-cooled absorption studies[2] of neighbouring transitions terminating on the (traditionally assigned) (5, 0, 1) and (6, 0, 0) excited-state levels show these to have the same rotational band types. A recent reanalysis[3] of the vibrational structure accompanying the Huggins band system seems to clear up this and other deficiencies in the previous assignments. Specifically, Joens[3] proposes that the Huggins bands arise from a forbidden $^1A_1$–$\tilde{X}\,^1A_1$ electronic transition, and that the observed vibronic bands all involve excited-state levels with an odd number of quanta in v_3. Thus all have $^1B_2$ vibronic symmetry, and gain transition strength by vibronic mixing with the strongly allowed $^1B_2$–$\tilde{X}\,^1A_1$ transition responsible for the Hartley band; Joens' reinterpretation[3] would label the two features identified in Fig. 3 of the work of Ball *et al.* as involving the (2, 0, 5) and (4, 0, 3) excited-state vibrational levels, respectively. I conclude with a general question: just how good is our understanding of the near-UV absorption spectrum of ozone?

1 D. H. Katayama, *J. Chem Phys.*, 1979, **71**, 815; 1986, **85**, 6809.
2 A. Sinha, D. Imre, J. H. Goble Jr. and J. L. Kinsey, *J. Chem Phys.*, 1986, **84**, 6108.
3 J. A. Joens, *J. Chem Phys.*, 1994, **101**, 5431.

Dr. Hancock responded: The recent questioning of the assignment of the structured features of the Huggins band certainly indicates that our understanding of the near-UV spectroscopy of ozone is incomplete. Fig. 3 of our paper shows measurements of the quantum yields of $O_2(a\,^1\Delta_g)$ over vibronic features which, if the reassignment is correct, need now to be labelled as (2, 0, 5) at 319.4 nm, and (4, 0, 3) at 320.0 nm. The reassignment does not change our major conclusion, that the $O_2(a\,^1\Delta_g)$ quantum yield shows no obvious correlation with this structure, and thus no dominant dependence of the dissociation mechanism leading to this fragment on the nature of the vibronic levels reached by absorption in this region.

Measurements of the quantum yield of the $O(^1D_2)$ fragment in this wavelength region are clearly of great importance for the needs of atmospheric modellers, but we should not be content with carrying out such pragmatic measurements without attempting to understand the spectroscopy and dissociation dynamics of ozone. One of the ways of finding information on the symmetries of the upper state(s) reached in the Huggins band is from TOF measurements currently underway in our laboratory. We find that the form of the TOF profile changes when the electric vector of the photolysis laser is rotated, and analyses of the effect shows that for both the spin-allowed (299 and 305 nm) and spin-forbidden (328 and the extreme peaks at 331 nm) dissociations, the transition is predominantly a parallel one. This is consistent with the upper levels having $^1B_2$ symmetry, but does not distinguish between direct dissociation following absorption to the $^1B_2$ surface, or absorption to the $^1B_2$ vibronic features of an electronically excited $^1A_1$ surface. More combinations of spectroscopic and dissociation dynamics experiments are needed as suggested by Prof. Ashfold's remark.

I point out one feature of the TOF data shown by Ball. At 331 nm there is a clear slow $O_2(a\,^1\Delta_g)$ peak between the fast forward and backward scattered features. We are, as yet, unsure of its origin, but an intriguing possibility is that it is from the photolysis of the small amount of vibrationally excited ozone in a spin-allowed step, with the forward and backward peaks resulting from spin-forbidden dissociation. However the relative magnitudes of these peaks only give a measure of the relative quantum yields into the specific quantum states of $O_2(a\,^1\Delta_g)$ probed at 331.5 nm. These are at low J in $v = 0$, and thus underestimate the true spin-forbidden process which, on energetic grounds, can form a wide range of internally excited states unobserved in this experiment. The data at 328 nm show fragments with lower kinetic energies than those found at the longer wavelength. This experiment was carried out with a single laser so that $O_2(a\,^1\Delta_g)$ fragments were formed and probed at the same wavelength, which, we believe, probes highly internally excited oxygen molecules. Under these conditions no slow spin-allowed fragments are seen. Experiments of this kind can be extended to probe the atomic oxygen fragments $O(^1D_2)$ and $O(^3P_{2,1,0})$ and can be used to unravel the complex dissociation dynamics (and hopefully the spectroscopy) of ozone in this region.

Prof. Zellner asked: What is the lifetime of O_3^* ($^1A_1{}^1B_1$) around and beyond 310 nm? Is it conceivable that the O_2 ($^1\Delta_g$) that you observe in your experiments results from collisional energy transfer from O_3^* to O_2 ($X\,^3\Sigma_g^-$)?

Dr. Hancock replied: Sinha *et al.*[1] have estimated an upper bound to the lifetime of the states reached by absorption near 325 nm as being 3.6 ps. This upper limit is greater than a rotational period (the low-J spectra support rotational structure). Our TOF measurements of the spin-forbidden channel near 331 nm (Fig. 1 of the comment by Ball), however, imply a smaller value, as the anisotropy of the dissociation is not washed out by rotation of the excited ozone molecule (although it should be emphasised that our observations are for the minor dissociation channel forming $O_2(a\,^1\Delta_g)$, and the major channel to form ground-state fragments may originate from states which have a longer lifetime). To produce a quantum yield of 0.2 by collisional energy transfer to O_2 by an excited ozone molecule whose lifetime is 3.6 ps requires, for our highest pressure measurements, a rate constant which is at least two orders of magnitude larger than gas kinetic, and seems unlikely. Measurements over a range of O_3/O_2 pressures and ratios produced no evidence of such collisional processes being of importance.

1 A. Sinha, D. Imre, J. H. Goble Jr. and J. L. Kinsey, *J. Chem. Phys.*, 1986, **84**, 6108.

Prof. Ravishankara remarked: When considering the quantum yields for various products in the dissociation of O_3 it is necessary to remember that at short wavelengths, *i.e.* $\lambda < 200$ nm, there is a channel for O_3 to dissociate into three O atoms.[1] This dissociation leads to a net ozone production. I wonder if this will have any impact on the upper stratospheric ozone production rate.

1 A. A. Turnipseed, G. L. Vaghjiani, T. Gierczak, J. E. Thompson and A. R. Ravishankara, *J. Chem. Phys.*, 1991, **95**, 3244.

Dr. Hancock responded: It would be interesting to look at the dissociation dynamics of ozone photolysis at 193 nm, where Turnipseed *et al.* find the quantum yield for oxygen atom production to be larger than unity. Velocity measurements of the ground-state O atom by TOF following REMPI at 226 nm may be able to show the existence of this fragmentation channel, and indicate whether it would compete as a net odd-oxygen source with photolysis of O_2 at these short wavelengths.

Some very recent measurements have provided further evidence for the existence of the long-wavelength tail in the quantum yield of $O(^1D_2)$ from the photolysis of ozone. Hofzumahaus and co-workers have made simultaneous measurements of solar UV flux and $O(^1D_2)$ formation rates over a range of ozone column heights.[1] They find that their data are explained better when the tail is included, and have shown how this will increase the calculated OH production rate in the troposphere by a factor of up to two. Armerding *et al.* have measured the $O(^1D_2)$ quantum yields between 300 and 330 nm in an indirect fashion by detecting the OH product of the reaction of the excited oxygen atom with H_2O.[2] Results are in good agreement with our observation of the $O_2(a\,^1\Delta_g)$ tail in this wavelength region. We are currently carrying out direct observations of the $O(^1D_2)$ product by REMPI and hope that this will lead to definitive values of the wavelength and temperature variation of its quantum yield in the near UV.

1 M. Müller, A. Kraus and A. Hofzumahaus, *Geophys. Res. Lett.*, 1995, **22**, 679.
2 W. Armerding, F. J. Comes and B. Schülke, *J. Phys., Chem.*, 1995, **99**, 3137.

Dr. Hernandez addressed Prof. Wodtke: I would like to present the results of theoretical work we have performed which is of direct relevance to the paper presented by Prof. Wodtke.

We have calculated, from first principles, the state-selected vibrational relaxation rates in collisions of two oxygen molecules with one molecule in a highly vibrationally excited state.[1] The method we use is based on high quality *ab initio* calculations of the potential-energy surface, including the explicit dependence on the O_2 vibrational coordinates, followed by quantum scattering computations of cross-sections and rate constants for the vibrational energy transfer. The scattering calculations were performed with the vibrationally close-coupling rotationally infinite-order sudden approximation (VCC-IOSA).[2]

In Fig. 2 we present a comparison of theoretical[1,3] and experimental[4–6] results for the state-selected vibrational relaxation rates. We obtain very good agreement with the three independent sets of experimental data for values of the vibrational quantum number, $v < 26$. These results confirm the conclusion that the relaxation rates are too fast for the autocatalytic ozone production cycle of Slanger[7–11] to be of relevance. At the higher values of v our calculations, which only treat the non-reactive dynamics, are unable to reproduce the sharp increase in rate constant observed by the Wodtke group.[4]

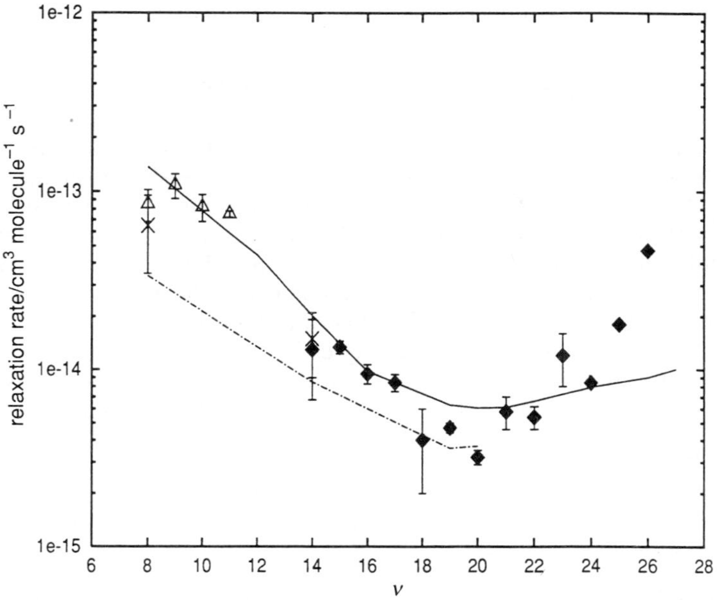

Fig. 2 Total state-to-state vibrational relaxation rates for $O_2 + O_2(v)$ at 295 K: comparison with previous experimental and theoretical work. All the experimental results include error bars. Present work is the solid line, ref. 1; – · –, ref. 3 (calculations by Billing and Kolesnick); ◆, ref. 4 (Wodtke and co-workers); ×, ref. 5 (Park and Slanger); △, ref. 6 (Smith and co-workers).

This is consistent with the opening of a reactive channel leading to ozone formation. We have started quantum scattering calculations of reactive rate constants to confirm that this is the new mechanism appearing for $v \geqslant 26$.

1 R. Hernández, R. Toumi and D. C. Clary, *J. Chem Phys.*, 1995, **102**, 9544.
2 D. C. Clary, *Chem. Phys. Lett.*, 1980, **74**, 454.
3 G. D. Billing and R. E. Kolesnick, *Chem. Phys. Lett.*, 1992, **200**, 382.
4 J. M. Price, J. A. Mack, C. A. Rogaski and A. M. Wodtke, *Chem Phys.*, 1993, **175**, 83.
5 H. Park and T. G. Slanger, *J. Chem Phys.*, 1994, **100**, 287.
6 M. Klatt, I. W. M. Smith, R. P. Tuckett and G. N. Ward, *Chem. Phys. Lett.*, 1994, **224**, 253.
7 T. G. Slanger, L. E. Jusinski, G. Black and G. E. Gadd, *Science*, 1988, **241**, 945.
8 R. Toumi, B. J. Kerridge and J. A. Pyle, *Nature (London)*, 1991, **351**, 217.
9 R. Toumi, *J. Atmos. Chem.*, **15**, 69 (1992).
10 T. G. Slanger, *Science*, 1994, **265**, 1817.
11 R. L. Miller, A. G. Suits, P. L. Houston, R. Toumi, J. A. Mack and A. M. Wodtke, *Science*, 1994, **265**, 1831.

Prof. Smith commented: I should like to give a few details of our experiments[1,2] on vibrationally excited O_2 referred to by Dr. Hernandez. In these experiments, we generate O_2 in vibrational levels $v \leqslant 11$ from the reaction between $O(^3P)$ atoms and NO_2.[3] This reaction is initiated by partial photolysis of NO_2 at 355 nm using radiation from a frequency-tripled Nd : YAG laser.

$$NO_2 + h\nu(\lambda = 355 \text{ nm}) \rightarrow NO + O(^3P)$$

$$O(^3P) + NO_2 \rightarrow NO + O_2(v \leqslant 11)$$

The kinetics of O_2 in individual vibrational levels ($8 \leqslant v \leqslant 11$) are then observed *via* laser-induced fluorescence as step-wise relaxation occurs in collisions with NO_2 and with other non-reactive relaxants (M) added to the gas mixture:

$$O_2(v) + NO_2 \rightarrow O_2(v-1) + NO_2$$

$$O_2(v) + M \rightarrow O_2(v-1) + M$$

Experiments have been performed with $M = O_2$, N_2, CO_2, N_2O, CH_4 and He. The rate constants for relaxation of $O_2(8 \leqslant v \leqslant 11)$ by O_2 are shown in the diagram presented by Dr. Hernandez. Our conclusions are: (a) that our results are in fair agreement with the calculations of Billing and Kolesniek;[4] (b) that they are in excellent agreement with the calculations of Hernandez *et al.*;[5] and (c) that they confirm Dr. Wodtke's conclusion that the relaxation of vibrationally excited O_2 in the atmosphere is too rapid for the 'Slanger mechanism'[6] to be a significant additional source of O_3.

I should also like to ask Dr. Wodtke a question regarding the proposed mechanism for additional formation of 'odd oxygen' in which O_2 molecules in levels $v \geqslant 26$ react with unexcited O_2. Given that (a) the triplet products of O_3 photolysis, *i.e.* $O_2(^3\Sigma_g^-) + O(^3P)$, constitute only *ca.* 10% of the whole, (b) the threshold for production of $O_2(^3\Sigma_g^-, v = 26) + O(^3P)$ lies at *ca.* 241 nm and (c) only a fraction of the photolysis leading to triplet products at short wavelengths yields O_2 in levels $v \geqslant 26$, is the overall quantum yield for the autocatalytic process sufficient to cause a significant change in the ozone production rate in the atmosphere?

1 M. Klatt, I. W. M. Smith, R. P. Tuckett and G. N. Ward, *Chem. Phys. Lett.*, 1994, **224**, 253.
2 I. W. M. Smith, A. C. Symonds, R. P. Tuckett and G. N. Ward, *J. Chem. Soc., Faraday Trans.*, in the press.
3 I. W. M. Smith, R. P. Tuckett and C. J. Ward, *Chem. Phys. Lett.*, 1992, **200**, 615.
4 G. D. Billing and R. E. Kolesniek, *Chem. Phys. Lett.*, 1992, **200**, 382.
5 R. Hernandez, R. Toumi and D. C. Clary, *J. Chem Phys.*, 1995, **102**, 9544.
6 T. G. Slanger, L. E. Jusinski, G. Black and G. E. Gadd, *Science*, 1988, **241**, 945.

Prof. Wodtke responded to both comments: I agree that this is another piece of indirect evidence for the existence of the reaction (1). Indeed, the many pieces of indirect evidence, and in particular this one, make the case for reaction (1) compelling. It should be stressed that these results are on a potential-energy surface that 'cannot react'. Dr. Clary and his collaborators are clearly in an excellent position to treat this problem from the point of view of pure *ab initio* theory including the reaction. The importance of that work is obviously great and I look forward to similar high-quality results on a potential-energy surface which allows for the possibility of reaction.

I would like to point out that our group has also studied a few other collision partners, notably: N_2, CO_2 and N_2O. Prof. Smith's work, together with our own, gives a remarkably complete picture of the vibrational-quantum-number dependence of the vibrational relaxation of highly vibrationally excited O_2, with a number of collision partners. This data set is, by far, the most complete experimental picture of such phenomena. It cries out for theoretical interpretation which is likely to influence significantly, our understanding of the collisional relaxation of hot molecules.

Regarding Prof. Smith's question concerning the atmospheric production of $O_2(X\,^3\Sigma_g^-, v \geqslant 26)$, in a recent paper of ours, two completely different kinds of experiments were carried out to try to understand the product-state distributions of ozone photolysis at 226 nm. Ion imaging experiments were carried out in the laboratory of Prof. Paul Houston at Cornell. In those experiments a markedly bimodal kinetic-energy distribution was observed for $O(^3P)$ products. Such bimodal distributions are not uncommon in translational-energy measurements of photolysis products and they are normally ascribed to the formation of two electronic states of the reaction products. However, there has never been any direct evidence that $O(^3P)$ can be formed with two different electronic states of O_2 at this photolysis wavelength. Only formation of $O_2(X\,^3\Sigma_g^-) + O(^3P)$ is considered important. This implied the counter-intuitive formation of a bimodal vibrational distribution $O_2(X\,^3\Sigma_g^-, v)$ at 226 nm. In the light of Houston's provocative results we set about the measurement of a conventional pump–probe experiment to try to find spectroscopic evidence of a bimodal vibrational distribution of $O_2(X\,^3\Sigma_g^-, v)$. The results of these experiments confirmed that, indeed, ozone photolysis at 226 nm results in a bimodal vibrational distribution of $O_2(X\,^3\Sigma_g^-, v)$ with a peak at $v = 27$ and another peak at *ca.* $v = 15$. Analysis of the ion imaging experiment allowed a quantitative determination of the quantum yield for formation of reactive states of O_2 and this information was included in a non-LTE 2D model by Dr. Ralf Toumi of Imperial College London. His modelling results showed that this mechanism can indeed have a significant impact on the stratospheric ozone budget, despite the low quantum yield for reactive states of O_2(*ca.* 1% at 226 nm). Further work is necessary before we close the book on this problem. In particular, the wavelength dependence of this quantum yield and the unambiguous detection of products of reaction (1) are clearly desirable and realizable experimental goals.

Dr. Pyle addressed Prof. Crutzen: I have a few observations on your interesting study with the HALOE data. First, it should be pointed out that there have always been quite large differences between models of the upper stratosphere. The ozone deficit calculated soon after the Crutzen and Schmailzl study ranged from nearly zero to perhaps 50% in a variety of models. Much of this must be due to model formulation, the calculation of photolysis rates *etc.* Intercomparisons of models are still required. We, as modellers, should be cautious about pointing to discrepancies between models and field data while there are large differences between the models.

Secondly, Paul has shown that the calculated ozone at around 45 km, is sensitive to H-radical reactions, and hence to the H_2O distribution used in the model. How well do the various H_2O measurements from UARS (HALOE, ISAMS, MLS *etc.*) agree?

Finally, Paul now indicates that the model calculations produce *ca.* 10% too much ozone at 45 km. Should we be worried about differences of this order or, given the complexity of the photochemical model at these heights, could this be regarded as satisfactory agreement?

Prof. Crutzen responded: I fully agree with your remarks. The main aim of our study, now published, was exactly to point out the problems you indicate. Furthermore, in recent years, it had become accepted that there was indeed an ozone deficit problem in the 40–70 km region and elaborate laboratory experiments are now devoted to this issue, such as those presented by Dr. Wodtke at this meeting. Although we find such investigations very interesting, giving detailed information on the photochemistry and spectroscopy of ozone, it should be known to the experimentalists that the problem may not exist.

For the future, we propose further analysis of satellite and *in situ* data, as well as model intercomparisons. A detailed field programme involving *in situ* measurements would also be very valuable. Regarding our discrepancy of the order of 10% at 45 km, we would like to point out that it grows with altitude, possibly indicating problems with the kinetic data in the HO_x reaction system. A check on the satellite measurements at higher altitudes is likewise recommended.

1 P. J. Crutzen, J-U. Grooss, C. Brühl, R. Müller and J. M. Russell III, *Science*, 1995, **268**, 705.

Dr. Tuck said: Direct observationally based papers in 1986–7[1,2] used balloon and satellite measurements to examine the O_3 balance in the upper stratosphere. Both concluded that any discrepancy between theory and experiment was 10–15% or less. Current stratospheric models do a less than perfect job of simulating the dynamics and radiation; as a result their 'tracer' distributions, *e.g.* CF_2Cl_2, N_2O, are incorrect to some degree. In order to test fast photochemistry, it is better to use observed tracer distributions to calculate the amount of reactive material, for example NO_y or Cl_y.

1 E. M. Weinstock, C. M. Schiller and J. G. Anderson, *J. Geophys. Res.*, 1986, **91**, 5237.
2 J. Austin, R. C. Pallister, J. A. Pyle, A. F. Tuck and A. M. Zavody, *Q. J. R. Meteorol. Soc.*, 1987, **113**, 361.

Prof. Crutzen replied: A discrepancy by 10–15% does not look much, but this is about the extra production of ozone that is being looked for. Furthermore, some of the earlier model studies underestimated the problem by allowing ozone concentrations to change (and not fixing them according to the observations), thus enhancing the penetration of ozone-forming radiation.

In answer to Dr. Tuck's second remark, our ozone budget analysis is basically independent of dynamics, as we take those compounds which are from the HALOE observations.

1 P. J. Crutzen, J-U. Grooss, C. Brühl, R. Müller and J. M. Russell III, *Science*, 1995, **268**, 705.

Prof. Gershenzon stated: You have found a fast reaction

$$O_2(v = 28) + O_2(v = 0) \rightarrow O_3 + O \qquad (1)$$

The important question is of course whether the binary-collision ozone source could remove the apparent deficiency in model ozone above 95 km. To answer this question we should take into account both reaction (1) and the back reaction

$$O_3 + O \rightarrow 2O_2(\text{vib}) \qquad (-1)$$

The role of the latter is two-fold. First, it should produce O_2(vib). The formation of O_2(vib) in the back reaction (−1) follows from your direct study of the reaction (1) and from our earlier indirect results.[1] On the other hand, we have shown indirectly in ref. 1 and directly by the CO_2 laser excitation of O_3 in ref. 2 and 3 that the reaction (−1) is effectively accelerated by vibrational excitation of O_3. Our work on the scheme[4]

$$O + O_2 + M - O_3(\text{vib}) + M$$

$$O + O_3(\text{vib}) = 2O_2$$

$$O_3(\text{vib}) + O, O_2 = O_3$$

brings us to the conclusion that vibrational excitation of ozone should increase the apparent deficiency in model ozone in the range 80 to 95 km. In reality, both vibrationally excited species O_2 and O_3 should be considered simultaneously but I am not sure that this will remove the deficiency in model ozone above 95 km.

1 Yu. M. Gershenzon and S. K. Chekin, *Kinet. Catal.*, 1977, **18**, 1126.
2 S. K. Chekin, Yu. M. Gershenzon, A. V. Konoplev and V. B. Rozenshtein, *Chem. Phys. Lett.*, 1979, **68**, 386.
3 S. D. Il'in, V. V. Selikhanovich, Yu. M. Gershenzon and V. B. Rozenshtein, *Khim. Fiz.*, 1991, **10**, 1060 (in Russian).
4 Yu. M. Gershenzon, V. M. Grigorieva, A. V. Konoplev and V. B. Rozenshtein *Khim. Fiz.*, 1985, **4**, 544 (in Russian).

Prof. Wodtke said in reply: I am not sure either. Conventional wisdom holds that the ratio of LTE to non-LTE ozone formation goes as:[1]

$$\frac{J_{O_2}[O_2]}{J_{O_3}[O_3]}$$

The numerator in this expression increases with altitude while the denominator decreases. Therefore, it may be that reaction (1) does not play a role in the mesosphere. On the other hand, one certainly expects the mesosphere to be full of non-LTE phenomena and it may be that there are many such interesting puzzles to solve before we can say we understand that part of the atmosphere.

1 C. A. Rogaski, J. M. Price, J. A. Mack and A. M. Wodtke, *Geophys. Res. Lett.*, 1993, **20**, 2885.

Prof. Zellner asked: Do you have any evidence that, in the collisional deactivation of $O_2(v'')$ by N_2, there may be a reactive channel producing N_2O? If this were the case, the impact of this process on stratospheric chemistry may be much more important than collisional/reactive deactivation of $O_2(v'')$ by O_2.

Prof. Wodtke replied: We do not have any evidence of the reaction you mention: $O_2(v) + N_2 \rightarrow N_2O + O$. However, we cannot rule it out based on our present results. I agree that observation of this reaction would make everything we have done so far look like small potatoes. Indeed, non-LTE formation of catalytic species is a very interesting area of investigation. It definitely deserves a closer look.

Prof. Crutzen asked: Are there any possibilities of producing electronically excited states such as $O_2(^3\Sigma_u{}^+)$ or $O_2(^1\Delta_u) \ldots$?

Prof. Wodtke replied: This channel cannot yet be ruled out with certainty, but I hasten to add that the chance that our observations are due to vibration–electronic (V–E) energy transfer is very small, especially in the light of the temperature-dependent results presented in this Discussion. These experiments were carried out specifically in order to try to rule out other channels than the reaction. I do not see how the observed temperature dependences and derived activation energies could be consistent with V–E energy transfer, whereas they offer quite strong, albeit indirect, evidence that the reaction is taking place. This question certainly shows the importance of studies which are designed to observe reaction products, studies which are technically very challenging.

Dr. Clary commented: Dr. Hernandez and I have calculated[1] Franck–Condon factors between several vibrational states in the ground electronic state of O_2 and corresponding near-resonant vibrational states for the three lowest excited electronic states of O_2. These Franck–Condon factors became large for $v > 15$. This implies that if V–E energy transfer is important, the effect on the vibrational relaxation rates would arise for much lower vibrational states than $v = 26$.

1 R. Hernandez, R. Toumi and D. C. Clary, *J. Chem. Phys.*, 1995, **102**, 9544.

Dr. Plane commented: The lower thermosphere is where Prof. Wodtke's proposed source of additional O_3 would be expected to have the greatest effect. This is because there is a comparatively high photon flux between 180 and 300 nm, and the mixing ratio of O_2 has not fallen significantly below 0.2. There is, in fact, strong evidence that an additional O_3 source operates in this region. During the Aladdin 74 programme, for instance, simultaneous daytime measurements of O, O_2, O_3 and N_2 were made by rocket-borne mass spectrometry between 90 and 110 km.[1] The measured O_3 levels were then shown to be significantly in excess of those predicted from simple atmospheric chemical models.[2] Although various explanations have been advanced,[3] this discrepancy has still not been satisfactorily resolved. Perhaps the data ought to be examined in the light of this new mechanism?

1 L. H. Weeks, R. E. Good, J. S. Randhawa and H. Trinks, *J. Geophys. Res.*, 1978, **83**, 978.

2 M. Allen, *J. Geophys. Res.*, 1986, **91**, 2844.
3 J. A. Joens, *J. Geophys. Res.*, 1986, **91**, 14533.

Prof. Wodtke responded: The point you make is very interesting and I think that it is quite likely that some non-LTE ozone-formation mechanism is responsible for the deviation between observation and model. However, I think it is not obvious that this particular mechanism is necessarily the correct candidate. As mentioned in an earlier comment, the ratio of LTE to non-LTE ozone formation goes as:

$$\frac{J_{O_2}[O_2]}{J_{O_3}[O_3]}$$

Therefore, it may be that reaction (1) does not play a role in the mesosphere. I would say this issue requires further thought.

1 R. L. Miller *et al.*, *Science*, 1994, **265**, 1831.

Dr. Whitehead moved the discussion on to the papers of Prof. Byers Brown and Prof. Molina: In your papers and earlier contributions, we have seen similar processes taking place in different phases: liquid droplets, clusters and ice films. We have also been given explanations both in terms of bulk macroscopic descriptions and in terms of microscopic accounts of dynamics and mechanism.

In the molecular simulations presented by Prof. Byers Brown, there was a marked reluctance of the species to leave the surface and become solvated in the conventional sense of being uniformly surrounded by water molecules, despite there being a sufficient number. Whilst the authors are cautious about the observation, I suggest that it might be quite common and may provide the intermediate surface state invoked previously in this Discussion. However, how do we reconcile this with Prof. Davidovits' model of 'clustering' prior to incorporation into the bulk of the droplet?

Is it not true that, in the heterogeneous processes, which we are trying to understand in atmospheric chemistry, the important processes take place only in a fairly fluid surface layer of a few molecular thicknesses which controls the chemistry irrespective of whether the layer is based on a 'substrate' which is a regular or supercooled liquid, a crystal or a solid film?

Prof. Hillier responded: Our simulation of the CT complex in the water droplet, where the solute migrates to the surface can be compared to the case of solvated Cl^- studied by Kollman[1] where migration to the surface also takes place. Thus, this result may be quite general.

1 J. W. Caldwell and P. A. Kollman, *J. Phys., Chem.*, 1992, **96**, 8249.

Prof. Molina added: Yes, we believe that the atmospheric reactions taking place on the surfaces of aerosols consisting of various ices (water ice, nitric acid trihydrate *etc.*) involve a high-mobility surface layer that leads to solvation and to reaction mechanisms that are ionic in nature. This point might be discussed again later, in connection with Dr. Clary's paper.

Dr. Herrmann addressed Prof. Byers Brown: You presented the absorption spectra of liquid water and of liquid water containing dissolved oxygen, both showing significant absorptions above 200 nm. From photochemical experiments our experience is that there should be no significant absorption of liquid water at wavelengths above 200 nm. Could you comment on the origin and the quality of the data presented which, I think, were taken from the literature?

Prof. Byers Brown replied: The figure on the overhead that I showed comparing the UV absorption spectrum of oxygenated liquid water with that of highly purified liquid water is Fig. 2 in a recent paper of mine.[1] The data used in the figure are from ref. 2 and 3. I cannot comment knowledgeably on the accuracy of the Quickenden and Irvin spectrum of pure water, except to say that the purification procedures described in their work appear to be impressively thorough, and led to lower absorption over the whole range down to 200 nm than that of all other workers quoted. We know of no more recent reference for the spectrum.

1 W. Byers Brown, *Chem. Phys. Lett.*, 1995, **235**, 94.
2 L. J. Heidt and L. Ekstrom, *J. Am. Chem. Soc.*, 1957, **79**, 1260.
3 T. I. Quickenden and J. A. Irvin, *J. Chem. Phys.*, 1980, **72**, 4416.

Dr. Tuck commented: The ER2 aircraft encountered at least 11 filaments of Concorde exhaust near New Zealand in the early afternoon of 8th October 1994, at about 100 hPa pressure. There are two problems with the analysis to which Prof. Byers Brown's paper may be relevant, given that the exhaust contains a lot of water vapour at low temperatures in the presence of ultraviolet radiation. (i) There are many more condensation nuclei (CN) in the exhaust plume than can be accounted for on the basis of sulfuric acid aerosol produced from the S in the fuel; (ii) There is a discrepancy in the value of J_{NO_2}, the photodissociation value of NO_2, which might be explicable by a high density of clusters in the exhaust filament. It would be interesting to see if there is any possibility that H_2O/O_2 photonucleation could explain these observations.

Prof. Byers Brown said in reply: Your report about the two unexpected measurements made last year in the lower stratosphere by a highly instrumented research aircraft are extremely interesting, and the Wilson mechanism for creating CNs seems to be, potentially, highly relevant.

In our paper we refer in passing to earlier measurements of CN concentrations by research aircraft in the troposphere at around 15 000 feet by Hoppel and Dinger 20 years ago (see ref. 3 of our paper). They suggested that the CN were produced by UV and cite earlier work (but not Wilson's papers). However, the instrumentation would be quite primitive compared with that in your case, and the pollutant levels were not measured.

Dr. Plane said: I have three questions on your paper. First, have any mechanisms been proposed for the production of Wilson clusters other than the formation of charge-transfer complexes? In particular, could trace impurities in the gases have led to aerosol formation? Secondly, Fig. 8 of the paper shows that the additional binding energy per molecule produced by the CT complex appears to increase with cluster size. Surely this goes against intuition? Thirdly, in the final section of the paper, where the rate of formation of Wilson aerosols in the atmosphere is estimated, what is the concentration of H_2O employed in the calculation?

Prof. Byers Brown replied: Trace impurities could certainly be responsible in principle for photonucleation, and to exclude them we are dependent on Wilson's reputation as an experimenter and his assertion that only H_2O and O_2 are required. However, further experimental confirmation is most desirable. With regard to other mechanisms involving oxygen, I would refer to my later reply to Prof. Vaida.

The water vapour concentration used in the calculation in Section 6 of the paper is implied, in the third paragraph, as that corresponding to saturated air at 0 °C. The partial pressure is therefore 4.5 Torr.

Dr. Masters communicated: For the assumed forms of the interaction potentials, the simulation results show an unexpectedly small binding energy for the CT complex. The statistical errors associated with this small quantity are too large, however, to permit one to seek a trend with cluster size. One can naturally question the model potentials used, but we clearly have simulated a highly polar molecule in a water cluster and yet it only binds weakly. This is in itself an intriguing result.

Dr. Dos Santos added: The variations in the potential energy of the clusters with the same aggregation number as shown in Fig. 8, show a general trend that replacing a water molecule with a CT complex marginally decreases the energy, whereas replacement with a hydrogen peroxide molecule increases the potential energy. The small energy differences between clusters of varying sizes cannot be readily compared and this is due to the limitations of our chosen water models. To our knowledge, there exists no published water model capable of giving good bulk and surface properties. Here, our aim was to observe whether these general trends were sensitive to the exact water model used. To study the stability of the clusters further, simulations involving the calculation of the Gibbs energy change in adding the two molecules of interest should be performed.

Prof. Okumura said: You consider only excitation to the charge-transfer triplet state of the $O_2 \cdot H_2O$ complex. This is an allowed transition which most likely has a very short radiative lifetime back to the ground state. One would expect that a longer-lived state or product is needed to induce nucleation. Have the authors considered one of the following possibilities? First, the excited triplet state could lead directly to reaction. The calculation clearly shows a bound potential-energy curve, but it appears that the calculation was constrained to C_{2v} symmetry. Have the authors looked for saddlepoints on this surface in lower symmetries which might correspond to transition states for reactions, or indeed is there a direct reaction possible if the symmetry constraint is released? Secondly, are there other triplet states below the charge-transfer band? Lastly, have the authors calculated the full potential-energy curve for the lowest-energy singlet state of the complex, analogous to the curve depicted in Fig. 4? Such a state could be prepared by a spin–orbit induced surface crossing or by a weak transition in the long-wavelength tail and could subsequently have a very long lifetime with respect to radiative decay.

Prof. Byers Brown responded: We have considered both possibilities raised in your interesting remarks, but have not yet completed computational investigation of either of them. The possibility that the triplet state of $H_2O^+O_2^-$ might react very rapidly with one more water molecules is mentioned in ref. 6 in our paper.

Dr. Palmer has begun to explore photochemical pathways from the CT complex leading to some of the possible photo-products outlined in Section 4 of the paper. This involves characterising not only topological features familiar in the study of adiabatic reaction coordinates but also regions where facile non-adiabatic transfer of electronic state occurs. These regions of Born–Oppenheimer violation allow competition of the photochemical pathways with radiative decay to the ground state. The possibility that a (long-lived) lowest singlet state of $H_2O^+ \cdot O_2^-$ could be prepared in one of the ways described in the question is intriguing, and we shall attempt to find it computationally.

With regard to both triplet and singlet states of the CT complex, the recent development and exploitation of optical–optical double resonance (OODR) spectroscopy may allow convenient observation and exploration of the ion-pair states of O_2 and H_2O (see for example ref. 1).

1 R. J. Donovan, K. P. Lawley, Z. Min, T. Ridley and A. J. Yarwood, *Chem. Phys. Lett.*, 1994, **226**, 525.

Prof. Vaida commented: The *ab initio* water–oxygen potential surfaces presented yield model absorption spectra for the CT complex, $H_2O^+O_2^-$, at higher energy

($\lambda < 180$ nm) than the observed oxygenated water. Based on our work with oxygen complexes,[1] notably the oxygen dimer, we suggest that the 200–300 nm absorption observed in oxygenated water could arise from enhancement of a forbidden O_2 band. In $(O_2)_2$, the $A'\,^3\Delta_u \leftarrow X\,^3\Sigma_g^-$ transition is enhanced three orders of magnitude compared with the same absorption observed in free O_2 and excitation at these wavelengths results in odd-oxygen formation.

1 V. Vaida, *Ber. Bunsen-Ges. Phys. Chem.*, 1995, **99**, 1.

Prof. Byers Brown responded: It is true that the *ab initio* computations of the van der Waals and CT complexes of H_2O and O_2 presented in our paper predict a maximum absorption with wavelength around 140 to 150 nm. However, we had to make a number of assumptions regarding the geometry, which are discussed in the paper, and a comparatively small change in the relative positions of the excited and ground potential-energy surfaces can lead to a large change in the calculated wavelength maximum. The main object of the calculations was to see if the theoretical CT spectrum was broad like the experimental one, and in roughly the right place.

Our views have undergone some development as a result of this Discussion. Our present position is that we believe that a transition to the CT state dominates down to a wavelength of about 220 nm, but that the absorption at longer wavelengths is due to the enhancement of the forbidden Herzberg continuum transition $^3\Sigma_u^+ \leftarrow {}^3\Sigma_g^-$, in keeping with the suggestion of Prof. Vaida. In particular, we believe this enhancement is caused by the interaction of the CT state $H_2O^+ \cdot O_2^-$.

This view is more or less in accord with a very interesting Russian paper[1] which is quoted in ref. 6 of our paper. The experimental spectra in the Russian paper have a change in slope of the oxygen–water absorption at about 220 nm which was missed by Heidt and co-workers (our ref. 8), whose results they criticize. It is very interesting that the photonucleation spectrum of Katz *et al.*, reproduced in our Fig. 1, shows a pronounced kink just below 220 nm. The absence of details in the short Russian paper renders it difficult to make quantitative use of their work, and further experimental studies of both kinds of spectra are desirable.

1 V. R. Blok, O. L. Lebedev and N. G. Mekhryakova, *Dokl. Akad. Nauk*, 1979, **249**, 633.

Dr. Roscoe said: At the start of your talk you made a tantalising reference to the implications of your new species for an outstanding and important unresolved issue in far-IR and microwave spectroscopy: the anomalous absorption by water vapour, far from the centres of its absorption lines. You did not mention the subject again, and it is not in your paper. What evidence is there for your suggestion, and have you calculated the far-IR spectrum of your new species?

Prof. Byers Brown replied: As you say, there is now no doubt about the reality and importance of the anomalous absorption of far-IR and millimetre waves by the atmosphere. Less well established is its enhancement in the presence of UV radiation, both in the atmosphere and in the laboratory (see references in the paper cited below).

In the absence of UV, laboratory experiments over the whole range of humidity suggest strongly that it is due to the ordinary aerosol population in equilibrium with the water vapour present at partial pressure, p. A simple phenomenological theory relating the absorbance to the relative humidity, p/p_0, through the Kohler equation has been proposed.[1] Despite attempts to identify the physical origin of the absorption by speculative modelling, in particular involving electromagnetically driven capillary oscillations of electrolyte aerosols, no convincing mechanism has yet emerged.

In the presence of solar or laboratory UV it seems likely that the greatly enhanced absorption is due to the extra 'Wilson' aerosol clusters containing the $H_2O^+ \cdot O_2^-$ complexes and/or their reaction products with water. We have calculated the far-IR spectrum of the new species itself, but it consists only of isolated absorption lines, none of which are in the millimetre-wave region where anomalous absorption is known to occur.

1 W. Byers Brown and H. A. Gebbie, *Infrared Phys.*, 1992, **33**, 359.

Prof. Byers Brown added in conclusion: I am grateful for the opportunity to conclude the discussion on our paper on the photonucleation of water vapour in the presence of oxygen and the absence of pollutants. I would like to do so by drawing attention to an old observation in the field, which we would no doubt now label as anecdotal, since the composition of the atmosphere was not specified in sufficient detail. However, it was made by the early pioneer of atmospheric science, John Aitken, who invented the first counter of pre-condensation nuclei, and who first pointed out the importance of pollutants of all kinds in a paper entitled '*The Sun as a Fog Producer*.' It therefore is worth repeating. It is quoted in *The Physics of Clouds*: 'As an example of the influence of solar radiation, one may recall Aitken's (1912) very interesting, but not yet fully elucidated, observation, that the nucleus concentration on a beach at low tide increased twentyfold when the sun came out.'

Assuming that Aitken was sufficiently aware of the effect of pollutants, so that their absence is implied, this historical observation can be added to that of C. T. R. Wilson, who observed that the sub-critical expansion of moist dust-free air which had been exposed to sunlight through a quartz window (but not window-glass) led to clouds of droplets [ref. 1(*b*) of our paper]. Wilson speculated: 'The nuclei which enable clouds to form may, in many cases, arise from this source. The upper clouds especially may owe their formation in this way to the action of sunlight.'

Despite the enormous advances in our knowledge and understanding of atmospheric photochemistry, this simple photochemical phenomenon awaits modern quantitative experimental confirmation and assessment.

1 J. Aitken, *Proc. R. Soc. Edinburgh*, 1912, **32**, 183.
2 J. B. Mason, *The Physics of Clouds*, Oxford University Press, 1971, p. 53.

Prof. Ravishankara addressed Prof. Molina: The question of whether it makes any difference if a heterogeneous reaction occurs close to the surface or in the bulk of the condensed phase came up. In the case of solids, it is now abundantly clear that the reactions are essentially confined to the surface. There are ways to think about where the reaction occurs and how to deal with the situation when the reaction is not taking place close to the surface, as one would expect for at least some reactions in liquid sulfuric acid. One such method is presented by Hanson *et al.*[1] If the square root of the ratio of the diffusion coefficient to the first-order rate coefficient for the reactive loss in the liquid is very small compared with the radius of the droplet, then the reaction can be envisioned to be taking place 'on' the surface and all that matters is the surface area of the droplet. In the opposite case, the volume of the droplet determines the processing rate. The differences between surface and bulk reaction does effect the rates of heterogeneous processes in the atmosphere. Some examples are discussed in the paper cited.

1 D. R. Hanson, A. R. Ravishankara and S. Solomon, *J. Geophys. Res.*, 1994, **99**, 3615.

Prof. Molina responded: A separate question, however, is whether heterogeneous reactions take place on the top surface layer of a solid (as is the case with refractory substrates such as silica or alumina), or else inside the first few surface layers, as we believe is the case with ice. The difference has mechanistic implications (*i.e.* involving solvation, rather than physical adsorption). In either case, however, the reaction rate should be proportional to the surface area (and not to the volume) of the aerosols.

Dr. Peter said: Fig. 3, below, shows measured reaction probabilities, γ, for HCl + $ClONO_2$ as presented in Fig. 5 of the paper in comparison with theoretical curves calculated from $\gamma = 4RTH_{ClONO_2}(D_l K^{II}[HCl])^{1/2}/\bar{c}$. Solubilities of HCl have been calculated using the model of Carslaw *et al.*[1] The liquid diffusion constant (D_l) is calculated from the cubic cell model.[2] The second-order rate coefficient (K^{II}) was calculated, assuming the reaction to be diffusion limited, and the Henry's law constant of chlorine nitrate was estimated as $H_{ClONO_2} = 10^3$ mol l^{-1} atm^{-1} at 202 K and assuming -6 kcal mol^{-1} for the enthalpy of solvation.[3] The good agreement between the measurements and the theoretical estimate corroborates the conclusion of the present paper that the reaction probability is governed by the solubility of HCl in the liquid aerosol. Furthermore, it supports the suggestion by Hanson and Ravishankara that $ClONO_2$ remains undissociated in the liquid, so that its interaction with the liquid involves hydrogen bonding only.

1 K. S. Carslaw, B. P. Luo, S. L. Clegg, Th. Peter and P. Brimblecombe, *Geophys. Res. Lett.*, 1994, **21**, 2479.
2 B. P. Luo, S. L. Clegg, Th. Peter, R. Müller and P. J. Crutzen, *Geophys. Res. Lett.*, 1994, **21**, 49.
3 D. Hanson and A. R. Ravishankara, *J. Phys., Chem.*, 1994, **98**, 5728.

Dr. K. Carslaw said: In Table 4 you give measured reaction probabilities (γ) for constant HCl concentrations ([HCl]), but for a range of H_2SO_4 concentrations. From the classical uptake description one would expect $\gamma \propto H^*_{ClONO_2} (k_1 D_1)^{1/2}$ where k_1 is the first-order rate constant and D_1 is the diffusion constant of the $ClONO_2$. If this is the case, then γ would fall with increasing H_2SO_4 owing both to the decreasing solubility ($H^*_{ClONO_2}$) and the decreasing diffusion coefficient. Could you explain this?

You show that γ appears to be independent of the amount of HNO_3 in the droplet. This is an important finding since it is now appreciated that liquid aerosols can absorb large amounts of HNO_3. However, a more important aspect is the increase in aerosol volume that accompanies the uptake of HNO_3, which can be a factor of 50 or so for typical conditions. This means that, although the γ value is unchanged, the actual rate of the heterogeneous processing will increase dramatically with HNO_3 uptake.

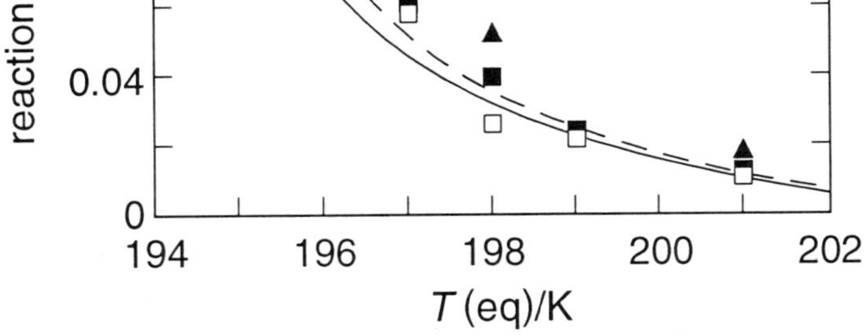

Fig. 3 Measured reaction probabilities and theoretical curves: solid line, H_2SO_4–H_2O; dashed line H_2SO_4–HNO_3–H_2O

Prof. Molina explained: The reaction probabilities listed in Table 4 were measured over a relatively narrow range of temperatures and H_2SO_4 concentrations; no clear trends are discernible given the experimental uncertainty. It appears that there are compensating factors; perhaps the rate constant itself increases as the concentration of H_2SO_4 increases; furthermore, there might be a surface component to the reaction rate. Clearly, more experiments need to be carried out over a wider range of conditions in order to establish better the mechansim and the reaction rate dependence on the various parameters.

Prof. Molina added in conclusion: To summarize, the rate for chlorine activation reactions on stratospheric aerosols depends primarily on the extent of HCl uptake which, in turn, depends on the relative humidity and hence is very sensitive to temperature. For liquid droplets the reaction probability increases rapidly as the HNO_3–H_2O dew point is approached, and for solid particles as the ice frost point is approached.

Faraday Discuss., 1995, **100**, 295–307

Dissociation Dynamics of $ClONO_2$ and Relative Cl and ClO Product Yields following Photoexcitation at 308 nm

Teresa A. Moore and Mitchio Okumura*

Arthur Amos Noyes Laboratory of Chemical Physics,† California Institute of Technology, Pasadena, California 91125, USA

Masahito Tagawa‡

Jet Propulsion Laboratory, California Institute of Technology, Pasadena, California 91109, USA

Timothy K. Minton*

Jet Propulsion Laboratory, California Institute of Technology, Pasadena, California 91109 and Center for Biofilm Engineering and Department of Chemistry and Biochemistry, Montana State University, Bozeman, Montana 59717, USA

Chlorine nitrate photolysis at 308 nm has been investigated with a molecular beam technique. Two primary decomposition pathways, leading to $Cl + NO_3$ and $ClO + NO_2$, were observed. The branching ratio between these two respective channels was determined to be $0.67 \pm 0.06 : 0.33 \pm 0.06$. This ratio is an upper limit because some of the ClO photoproducts may have undergone secondary photodissociation. The angular distributions of the photoproducts with respect to the direction of polarization of the exciting light were anisotropic. The anisotropy parameters were $\beta = 0.5 \pm 0.2$ for the $Cl + NO_3$ channel and $\beta = 1.1 \pm 0.2$ for the $ClO + NO_2$ channel, indicating that dissociation of $ClONO_2$ by either pathway occurs within a rotational period. Weak signal at mass-to-charge ratios of 35 and 51, arising from products with laboratory velocities close to the beam velocity, was observed. While this signal could result from statistical dissociation channels with a total relative yield of 0.07 or less, it is more likely attributable to products from ClO secondary photodissociation or from dissociation of clusters.

Chlorine nitrate ($ClONO_2$) plays a key role in stratospheric chemistry as a reservoir for reactive chlorine and nitrogen oxides.[1,2] Photolysis is a major decomposition process for $ClONO_2$,[3] and has been the subject of many studies.[4–13] At wavelengths <250 nm, photodecomposition follows two primary channels:[9]

$$ClONO_2 + h\nu \rightarrow ClO + NO_2; \qquad \Delta H = 26 \text{ kcal mol}^{-1} \tag{1}$$

$$\rightarrow Cl + NO_3; \qquad \Delta H = 40 \text{ kcal mol}^{-1} \tag{2}$$

† Contribution no. 9024.

‡ Visiting scientist from Department of Precision Engineering, Faculty of Engineering, Osaka University, 2-1 Yamada-oka, Suita, Osaka 565, Japan.

Channel (2) can lead directly to loss of two ozone molecules *via* the following reactions:[13]

$$NO_3 + h\nu \rightarrow NO + O_2 \qquad (3a)$$

$$NO + O_3 \rightarrow NO_2 + O_2 \qquad (3b)$$

$$Cl + O_3 \rightarrow ClO + O_2 . \qquad (3c)$$

Toumi *et al.* have recently shown in model studies that this catalytic cycle[13] including steps (2) and (3*a*)–(3*c*) becomes the dominant ozone-loss mechanism in the latter stages of springtime Arctic ozone depletion.[14] In this model, the extent of ozone loss depends upon the quantum yield of channel (2), especially at $\lambda > 300$ nm. Toumi *et al.* used the value of 0.9 for this quantum yield, which was recommended by the 1992 JPL evaluation of kinetic data.[15] However, results from our laboratory at shorter wavelengths (193 and 248 nm) showed the Cl + NO$_3$ product yield to be significantly lower,[9] a conclusion later corroborated by other groups.[10–12] A lower yield at longer wavelengths would diminish the predicted contribution of the ClONO$_2$ catalytic cycle to Arctic ozone depletion and reduce the predicted magnitude and duration of the ozone-loss episode.

Sander and co-workers have recently obtained evidence from flow tube studies that excited ClONO$_2$ formed by wavelengths >300 nm lives long enough to be quenched under stratospheric conditions.[12] Quenching of ClONO$_2$ photolysis would profoundly affect the partitioning of the Cl and NO$_x$ budgets and would largely eliminate the ClON$_2$ catalytic cycle as a candidate loss mechanism for stratospheric ozone. In their experiment, Sander and co-workers observed products from channels (1) and (2) when ClONO$_2$ was excited by a broadband Xe lamp; however, at wavelengths >300 nm they found that the total quantum yield was significantly quenched at N$_2$ buffer gas pressures >50 Torr. Under their conditions (*ca.* 50 Torr and 200 K), the mean time between gas-kinetic collisions is 3 ns, so an excited molecule must have a lifetime of the order of nanoseconds or longer to be quenched rather than dissociated.

Previous flow-tube studies revealed no direct evidence of a relatively long-lived state following UV excitation of ClONO$_2$.[4,8,10,11] However, recent *ab initio* calculations by Graña *et al.* do predict the existence of a metastable state.[16] These authors performed high-level configuration interaction (CI) and coupled-cluster calculations to predict the vertical excitation energies and oscillator strengths for transitions to eight singlet and eight triplet states of ClONO$_2$ between 400 and 160 nm. They also found a bound, metastable triplet state, with a non-planar geometry, lying 62 kcal mol^{-1} above the ground-state equilibrium configuration. If ClONO$_2$ were formed in this state either by direct excitation or by relaxation from an initially prepared state, the excited ClONO$_2$ could conceivably persist for nanoseconds or longer.

The fate of chlorine nitrate excited at wavelengths >300 nm has important implications for stratospheric ozone depletion. To address questions concerning the dissociation pathways and the possibility of a long-lived excited state, we extended our previous photodissociation studies of ClONO$_2$ to a longer wavelength. With the use of molecular beam methods, we can directly detect photoproducts, measure their relative yields, and determine angular and translational energy distributions in the absence of collisions. Furthermore, we can infer the timescale for dissociation from the anisotropy of the product angular distribution. In our earlier experiments, we observed comparable yields of Cl and ClO products from ClONO$_2$ photolysis at both 193 and 248 nm.[9] We calibrated the relative detector sensitivity for Cl and ClO by performing an experiment on Cl$_2$O photolysis at 308 nm and found Cl : ClO branching ratios of 0.57 : 0.43 and 0.67 : 0.33 for ClONO$_2$ photolysis at 248 and 193 nm, respectively. Furthermore, we observed anisotropic angular distributions for the two channels and inferred that the dissociation is prompt, occurring within a rotational period ($\tau_{rot} \approx 9$ ps at the rotational temperature of the beam). Here we present results from a similar investigation of $ClONO_2$ photolysis at 308 nm.

Experimental

The molecular beam apparatus and methods have been described in detail elsewhere.[9,17] Briefly, a molecular beam was crossed with a laser beam in a low-pressure region at the centre of rotation of a rotatable mass spectrometer detector[18] (Fig. 1). Collisionless conditions allowed for detection of primary dissociation products recoiling away from the beam direction. Centre-of-mass (cm) product translational energy and angular distributions were derived from the arrival-time distributions of the neutral photoproducts over a calibrated flight length as a function of detector angle away from the beam direction.

Continuous molecular beams were formed by passing H_2 over $ClONO_2$ at $-50\,°C$ or He over Cl_2O at $-74\,°C$ and expanding the mixture through a 0.1 mm diameter quartz nozzle maintained at $+5\,°C$. $ClONO_2$ was seeded in hydrogen rather than helium in order to obtain a faster beam, which enhanced the signal through kinematic focussing. Stagnation pressures of the gas mixtures were 300 Torr for the $ClONO_2$ experiment and 200 Torr for the Cl_2O experiment. The $ClONO_2$ data were obtained in two sets, with corresponding beam velocities of 1350 and 1580 m s^{-1}, as determined by beam time-of-flight (TOF) distributions taken with the aid of a chopper wheel. The Cl_2O beam velocity (1480 m s^{-1}) was determined by examining the arrival time of the laser-induced depletion of signal at a mass-to-charge ratio (m/z) of 37 with the detector directly viewing the molecular beam.[19,20]

Cl_2O was synthesized by oxidation of Cl_2 on HgO and was purified by pumping on the sample at $-118\,°C$ (ethanol slush bath).[21] $ClONO_2$ was synthesized from the reaction of Cl_2O with N_2O_5 and purified first by cannulation and then by pumping on the sample at $-95\,°C$ (methanol slush bath).[22] Purity was checked by a survey of beam TOF distributions at all masses where signal was observed. The only impurities in either

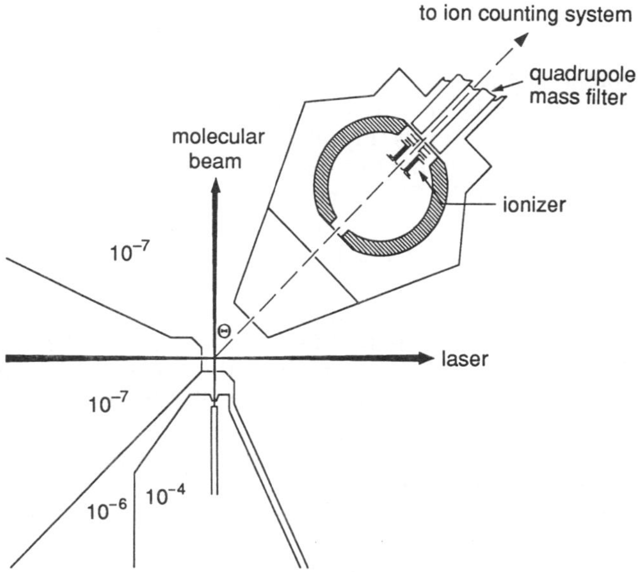

Fig. 1 Schematic diagram of the molecular beam photodissociation experiment. The numbers correspond to pressures in Torr for the various regions. Θ is the detector angle.

molecular beam could be attributed to dimers and/or clusters. A small peak in the beam TOF at $m/z = 146$ ($ClONO_2 \cdot NO_2^+$) indicated that clusters were present in the $ClONO_2$ beam; however, the product signals observed were insensitive to the relative magnitudes of the cluster signals.

The molecular beam was crossed in the interaction region with a laser beam at 308 nm from a Lambda Physik EMG103 excimer laser operating at 200 Hz. The laser beam was randomly polarized in a plane perpendicular to the scattering plane and parallel to the molecular beam axis. The laser beam was focussed to a spot size of 2.5 mm × 2.5 mm, where it intersected the molecular beam. The laser pulse energy used for the $ClONO_2$ experiment was 170 mJ. The consequent laser fluence was high enough that secondary photodissociation of the primary ClO products was a concern; however, the low absorption cross-section[15] of $ClONO_2$ at 308 nm ($\sigma = 1.6 \times 10^{-20}$ cm^2) required the use of relatively high laser fluences to make the primary photoproduct signals observable. For the Cl_2O experiment, a pulse energy of 126 mJ was used for data collection at $m/z = 51$, and the laser was attenuated with a stack of quartz plates to 36 mJ per pulse for data collection at $m/z = 35$ in order to minimize secondary photodissociation of the ClO product. TOF distributions were collected at $m/z = 35$ (Cl^+), 51 (ClO^+), 30 (NO^+), 46 (NO_2^+), 16 (O^+) and 81 ($ClONO^+$), and angular distributions of product signals detected at $m/z = 35$, 51 and 46 were measured.

Results and Analysis

Analysis of TOF and angular distributions was carried out by a forward convolution technique.[20,23–25] For each channel, trial cm translational energy $P(E_T)$ and angular $\omega(\theta)$ distributions were transformed to the laboratory frame of reference to obtain a predicted number density TOF distribution $N(t)$ that could be compared with the experimental data. The $P(E_T)$ and $\omega(\theta)$ distributions were then adjusted iteratively until the predicted TOF distributions agreed with the observed distributions at all angles.

The data were similar to our earlier results at 193 and 248 nm.[9] A single peak was prominent at $m/z = 51$ (see Figs. 2 and 3), which we assigned to ClO from the ClO + NO_2 channel by analogy with our previous results. The $m/z = 35$ TOF distribution contained two main components, the slower of which matched the peak at $m/z = 51$, indicating that it must arise from fragmentation of ClO to Cl^+ in the ionizer.† We assigned the fast component at $m/z = 35$ to Cl from the Cl + NO_3 channel. We used the TOF spectra taken at $m/z = 35$ and 51 to derive the cm translational energy distributions for the two dissociation channels (Fig. 4).

We verified our assignments of the two dissociation channels by using the $P(E_T)$ and $\omega(\theta)$ distributions derived from the ClO^+ and Cl^+ data to predict the TOF distributions for the respective counterfragments, NO_2 and NO_3 (Fig. 5). No signal was observed at $m/z = 62$, the NO_3 parent mass, but the TOF spectrum collected at $m/z = 46$ (NO_2^+) fitted well with the distribution predicted for the NO_3 fragment by the $P(E_T)$ and $\omega(\theta)$ distributions for the Cl + NO_3 channel. Apparently, the NO_2^+ signal comes not from NO_2, but from NO_3 which fragments to NO_2^+ in the ionizer. The bimodal TOF distribution at $m/z = 30$ (NO^+) could not be fit from the $P(E_T)$ and $\omega(\theta)$ distributions for either the Cl + NO_3 or ClO + NO_2 channels alone, but it could be fit with a sum of contributions from the predicted TOF distributions for both NO_3 and NO_2 fragments. Hence, both NO_3 and NO_2 photofragments lead to NO^+ in the ionizer; similar ionizer fragmentation patterns were observed in our previous study.[9] The ability to fit both counterfragments of each dissociation channel with the use of the same $P(E_T)$ and $\omega(\theta)$

† The 4 μs difference in peak positions of the ClO signals detected at $m/z = 51$ and $m/z = 35$ reflects the mass dependence of the ion flight time from the ionizer of the mass spectrometer to the Daly-type ion counter. The ion flight time was determined to be $t = 3.5\,(m/z)^{1/2}$ μs.

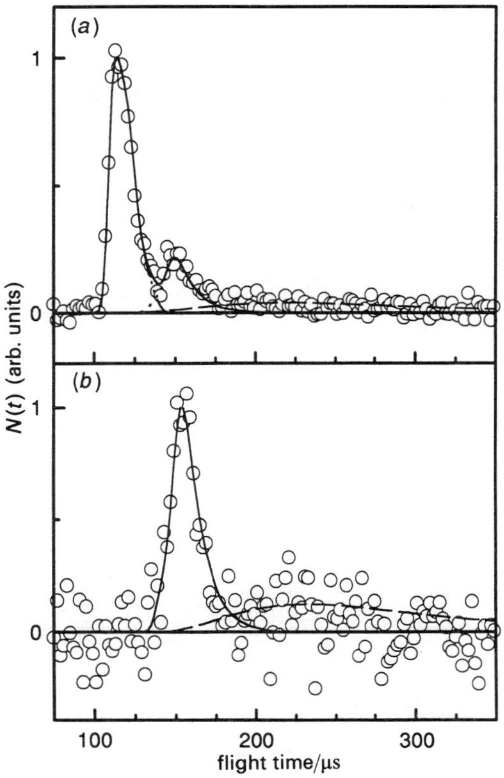

Fig. 2 TOF distributions $N(t)$ collected at a detector angle of 30° for (*a*) Cl^+ and (*b*) ClO^+ fragments. The molecular beam velocity was 1580 m s^{-1}. Fits to the dominant product channels, $Cl + NO_3$ and $ClO + NO_2$, are calculated with optimized centre-of-mass translational energy $P(E_T)$ and angular $\omega(\theta)$ distributions (see Fig. 4 and 6). The long-dashed line at slow times in (*b*) is a hypothetical fit based on the assumption that ground electronic state ClO and NO_2 products are formed by a statistical (RRKM) process. The long-dashed line at slow times in (*a*) is also a hypothetical fit based on statistical (RRKM) decay that leads to ground electronic state products. This 'fit' includes contributions to a Cl^+ signal from both $ClO + NO_2$ and $Cl + NO_3$ channels. The upper limit on the total yield from both channels is determined from the fits in (*a*) to be 0.07. (*a*) (·—·) $Cl + NO_3$, (····) $ClO + NO_2$; (*b*) (——) $ClO + NO_2$.

distributions thus confirms our original assignments of the two primary photodissociation channels.

Laboratory angular distributions of ClO, Cl and NO_3 were obtained from the integrated TOF data detected at $m/z = 51$, 35 and 46, respectively. At $m/z = 51$ and 46, where the TOF distributions corresponded only to a single photoproduct, the signal for each detector angle was simply integrated. The Cl angular distribution was extracted from the $m/z = 35$ TOF distributions by integrating the calculated TOF distributions for the Cl^+ signal arising from the $Cl + NO_3$ channel [obtained from the optimized $P(E_T)$ and $\omega(\theta)$ distributions].

The cm angular distribution must have the form $\omega(\theta) \propto 1 + \beta P_2(\cos\theta)$, where P_2 denotes the second-order Legendre polynomial and θ the angle between the electric field vector of the excitation photon and the product recoil direction in the cm reference frame.[23,26] β is the anisotropy parameter and must lie in the range $-1 \leqslant \beta \leqslant 2$. We obtained $\beta = 1.1 \pm 0.2$ for the $ClO + NO_2$ channel from the ClO angular distribution and $\beta = 0.5 \pm 0.2$ for the $Cl + NO_3$ channel from both the Cl and NO_3 angular distributions. The fits used to determine the anisotropy parameters are shown in Fig. 6.

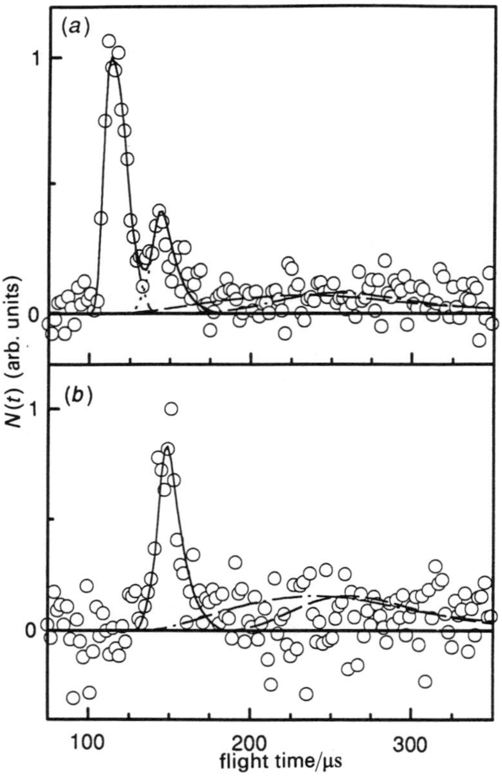

Fig. 3 TOF distributions collected at a detector angle of 15° for (*a*) Cl^+ and (*b*) ClO^+ fragments. The molecular beam velocity was 1350 m s^{-1}. Fits to the main peaks are calculated with the same $P(E_T)$ and $\omega(\theta)$ distributions used to achieve the analogous fits in Fig. 2. Two kinds of hypothetical fits are shown at longer flight times. One is based on the assumption of statistical decay to ground-state products and the other is based on the assumption of statistical decay to excited-state products. As in Fig. 2, the hypothetical fits at $m/z = 35$ include dissociation along both $ClO + NO_2$ and $Cl + NO_3$ product channels. Statistical decay to excited state products *via* channels (4) and (5) leads to low product translational energies and thus a smaller possible contribution to the signal than dissociation to ground-state products. The upper limit on the total yield of ground-state products is 0.07, while the upper limit on the total yield of excited-state products at this detector angle is 0.03. The total yield of all slow products cannot exceed 0.07. At detector angles of 30° and larger, the excited-state products cannot be detected because their centre-of-mass velocities are too low. (*a*) (·—·) $Cl + NO_3$, (···) $ClO + NO_2$, (—·—) RRKM (ground-state products), (———) RRKM (excited-state products); (*b*) (——) $ClO + NO_2$, (—·—) RRKM (ground-state products), (——) RRKM (excited-state products).

The broad, slow signal at $m/z = 35$ may be explained by a number of possible mechanisms: spontaneous secondary dissociation, dissociation of clusters formed in the beam expansion, secondary photodissociation of primary ClO photoproducts, or a third primary channel leading to products with low cm translational energies. We can rule out spontaneous secondary dissociation because the $P(E_T)$ distribution [Fig. 4(*b*)] implies that no ClO photoproducts have enough internal energy to dissociate. Clusters were present in the beam, but we expect the contribution from dissociation of clusters to the $m/z = 35$ TOF to be small because changes in the stagnation pressure did not significantly alter the magnitude of the slow signal relative to the fast peaks. The most likely explanation for the tail at $m/z = 35$ is secondary photodissociation of ClO. The ClO absorption cross-section is larger than that of $ClONO_2$ at 308 nm, and at the high laser

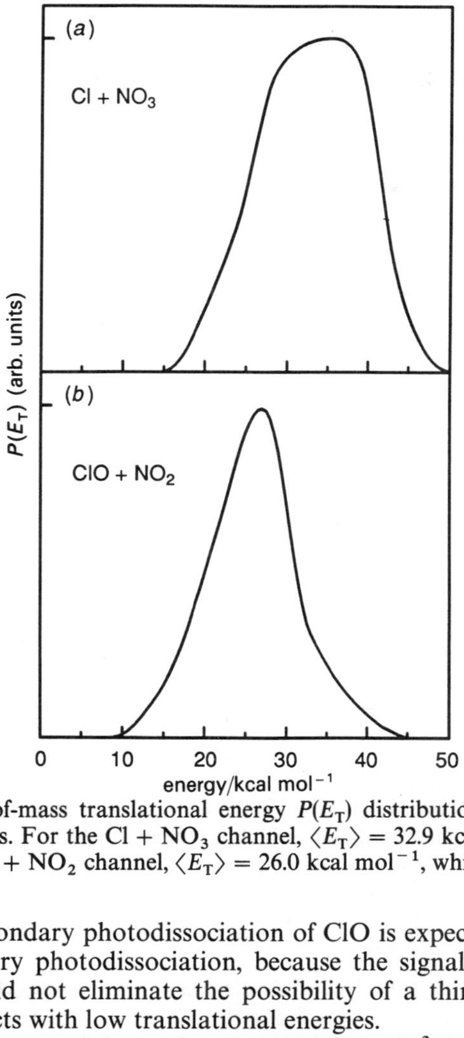

Fig. 4 Optimized centre-of-mass translational energy $P(E_T)$ distributions for the two dominant photodissociation channels. For the Cl + NO_3 channel, $\langle E_T \rangle = 32.9$ kcal mol^{-1}, which is 62% of E_{avail}. For the ClO + NO_2 channel, $\langle E_T \rangle = 26.0$ kcal mol^{-1}, which is 39% of E_{avail}.

fluences used, some secondary photodissociation of ClO is expected. However, we could not model the secondary photodissociation, because the signal-to-noise ratio was too low. We therefore could not eliminate the possibility of a third primary dissociation channel forming products with low translational energies.

Direct dissociation on a triplet surface to ClONO + $O(^3P)$ could conceivably contribute to a slow signal at $m/z = 51$ and 35 through fragmentation to ClO^+ or Cl^+ in the ionizer. ClONO might also fragment to NO_2^+ or NO^+ and give rise to a slow signal at $m/z = 46$ and 30. The energy available for translation after breaking the O—N bond is only *ca.* 26 kcal mol^{-1}, and the ClONO fragment will recoil slowly from the relatively light O-atom counterfragment. Therefore, the minimum flight time possible at a detector angle of 30° is 133 μs (not including ion flight time), and any signal present should manifest itself as a peak in the slow tail of the product TOF distributions. The only TOF distributions which showed an obvious hint of slow signal were the Cl^+ distributions. If all slow signal at $m/z = 35$ not assigned to the dominant channels did come from ClONO, then the ClONO + O yield would be *ca.* 0.07. Even this small yield is suspect because (1) no photofragment signal was detected at the ClONO parent mass ($m/z = 81$), (2) no obvious slow signal was seen at other masses and (3) ClONO can only be formed by a spin-forbidden process at an excitation wavelength of 308 nm.

Statistical dissociation of $ClONO_2$ could also lead to products with low translational energies. The cm translational energy distributions for products of a statistical dissociation process have a maximum near zero and decay rapidly with E_T; therefore,

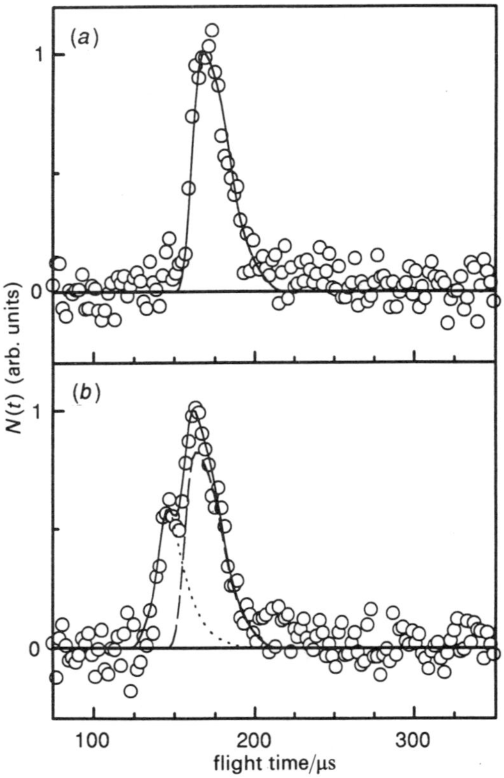

Fig. 5 TOF distributions collected at 30° for (*a*) NO_2^+ and (*b*) NO^+ fragments. The peak in (*a*) is fitted with the $P(E_T)$ and $\omega(\theta)$ distributions for the Cl + NO_3 channel under the assumption that all the signal at $m/z = 46$ arises from NO_3 which fragments to NO_2^+ in the ionizer. The bimodal distribution in (*b*) apparently contains signal from both NO_3 and NO_2 products which fragment to NO^+ in the ionizer; this distribution is fitted with two components corresponding to both Cl + NO_3 (— —) and ClO + NO_2 (---) channels.

the observed signal would have low cm velocities. At the smallest detector angle used here, 15°, we can detect Cl products from the Cl + NO_3 channel with total cm translational energies $E_T \geqslant 0.7$ kcal mol^{-1} and ClO products from the ClO + NO_2 channel with $E_T \geqslant 1.4$ kcal mol^{-1}. The experiment should therefore be sensitive to statistical dissociation channels with available energies of a few kcal mol^{-1} above threshold.

We can estimate the maximum possible yield from statistical dissociation by fitting the slow component to a statistical model, but the modeling will depend upon the final dissociation channel. Several dissociation channels are energetically available to $ClONO_2$ with 92.8 kcal mol^{-1} of internal energy (*i.e.* the photon energy of 308 nm light). While statistical dissociation of $ClONO_2$ on the ground electronic state potential-energy surface will lead predominantly to ClO and NO_2 in their ground electronic states, an excited state of $ClONO_2$ may correlate to a higher energy asymptote and dissociate to form excited-state products. The statistical decay rate will be slower and the translational energy release smaller for channels with lower available energy. We therefore considered dissociation *via* two excited-state channels which consume most of the 308 nm photon energy:

$$ClONO_2 \rightarrow Cl(^2P) + NO_3(\tilde{A}\,^2E'); \qquad \Delta H = 83 \text{ kcal mol}^{-1} \qquad (4)$$

$$\rightarrow ClO(a\,^4\Sigma^-) + NO_2(\tilde{X}\,^2A_1); \qquad \Delta H \approx 77 \text{ kcal mol}^{-1} \qquad (5)$$

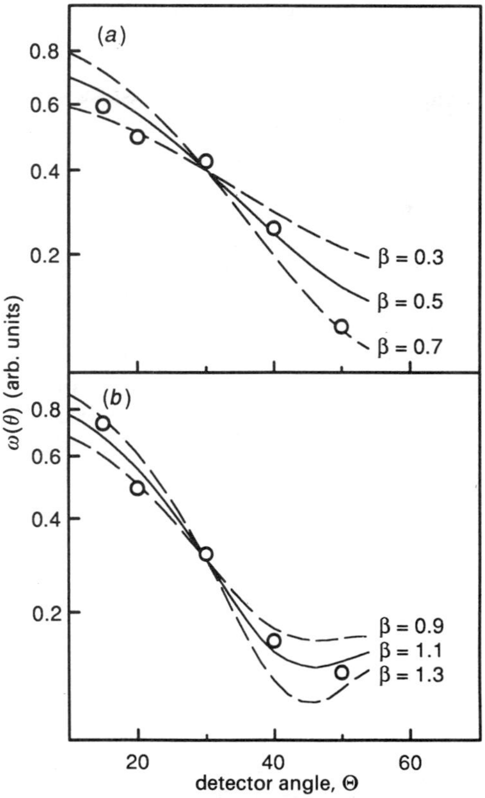

Fig. 6 Semi-logarithmic plot of the laboratory angular distributions for the two photodissociation channels observed: (*a*) $Cl + NO_3$, (*b*) $ClO + NO_2$. The circles are the experimental data, and the lines are the calculated distributions using values of the anisotropy parameter β shown. The solid lines are the best fit to the data; the dashed lines are calculated using upper and lower limits for the parameter β.

We proposed the existence of a bound ClO quartet state in a previous study of photoproducts from Cl_2O dissociation,[17] and this hypothesis was confirmed theoretically by Langhoff, who computed the energy and geometry of this state.[27] In order to estimate the maximum possible yield of these channels, we first performed RRKM calculations to obtain a cm translational energy distribution for each channel.[28] We then compared the slow component of the TOF distribution at the relevant m/z to the TOF distribution predicted by the RRKM $P(E_T)$ distribution. We obtained upper limits on the yields by assuming that the signal in the slow components arose solely from statistical dissociation (see Fig. 2 and 3). We estimated a yield of 0.05 for dissociation to ground-state ClO and NO_2 products, based on the signal at $m/z = 51$. We then determined the ClO contribution to the slow component at $m/z = 35$ by using the same ClO ionizer fragmentation pattern observed for the main peaks. We assumed that the remainder of the slow signal at $m/z = 35$ corresponded to Cl products from statistical decay to ground-state Cl and NO_3 products and obtained a maximum yield of 0.02 for this channel. Thus, the maximum total relative yield for statistical dissociation to ground-state products *via* either channel is 0.07. For the excited-state products, we found a maximum possible total yield of 0.03 based on the data collected at 15° (Fig. 3). However, the RRKM $P(E_T)$ distributions for the excited-state product channels predict negligible contributions to the slow signal at larger detector angles, because the $P(E_T)$ distributions fall off quickly with E_T. Our inability to achieve a consistent fit of all the data under the assumption of excited-state products makes it highly unlikely that channels (4) or (5) play any role in $ClONO_2$ photodissociation at 308 nm. In any case, these yields are upper limits because it is improbable that all of the slow signal arises from statistical dissociation. Secondary photodissociation, observed unambiguously in our previous experiments at shorter wavelengths, may even account for all of this broad underlying signal at $m/z = 35$. Regardless of the actual product channel assumed, we conclude that any statistical dissociation process is at most a minor channel.

We performed a photodissociation experiment on Cl_2O at 308 nm in order to calibrate the relative detection efficiency for Cl and ClO and thereby determine the branching ratio between the two channels in the photodissociation of $ClONO_2$. The Cl_2O data (not shown) were analogous to previous observations and were fit with the same $P(E_T)$ and $\omega(\theta)$ distributions as before.[17] The relative yields for the two photodissociation channels of $ClONO_2$ following excitation at 308 nm were determined to be 0.67 ± 0.06 for the $Cl + NO_3$ channel and 0.33 ± 0.06 for the $ClO + NO_2$ channel. Our results may underestimate the ClO yield because the primary ClO product has a significant probability of undergoing secondary photodissociation during the laser pulse; therefore the Cl : ClO branching ratio of 0.67 : 0.33 is an upper limit.

Discussion

Excited-state Lifetime

The anisotropy of the angular distributions allows us to determine whether the main products associated with the dominant peaks were formed by prompt dissociation, *i.e.* occurring in less than a rotational period, τ_{rot}. If the excited molecule has time to rotate before dissociating, then the molecule will change orientation, causing the angular distribution to become more isotropic. Rotational averaging thus reduces the absolute magnitude of β, though not necessarily exactly to zero. We can estimate upper limits on the minimum values of β for the two channels with the use of expressions that Yang and Bersohn have derived for calculating β when the dissociating molecule is long-lived.[29] We assume (1) that the transition dipole moments are parallel to the axes of the respective dissociating bonds, (2) that the geometry of $ClONO_2$ does not change on going to the dissociative state and (3) that $ClONO_2$ can be approximated as a symmetric top ($\kappa = -0.89$). We find that the upper limits on the anisotropy parameters for a long-lived state of $ClONO_2$ are $\beta = 0.07$ for dissociation to $ClO + NO_2$ and $\beta = 0.13$ for dissociation to $Cl + NO_3$. Effects such as vibrational averaging and breakdown of the symmetric-top assumption will reduce β further. The calculated values lie below the uncertainties in the anisotropy parameters for the two main channels, $\beta = 1.1 \pm 0.2$ for $ClO + NO_2$ and $\beta = 0.5 \pm 0.2$ for $Cl + NO_3$. The observed anisotropies thus indicate that dissociation *via* either channel occurs within τ_{rot}. Assuming a typical rotational temperature of *ca.* 50 K in the molecular beam, τ_{rot} would be *ca.* 9 ps. We conclude that the main products observed in this experiment are formed in $\lesssim 9$ ps, which is shorter than the mean collision time (*ca.* 5 ns) under stratospheric conditions, and thus cannot arise from a long-lived excited state.

We could not rule out assignment of the slow signal at $m/z = 35$ and 51 to a third dissociation channel, such as statistical decomposition, with low cm translational energies; however, the contribution from such a process will be small. Dissociation *via* a statistical process should be dominated by fission of the weakest bond, making ground state ClO and NO_2 the most likely major products if $ClONO_2$ dissociates on its ground electronic state surface. We derived an upper limit of 0.05 for the yield of this channel from the $m/z = 51$ data (Fig. 1 and 2). However, the RRKM lifetime for ground-state $ClONO_2$ with 92.8 kcal mol^{-1} of internal energy will be on the subpicosecond timescale. To account for a nanosecond lifetime, we would have to invoke a rate-limiting step, such as internal conversion or intersystem crossing from an initially prepared state. Statistical dissociation of metastable $ClONO_2$ could occur on a nanosecond timescale only if the available energy is small (on the order of 10 kcal mol^{-1} or less for $ClONO_2$), *e.g.* dissociation to a channel near threshold leading to excited-state products. Dissociation to channels (4) and (5) may have lifetimes as long as nanoseconds, but we did not have enough information to calculate accurate decay rates for processes so close to threshold. The bound triplet state predicted by Graña *et al.* is a candidate for this metastable state, but there is no information on its possible decomposition pathways. Although the translational energy release would be even lower than for dissociation to ground-state products, our experiment still places an upper limit of 0.03 on the total relative yield for the plausible product channels (4) and (5) near 93 kcal mol^{-1} (see Fig. 2). Thus, the slow product signal could account for only a small fraction of the total observed products; furthermore, all of these estimates are upper bounds, because secondary photodissociation of ClO will likely contribute substantially to the observed slow signal at $m/z = 35$.

Given our beam velocity, laser spot size and the solid angle viewed by our detector, we can only detect fragments from excited $ClONO_2$ molecules that dissociate within 2.5 μs. Our sensitivity allows us to detect molecules in states with dissociation lifetimes $\tau \leqslant 10$ μs; metastable states which live longer would go unobserved in our experiment. Such lifetimes are significantly longer than typical mean collision times of 1–10 ns in flow tube experiments, where unit quantum yields are observed for $ClONO_2$ photolysis at buffer gas pressures of 5–10 Torr.[7,10,11] If dissociation occurs through a state with a lifetime of 10 μs or longer, then the metastable $ClONO_2$ must undergo $>10^3$ gas kinetic collisions without being quenched at these pressures. Thus, the molecular beam and flow tube studies are together consistent with a long-lived metastable only if that state has a lifetime $\tau > 10$ μs and is very inefficiently quenched.

Our results would also be consistent with the experiment of Sander and co-workers if a metastable state of $ClONO_2$ is prepared by excitation at wavelengths >308 nm. Sander and co-workers photolysed chlorine nitrate with a broad-band, filtered Xe lamp whose output extended from *ca.* 300 nm to visible wavelengths.[12] Such long-wavelength radiation would be absorbed in the 370–380 nm region, where Molina and Molina[3] observed a weak absorption band that was distinct from absorption bands deeper in the ultraviolet. This band lies close to 362 nm, the wavelength that Graña *et al.* predict for the lowest vertical triplet excitation.[16] This state is predicted to be directly dissociative, but may lead by curve crossing to the bound metastable state. Additional experiments at longer wavelengths are necessary to examine the possiblility that this band involves a transition leading to formation of a long-lived state; however, the small absorption cross-section, nine times lower at 362 nm than at 308 nm, will make these measurements difficult.

Product Yields

We found channels (1) and (2) to dominate $ClONO_2$ photochemistry at 308 nm; the total yield of all other possible channels must be $\leqslant 0.07$. The Cl : ClO product branching ratio was determined to be 0.67 : 0.33. The relative yield of 0.33 for ClO is a lower limit, because some of the ClO products may have undergone secondary photodissociation. This branching ratio does not agree with a number of earlier studies performed at $\lambda > 300$ nm by Smith *et al.*,[4] Chang *et al.*,[13] Knauth and Schindler[6] and Margitan.[7] The last three experiments all found near unit quantum yields of Cl atoms and failed to detect ClO products, even for photolysis at wavelengths <300 nm. The JPL evaluation,[15] which recommends quantum yields of 0.9 ± 0.1 for the Cl + NO_3 channel and 0.0 for the ClO + NO_2 channel, was based primarily on the results of the most direct study, that of Margitan. However, our results are more consistent with recent unpublished flow-tube studies[10–12] which detect both channels with similar relative yields.

The decreased yield for dissociation of the weakest bond, $ClO{-}NO_2$, in the present study relative to our earlier results suggests that photons at 308 nm excite transitions favouring fission of the Cl—O bond. Graña *et al.* calculate that the lowest singlet state $1\,^1A''$ possesses ClO $n \rightarrow \sigma^*$ character and has a vertical excitation wavelength of 284 nm with an oscillator strength *ca.* 1% of the band strength at 190 nm. The next highest state, the $2\,^1A''$ state, is predicted to be a $\pi \rightarrow \pi^*$ transition localized on the NO_2 group at 252 nm but with zero oscillator strength. While excitation to the $1\,^1A''$ state could explain the higher Cl yield at 308 nm relative to 248 nm, it is not consistent with the observed angular distributions which are indicative of a parallel transition. To excite a state of A″ symmetry, the transition moment must be out of plane and perpendicular to either dissociation coordinate in the ground-state geometry. A perpendicular transition would result in a negative value of β unless the dissociating bond rotates significantly out of plane prior to dissociation. There are many states of both symmetries at slightly higher energies, and excitation at 308 nm could instead readily excite the Franck–Condon wing of a dissociative A′ state. It is possible that there are simultaneous parallel and perpendicular transitions to multiple excited states or a transition to a mixed state with the dipole transition moment aligned midway between the two bond axes, making the net β small and positive. The complexity of the $ClONO_2$ excited-state manifold suggests that photodissociation of $ClONO_2$ cannot be described by a simple bond-selective picture but rather that curve crossings and the dynamics of nuclear motion following excitation may largely determine the final product branching ratio.

The dominance of direct dissociation processes at excitation wavelengths as long as 308 nm implies that it is important to consider a catalytic cycle involving $ClONO_2$ photolysis in models of stratospheric ozone depletion. However, our finding of a significant yield for the $ClO + NO_2$ channel will lead to a decrease in the ozone-depleting efficiency of the $ClONO_2$ cycle from what has been predicted. At mid-latitudes, where this cycle contributes only 3–5% of the ozone loss,[2] the new product yield ratio will not affect the predicted ozone depletion. The impact will be greater on model predictions of ozone loss in the Arctic polar spring as well as in the Antarctic collar region, where Toumi *et al.* have shown that this cycle plays a major role.[14]

The experiments were performed at the Jet Propulsion Laboratory, California Institute of Technology, under contract with the National Aeronautics and Space Administration (NASA), and were supported by the NASA Upper Atmosphere Research Program under contract NAGW-3893 and by E. I. duPont de Nemours and Co., Inc. We acknowledge support for T.A.M. from a NASA Graduate Research Fellowship and for M.T. from the Murata Overseas Scholarship Foundation.

References

1 F. S. Rowland, J. E. Spencer and M. J. Molina, *J. Phys. Chem.*, 1976, **80**, 2711; 2713.

2 P. O. Wennberg, R. C. Cohen, R. M. Stimpfle, J. P. Koplow, J. G. Anderson, R. J. Salawitch, D. W. Fahey, E. L. Woodbridge, E. R. Keim, R. S. Gao, C. R. Webster, R. D. May, D. W. Toohey, L. M. Avallone, M. H. Proffit, M. Loewenstein, J. R. Podolske, K. R. Chan and S. C. Wofsy, *Science*, 1994, **266**, 398.

3 M. Molina and L. T. Molina, *J. Photochem.*, 1979, **11**, 139.

4 W. S. Smith, C. C. Chou and F. S. Rowland, *Geophys. Res. Lett.*, 1977, **4**, 517.

5 S. M. Adler-Golden and J. R. Wiesenfeld, *Chem. Phys. Lett.*, 1981, **82**, 281.

6 H. D. Knauth and R. N. Schindler, *Z. Naturforsch., Teil A*, 1983, **38**, 893.

7 J. J. Margitan, *J. Phys. Chem.*, 1983, **87**, 674.

8 J. P. Burrows, G. S. Tyndall and G. K. Moortgat, *J. Phys. Chem.*, 1988, **92**, 4340. These authors proposed a hypothetical metastable state of $ClONO_2$ excited at 254 nm, because $Cl + NO_3$ products could not account for all the photolysis yield. It is likely that the missing product is in the $ClO + NO_2$ channel, which they did not detect.

9 T. K. Minton, C. M. Nelson, T. A. Moore and M. Okumura, *Science*, 1992, **258**, 1342.
10 A. R. Ravishankara, personal communication.
11 J. J. Orlando, C. S. Kegley-Owen, G. S. Tyndall and J. G. Calvert, XXIst Informal Conference on Photochemistry, Toronto, Ontario, Canada, May 1994.
12 S. L. Nickolaisen, S. P. Sander, R. R. Friedl, L. Jaegle, M. A. Allen, Y. L. Yung, C. R. Webster, R. D. May, G. C. Toon, B. Sen and J-F. Blavier, XXIst Informal Conference on Photochemistry, Toronto, Ontario, Canada, May 1994.
13 J. S. Chang, J. R. Barker, J. E. Davenport and D. M. Golden, *Chem. Phys. Lett.*, 1979, **60**, 385.
14 R. Toumi, R. L. Jones and J. A. Pyle, *Nature* (*London*), 1993, **365**, 37.
15 W. B. DeMore, S. P. Sander, D. M. Golden, R. F. Hampson, M. J. Kurlyo, C. J. Howard, A. R. Ravishankara, C. E. Kolb and M. J. Molina, *Chemical Kinetics and Photochemical Data for use in Stratospheric Modeling*, JPL Publication 92-20, 1992.
16 A. Graña, T. J. Lee and M. Head-Gordon, *J. Phys. Chem.*, 1995, **99**, 3493.
17 C. M. Nelson, T. A. Moore, M. Okumura and T. K. Minton, *J. Chem. Phys.*, 1994, **100**, 8055.
18 Y. T. Lee, J. D. McDonald, P. R. LeBreton and D. R. Herschbach, *Rev. Sci. Instrum.*, 1969, **40**, 1402.
19 T. K. Minton, P. Felder, R. J. Brudzynski and Y. T. Lee, *J. Chem. Phys.*, 1984, **81**, 1759.
20 P. Felder, Habilitationsschrift, University of Zürich, Switzerland, 1993.
21 G. H. Cady, *Inorg. Synth.*, 1957, **5**, 156.
22 M. Schmeisser, *Inorg. Synth.*, 1967, **9**, 127.
23 G. E. Busch and K. R. Wilson, *J. Chem. Phys.*, 1972, **56**, 3626; 3638.
24 T. K. Minton, G. M. Nathanson and Y. T. Lee, *J. Chem. Phys.*, 1987, **86**, 1991.
25 A. M. Wodtke and Y. T. Lee, *J. Phys. Chem.*, 1985, **89**, 4744.
26 R. N. Zare, *Mol. Photochem.*, 1972, **4**, 1.
27 S. R. Langhoff, unpublished results.
28 P. J. Robinson and J. A. Holbrook, *Unimolecular Reactions*, Wiley-Interscience, New York, 1972.
29 S-C. Yang and R. Bersohn, *J. Chem. Phys.*, 1974, **61**, 4400.

Paper 5/00231A; *Received* 13*th January*, 1995

Faraday Discuss., 1995, **100**, 309–320

Solvation of Hydrogen Halides on the Surface of Ice

Struan H. Robertson and David C. Clary*
Department of Chemistry, University of Cambridge, Lensfield Road, Cambridge, UK CB2 1EW

Heterogeneous processes on the surface of ice crystals in polar stratospheric clouds are thought to play a crucial role in the catalysis of reactions such as HCl + $ClONO_2$. An important consideration is the mechanism of solvation of the HCl molecules on the ice surface. A molecular dynamics simulation of the solvation of HCl, HF and HBr by an ice surface at 190 K is reported. The model allows for the solvation of the hydrogen halide by water molecules on the ice surface to form solvated ions. The simulations suggest that the ionic solvation process is energetically feasible for HCl and HBr, but not for HF.

1. Introduction

The depletion of ozone in the Antarctic stratosphere was first documented in 1985[1] and has since generated numerous investigations. The magnitude and dynamics of the depletion has prompted speculation that mechanisms peculiar to Antarctic stratospheric conditions are responsible for the observed depletions.[2] It has been proposed that heterogeneous mechanisms on the surface of ice crystals play a crucial role in the catalysis of the reactions,[3,4]

$$HCl + ClONO_2 \rightarrow HNO_3 + Cl_2 \quad (1)$$

$$HCl + ClOH \rightarrow H_2O + Cl_2 \quad (2)$$

The main consequence of such heterogeneous catalysis is that the species HCl and $ClONO_2$, which had hitherto been regarded as inert reservoir species for chlorine, may be important in Cl_2 production. It has been demonstrated experimentally that under gas-phase conditions equivalent to the Antarctic stratosphere, reaction (1) proceeds at negligible rate.[5] There is, however, a growing body of evidence that suggests that these reactions are catalysed by ice particles that are present in polar stratospheric clouds (PSCs),[3,6–8] and that this mechanism is the main source of diatomic chlorine, Cl_2. It is this production of Cl_2 during the Antarctic winter followed by photolysis to give reactive Cl atoms that leads ultimately to the catalytically destructive ClO_x species.[2]

There have been a number of fundamental experimental investigations of these mechanisms with emphasis on determining the degree of adsorption of HCl at an ice surface.[9] This work suggests that there is close to monolayer coverage of the ice surface by HCl under stratospheric conditions. Horn *et al.*[10] have recently reported experiments in which they examined the spectra of various species adsorbed onto heavy-water ice and in particular DCl at a temperature of 110 K. The observed spectrum appears to be lacking the DCl stretch frequency and this has been interpreted as evidence for the formation of an ion pair at the surface, though it should perhaps be noted that the conditions of these experiments are somewhat removed from those found in the polar stratosphere. More recently Delzeit *et al.*[11] have reported experiments in which the IR spectra of HCl at an ice surface for a series of temperatures was monitored. These spectra indicated that there was significant ionisation at temperatures greater than 60 K. Graham and Roberts[12] have also recently reported experimental measurements on the desorption of HCl from ice layers.

In a previous theoretical study[13,14] the adsorption of HCl on ice crystals was examined using the molecular dynamics (MD) method.[15] The results of this study demonstrated that HCl molecules did indeed stick to an ice surface but the residence time at the surface was so small that the resulting surface coverage was in disagreement with existing experimental[9] results by orders of magnitude.

A difficulty is that the theoretical model did not allow for ionic dissociation of HCl on the ice surface. Previous calculations[16] demonstrate that the ice, while not a liquid, does exhibit some fluid characteristics and recent work by Molina and co-workers[9b] suggest that ionic solvation of HCl is a crucial step in the catalysis of reactions (1) and (2). Experimental evidence for an ionic mechanism which is energetically feasible has been found by Nelson and Okumura.[17] In the gas phase HCl does not form an ion pair with a single water molecule but it does dissociate into ions when solvated in liquid water. A crucial feature is whether there are enough moving water molecules on the surface of ice to solvate the HCl and thus allow ionic solvation to occur.

The HCl-adsorption experiments[9] have all been performed on polycrystalline ice. Type II PSCs are generally thought to be single crystals, and it is not yet completely clear whether the experimental results obtained for polycrystalline ice are applicable to the single-crystal case. It should be noted that HCl is much more soluble in polycrystalline ice than in single-crystal ice, owing to the high affinity for HCl of the triple junctions between grain boundaries.[18]

Work has tended to focus on HCl studies but there is increasing concern that other hydrogen halides also contribute to ozone depletion. In particular, the level of HBr is a cause for concern. Volatile brominated products are generated by industry and through deforestation activities. HBr can be generated *via* similar tropospheric cycles as those that generate HCl. HBr is potentially more destructive because of its weaker bond strength. In recent experiments, Hanson and Ravishankara[19] demonstrated that HBr was adsorbed at least as efficiently as HCl was on an ice surface. Interestingly, they also demonstrated that HF was hardly adsorbed at all. These observations were rationalised in terms of relative acid strengths.

It is clearly desirable to extend the MD calculations presented earlier by allowing the possibility of ionic solvation of hydrogen halides (HX) on the ice surface. Two questions need to be addressed: the first concerns whether the ionic solvation is energetically feasible; the second concerns whether the process is kinetically feasible. This work is concerned with the first of these questions, to determine if the ionic solvation is an energetically feasible process. The MD methods to do this are adapted from those developed by Hynes, Borgis and co-workers[20] for proton transfer reactions in solution. It should be emphasised that the kinetic aspects of the problem are not considered here and will be the subject of future work.

In Section 2 the method used to investigate the energetic feasibility of ionic solvation is outlined. In Section 3 results are discussed. Concluding remarks are collected in Section 4.

2. Method

The reaction of primary interest in the solvation of HX at an ice surface is

$$HX_{(g)} + H_2O_{(l)} \rightarrow X_{(l)}^{-} + H_3O_{(l)}^{+} \tag{3}$$

where the subscript (l) implies the existence of a layer at the surface of the ice, which, while not liquid has some fluid characteristics. From this it is seen that the principal step is the transfer of a proton, and so this section begins with a discussion of the general aspects of proton transfer.

2.1 Proton Transfer in Solution

There have been many experimental[21–23] and theoretical[20,24,25] investigations of proton transfer in solvated environments. The general process can be written as

$$\mathrm{A\cdots H{-}D \rightarrow A{-}H\cdots D} \tag{4}$$

where A is the proton-accepting species and D the proton-donating species. Fig. 1(*a*) is a schematic representation of the potential-energy surface on which the process occurs.[20] The coördinate, q, is the displacement of the proton from the donor (D) in the direction of the acceptor (A). The distance between the donor and acceptor sites is denoted Q. The first minimum at the point $q = \mathrm{R}$ corresponds to the equilibrium position in the local mode stretch of the D—H bond. The second minimum at point $q = \mathrm{P}$, likewise, corresponds to the minimum in the local mode stretch of the A—H bond. Implicit in the schematic representation of Fig. 1(*a*), is that the process occurs on an electronically adiabatic surface, *i.e.* the local modes are coupled, resulting in an overall surface where the two wells are separated by a barrier of finite width and height. The centre of this barrier occurs at values of q between R and P. In the absence of tunnelling the proton would be confined to the donor well and transition to the other well could only be made by a series of activation and deactivation steps. This route will become more likely at high temperatures when the population of higher vibrational states is possible. PSCs exist at temperatures of around 190 K so it is assumed that in either the acceptor or donor site the proton will be in the ground vibrational state, and that there is very little activation to any higher vibrational states.

The proton can tunnel through the barrier between the two sites and in the absence of an activation/deactivation process this mechanism will be the dominant one. The

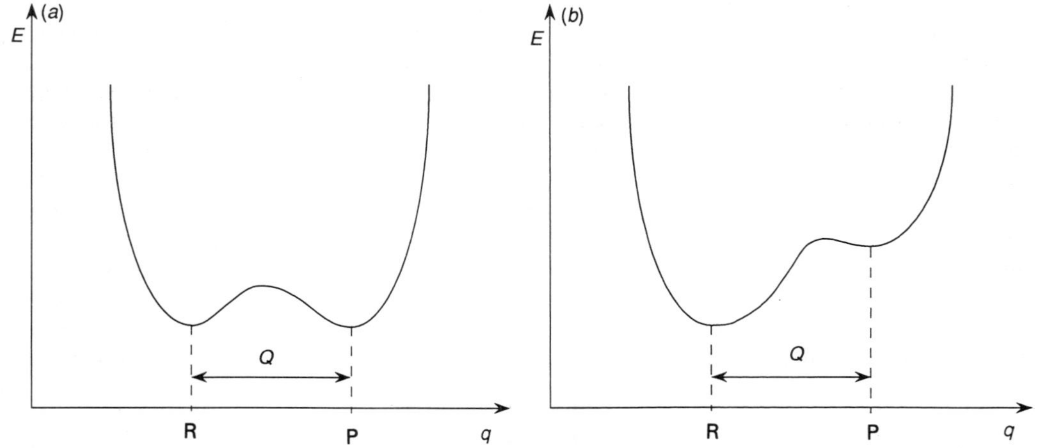

Fig. 1 (*a*) Schematic of the proton-transfer potential-energy surface showing the q and Q coordinates. Note that the minima are at the same energy thus allowing proton transfer. (*b*) The same as (*a*) but this time the minima are now at different energies so that tunnelling is not possible.

transfer of a proton in a solvated environment is determined by two main factors: (i) In common with gas-phase reactions that involve tunnelling, the probability that a proton tunnels through the barrier is dependent on the width and the height of the barrier in the effective potential.[23,26] (ii) The other main effect is the interaction of the proton-transfer system with the solvent which can alter the potential-energy surface on which the process occurs. The solvent tends to arrange itself around the proton-transfer system so as to make one configuration (*e.g.* A···H—D) energetically more favourable than the conjugate configuration (A—H···D). This is represented schematically in Fig. 1(*b*), where the H—D well lies lower than the H—A well. Whilst this state of affairs persists the proton cannot tunnel as there is no state of comparable energy on the other side of the barrier. However, the solvent is not static but is subject to fluctuations in configuration during which the solvation of the proton-transfer system may alter such that the energy levels of the local mode vibrations become equal as depicted in Fig. 1(*b*), thus allowing the ionic dissociation to be energetically feasible.

The analysis of Hynes, Borgis and co-workers[20] considered other, more subtle points *e.g.* it was noted that the magnitude of the preferential stabilisation of one site with respect to the other was a function of Q in that the more extended Q the more likely that solvent could enter into the transfer region thereby stabilising the protonated site. For the present case this and other effects are likely to be less marked as the proton-transfer system is not actually immersed in the ice surface.

Strictly speaking the Hynes–Borgis analysis is only applicable to situations where the donor–acceptor moiety is immersed in a polar aprotic solvent. For HX adsorbed at the surface of ice this is clearly not the case. However, it will be assumed that the difficulties introduced by protic solvents are at least partially mitigated by the fact that the donor–acceptor moiety lies on top of the solvent. The surface must also exhibit fluid characteristics for the theory to be applicable at all. Previous calculations showed this to be the case.[16]

The point of view adopted here is that the transfer of a proton from HX to the water molecules moving on the ice surface is determined by a number of competing factors, each of which could be rate limiting. For simplicity we break the process up into three steps, each with a characteristic time period: (i) The physisorption of HX at the ice surface (with mean period τ_p). (ii) Fluctuation of surface water molecules and the HX molecule to produce a configuration which would lead to energetically favourable proton transfer (with mean period τ_f). (iii) Tunnelling of the proton through the potential barrier (with mean period τ_t). This division excludes any correlation between solvent fluctuation and tunnelling, and thus the treatment diverges from that of Hynes, Borgis and co-workers[20] where these effects are taken into account explicitly.

The first step has been investigated in the previous work[14] where it was estimated that the physisorption lifetime was of the order of 10^{-7} s. The second step is the focus of the present work. The simplest considerations suggest that solvation will occur if

$$\tau_p > \tau_f > \tau_t \qquad (\mathrm{I}a)$$

However, eqn. (I*a*) is too restrictive if the system's fluctuations produce a reactive configuration during a large proportion of the physisorption time. Therefore, tunnelling will occur if the average total time spent in a reactive configuration while the molecule is physisorbed (considering the tunnelling channel to be closed) exceeds the tunnelling time, *i.e.* if

$$P_r \tau_p > \tau_t \qquad (\mathrm{I}b)$$

where P_r is the probability of the system being in a reactive configuration while physisorbed. A condition for eqn. (I*b*) to hold is that the tunnelling process is statistical, in the sense that we should have $\tau_f > \tau_v$ where τ_v is the vibrational period of the HX bond (11 fs for HCl).

2.2 Energetic Considerations

An approximate idea of the energies involved in the solvation of HCl at an ice surface can be obtained from the Born–Haber cycle shown in Fig. 2. This cycle is for the purely gas-phase reaction of transferring a proton from within an H_2O–HCl complex. Briefly the steps involved are: 1. The dissociation of the H-bonded complex. 2. The dissociation of the HCl molecule. 3. The ionisation of the H atom. 4. The formation of the chloride ion (electron affinity of chlorine). 5. The formation of the hydroxonium ion (proton affinity of water). 6. The electrostatic interaction of the chloride and hydroxonium ions.

As can be seen this gas-phase process is endothermic by *ca.* 169 kJ mol^{-1}.[27,28] Thus, if ionic solvation is to occur then it must overcome this enthalpy deficit. In the case of dissolution of HCl in liquid water under standard conditions the overall enthalpy change is exothermic by *ca.* 75 kJ mol^{-1},[27] hence solvation contributes around 244 kJ mol^{-1} to the process.

Similar cycles can be constructed for the HF–water and HBr–water complexes, the major difference being the bond dissociation energies. As for the HCl–water complex, transfer of a proton is an endothermic process: *ca.* 190 kJ mol^{-1} for the HBr–water complex and 280 kJ mol^{-1} for the HF–water complex.

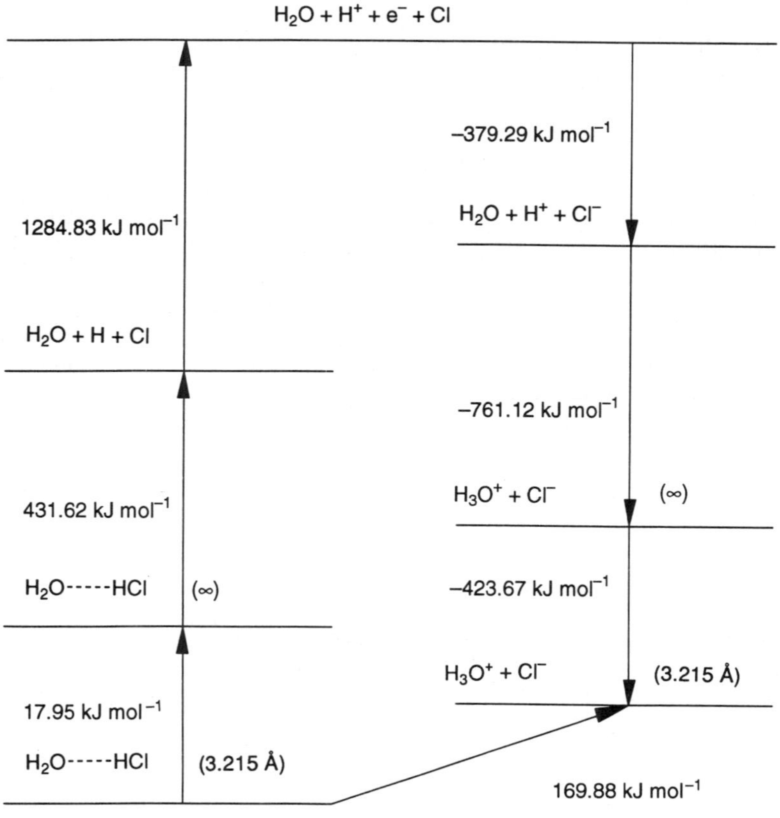

Fig. 2 Born–Haber cycle for the transfer of a proton from HCl to H_2O in the isolated HCl–H_2O complex

2.3 Details of Simulation

The basic model is an extension of that presented previously.[13,14] Briefly, an MD[15] model was constructed consisting of 360 water molecules of fixed geometry and interacting *via* the TIP4P[29] potential. The molecules are initially arranged in 12 layers of 30 molecules each. The molecules in the bottom four layers are held fixed to mimic the properties of an infinite crystal lattice. The top eight layers are allowed to move. Translation of the molecular centre of mass was performed using a standard leapfrog algorithm.[15] Rotation about the centre of mass was performed using the implicit quaternion algorithm devised by Fincham.[30] A damping factor was placed on the long-range component of the TIP4P potential to facilitate rapid calculation.[14] As before, the initial configuration was integrated for a period during which the velocities were rescaled at each time-step so as to produce a surface close to the specified temperature[31] of 190 K. The rescaling was then switched off and the simulation run on to give an equilibrated surface and to check that energy was conserved. Once a satisfactory ice surface was obtained, a scattering simulation was performed. An HX molecule was introduced with a velocity component such that the centre of mass was moving toward the surface. The interaction potential of the HCl with the ice surface was the same as that used in previous work.[14]

Two new potentials were developed for the HBr and HF systems. The basis of these potentials was that of the HCl–ice system. The structure and energy of a number of hydrogen halide–water molecule complexes has been the subject of a recent *ab initio* study by Hannachi *et al.*[32] The parameters of the HCl–ice model were altered to values appropriate to HBr and HF so that the structures of the HBr–water and HF–water complexes calculated by Hannachi *et al.*[32] were reproduced. As with the HCl–water complex the main contribution to these potentials was from the coulombic interaction terms. The charges and sites for each of the three hydrogen halides are listed in Table 1.

For this particular study, initial conditions were chosen such that a sticking collision would result. Once the HX molecule had become attached to the surface the simulation was integrated for a period of 10 ps, to ensure that it was properly physisorbed.

In the previous study of HCl on an ice surface, minimisation of the HCl–ice potential was performed, the aim being to ascertain the arrangement of the HCl on the ice surface. From this study an esitmate of the surface binding energy can be obtained and, together with the frequency of vibration of the hydrogen halide on the surface of the ice, an adsorption time can be calculated. Table 2 gives estimates of the surface binding energies for the three hydrogen halides considered. It is noted that the adsorption energies are of very similar magnitude and that there is no discernible trend. Also reported are the surface vibrational frequencies obtained from trajectory calculations which are also

Table 1 Three point model charges for hydrogen halides

molecule	site	coordinate	charge
HF	H	0.8712	0.707
	F	0.1590	−1.414
	M	−0.0459	0.707
HCl	H	1.2392	0.403
	Cl	−0.0354	−0.909
	M	−0.5646	0.506
HBr	H	1.3964	0.358
	Br	−0.0176	−0.700
	M	−1.0000	0.342

Table 2 Surface characteristics of hydrogen halides on ice

molecule	surface period/ps	minimised energy/kJ mol^{-1}	surface lifetime/ns
HF	0.5	−20 to −27	96
HCl	0.5	−22 to −29	130
HBr	0.25–0.5	−17 to −24	30

rather similar and, as a consequence, the calculated physisorbed lifetimes are close in magnitude. It follows that the corresponding surface coverage for HBr and HF would be similar to that obtained in the previous simulations[14] on HCl and this is considerably less than unit monolayer coverage.

Addressing the question of energetic feasibility we follow Hynes, Borgis and co-workers[20] and define ΔE to be the potential-energy difference of the proton in the donor and acceptor sites. To calculate the potential-energy difference it is necessary to calculate the potential when the proton is in the two different sites. The calculation of the potential in the first site is straightfoward as this is simply the interaction energy of the HX molecules with the rest of the surface and is necessarily calculated within each integration cycle of a normal MD execution.

The second site of the proton is when it is bonded to an oxygen atom of a water molecule. In previous work[14] it was found that the HX, once stuck to the surface, became hydrogen bonded to one particular water molecule. In the present calculation it is assumed that the oxygen atom of this water molecule is the acceptor site and that the probability of the proton tunnelling to another oxygen atom is negligible because the distance is much greater. Hence the oxygen atom of this water molecule and the halide atom of the HX molecule are taken to be the two heavy atoms between which the proton is transferred, the distance between these two being the coordinate Q.

Having established on to which molecule the proton could jump, the next task is to determine the position of the proton. In common with other hydrogen-bonded complexes we assume that the hydrogen bond is linear and that the proton moves along the O—X vector. The HCl and HO bond lengths were taken to be 1.27 Å and 0.96 Å, respectively.[14] The experimental O—Cl distance in the gas phase H_2O–HCl complex is[33] 3.2 Å and so the distance that the proton has to tunnel in this complex is 0.97 Å.

Next, an interaction potential was devised for the ionic system. This was done in four parts: (i) The interaction of bulk water molecules with each other is straightforward to obtain since it is simply the previous bulk water interaction potential less terms that are due to the acceptor water molecule. (ii) The interaction of the X^- ion with bulk water molecules has two components, the first of these being electrostatic interaction with the charges of the TIP4P model and the second an induced dipole term:[34]

$$V = -\sum_{i,j} \frac{q_{i,j}}{r_{i,j}} - \sum_{i} \frac{\alpha}{2R_i^4} \tag{II}$$

where $q_{i,j}$ is the charge at the jth charge site of the TIP4P potential on the ith molecule and $r_{i,j}$ the distance between the X^- ion and this charge site. α is the polarisability of the H_2O molecule and was taken to be[35] 1.45 $Å^3$, and R_i is the centre of mass distance between the X^- ion and the ith water molecule. (iii) The interaction of the H_3O^+ with bulk water molecules has the difficulty that H_3O^+ clearly cannot be treated as a point charge. Matters are exacerbated by the closer proximity of the water molecules with which it interacts. For the sake of computational convenience a model based on that of eqn. (II) was adopted. The charge was distributed about the ion such that there was a third on each proton and zero charge on the central oxygen atom. The charges on the protons interacted with the water molecules *via* both electrostatic and induced dipole

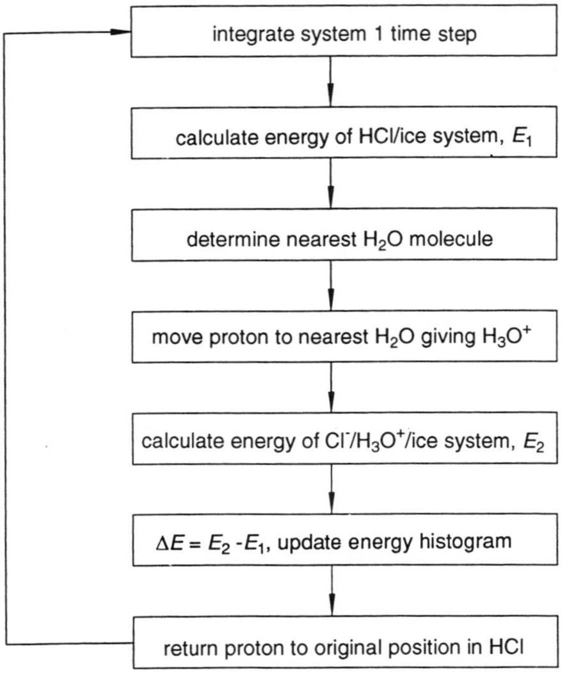

Fig. 3 Flow chart describing the computer simulation

terms, the potential due to one of these charges being,

$$V = \frac{1}{3}\sum_{i,j}\frac{q_{i,j}}{r_{i,j}} - \frac{1}{9}\sum_{i}\frac{\alpha}{2R_i^4} \qquad \text{(III)}$$

The Lennard-Jones term of the TIP4P model was also retained, though in the calculation of the difference in energy of the two sites this term will cancel. (iv) The interaction of the X^- ion with the H_3O^+ ion was assumed to be purely electrostatic.

Having constructed a potential, the calculation of ΔE can now proceed much along the same lines as devised by Hynes, Borgis and co-workers.[20] An integration cycle is performed. The potential energy of the HX–ice system is then calculated, and the proton is moved to form the ion pair. The potential energy for this system is calculated again and the proton is then replaced to its HCl position. The system is then integrated another step and the sequence is repeated. The potential-energy differences of the proton in the two different sites, ΔE, were collected together and used to construct a histogram, from which the mean energy for the formation of ions was calculated together with the proportion of configurations that exceed the endothermicity of the HCl bond breakage. The procedure is summarised in Fig. 3.

3. Results and Discussion

Fig. 4 shows the ΔE histogram for the HCl–ice system at 190 K. These data were obtained from a run with 2×10^5 time-steps of length of 1 fs. The mean value for ΔE is 63.2 kJ mol^{-1}. The origin corresponds to the situation where the local mode oscillators are in resonance [as in Fig. 1(*a*)] and so ionic solvation is energetically feasible. It is assumed that any configuration for which $\Delta E \leqslant 0$ can lead to reaction. From the histogram, the probability of the system being in a reactive configuration, P_r, is 0.264.

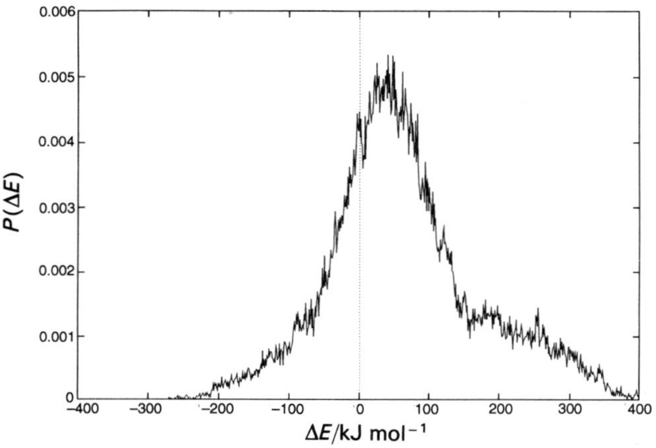

Fig. 4 Histogram of ΔE, the potential-energy difference between the proton in the acceptor and donor sites for the HCl–ice system

The histogram demonstrates that there are configurations for which the transfer of a proton from the HCl to the bulk is exothermic and therefore an energetically allowed process. It says little about the kinetics of the process. If the rate of transfer takes place on a timescale greater than the 10^{-7} s physisorption timescale found previously then clearly the molecule will have desorbed before tunnelling can have occurred.

As stated above, the kinetics of this process are a complex interplay of three factors any of which could be rate determining. The first step in the above sequence, the physisorption/desorption of HCl onto ice, has already been considered.[14] A very approximate analysis of the second step, the formation and decay of reactive configurations, can be performed using an auto-correlation formalism in conjunction with the histogram of Fig. 4.

Let the energy axis of Fig. 4 be divided into two intervals at $\Delta E = 0$, those configurations lying to the right are unreactive configurations, U, and those lying to the left are reactive, R. There exists an equilibrium between these two states, for which the equilibrium constant, K, is simply the ratio of the two proportions outline above. The equilibrium is a dynamic one, and if we represent it as,

$$\mathrm{R} \rightleftharpoons \mathrm{U} \tag{5}$$

then

$$K = k_f/k_r \tag{IV}$$

The quantity of interest here is k_f as this dictates the rate of decay of the reactive R states. If this rate coefficient is less than that for the tunnelling process then transfer of the proton is kinetically feasible. To estimate k_f the indicator function I is defined as,

$$I[\Delta E(t)] = \begin{cases} 1; & \text{if } \Delta E[t] \leqslant 0 \\ 0; & \text{otherwise} \end{cases} \tag{V}$$

and the auto-correlation function,[36]

$$C(t) = \langle I(t)I(0) \rangle \tag{VI}$$

is calculated using the standard techniques as outlined in ref. 15. This function can be interpreted as the unnormalised probability that the system is in a reactive configuration at time t, given that it was in a reactive configuration at $t = 0$. The rate coefficient for

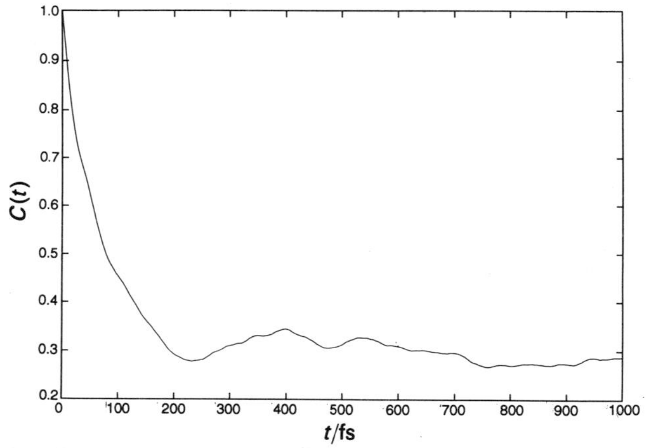

Fig. 5 Autocorrelation function $C(t)$ for the decay of reactive configurations for the HCl–ice system

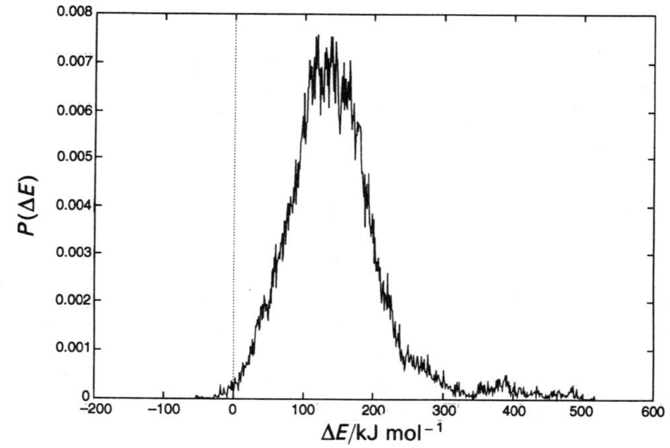

Fig. 6 As for Fig. 4 but for the HF–ice system

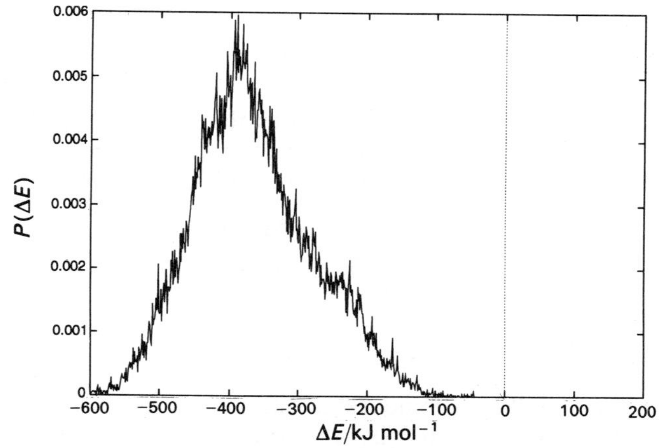

Fig. 7 As for Fig. 4 but for the HBr–ice system

the decay of R states can be obtained from,

$$\frac{\langle I(t)I(0)\rangle - \langle I\rangle^2}{\langle I^2\rangle - \langle I\rangle^2} = \exp(-k_f t) \tag{VII}$$

where $\langle I\rangle$ is the mean value of the indicator function and, since $I^2 = I$, it follows that $\langle I^2\rangle = \langle I\rangle$.

Fig. 5 shows such a plot of $C(t)$ normalised to one at time $t = 0$. As can be seen there is a rapid decline of $C(t)$, as the initial R state decays. At long times the $C(t)$ approaches an asymptotic value, which is the equilibrium proportion of R since at long times the system becomes uncorrelated with the initial state. From eqn. (VII) and the data presented in Fig. 5, the rate coefficient for the decay of reactive states was found by least-squares fitting. The value of the rate coefficient obtained was 1.38×10^{13} s^{-1}, or a lifetime, τ_f, of 0.725×10^{-13} s, which is considerably less than the physisorption lifetime. This lifetime, together with the ratio of reactive to unreactive states, would indicate that the system makes frequent excursions into the set of reactive configurations before the HCl molecule desorbs. In previous work[14] the period of oscillation of the centre of mass of the HCl molecule on the surface was shown to be of the order of 0.5 ps. It would thus seem likely that the lifetime of the reactive state is directly determined by the oscillation of the HCl on the surface.

Similar calculations and analysis were performed for the HF–ice and HBr–ice systems. Fig. 6 and 7 show histograms for the ΔE of proton transfer for the HF–ice and HBr–ice systems, respectively. It is seen from Fig. 6 that there are very few configurations of the HF–ice system for which tunnelling would be energetically favourable. The number is so small that the chances of being in an energetically feasible configuration for long enough for tunnelling to occur are negligible. Fig. 7 shows that the situation is rather different for the HBr–ice system. It is seen that all the configurations are energetically feasible to tunnelling.

The single most important factor determining the histogram distributions is the bond strength of the HX molecule. Previously Hanson and Ravishankara[19] have pointed out that the adsorption was correlated to the acid strength of the HX molecules and the present calculation confirms this observation and consequently indicates that the mechanism of adsorption involves ionic solvation.

4. Conclusions

There are two main conclusions that we make from our computer simulations on the solvation of hydrogen halides on ice surfaces. First, energy minimisations show that covalent HCl, HBr and HF have very similar properties at the surface of ice that lead to surface coverage that is considerably less than the monolayer observed experimentally for HBr and HCl. Secondly, the MD simulations indicate that solvation of the HX molecules to form ions on the ice surface is a thermodynamically feasible process for HCl and HBr, but not for HF. This conclusion is subject to a number of qualifications. First, the potential-energy surface is very approximate. In particular, better treatments of induced dipole effects are needed but it is expected that these are, in general, negative quantities that will lead to a more exothermic solvation process that will only enhance the above effects. Secondly, the thermodynamic feasibility has been assessed by consideration of enthalpy only. The free energy of proton transfer should be calculated and used to determine the position of the equilibrium.

This paper has not been concerned with a treatment of the kinetics of the proton tunnelling that is needed for the ionic solvation of HX. A study of this aspect of the problem will require a more sophisticated treatment of the potential-energy surface and solvation effects. Furthermore, a more advanced MD model is needed to treat the tunnelling dynamics. Work in both of these directions is currently going ahead.

The results provide strong evidence to suggest that HCl and HBr are both solvated to form ions on the surface of type II PSCs at 190 K. However, HF is not solvated in this way and will rapidly desorb. Therefore, it is likely that the reactions on type II PSCs involving HCl and HBr must involve an ionic mechanism, similar to reactions in solution, and not a mechanism involving neutral molecules.

This work has been supported by the EC, EPSRC and NERC. We are grateful to Geert-Jan Kroes and Ruth Lynden-Bell for useful suggestions and discussions.

References

1 J. C. Farman, B. G. Gardiner and J. D. Shanklin, *Nature (London)*, 1985, **315**, 825.
2 R. P. Wayne, Chemistry of Atmospheres, Oxford University Press, Oxford, 2nd edn., 1991.
3 M. J. Molina, T-L. Tso, L. T. Molina and F. C-Y. Wang, *Science*, 1987, **238**, 1253.
4 S. Solomon, R. R. Garcia, F. S. Rowland and D. J. Wuebbles *Nature (London)*, 1986, **321**, 755.
5 L. T. Molina, M. J. Molina, R. A. Stachnik and R. D. Tom, *J. Chem. Phys.*, 1985, **89**, 3779.
6 M. A. Tolbert, M. J. Rossi, R. Malhotra and D. M. Golden, *Science*, 1987, **238**, 1258.
7 M-T. Leu, *Geophys. Res. Lett.*, 1988, **15**, 17.
8 D. R. Hanson and A. R. Ravishankara, *J. Geophys. Res.*, 1991, **96**, 5081; J. P. D. Abbatt and M. J. Molina, *J. Phys. Chem.*, 1992, **96**, 7674.
9 (*a*) D. R. Hanson and A. R. Ravishankara, *J. Phys. Chem.*, 1992, **96**, 2682; (*b*) J. P. D. Abbatt, K. D. Beyer, A. F. Fucaloro, J. R. McMahon, P. J. Wooldridge, R. Zhang and M. J. Molina, *J. Geophys. Res.*, 1992, **97**, 15 819.
10 A. B. Horn, M. A. Chesters, M. R. S. McCoustra and J. R. Sodeau, *J. Chem. Soc., Faraday Trans.*, 1992, **88**, 1077.
11 L. Delzeit, B. Rowland and J. P. Devlin, *J. Phys. Chem.*, 1993, **97**, 10 312.
12 J. D. Graham and J. T. Roberts, *J. Phys. Chem.*, 1994, **98**, 5974.
13 G-J. Kroes and D. C. Clary, *Geophys. Res. Lett.*, 1992, **19**, 1355.
14 G-J. Kroes and D. C. Clary, *J. Phys. Chem.*, 1992, **96**, 7079.
15 M. P. Allen and D. J. Tildesley, *Computer Simulation of Liquids*, Oxford University Press, Oxford, 1989.
16 G-J. Kroes, *Surf. Sci.*, 1992, **275**, 365.
17 C. M. Nelson and M. Okumura, *J. Phys. Chem.*, 1992, **96**, 6112.
18 E. W. Wolff, R. Mulvaney and K. Oates, *Geophys. Res. Lett.*, 1989, **16**, 487.
19 D. R. Hanson and A. R. Ravishankara, *J. Phys. Chem.*, 1992, **96**, 9441.
20 D. Borgis and J. T. Hynes, in *The Enzyme Catalysis Process*, ed. A. Cooper, J. Houben and L. Chien, Plenum, New York, 1989, 293; D. Borgis, S. Lee and J. T. Hynes, *Chem. Phys. Lett.*, 1989, **162**, 19; D. Borgis and J. T. Hynes, *J. Chem. Phys.*, 1991, **94**, 3619.
21 R. P. Bell, *The Proton in Chemistry*, Chapman and Hall, London, 1973.
22 R. Knochenmuss and S. Leutwyler, *J. Chem. Phys.*, 1989, **91**, 1268.
23 E. R. Bernstein, *J. Phys. Chem.*, 1992, **96**, 10105.
24 R. A. Marcus, *J. Phys. Chem.*, 1968, **72**, 892.
25 A. Warshel, J. K. Hwang and J. Åqvist, *Faraday Discuss.*, 1992, **93**, 225.
26 L. I. Schiff, *Quantum Mechanics*, McGraw-Hill, London, 3rd edn., 1968.
27 *CRC Handbook of Chemistry and Physics*, CRC Press, Boca Raton, FA, 72 edn., 1991.
28 B. E. Conway, *Ionic Hydration in Chemistry and Biophysics*, Elsevier, Amsterdam, 1981.
29 W. L. Jorgensen, J. Chandrasekhar, J. D. Madura, R. W. Impey and M. L. Klein, *J. Chem. Phys.*, 1983, **79**, 926.
30 D. Fincham, *Mol. Simul.*, 1992, **8**, 165.
31 H. F. C. Berendsen, J. P. M. Postma, W. F. Van Gunsteren, A. DiNola and J. R. Haak, *J. Chem. Phys.*, 1984, **81**, 3684.
32 Y. Hannachi, B. Silvi and Y. Bouteiller, *J. Chem. Phys.*, 1991, **94**, 2915.
33 A. C. Legon and L. C. Willoughby, *Chem. Phys. Lett.*, 1983, **95**, 449.
34 A. D. Buckingham, *Quart. Rev. Chem. Soc.*, 1959, **13**, 183.
35 T. Stoecklin, D. C. Clary and A. Palma, *J. Chem. Soc., Faraday Trans.*, 1992, **88**, 901.
36 D. Chandler, *Satistical Mechanics*, Oxford University Press, Oxford, 1987.

Paper 5/00172B; Received 5th January, 1995

Faraday Discuss., 1995, **100**, 321–332

Ionisation and Solvation of Stratospherically Relevant Molecules on Ice Films

Sally F. Banham, Andrew B. Horn,* Thomas G. Koch and John R. Sodeau*
School of Chemical Sciences, University of East Anglia, Norwich, UK NR4 7TJ

Heterogeneous reactions of stratospheric reservoir species such as HCl, N_2O_5 and $ClONO_2$ on type I (nitric acid hydrates) and type II (water-ice) polar stratospheric cloud (PSC) particles are believed to play an important role in the extremely large losses of stratospheric ozone observed during the Antarctic spring. Laboratory studies of such processes, using thin ice films as PSC particle surface mimics, can provide mechanistic information about individual elementary steps and overall reaction schemes. IR and mass spectrometry have been used to identify reaction products and intermediates and it has been shown that the primary step in all the studied reaction schemes involves the formation of ionic surface species. Furthermore, the nature and stability of the ions are found to be inextricably linked with the amount of 'free' water available to solvate them. The following questions are also addressed: Are adsorbed ions (*e.g.* nitrate, chloride and oxonium) reactive on ice surfaces? If so, are the products formed the same as those suggested when their corresponding molecular parents are involved? The answers, which have important implications for understanding stratospheric heterogeneous chemistry, are discussed in terms of ionic reaction mechanisms and novel surface intermediates.

The heterogeneous chemistry of the atmosphere is currently the subject of detailed scrutiny by experimentalists in the laboratory, measurers in the field and modellers at their screens owing to the newly established influence of cold particle surfaces on stratospheric ozone. This low-temperature phenomenon is known to reactivate chlorine sink compounds into active, ozone-depleting species in Austral spring over Antarctica and leads to its infamous stratospheric 'hole'.[1] The principal effect can be resolved into two components: first, the conversion of inert, reservoir species such as chlorine nitrate ($ClONO_2$) and hydrogen chloride (HCl) into photolytically labile species such as Cl_2 and secondly the removal of nitrogen oxides, species which, in the gas phase are able to sequester catalytically active chlorine radicals. In the winter Antarctic stratosphere, the most common forms of low-temperature solid material found are the particles which form polar stratospheric clouds (PSCs). Two main types of solid are thought to be involved: type I, consisting of sub-μm particles of nitric acid trihydrate (NAT) and type II, which are essentially water-ice condensed around an NAT nucleus.[2] Unsurprisingly, *in situ* spectroscopic measurements of reactions on actual PSCs at altitudes of 15–20 km have not yet been attempted. Therefore, laboratory studies utilising ice-substrate mimics of PSC particles, are currently the only feasible way of obtaining relevant mechanistic information.

Previous laboratory measurements using FTIR spectroscopy have successfully monitored the reactions of gas-phase chlorine nitrate, hydrogen chloride and dinitrogen pentoxide with solid ice films at 140–200 K.[3,4] In each case, there is compelling spectroscopic evidence for the formation of surface ions in the stratospherically relevent, low-exposure regime, *i.e.* experiments in which the reactant partial pressures are significantly lower than that of water (*ca.* 10^{-4} mbar). These conditions result in an excess of water at the substrate surface because of the high sticking probability of water at relevant temperatures. Care, of course, must be taken to avoid high reactant pressures. Such mimics may represent significantly different reaction systems from those present in the atmosphere because of the controlling influence of phase equilibria. For example, in the HCl–water system a pseudo-liquid layer on an ice film may be produced by adsorbate-induced surface-localised melting.

FTIR studies are not the only way by which 'ice-reactions' can be followed, indeed mass spectrometric (MS) techniques are essential for determining the nature of any products being evolved from the surface. The information obtained by the two techniques is, in fact, complementary because the MS-based experiments on their own do not provide a direct analysis of the surface species. These fundamental differences may explain current disagreement about the assignment of HCl adsorbed onto a water-ice surface as being molecular or ionic in nature.[5,6] The lesson here is that spectroscopic analysis of the ice surfaces should be used in conjunction with temperature-programmed (TP) MS to obtain the full picture. The use of both techniques also allows the following questions to be addressed: Are adsorbed ions reactive on ice surfaces? If so, are the products formed the same as those suggested when their corresponding molecular parents are involved? The methodology required for answers to these questions requires many potential elementary steps to be probed but can eventually lead to reaction schemes which effectively map the transition-state surface. This approach is the one taken in this study, which details the interactions of $ClONO_2$, HCl, Cl_2 and HOCl with water-ice surfaces. These compounds were chosen because of their central importance in the chlorine-reactivating, heterogeneous process shown as reaction (1):[7a,b]

$$ClONO_2 + HCl \rightarrow Cl_2 + HONO_2 \quad (1)$$

The transformation is clearly not a simple, one-step process and the following reactions, (2*a*) and (2*b*) have been proposed to explain the chemistry:

$$ClONO_2 + H_2O \rightarrow HOCl + HONO_2 \quad (2a)$$

$$HOCl + HCl \rightarrow Cl_2 + H_2O \quad (2b)$$

A recent FTIR–TPMS study by us on reaction (2*a*) in conditions of restricted water has provided clear evidence that the heterogeneous interaction of chlorine nitrate with water-ice does not directly produce HOCl and $HONO_2$ but in fact produces the $(H_2OCl)^+$ ion and a nitrate ion contained in an unsymmetric environment.[3] The results are summarised as reaction (3):

$$ClONO_2 + H_2O \rightarrow (H_2OCl)^+ + (ONO_2)^- \quad (3)$$

In reality, reaction (2*b*) involves the interaction of both HCl and HOCl with water-ice. Recent reflection–absorption (RA) IR measurements indicate that molecular HCl does not exist in high concentrations on water-ice surfaces at any stratospherically relevant temperature or pressure. The spectra show that it is adsorbed dissociatively to form $H_3O^+Cl^-$ and its hydrates, as shown in reaction (4):[6,8,9,26]

$$HCl + nH_2O \rightarrow H_3O^+ \cdot (H_2O)_{n-1} + Cl^- \quad (4)$$

The present study has two main objectives: (i) to focus on the interactions between water-ice and the other two chlorine-containing species involved in reaction (2*b*), namely Cl_2 and HOCl; (ii) to investigate the reaction between $ClONO_2$ and HCl on water-ice.

The complete mapping of reaction (1) allows the following key questions to be discussed: Are the ions formed in reactions (3) and (4) able to produce neutral species as predicted by reaction (1)? If so, can the products they form be released into the gas phase? In other words, the purpose of our experiments is to test the following alternate possibilities: If the relevant species, ionic or molecular, for the generation of molecular chlorine (*i.e.* HOCl, Cl^-, $[H_2OCl]^+$, H_2O, H_3O^+) are present upon the surface of a water-ice film, can they combine to eject it into the gas phase or are the adsorbed ions subject to association and desorption barriers which retain them on the surface under stratospheric conditions?

Experimental

Gas Dosing and Ice-film Generation

Ice films were deposited from a gaseous water source onto a variable-temperature gold foil substrate held *in vacuo*. Temperatures between 80 and 300 K were maintained at ± 2 K for the duration of an experiment. Reactants were dosed into the chamber by controlled leaks from three isolated glass gas-handling lines, which ensured that no mixing could occur prior to arrival at the surface. Co-deposition experiments were performed by setting the partial pressure of water vapour to the required value and then co-dosing the second reactant. Pressure measurement inside the chamber was made using a trigger Penning gauge. A quadrupole mass spectrometer (VG QX200) coupled to a PC-based data acquisition system was employed for measurements of gas purity and in the temperature-programmed experiments.

Generation of Reactants

Chlorine nitrate was prepared from dinitrogen pentoxide (N_2O_5) and dichlorine monoxide (Cl_2O) based on the method of Schmeisser.[10] Cl_2O was synthesised according to reaction (5):[11]

$$2Cl_2 + (n + 1)HgO \rightarrow Cl_2O + HgCl_2 \cdot nHgO \tag{5}$$

Chlorine (Air Products, 99.9%) in a stream of nitrogen (BOC zero grade, 99.995%) was passed through a Pyrex-glass column packed with 200 g of dry, yellow mercury (II) oxide (Aldrich, 99+%) dispersed on a bed of glass balls at room temperature. A mixing ratio of 10% v/v at a total flow rate of 200 cm^3 min^{-1} was maintained throughout the preparation. The Cl_2–N_2 gas mixture and the Cl_2O product were dried by passage through P_2O_5 (Aldrich, 97%) at either end of the column. The cherry-red liquid product Cl_2O was collected at 195 K in a cardice–acetone trap and purified by degassing at 195 K followed by gradual warming of the product to *ca.* 250 K to remove the residual N_2–Cl_2.

Chamber Conditions

Experiments were performed over a range of temperatures and pressures using two different methods: (i) ice-film surface exposure and (ii) co-deposition. Ice films were grown using a reproducible method described by us elsewhere[12] and were supported in a vacuum of 5×10^{-9} mbar in the temperature range, 80–160 K. Co-deposition experiments were performed by simultaneous gas-phase deposition of two components directly onto the gold foil support held at 180 K.

Results and Discussion

Ionised Chlorine Nitrate

Solid films of chlorine nitrate and water were deposited onto the gold substrate at 80 K in order to check purity and to provide reference spectra. Gaseous $ClONO_2$ at a pressure of 2×10^{-8} mbar was admitted to the chamber for 30 s resulting in the reflection–absorption IR (RAIR) spectrum shown in Fig. 1(*a*). In another experiment, $ClONO_2$ (2×10^{-8} mbar) and H_2O (2×10^{7} mbar) were co-deposited at 80 K for 30 s to produce

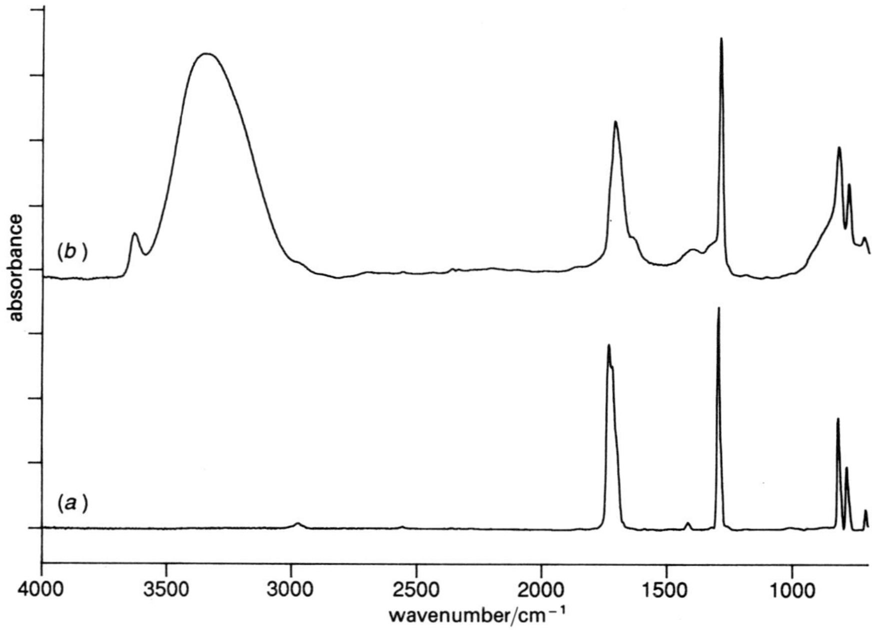

Fig. 1 RAIR spectra of thin films deposited on a gold foil substrate at 80 K. (*a*) pure chlorine nitrate, (*b*) co-deposited chlorine nitrate and water.

a mixed, unreacted film, the spectrum of which is shown in Fig. 1(*b*). Comparison with matrix isolation data[13] indicates that both traces are characteristic of molecular chlorine nitrate with C_s symmetry. The IR detector used was not sensitive in the far-IR region and therefore only five of the nine fundementals could be observed, namely the NO_2 asymmetric stretch at 1734 cm^{-1}, the symmetric stretch at 1300 cm^{-1}, the ClO stretch at 820 cm^{-1}, the ONO scissors mode at 786 cm^{-1} and the NO_2 wagging mode at 709 cm^{-1}. The chlorine nitrate fundamentals observed for the co-deposited $ClONO_2$–H_2O film at 80 K are qualitatively similar to the spectrum of the pure solid film. These observations indicate that the $ClONO_2$ molecules retain their symmetry in the water-matrix environment and little or no reaction occurs at this low temperature. Such unreactivity is not surprising because at 80 K the activation barrier for solvation is unfavourable and the mobility of water molecules is limited. The broad OH stretching band observed near 3300 cm^{-1} in the co-deposited film is typical of amorphous, unreacted water-ice. The sharper feature at 3632 cm^{-1} can be assigned to the free, non-hydrogen-bonded OH stretch of a relatively unperturbed water molecule often termed the 'dangling bond'.[6]

In order to investigate the reaction between chlorine nitrate and water-ice at a more relevant stratospheric temperature, similar experiments were carried out at 180 K. Chlorine nitrate was dosed onto the substrate at a pressure of 5 × 10^{6} mbar resulting in the RAIR spectrum shown in Fig. 2(*a*). The spectrum given as Fig. 2(*b*) was obtained after simultaneous exposure of the cold substrate to partial pressure of 1.5 × 10^{-7} mbar H_2O and 1.5 × 10^{-6} mbar $ClONO_2$. These partial pressures are sufficiently low to prevent the condensation of a neat film of either compound at 180 K. In both experiments the resulting spectra exhibit the characteristics of ionic nitrates.[14,15] To our knowledge, the IR spectrum of an ionic form of chlorine nitrate has not been reported previously. However, ionisation is well established for another stratospheric NO_y reservoir species, N_2O_5. Dinitrogen pentoxide exists in the gas phase only as a covalent molecule with near C_{2v} symmetry but in the solid form it is most stable as nitronium nitrate,

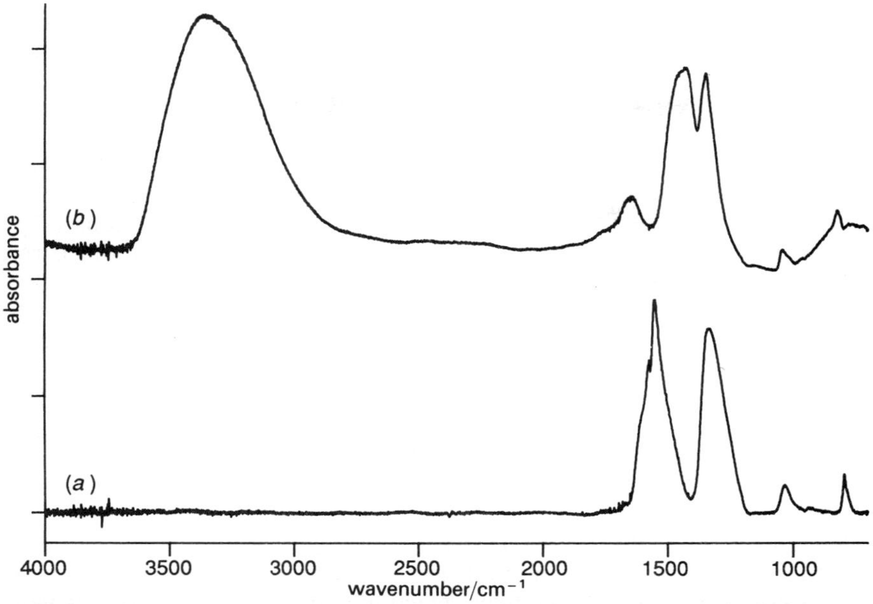

Fig. 2 RAIR spectra of thin films deposited on a gold foil substrate at 180 K. (*a*) pure chlorine nitrate, (*b*) co-deposited chlorine nitrate and water.

$NO_2^+NO_3^-$.[15] The published FTIR spectrum of this species shows two absorption bands: one at 2386 cm^{-1}, assigned to the asymmetric stretch of NO_2^+ and the other at 1420 cm^{-1} corresponding to the doubly degenerate, asymmetric stetch of $D_{3h}NO_3^-$.

Fig. 2(*a*) shows that for $ClONO_2$ at 180 K, three bands are found in the 750–1450 cm^{-1} region and a fourth observed at 1555 cm^{-1}. This pattern indicates that the nitrate ion formed with chlorine nitrate dissociation is not as symmetric as that made from dinitrogen pentoxide. Furthermore, the relatively high energy value associated with the absorption band at 1555 cm^{-1} is characteristic of a ν(N=O) stretch. The observation of a double-bond stretching vibration for the ion can be most easily rationalised by proposing that the NO_3^- group interacts with a chlorine counter-ion *via* two of its oxygen ions to give an $O{=}NO_2^-Cl^+$ complex. Such a pairing would lead to an overall C_{2v} symmetry for the nitrate ion as observed for the bidentate nitrato complexes formed by metals.[14] The absorption bands observed at 1335, 1035 and 790 cm^{-1} may then be assigned to the asymmetric NO_2 stretch, symmetric NO_2 stretch and NO_2 bending mode of $C_{2v}NO_3^-$, respectively. An alternate model, which may also lead to the strengthening of one particular NO stretching mode relies on the formation of cyclic $ClONO_2$ dimers. These eight-membered ring structures have been reported to exist in low-temperature solids for certain metal nitrates[14] as well as molecular nitric acid.[16]

In the absence of crystallographic data the above assignment to the C_{2v} structure of ionic $ClONO_2$ must remain tentative. However the presented IR evidence leaves little doubt that literally, chlorine nitrate, $Cl^+NO_3^-$, exists.

Analysis of the spectrum given in Fig. 2(*b*) shows that, at 180 K, incident chlorine nitrate molecules react with water, resulting in the formation of a number of new absorption bands, which are clearly different from those reported for crystalline nitric acid hydrates.[17] One of the products appears to be due to a nitrate ion sited in an unsymmetrical environment. This conclusion is evidenced by the following observations: (i) the degeneracy of the asymmetric stretching mode of free NO_3^- is lifted to produce two components at 1430 and 1350 cm^{-1}; (ii) the symmetric stretching mode, which is formally forbidden for a planar ion when symmetrically surrounded, is present as a weak band at 1040 cm^{-1}.[18] Additional spectral features due to another species are also observed: (i) a broad band, characteristic of strongly hydrogen-bonded OH-containing species centred at 3300 cm^{-1} and (ii) a sharper feature, at 1650 cm^{-1}. By comparison with the isoelectronic NH_2Cl molecule and also the H_3O^+ ion, which exhibits a broad, doubly degenerate deformation mode near 1750 cm^{-1}, we have previously assigned the two bands to the OH stretching and H_2O bending modes of the ion $(H_2OCl)^+$.[3] Although this rather unusual species is only known to exist from MS experiments, semi-empirical calculations suggest that, similar to other hypohalous acidium ions, $(H_2OCl)^+$ is a covalent, non-planar molecule with the electronegative chlorine atom carrying a significant part of the positive charge ($\rho Cl = 0.34$).[19] With an appreciable charge on the chlorine atom, it is reasonable to expect that the association between the $(H_2OCl)^+$ ion and nitrate ion will be close. Consequently, this effective complexation is likely to be responsible for the observed asymmetry of the nitrate ion.

The most likely mechanism for the formation of the two products is the nucleophilic attack of the δ^- oxygen atom of the water molecule on the δ^+ chlorine atom of chlorine nitrate forming a protonated acid intermediate and a stable anion, as follows:

$$H_2O + ClONO_2 - > [H_2OCl^+ \cdots {}^-ONO_2] \qquad (3^*)$$

Recent *ab initio* studies of the molecular structure of chlorine nitrate appear to confirm the importance of this type of interaction for $ClONO_2$ with water molecules. In the report, it was also suggested that the likely reaction coordinate for hydrolysis would be along the Cl---O internuclear distance.[20] In fact, as detailed elsewhere, $(H_2OCl)^+$ is quickly hydrolysed in the presence of excess water to produce a chlorine-containing compound, presumably molecular HOCl, which is desorbed at 180 K.[3] The step is shown as reaction (6):

$$H_2OCl^+ + H_2O - > H_3O^+ + HOCl \qquad (6)$$

Chlorine Nitrate and Hydrochloric Acid on Ice

In order to study the direct reaction of chlorine nitrate with HCl we exposed the $[H_2OCl^+ \cdots {}^-ONO_2]$ film as respresented by Fig. 2(*b*), to molecular HCl at 180 K. However, unlike the water-dosing experiment described above, neither a change in the IR spectrum was observed nor were measurable quantities of chlorine-containing compounds released from the surface. Such an observation can be explained if solvation of HCl is a critical feature of the overall reaction mechanism. To test this hypothesis we investigated the chemistry of the HCl–HOCl–H_2O system.

Hypochlorous and Hydrochloric Acids on Ice

Hypochlorous acid, HOCl, can be produced by the reaction of Cl_2O with water. In the gas phase it is generated as an equilibrium mixture with the reactants according to reaction (7):

$$H_2O + Cl_2O \Leftrightarrow 2HOCl \qquad (7)$$

The equilibrium constant for this reaction is 0.1.[21] In our experiment, it was determined that an unequal mixture containing a small excess of Cl_2O with a total pressure of 3 mbar produced sufficient HOCl for our deposition purposes. Initially, several experiments were carried out by depositing neat H_2O and Cl_2O directly without premixing.

When pure Cl_2O and water were co-deposited from separate sources at 80 K, a composite spectrum was obtained, as shown in Fig. 3(*a*). The broad bands near 3400 and 800 cm^{-1} arise from amorphous water. The sharp absorptions at 670 and 629 cm^{-1} can be assigned to the asymmetric and symmetric stretching modes of pure Cl_2O,

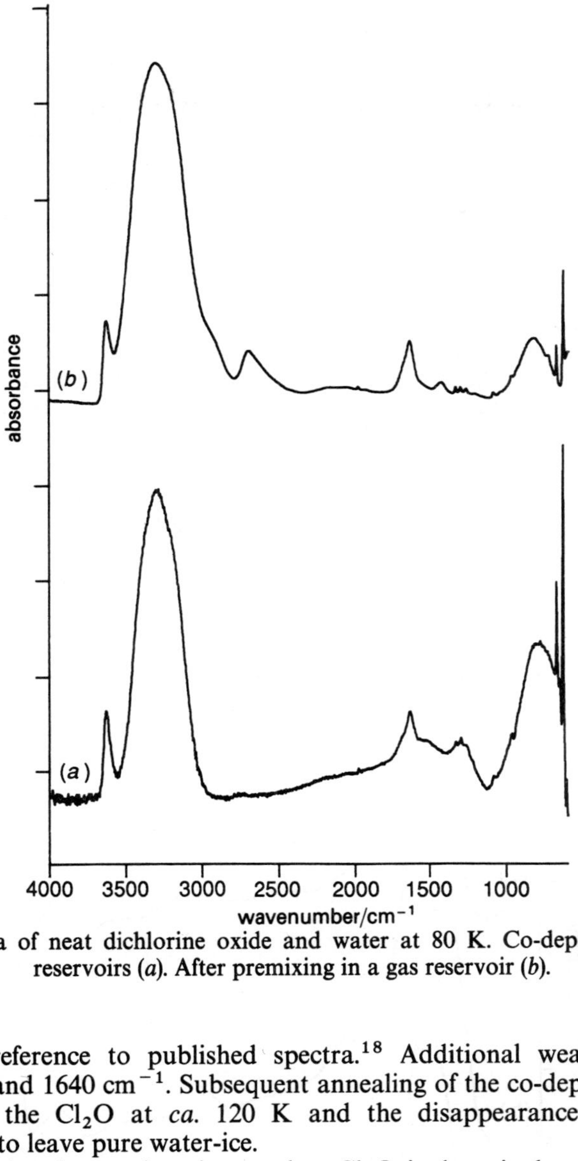

Fig. 3 RAIR spectra of neat dichlorine oxide and water at 80 K. Co-deposited from separate reservoirs (*a*). After premixing in a gas reservoir (*b*).

respectively, by reference to published spectra.[18] Additional weak bands are also observed at 3650 and 1640 cm^{-1}. Subsequent annealing of the co-deposited layer results in desorption of the Cl_2O at *ca.* 120 K and the disappearance of the 3650 and 1640 cm^{-1} bands to leave pure water-ice.

The two weak bands are also absent when Cl_2O is deposited onto the surface of a neat water-ice film at 80 K, which otherwise produces a similar spectrum to that shown in Fig. 3(*a*). The 3650 and 1640 cm^{-1} bands may, therefore, be attributed to interactions which exist in the bulk solid phase between the two components. By analogy with the results from the $ClONO_2$–H_2O co-deposit at 80 K, the 3650 cm^{-1} band can be assigned to the well established OH 'dangling bond' and the 1640 cm^{-1} band is likely to be due to the bending mode of H_2O complexed to the chlorine atom of a partner Cl_2O molecule.

Exposure of the metal substrate to the equilibrium H_2O–Cl_2O–HOCl gas mixture at 80 K resulted in the spectrum shown in Fig. 3(*b*). In addition to the bands due to the adsorption of unreacted water and Cl_2O, new absorption features at 2710, 1410 and 721 cm^{-1} were detected, which can be assigned to the OH stretch, HOCl bending mode and OCl stretch, respectively, of molecular hypochlorous acid. The three bands

Table 1 IR band positions for Cl_2O, H_2O, HOCl (in cm^{-1})

Cl_2O	H_2O	HOCl	H_2O–Cl_2O	assignment
	3500–2800			OH stretch
		2710		H-bonded OH
	1650 (br)		1640	H_2O deformation
		1410		HO deformation
670		712		Cl—O stretch
630				

disappear simultaneously upon annealing to 175 K. At the same time a species with $m/z = 52$ is released in the thermal desorption mass spectrum verifying the assignment of the material to HOCl. The fact that the OH stretching and bending absorption band wavenumbers are quite different from the corresponding Ar-matrix isolation values of 3587 and 1240 cm^{-1}, respectively,[22] is expected because, in the water-ice experiments, molecular HOCl should exist with a large degree of hydrogen bonding. In contrast, the ClO stretch is little perturbed (721 cm^{-1} for water-ice of 729 cm^{-1} for the Ar-matrix). All relevant IR band positions are summarised in Table 1.

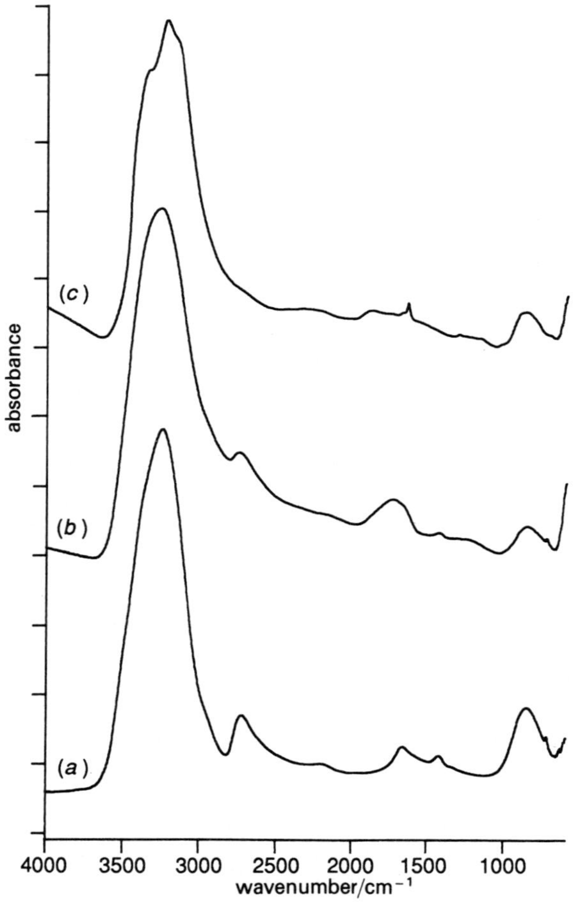

Fig. 4 RAIR spectra of (*a*) an HOCl–H_2O mixture at 140 K; (*b*) after exposure of (*a*) to HCl; (*c*) after annealing (*b*) to 165 K

The reaction between HCl and HOCl was investigated by exposing an $HOCl–H_2O$ film to gas-phase HCl at 140 K.

Fig. 4(*a*) shows the IR spectrum resulting from the co-deposition of a $Cl_2O–HOCl–H_2O$ mixture at 80 K followed by heating to 140 K to remove any residual Cl_2O. Exposure of this adsorbed HOCl layer to HCl at 140 K results in the formation of additional bands previously assigned to the ionic, solvated species $H_3O^+ \cdot (H_2O)_n$ [see reaction (4)]. The process is illustrated in Fig. 4(*b*) and shows the formation of a new broad band at *ca.* 1750 cm^{-1} (assigned to H_3O^+) whilst the features at 2710, 1410 and 721 cm^{-1} are reduced in intensity. A very small reduction in the water absorption bands is also observed.

Subsequent heating to 165 K results in the loss of the IR features due to H_3O^+ and surface HOCl, as shown in Fig. 4(*c*). The transformation is accompanied by the release of molecular chlorine ($m/z = 70$) in the thermal desorption mass spectrum over the range 145–160 K. Furthermore, no HOCl (monitored as $m/z = 51$) is released into the gas phase between the same temperature limits. These results are illustrated in Fig. 5. The overall process can be summarised as follows:

$$HOCl + H_3O^+ \cdot (H_2O)_nCl^- \rightarrow Cl_2(g) + (n + 2)H_2O(s) \quad (8)$$

In the higher-temperature IR spectrum shown in Fig. 4(*c*), immediately prior to complete film desorption, there is evidence for the production of a sharp band near 1640 cm^{-1}. This band can be assigned to a crystalline HCl hydrate $H_3O^+Cl^-$ being in excess in these experiments.

Mechanistically, it is not possible with our equipment to establish whether or not the action occurs directly between $HOCl + Cl^-$ or *via* the intermediary of the $[H_2OCl]^+$ ion. Unfortunately, it is not possible to monitor the signal due to Cl_2 during HCl deposition with our current MS equipment. Therefore we are only able to infer from the loss of HOCl features upon HCl exposure to the film that molecular chlorine is also generated directly at 140 K. Any Cl_2 which is formed would be expected to be desorbed since we have determined in a separate experiment that its desorption temperature from water-ice is *ca.* 110 K.

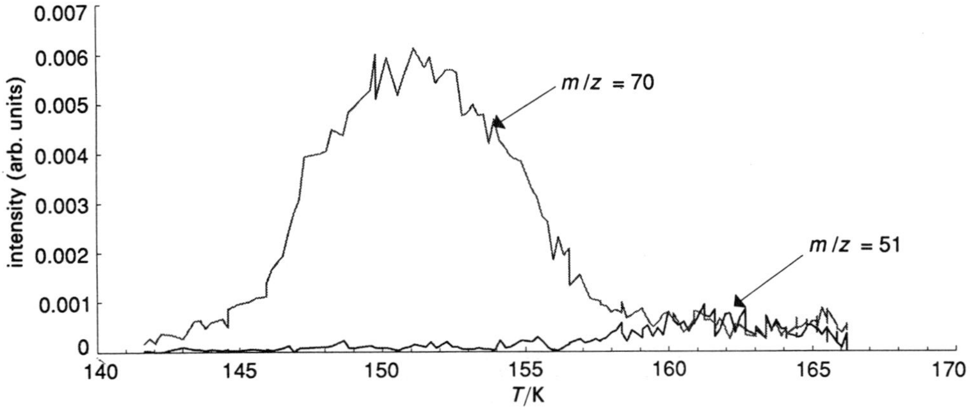

Fig. 5 MS signals at m/z 70 and m/z 51 during the annealing of the $HOCl–HCl–H_2O$ film shown in Fig. 4

Molecular Chlorine on Ice

To obtain a more complete picture of the release of chlorine from water at higher temperatures, we have studied the reverse of the reactions leading to the production of Cl_2 by co-depositing molecular chlorine and water at 180 K.

Spectra were obtained for two different chlorine : water ratios in order to determine the effect of water availability upon any reaction products. Fig. 6(*a*) shows the spectrum resulting from the co-deposition of 2×10^{-6} mbar chlorine and 2×10^{-8} mbar water at 180 K. Single-component deposition at this temperature does not occur and therefore it can be concluded that the resulting film consists of reaction products alone. The strongest features in the spectrum are a series of broad, overlapping bands from 3600–2700 cm^{-1} and a relatively sharp band at 1608 cm^{-1}. Co-deposition with a greater amount of water *i.e.* 5×10^{-7} mbar results in a thicker film, the spectrum of which is shown in Fig. 6(*b*). As before, a series of broad, overlapping bands are observed in the 3600–2700 cm^{-1} region. However, in this case a strong, sharp band at 3559 cm^{-1} is also present and the sharp band in the 1600 cm^{-1} region is shifted to 1626 cm^{-1}.

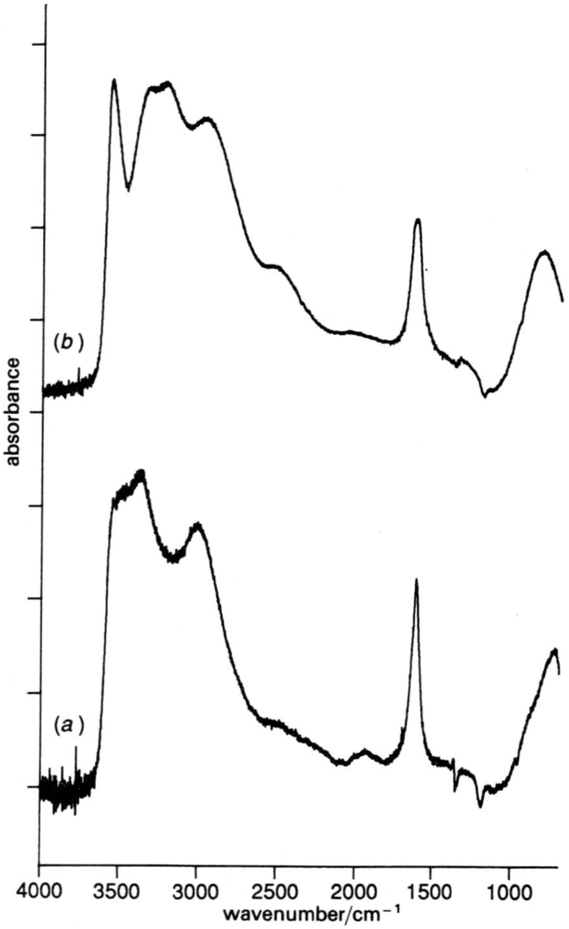

Fig. 6 RAIR spectrum of Cl_2–H_2O co-deposited at 180 K. (*a*) 2×10^{-6} Cl_2–2×10^{-8} H_2O mbar, (*b*) 2×10^{-6} Cl_2–5×10^{-7} H_2O mbar.

These chlorine–water co-deposited films show similar spectral behaviour in the OH stretching region to crystalline NAD (nitric acid dihydrate), NAT and hydrochloric acid hydrate films at 180 K.[23,24] Thus, in all experiments where more water is available, a strong, sharp absorption band, blue-shifted from that due to water-ice, appears. The feature has previously been assigned to 'water of hydration'; it is not due to a 'dangling bond' because such features do not appear at temperatures above *ca.* 140 K. The lower wavenumber band, which shifts from 1608 to 1626 cm^{-1} with increasing water content can have few origins in such a simple chemical system. Their close proximity to the band position assigned to the $[H_2OCl]^+$ ion discussed above leads us to speculate that in molecular chlorine–water-ice films, the species $(H_2OCl)^+(Cl)^-$ and its hydrates can be formed.

A further noteworthy feature of the experiments is that the partial pressure of chlorine admitted to the chamber is relatively high compared with the water pressure. If this pressure is lowered, the rate of formation of the film is much slower. This observation suggests that the sticking probability for chlorine is not high.

Although a six to one excess of chlorine to water is not at all representative of stratospheric conditions, the experiments show that ionisation and solvation of molecular chlorine in water-ice is possible.

Conclusion

Are adsorbed ions reactive on ice surfaces? If so, are the products formed the same as those suggested when their corresponding molecular parents are involved? The answers to our original questions are clear for the case of $ClONO_2$, HCl and HOCl on water-ice: if the relevant ions are present, they can react to evolve the expected gas-phase products at an appropriate temperature. Thus, by combining reactions (3), (4), (6) and (8), the following overall description is obtained:

$$ClONO_2 + HCl + nH_2O \rightarrow \{[H_2OCl]^+ + ONO_2^- + H_3O^+ + (H_2O)_{n-2} + Cl^-(+HOCl?)\} \rightarrow Cl_2(g) + H_3O^+ \cdot (H_2O)_{n-1}ONO_2^-$$

The intermediacy of ionic species can be used to explain a number of features of the kinetics of these reactions, as measured in flow-tube experiments. For example, the formation and stability of ionic species will be strongly influenced by the presence of other ions. Ionic intermediacy therefore provides a reason why the rates of many reactions studied on type I PSC mimics (NAT) are in general much slower than on pure water-ice and increase in rate as the NAT becomes water rich.[25] It also explains the observed dependence upon the amount of available water: an ionic reaction intermediate is initially stabilised by solvation although it may then be hydrolysed in the presence of a large excess of water, as in the stratosphere. The phenomenon may affect the release of chlorine to the gas phase because we have demonstrated in this paper that Cl_2 can be both adsorbed dissociatively in an ionic form and desorbed associatively by a surface reaction sequence involving ionic and molecular species. Finally, it should not be forgotten that ionic solvation depends upon the dynamic equilibrium between surface and gas-phase water molecules. At stratospheric temperatures, reaction products formed on the 'ices' may effectively be transported into the bulk solid when the surface is replenished with water: laboratory experiments are not yet sophisticated enough to mimic such complicated behaviour.

This work was supported by the SERC and the Commission of the European Community STEP programme (CT90-0071).

References

1 R. F. Pueschel, G. V. Ferry, G. V. Snetsinger, K. G. Goodman, J. Dye, J. E. Baumgardner and B. W. Gandred, *J. Geophys. Res.*, 1992, **97**, 8105.
2 O. B. Toon, P. Hamill, R. P. Turco and J. Pinto, *Geophys. Res. Lett.*, 1990, **17**, 393.
3 J. R. Sodeau, S. F. Banham, A. B. Horn and T. G. Koch, *J. Phys. Chem.*, 1995, **99**, 6258.
4 M. J. Molina, T-L. Tso, L. T. Molina and F. C-Y. Wang, *Science*, 1987, **238**, 1253.
5 J. T. Roberts and J. D. Graham, *J. Phys. Chem.*, 1994, **98**, 5974.
6 A. B. Horn, M. A. Chesters, M. R. S. McCoustra and J. R. Sodeau, *J. Chem. Soc., Faraday Trans.*, 1992, **88**, 1077.
7 (*a*) S. Solomon, R. R. Garcia, F. S. Rowland and D. J. Wuebbles, *Nature* (*London*), 1986, **321**, 755; (*b*) M. B. McElroy, R. J. Salawitch, S. C. Wofsy and J. A. Logan, *Nature* (*London*), 1986, **321**, 759.
8 L. Delzeit, B. Rowland and J. P. Devlin, *J. Phys. Chem.*, 1993, **97**, 10 312.
9 M. R. S. McCoustra and A. B. Horn, *Chem. Soc. Rev.*, 1994, 195.
10 M. Schmeisser, *Inorg. Synth.*, 1967, **9**, 127.
11 G. H. Cady, *Inorg. Synth.*, 1957, **5**, 156.
12 A. B. Horn, T. G. Koch, M. A. Chesters, M. R. S. McCoustra and J. R. Sodeau, *J. Phys. Chem.*, 1994, **98**, 946.
13 A. M. De Saxce and L. Schriver, *Chem. Phys. Lett.*, 1992, **199**, 596.
14 J. Laane and J. R. Ohlson, *Prog. Inorg. Chem.*, 1980, **27**, 465.
15 T. G. Koch, A. B. Horn, M. A. Chesters, M. R. S. McCoustra and J. R. Sodeau, *J. Phys. Chem.*, 1995, **99**, 8362.
16 W. A. Guillory and M. L. Bernstein, *J. Phys. Chem.*, 1975, **62**, 1058.
17 M. A. Tolbert and A. M. Middlebrook, *J. Geophys. Res.*, 1990, **95**, 22 423.
18 K. Nakamoto, *Infrared Spectra of Inorganic and Coordination Compounds*, Wiley, New York, 1963.
19 H. Egsgaard and L. Carlsen, *Int. J. Mass Spectrom. Ion Processes.*, 1992, **113**, 233.
20 G. La Manna, *J. Mol. Struct.*, 1994, **309**, 31.
21 J. B. Burkholder, *J. Phys. Chem.*, 1993, **98**, 2963.
22 K. Johnsson, A. Engdahl, P. Ouis and B. Nelander, *J. Phys. Chem.*, 1992, **962**, 5778.
23 J. R. Sodeau, in *Spectroscopy in Environmental Sciences*, Wiley, New York, 1995, 349.
24 G. Ritzhaupt and J. P. Devlin, *J. Phys. Chem.*, 1991, **95**, 90.
25 J. P. D. Abbatt and M. J. Molina, *J. Phys. Chem.*, 1992, **96**, 7674.
26 M. J. Molina, R. Zhang, P. J. Wooldridge, J. R. McMahon, J. E. Kim, H. Y. Chang and K. D. Beyer, *Science*, 1993, **261**, 1418.

Paper 5/00270B; *Received* 13*th January*, 1995

Faraday Discuss., 1995, **100**, 333–351

GENERAL DISCUSSION

Prof. Wodtke opened the discussion of Prof. Minton's paper: In your paper you discussed the possible atmospheric role of a metastable excited state of $ClONO_2$. The photofragmentation translational spectroscopy method is known to suffer in attempts at detecting metastable dissociation when the lifetime of the metastable state is much longer than the viewing time between photoexcitation and the time at which the centre-of-mass flies out of the viewing window of the detector. In the light of this, is it possible that all experiments are consistent with the formation of a metastable state of $ClONO_2$ being the dominant photochemical pathway at 308 nm?

Prof. Okumura responded: The solid angle viewed by our detector and the velocity of the parent beam do limit the viewing time of our instrument. We can only detect dissociation events occurring within 2.5 μs of the laser pulse. Given our sensitivity, we should be able to detect metastable molecules having lifetimes as long as 10 μs with respect to dissociation.

Sander and co-workers[1] observe quenching of the photolysis when the gas pressure in their cell is 50–100 Torr; they do not observe quenching at pressures of a few Torr and lower. If the rate constant for quenching the proposed excited state of $ClONO_2$ is gas-kinetic, then the observation of a reduced photolysis quantum yield in the 50–100 Torr pressure range would imply that the excited-state lifetime is of the order of nanoseconds. We can readily rule out dissociation lifetimes in the nanosecond range in our experiment. If, however, the proposed excited state is quenched with a rate constant that is $<10^{-3}$ of the gas-kinetic collision limit, then it would have a lifetime greater than 10 μs. Thus, the two experiments would be consistent only if the metastable state of $ClONO_2$ were quenched very inefficiently.

1 S. L. Nickolaisen, S. P. Sander and R. R. Friedl, personal communication.

Prof. Vaida said: $ClONO_2$ photochemical decomposition was shown to occur by several pathways. The branching ratio between the different products needs to be known accurately to estimate the effect of this reaction in ozone depletion. Our work on OClO[1] provides another example of such complicated excited-state reactions with formation of both ClO + O and Cl + O_2. The analogy with OClO suggests two questions:

(1) As multiple potential-energy surfaces are involved in reaction, the photoproduct branching ratio is expected to be very sensitive to pressure, temperature and complex formation. Have such effects been observed? Might they account for the differences in the product branching ratios of $ClONO_2$ photolysis observed in different laboratories?

(2) OClO photolysis leads to formation of highly vibrationally excited ClO[2] with sufficient internal energy to exceed the activation energy for reaction of ClO(v) with O_3. Is the ClO product formed on photolysis of $ClONO_2$ vibrationally excited?

1 V. Vaida, S. Solomon, E. C. Richard, E. Ruhl and A. Jefferson, *Nature* (*London*), 1989, **342**, 405.
2 V. Vaida and J. D. Simon, *Science*, 1995, **268**, 1443.

Prof. Okumura said: (1) As we conclude in our paper, the dynamics of $ClONO_2$ dissociation may well involve multiple potential-energy surfaces, especially when one considers the large number of excited singlet and triplet surfaces that are predicted by Graña *et al.* One of our primary motivations in undertaking this study was to examine the possibility of relaxation to a long-lived state (possibly a bound triplet state predicted by Graña *et al.*) raised by Sander and co-workers at the Jet Propulsion Laboratory.

We find that photodissociation at 308 nm is so rapid (on the picosecond timescale or faster) compared with mean collision times that no pressure effects are possible below one atmosphere. Furthermore, our relative yield measurements are in fair agreement with recent, unpublished quantum yields measured in flow-tube studies at NOAA, NCAR and JPL. It is likely that many of the discrepancies in previous experiments arose from measurement errors or misinterpretation.

In addition to the main features described in our paper, we have observed a pressure-dependent signal appearing at low centre-of-mass translational energies in our TOF spectra, which we assign to products from the photodissociation of clusters. We have always reduced source stagnation pressures and $ClONO_2$ concentrations to minimize or eliminate the cluster signal. We have not attempted to characterize this signal, although the low E_T of these products is consistent with radiationless decay to the ground state followed by statistical dissociation.

We have not studied the temperature dependence of the product yields; given our source conditions, the vibrational temperature is likely to be near 200 K, comparable to stratospheric temperatures.

(2) Our experiments provide us with the distribution of total internal energies in both the $ClO + NO_2$ and $Cl + NO_3$ channels. These are obtained from conservation of energy, $E_{int} = E_{avl} - E_T$. We find that in the $ClO + NO_2$ channel, the average total internal energy of the photoproducts is 67 kcal mol^{-1}, or 61% of the available energy. However, we cannot determine how this internal energy is partitioned between the two fragments.

Prof. Simons commented: Interpretation of the magnitude and sign of the photofragment anisotropy β, when it lies within the limiting values $-1 \leqslant \beta \leqslant 2$ is a subtle affair. While values may reflect the parent molecular rotation and hence the lifetime prior to fragment separation, they may also indicate changes in molecular shape (*e.g.* planar → pyramidal or other out-of-plane motions) or overlapping dissociation continua. The lifetime of the excited $ClONO_2$ is assuredly of short duration, but the intermediate magnitudes of β most probably reflect a spiral trajectory associated with the out-of-plane motion(s); *cf.* the photodissociation of $HONO_2$.[1] In this case the sign of β may not rigorously establish the alignment of the transition moment, in, or perpendicular to, the molecular plane.

1 M. Brouard, S. Duxon, P. A. Enriquez and J. P. Simons, *J. Chem. Soc., Faraday Trans.*, 1993, **89**, 1435.

Prof. Okumura replied: Our chief concern in measuring the anisotropies of the product angular distributions was in determining whether the dissociation occurs promptly, within one rotational period. The non-zero values of the anisotropy parameter β for both channels unambiguously confirm that the main products are formed within a few picoseconds.

In addition, we have inferred the orientation of the transition dipole relative to the plane of the molecule for excitation leading to the $Cl + NO_3$ channel from the observed anisotropy parameter β. You correctly point out the possible ambiguity in determining the direction of the transition dipole moment from this parameter alone when β is intermediate between the limits of 2 and -1. However, the translational energy distribution observed for this channel provides us with additional insight into the geometry of the molecule during dissociation.

A large fraction of the available energy is partitioned into product translation in the $Cl + NO_3$ channel. The average translational energy release is $\langle E_T \rangle = 32.9$ kcal mol^{-1}, or 62% of the available energy, and the maximum in E_T is 50 kcal mol^{-1}, or 94% of the available energy. Conservation of momentum and energy places an upper bound on the energy that can be released in translation for a given geometry. If $ClONO_2$ dissociates from the ground-state equilibrium geometry, a simple rigid radical model predicts that, at most, 69% of the available energy will be partitioned into translation. The high translational energy release that is observed can only occur if the recoil direction (the Cl—O bond) is nearly collinear with the line-of-centres connecting the Cl atom and the NO_3 centre-of-mass, *i.e.* the dissociation must have a small impact parameter. Furthermore, the NO_3 moiety must be close to its planar equilibrium geometry during dissociation, or a large fraction of the energy would remain in the NO_3 bending mode. Thus, the recoil direction lies in or close to the plane of the molecule. From the observed anisotropy parameter β, we conclude that the transition dipole must lie within 40° of the plane and, if a single state is excited, that state must be of A′ symmetry.

Your arguments are more relevant to our results on the ClO + NO_2 dissociation channel. Although we measure a larger anisotropy parameter ($\beta = 1.1$), on average less than half of the available energy is partitioned into product translation. We have therefore not drawn any inferences concerning the transition dipole orientation or other vector correlations in this channel. The ClO + NO_2 channel is the analogue of the OH + NO_2 channel observed in HNO_3 photolysis at 280 nm. While it is possible that the two molecules possess similar dissociation dynamics, our measurements cannot directly tell us if the $ClONO_2$ photodissociation dynamics for this channel involve analogous out-of-plane distortion in the exit channel and consequent spiral motion of the ClO fragment. Furthermore, *ab initio* theory indicates that additional excited states of $ClONO_2$ are present in this wavelength region. A study of ClO product rotation–translation correlations similar to the elegant investigation by Brouard *et al.* of OH product distributions following HNO_3 photolysis could shed light on this issue.

Prof. Molina asked: We have measured the quantum yield for the destruction of both Cl_2O and $ClONO_2$ using steady-state photolysis coupled to FTIR analysis of the gaseous mixture. Around 300 nm, the quantum yield is unity even at atmospheric pressure. At longer wavelengths (*ca.* 350 nm) the results are not as definitive, but still indicate unit quantum yield. Is there any precedent for radiationless transitions occurring in molecules as small as Cl_2O on the very short timescales required to explain the formation of a metastable triplet?

Prof. Ravishankara commented: We have measured the quantum yields for various products in the photodissociation of $ClONO_2$ at various wavelengths. The results from this, as yet unpublished, study are shown in Table 1 below. The main point to note is that we see both ClO and Cl, in agreement with the findings of Minton *et al.* The other major noteworthy point is that our experiments were carried out at high pressures of N_2 and/or He and we do not see any evidence for the decrease in quantum yields with pressure for any of the products, in contrast to the observations of Sander and co-

Table 1 Quantum yields for various products in the photodissociation of $ClONO_2$

wavelength/nm	$Cl(^2P)$	$O(^3P)$	$O(^1D)$	NO_3	ClO
193	0.6	0.4	<0.05	0.2	0.2
222	0.5	0.2			~0.5
248	0.5	0.1		0.6	0.4
308	0.8	<0.05		0.7	0.2
351				0.9	

These measurements were made using different detection techniques such as atomic resonance fluorescence, tunable diode laser absorption and diode array spectroscopy.

workers, which were presented and discussed by Prof. Minton and Prof. Okumura. Also, we see that the quantum yield for NO_3 is essentially unity at 351 nm. Based on these observations we suggest that, in the atmosphere, the primary photodissociation products are Cl and NO_3 and their quantum yields do not change with pressure. It is important to remember that the *J* value for $ClONO_2$ photolysis in the stratosphere peaks around 350 nm and the quantum yields of interest are those at wavelengths greater than 308 nm.

Prof. Okumura responded: The central question addressed by our paper and the comments of Prof. Molina and Prof. Ravishankara is whether a metastable state of $ClONO_2$ is prepared by long-wavelength excitation ($\lambda > 300$ nm). Such concerns were raised by recent flow-cell studies performed by Sander *et al.* at the Jet Propulsion Laboratory (JPL), who found evidence for quenching of the photolysis of both Cl_2O and $ClONO_2$ at high pressures.

The results of Prof. Molina and of Prof. Ravishankara corroborate our finding that no metastable state is formed at 308 nm. The branching ratios reported by Prof. Ravishankara at 308, 248 and 193 nm are also in fair agreement with the relative yields measured in our apparatus. We concluded that metastable formation could only occur at longer wavelengths, but the results at *ca.* 350 nm mentioned above would appear to eliminate this possibility.

While the results reported at this meeting contradict the observations of Sander *et al.*, the JPL group has not found any reasons for the discrepancies. In brief, the JPL experiment was performed in flowing flash photolysis cells with photolysis radiation limited to wavelengths longer than 200 nm (quartz cell) and 300 nm (Pyrex cell). For the experiments in the quartz cell, the yields of NO_3 and ClO were independent of the bath-gas pressure, while for the Pyrex cell, these yields and the disappearance of $ClONO_2$ were dependent on the total pressure, decreasing by about an order of magnitude between 10 and 100 Torr of N_2.[1]

In the case of Cl_2O, these authors directly observed a transient absorption spectrum, which they have assigned, with the aid of *ab initio* calculations, to a bound-to-bound transition from an excited triplet state.[2] Although no transient absorption spectrum was observed for $ClONO_2$, it exhibited similar kinetic behaviour. The peak in the photodissociation efficiency (the absorption cross-section convoluted with the lamp flux) of $ClONO_2$ was *ca.* 330 nm, suggesting that metastable formation should have been observed in the Molina and Ravishankara experiments.

Spin–orbit-induced radiationless transitions are of course well known, even in small molecules. Sander's observation of a transient absorption spectrum in Cl_2O provides evidence for such processes. Even in the diatomic NH radical, the A $^3\Pi$ state predissociates with a lifetime of 0.5–1 μs owing to spin–orbit coupling at a curve crossing with a repulsive $^5\Sigma$ state,[3,4] and one expects more rapid processes in chlorine-containing species that would have spin–orbit matrix elements an order of magnitude larger.[5] The theoretical work of Lengsfield and Yarkony suggests that spin-forbidden processes due to non-adiabatic coupling at crossing seams should be quite frequent even in small polyatomic molecules.[6]

1 S. P. Sander, personal communication.
2 S. L. Nickolaisen, C. E. Miller, S. P. Sander, M. R. Head, I. H. Williams and J. S. Francisco, to be published.
3 N. L. Garland and D. R. Crosley, *J. Chem. Phys.*, 1989, **90**, 3566.
4 D. Patel-Misra, G. Parlant, D. G. Sauder, D. R. Yarkony and P. J. Dagdigian, *J. Chem. Phys.*, 1991, **94**, 1913.
5 W. G. Richards, H. P. Trivedi and D. L. Cooper, *Spin Orbit Coupling in Molecules*, Clarendon Press, Oxford, 1981.
6 B. H. Lengsfield and D. R. Yarkony, *Adv. Chem. Phys.*, 1992, **81**, 1.

Prof. Minton concluded: Although a great deal of effort was spent placing an upper limit of 0.07 on the fractional yield of products from RRKM-like dissociation of $ClONO_2$, a point not to be overlooked is that the slow signal in the product TOF distributions most likely comes from photodissociation of primary ClO products. We observed a laser-intensity dependence of an analogous signal in our earlier study of $ClONO_2$ photodissociation at 248 nm that indicated a two-photon process. Furthermore, in the current experiment, the only unambiguous slow signal that could be ascribed to a statistical process was detected at $m/z = 35$ (Cl^+). If some fraction of excited $ClONO_2$ really does dissociate at 308 nm *via* a statistical process, then we should have observed a similar slow signal in the time-of-flight distributions collected at $m/z = 51$ (ClO^+), 46 (NO_2^+) and 30 (NO^+), but no such signal was distinct from the noise at these masses. The most likely assignment of the slow signal in the $m/z = 35$ TOF distribution as secondary photodissociation of ClO means that our reported relative yield (0.33) for the $ClO + NO_2$ channel is a lower limit. This consideration should be borne in mind when attempting to draw conclusions about the wavelength dependence of the photoproduct yields.

Prof. Vaida opened the discussion of Dr. Clary's paper: It could prove to be important to find the correct sites for any potential ionization. In recent unpublished work, Gertner and Hynes[1] have used the evident dynamic character of the ice surface[2] to argue that HCl is initially molecularly incorporated in, rather than atop, the ice lattice, although it is not incorporated in the bulk. Molecular dynamics simulations at 190 K of acid ionization from these sites to produce a contact ion pair find Gibbs energy changes of *ca.* -8 kcal mol^{-1}. This exothermic character, together with the estimate of no more than a small kinetic barrier for the reaction, strongly indicates significant HCl ionization at the ice surface, and thus suggests the importance of ionic mechanisms for ozone depletion.

1 B. J. Gertner and J. T. Hynes, *Science*. submitted.
2 D. R. Haynes, N. J. Tro and S. M. George, *J. Phys. Chem.*, 1992, **96**, 8502.

Prof. Thrush addressed both Dr. Clary and Dr. Horn: Both of your papers discuss ions on the surface of ice. In liquid water, *ca.* 3 Å depth of the surface layer is depleted of ions because of the loss of solvation energy as an ion approaches the surface. This energy would be particularly significant at low temperatures. Would the ions mentioned by these authors have a 'bump' of water molecules above an ion on the surface or would these ions be, say, *ca.* 3–4 Å below the surface?

Dr. Horn replied: In a liquid, ion mobility is quite high and they may diffuse readily into the bulk to achieve complete solvation. In an ice, reduced mobility is likely to prevent this, although there is still a potential forcing the system towards higher solvation. This potential may drive the solvation in two ways: by accumulating water molecules over the ion as suggested by Prof. Davidovits or by driving the ions sub-surface. Our experiments cannot directly distinguish between these cases. However, as Prof. Molina commented earlier, the rate of most of the surface reactions studied to date is incommensurate with bulk-to-surface diffusion processes, pointing strongly towards surface localization for most species.

Dr. Clary replied to Prof. Thrush: The ions formed in our model at 190 K are close to the surface as the second and lower layers of water molecules in ice at this temperature are almost rigid. To look in detail at the dissociation of HCl, we have recently[1] carried out a systematic study of HCl $(H_2O)_n$, $n = 1–3$, clusters using *ab initio* calculations at the MP2 level. We find a ring structure with the HCl bond length increasing significantly with n and the charge on the H of HCl increasing also. An example of such a structure is shown in Fig. 1 for $HCl(H_2O)_3$.

1 M. J. Packer and D. C. Clary, *J. Phys. Chem.*, in the press.

Prof. Davidovits said: I think there is reason to believe that under certain circumstances the surface of ice is not much different from the surface of liquid water, except that perhaps the surface mobility is reduced. Therefore, processes that have been observed to occur on liquid surfaces are likewise likely to occur on ice surfaces. For example, we have found that at the gas/liquid interface chlorine reacts much more rapidly with a halide ion than it does in the bulk. One such surface-enhanced reaction is $Cl_2 + I^- \rightarrow Cl^- + ICl$. Because the Cl_2 molecule is not hydrophilic, in the absence of the ion, one would expect the mass accommodation of chlorine to be rather low. However, in the presence of the ion the chlorine at the surface becomes polarized by the ion near the interface and the polarized Cl_2 can serve as a centre for nucleation (see our paper). The uptake is now greatly enhanced leading to a surface-enhanced reaction and uptake into the bulk. Ice surfaces are likely to exhibit similar surface-enhanced reactivity.

Dr. Horn responded: Your observations concerning the mechanism of the Cl_2–I^- interaction may also be relevant to the processes occurring upon an ice surface. The presence at an ice surface of water molecules of restricted mobility may lead to the polarization of the incoming chlorine molecule, acting as a precursor to dissociative adsorption in the form $[H_2OCl]^+Cl^-$. This is substantiated by the observation that the reaction products formed from $ClONO_2$, N_2O_5 and HCl impinging upon an ice surface follow a mechanistic scheme in which the most accessible electrophilic site appears to interact with the oxygen of a surface water molecule.

Prof. Molina commented: We have investigated recently the uptake of HCl vapour by μm-sized aerosols consisting of ice crystals. We estimate the surface coverage to be about 10^{13} molecule cm^{-2} at HCl partial pressures and temperatures corresponding to polar stratospheric conditions. We obtain similar results using bulk ice made by freezing liquid water and consisting of crystals with sizes of the order of a centimetre. Earlier uptake experiments had been carried out with ice films deposited from the gas phase, which are often rather porous; it was not possible to rule out in those experiments that the HCl uptake was taking place predominantly at grain boundaries or on surface imperfections.

We have also carried out isotope exchange experiments: DCl vapour yields HCl very rapidly in the presence of H_2O ice, a result which is consistent with an uptake mechanism involving solvation.

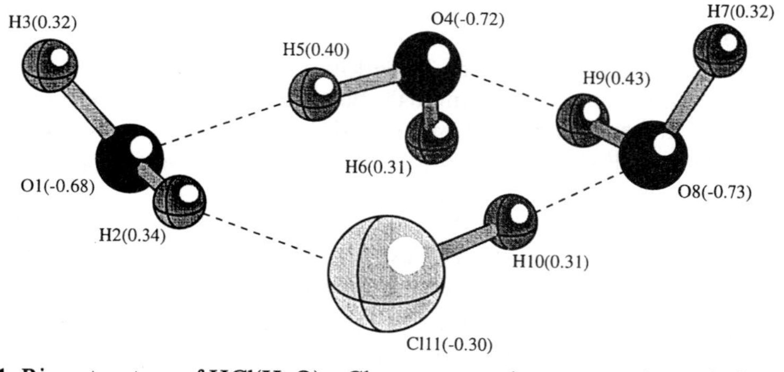

Fig. 1 Ring structure of $HCl(H_2O)_3$. Charges on each atom are shown in brackets.

Dr. Clary asked: Have you managed to measure the temperature dependence of the rate for the exchange reaction $HCl + D_2O \rightarrow DCl + HDO$?

Prof. Molina responded: Considering the $HCl–H_2O$ phase diagram, we expect that for temperature and HCl partial pressure conditions inside the ice stability regime and sufficiently removed from the liquid/solid coexistence line (perhaps by 20 K or more) the HCl will no longer solvate on the surface, and that the DCl–HCl exchange will not proceed rapidly. We have not yet carried out experiments under such conditions.

Prof. Ravishankara said: This comment is not only for Dr. Horn and co-workers, but also for many of the people doing surface work who presented posters and talks at this meeting.

There is no doubt that an understanding of the mechanisms of the heterogeneous reactions is needed to enhance our confidence in modelling atmospheric processes. However, it is also very important to remember that much of the information needed for atmospheric purposes is already known. Just to make this point, let me present a figure from our 1992 paper[1] dealing with the reactions of $ClONO_2$ and HOCl with HCl on HNO_3-treated ice (NAT). This figure (Fig. 2) shows that, with atmospheric levels of HCl, the loss of $ClONO_2$ is first order in $ClONO_2$. Indeed, it is first order even when the HCl abundances are lower than that of $ClONO_2$. When the gas-phase concentration of HCl is much less than that of $ClONO_2$, HCl loss is first order in HCl. Cl_2 is produced under all these conditions. Therefore, for atmospheric purposes, we already know how this reaction occurs and how to deal with it! Also, in coming up with detailed mechanisms of these reactions, one should not forget to explain these 'macroscopic' observations.

I am a little surprised at the stability of the H_2OCl^+ species on ice. Eigen's work showed that H_2OCl^+ would react with H_2O to give HOCl and H_3O^+. Also, we have noted that the rate coefficients for the HOCl + HCl reaction in sulfuric acid[2] is only a

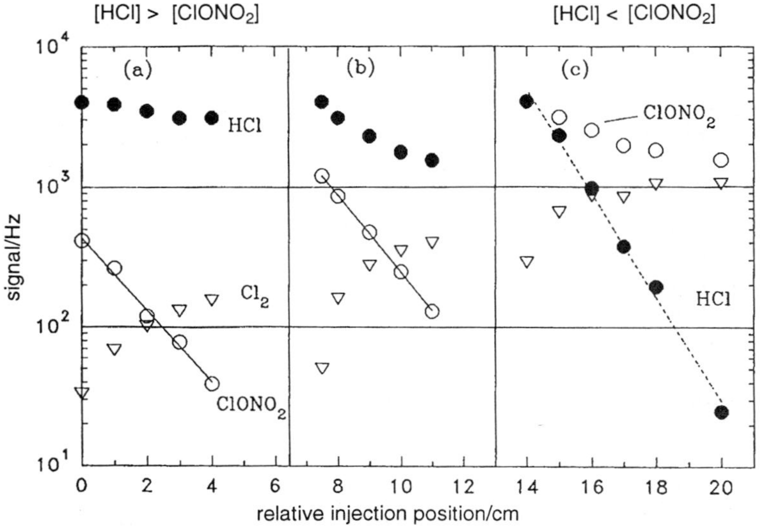

Fig. 2 Concentrations of HCl, $ClONO_2$, and Cl_2 as a function of injector position where the reactants, HCl and $ClONO_2$, are exposed to ice coated on the inside wall of a flow-tube reactor. The injector position translates into reaction time, *i.e.* time for which the reactants are in contact. The signals on the *y*-axis are proportional to the gas phase concentrations of the species. The two right panels, (b) and (c), have been moved up in injector position by 6.5 and 13 cm, respectively. Filled circles, open circles, and open triangles represent HCl, $ClONO_2$ and Cl_2, respectively. A figure similar to this has been published in ref. 1.

factor of 10 higher than that measured in very dilute acids by Eigen and Kustin.[3] Could you comment on why you see this species in your experiments? Also, in the atmosphere, the surface coverage of this species will have to be very low. So, do you expect this to be the intermediate which reacts with HCl in the stratosphere? In other words, would not H_2OCl^+ be converted to HOCl before it reacts with HCl?

1 D. R. Hanson and A. R. Ravishankara, *J. Phys. Chem.*, 1992, **96**, 2682.
2 D. R. Hanson and A. R. Ravishankara, *J. Phys. Chem.*, 1993, **97**, 12319.
3 M. Eigen and K. Kustin, *J. Am. Chem. Soc.*, 1962, **84**, 1355.

Dr. Horn replied: I strongly agree with your comments concerning the relationship between kinetic and compositional measurements in flow tubes and detailed mechanistic studies. Much of the available material concerning the rates of reaction can be used to infer mechanisms: conversely, mechanistic data can be used to explain kinetic observations. However, the gulf between macroscopic kinetics and microscopic mechanisms is still large and merits further research, the goal being to achieve a uniform description of low-temperature heterogeneous chemistry on all solid and liquid particle surfaces found in the atmosphere.

Dr. Rieley said: We wish to comment on the relationship of the conditions under which experimental observations on ice films are made to those which exist in the regions of the stratosphere where PSCs form. We note that the partial pressures of reagent species used in this study are rather high compared with those which occur in the stratosphere. These elevated pressures appear to be necessary in order to observe evidence for the existence of the $[H_2OCl]^+ \cdot Cl^-$ intermediate, presumably to obtain a detectable FTIR absorption signature within a reasonable timescale. The emphasis of this work and other low-temperature ice film work (including our own) is on the determination of chemical mechanisms through the isolation and characterization of reagents, products and key intermediates, with the implicit assumption that under elevated conditions of temperature and pressure the same intermediates and chemical mechanisms will also prevail. There is some question as to the validity of this assumption in the case of 'ionic' mechanisms. As a case in point, we mention the uptake of HCl on ice surfaces, which forms an 'ionic' or hydrated species in the adsorbed state. Previous work by some of the authors of this paper,[1] our own work,[2] that of Graham and Roberts[3] and others indicates that the degree of hydration (*i.e.* the prevalence of various ionization channels) depends on the partial pressure of the HCl gas above the surface of the ice film. In fact, as can be seen readily from an appropriate phase diagram, *e.g.* that of Hanson and Mauersberger,[4] the degree of hydration in the relevant temperature range decreases from hexahydrate through to trihydrate as the partial pressure of HCl is increased from levels typical of the stratosphere (*ca.* 10^{-8} Torr) to those typical of some of the laboratory experiments carried out (*ca.* 10^{-4} Torr). Clearly, the lower hydrates of HCl are interesting ionic intermediates, but they are unlikely to play a significant role under stratospheric conditions for which the hexahydrate is the thermodynamically most stable form.

We further comment that the importance or otherwise of ionic species will depend not only on the surface water available for hydration of incident species, but also on the relative sticking coefficients of the various species involved. This will be particularly important in reactive situations such as in the heterogeneous reaction: $N_2O_5 + HCl$. As mentioned above in a previous comment, we have measured the sticking coefficients of HCl and N_2O_5 on a pure water ice film at *ca.* 100 K and determine values of 0.95 and $\leqslant 0.1$, respectively. It is clear, in this case, that the important reaction to consider is:

$$N_2O_5(g) + [(H_2O)_n \cdot H_3O]^+ \cdot Cl^- \rightarrow ClNO_2(g) + HNO_3(s) + (n+1)H_2O(s) \quad (1)$$

rather than

$$HCl(g) + [(H_2O)_m \cdot H_3O]^+ \cdot NO_3^- \rightarrow ClNO_2(g) + HNO_3(s) + (m + 1)H_2O(s) \quad (2)$$

where $n = 5$ and $m = 2$ under typical stratospheric conditions. In the first instance hydrated HCl will be much more abundant on the surface and secondly, reaction (1) naturally produces an ionic, yet highly hydrated, surface.

We note that, in the paper presented, the reaction $HCl(g) + [H_2OCl^+ \cdots NO_3^-]$ did not appear to occur (no Cl_2 was observed). In the light of the above comments on the reaction of HCl with N_2O_5, it seems likely that the more important channel under stratospheric conditions, and the one which would yield Cl_2 on an ice film also, is the reaction $ClONO_2(g) + [(H_2O)_n \cdot H_3O]^+ \cdot Cl^-$.

Following on from these two comments we pose the following associated questions:

(1) Is the ionic intermediate, $[H_2OCl^+ \cdot NO_3^-]$, identified in this work, likely to be an important intermediate under stratospheric conditions?

(2) Given the comments on the importance of sticking coefficient and water of hydration, have the authors attempted to observe the heterogeneous reaction of $ClONO_2$ and HCl through reaction of gas-phase $ClONO_2$ and HCl hexahydrate?

1 A. B. Horn, M. A. Chesters, M. R. S. McCoustra and J. R. Sodeau, *J. Chem. Soc., Faraday Trans.*, 1992, **77**, 261.
2 H. Rieley, H. D. Aslin and S. Haq, *J. Chem. Soc., Faraday Trans.*, 1995, **91**, 2349.
3 J. D. Graham and J. T. Roberts, *J. Phys. Chem.*, 1994, **98**, 5974.
4 D. R. Hanson and K. Mauersberger, *J. Phys. Chem.*, 1990, **94**, 4700.

Dr. Horn replied: The reaction of N_2O_5 with HCl is a subject of current controversy. The WMO report for 1994 suggests that the reaction may be unimportant because it is slow, but it has not yet been well studied. In our investigations, the reaction does not appear to produce $ClNO_2$ in any appreciable quantity. The reaction $ClONO_2 + H_3O^+Cl^-$ is currently under study.

Dr. Rieley responded: I understand, and fully agree with, the notion that an important factor in the viability of heterogeneous processes *via* ionic intermediates is the available water of hydration. The point I wish to emphasize is that the 'natural' state of hydrated HCl under stratospheric conditions is a highly hydrated form. Thus under stratospheric conditions, given the relative abundances and sticking coefficients of HCl and potential co-reactants such as N_2O_5 or $ClONO_2$, it is likely that PSC surfaces will be coated with highly hydrated HCl which naturally allows for the hydration of, and subsequent reaction with, these other species.

Dr. Whitehead said (**Dr. Ahmed**, **Miss Apps**, **Miss Hughes** and **Mr. Watt** in part communicated): Dr. Clary has posed the question as to what happens when molecules stick or are adsorbed onto an ice crystal. I would like to present data that show that a facile reaction can occur when N_2O_5 is adsorbed onto the surface of a nm-sized ice crystal. We produce ice crystals by a supersonic molecular beam expansion of pure water vapour at 1.2 bar giving clusters with a mean size of a few hundred water molecules. Under these conditions, the ice crystals have an amorphous crystalline structure and a limiting temperature of 180 K. The N_2O_5 is attached onto the water cluster using the 'pick-up' method to spray N_2O_5 at pressures up to 7 mbar onto preformed clusters. This results in a relatively low coverage of N_2O_5 molecules with less than five molecules being attached to each cluster. The resulting cluster is detected after 0.5 ms by mass spectroscopy. Fig. 3 shows that there has been a clear change from the fragmentation pattern associated with pure, unaccommodated N_2O_5 (upper panel) upon adsorption of N_2O_5 onto the ice crystal (middle panel). Specifically, there is a reversal of the ordering of the relative intensities of the NO to NO_2^+ ion peaks following accommodation to

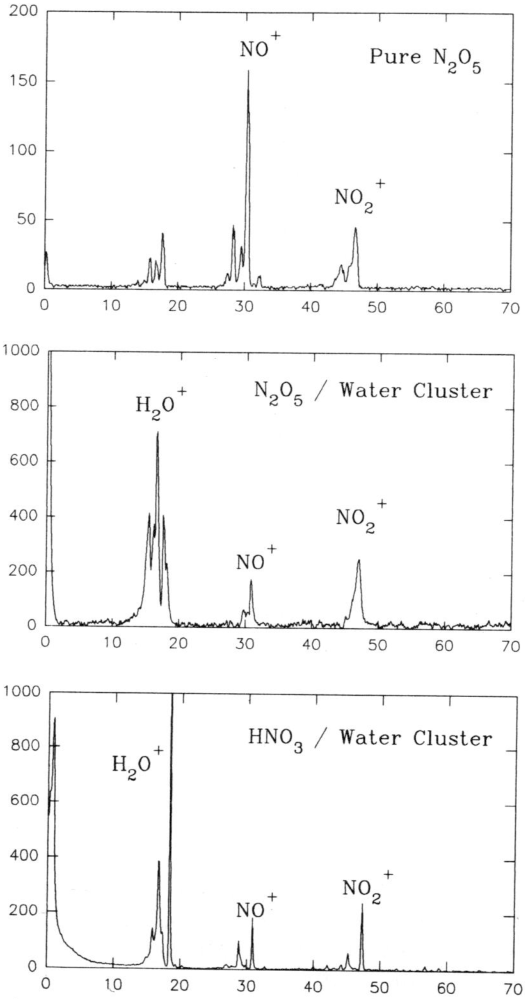

Fig. 3 Comparison of the mass spectrum of gaseous N_2O_5 with N_2O_5 adsorbed onto a water cluster and HNO_3 adsorbed onto a water cluster.

give a fragmentation pattern which is essentially identical to that resulting from the direct accommodation of HNO_3 onto a water cluster (lower panel). We estimate that about 80% of the N_2O_5 has been converted into HNO_3 *via* the reaction

$$N_2O_5 + H_2O_{(s)} \rightarrow 2HNO_3$$

Like Prof. Davidovits, we believe that local melting may take place at the surface, owing to the heat released by adsorption and the exothermicity of the reaction, and that the reaction probably takes place in a localized solution phase.

Prof. Molina commented: Fundamental aspects of atmospheric processes can often be investigated with experiments carried out under conditions well outside those applicable to the atmosphere. However, the identity of various solid phases generated in the laboratory may be quite different from that found in the atmosphere. An example is HCl deposited on ice; at sufficiently low temperatures, *i.e.* below 160 K, partial pressures of HCl in the 10^{-8} Torr range may lead to the formation of liquid or solid solutions

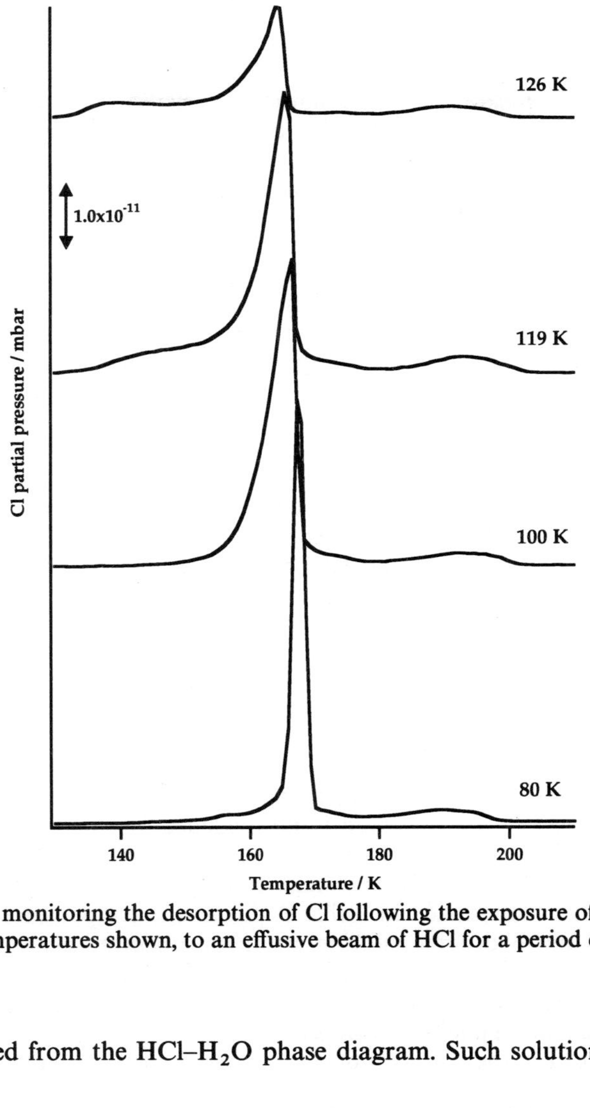

Fig. 4 TPD spectra monitoring the desorption of Cl following the exposure of pure ice films, held at the temperatures shown, to an effusive beam of HCl for a period of 10 min

(glasses), as inferred from the HCl–H_2O phase diagram. Such solutions do not form in the atmosphere.

Prof. Ravishankara addressed Dr. Horn: I agree with your comment that the rates of reactions such as those of $ClONO_2$ and HOCl with HCl will depend on the availability of water on the surfaces. There is clear evidence for the dependence on the availability of water. It has been shown that the reactive uptake due to the reactions of $ClONO_2$ and HOCl with HCl is essentially independent of the surface (*i.e.* NAT, frozen SAT, glass *etc.*) but depends on the relative humidity, which controls the amount of water available at the surface. (See ref. 1–3).

1 D. R. Hanson and A. R. Ravishankara, *J. Geophys. Res.*, 1993, **98**, 22931.
2 J. P. D. Abbatt and M. J. Molina, *J. Phys. Chem.*, 1992, **96**, 7674.
3 J. P. D. Abbatt and M. J. Molina, *Geophys. Res. Lett.*, 1992, **19**, 461.

Dr. Horn responded: The discrepancy between the conditions prevailing in the stratosphere and those in the laboratory is large. In our experiments, we restrict the amount of water available in order to investigate the effect of relative humidity. This has the effect of trapping intermediates such as $[H_2OCl]^+$ for a sufficient timescale for observation. The formation of this intermediate does not necessarily imply that it is a long-lived, stratospheric component: it is observed to react to produce HOCl and H_3O^+ in the presence of excess water (conditions expected in the stratosphere) and it is not expected to be present in sufficient quantities for direct reaction with, say, $H_3O^+Cl^-$. Its formation does, however, give an insight into the mechanism of the heterogeneous hydrolysis of stratospheric species and the key role played by water.

Dr. Rieley commented: We wish to present some recent experimental results which concur with the theoretical findings of Robertson and Clary, that the ionic solvation process is feasible (and expected) for HCl and HBr. We have undertaken the measurement of sticking coefficients of species such as HCl and HBr on low-temperature water ices (Type II PSC mimic). The methodology adopted is one which employs a thermal molecular beam source in conjunction with grazing-incidence FTIR in a high-vacuum environment. The details of these sticking experiments, which were conducted at a surface temperature of *ca.* 100 K, appear elsewhere.[1] We have also studied the uptake of HBr and HCl at various temperatures in the region 80–130 K and provide here a sample result from this work. Fig. 4 shows temperature-programmed desorption (TPD) spectra of Cl (as monitored at $m/z = 35$) after 10 min exposures of ice samples held, in turn, at a different temperature, to an HCl beam (the flux was the same in each case). The integrated area of each trace is proportional to the amount of hydrogen chloride uptake in each case. Each feature in a TPD trace, in principal, represents the existence of a different environment for chlorine at the surface. We ascribe the main desorption feature at each surface temperature to chlorine which is present as a surface hydrate. Qualitatively, it can be seen that the integrated uptake increases as a function of temperature to around 120 K and then starts to decrease; a more rigorous analysis will appear elsewhere.[2]

The initial conclusions from our experimental studies of the sticking of hydrogen halides on ice are as follows: (i) There is no evidence for the existence of molecular (covalent) HBr or HCl on the ice surface: each is adsorbed as a highly hydrated form. (ii) The initial sticking coefficients for HCl and HBr on low-temperature ice have been established: these are summarized in Table 2. (iii) In the temperature range 80–130 K, hydrogen bromide and hydrogen chloride are taken up primarily as surface hydrates: each anion being associated with perhaps four or more water molecules. (iv) As a function of temperature, the saturated uptake of HCl and HBr on the ice films increases in the range 80–120 K and then falls above 120 K.

Finally, we note that the surface uptake suggested by our analysis naturally leaves water of hydration available for incoming reagents, *i.e.* there is effective activation of the surface. This assumes, of course, that surface hydration/adsorption is a necessary part of

Table 2 Sticking coefficients of HCl and HBr on an ice film at 100 K

reaction studied[a]	sticking coefficient
$HCl(g) + H_2O(s) \rightarrow HCl(a) + H_2O(s)$	0
$HBr(g) + H_2O(s) \rightarrow HBr(a) + H_2O(s)$	0
$HCl(g) + (n + 1)H_2O(s) \rightarrow [(H_2O)_n \cdot H_3O]^+ \cdot Cl^-$	0.95
$HBr(g) + (n + 1)H_2O(s) \rightarrow [(H_2O)_n \cdot H_3O]^+ \cdot Br^-$	1.0

[a] There is some uncertainty in the value of n, but we believe that $n \geqslant 3$.

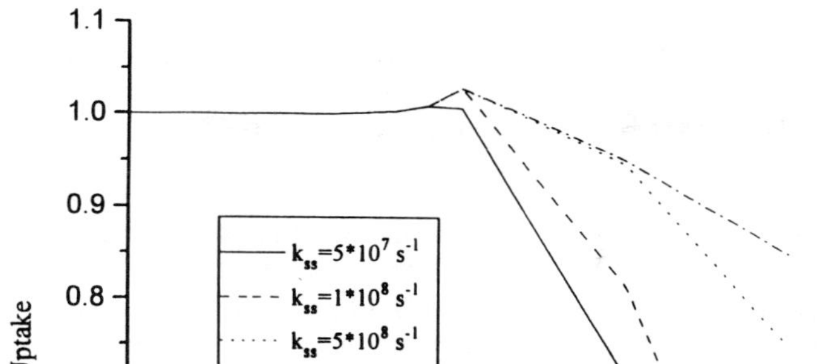

Fig. 5 Variation of relative initial rate of H_3O^+ uptake with surface temperature and rate of surface solvation for the interaction of HCl at a partial pressure of 10^{-7} mbar with an ice surface calculated using the model described in the text

the heterogeneous reaction mechanism between two different species which initially existed in the gas phase. It seems that an issue which we should further address is the fundamental mechanism of the heterogeneous reactions on PSC and other aerosols. The implicit assumption (perhaps rightly) is that heterogeneous reactions are dominated by Langmuir–Hinshelwood type behaviour. Such a reaction between two species X and Y would be represented as follows:

$$X(g) + Y(g) + S \rightarrow X(a) + Y(a) \rightarrow XY(g) + S$$

where S is a suitable (ice) surface and (a) implies that the species is adsorbed in some form at the surface. However, we have some evidence for the existence of an Eley–Rideal mechanism in certain circumstances, *e.g.* the reaction of HCl(g) with $NO_2^+ \cdot NO_3^-(a)$ mentioned in an earlier comment in this discussion. A reaction of this type between X and Y would be written:

$$X(g) + Y(g) + S \rightarrow X(a) + Y(g) \rightarrow XY(g) + S$$

i.e. gas-phase Y reacts with adsorbed-phase X to yield XY(g). The determination of the relative importance of these mechanisms awaits a more detailed mechanistic and kinetic study.

1 H. Rieley, H. D. Aslin and S. Haq, *J. Chem. Soc., Faraday Trans.*, 1995, **91**, 2349.
2 H. Rieley, H. D. Aslin and S. Haq, to be published.

Dr. McCoustra communicated to Dr. Rieley: Your observation that the rate of uptake of HCl and HBr appears to decrease with increasing surface temperature is indicative of a precursor-mediated adsorption mechanism. I have carried out some simple kinetic calculations based around the mechanism

$$HX \rightleftharpoons HX(ads) \rightarrow H_3O^+ \cdot X^-(ads) \rightarrow H_3O^+ \cdot X^-(abs)$$

which is based on the precursor-mediated adsorption of HX, but is extended to include diffusion of surface-adsorbed $H_3O^+ \cdot X^-$ ion pairs [$H_3O^+ \cdot X^-(ads)$] into the bulk of the ice substrate yielding absorbed ion pairs [$H_3O^+ \cdot X^-(abs)$]. Using realistic estimates for the rates of the forward and reverse steps in the pre-equilibrium and for the surface to bulk diffusion, calculations were carried out on the uptake of HCl by ice at a range of surface temperatures using a number of different values for the rate coefficient of the surface solvation process.

$$HX(ads) \xrightarrow{k_{ss}} H_3O^+ \cdot X^-(ads)$$

The results of these calculations are shown in Fig. 5 and 6.

Fig. 5 shows clearly the behaviour you observed. The relative initial rate of HX uptake (expressed as H_3O^+ uptake) is observed to fall at high temperatures reflecting the increased rate of precursor desorption at these temperatures. In contrast, from Fig. 6, it can be seen that the substrate uptake increases with increasing temperature as material is more readily absorbed into the bulk.

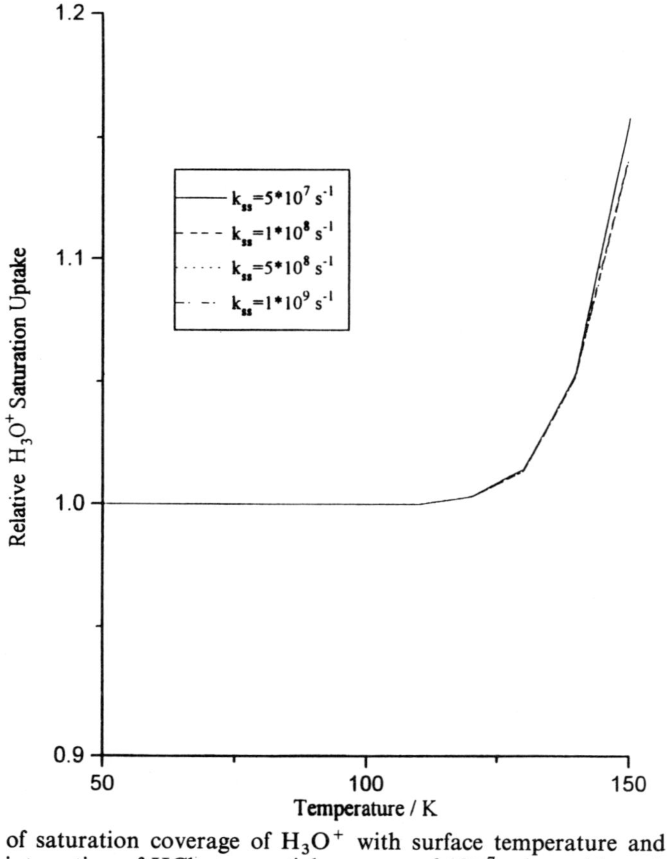

Fig. 6 Variation of saturation coverage of H_3O^+ with surface temperature and rate of surface solvation for the interaction of HCl at a partial pressure of 10^{-7} mbar with an ice surface calculated using the model described in the text

Dr. Rieley communicated in response: You provide an intuitively simple adsorption scheme for HCl on ice, which can be usefully compared with our observations. Unfortunately, one important aspect not provided is the effect of HCl partial pressure on the uptake behaviour as a function of temperature. For the experimental partial pressures used, in the temperature range studied, we see no change in the initial rate of uptake (as given by the initial sticking coefficient); the duration of high sticking does seem to shorten with temperature. In addition, under these same conditions the total HCl uptake seems to increase with temperature to *ca.* 120 K and then decrease, rather than continue to increase above 120 K as suggested by your Fig. 6. However, it seems that, consistent with the HCl/H_2O phase diagram,[1] surface uptake is likely to be the dominant process in our experiments. At higher HCl partial pressures the behaviour suggested by McCoustra's analysis is more likely to be followed.

1 M. J. Molina, in *The Chemistry of the Atmosphere: Its Impact on Global Change*, ed. J. G. Calvert, Blackwell Scientific, Oxford, 1994, p. 27.

Prof. Okumura addressed Dr. Clary: Your paper is very exciting, because it provides a first theoretical step towards addressing the issue of HCl dissociative ionization on ice surfaces. The largest contribution to the stability of $H^+ + Cl^-$ will most likely be from the three ionic hydrogen bonds of H_2O molecules to H_3O^+ in the first solvation shell of the cation. How accurately does the model potential used in this simulation account for the solvation energetics of H_3O^+?

Dr. Clary responded: You are correct that the formation of H_3O^+ is the most significant contribution to the dissociative ionization of HCl. However, as yet we have not used our simple model potential to simulate the solvation energetics of H_3O^+.

Dr. McCoustra communicated a question addressed to Dr. Clary: In your paper you characterize the dissociative ionization of an HX molecule on an ice surface in terms of three steps: (i) The physisorption of the molecule on the surface, (ii) the generation of a suitable active site on the surface by fluctuations of the configuration of surface molecules and (iii) the tunnelling of the proton from HX to the surface. Each of these was characterized by an appropriate timescale (τ_p, τ_f or τ_t). You have given in your paper estimates for τ_p (*ca.* 10^{-7} s) and τ_f (*ca.* 10^{-12} s). Do you have any estimate of the remaining timescale, τ_t?

Dr. Clary communicated in response: As yet we have no reasonable estimate of the tunnelling time. That requires more sophisticated calculations than those reported in our paper.

Prof. Okumura addressed Dr. Horn: The surface experiments described in your paper and in Dr. Rieley's remarks provide a very elegant approach to examining the properties of these model PSC film surfaces. Is there any evidence from TPD data in the HCl/ice surface experiments for a quasi-liquid layer as proposed by Prof. Molina? Specifically, does adsorption of even a sub-monolayer of HCl induce a change in the H_2O TPD spectrum of water which might indicate a change in the structure or binding of H_2O at the surface upon dissolution of HCl?

Dr Horn replied: The TPD spectra of pure water-ice films and those exposed to various amounts of HCl show a number of differences. The addition of HCl seems in general to increase the temperature at which complete film desorption occurs. Additionally, there are a number of HCl desorption features which can be attributed to changes in hydrate stoichiometry (measured spectroscopically by Delzeit *et al.*[1]). At high

HCl pressures, features appear in both the water and HCl desorption traces which may be indicative of liquid-like layers formed by adsorbate-induced, surface-localized melting.

1 L. Delzeit, B. Rowland and J. P. Devlin, *J. Phys. Chem.*, 1993, **97**, 10312.

Dr. Rieley added: In response to the question of the comparison between our work and the work of Graham and Roberts[1] and the present and previous work of Horn *et al.*[2] in which the former suggest the existence of a weakly adsorbed molecular state of HCl on ice, we comment that the pressures used in our experiments are somewhat lower than those of the aforementioned groups and are somewhat closer to the partial pressures of HCl which exist in the stratosphere. At no time in any of our experiments did we see any evidence for the existence of a molecular form of HCl. It is our belief that had a molecular form existed, even at low sub-monolayer levels, we would have observed this species; both gaseous[3] and crystalline[4] forms of HCl have distinct spectroscopic signatures in the IR region to which we are sensitive.

Under our conditions of low HCl partial pressure above the surface, we did not observe any significant change in the desorption temperature of water from the film following exposure to the HCl. Furthermore, the total amount of HCl taken up by the film was of the same order of magnitude over the temperature range studied (80–130 K).

1 J. D. Graham and J. T. Roberts, *J. Phys. Chem.*, 1994, **98**, 5974.
2 A. B. Horn, M. A. Chesters, M. R. S. McCoustra and J. R. Sodeau, *J. Chem. Soc., Faraday Trans.*, 1992, **77**, 261.
3 G. Herzberg, *Molecular Spectra and Molecular Structure*, Krieger, Florida, 2nd edn., 1989, vol. I.
4 L. Delzeit, B. Rowland and J. P. Devlin, *J. Phys. Chem.*, 1993, **97**, 10312.

Dr. Wolff made a general comment: I would like to make a comment to all the people looking at uptake and reactivity on ice surfaces; please do not stop your thinking, and your experiments, at stratospheric temperatures. There are two very important applications where uptake, and maybe reactions, on ice at polar tropospheric temperatures (*ca.* 200–270 K) are important. (1) In the Antarctic and Arctic tropospheres, there are ice surfaces in clouds and elsewhere, and heterogeneous reactions may take place on them. (2) Ice cores have the ability to reveal not just past climate, but also past atmospheric composition. We can measure, in ice cores, species such as HNO_3 and H_2O_2, that certainly contain information about crucial tropospheric concentrations. However, our ability to interpret these measurements is severely hampered by a lack of understanding of the relationship between air and ice concentrations. Data on the uptake (or loss) of *e.g.* HNO_3, HCl, H_2O_2 and other species on ice surfaces at relevant temperatures would greatly help, and might allow us to deduce many aspects of the state of the pre-industrial atmosphere, against which modern perturbations can be judged.

Dr. Donsig communicated: Recently, we at UMIST (Vickerman and Donsig) began a collaborative project with Sodeau, Horn and co-workers at UEA. Our contribution has been the analysis of some ice–gas-phase reactant systems using static secondary ion mass spectrometry (SSIMS). As a surface-sensitive mass spectrometric technique, SSIMS should be suitable for probing the chemical nature of the proposed ionic intermediates found by the team at UEA. In addition, using the depth-profiling capability of SIMS it should be possible to study the diffusion of intermediates into the bulk.

Our experiments have been deliberately designed to mimic those performed at UEA. We condense the reactants onto a cooled gold substrate, through isolated gas-inlet lines in a dosing system similar to the one at UEA. As little is known about the SSIMS of reactive species adsorbed on condensed ice films, we have begun by acquiring a set of 'standard' SSIMS spectra of water condensed at 90 to 100 K, of Cl_2 condensed on water at the same temperature, and similarly for HCl on water. The spectra were all acquired at temperatures in the range of 90 to 100 K.

We have measured the SIMS spectra of water both 'as dosed' (presumably an amorphous solid), and after thermal annealing to 150 K followed by cooling back to 90 K. Fig. 7 shows a characteristic positive ion SSIMS spectrum of annealed water. Note that there are two major 'families' of peaks: one of relatively low-intensity peaks from water clusters of the form $(H_2O)_{n'}^+$, and another of relatively high-intensity peaks from protonated water clusters of the form $(H_2O)_nH^+$. Both sets of peaks also appear in the positive-ion SSIMS spectra of unannealed water. The negative-ion spectra are much simpler, the major peaks being from O^- and OH^-.

Both the positive- and negative-ion spectra of Cl_2 condensed on water [Fig. 8(*a*) and (*b*)] show $Cl_m^{+/-}$ clusters, up to $m = 5$ and beyond. These appear in addition to the characteristic water peaks. In the negative-ion SSIMS spectrum of Cl_2 condensed directly onto the gold substrate, the major peaks are from Cl_m^- clusters. Thus a spectrum from the Cl_2 on water system looks like a spectrum of condensed Cl_2 superimposed on a spectrum of water, indicating no significant degree of interaction between the two species at 90 K. This is consistent with the results presented by Sodeau *et al.*, where reaction did occur, but at the much higher temperature of 180 K. Having established reference spectra for the low-temperature behaviour of the system, we will now move on to higher temperatures, where we should see evidence of a water–Cl_2 interaction.

With HCl on water, even at 90 K, the situation is somewhat different. Qualitatively, the positive-ion spectrum of HCl on water is not radically different from a spectrum of water alone, although the spectra are complicated by overlapping signals from H_2O ($m/z = 18$) and $H\,^{35}Cl$ ($m/z = 36$) clusters. [Mixed clusters of the form $(HCl)_k(H_2O)_n^+$ or $(HCl)_k(H_2O)_nH^+$ are, of course, possible.] Two significant features can be readily observed. The first is the absence of $(HCl)_{k'}^+$ or $(HCl)_kH^+$ clusters which, if present, should be identifiable from cluster peaks containing at least one ^{37}Cl atom. Significant quantities of such clusters are not observed. The second noteworthy feature is a dramatic increase in the absolute and relative intensities of the protonated *vs.* unprotonated water cluster peaks. This is illustrated in Fig. 9 which shows the relative intensities as a function of 'time' (spectra run sequentially). The most dramatic change occurs with the

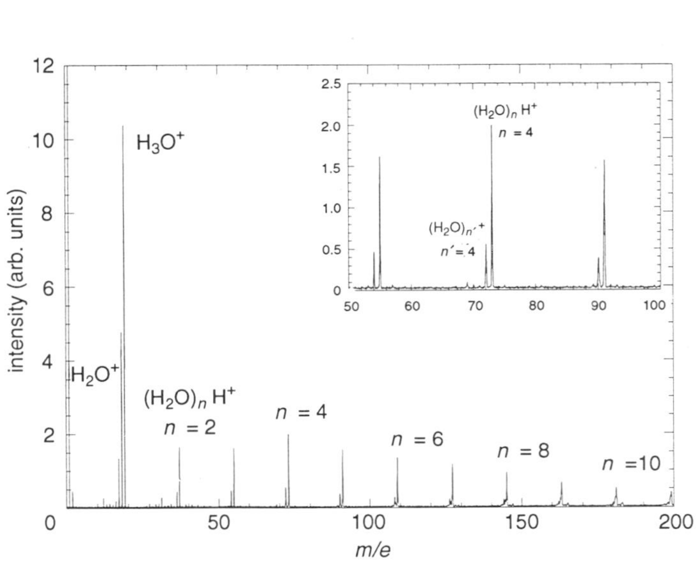

Fig. 7 Positive ion SIMS spectrum of annealed water

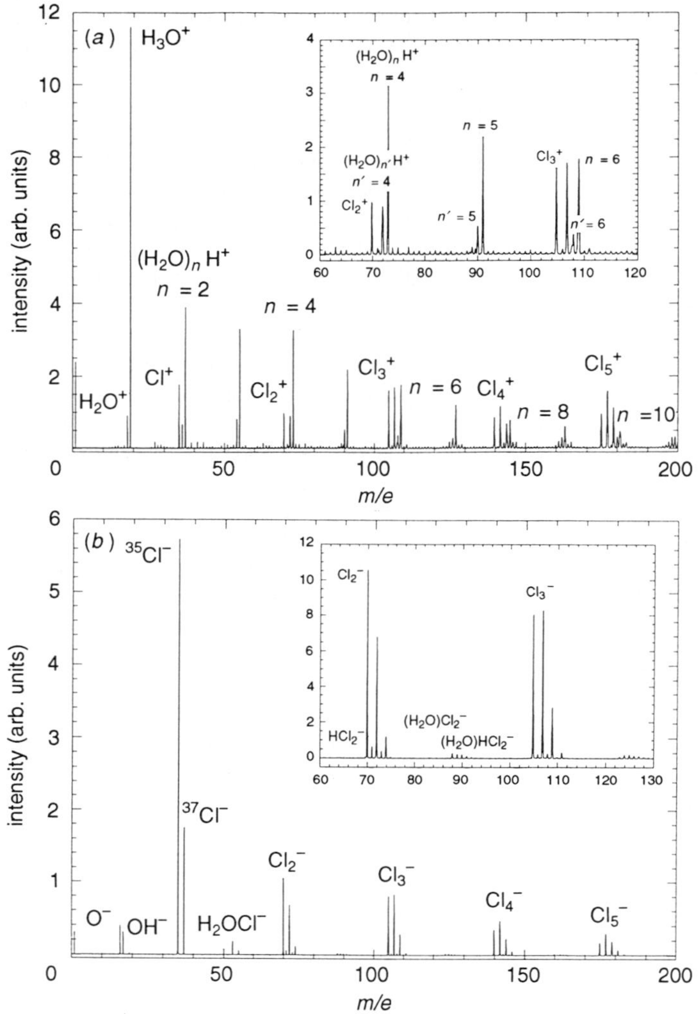

Fig. 8 (*a*) Positive ion and (*b*) negative ion SIMS spectra of Cl_2 on water

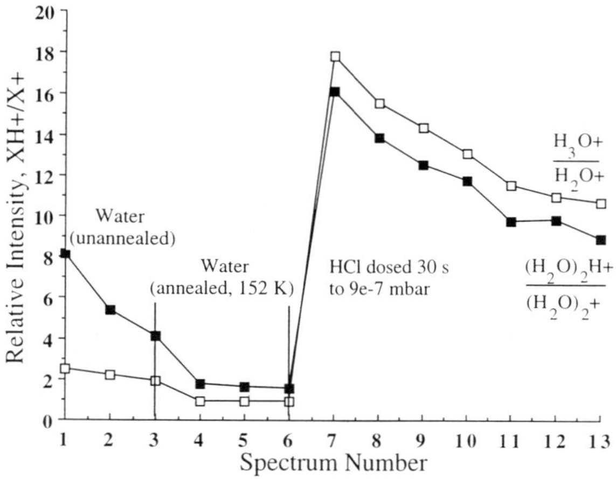

Fig. 9 Change in relative peak intensities during HCl–water experiment

addition of HCl, where the XH^+/X^+ ratio jumps by a factor of 15 or so. Although not clear from the graph, this increase is due almost entirely to an increase in the absolute intensity of the protonated species. This leads us to the unsurprising conclusion that ions are formed in the interaction between water and HCl, even at 90 K. The absence of HCl clusters in the positive-ion spectrum indicates an absence of condensed molecular HCl. In the negative-ion spectrum (not shown), the absence of the Cl_m^- clusters observed for chlorine verifies that the nature of the Cl species in the HCl–ice interaction differs from that in the Cl_2–ice system.

The results from the SIMS experiments thus far look promising. We have detected clear differences in the spectra of Cl_2 on water *vs.* HCl on water. The interpretation of those results, that Cl_2 is unreactive at 90 K while HCl dissociates to form ions, is reasonable and agrees with the interpretation of the results of Sodeau *et al.* from IR spectroscopy. We are now in a position to use SSIMS to examine the reactive behaviour of the Cl_2–water and HCl–water systems at higher temperatures.

Dr. Clary concluded the discussion: I would like to suggest an experiment which would be very important in understanding the interaction of HCl with water droplets and ice surfaces. This would be the measurement of the IR spectrum of the size-selected clusters $HCl(H_2O)_n$ as a function of n. In Fig. 10 I show *ab initio* predictions that we[1] have made of this spectrum. A strong shift of the HCl vibration is seen as n increases. Eventually, the HCl peak will disappear when the HCl is dissociated into ions and such an experiment will give the value of n that is needed for this to happen.

1 M. J. Packer and D. C. Clary, *J. Phys. Chem.*, in the press.

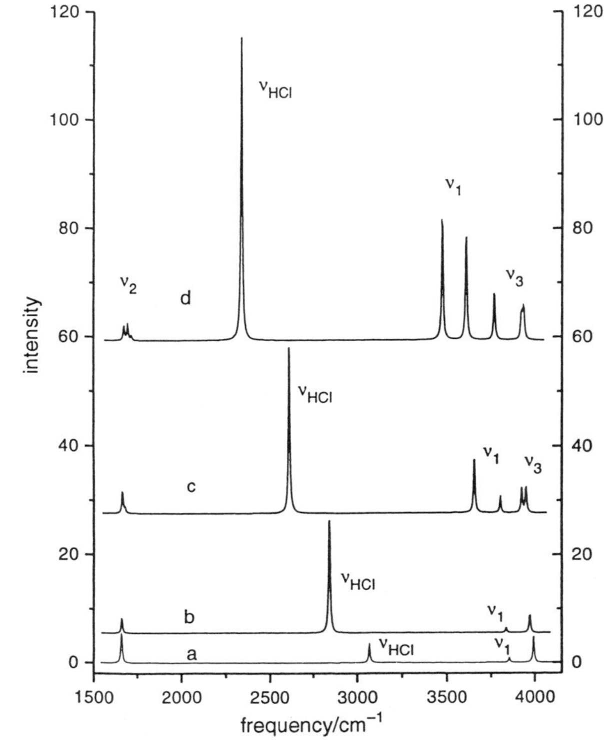

Fig. 10 Spectra of the $(H_2O)_nHCl$ clusters, assuming a Lorentzian line shape: a, H_2O and HCl; b, H_2O–HCl; c, $(H_2O)_2HCl$; d, $(H_2O)_3HCl$. Harmonic frequencies calculated at MP2 : 6–31g(2dp) level.

Faraday Discuss., 1995, **100**, 353–369

Global Atmospheric Chemistry from Satellites: Results from UARS/ISAMS

F. W. Taylor, C. D. Rodgers, J. J. Remedios, R. G. Grainger, A. Lambert, M. Lopez-Valverde, M. Goss-Custard and J. Reburn
Atmospheric, Oceanic and Planetary Physics, Department of Physics, University of Oxford, Clarendon Laboratory, Parks Road, Oxford, UK OX1 3PU

Satellites are a relatively new tool for investigating atmospheric chemical processes and their products. The precision and, in particular, the space and time coverage, now being achieved is immensely useful, and often essential, where global-scale phenomena like the variability of the stratospheric ozone layer are under study. The IR remote-sensing method using pressure modulator radiometry has recently been extended to the mapping of trace species. The technique is described, and its capabilities demonstrated with some selected new results from the improved stratospheric and mesopheric sounder (ISAMS) on the upper atmosphere research satellite (UARS). ISAMS was designed to study nitrogen-catalysed ozone chemistry, plus transport and other processes involving water vapour, methane and carbon monoxide, in the middle atmosphere. Its primary products are fields of temperature and composition, including all of the important members of the active nitrogen family, water, methane, carbon monoxide, ozone, aerosols and ice clouds. In the new data, all of these show large variations with time, height and latitude. The results have not yet been fully analysed, but early indications are that they confirm some models based on theoretical predictions and earlier, more limited, data, while in other cases they show that the situation is more complex than had been assumed. Sophisticated studies of the radiative–chemical–dynamical system in the stratosphere using satellite data and computer models are now in progress. In the future, even more advanced sensors now being built will obtain improved spatial resolution in the stratosphere, and map the global budgets of tropospheric trace gases as well.

Many of the currently most interesting problems of atmospheric chemistry are global in nature. The origins, conversion, evolution and eventual fate of a range of tropospheric pollutants, and the details of stratospheric ozone balance and its secular change, are just two topical examples. To understand such problems requires, along with careful laboratory work and regional sampling of composition by surface stations and aircraft, observations of the whole planet with high sensitivity to a range of minor and trace constituents, and with good vertical and time resolution. At present the only technique which is a serious candidate for satisfying the global requirement is spectroscopic measurements from satellites.

Unfortunately, many of the species of interest are present only in minute amounts and have proved difficult to observe by any technique, including *in situ* sampling. Also, the remote sensing methods of most promise are still, by and large, in their infancy. Those instruments which do offer a potentially useful performance tend to be large, complex and expensive. For example, large-aperture optics coupled to sensitive detectors cooled to very low temperatures are often required. Flight opportunities on satellite platforms which can accommodate such sensors are in extremely short supply, and space on them is obtained only after very arduous competition with other would-be users.

It follows that a significant measure of progress is essential from each major space-based atmospheric chemistry project, as it occurs, if the evolution which is essential to reach certain well defined goals is to continue to occur. Recently, the authors' department at Oxford developed, with collaborators in government laboratories and in industry, an advanced and comprehensive global atmospheric chemistry experiment called the 'improved stratospheric and mesopheric sounder'. ISAMS was part of an American project, the upper atmosphere research satellite (UARS), which itself has been designated the first component of the 'Mission to Planet Earth' programme designed, *inter alia*, to address global atmospheric chemistry problems over the next two decades or so.

Data gathering by ISAMS took place in the period from September 1991 to July 1992. A longer term, covering two rather than a single northern winter had been planned, but could not be achieved owing to a failure in one of the electro-mechanical devices in the instrument. Nevertheless, the period of data analysis which followed, and of interpretation which is just beginning, has shown dramatically how the new measurements open up the global details of problems which previously could only be glimpsed. They also show the limitations of the data, and the progress which will be required in the future. Three new instruments, destined for future 'Mission to Planet Earth' satellites, now called the Earth observing system, are already under development and are expected to make further important progress, including the extension to tropospheric measurements.

Here, we present some of the new results from ISAMS in selected areas related to the original goals, which were to elucidate the role of oxides of nitrogen in the stratospheric ozone depletion, to investigate relationships between chemistry and dynamics in the middle atmosphere, and to investigate the budgets of water vapour, methane and carbon monoxide, including their inter-relationship. A fourth category was added later, after the importance of heterogeneous chemistry in the stratosphere was realised; ISAMS can make sophisticated observations of the variability and physical properties of sulfuric aerosols and polar stratospheric clouds. It should be stressed that these measurements are preliminary, since the reduction of the raw spectral radiance data to temperatures and molecular abundances is not yet complete, nor likely to be for some time. This so-called retrieval process is complicated by phenomena such as departures from local thermodynamic equilibrium, in which radiative emissions due to non-thermal physical and chemical excitation of molecular energy levels are observed, and stratospheric aerosol emission, which was greatly enhanced during the ISAMS observing period because of the eruption of Mt. Pinatubo in the Philippines, which took place shortly before the launch of UARS.

The goal of the paper is, therefore, to show how these new data reveal details of important atmospheric chemical processes, and how this information is being brought to bear on a range of current problems. Final conclusions are still some years away, and will rely on a careful blend of satellite data with observations from balloons and aircraft, with laboratory data and with detailed numerical models.

Experimental

ISAMS is a limb-viewing, multispectral instrument[1] measuring thermal IR emission from UARS, which is in a moderately high (52°) inclination orbit. Limb-viewing extends latitude coverage to within 10° of the pole, and permits a high vertical resolution (compared with nadir-viewing sensors, like those on most weather satellites, for example), and global, day and night coverage. The spacing between profiles is *ca.*

200 km along the orbit track. Perpendicular to the track it varies between overlap at the poles and about 2200 km at the equator.

ISAMS measures thermal emission at 14 wavelengths between 4.5 and 16.3 μm, in 14 wide-band and 10 pressure-modulated channels. The pressure-modulator (PM) technique involves the use of a cell containing the gas being observed, as an additional selective filter, whereas 'wide-band' refers to the multilayer dielectric interference filters which select a pass band which transmits the emission lines of all species that are active in that wavelength range. Despite the terminology, which is relative to the extremely high resolution of the pressure-modulated channel, this pass band is quite narrow, being close to the limits of filter technology at around 2% of the central wavelength (see Table 1). Over this limited range, the Planck function for thermal emission at typical stratospheric temperatures varies only slightly with wavelength. The measured intensity is, therefore, a function primarily of temperature and absorber number density along the line of sight. Both PM and wide-band signals are measured simultaneously at the same detector by using phase-sensitive detectors tuned to the two different modulation frequencies. The difference in number between wide-band and pressure-modulated channels arises because some of the channels do not include pressure modulators, usually because the species in question are too chemically unstable even when ultra-clean techniques are used (O_3 and HNO_3). In two other cases (aerosol and N_2O_5) the technique is not applicable anyway because the absorption/emission spectrum in each case is a continuum. Moderately reactive species such as NO and NO_2 have been successfully contained, although the actual amount in the modulator and cell changes slowly after filling and has to be determined during flight as part of the data analysis. This is not a particularly difficult problem in principle, provided sufficient gas remains to produce an adequate signal, and the products of any dissociation which takes place do not themselves have strong spectral lines in the pass band.

The spectral positions and widths of the channels, listed in Table 1, were selected to optimise the measurements of temperature, pressure, aerosol opacity and minor constituent abundances. The channel at 12.1 μm was positioned in a spectral 'window', *i.e.* a region free of strong gaseous absorption, and therefore ideal for aerosol detection and measurement. However, since aerosol emission is a continuum across the spectrum, the other channels are also sensitive to it, the wide band more so than the pressure-

Table 1 ISAMS spectral channel designations, central wavelengths and bandwidths, and target species

channel number	wavelength /μm	spectral bandwidth, $\Delta T_{50\%}/\text{cm}^{-1}$	species observed
0	4.5	81	CO
1	6.8	71	H_2O
2	7.8	36	N_2O
3.0	15.6	27	CO_2
3.1	16.3	19	CO_2
3.2	11.4	38	HNO_3
3.3	10.0	20	O_3
4	5.3	76	NO
5	6.2	34	NO_2
6	7.4	52	CH_4
7.0	15.6	27	CO_2
7.1	16.3	20	CO_2
7.2	8.1	25	N_2O_5
7.3	12.1	15	aerosol

modulated channels. The ISAMS retrieval procedures[2] were designed to separate the opacity of the principal species in each channel from other emitters, including aerosol. At the same time crude spectra of the aerosol can be extracted, and it has been shown[3] that these readily distinguish between ice, nitric acid trihydrate and H_2SO_4–H_2O particles.

The vertical resolution of ISAMS at the limb is about 2.5 km; four layers of this thickness are observed simultaneously by a mechanically cooled detector array whose field-of-view is scanned vertically to obtain a complete profile from 10 to 100 km altitude every 16 s. Special long-life, high-efficiency coolers were developed at Oxford to achieve the necessary cooling with low mass and power requirements. By means of a system of mirrors, ISAMS views the atmosphere at the limb from either side of the UARS spacecraft, and the orbital parameters are such that with this the Earth can be covered from 80°S to 80°N latitude, that is, over more than 98% of its surface, every operating day.

Some Early Results

It is not possible to cover in useful depth all of the preliminary results from ISAMS in a single paper of this length. We discuss instead some previously unpublished highlights selected for their potential interest to the chemistry community, and representative of the more extensive data set which is gradually being released elsewhere in publications[4] and databases.†

Temperature and Dynamics

Chemical reactions in the stratosphere are temperature dependent and the reactants are often dynamically transported over considerable distances by thermally induced pressure gradients. The thermal structure of the atmosphere must, therefore, be investigated as an essential part of understanding its composition variations. It has already been noted above that it is necessary to know the kinetic temperature accurately in order to retrieve abundances from radiance measurements, and quite small uncertainties in temperature can dominate the error in a composition measurement. In general, an error of 1 or 2 °C in the former maps into a 10–20% error in the retrieved mixing ratio. Temperature is measured by observing the intensity of the emission from atmospheric carbon dioxide, on the assumption that the mixing ratio of this constituent is constant at a known value (which is, in fact, only approximately true, one of many details which complicates the definitive analysis of the data). ISAMS had two independent pressure-modulated channels, with their associated wide bands, positioned in strongly and weakly absorbing parts of the CO_2 15 μm band, in order to optimise the temperature determination over a wide vertical range. The current data extend from *ca.* 100 to *ca.* 0.01 mbar (*ca.* 15 to 80 km), and agree to within about 1 K with more conventional temperature measurements such as radiosondes, ISAMS measurements being slightly lower overall.[5]

The most dramatic event during the seasonal cycle of atmospheric temperatures is the stratospheric sudden warming, a phenomenon which occurs in the northern winter, and which is associated with large-scale changes in the dynamical regime. Satellite sensors can observe this dramatic phenomenon, and its consequent effect on the dynamics of the vortex, and material transport within and around it, as it evolves, and ISAMS did so with unprecedented detail. Plate 1 shows a temperature cross-section, from low latitudes on one side of the Earth to the other, passing close to the north pole,

† The complete ISAMS data set is available to qualified users from the British Atmospheric Data Centre at Rutherford Appleton Laboratory, Chilton, Didcot, Oxon., UK OX11 0QX.

on 9 January 1992 during the occurrence of such a warming. The maximum temperature near the pole on this day approaches 300 K, more than 60 K higher than the climatological mean for that time of year.

The onset and evolution of the warming can be studied using polar stereographic projections on pressure surfaces; the ISAMS vertical resolution is such that more than 20 different height levels are observed simultaneously. From these it is found that the warming is in the form of a single wave disturbance surrounding the pole, with the phase of the wave varying smoothly in the vertical direction, and changing by 180° in about 30 km, so that the maximum near 50 km overlies the minimum at 20 km. This can also be seen in Plate 1. Long sequences or 'movies' showing the variations of the temperature field throughout the winter reveal that several smaller warmings took place before the large event in January. Elevated temperatures appear first at higher levels and then move downwards, and the main, cyclonic polar vortex is pushed from its usual position over the pole to lower latitudes, while a secondary, anticyclonic vortex (which occurs annually and is known as the 'Aleutian high') builds up until its strength is nearly the same as that of the main vortex. This pattern begins to break up at higher levels after the peak of the warming.[6] The chemistry that occurred at different times and places as part of this phenomenon is reflected in compositional variations which are detected in the ISAMS data, some of which are discussed below.

Budgets of Minor Constituents

General Considerations

Virtually all of the ozone in the stratosphere is produced *in situ*, but many of the other minor constituents present, including those which act as source gases for species which react with ozone, originate below in the troposphere or above in the mesosphere. Table 2 shows approximate source–sink relationships responsible for the gross vertical budgets of five of the key species measured by ISAMS. It is striking that the patterns are different in every case, and almost every possible combination is present. We therefore expect, and observe, that the general characteristics of the vertical distributions of each will be qualitatively different. Of course, the sources and sinks vary horizontally, especially with latitude, as well as vertically, and also with time, diurnally, seasonally, secularly and quasi-randomly. In data from satellites these variables are covered, although never perfectly since a single satellite can only be in one place, viewing a limited region, at a given time. The production/destruction processes themselves can be quantified and understood by careful analysis, often by employing models to help separate the many variables, and in particular to take account of simultaneous chemistry and transport.

Oxides of Nitrogen

The basic nitrogen chemistry cycle in the stratosphere has two major characteristics: (i) the mid- and low-latitude diurnal variation in which NO_2 is converted by day into NO,

Table 2 One-dimensional budgets of some important stratospheric minor constituents measured by ISAMS (the labels refer to whether or not each region is a net contributor to the abundance of the relevant species in the stratosphere)

	CO	O_3	H_2O	CH_4, N_2O	NO_x
mesosphere and thermosphere	source	sink	sink	sink	source
stratosphere	sink (+source)	source (+sink)	source	sink	source (+sink)
troposphere	source	sink	source	source	sink

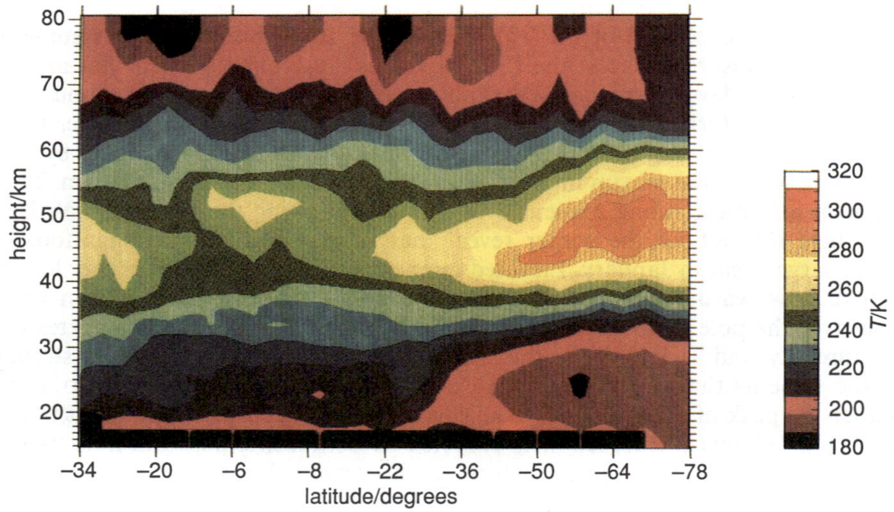

Plate 1 Temperature cross-section from low southern latitudes through the north polar region, obtained by ISAMS during part of one orbit on 9 January 1992

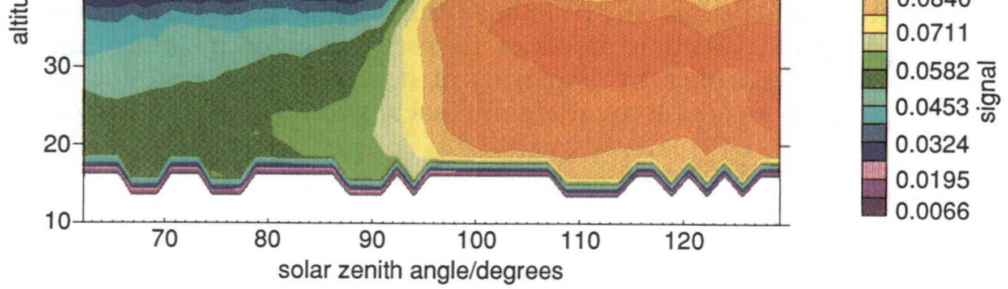

Plate 2 ISAMS NO_2 radiance signal as a function of height and solar zenith angle across the terminator region

with the reverse occurring at night; and (ii) the conversion between these chemically active species and the more stable 'reservoirs', principally N_2O_5 and HNO_3, which is particularly prevalent in the long polar nights. ISAMS data, in the form of a vertical section of the NO_2 radiance signal as a function of solar zenith angle across the terminator region, reveal the diurnal cycle in dramatic detail (Plate 2). Of particular interest are the way in which the amount of NO_2 builds up as the Sun sets, at a rate which is clearly very height dependent, and the vertical structure itself, which shows that the high NO_2 content is confined to a region of depth *ca.* 20 km. This region is, of course, also that which contains the greatest concentration of ozone. Model intercomparisons of many measurement fields similar to the example in Plate 2 will lead to a refined understanding of this aspect of the NO_x cycle.

The composition of the atmosphere inside and outside the winter polar vortices in both hemispheres shows, as expected, the largest contrasts between active and inactive nitrogen abundances. For instance, Plate 3 shows the NO_2 abundance on the 10 mbar constant-pressure surface (about 30 km altitude) on 9 January 1992. Here both NO and NO_2 exhibit low values in the night-time, because the polar night is so long that extensive conversion of NO to NO_2, and NO_2 to the 'reservoirs', N_2O_5 and HNO_3, has taken place. From the vantage point of a satellite, it can be clearly seen how a 'cliff' forms in the NO_2 profile across North America in winter. This phenomenon had been detected previously by individual profile sampling from the ground, but obviously cannot be explained completely without seeing the global, time-dependent picture. In ISAMS maps, the active and reservoir species are found to be spatially and temporally quite variable, with, for example, large masses of air from lower latitudes, characterised by their minor constituent composition, seen to be mixing with the northern vortex, after which the composition of the mass is observed to evolve. Filaments from the vortex also separate and mix with mid-latitude air. The relationship between the amounts of N_2O_5 and HNO_3 present, in particular, during this sort of process, is complicated and is still under study. In fact, it is already clear that the horizontal spatial resolution obtained by ISAMS, although good by contemporary standards, is not high enough to resolve the filamentary structure in the NO_x composition fields in the regions near the edge of the polar vortex, to elucidate completely the relationship between the chemistry and dynamics there. Future instruments will need to address this point (see below).

It would be valuable to know how much of the stratospheric active nitrogen originates in the troposphere (primarily as N_2O which is then photochemically converted in the stratosphere), and how much is propagated downwards from the thermosphere, perhaps mainly as NO. The latter is formed by the action of energetic particles, and possibly also X-rays, mainly of solar origin, on N_2. This process is, therefore, expected to be very dependent on solar activity, and the ISAMS observations of thermospheric NO profiles during a period of high geomagnetic activity following major solar flares (Fig. 1) show that this is indeed the case.[7] They also show that the enhancements in NO in the atmosphere are very spatially focused, and quite variable. In the particular event illustrated in Fig. 1, the maximum NO contents were observed between 30° and 40°N latitude, and 0° and 30°E longitude. Anomalies such as these, and probably the background levels of NO_x produced during more typical conditions, make a contribution to the stratospheric burden of ozone-depleting molecules, which is probably significant but which remains to be quantified accurately using ISAMS and other data.

Water, Methane and Carbon Monoxide

These species all have intermediate lifetimes in the stratosphere and hence distributions which tend to reflect the mean circulation and its seasonal variations (but in different ways because of the different source–sink relationships discussed earlier). The principal source of stratospheric water vapour is that which is transported through the tropo-

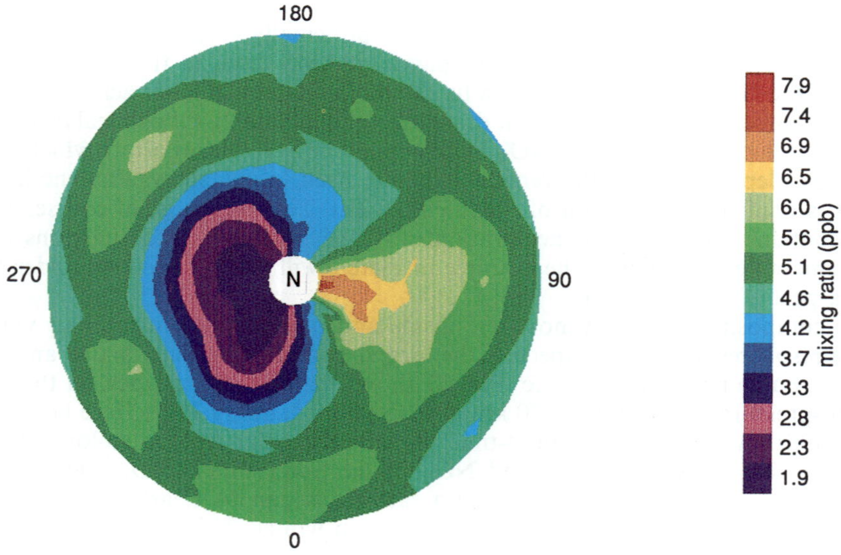

Plate 3 NO_2 mixing ratio on the 10 mbar pressure surface (about 30 km altitude) on 9 January 1992

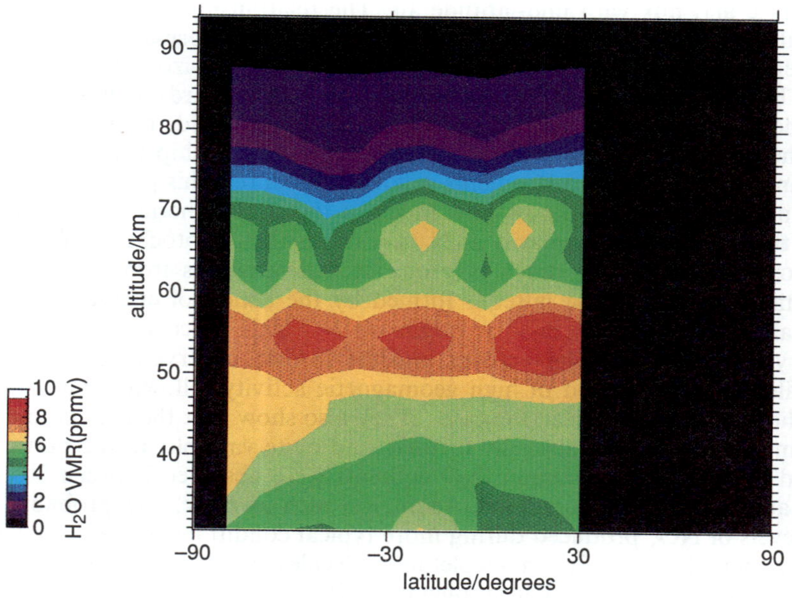

Plate 4 Zonal mean values of water vapour mixing ratio *vs.* latitude and height on 29 March 1992, derived from night-time observations only

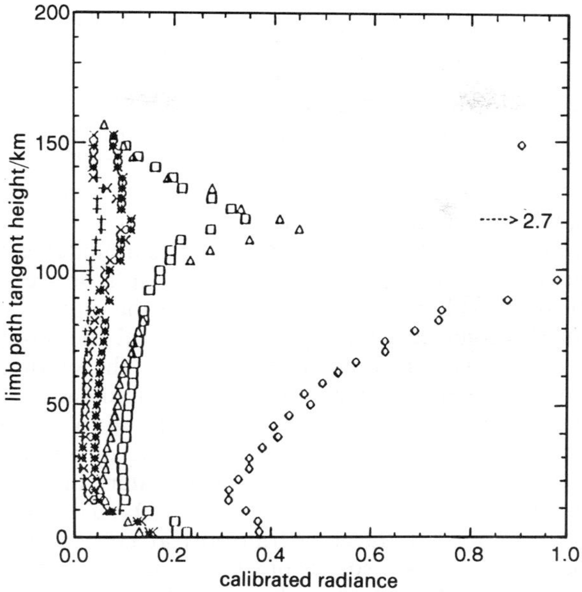

Fig. 1 Nitric oxide profiles at 35°N, 15°W for three successive days in November, 1991, showing the respone of the thermospheric abundance of NO to a period of strong solar flare activity. (+) 8 Nov 1991 am; (*) 8 Nov 1991 pm; (◇) 9 Nov 1991 am; (△) 9 Nov 1991 pm; (□) 10 Nov 1991 am; (×) 10 Nov 1991 pm. The density scale is very preliminary (after Ballard *et al.*[7]).

pause, while the main sink is in the upper mesosphere and thermosphere where photolysis by solar Lyman α radiation proceeds rapidly during the daytime. ISAMS water vapour fields (Plate 4) show the abundance falling off rapidly above 70 km in the region where this effect is expected to be significant. In due course, a complete analysis of the water vapour mixing ratio in this region will shed light on the vertical eddy diffusion rate in this little-explored region and also quantify better the rate at which water vapour is being lost from the atmosphere. It had been hoped that the new data would also allow progress to be made on understanding the global variability of the stratospheric–tropospheric exchange process, but prospects for this are much reduced owing to the very heavy aerosol loading in the lower stratosphere during the observing period. This limits the lower height limit in the currently available water vapour maps to about 10 mbar (30 km), probably too high to be sensitive to source fluctuations at the tropopause. Later improvements in the retrieval method should improve the lower limit for all species to around 100 mbar (15 km) when a fresh attempt on this problem will be possible. Future instruments will use higher spectral resolution to penetrate right into the troposphere (see below).

The ISAMS water vapour data also show a distinct maximum in the mixing ratio near the stratopause, and a weaker maximum at higher latitudes in the stratosphere. The former is believed to be due to the photodissociation of methane, and subsequent oxidation of its components, an hypothesis which can be tested by using the coincident ISAMS H_2O and CH_4 measurements to calculate fields of the sum $2[CH_4] + [H_2O]$, which should be constant if $CH_4 \rightarrow 2H_2O + CO_2$ is the net conversion taking place. The sum is indeed found to be very nearly constant, with a preliminary value[8] of about 8 ppmv, except at high altitudes and high latitudes where lower values are found. These are probably due to the downward moving of mesospheric air rich in H_2, although direct measurements of molecular hydrogen abundance at high latitudes and high altitudes, by ISAMS or any other method, are not available to verify this at present.

It is also not possible at present to interpret ISAMS water vapour measurements made during the day. The reason for this is the strong non-thermal component to the emission, caused by solar pumping of the ν_2 vibrational levels which give rise to the transitions observed in ISAMS channel 1. Enough is now known about the processes involved [basically, absorption in the shorter-wavelength bands of H_2O near 1.9 and 2.7 μm, where the solar intensity is greater, followed by cascading down to the $\nu_2 = 1$ or (0, 1, 0) level, plus excitation resulting from the photodissociation of O_3 and subsequent excitation of $O_2(1)$, which is in near-resonance with $H_2O(0, 1, 0)$] to make corrections in the retrievals accurate enough to produce useful daytime abundances.[9]

A similar problem applies to carbon monoxide, which fluoresces brightly in sunlight, but in this case the simplified energy-level structure of the diatomic molecular is easier to treat, in spite of a resonance with N_2, and non-local thermodynamic equilibrium retrievals have been possible for some time.[10] In fact, the extra contribution to the source function, coupled with the high mesospheric abundance of CO, means that daytime ISAMS measurements of this species can extend up to 100 km or higher. This is fortunate since the source–sink relationship, and hence direction of propogation, of carbon monoxide is broadly the opposite to that of water vapour, in that production occurs mainly in the thermosphere, by dissociation of CO_2. The species then propagates downwards, and is an important tracer of transport at mesospheric levels. The troposphere is also probably a significant source of CO, as may be chemical production in an intermediate stage of the photolysis of methane in the stratosphere. The ISAMS data allow us, in principle, to assess the relative contributions of each of these sources, and perhaps to identify the major sinks, but the picture is complicated by dynamics, horizontal as well as vertical (Plate 5). One of the earliest results from ISAMS was the first experimental evidence,[11] from maps of the CO abundance, for mesospheric air descending strongly into the stratosphere over the winter pole, first in one hemisphere and then in the other as the seasons reverse. This lends indirect support to the explanation of the water vapour maps advanced above, in that the CO-rich descending air is likely to be H_2-rich also.

Methane and nitrous oxide both originate in the troposphere and have long lives, measured in months, in the stratosphere before oxidation to H_2O, CO and CO_2 and conversion to NO_x, respectively. Injection into the stratosphere occurs mainly at low latitudes, as can be seen (Plate 6) from the large gradients from equator to pole in each hemisphere, as well as in the vertical direction. In addition to the chemical budget studies already mentioned, the new measurements are important for high-resolution tracer studies of global transport in the stratosphere. For example, the double-peaked structure apparent in Plate 6 arises because of the basic equator-to-pole overturning motion which characterises the general circulation of each hemisphere, outside the polar vortices. Time series of similar data show the peak in the summer hemisphere strengthen with the advancing season, and then weaken during the winter; the roughly symmetric pattern shown in Plate 6 is characteristic of the period around the equinoxes.

Ozone

The strong diurnal and latitudinal variability of ozone is well captured in satellite data. Plate 7 shows the ozone concentration near the stratopause (on the 1 mbar surface, or at about 50 km altitude) in the northern hemisphere in early October, 1991. Note that the highest concentration is over the pole; although production is a maximum at the equator. Clearly, ozone destruction is also latitude-dependent, and meridional transport must be efficient enough to achieve the observed result. In principle, this balance can be worked out in detail at every level using satellite-derived abundances, temperatures and winds in a model scheme.

Much current interest is focused on how variations in ozone density may correlate with the distributions in chemically active minor constituents, especially the nitrogen

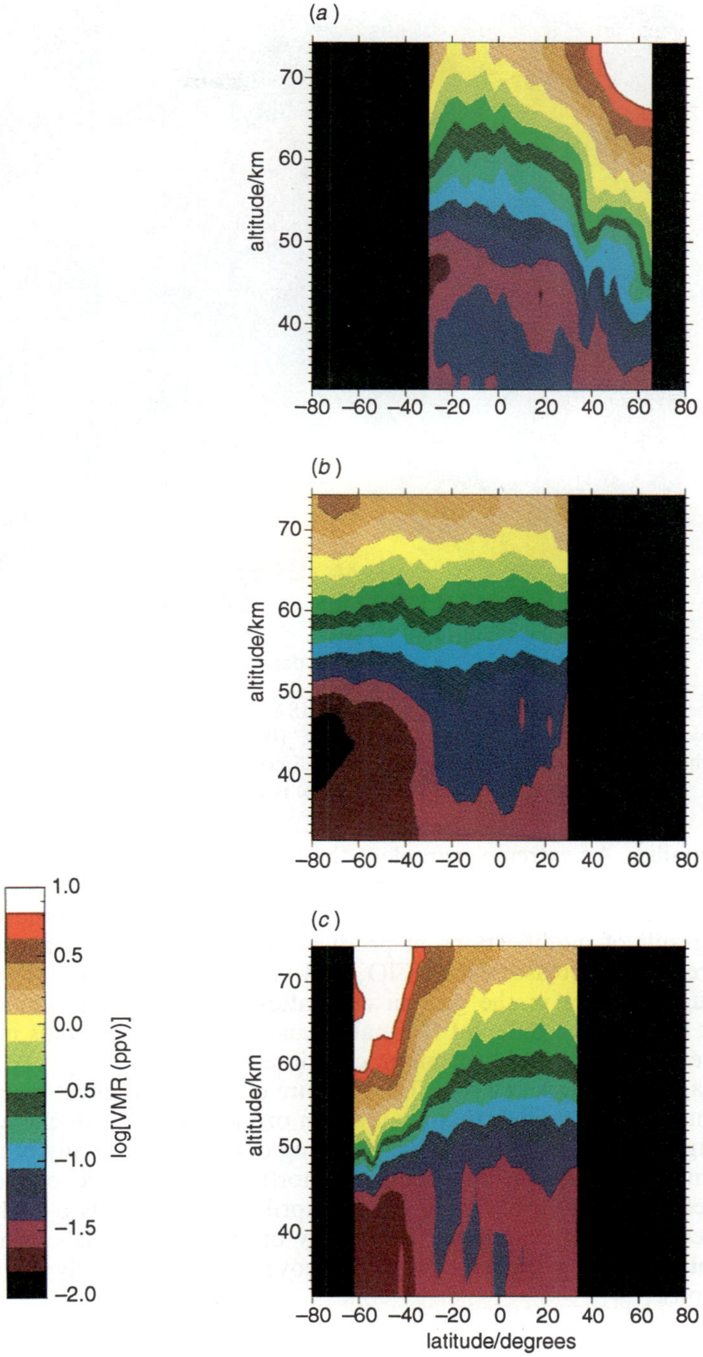

Plate 5 Carbon monoxide zonal mean mixing ratio *vs*. height in (*a*) northern-hemisphere winter, (*c*) southern-hemisphere winter and (*b*) near the spring equinox

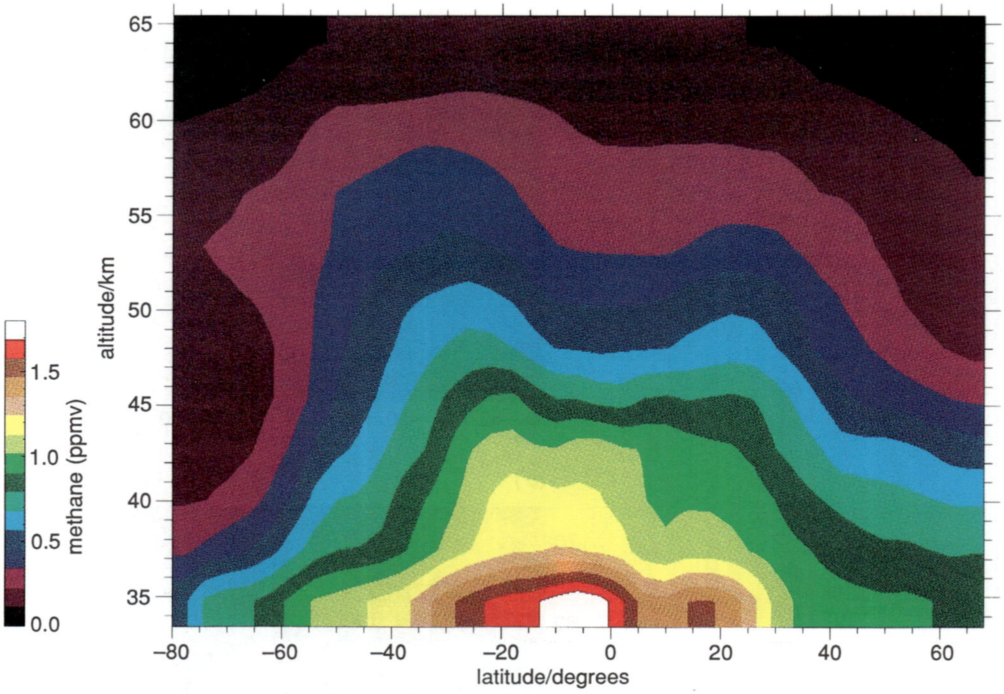

Plate 6 Zonal mean methane abundance as measured by ISAMS on 18 April 1992. Measurements were made well below the range shown, in fact down to *ca.* 100 mbar (*ca.* 15 km), but the retrieval of composition in the lower part of the stratosphere was still complicated by massive amounts of aerosol from Mt. Pinatubo at this time, and further work is required to obtain reliable results.

and halogen families. An extreme example is the Antarctic ozone hole which develops in the austral spring, and which is believed to be caused by heterogeneous chemistry (see below). Even more crucial, however, are subtle or gradual changes in the global ozone balance as a result of modifications in production, destruction or transport, whether directly as a result of increased NO_x, ClO_x and background aerosol, or indirectly, for example through changes in the general circulation as a result of a modified global energy budget leading to a different thermal balance. The latter depends mainly on the longer-lived, less chemically active minor constituents such as CO_2, H_2O and CH_4, with aerosols again having a key role. Temperature changes also affect reaction rates; it has been estimated using models that equilibrium ozone amounts can change by 1% for each 1 K change in temperature, and that doubling of CO_2 without any other change in composition may be sufficient to produce a northern hemisphere ozone hole, with almost complete destruction of ozone in mid-April.[12] Clearly, it is a high priority to verify such predictions by examining the validity of the latest models using data from ISAMS and other satellite instruments, which cover comprehensively the current seasonal fluctuations.

Heterogeneous Chemistry: Sulfate Aerosols and Polar Stratospheric Clouds

Aerosols of various types in the middle atmosphere are of interest in their own right, but recently have drawn increased attention for their potentially important role in altering the albedo of the Earth and therefore giving rise to the greenhouse effect. Of more immediate interest is their effect on the abundances of gaseous species by acting as

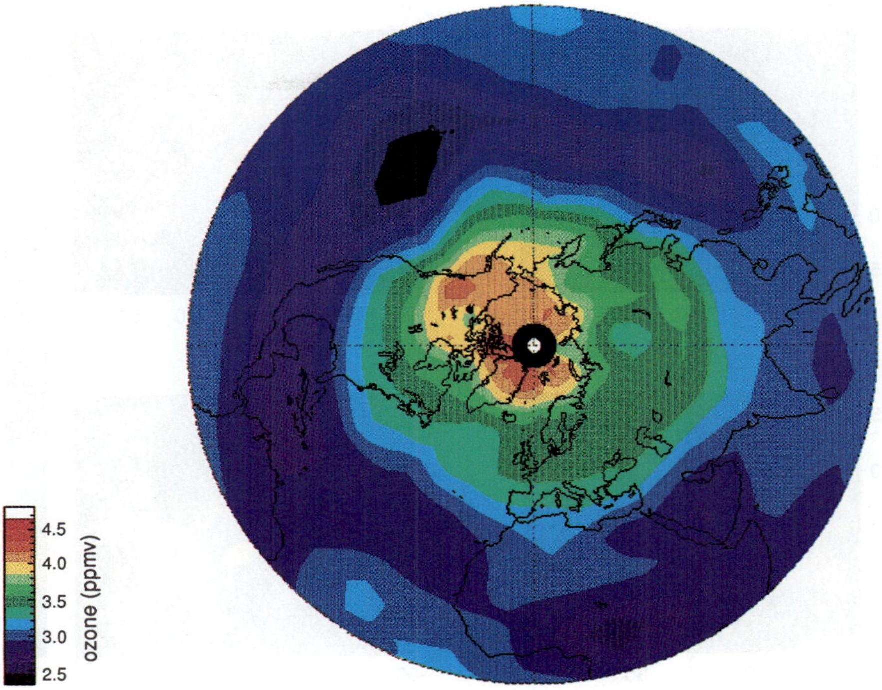

Plate 7 Ozone abundances over the northern hemisphere on 6 October 1991

reservoirs of sulfur, nitrogen and chlorine compounds, and as surfaces on which heterogeneous chemical reactions can take place.

The ISAMS spectral data discriminate readily between aerosols composed primarily of sulfuric acid (generally of volcanic origin) and those of nitric acid (polar stratospheric clouds), but the global maps of each have shown that the two are physically interrelated in the important winter polar regions, with the former providing possible condensation partners for the latter. ISAMS observed the Pinatubo aerosol cloud as it travelled vertically and polewards for many months after the eruption (Plate 8). It was also able to monitor changes in the droplet size and acid concentration. A year after eruption, large amounts of aerosol had migrated to high northern latitudes, and a smaller but still very significant amount to the Antarctic; the total mass of sulfate aerosol injected into the stratosphere has been estimated to be around 20 million tonnes.[13] It is to be expected that such large quantities of material will significantly affect the chemical balance in the ozone layer, and work is underway using trajectory-following methods to identify and quantify this effect in the ISAMS data.

Polar stratospheric clouds, or PSCs, particularly in the northern hemisphere where they are rarer and more ephemeral owing to higher mean temperatures, do not seem to form automatically when the temperature falls below the condensation temperature for nitric acid tryhydrate (NAT), as models commonly assume. This may be because they are not, in fact, NAT at all, but rather drops of a ternary compound of HNO_3–H_2SO_4–H_2O, as proposed by Drdla *et al.*[14] In any case, it is important to elucidate the processes by which they form (and dissipate), and this requires good coverage and resolution in space and time like that provided by ISAMS. Wide spectral coverage is also useful, not only to delineate between aerosol types but also to allow the optical properties to be determined and the microphysics (particle size, *etc.*) to be inferred.

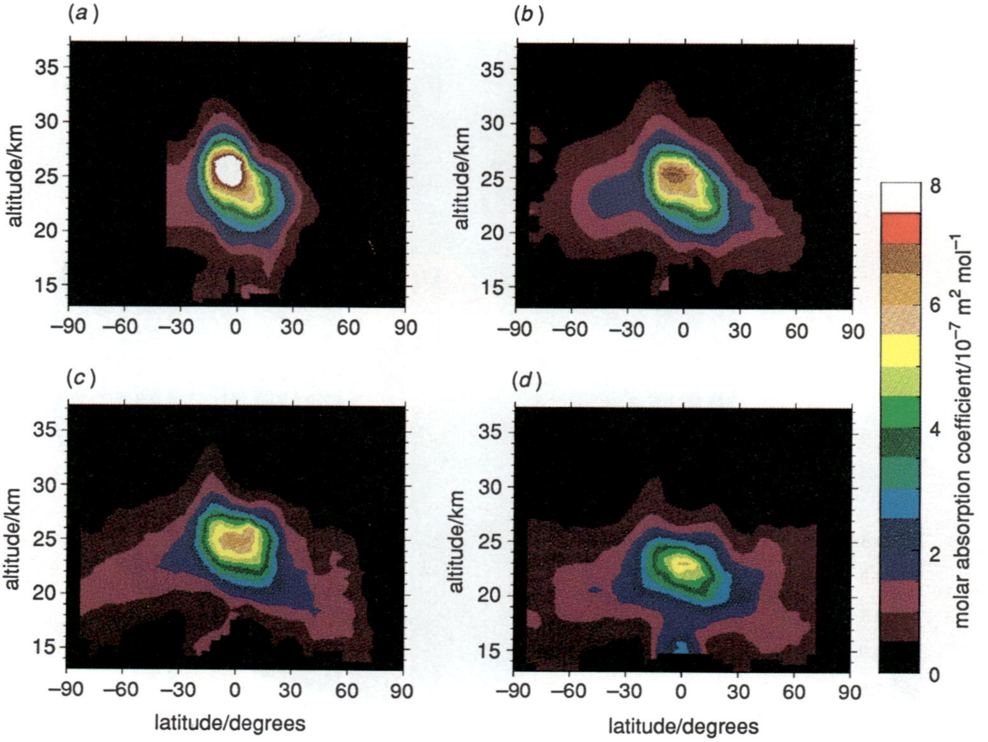

Plate 8 Evolution of the Mt. Pinatubo aerosol cloud as observed by ISAMS at a wavelength of 12.1 μm, an atmospheric 'window'. (*a*) 7 October 1991; (*b*) 14 November 1991; (*c*) 15 January 1992; (*d*) 2 April 1992.

The PSC in Plate 9 formed over northern Europe on 27–28 December 1991 at a height of 20 km, and persisted for only about three days. It was one of only two substantial PSC episodes in the winter of 1991/2, a much lower rate of occurrence than has been inferred for previous northern winters.[15] This is possibly a result of a greatly increased amount of H_2SO_4 present following the Pinatubo eruption, depressing the freezing point of the mixture, which has been shown[16] to be affected strongly by composition. When PSCs did occur, major changes in the chemistry (including, in this case, the release of active chlorine as the oxide, ClO, and subsequent depletion of ozone) were observed.[17]

Further Developments

Detailed Coupling between Dynamics and Chemistry

The analysis of the ISAMS data is showing increasingly that dynamics and chemistry are closely coupled in the stratosphere, not only through the global transport of constituents, but also because structure is sometimes present in the composition fields (at the edge of the polar vortex, for example) down to, and probably beyond, the smallest observable scales. Then, the rate at which a reaction between two constituents can proceed can no longer be estimated from the product of the mean measured abundances observed. It is already clear from the new UARS data that mixing on smaller scales than has been observed so far is crucial in understanding chemical budgets. This interesting problem implies a requirement for improving the horizontal resolution of future global

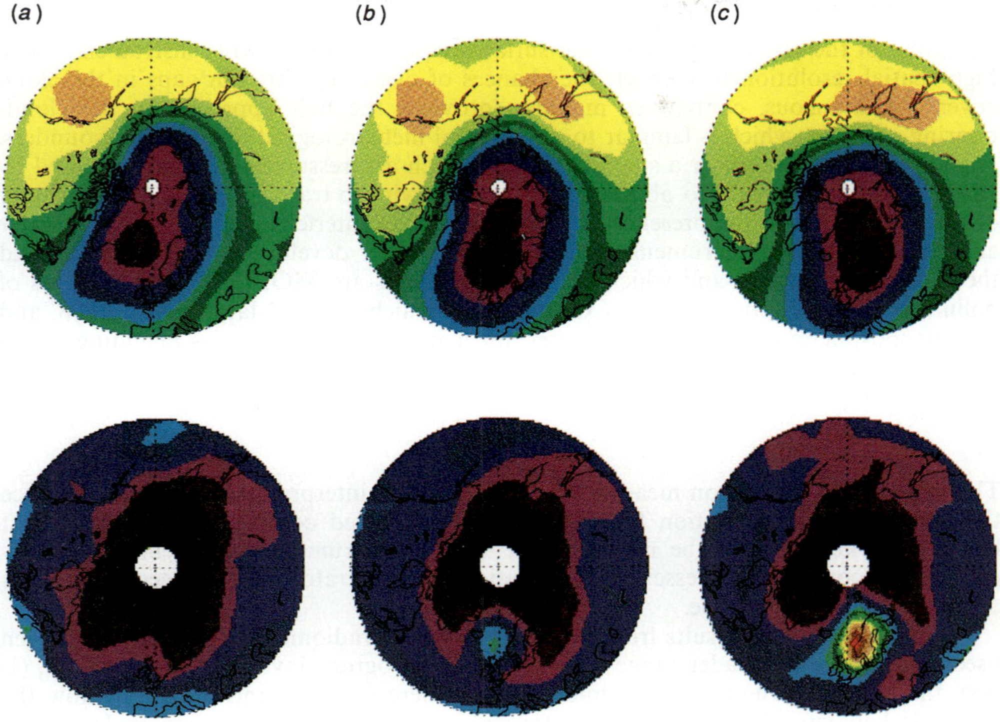

Plate 9 Polar stereographic plots of a polar stratospheric cloud forming over northern Europe in December, 1991: (*a*) 26th, (*b*) 27th and (*c*) 28th. The maps are polar stereographic projections on the 46 mbar surface, and show temperature (above) and the molar absorption coefficient at 12.1 μm (below).

measurements, as well as maintaining (and possibly slightly improving, although ISAMS is close to the theoretical limit even for limb sounding instruments) the vertical resolution. The next generation of satellite instruments, under development now, includes the high-resolution Infrared Dynamics Limb Sounder (HIRDLS) which achieves this by using a very fast integration time and an azimuth scanning technique, *i.e.* one which sweeps the field of view from side to side of the orbital track. Improvements in detector technology also mean that larger detector arrays can be used to make measurements in several directions (for example over the complete vertical range) simultaneously, reducing the need for mechanical scanning.

Chemical Coupling between Atmospheric Regions

Another limitation of current limb-scanners like ISAMS is their very restricted ability to sound in the region of the tropopause and below. Because of this, a number of key questions such as how water, and nitrogen- and chlorine-containing pollutants reach the stratosphere from below, and the identification of the tropospheric sinks for stratospheric pollution, can only be addressed indirectly. Future instruments, including HIRDLS, will use higher spectral resolution to penetrate below the tropopause whenever high clouds or dense aerosols are absent, employing 'microwindows' in the atmospheric spectrum. Calculations[18] show that measurements of several species, including water, methane, nitrous oxide and the simpler chlorofluorocarbons, can extend down to as low as 6 to 8 km above the surface under favourable conditions.

Tropospheric Chemistry from Space

Sounding of the composition near the surface requires nadir-viewing instruments, with high spatial resolution to increase the chances of observing through gaps in the cloud cover. Simultaneous, contiguous mapping provides the best approach to the cloud-clearing problem, which is familiar to operational meteorological temperature sounders. High spectral resolution or a spatial technique such as pressure modulation, as used by ISAMS, is also necessary to obtain adequate sensitivity to trace species in the proximity of the surface and in the presence of large amounts of interfering species such as water and carbon dioxide. Instruments which are now under development for flight around the turn of the century and which aim to achieve this are MOPITT (measurements of pollution in the troposphere, for CO and CH_4) which is to be launched in 1998, and TES (tropospheric emission spectrometer, for a wider range of species including ozone and fluorocarbons) which begins observing around 2002.

Conclusion

This paper concentrates on measurements, rather than interpretation, particularly since the extraction of information from the ISAMS-measured composition fields has just begun. The key point to be made is that suitable instruments on satellites are now providing a cornucopia of essential information on key stratospheric processes, and this can be expected to continue.

Some selected new results from the ISAMS spectroradiometer on UARS have been used to illustrate a broader range of work now in progress, involving, in particular, (1) how the thermal structure of the middle atmosphere changes, and from that how the region is behaving dynamically; (2) details of the stratospheric nitrogen cycle and its variation in space and time, species by species; (3) the chemistry and physics of the stratospheric ozone balance, leading to better predictions of possible future ozone depletion; (4) the global stratospheric budgets of four related, key minor constituents, namely water vapour, methane, nitrous oxide and carbon monoxide; and (5) the evolution of stratospheric aerosol and condensate clouds, and their roles in the preceding four scenarios through their radiative effects, their role as reservoirs of gaseous species, and heterogeneous chemistry.

The full scientific return from ISAMS will continue to accumulate in the years ahead. Meanwhile, new instruments using alternative or improved techniques are being developed which will further extend the scope and value of measurements from space in the future.

Thanks are due to the many members of the science and spacecraft teams in the UK and the USA who made the ISAMS experiment possible.

References

1 F. W. Taylor, C. D. Rodgers, J. G. Whitney, S. T. Werrett, J. J. Barnett, G. D. Peskett, P. Venters, J. Ballard, C. W. P. Palmer, R. J. Knight, P. Morris, T. Nightingale and A. Dudhia, *J. Geophys. Res.*, 1993, **10**, 799.
2 C. D. Rodgers, R. G. Grainger and R. J. Wells, *J. Geophys. Res.*, submitted.
3 F. W. Taylor, A. Lambert, R. G. Grainger and C. D. Rodgers, *J. Atmos. Sci.*, 1994, **51**, 3019.
4 See, for example, the special UARS issues of *Geophys. Res. Lett.*, 18 June 1993, and *J. Atmos. Sci.*, 15 October 1994.
5 A. Dudhia and N. Livesey, *J. Geophys. Res.*, submitted.
6 S. M. Rosier, B. N. Lawrence, D. G. Andrews and F. W. Taylor, *J. Atmos. Sci.*, 1994, **15**, 2783.
7 J. Ballard, B. J. Kerridge, P. Morris and F. W. Taylor, *Geophys. Res. Lett.*, 1993, **20**, 1311.
8 M. Goss-Custard, D.Phil. Thesis, University of Oxford, 1994.
9 M. Lopez-Puertas, G. Zaragova, B. J. Kerridge and F. W. Taylor, *J. Geophys. Res.*, in the press.

10 M. Lopez-Puertas, M. A. Lopez-Valverde, D. P. Edwards and F. W. Taylor, *J. Geophys. Res.*, 1993, **98**, 8933.
11 M. A. Lopez-Valverde, M. Lopez-Puertas, C. J. Marks and F. W. Taylor, *Geophys. Res. Lett.*, 1993, **20**(12), 1247.
12 J. Austin, N. Buchart and K. P. Shine, *Nature* (*London*), 1992, **360**, 221.
13 A. Lambert, R. G. Grainger, J. J. Remedios, C. D. Rodgers, M. Corney and F. W. Taylor, *Geophys. Res. Lett.*, 1993, **20**, 12, 1287.
14 K. Drdla, A. Tabazadeh, R. P. Turco, M. Z. Jacobson, J. E. Dye, C. Twohy and D. Baumgardner, *Geophys. Res. Lett.*, 1994, **21**, 2475.
15 M. P. McCormick and C. R. Trepte, *J. Geophys. Res.*, 1987, **82**, 4297.
16 N. Song, *Geophys. Res. Lett.*, 1994, **21**, 2709.
17 J. W. Waters, L. Froidevaux, W. G. Read, G. L. Manney, L. S. Elson, D. A. Flower, R. F. Jarnot and R. S. Harwood, *Nature* (*London*), 1993, **362**, 597.
18 J. J. Barnett, personal communication.

Paper 5/00901D; *Received* 14*th February*, 1995

Faraday Discuss., 1995, **100**, 371–387

Early Modelling Results from the SESAME and ASHOE Campaigns

John A. Pyle, Martyn P. Chipperfield, Iarla Kilbane-Dawe and Adrian M. Lee
Centre for Atmospheric Science, Department of Chemistry, University of Cambridge, Lensfield Road, Cambridge, UK CB2 1EW

Richard M. Stimpfle and Daniel Kohn
Harvard University, Cambridge, USA

Wolfgang Renger
Deutsche Forschungsanstalt für Luft und Raumfahrt, Oberfaffenhofen, Germany

Joe W. Waters
Jet Propulsion Laboratory, Pasadena, USA

Data from the second European stratospheric arctic and middle latitude experiment (SESAME) and the airborne southern hemisphere ozone experiment (ASHOE) (recent campaigns in both hemispheres), and from the upper atmosphere research satellite (UARS) provide information about the movement, and possible mixing, of stratospheric air from the vortex and vortex edge into middle latitudes. Model studies reported here reproduce the observed features and may provide insight into their larger-scale structure. It is now clear that filaments of activated air, removed from the vortex edge, contribute to the observed decline of ozone in middle latitudes.

Since the discovery of the Antarctic ozone hole just ten years ago[1] there has been tremendous progress in our understanding, both qualitative and quantitative, of the fundamental processes responsible for the ozone loss (see, for example, ref. 2). It is now also apparent that levels of ozone away from polar regions have declined during the last ten or more years.[2–4] This decline is not especially well understood and, in particular, the quantitative understanding of middle-latitude ozone decline is not satisfactory.

A number of different ideas have been put forward to explain the mid-latitude ozone loss (see ref. 2 and 5 for more detailed reviews). Three particular ideas should be mentioned. First, it could be that air, primed for ozone depletion in the cold polar lower stratosphere, is carried to middle latitudes, and possibly mixed with mid-latitude air, followed by local ozone loss. If there is a large throughput of air from the polar vortex to middle latitudes[6,7] so that the vortex can be thought of as a 'flowing processor', then substantial ozone loss could occur at lower latitudes, probably increasing significantly with years as the amount of chlorine in the stratosphere continues to increase. An alternative idea, connecting polar processes to middle latitudes, is that polar air, depleted of ozone, is mixed into middle latitudes causing a general 'dilution' of ozone there. Yet another possibility is that the middle latitude decline arises from *in situ* processes when, for example, the ozone-destroying chlorine radicals are activated by reactions occurring on the lower stratospheric sulfate aerosol layer.

It is likely that the explanation for mid-latitude ozone loss will involve all three processes. Dilution will certainly occur when the polar vortex breaks down at the end of the winter. There is both direct[8] and indirect[9] evidence for activation of chlorine outside the polar vortex. A significant number of studies have shown that polar vortex air can be found in middle latitudes.

It is the objective of this paper to use field data and numerical models to investigate further, through a number of case studies, examples of polar air being removed (possibly permanently) to middle latitudes. Tuck *et al.*[6] compared lower stratospheric potential vorticity (PV) fields, obtained from the European centre for medium-range weather forecasts (ECMWF), with data collected during the airborne arctic stratospheric experiment (AASE) in 1989 on a flight (in an ER-2) from Stavanger to North America. The flight crossed the polar vortex into middle-latitude air. The PV fields, in addition to revealing the large-scale structure of the lower stratosphere, show a considerable degree of structure, with many 'blobs' of high PV, characteristic of the polar vortex, seen in middle latitudes. Coincident with some of these 'blobs', instruments onboard the ER-2 measured concentrations of species which were indicative of the polar lower-stratospheric environment. There is some debate about whether the small-scale structure in the PV maps should be believed, and Knudsen[10] shows that they can be an artefact of the analysis procedure, but what is not in dispute is that the chemical measurements indicate that polar air does intertwine with air characteristic of lower latitudes.

Other studies have also shown similar results. For example, Pyle *et al.*[11] ran a large number of trajectory calculations, calculated from ECMWF data, during January 1992, when the polar vortex became highly distorted. They showed that a large parcel of air became detached from the main vortex and moved to lower latitudes. Chemical calculations[12] indicated an ozone loss in this parcel of about 1% per day. During the same period, contour advection studies[13,14] showed that a large number of filaments of vortex air were stripped off the vortex edge. These filaments become of increasingly fine scale with time and presumably eventually mix into middle latitudes. In contrast, at least part of the large 'blob' discussed by Pyle *et al.*[11] may have eventually returned to the polar vortex. Nevertheless, this event, and indeed the several large distortions of the polar vortex during that winter, would have contributed to the observed middle-latitude ozone depletion.

The contour advection calculations[13,14] are effectively run at very high spatial resolution. Note that a very similar transfer of air from polar to middle latitudes was reproduced in a relatively low-resolution (*ca.* $6^\circ \times 6^\circ$) chemical transport model.[15]

It must be an objective of future studies to quantify the amount of air transferred out of the polar vortex into middle latitudes during the winter months. This study aims rather more simply to investigate a number of cases where observational evidence indicates that air has been stripped off the polar vortex, either as a large 'blob' or in a fine filament. It will be our object to see to what extent current numerical models can explain the observed phenomena. Without the ability to explain these features, there is no possibility of realistic prognostic modelling to study how the stratosphere of the future may evolve.

In the next section the sources of data will be introduced briefly, followed by a description of the various numerical models which will be used. Then, four particular cases will be examined, two from the northern winter in January and February 1994 and two from June and October 1994 during the southern winter.

Data Sets

During 1994 and 1995 two campaigns have been mounted to study polar processes and their impact on middle latitudes. SESAME began in January 1994 and continued until spring 1995. The campaign was designed to make measurements over a long period, so that the build up and decay of the Arctic polar vortex could be observed, and over a wide geographical area, so that a nearly hemispheric picture could be developed. Measurements from the ground, from ozone sondes, from large stratospheric balloons and from remote sensing instruments on aircraft all contribute to SESAME. Here aerosol data from the DLR lidar on board a Transall aircraft will be used.

ASHOE was designed around the high-quality *in situ* measurements that are made when the ER-2 flies in the lower stratosphere. Thus, in contrast to SESAME, ASHOE provides a more local picture, but at very high resolution. Of course, to put either data set into context requires both numerical models and the best meteorological data. The constituent data from ASHOE to be used in this study are the measurements of ClO by the Harvard University resonance fluorescence instrument.

In addition to data from these two campaigns, measurements from UARS were taken throughout this period. ClO data from the UARS microwave limb sounder (MLS) are also used here.

Models

A hierarchy of numerical models has been developed as part of the UK Univerisities global atmospheric modelling programme (UGAMP). At Cambridge University emphasis has been placed on building models to study chemistry and transport. A particularly important member of the hierarchy is TOMCAT, a chemical transport model in which the fields are taken from meteorological analyses. Thus, in the case described below, successive ECMWF analyses are incorporated into the model to give the best possible description of transport. A full stratospheric chemistry scheme, for both homogeneous and heterogeneous chemistry, is included in the model. The model can be run at a variety of horizontal and vertical resolutions. The use of meteorological analyses provides an ideal framework for comparison between model and experimental data. Earlier studies with the model have, for example, been able to explain subtle changes in the concentration of $ClONO_2$ around the edge of the polar vortex.[16]

In TOMCAT, the vertical levels in the stratosphere are pressure surfaces. Isentropic surfaces offer an advantage in studies of tracers, since on short timescales tracers will move on such surfaces. Transport across isentropic surfaces depends on the diabatic terms, radiative heating and cooling. A 3D isentropic model (SLIMCAT) has also been developed and results are presented here for the first time. In essence, the model consists of several stacked isentropic surfaces in the lower stratosphere. Horizontal transport on each isentropic surface is derived from meteorological analyses, just as with TOMCAT. Transport across the isentropic surfaces is calculated using the MIDRAD radiation scheme developed by Shine.[17] A zero-flux boundary condition is applied to the top and bottom levels of the model. The isentropic and cross-isentropic transports, calculated from the analyses and the diabatic heating, respectively, do not necessarily balance. Accordingly, mass adjustment is applied continuously to ensure that the mass of a grid box remains correct.

Finally, we have also studied the particular cases using very high resolution trajectory calculations by the method of contour advection, using code developed by Norton.[14] Data from the ECMWF are again used to derive the contours. These provide a purely meteorological picture; chemistry schemes have not been coupled into the calculations.

Results and Discussion

28 January 1994

On 28 January 1994 the German Transall aircraft flew from Kiruna in northern Sweden to southern Germany. The polar vortex was displaced towards northern Europe so that the initial part of the flight was within the vortex while the lower stratosphere above Germany was clearly outside the vortex. Measurements made with the DLR lidar at 532 nm are shown in Plate 1. The quantity plotted is backscatter ratio; lower values indicate 'clean' air with fewer scatters. Thus, soon after take-off, the lidar returns indicated a very clean lower stratosphere above *ca.* 425 K. The signal from lower potential temperature surfaces is from aerosol while the very low backscatter ratios indicate the cleaner air which has descended from higher in the stratosphere to deep inside the vortex. In contrast, over Germany the backscatter ratio remains high to potential temperatures of 650 K, consistent with a higher altitude of the stratospheric aerosol layer outside the vortex.

The most interesting features in Plate 1 occur on the vortex edge (where the potential vorticity is changing most rapidly with latitude). Leaving the vortex and moving southward, the lidar first detects the edge of vortex air, characterised by the increased backscatter ratio, at about 67°N. However, slightly further south, at 58°N, very clean air characteristic of the polar vortex is encountered in a band about one degree of latitude wide. The band of clean air is remarkable for its coherence througout a substantial depth of the stratosphere. OClO measurements made at the same time also seem to show some structure at this point (Pfeilsticker, personal communication) but a detailed interpretation of these data has not been attempted here. Finally, as the aircraft flew southward, the backscatter ratio increased again.

Notice that the clean structure is not correlated with the large scale PV values also plotted on Plate 1. Of course, the analysis procedure, which depends on a combination of new atmospheric observations being incorporated into the numerical weather-forecasting model, is unlikely to maintain even extensive features if some of the spatial

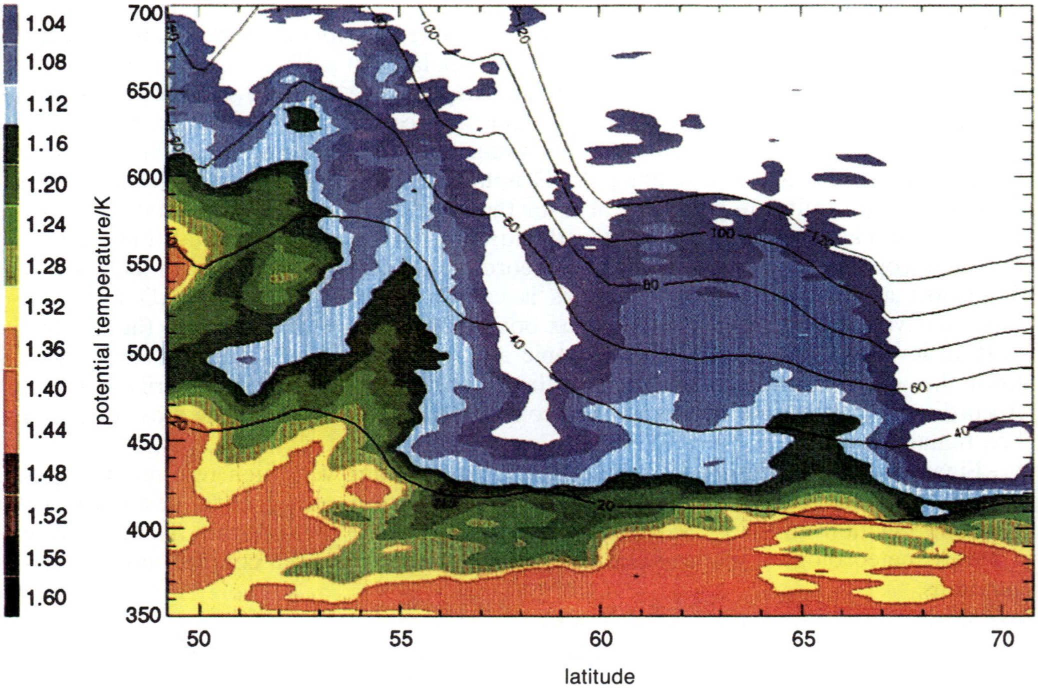

Plate 1 Backscatter ratio, indicated by the colour scale, as measured on a southbound flight of the TRANSALL on 28 January 1994. The data were collected at 532 nm by the DLR lidar. Also shown are contours of PV calculated from ECMWF analyses.

dimensions are less than either the spatial resolution of the model or of the observations. Furthermore, PV is not an advected quantity in the model (or the analysis) but is instead derived following differentiation of other meteorological fields. It is interesting, therefore, to ask if the aircraft has sampled a filament being removed from the vortex edge but not picked up in the analysis. If so, what is its structure? To address this question the SLIMCAT model was run. The model was initialised on 1 January 1994 and run on five stratospheric levels between 400 and 650 K for the entire month at a horizontal resolution of 1° × 1°. No chemistry was included but a passive N_2O-like tracer was also initialised on 1 January, using an equivalent latitude-potential temperature initialisation[18] with N_2O data taken from the Cambridge 2D model. The N_2O distribution is characterised by very low values inside the polar vortex with much higher values outside. This model tracer is therefore a good marker of the origin of air.

Plate 2 shows the calculated distribution of the N_2O tracer on 28 January. The main dark-blue region marks the polar vortex while lighter colours (higher N_2O) indicate lower latitude air of lower potential vorticity. The model clearly indicates that, on this day, northern Sweden lay beneath the polar vortex while the lower stratosphere above Germany was obviously outside the vortex. There is some very interesting structure around the vortex edge. For example, starting inside the vortex and moving southwards along the Greenwich meridian (essentially the Transall flight track), higher N_2O values are encountered at 60°N. At a slightly lower latitude the model produces a filament of vortex air with low N_2O being pulled out of the vortex. Then, moving further south, the values of N_2O increase as expected.

It seems clear that the model has reproduced the structure seen in the lidar data. Furthermore, it is evident from Plate 2 that we are seeing a filament, with the vortex (and vortex-edge) air being stretched right around the vortex and being preferentially moved to lower latitudes at about 180°E. As the filament continues to be stretched, its final fate will be to mix with middle-latitude air.

While the model seems to have captured the observed structure at 550 K, it does not reproduce the remarkable depth of the structure seen in the observations. The reason for this is not clear but it is perhaps not too surprising given that the model includes only five widely spaced levels at about the same coarse vertical resolution as the ECMWF winds. The number of levels chosen was dictated to some extent by computing time and was a compromise between very high resolution on the isentropic surfaces and much poorer vertical resolution. It would obviously be very interesting to repeat the calculation at the same horizontal resolution but with many more levels in the vertical direction.

5 February 1994

During the days following the event described above the polar vortex became highly disturbed, elongating and then, by 5 February, forming two centres, at *ca.* 90°E and 270°E. Fig. 1 shows the ECMWF potential vorticity map at 475 K showing the structure described. On the same day the UARS satellite made its yaw procedure so that the high latitude northern hemisphere was again visible to the MLS instrument for the first time in about a month. The observed ClO map on the 465 K surface is shown in Plate 3. The two-centred, dumb-bell structure evident in the PV field is mirrored in the ClO observations.[19] This is not surprising since temperatures within the polar vortex had previously fallen low enough for polar stratospheric clouds (PSCs) to form and, thus, for chlorine to be turned into its active form.

One feature in Plate 3 is surprising: the large 'blob' of elevated ClO at 55°N, 180°E. No similar feature is evident in the PV map. Again, it could be that the feature is lost during the meteorological analysis procedure. There is some evidence that the 'blob' persists for a few days, while shrinking. There are unfortunately no observations on the

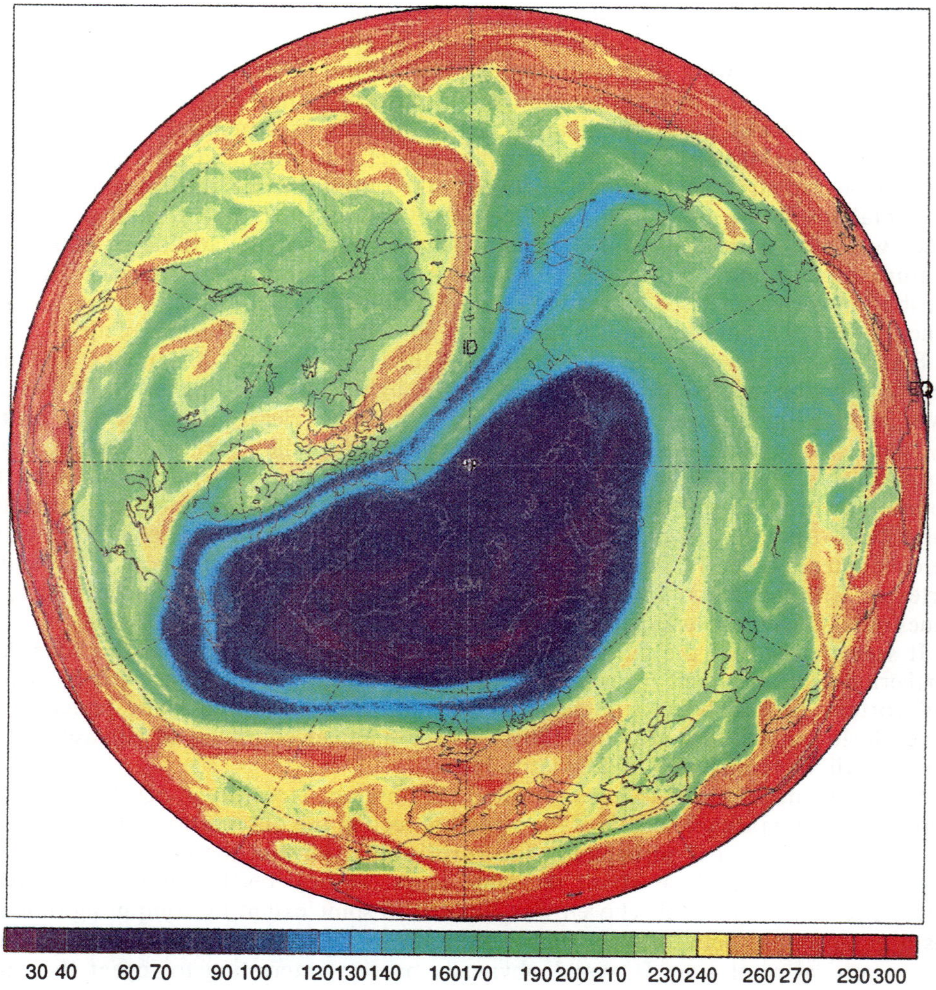

Plate 2 Calculated distribution on the 550 K surface of an inert N_2O-like tracer for 28 January 1994, from a run of the SLIMCAT model. Mixing ratio, in ppb by volume, is indicated by the colour scale.

preceding days when the MLS was looking southward. Waters *et al.*[20] have discussed the small-scale high ClO features found outside the polar vortex. They conclude that they may be unreal, with the instrument performing at its signal-to-noise limit in the middle-latitude lower stratosphere. Schoeberl *et al.*[21] studied these small features in a theoretical investigation. They too concluded that most features were unlikely to be real.

In order to investigate whether the 'blob' may be real we have carried out a number of separate modelling studies. Results from contour advection studies, from the SLIMCAT model run (as described in the previous section) and from lower-resolution studies with the chemical transport model, TOMCAT, are presented here.

A contour advection experiment was run in which PV calculated from ECMWF analyses for 25 January 1994 was advected using six hourly ECMWF wind analyses. An effective length scale of 31 km was used. The initial contours described the outline of an approximately triangular vortex edge and centre. By 2 February (experiment day eight) the vortex core had elongated and a large thick filament extended from about 65°N,

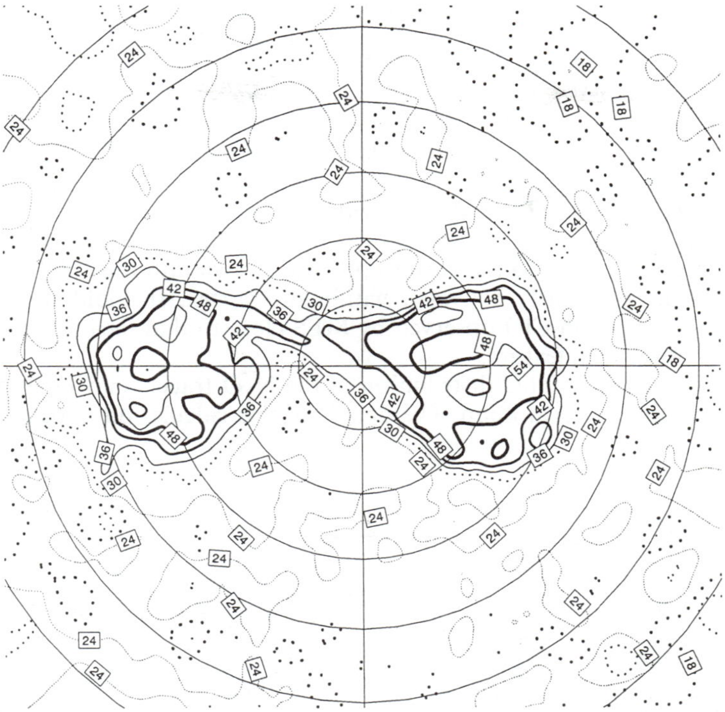

Fig. 1 Potential vorticity on the 475 K surface on 5 February 1994 calculated from ECMWF analyses

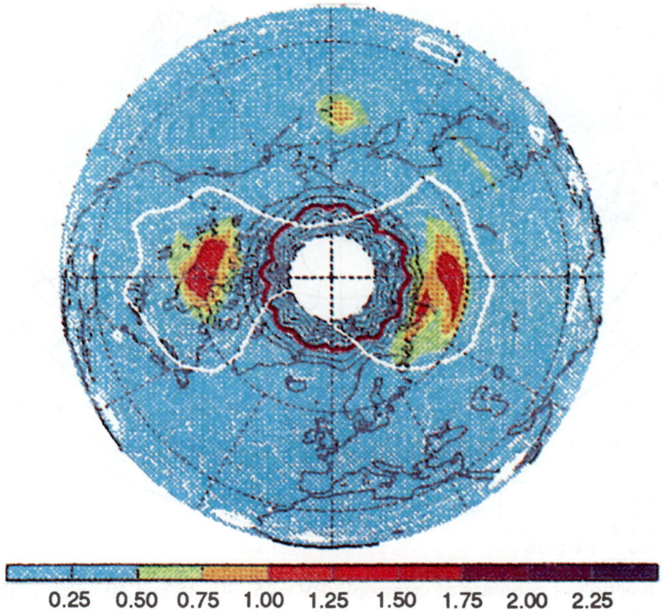

Plate 3 Northern hemisphere ClO mixing ratio at 465 K measured on 5 February 1994 by the MLS instrument on the UARS satellite. The colour scale is in ppb by volume.

120°E clockwise around two sides of the vortex. The experiment indicated that while the filament was being drawn out a large 'blob' of formerly high PV air (30–40 PV units) remained coherent within the main body of the filament. One day later both ends of the elongated vortex core had wound up to form two distinct vortices.

Plate 4 shows the final result of this contour advection experiment. The two-centred structure of the vortex is again evident while the higher resolution of the contour advection calculation clearly shows the filament extending from the western vortex centre. Over the next two days the filament split, with one end following the vortex edge while the other rolled into a 'blob' at *ca.* 50°N, 170°E as shown. Some of the air in the filament, which was formerly in the vortex core, was drawn into this 'blob' and may well have contained air previously activated within the vortex. It could well be that this would account for the high ClO observed by MLS close to that location, indicating that the observation was a real feature.

Plate 5 shows the calculated distribution of the N_2O tracer on 5 February at 00h00 GMT, corresponding to local noon. The details of this calculation appear to be qualit-

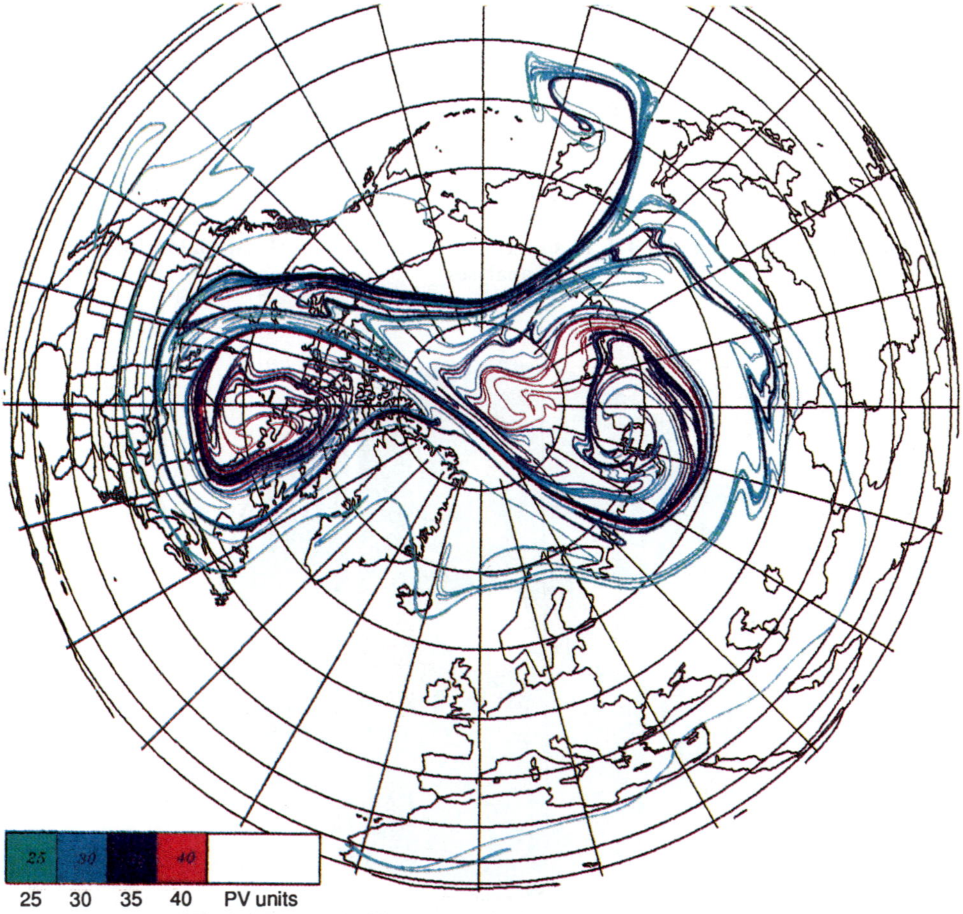

Plate 4 Result of advection of contours of isentropic PV=25, 30, 35 and 40 units on the 465 K surface, calculated from ECMWF analyses on 25 January 1994 at 00h00 GMT. Advection was carried out using 465 K isentropic winds derived from ECMWF analyses. The plot shows the result for 5 February 1994, 00h00 GMT, experiment day 11. Calculations were performed using a length scale of 31 km.

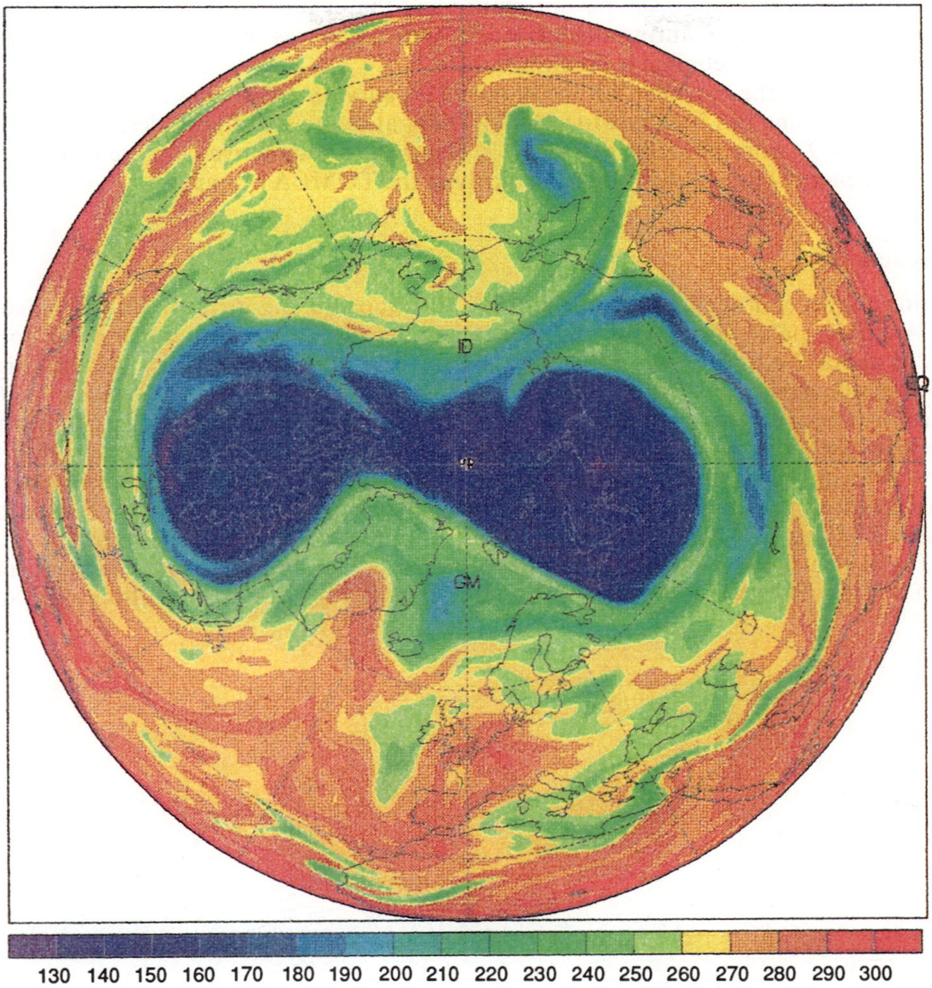

Plate 5 Calculated distribution on the 465 K surface of an inert N_2O-like tracer for 5 February 1994, from a run of the SLIMCAT model

atively very similar to the contour advection calculation. The blue region at *ca.* 50°N, 170°E is collocated with the 30 PV units contour in the contour advection calculation. The splitting of the broad filament is also well represented.

The TOMCAT model with full chemistry was run for the entire winter of 1993–94 at low horizontal resolution (6° × 6°). The model did not therefore reproduce the fine filament discussed in the previous section but, as is shown in Fig. 2 and 3, its results may be relevant to the ClO 'blob' of 5 February.

Fig. 2 shows the ClO_y levels ($Cl + ClO + 2Cl_2O_2 + HCl + ClONO_2 + HOCl$) calculated in the model. High ClO (not shown) is confined to the region of high PV within the vortex. The field of high ClO_y shows in addition a region of elevated inorganic chlorine outside the polar vortex although the initial conditions placed high ClO_y only within the vortex. Mixing into middle latitudes has occurred. The high ClO_y outside the vortex is not displaced equatorward as seen in the MLS observations, or as would be suggested by the contour advection calculations. Nevertheless, it does appear that the structures in all three calculations are similar and that a substantial volume of activated chlorine left the vortex during this period. Whether this was what the MLS instrument observed cannot be definitively confirmed.

Fig. 3 shows the modelled $ClONO_2$ on 5 February. It is clear that, in the model, the activated chlorine has been returned to $ClONO_2$ at lower sunlit latitudes, following the release of NO_2 from photolysis of HNO_3. This recovery of active chlorine to $ClONO_2$ appears generally to occur a little too quickly in the TOMCAT model.[16]

Early June 1994

During the second deployment of the ASHOE campaign, flights from New Zealand to the Antarctic polar vortex took place on 1 and 3 June. In both cases the polar vortex was at about the limit of the ER-2 duration for a return flight to New Zealand. Fig. 4 shows preliminary ClO mixing ratios as measured by the Harvard resonance fluorescence instrument. Levels of ClO were very low during the night-time flight southward. Elevated ClO was encountered when the edge of the polar vortex was reached (at 80×10^3 s in Fig. 4). Shortly afterwards, the ER-2 executed a dive when observed ClO concentrations rose even higher to about 0.4 ppb by volume. On the flight back, this time in daylight, ClO concentrations of about 50 ppt by volume were measured.

Fig. 4 also shows modelled ClO, interpolated in space and time onto the flight track from a run with the TOMCAT model, in this case at a horizontal resolution of $3° \times 3°$.

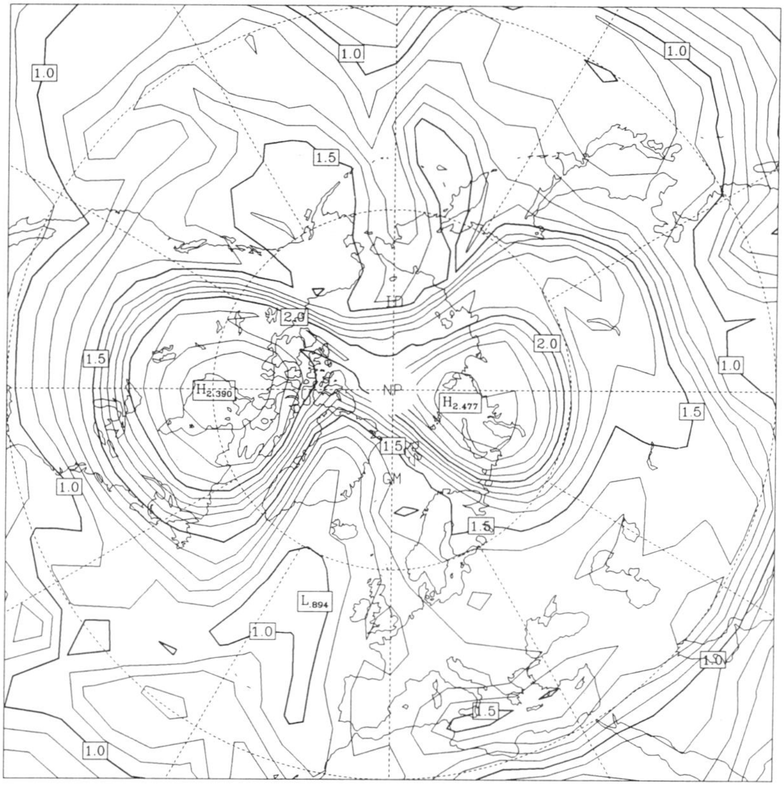

Fig. 2 ClO_y calculated on the 475 K surface, 5 February 1994, from the TOMCAT model. Contours are of volume mixing ratio in ppb, from 0.4 to 2.4 in 0.1 steps.

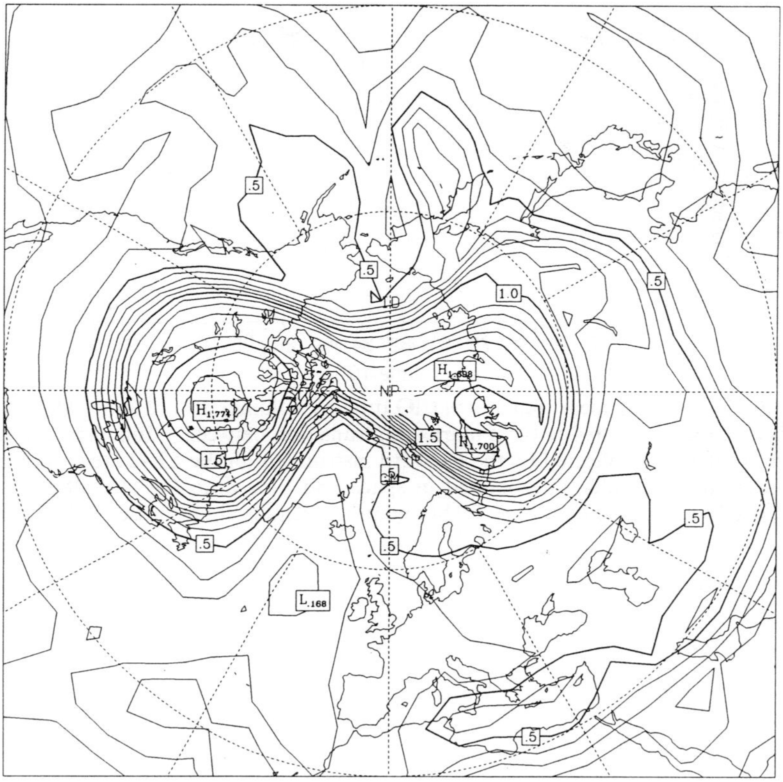

Fig. 3 As Fig. 2 for $ClONO_2$ with contours from 0.1 to 1.7

The model reproduces the main observed features, with the rise at the vortex edge modelled well. During the aircraft dive the modelled ClO stays reasonably constant and does not reproduce the observed increase. The elevated ClO levels modelled at the edge of the polar vortex are too high in comparison with the data. This highlights the problem of the coarse vertical grid. Even so the model run does show the two-peak structure during the dive, with a maximum in the ClO at the bottom of the dive. The model indicates that the maximum ClO mixing ratios occur on the 430 K surface.

Fig. 5 shows the modelled ClO field on the 400 K surface, along with the ER-2 flight track. (Note that the flights to and from the vortex are at and above 460 K). Elevated ClO is found within the vortex. A filament of activated chlorine is being stripped from the vortex edge, explaining the high ClO seen in the dive when the ER-2 descends to this altitude. Note that even a small change in the modelled location of the filament could have caused the modelled ClO in the dive to be significantly higher, indicating the difficulty of getting exact agreement between model and observations even in a model whose transport is constrained by the meterological analyses.

Two days later the model shows that the filament is still intact and at its northernmost extension has crossed above New Zealand (see Fig. 6). The filament is quite broad in the model run at this resolution. In runs, not shown, at higher resolution a much thinner filament is seen.

Fig. 7 shows the observed and modelled ClO along the aircraft flight track for the flight of 3–4 June 1994. Again, the model reproduces the increase across the vortex edge. In this case when the aircraft dives (at about 87×10^3 s flight duration) very low ClO is

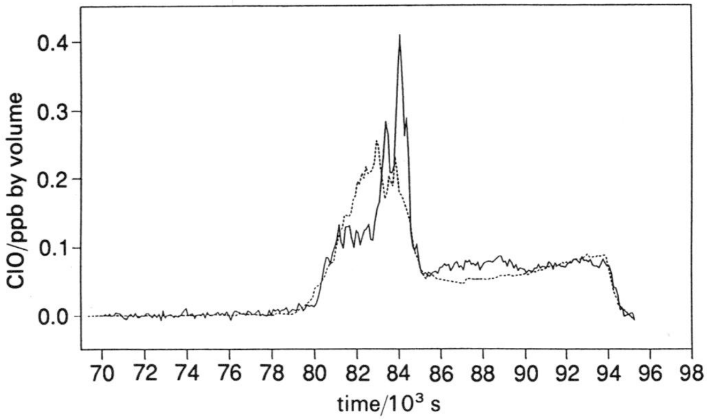

Fig. 4 Modelled (– – –) and measured (——) ClO (ppb by volume) on 1–2 June 1994. The horizontal axis indicates time in s. 86.4×10^3 s marks the transition between 1 June and 2 June. Preliminary analyses of the data collected by the Harvard instrument on the ER-2 are shown by the solid line. TOMCAT model results, interpolated in space and time onto the aircraft flight track, are shown by the dashed line. The flight and model details are discussed in the text.

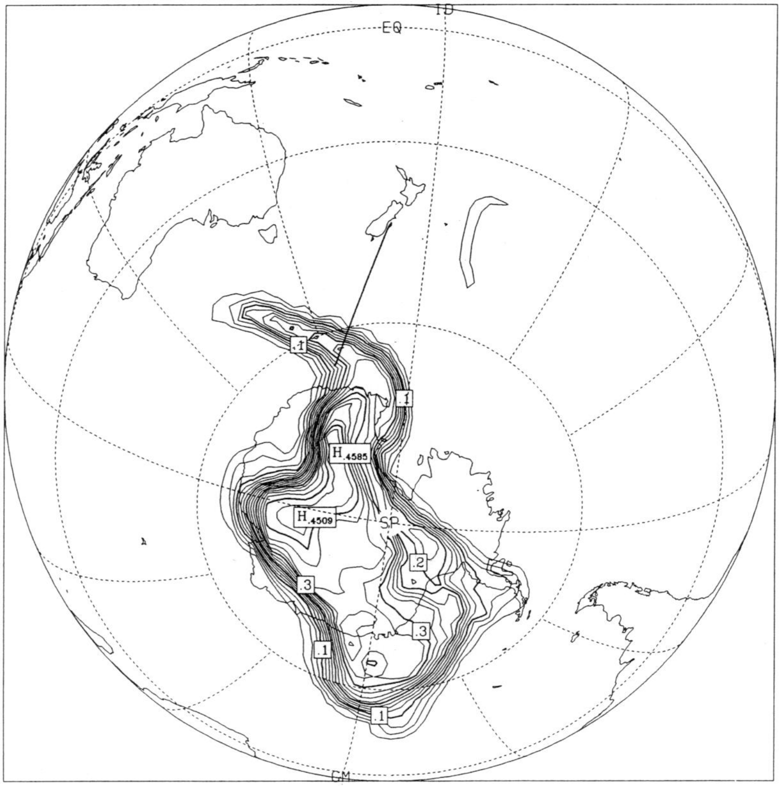

Fig. 5 Modelled ClO mixing ratio (ppb by volume) on the 400 K surface at 00h00 GMT, 2 June 1994. Contours are from 0.025 to 0.45 in steps of 0.025.

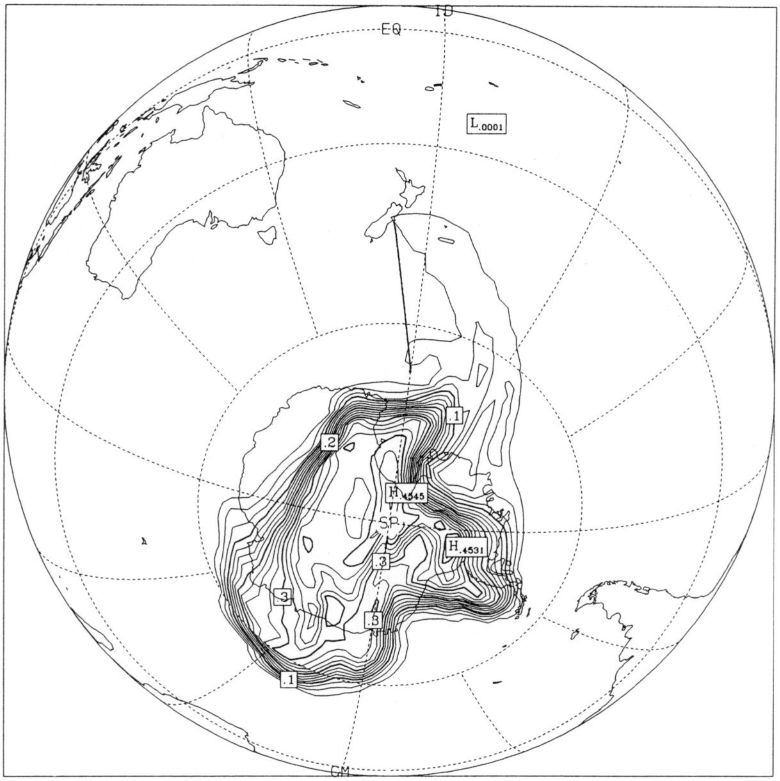

Fig. 6 As Fig. 5, for 4 June 1994

found. Note in Fig. 6 that the dive to 400 K occurs well outside the vortex at that level. The particularly interesting feature of this flight is the very high ClO measured soon after take-off. The model does not reproduce this but Fig. 6 indicates that the early part of the flight possibly passed through the activated filament which had been seen closer to the vortex edge a few days earlier. Given that the model has a tendency to return activated chlorine to $ClONO_2$ too quickly (see above), and that the modelled filament over New Zealand contains elevated $ClONO_2$, it does appear that the model can qualit-

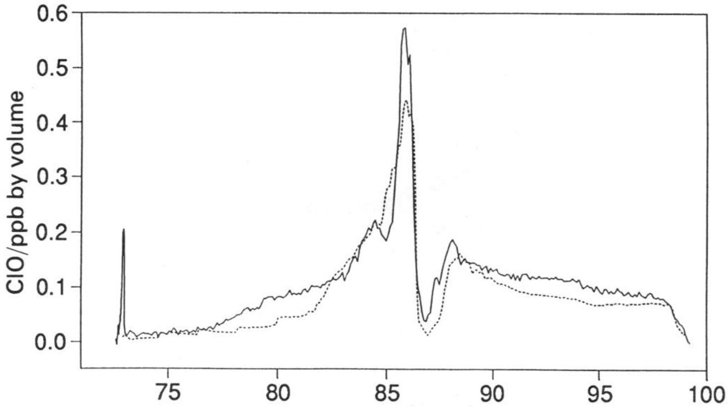

Fig. 7 As Fig. 4, for 3–4 June 1994

atively explain the high ClO observed: the ER-2 at take-off has sampled air of recent vortex origin.

To confirm this, a contour surrounding a parcel of air of 85 km radius was run backwards on the 375 K surface from the location at which the high ClO was measured. This was driven by winds derived from ECMWF analyses. Fig. 8 shows the calculated path of the air parcel. It shows that the air intersected by the ER-2 flightpath was drawn from the edge of the vortex during 30–31 May. While the air in this small parcel was not directly exposed to temperatures likely to form PSCs, a larger parcel (650 km radius) run backward from the same location was exposed to PSC-forming temperatures and underwent considerable stirring. A forward advection of vortex centre PV contours (inset Fig. 8) also places vortex centre air close to the measurement site. The calculations show that air from both the vortex edge and the vortex core was placed near to the ER-2 measurement close to New Zealand on 3 June.

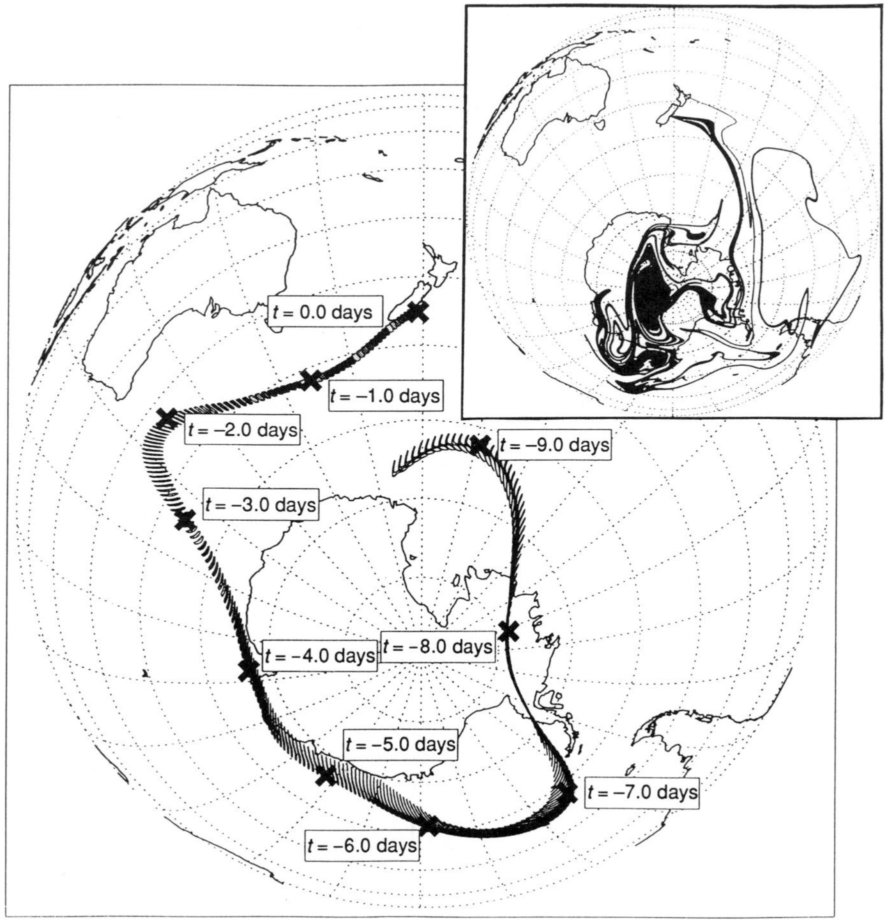

Fig. 8 Backward isentropic advection of contour surrounding an 85 km radius air parcel centred on 45.4°S, 173.1°E starting at 20h28 GMT June 3 1994 at 375 K. Advected using isentropic winds derived from ECMWF analyses. Timestep between points is 1 h. Inset shows the calculated position of contours of vortex centre PV on 00h00 25 May advected forward to 00h00 GMT June 4 1994 on the 375 K surface using similar winds.

11 October 1994

A final example of filamentary structure is taken from the fourth ASHOE deployment in October 1994. The modelled and observed flight tracks are shown in Fig. 9. A feature of particular interest occurs at about 81×10^3 s, when a sharp, elevated peak in ClO was measured. This particular peak was measured during level flight at about 480 K. This period of the campaign was modelled using SLIMCAT run on 5 vertical levels between 375 and 550 K. The peak is seen in SLIMCAT, but even at a resolution of $1.5° \times 1.5°$, it is much broader than observed and lower in magnitude. Fig. 10 shows the modelled ClO field for the same day on the 475 K surface. A filament stretching towards New Zealand is a prominent feature. The model calculation shows that this filament is present on all levels from 375 to 475 K, sloping steeply with altitude. The smaller peak in the observed ClO at about 77×10^3 s in Fig. 9 appears to be the same filament as is shown in Fig. 10 on the 475 surface, but on the 400 K surface. Again, the model does not reproduce the magnitude of the peak in the ClO, although evidence for the peak is seen in the modelled HCl and $ClONO_2$ (not shown). Once more the model can only qualitatively explain the high early morning measurements of ClO.

Conclusions

A combination of numerical model calculations and data from two field campaigns, SESAME and ASHOE, has been used to explore cases when filaments of air from the vortex and the vortex edge have been moved to lower latitudes. A case in the northern hemisphere winter was studied at very high resolution with a model including a passive tracer. The observations and model agreed extremely well on the 550 K surface, but less well on other levels. It is clear that the airborne lidar did measure a filament whose horizontal extent was, according to the model, large and which certainly led to mixing with lower latitude air.

In the southern hemisphere, two examples were given where, again, the combination of data and model indicated that air, sampled in middle latitudes, had previously been activated in polar latitudes. This air must have contributed to middle-latitude ozone loss.

A further case was considered in the northern hemisphere when, during a period in which the polar vortex became particularly disturbed, a region of high ClO was measured well outside the vortex by the MLS instrument on UARS. Model calculations suggested that a 'blob' of activated air did indeed leave the vortex although the posi-

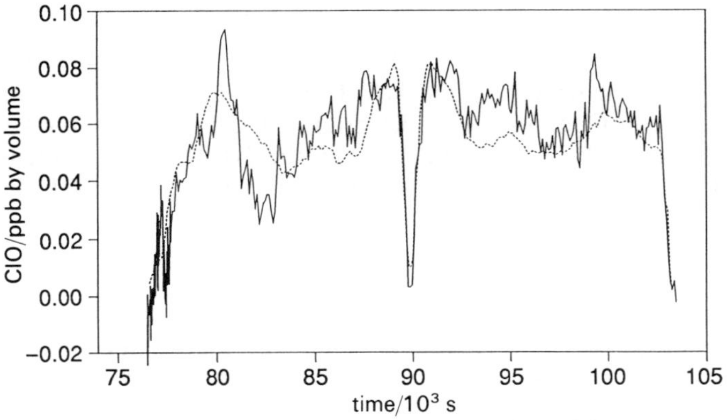

Fig. 9 As Fig. 4, for 10–11 October 1994

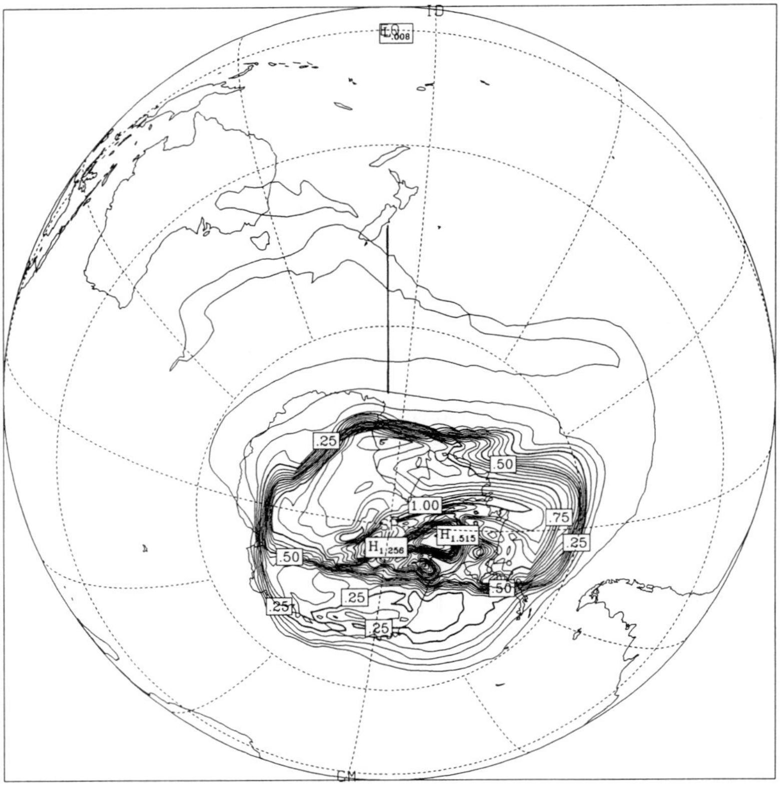

Fig. 10 As Fig. 5, for 11 October 1994 and for the 475 K surface, with contours from 0.05 to 1.5 in 0.05 steps

tions of the observed and modelled features differed a little. Furthermore, in this case, the 'blob' seems to have resulted when a filament, detached from the vortex, wound around itself to reform a large feature.

In this paper, the existence of filamentary structures, seen in earlier model calculations, has been demonstrated by comparison between observations and models. It is clear that at least some of these filaments must mix into middle-latitude air and contribute to the ozone decline recorded there. The next task will be to quantify the importance of these events.

This work forms part of NERC's UGAMP programme. We are happy to acknowledge the additional support provided by the UK Department of the Environment and by DGXII of the European Commission under grant CT93-0351. I.K-D. was supported by an HCMP fellowship of the EC. A.M.L. thanks the NERC for a studentship. We are indebted to Dr Paul Berrisford for his assistance in accessing ECMWF data. We thank ECMWF for the use of their data.

References

1 J. C. Farman, B. G. Gardiner and J. D. Shanklin, *Nature* (*London*), 1995, **315**, 207.
2 World Meteorological Organisation (WMO), *Scientific Assessment of Atmospheric Ozone*, 1995, **37**, in the press.
3 R. S. Stolarski, P. Bloomfield, R. D. McPeters and J. R. Herman, *Geophys. Res. Lett.*, 1991, **18**, 1015.

4 J. F. Gleason, P. K. Bhartia, J. R. Herman, R. D. McPeters, P. Newman, R. S. Stolarski, L. Flynn, G. Labow, D. Larko, C. Seftor, C. Wellemeyer, W. D. Komhyr, A. J. Miller and W. Planet, *Science*, 1993, **260**, 523.
5 SORG, *Stratospheric Ozone 1993*, 1993.
6 A. F. Tuck, T. Davies, S. J. Hovde, A. Noguer-Alba, D. W. Fahey, S. R. Kawa, K. K. Kelly, D. M. Murphy, M. R. Proffitt, J. J. Margitan, M. Loewenstein, J. R. Podolske, S. E. Strahan and C. R. Chan, *J. Geophys. Res.*, 1992, **97**(D8), 7883.
7 A. F. Tuck, J. M. Russell and J. E. Harries, *Geophys. Res. Lett.*, 1993, **20**, 1227.
8 D. W. Fahey, S. R. Kawa, E. L. Woodbridge, P. Tin, J. C. Wilson, H. H. Jonsson, J. E. Dye, D. Baumgarten, S. Borrmann, D. W. Toohey, L. M. Avallone, M. H. Proffit, J. J. Margitan, M. Loewenstein, J. R. Podolske, R. J. Salawitch, S. C. Wofsy, M. K. W. Ko, D. E. Anderson, M. R. Schoeberl and K. R. Chan, *Nature (London)*, 1993, **363**, 509.
9 D. Perner, A. Roth and T. Klüpfel, *Geophys. Res. Lett.*, 1994, **21**, 1367.
10 B. M. Knudsen, PhD Thesis, University of Copenhagen, 1994.
11 J. A. Pyle, G. D. Carver and U. Schmidt, *Geophys. Res. Lett.*, 1994, **21**, 1431.
12 L. Grenfell, PhD Thesis, University of Cambridge, 1994.
13 D. W. Waugh and R. A. Plumb, *J. Atmos. Sci.*, 1994, **51**, 530.
14 W. A. Norton, *J. Atmos. Sci.*, 1994, **51**, 654.
15 M. P. Chipperfield, D. Cariolle and P. Simon, *Geophys. Res. Lett.*, 1994, **21**, 1467.
16 M. P. Chipperfield, J. A. Pyle, C. E. Blom, N. Glatthor, M. Höpfner, T. Gulde, Ch. Piesch and P. Simon, *J. Geophys. Res.*, in the press.
17 K. P. Shine, *Quart. J. Roy. Met. Soc.*, 1987, **113**, 603.
18 D. J. Lary, M. P. Chipperfield, J. A. Pyle, W. A. Norton and L. P. Riishojgaard, *Quart. J. Roy. Met. Soc.*, 1995, **121**, 187.
19 J. W. Waters, G. L. Manney, W. G. Read, L. Froidevaux, D. A. Flower and R. F. Jarnot, *Geophys. Res. Lett.*, 1995, in the press.
20 J. W. Waters, L. Froidevaux., W. G. Read, G. L. Manney, L. S. Elson, D. A. Flower, R. F. Jarnot and R. S. Harwood, *Nature (London)*, 1993, **362**, 597.
21 M. R. Schoeberl, R. S. Stolarski, A. R. Douglass, P. A. Newman, L. R. Lait, J. W. Waters, L. Froidevaux and W. G. Ready, *Geophys. Res. Lett.*, 1993, **20**, 2861.

Paper 5/00298B; *Received* 17*th January*, 1995

Faraday Discuss., 1995, **100**, 389–410

Time and Temperature Dependences of Fractional HCl Abundances from Airborne Data in the Southern Hemisphere during 1994

Adrian F. Tuck, Christopher R. Webster, Randy D. May, David C. Scott, Susan J. Hovde, James W. Elkins and K. Roland Chan
NOAA Aeronomy Laboratory, 325 Broadway, Boulder, CO, 80303-3328, USA

Measurements of HCl and CH_4 taken by the aircraft laser infrared absorption spectrometer (ALIAS) on the ER-2 high-altitude research aircraft during the Southern Hemisphere winter of 1994 have been used to examine the abundance of HCl as a fraction of total inorganic chlorine. The fractional abundance of HCl shows a threshold behaviour as a function of temperature history; on a 10 day timescale, the abundance dropped sharply in those air parcels experiencing a temperature <195 K, but little or no change was seen in parcels which stayed warmer than this temperature. The behaviour mirrors well the temperature behaviour calculated for the transformation of HCl into reactive forms (Cl_2, HOCl) from laboratory studies of sulfate aerosols and polar stratospheric clouds. During the course of the winter, the fractional abundance of HCl outside the vortex decreased from its values in late May by about a third, while inside it dropped to near zero by early August. Some recovery was evident in October. Examples of the peel-off of low-HCl air equatorward of the wind maximum were evident in early June. Meteorological trajectories are used to show, in a case study of a flight in early August, that air parcels which experienced temperatures of <195 K, and as a result had low fractional HCl abundances, did so largely poleward of the maximum in the polar night jet stream. Encountering temperatures of <195 K during the previous 10 days was a necessary and sufficient condition for the transformation of HCl into reactive forms by heterogeneous reactions. The trajectories further showed that air arriving from sub-tropical latitudes had higher fractional HCl abundances than the air in the middle latitudes, and much higher fractions than the air at high latitudes. The resulting picture is one in which the fractional abundance of HCl in air at mid latitudes was the result of mixing of air from sub-tropical latitudes with air mainly from polward of the jet stream core which has experienced temperatures <195 K. The sensitivity of the fractional abundance of HCl to the assumption that no HCl enters the stratosphere *via* the tropical tropopause is examined in the light of an observed profile near the equator with a volume fraction of 0.4 ppb HCl, zero ClO and tropospheric mixing ratios of CFCs at the tropical tropopause.

Ozone losses since 1979 have averaged about 1% of the overhead column per year in mid latitudes, with the largest effect in winter.[1] Height-resolved observations demonstrate that this loss is primarily located in the lower stratosphere,[2] below *ca.* 20 km. Mechanisms which have been proposed to explain this include the conversion of HCl and $ClONO_2$ to Cl_2 and HOCl and then to Cl, ClO and its dimer on *in situ* mid-latitude sulfuric acid aerosol,[3–5] a mechanism which might be expected to accelerate in the presence of volcanic emissions.[6,7] Another mechanism is the fluid mechanical transport equatorward of air exposed to temperatures below 195 K in the lower stratospheric winter vortex,[8–10] where polar stratospheric cloud[11–14] or ternary super-cooled aerosol droplets of sulfuric acid, nitric acid and water[15–17] transform HCl and $ClONO_2$ into Cl_2 and thence Cl, ClO and Cl_2O_2.

The ER-2 was equipped with a number of instruments which made measurements bearing directly upon these above questions, including those of ClO,[18] the organic chlorine budget,[19] aerosol particle-size distributions[20,21] and HCl.[22] In this work, we focus on the measurements of HCl made by the ALIAS; results from this instrument have been used to examine the 1992–1993 Northern Hemisphere chlorine partitioning in the presence of significant amounts of volcanic aerosol, by Webster *et al.*[23,24] A significant discrepancy existed between the HCl amount calculated from the photochemical balance between ClO and NO_2 measurements for an observationally constrained[25] total amount of inorganic chlorine, Cl_y, and that measured directly. Here we report an analysis of the temperature and time dependence of the HCl abundance as a fraction of the total inorganic chlorine during the Southern Hemisphere winter of 1994 (March–November); the observations were made from the NASA ER-2 high-altitude research aircraft during the airborne Southern Hemisphere ozone experiment/measurements for the assessment of the effects of stratospheric aircraft (ASHOE/MAESA). We note that in an analysis of the temperature distributions in the Southern Hemisphere lower stratosphere during the winter of 1987, the year of the airborne antarctic ozone experiment (AAOE), there was a significant frequency of occurrence of air with temperatures <196 K equatorward of the wind maximum, *i.e.* outside the dynamically defined vortex.[26] A further important point is that during ASHOE/MAESA the abundance of aerosol was generally at one to two times that measured from the ER-2 before the eruption of Mount Pinatubo in June 1991, or at least an order of magnitude less than that encountered during the Northern Hemisphere missions of late 1991 through to spring 1993 (J. C. Wilson, H. Jonsson and C. A. Brock, personal communication).

Method

The formulae of Hanson and Ravishankara[5] were used, with typical measured values of HCl mixing ratio, water vapour mixing ratio, aerosol droplet radius, pressure and temperature measured from the ER-2 in austral mid-winter 1994, to calculate the values of the reactive uptake coefficients, γ, for the following reactions on sulfuric acid aerosol:

$$ClONO_2 + H_2O \rightarrow HOCl + HNO_3 \quad (1)$$

$$ClONO_2 + HCl \rightarrow Cl_2 + HNO_3 \quad (2)$$

$$HOCl + HCl \rightarrow Cl_2 + H_2O \quad (3)$$

The results are plotted in Fig. 1; note that the values of the observed variables listed above are given in the figure. The values of γ for reaction (1) are considerably less than those for reactions (2) and (3), and contribute very little to the curve of the sum $\gamma_t = (\gamma_1 + \gamma_2)$. Since (1) is faster than any gas-phase source of HOCl at ER-2 altitudes and must therefore precede (3), the indication is that reaction (2) will be the most effective. We seek to compare the observed fractional HCl abundance, H, where

$$H = HCl/Cl_y,$$

with γ_t as a function of temperature and time.

The electron capture gas chromatographic instrument from whose observations the total organic chlorine budget was available[19,25] flew only on the full payload flights; along with the HO_x, ClO/BrO and NO_y instruments it did not accompany ALIAS on the dynamics and radiation flights. To maximize the number of flights which can be

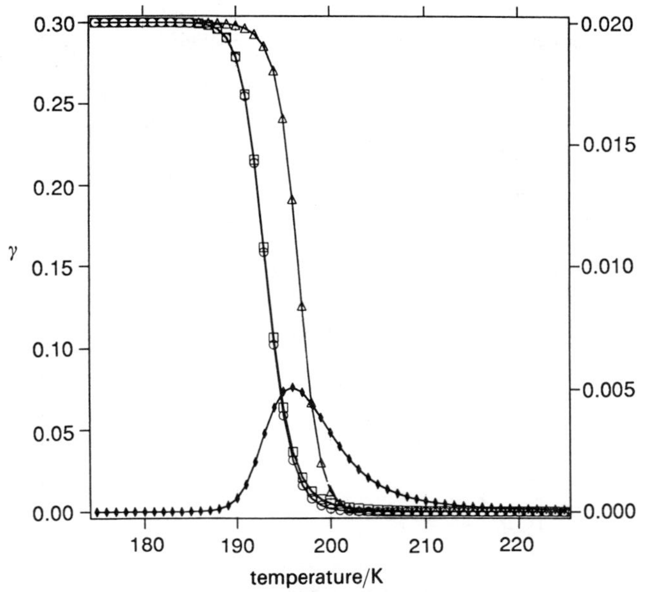

Fig. 1 Curves of γ, the coefficient for reactions (1) and (2) on sulfuric acid aerosol, using the formulae in ref. 3 and 5. The conditions are representative of those experienced along the ER-2 flight tracks in mid-winter as regards pressure, water vapour mixing ratio and aerosol particle radius. (◆), reaction (1); (○), reaction (2); (△) reaction (3). Reaction (1) plus reaction (3) is equivalent to reaction (2), but reaction (1) is rate-limiting in the sequence (1) followed by (3). The solid curve with open squares is the sum of (1) and (2), and its values and those of (3) are on the left ordinate. The right ordinate applies to (1).

used in this analysis, we employ a proxy for the total inorganic chlorine, Cl_y: the quantity $(1.7 - CH_4)$ ppm by volume. The CH_4 is measured by ALIAS, in some cases using the same laser and optics as is used to measure the HCl; the value 1.7 is the methane mixing ratio measured in the troposphere and at the upper tropical tropopause. The value of $(1.7 - CH_4)$ is the amount of methane lost from the observed air since it entered the stratosphere; in the ER-2 envelope, it is linearly related to Cl_y (itself obtained[25] by subtracting the observed amount of organic chlorine, CCl_y, from that in the air at the tropical tropopause upon entry to the stratosphere). Fig. 2 shows the result of regressing $(1.7 - CH_4)$ against Cl_y for a particular flight, that of 940806 (dates are in yymmdd format: read 6 August, 1994). We shall define $H_p = HCl/(1.7 - CH_4)$.

In analysing the observations, we shall not be concerned solely with the temperature measured along the ER-2 flight track, but with the temperature history of that air. This may be calculated with useful accuracy along Lagrangian back trajectories for up to 10 days.[27–29] Here, we employ three-dimensional (3D) trajectory calculations using the archived velocity components from the European centre for medium-range weather forecasts (ECMWF), in a calculation technique based on that of Kållberg,[30] which we have used previously[10] in the Northern Hemisphere. The meteorological data were obtained from ECMWF on a grid of 1.0° latitude × 1.0° longitude, on pressure levels of 150, 100, 70, 50, 30 and 10 hPa. One quantity obtained for each trajectory which is used here is minimum temperature of the air parcel during the last ten days, T_{min}. Also calculated from the ECMWF analyses, which were available every 12 h, at 00 and 12 h universal time convention (UTC), is the distribution of potential vorticity, PV.[10] Maps of PV can be used as a proxy for tracer distributions; it is calculable from routine meteorological analyses and is useful in the larger scale characterization of the vortex on any particular day. The ER-2 was limited to eight-hour flights, giving a highest attainable latitude

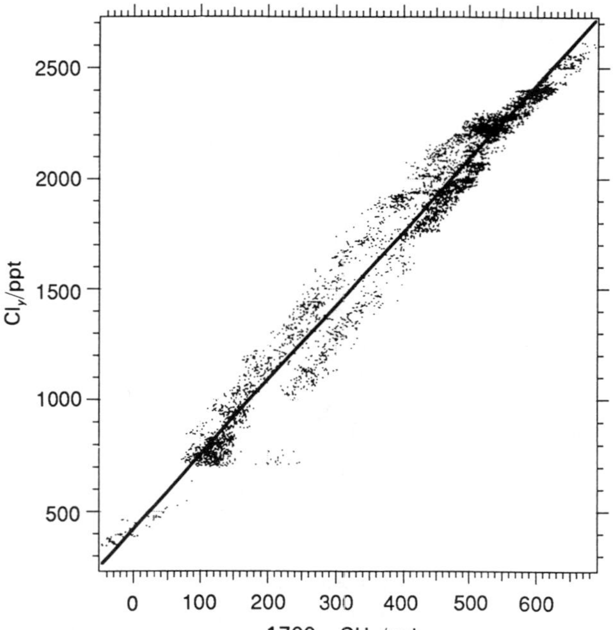

Fig. 2 Total inorganic chlorine, Cl_y, calculated from the formula of Woodbridge[25] regressed against (1.7 − CH_4) for the data of 940806.

of 70°S when flying due south; in practice this was achieved only twice as a result of a combination of cross winds, temperatures and planned deviations from due south arising from the forecast shape of the vortex. The usual flight profile was to fly an approximately isentropic leg southbound, turn, descent from *ca.* 19 to 15 km, ascent back to altitude within 2 to 3 degrees of latitude, and continue to climb northward, reaching *ca.* 20 km before descent into Christchurch. The objective of the isentropic leg

Table 1 Southbound ER-2 flights used in this paper

flight date (yymmdd)	maximum latitude (°S)	latitude of maximum wind (°S)	min. *T* near southern end /K
940413	68	50	209
940524	68	66	220
940601	65	62.5	206
940603	67	66	198
940605	69	69	197
940728	68	66	190
940730	68	67	191
940806	68	64	201
940808	68.5	57	200
940810	67.5	65.5	203
941005	65.5	64	202
941010	70	69	213
941013	68.5	66	217
941016	69.5	69	231
941020	64	63	218

was to search for filaments of vortex or tropical air, whose interpretation is simplified since air moves approximately adiabatically on a timescale of a few days in the lower stratosphere. The attainable value of the potential temperature, θ, where

$$\theta = T(1000/P)^{(\gamma - 1)/\gamma}$$

and T is temperature in Kelvin, P is pressure in hPa and γ is the ratio of specific heats, C_p/C_v, of air, varied from flight to flight but was in the range 450 to 520 K during cruise, and reached 380 K at 15 km. The entropy of an air parcel is proportional to $C_p \ln\theta$.

Table 1 lists the flights used here, together with the latitude of the wind maximum, the minimum temperature and the maximum latitude.

Errors for the ALIAS instrument are quoted at the 2σ level as 15% for HCl and 10% for CH_4. The random error is dependent upon the detected abundance, but is in all cases much smaller than the total error limits given above. There is a small number of flights on which fringing in the HCl channel resulted in error bars considerably larger than those quoted; examples are the flights on 940413, 940728, 940730 and 941029. The preliminary error estimates in volume fraction on a flight-by-flight basis are as follows: HCl, flights 940413 to 940605, ± 0.15 ppb; flights 940728, 940730 and 941029, ± 0.2 ppb; flights 940801 to 941022, ± 0.1 ppb; CH_4, for all flights used, 6% to 10%.

Results

The course of the minimum temperatures, and the geographical distribution of temperatures < 195 K during the austral winter of 1994 are central to the investigation of the loss of HCl as a result of heterogeneous reactions on sulfuric acid aerosol or on polar stratospheric clouds (PSCs), as is evident from Fig. 1 and ref. 3–16. The history of the minimum temperatures at 50 hPa poleward of 20°S is shown in Fig. 3 for the four periods of ER-2 flights and the 10 days preceding them. Note that in some instances PSCs have occurred where predicted assuming a composition of nitric acid trihydrate,[15] but in others substantial supersaturations have been observed.[31] Fig. 3(*a*) shows that at 50 hPa, for example, the minimum temperature dropped below the PSC type I threshold, 195 K, in early May and remained there until mid-October. Inspection of the Meteorological Office 50 hPa data for each day from 940401 to 941031 (not shown) confirms that this was also so for the fallow periods between phases 2, 3 and 4. The location of these minimum temperatures was poleward of the maximum wind speed except for episodic occurrences of a few per cent, *i.e.* they were largely inside the vortex. The distribution of the contours of the 195 K isotherm with respect to the wind maximum is also important;[25] it determines how much, if any, chlorine can be converted by heterogeneous chemistry outside the vortex. Fig. 3(*b*) shows the percentage of the area bounded by the 195 K isotherm that was equatorward of the wind speed maximum, plotted every 12 h from 940401 to 941031 for the four phases and the 10 days preceding them. During phase 1 there was no air cooler than 195 K; during phase 4 all such air was inside the vortex. While nearly all the air cooler than 195 K was poleward of the jet-stream core, the 195 K contour was episodically equatorward of the wind maximum at some longitudes during phases 2 and 3, to the extent shown in Fig. 3(*b*). Such proximity put air in which chlorine had probably been activated within the $\pm 5°$ latitude wide belt centred on the wind maximum where it could mix, by fluid mechanical filamentation ('chaotic advection'), across the jet stream.[26–29,32] Note, however, that occurrences of air with temperatures < 195 K equatorward of the jet-stream maximum in 1994 were fewer[26] than during 1987, the previous year with an airborne experiment in the Southern Hemisphere.

There were four phases during the 1994 austral winter when the ER-2 made flights toward Antarctica from Christchurch (43.5°S, 172.5°E). The eight-hour flights during which the ALIAS instrument measured HCl are: 940413, 940524, 940528, 940601,

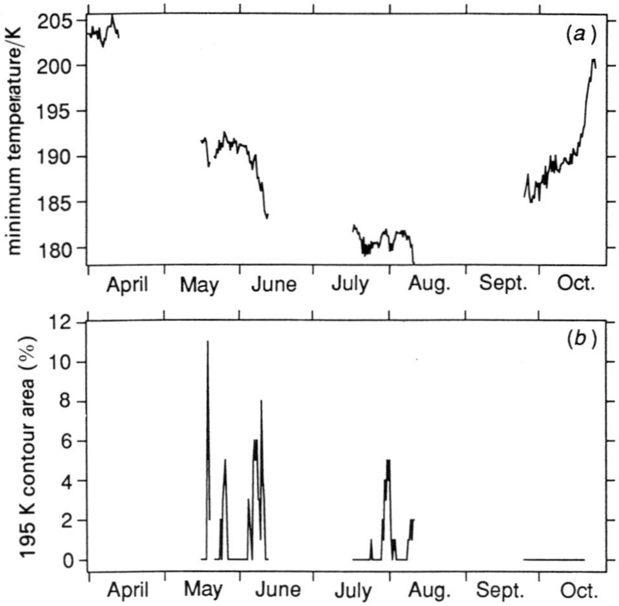

Fig. 3 (*a*) Minimum temperatures poleward of 20°S at 50 hPa from ECMWF analyses, 940401–941031. (*b*) Percentage area of that bounded by the 195 K contour which lies equatorward of the wind speed maximum, *i.e.* outside the vortex. Note that the data cover only the flight phases and the 10 days prior. In (*a*), the temperatures dropped below 195 K in very early May. In (*b*), note that there were no temperatures below 195 K in phase 1, and in phase 4 all such temperatures were poleward of the wind maximum.

940603, 940605, 940728, 940730, 940801, 940806, 940808, 940810, 941005, 941010, 941013, 941016 and 941020. We will discuss the results from each phase in turn. The results will be presented as plots of H_p, the proxy fractional abundance of HCl, as a function of latitude along the approximately isentropic leg of each flight, with the points inside the vortex (poleward of the wind maximum) distinguished from those outside the vortex.

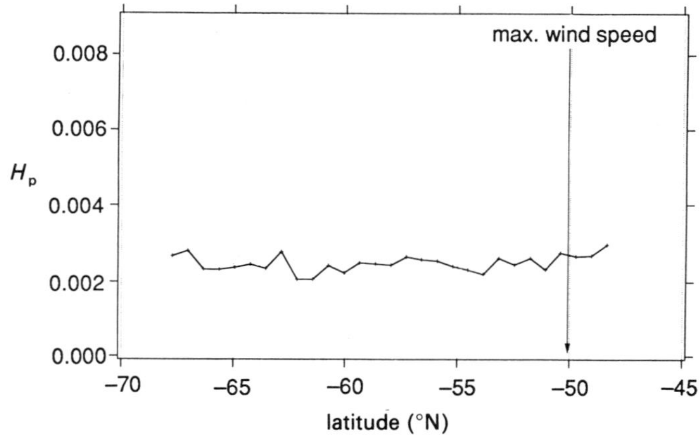

Fig. 4 H_p *vs.* latitude, southbound leg, 940413. Note that the wind speed maximum was at 50°S (Table 1).

Fig. 4 shows the values of H_p in this manner for the one available flight during phase 1, 940413. There is no significant difference between the air inside and outside; the mean value of H_p is close to 2.5×10^{-3}. The constancy is consistent with the absence of temperatures lower than 195 K anywhere south of 20°S by the flight data [Fig. 3(*a*)].

There were four flights during phase 2 which produced HCl measurements poleward of New Zealand, see Table 1. The first flight, on 940524, shows values of H_p ranging from $(3–4) \times 10^{-3}$ with little or no depression in the values poleward of the wind speed maximum, but the remaining three, 940601, 940603 and 940605, show a clear drop in the value of H_p not only inside the vortex but also outside it (Fig. 5). The values outside the vortex with low H_p are correlated with low values of tracer; this is demonstrated by

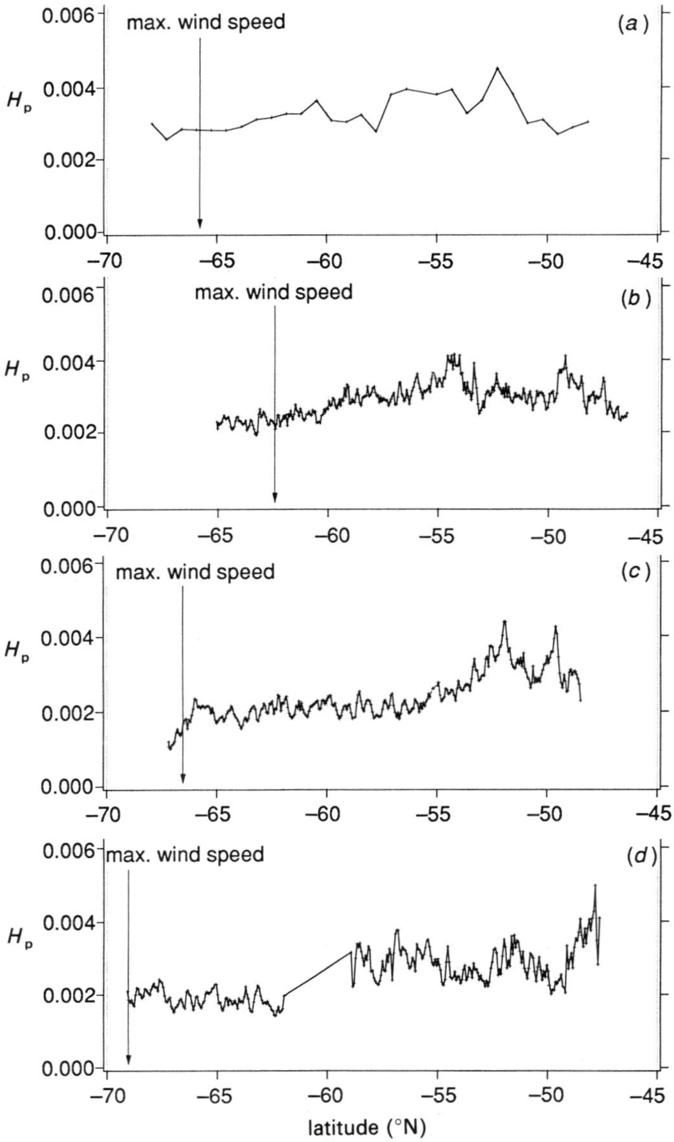

Fig. 5 H_p *vs.* latitude, southbound legs; (*a*) 940524, (*b*) 940601, (*c*) 940603 and (*d*) 940605. Note the drop in H_p equatorward of the wind maximum between 940524 and 940601.

comparing Fig. 5(*c*) with Fig. 6, for 940603, in which the CH_4, HCl and windspeed traces are shown. The stretches of flight track with low H_p equatorward of the wind maximum have low CH_4, characteristic of vortex air originating from the poleward side of the wind maximum. Note that such air is found between 63°S and 65.5°S, between 62°S and 56°S, and in a diluted form at 51°S and equatorward of 48°S. The spikes of high H_p, exceeding 4×10^{-3} at 49°S and 52°S, are correlated with high CH_4 values

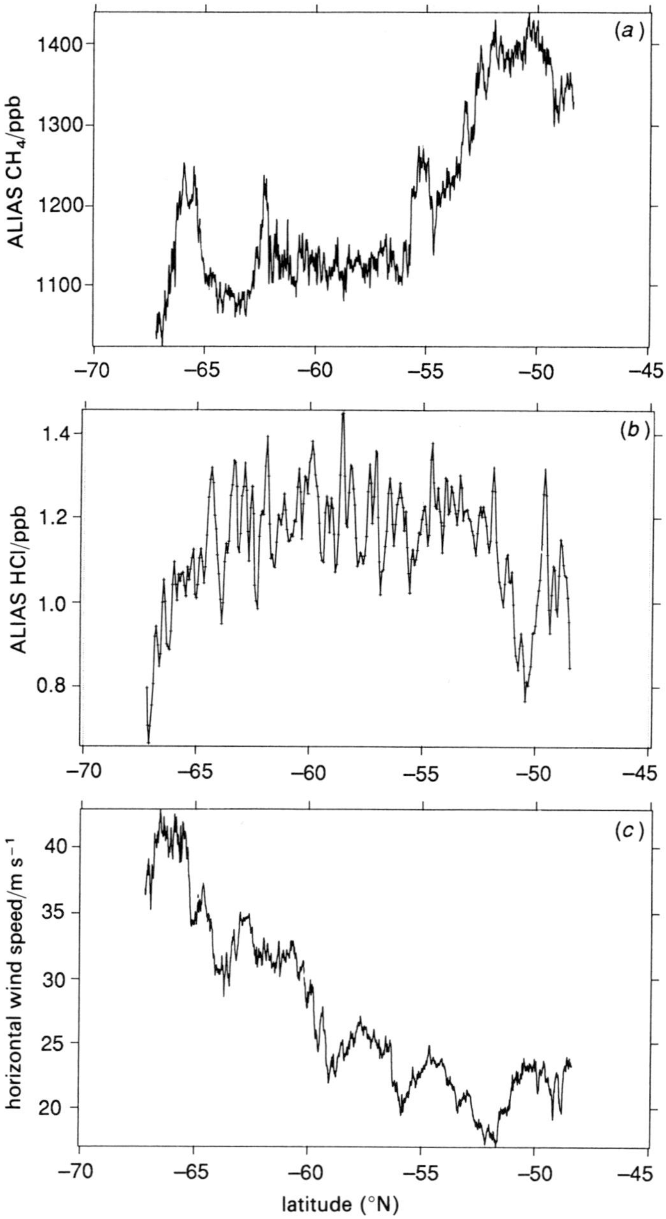

Fig. 6 (*a*) CH_4, (*b*) HCl and (*c*) wind speed *vs.* latitude, southbound leg, 940603. Note the low CH_4 air, characteristic of the poleward side of the jet stream, which has peeled-off into mid-latitudes, bringing with it low values of H_p [Fig. 5(*c*)].

characteristic of lower latitudes. There is a suggestion of slightly higher values of H_p at 62°S and 66°S, correlated with relatively higher values of CH_4. Each of the last three flights shows a depressed value of H_p in regions of relatively low CH_4; the values are generally lower by 20–30% than those of 940524.

During phase 3, there were five relevant flights, see Table 1. The data for 940728 and 940730 have sufficiently low signal-to-noise ratio that no analysis will be attempted here. The flights of 940806, 940808 and 940810 are shown in Fig. 7. The flight of 940806

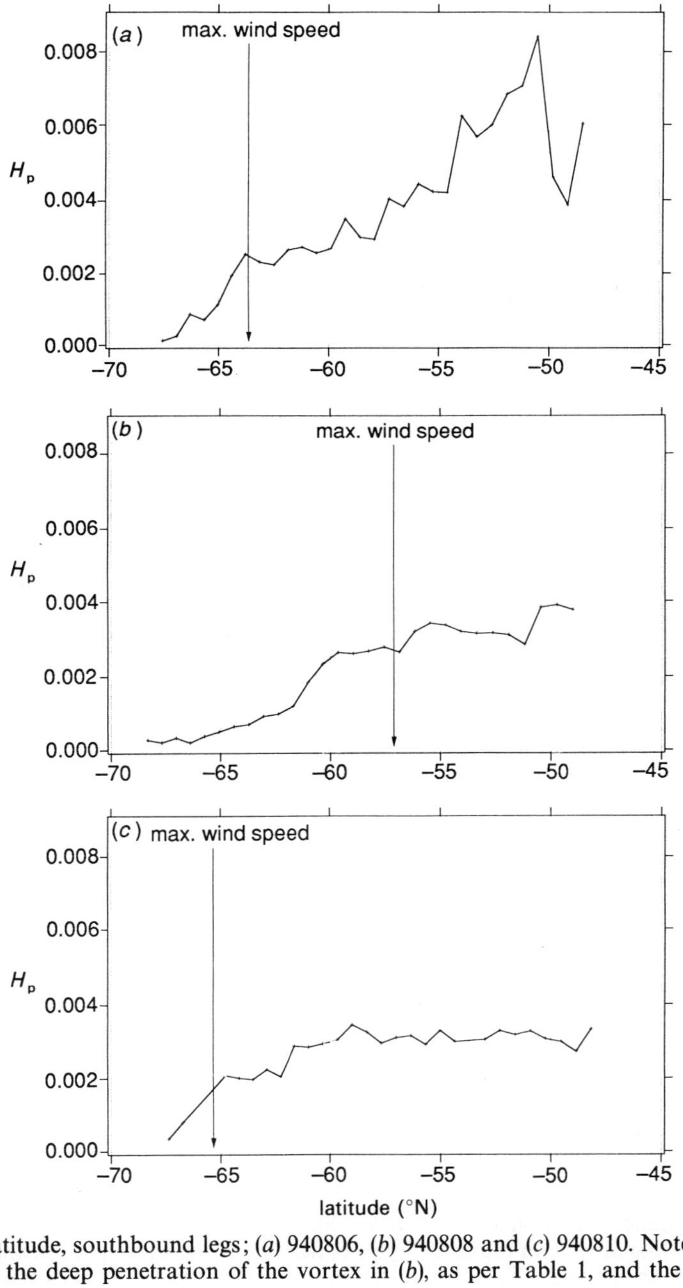

Fig. 7 H_p *vs.* latitude, southbound legs; (*a*) 940806, (*b*) 940808 and (*c*) 940810. Note the meridional gradient in (*a*), the deep penetration of the vortex in (*b*), as per Table 1, and the low H_p in each case.

shows a steep meridional gradient in H_p, with very low values at the highest latitudes. Note that there is air with relatively low H_p at 48.5°S to 50.5°S, correlated with low CH_4, separated from the low H_p air poleward of 60°S by air with high H_p, reaching as high as 8×10^{-3}. The high H_p air is all associated with high tracer (CH_4) values. We shall return to a detailed study of this flight later. The following flight, 940808, shows a

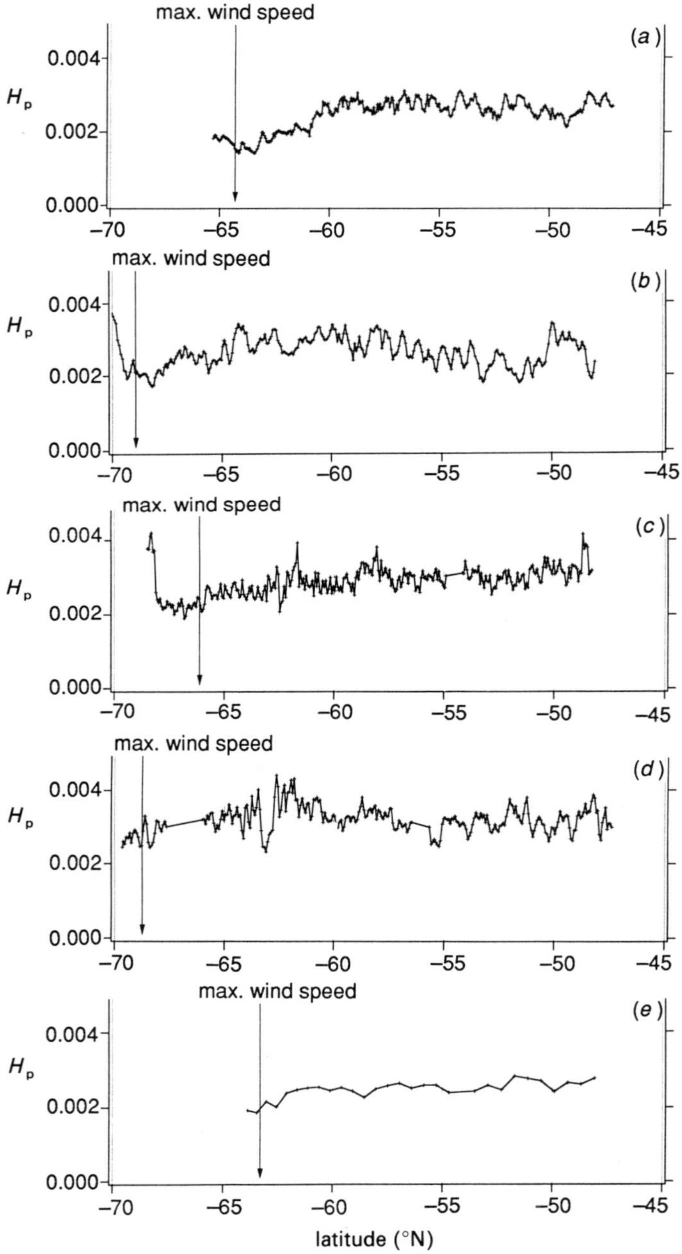

Fig. 8 H_p *vs.* latitude, southbound legs; (*a*) 941005, (*b*) 941010, (*c*) 941013, (*d*) 941016 and (*e*) 941020. Note the loss of HCl on the early flights either side of the wind speed maximum (Table 1), and the recovery on the later flights inside the vortex in (*b*) and (*c*), which occurs at the lower altitudes during the vertical profile.

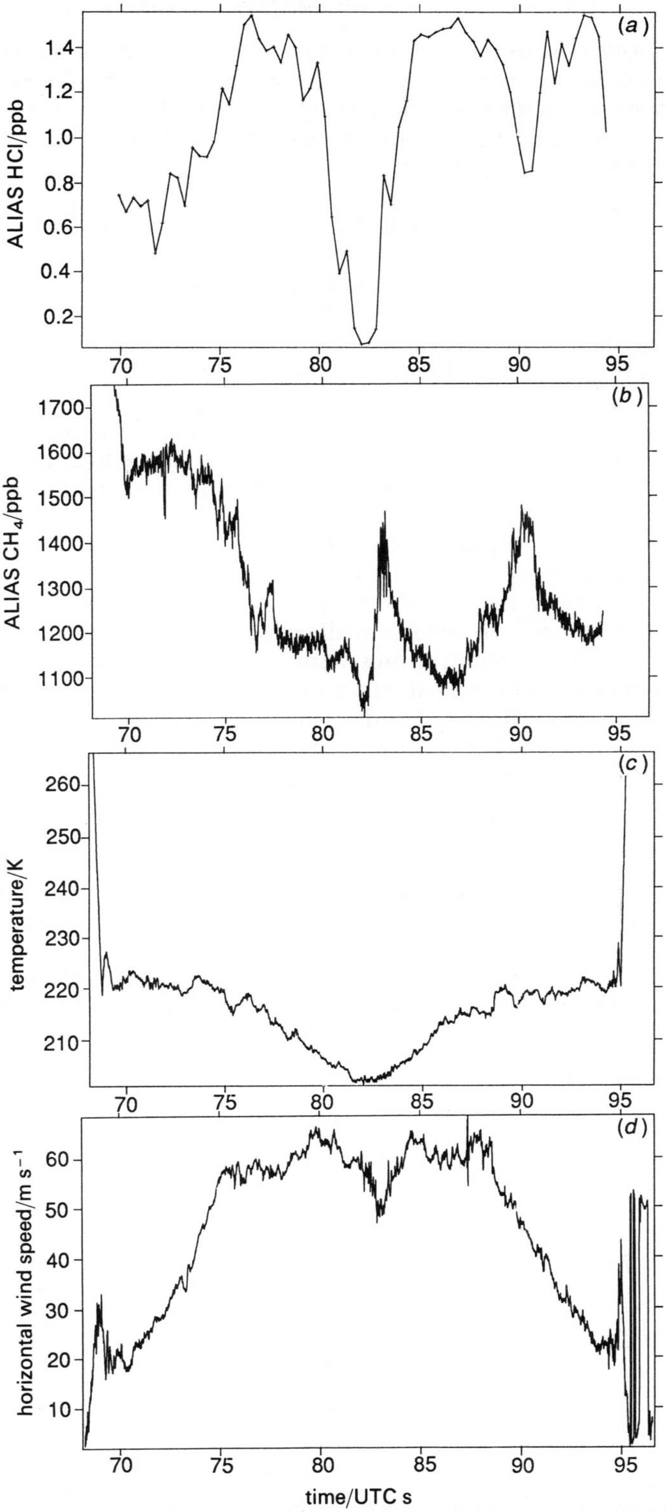

Fig. 9 The flight of 940806, plots *vs.* time along the entire track. (*a*) HCl, (*b*) CH_4, (*c*) temperature and (*d*) wind speed.

steep gradient inside the vortex, with low values of H_p at the highest latitudes increasing toward the wind speed maximum at 57°S. Equatorward of the wind maximum, there is a shallower gradient out to 45°S, but in the same sense as that inside the vortex. The last flight, 940810, shows similar behaviour to 940808 but in the context of a less deep penetration of the vortex by the ER-2.

During phase 4, there were five flights with HCl data at latitudes south of New Zealand. The values of H_p are shown in Fig. 8; on the first three flights there was still a decrease towards the pole at latitudes equatorward of the wind maximum, but the values had recovered substantially at latitudes poleward of the wind maximum since the midwinter flights of phase 3. The flight of 941010 had low values of H_p between 45°S and 55°S. The second and third flights, 941010 and 941013, show higher values inside and near the vortex than existed even in phases 1 and 2. The last flight, 941020, has lower values inside the vortex than outside. The interpretation of the observations on this fourth phase is complicated by the production of HCl inside the vortex by reaction by Cl atoms with CH_4 as ozone and ClO drop to zero in the final phases of the ozone hole; this evolution of photochemistry is being dealt with in a separate paper.[33]

We return now to the flight of 940806 to make a detailed study of the temperature history of the airmasses sampled by the ER-2 instruments along the flight track. The reason for so doing is suggested by the behaviour of H_p as a function of latitude and tracer mixing ratio during the first three phases, see Fig. 4–8. The temperatures sufficiently cold to activate HCl on sulfuric acid aerosol, on the basis of laboratory studies, are shown in Fig. 1. These temperatures were in general poleward, or occasionally at and slightly equatorward, of the jet-stream core (the wind speed maximum), see Fig. 3(*b*). The 940806 flight includes airmasses with high values of tracer and H_p juxtaposed with air near and in the vortex having low values. We start by displaying in Fig. 9 the observations of HCl, CH_4, temperature and windspeed *vs.* time along the aircraft track. The air encountered from 70 000 to 75 000 s UTC (*ca.* 45°S to 55°S) had high CH_4 mixing ratios, characteristic of low-latitude air. The wind maximum occurred close to 80 000 s UTC (*ca.* 64°S); the section of track from there to about 84 000 s, which includes the vertical profile, is therefore inside the vortex. The variation of H_p along this flight track is shown in Fig. 10. There are high fractional abundances in the high-methane air, and low ones at high latitudes, particularly poleward of the wind maximum. We will use back-trajectory analysis to examine where the air sampled along the flight path originated; such analysis also permits examination of the temperature

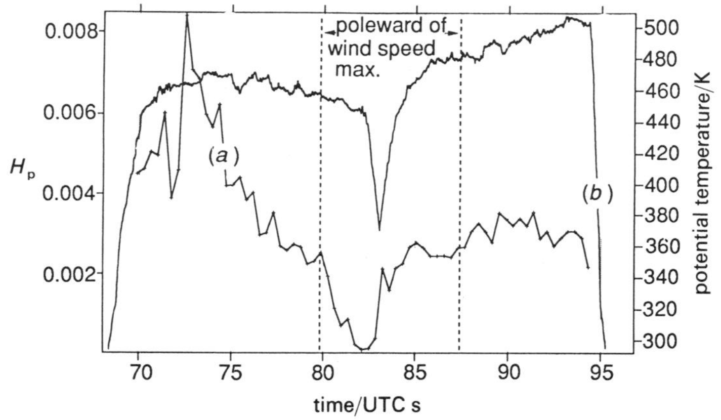

Fig. 10 (*a*) H_p and (*b*) potential temperature θ, *vs.* time along the ER-2 flight track, 940806. Note that, on average, higher θ values were encountered on the northbound leg than on the southbound leg.

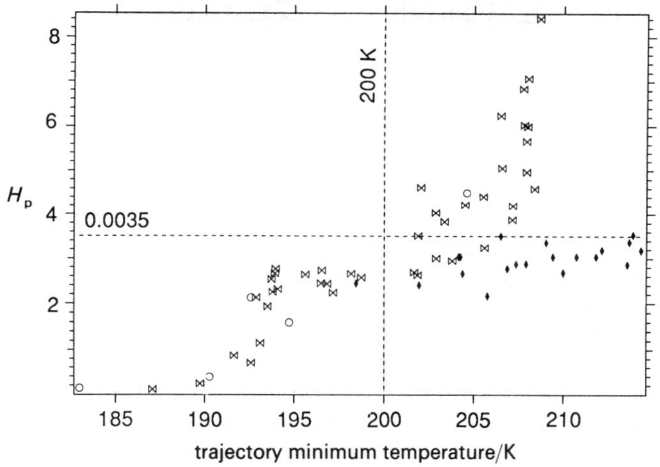

Fig. 11 H_p *vs.* minimum temperature of the sampled air parcels during the last 10 days, 940806: ○, $\theta < 440$; ⋈, $440 \leqslant \theta < 480$; ◆, $\theta \geqslant 480$ K. Note the threshold behaviour centred at 194 K (*cf.* Fig. 1).

history of each air parcel over the 10 days prior to 940806. Fig. 11 plots the proxy fractional HCl abundance, H_p, against the minimum temperature, T_{min}, experienced by the air parcel during the 10 days prior to 940806. The most striking result is that for those parcels whose minimum temperature dropped below 200 K, the value of H_p is

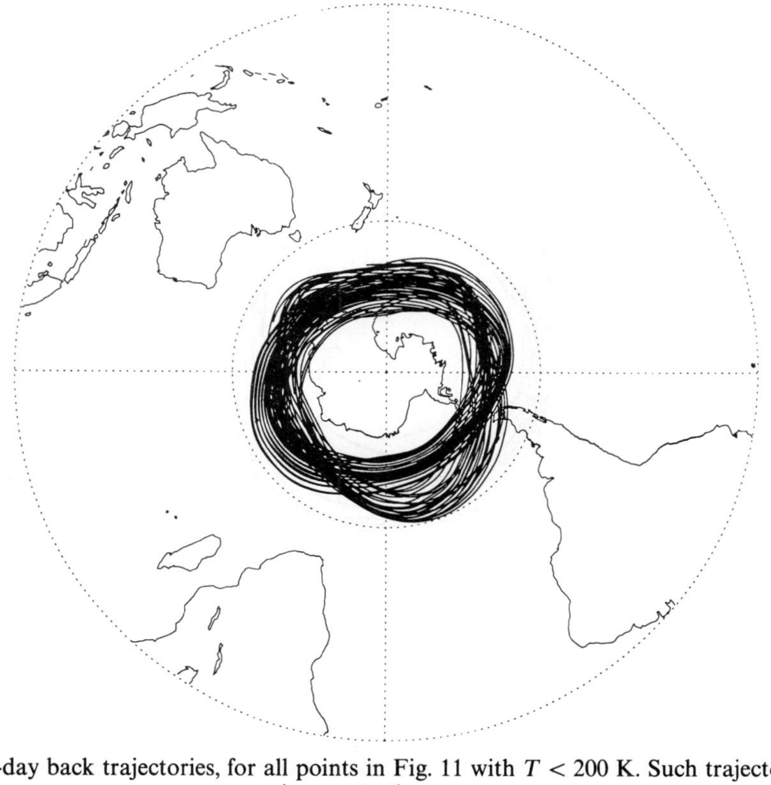

Fig. 12 10-day back trajectories, for all points in Fig. 11 with T < 200 K. Such trajectories are all in or near the vortex.

constant as a function of T_{min} from 200 K to 195 K. At temperatures below 194 K, there is a precipitous drop, until at temperatures below 190 K, H_p is near zero. There is a population almost parallel to the temperature axis which extends the curve from 200 to 215 K; for the totality of these air parcels with $H_p < 3.5 \times 10^{-3}$, it can be concluded that on a 10-day timescale, exposure to temperatures $\leqslant$194 K is necessary to remove HCl, and is also sufficient. There is a separate population in Fig. 4 with values of $H_p >$ 3.5×10^{-3} and whose minimum temperatures during the 10 days ranged between 200 and 208 K. The values of H_p range as high as 8×10^{-3}; all these points were encountered on the southbound leg, and are equatorward of 60°S.

We examine back trajectories for the three populations. Fig. 12 shows back trajectories for all points having experienced $T_{min} < 200$ K. All originated over or near Antarctica. Fig. 13 shows the back trajectories for those air parcels having $H_p < 3.5 \times 10^{-3}$ and with $T_{min} > 200$ K; all were confined to the latitude belt between 60°S and 45°S during the 10 days, with a few originating from near 35°S in the South American sector. Fig. 14 shows the back trajectories for air parcels having $H_p > 3.5 \times 10^{-3}$. Most originated equatorward of 45°S, with many coming from the South Atlantic between 30°S and 40°S.

We will pursue further the temperature dependence and its relationship to meridional gradients for the flight of 940806. To examine the structure of Fig. 11, and in the light of Fig. 12–14, we display in Fig. 15 the meridional gradient in H_p measured for the three northbound flights from Christchurch: 940528, 940801 and 941022. The data suggest that at ER-2 flight altitudes, H_p decreased from *ca.* 40°S to 15°S, particularly

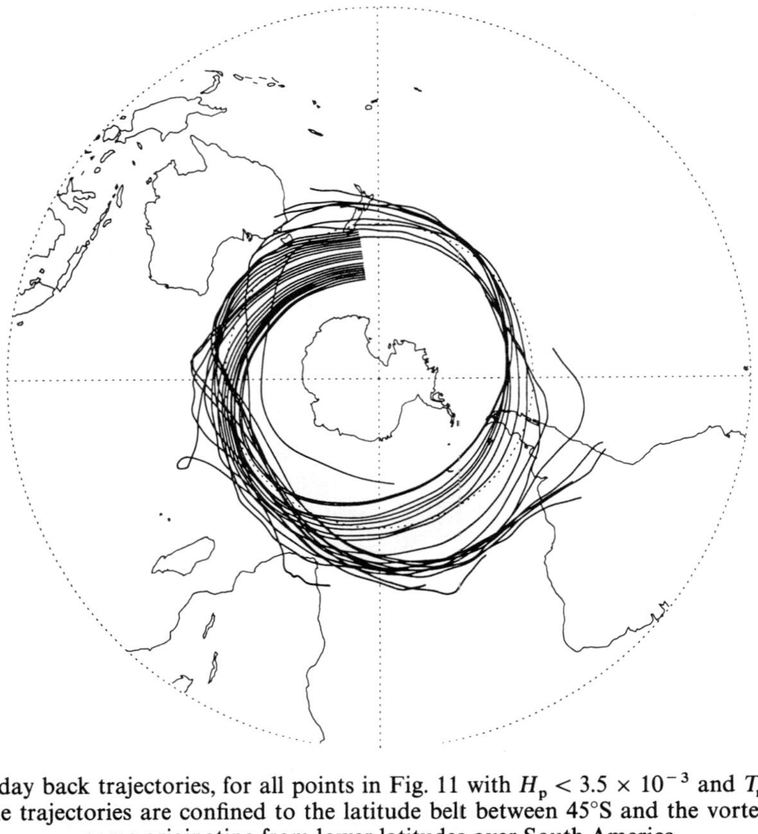

Fig. 13 10-day back trajectories, for all points in Fig. 11 with $H_p < 3.5 \times 10^{-3}$ and $T_{min} > 200$ K. Many of the trajectories are confined to the latitude belt between 45°S and the vortex edge, with some originating from lower latitudes over South America.

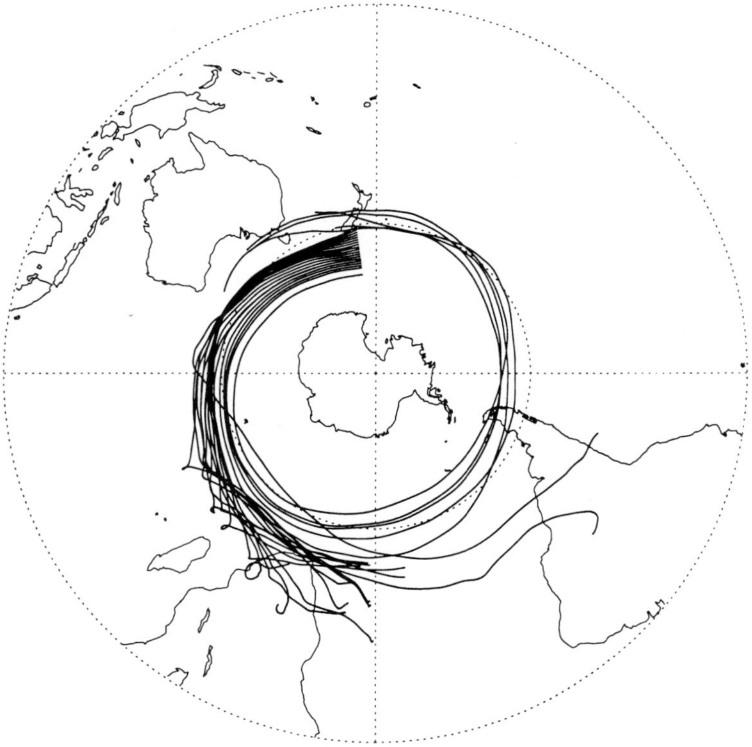

Fig. 14 10-day back trajectories, for all points in Fig. 11 with $H_p > 3.5 \times 10^{-3}$. Such trajectories mainly originate equatorward of 45°S, with those having the highest H_p coming from near 30°S–35°S over the South Atlantic.

during the second and fourth phases. We can now offer insight into the population on Fig. 11 with $H_p > 3.5 \times 10^{-3}$; the back trajectories on Fig. 14 show them to originate from between 30°S and 45°S, which Fig. 15 shows to be the location of the maximum values of H_p. The variability in H_p, evident in Fig. 4, 5, 7 and 8, and the presence of a maximum in subtropical latitudes in late May and late October appear to be difficult to reconcile with the possibility of a slow, pervasive and ubiquitous heterogeneous process which converts HCl to Cl and ClO *via* Cl_2 and HOCl at temperatures >195 K.

The question of how many of the points which experienced $T_{min} < 195$ K in Fig. 11 did so outside the vortex is of interest. Fig. 16 shows that virtually none of them did so; the temperatures needed for conversion of HCl into active chlorine were largely experienced poleward of the wind maximum, *i.e.* inside the vortex.

We will examine a portion of one other flight here, since it bears on the calculation of total chlorine and its partitioning. The ER-2 made three vertical profiles near the equator on its way back to California during late October. The HCl at the tropical tropopause was variable in the three profiles; on 941024, 941026 and 941029 it was, respectively, 0.04, 0.07 and 0.40 ppb by volume. The first two have little impact on H or Cl_y, but a volume fraction of 0.4 ppb HCl at the tropical tropopause would be very significant, particularly for the analysis of airmasses with low Cl_y. The two-hour section of flight track centred on the bottom of the vertical profile is shown in Fig. 17, for HCl, ClO, $CFCl_3$ and temperature on 941029. The ClO is zero at the tropical tropopause (16.4 km); there is no obvious loss of $CFCl_3$, and the volume fraction of HCl is 0.4 ppb. The implication is that a soluble molecule like HCl may reach the tropical tropopause;

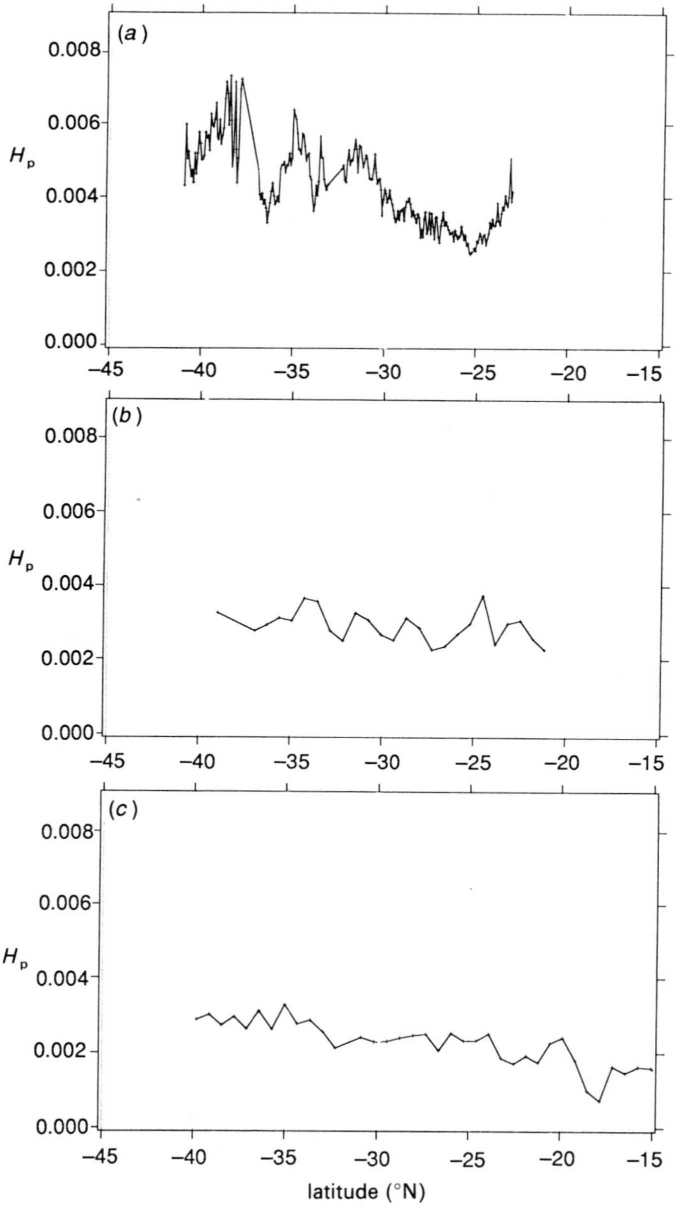

Fig. 15 H_p *vs.* latitude, 15°S–45°S; (*a*) 940528, (*b*) 940801 and (*c*) 941022. Note the tendency for H_p to decrease toward the equator. *Cf.* Fig. 5, 7 and 8; H_p tends to have maximum values between 30°S and 40°S, particularly in phases 2 and 4, corresponding to (*a*) and (*c*).

another possibility is production there *in situ*, possibly by photochemical reaction of labile chlorine containing molecules not normally counted in the stratospheric chlorine budget (for example C_2Cl_4). Another possibility is the production of HCl by reaction of nitric acid with sodium chloride in marine aerosol. In any event, the sensitivity of the analysis of chlorine partitioning[25] to an increment of 0.4 ppb in Cl_y is examined in Fig. 18, where Fig. 11 is repeated with HCl/Cl_y and $HCl/(Cl_y + 0.4)$, *i.e.* with H instead of H_p. The major impact of including the extra 0.4 ppb of chlorine is upon the lower

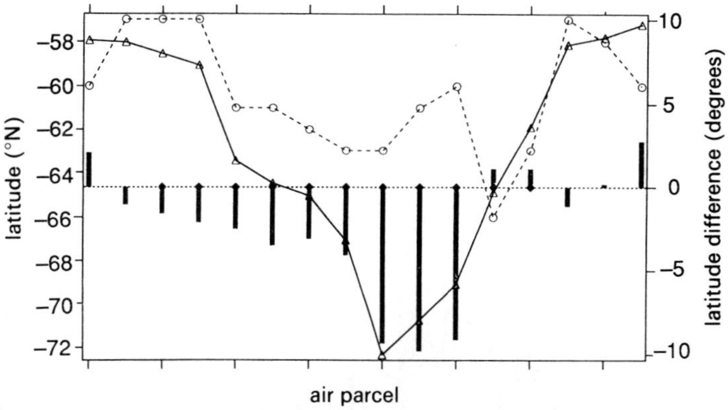

Fig. 16 Location of the occurrence of the values of $T_{min} < 195$ K, for those air parcels in Fig. 11 which experienced temperatures below this threshold during the 10 days prior to 940806, with respect to the wind speed maximum at the time of the occurrence of the minimum temperature. The great majority of the occurrences were poleward of the wind speed maximum. The curves are referenced to the left ordinate, the shaded vertical difference bars are referenced to the right ordinate. △, latitude of air parcel minimum temperature; ○, latitude of wind speed maximum; ▮, T_{min} latitude − wind maximum latitude; ◇, air parcel poleward of wind maximum when measured.

latitude, higher tracer air which has relatively high values of H_p, see Fig. 11. These points have their value of H reduced, from a range of 70–112%, to <75%, approximately equivalent to dropping H_p from 3.5×10^{-3}–8×10^{-3} to $<3.7 \times 10^{-3}$. The occurrence of some points with $H > 1$ does suggest that Cl_y has been underestimated by the procedure used to estimate it from the gas chromatographic measurements of organic chlorine,[25] perhaps because the possibility of a non-zero contribution from HCl at the tropical tropopause was not considered. The points which experienced temperatures less than 200 K were less sensitive, presumably because they represent older air with higher Cl_y.

Discussion and Conclusions

Joint consideration of Fig. 1, 11 and 18 shows that on a 10-day timescale temperatures below 195 K are a necessary and sufficient condition for the loss of HCl. The threshold behaviour predicted by laboratory studies is observed in the real stratosphere. Whether PSCs are nitric acid trihydrate or a ternary H_2O–HNO_3–H_2SO_4 solution is a question that is not resolvable by this analysis. The behaviour of those air parcels in Fig. 11 and 18 whose minimum temperature did not drop below 200 K in the 10 days 940727–940806 suggests that little or no loss of HCl occurs at temperatures above 194 K, irrespective of altitude. For air parcels whose trajectories were entirely poleward of 45°S, these conclusions extend to include the range $200 < T_{min}/\mathrm{K} < 215$. Air parcels in, or originating recently in latitudes 30°S–40°S exhibited fractional HCl abundances up to a factor of 2.5 higher than those poleward of 45°S during the second and fourth phases. In photochemical models, the tendency of air parcels in the Southern Hemisphere in August–September, away from temperatures less than 195 K, is to increase HCl at about 1% day^{-1},[34] *via* reaction of Cl with CH_4. The value of H_p of *ca.* 3×10^{-3} seen in most parcels poleward of 45°S which have not recently been below 195 K is probably maintained by mixing between the air at and just poleward of the wind maximum, which has recently been below 195 K, with air from equatorward of 40°S having high values of H_p: the trajectories of Fig. 13 have H_p values maintained by mixing air parcels from Fig. 12

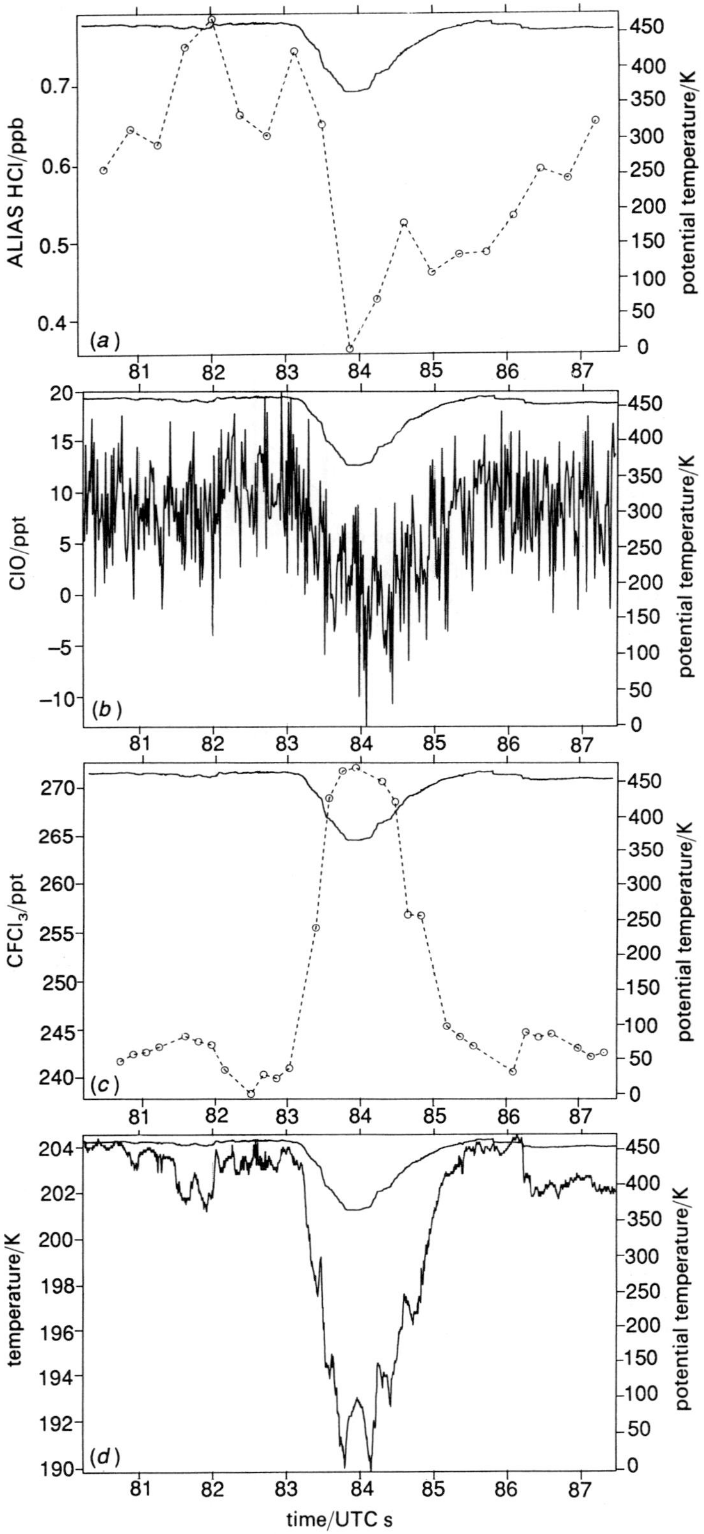

Fig. 17 Observations for the two-hour period centred on vertical profiles of (*a*) HCl, (*b*) ClO, (*c*) $CFCl_3$ and (*d*) temperature as a function of UTC time (s), 941029. In each case, the upper solid curve represents θ. The profile was near 1.5°S; the tropopause is located at the twin temperature minima either side of 84 000 s UTC, at a potential temperature of *ca.* 380 K. Note that at the tropopause, $ClO \approx 0$, there is essentially no loss of $CFCl_3$ compared with the value at potential temperatures less than 380 K, and the volume fraction of HCl is *ca.* 0.4 ppb.

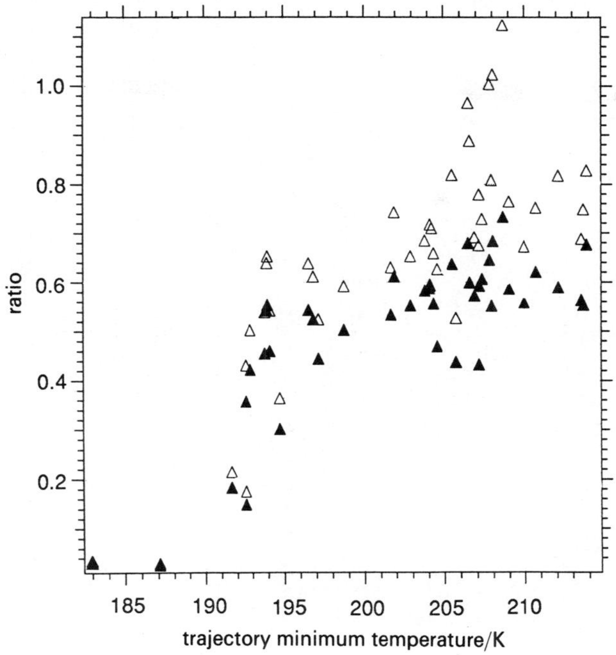

Fig. 18 △, HCl/Cl_y *vs.* T_{min}, 940806; ▲, $HCl/(0.4 + Cl_y)$ *vs.* T_{min}, 940806. The difference shows the effect of including an extra 0.4 ppb of chlorine in the denominator, as implied by the HCl at the tropical tropopause in Fig. 17. The low Cl_y air having high values of the ratio H is affected most; there are open triangles concealed beneath the solid triangles at the lowest values of T_{min}.

and 14. The agency for maintaining the near constant values of H_p between the wind maximum in the polar night jet stream and the subtropics is the shear in the horizontal wind as suggested by Kelly *et al.*[35] on observational grounds, and characterized[36,37] as the 'surf zone' in which chaotic advection[27] tends to flatten large scale horizontal gradients by the mixing action arising from the continual distortion and stretching of interleaved filaments of air having disparate origins. The reality of such processes is exemplified in Fig. 6, where filaments of air originating from poleward of the wind maximum are found in mid latitudes on 940603. The PV maps for 940806 at $\theta = 450$ K (southbound leg) and $\theta = 500$ K (northbound leg) are shown in Plate 1. The peel-off of vortex air at the edge is clearly visible; the structure at 450 K encountered by the ER-2 on its southbound leg (a filament of vortex edge air), followed by a mass of low PV (recently subtropical) air, followed in turn by vortex air is reproduced approximately, although the filament's location on the PV map is well north of that actually encountered by the ER-2.

The value of H_p, and by extension HCl/Cl_y, was very variable in mid-latitudes once temperatures had dropped below 195 K near the wind speed maximum in the polar night jet stream; the value of H_p showed little or no difference poleward or equatorward of the jet stream during the flights of 940413 and 940524, but did during every succeeding flight, from 940601 to 941020. This high variability is probably the result of two variable meteorological processes: the peel-off of air from near the jet-stream core, and the effect of rapid small scale adiabatic temperature fluctuations upon non-linear threshold processes, such as PSC formation or the heterogeneous reactions typified by Fig. 1.[38] The latter effect can arise from the cooling rates experienced by an air parcel in the real atmosphere being the net effect of many rapid small scale fluctuations, whereas the temperatures taken from global meteorological analyses necessarily smooth those out.

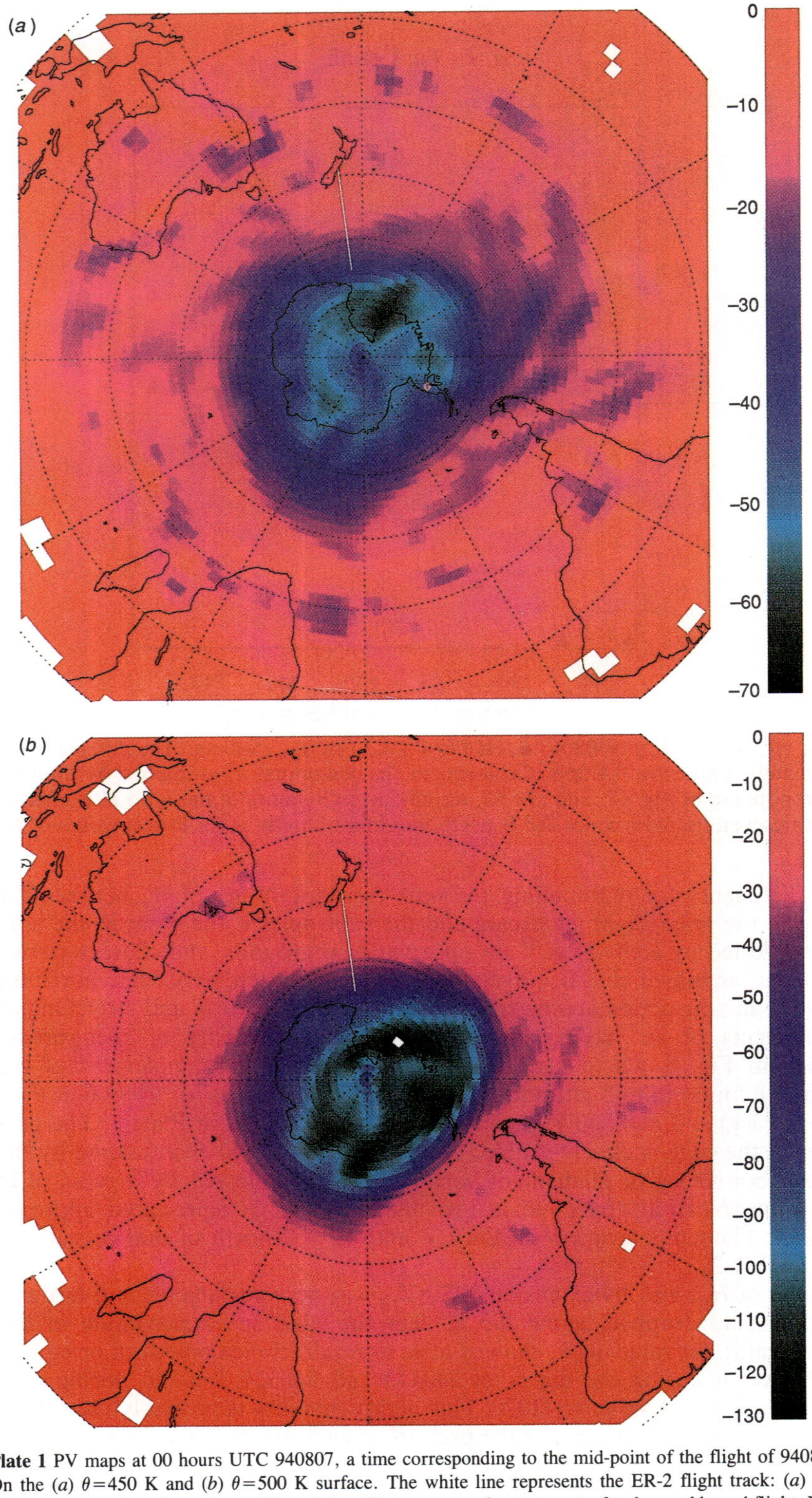

Plate 1 PV maps at 00 hours UTC 940807, a time corresponding to the mid-point of the flight of 940806. On the (*a*) θ=450 K and (*b*) θ=500 K surface. The white line represents the ER-2 flight track: (*a*) corresponds approximately to the southbound leg, (*b*) is an approximate average for the northbound flight. Note the filamentation of vortex edge air to mid-latitudes, and the rough correspondence with the features in CH_4 and H_p in Fig. 9 and 10, although the PV maps have the blue vortex edge filament too far north at the longitudes of the flight track.

The net effect[38] would shift the curves in Fig. 11 and 18 by *ca.* 2 K towards warmer temperatures.

The flights in late May and early June provide direct evidence for the lowering of HCl fractional abundance equatorward of the wind maximum, as a result of temperatures dropping below 195 K poleward of and at this jet-stream core followed by peel-off from the vortex to mid-latitudes (Fig. 3 and 5).

We conclude that temperatures below 195 K are a necessary and sufficient condition for the rapid loss of HCl; this result agrees well with laboratory studies.[3–5] We further observe the fraction of Cl_y residing in HCl in middle and subpolar latitudes to be very variable, and that such variability results from the combined effects of filamentation of air from the polar night jet stream into middle latitudes, and very likely the effects of small-scale temperature fluctuations upon a threshold process. The fractional abundance of HCl in air recently from subtropical latitudes was sensitive to the inclusion of an extra volume fraction of 0.4 ppb Cl_y, a test suggested by the observation of 0.4 ppb HCl at the tropical tropopause in the presence of zero ClO and no loss of $CFCl_3$. The above considerations taken together suggest that the observed lower stratospheric loss of ozone in austral mid-latitudes, peaking during winter and spring, are caused in whole or in part by the heterogeneous activation of the chlorine in air experiencing temperatures below 195 K at and poleward of the core of the polar night jet stream.

We thank Richard Stimpfle for the ClO data in Fig. 17(*b*). Part of the research described in this paper was carried out by the Jet Propulsion Laboratory, under a contract with NASA. The mission was in large part funded by NASA's Upper Atmosphere Research Program, under the leadership of Michael Kurylo, with substantial assistance from the Atmospheric Effects of Aviation Program. We are grateful to the ECMWF and the Meteorological Office for the provision of analyses and forecasts. The ASHOE/MAESA was a team effort, involving over 100 people; this work would not have been possible without them. A special thank you is owed to the ER-2 pilots: Jim Barrilleaux, Ken Broda, Bill Collette, Doyle Krumrey, Jan Nystrom and Ron Williams. We also thank William Brune, Akkihebbal Ravishankara, Ross Salawitch, and Susan Solomon for constructive discussions.

References

1 R. S. Stolarski, P. Bloomfield, R. D. McPeters and J. R. Herman, *Geophys. Res. Lett.*, 1991, **18**, 1015.
2 *Scientific Assessment of Ozone Depletion*, World Meteorological Organization, Global Ozone Research and Monitoring Project Report, 1992, **25**, Geneva, p. 2.27.
3 D. R. Hanson and A. R. Ravishankara, *J. Phys. Chem.*, 1992, **96**, 2682.
4 M. A. Tolbert, M. J. Rossi and D. M. Golden, *Geophys. Res. Lett.*, 1988, **15**, 847.
5 D. R. Hanson and A. R. Ravishankara, *J. Phys. Chem.*, 1994, **98**, 5728.
6 D. J. Hofmann and S. Solomon, *J. Geophys. Res.*, 1989, **94**, 5029.
7 D. J. Hofmann, S. J. Oltmans, W. D. Komhyr, J. M. Harris, J. A. Lathrop, A. O. Langford, T. Deshler, B. J. Johnson, A. Torres and W. A. Matthews, *Geophys. Res. Lett.*, 1994, **21**, 65.
8 A. F. Tuck, *J. Geophys. Res.*, 1989, **94**, 11687.
9 M. H. Proffitt, D. W. Fahey, K. K. Kelly and A. F. Tuck, *Nature (London)*, 1989, **342**, 233.
10 A. F. Tuck, T. Davies, M. Noguer-Alba, D. W. Fahey, S. R. Kawa, K. K. Kelly, D. M. Murphy, M. H. Proffitt, J. J. Margitan, M. Loewenstein, J. R. Podolske, S. E. Strahan and K. R. Chan, *J. Geophys. Res.*, 1992, **97**, 7883.
11 M. P. McCormick, H. M. Steele, P. Hamill, W. P. Chu and T. J. Swissler, *J. Atmos. Sci.*, 1982, **39**, 1387.
12 S. Solomon, R. R. Garcia, F. S. Rowland and D. J. Wuebbles, *Nature (London)*, 1986, **321**, 755.
13 O. B. Toon, P. Hamill, R. P. Turco and J. Pinto, *Geophys. Res. Lett.*, 1986, **13**, 1284.
14 D. W. Fahey, K. K. Kelly, G. V. Ferry, L. R. Poole, J. C. Wilson, D. M. Murphy, M. Loewenstein and K. R. Chan, *J. Geophys. Res.*, 1989, **94**, 11299.
15 R. A. Cox, A. R. MacKenzie, R. H. Müller, Th. Peter and P. J. Crutzen, *Geophys. Res. Lett.*, 1994, **21**, 1439.
16 K. Drdla, A. Tabazadeh, R. P. Turco, M. Z. Jacobson, J. E. Dye, C. Twohy and D. Baumgardner, *Geophys. Res. Lett.*, 1994, **21**, 2475.

17 K. S. Carslaw, B. P. Luo, S. L. Clegg, Th. Peter, P. Brimblecombe and P. J. Crutzen, *Geophys. Res. Lett.*, 1994, **21**, 2479.
18 W. H. Brune and J. G. Anderson, *Geophys. Res. Lett.*, 1986, **13**, 1391.
19 J. W. Elkins, in *Summary Report 1991, No. 20*, NOAA Climate and Diagnostics Laboratory, U.S. Department of Commerce, Boulder, 1992, p. 77.
20 D. Baumgardner, J. E. Dye and B. W. Gandrud, in *Airborne Southern Hemisphere Ozone Experiment/Measurements for Assessing the Effects of Stratospheric Aircraft*, National Aeronautics and Space Administration, Moffett Field, 1994, p. 12.
21 J. C. Wilson, H. H. Jonsson, C. A. Brock, D. W. Toohey, L. M. Avallone, D. Baumgardner, J. E. Dye, L. R. Poole, D. C. Woods, R. J. DeCoursey, M. Osborn, M. C. Pitts, K. K. Kelly, K. R. Chan, G. V. Ferry, M. Loewenstein, J. R. Podolske and A. Weaver, *Science*, 1993, **261**, 1140.
22 C. R. Webster, R. D. May, C. A. Trimble, R G. Chave and J. Kendall, *Appl. Opt.*, 1994, **33**, 454.
23 C. R. Webster, R. D. May, D. W. Toohey, L. M. Avallone, J. G. Anderson, P. Newman, L. Lait, M. R. Schoeberl, J. W. Elkins and K. R. Chan, *Science*, 1993, **261**, 1130.
24 C. R. Webster, R. D. May, L. Jaeglé, H. Hu, S. P. Sander, M. R. Gunson, G. C. Toon, J. M. Russell III, R. M. Stimpfle, J. P. Koplow, R. J. Salawitch and H. A. Michelsen, *Geophys. Res. Lett.*, 1994, **21**, 2575.
25 E. L. Woodbridge, J. W. Elkins, D. W. Fahey, L. E. Heidt, S. Solomon, T. J. Baring, T. M. Gilpin, W. H. Pollock, S. M. Schauffler, E. L. Atlas, M. Loewenstein, J. R. Podolske, C. R. Webster, R. D. May, J. M. Gilligan, S. A. Montzka and R. J. Salawitch, *J. Geophys. Res.*, 1995, **100**, in the press.
26 X. Tao and A. F. Tuck, *J. Geophys. Res.*, 1994, **99**, 3431.
27 R. B. Pierce and T. D. A. Fairlie, *J. Geophys. Res.*, 1993, **98**, 18589.
28 R. B. Pierce, W. L. Grose, J. M. Russell III, A. F. Tuck, R. Swinbank and A. O'Neill, *J. Atmos. Sci.*, 1994, **51**, 2931.
29 R. B. Pierce, T. D. A. Fairlie, W. L. Grose, R. Swinbank and A. O'Neill, *J. Atmos. Sci.*, 1994, **51**, 2957.
30 P. Kållberg, *SMHI R & D Note, No. 37*, Swedish Meteorological Institute, Stockholm, 1984.
31 J. E. Dye, D. Baumgardner, B. W. Gandrud, S. R. Kawa, K. K. Kelly, M. Loewenstein, G. V. Ferry, K. R. Chan and B. L. Gary, *J. Geophys. Res.*, 1992, **97**, 8015.
32 M. Bithell, L. J. Gray, J. E. Harries, J. M. Russell III and A. F. Tuck, *J. Atmos. Sci.*, 1994, **51**, 2942.
33 C. R. Webster, R. D. May, D. C. Scott, H. Hu, L. Jaeglé, R. Stimpfle, D. W. Fahey, M. H. Proffitt and R. J. Salawitch, in preparation.
34 R. L. Jones, J. Austin, D. S. McKenna, J. G. Anderson, D. W. Fahey, C. B. Farmer, L. E. Heidt, K. K. Kelly, D. M. Murphy, M. H. Proffitt, A. F. Tuck, and J. F. Vedder, *J. Geophys. Res.*, 1989, **94**, 11529.
35 K. K. Kelly, A. F. Tuck, D. M. Murphy, M. H. Proffitt, D. W. Fahey, R. L. Jones, D. S. McKenna, M. Loewenstein, J. R. Podolske, S. E. Strahan, G. V. Ferry, K. R. Chan, J. F. Vedder, G. L. Gregory, W. D. Hypes, M. P. McCormick, E. V. Browell and L. E. Heidt, *J. Geophys. Res.*, 1989, **94**, 11317.
36 M. E. McIntyre and T. N. Palmer, *J. Atmos. Terr. Phys.*, 1984, **46**, 825.
37 M. N. Juckes and M. E. McIntyre, *Nature (London)*, 1987, **328**, 590.
38 D. M. Murphy and B. L. Gary, *J. Atmos. Sci.*, 1995, **52**, in the press.

Paper 5/00175G; *Received* 10*th January*, 1995

Faraday Discuss., 1995, **100**, 411–430

Laboratory Study of the Reactions $Mg + O_3$ and $MgO + O_3$

Implications for the Chemistry of Magnesium in the Upper Atmosphere

J. M. C. Plane and M. Helmer
School of Environmental Sciences, University of East Anglia, Norwich, UK, NR4 7TJ

The reaction $Mg + O_3 \rightarrow MgO + O_2$ has been studied by the pulsed photodissociation at 193.3 nm of magnesium acetyl acetonate $[Mg(C_5H_7O_2)_2]$ vapour to produce Mg atoms in an excess of O_3 and N_2 bath gas, followed by time-resolved laser-induced fluorescence (LIF) spectroscopy of atomic Mg at 285.2 nm $[Mg(3^1P_1–3^1S_0)]$. The resulting rate coefficient is given in the Arrhenius form by $k(196 < T/K < 368) = (2.28 \pm 0.74) \times 10^{-10} \exp[-(139 \pm 84\ K)/T]\ cm^3\ molecule^{-1}\ s^{-1}$. This reaction is therefore the most rapid oxidation process of atomic Mg in the atmosphere between 65 and 95 km. The reaction $MgO + O_3 \rightarrow MgO_2 + O_2$ was investigated by the pulsed photodissociation of MgO_3 coupled with time-resolved LIF by pumping the $MgO(B\,^1\Sigma^+–X\,^1\Sigma^+, \Delta v = 0)$ transition at 499.4 nm and monitoring emission from the $MgO(B\,^1\Sigma^+–A\,^1\Sigma^+, \Delta v \leqslant 0)$ transition at wavelengths greater than 600 nm. This yields $k(217 < T < 366\ K) = (2.19 \pm 1.8) \times 10^{-10} \exp[-(548 \pm 271\ K)/T]\ cm^3\ molecule^{-1}\ s^{-1}$. *Ab initio* quantum calculations were used to show that the most stable form of magnesium in the upper atmosphere is $Mg(OH)_2$, formed from a rapid recombination reaction between MgO and H_2O. A one-dimensional (1D) model of magnesium was then constructed by using observed Mg^+ profiles above 85 km to constrain both the chemical reaction scheme and the meteoric input flux of the metal. The atomic Mg layer is predicted to occur at the same height as the Na layer (*ca.* 90 km), but with a peak concentration that is smaller by a factor of *ca.* 5. The model indicates that the ratio of the Mg/Na flux from meteoric ablation is about 0.5. This is strong evidence that the available mass of meteoroids does not ablate completely and that the residual fraction is enriched in magnesium.

Magnesium is the most abundant metallic constituent in meteorites, and about 6 t per day enters the atmosphere from space. This is expected to give rise to a layer of atomic Mg in the upper mesosphere, as in the case of other meteoric metals. However, such a layer cannot be observed by ground-based optical techniques because there is not a suitable spectroscopic transition of atomic Mg at wavelengths between 330 and 800 nm. Furthermore, there is almost no information about the gas-phase chemistry of neutral magnesium which would be required to predict the characteristic features of such a layer.

The presence of free metal atoms in the upper atmosphere has been known for more than 50 years.[1,2] Quantitative observations of metal atoms were first made in the 1950s by ground-based photometric measurements of the resonance fluorescence from the spectroscopic transitions of metal atoms such as Na, K, Li and Fe, excited by solar radiation during twilight.[1,3] Photometry was superseded in the 1970s when the discovery of pulsed tunable laser sources allowed the development of the lidar technique.[1,4]

Lidar measurements have shown that these metals occur globally in thin layers at altitudes between 85 and 95 km, *i.e.* in the upper mesosphere and lower thermosphere.[1,2] In the case of atomic Na, which is the most easily detected metal, a profile of the Na layer can be obtained with a modern lidar by signal-averaging for less than a minute, with a height resolution of <50 m.[5] This impressive temporal and spatial resolution has enabled the metal layers to be employed as tracers of dynamical processes such as gravity waves and tidal forces,[6] in a region of the atmosphere that would otherwise be difficult to explore. The only other metallic species that can be observed in the upper atmosphere are positive ions, measured *in situ* by mass spectrometers flown on rockets.[7–9]

The evidence that meteoric ablation is the major source of these metals includes the good correlation between the relative abundances of metallic ions observed in the lower thermosphere and their elemental abundances in chondritic meteorites,[8] sporadic enhancements in the total column abundance of Na during meteor showers,[10,11] and the direct observation by a high-resolution lidar of metal deposition in short-lived meteor trails.[5] The daily input of extraterrestrial material into the Earth's atmosphere has been estimated from meteor radar observations to be about 44 t.[12,13] It has been postulated[14] that most of the incoming meteoric mass is in the form of interplanetary dust particles in the mass range 1–1000 μg (median mass, 10 μg), with a radius range of 50–500 μm (median radius, 100 μm). The mean entry velocity of these particles lies between 14.5 and 17 km s^{-1}, so that frictional heating to well over 2000 K causes most of the meteoric material entering the atmosphere to ablate in the 80–100 km region, providing a direct source of metal atoms and ions.[14]

The major metallic constituents of meteorites by weight are: Mg 12.5%, Fe 11.5%, Al 1.7%, Ni 1.5%, Ca 1.0% and Na 0.6%.[15] Magnesium is therefore the most abundant metallic species, and large quantities of Mg^+ ions have been detected in the upper atmosphere by rocket-borne mass spectometry,[7–9] and by rocket[16] and satellite[17] photometric observations of the $Mg^+(3\,^2P_J{-}3\,^2S)$ doublet at 279.6–280.3 nm in the dayglow. Indeed, sporadic E layers (the thin, highly concentrated layers of ions which form between 90 and 105 km and can severely affect short-wave radio communications through the ionosphere) are primarily composed of Mg^+ and Fe^+ ions.[18] However, no observations appear to have been reported of neutral atomic Mg. This is because measurements of atomic Mg in the upper atmosphere by ground-based photometry or lidar would utilise the $Mg(3\,^1P_1{-}3\,^1S_0)$ transition at 285.2 nm, but the transmission through the troposphere and stratosphere of wavelengths shorter than 320 nm is prevented by absorption in the Huggins and Hartley bands of ozone.

The objective of this paper is to predict, through a combination of laboratory kinetic studies and modelling, the appearance of the atomic Mg layer in the upper atmosphere. A sensible starting-point for this exercise is to consider the Na layer: over the past decade, experimental work in several laboratories has revealed a substantial amount about the atmospheric chemistry of sodium species,[1,2] enabling a model to be developed which satisfactorily describes the observed diurnal and seasonal behaviour of the Na layer as a function of latitude.[19] Above 90 km, the ion chemistry of the metal predominates. This is because of the enhanced rate of metal-ion formation through charge transfer between the metal atom, with its low ionization energy, and the increasing concentrations of the major ions, NO^+ and O_2^+. Below 90 km, neutral atomic Na is oxidized by O_3 and O_2 to form a variety of oxides, hydroxides and carbonates.

In the case of magnesium, many of the reactions pertinent to its ion chemistry in the atmosphere have been studied by Rowe *et al.*[20] using the flowing afterglow technique. By contrast, the only relevant reaction of neutral Mg that has been investigated is recombination with O_2:[21]

$$Mg + O_2 + N_2 \rightarrow MgO_2 + N_2 \quad (1)$$

This reaction has a large positive activation energy (23.2 kJ mol^{-1}), so that the lifetime of an Mg atom at an altitude of 85 km would be about 1.5 years if reaction (1) were the only removal pathway. By analogy with the chemistry of Na,[19] the most rapid reaction likely to be undergone by freshly ablated Mg atoms in the mesosphere is:

$$Mg + O_3 \rightarrow MgO + O_2 \qquad (2)$$

This reaction is highly exothermic ($\Delta H^\circ_0(2) = -165$ kJ mol^{-1}),[22] and has in fact been observed to occur in an inert gas matrix.[23] Therefore, we will measure $k_2(T)$ over a significant temperature range, including temperatures applicable to the upper atmosphere, and test the results against the predictions of theoretical formalisms for calculating the rate coefficients of fast reactions governed by long-range attractive forces between neutral species.

This will be followed by a kinetic study of the reaction

$$MgO + O_3 \rightarrow MgO_2 + O_2 \qquad (3)$$

However, MgO_2 is not very strongly bound [$D_0(Mg{-}O_2) = 95$ kJ mol^{-1}],[24] so that more stable sinks for magnesium are likely to be provided through the recombination of MgO with H_2O and CO_2:

$$MgO + H_2O + N_2 \rightarrow Mg(OH)_2 + N_2 \qquad (4)$$

$$MgO + CO_2 + N_2 \rightarrow MgCO_3 + N_2 \qquad (5)$$

These reactions will be investigated through *ab initio* calculations on $Mg(OH)_2$ and $MgCO_3$ and the application of simplified RRKM theory.[25] Finally, all of these results will be used to construct a 1D model of magnesium that includes both chemistry and transport in the upper atmosphere.

Experimental Method

Reaction (2) was investigated by the pulsed photolysis of magnesium acetyl acetonate [MgAcAc, $Mg(C_5H_7O_2)_2$] to produce atomic Mg in an excess of O_3 and N_2 bath gas. The reaction was then monitored by time-resolved LIF spectroscopy of the atomic Mg. Fig. 1 is a schematic drawing of the experimental system employed. The reaction was initiated and then monitored spectroscopically in the central cylindrical chamber of the stainless-steel reactor. Four horizontal side-arms attached to this chamber provided the optical coupling for the lasers into and out of the central chamber. One of these side-arms was independently heated to act as a heat-pipe source of MgAcAc vapour, and two of the other side-arms carried the flows of O_3, diluted in N_2, into the chamber. A fifth, vertical, side-arm provided the optical coupling to the photomultiplier tube which monitored the LIF signal.

Powdered MgAcAc was placed in a tantalum boat in the heat pipe. The heat-pipe temperature was set to between 383 and 413 K, depending on the concentration of MgAcAc vapour required, and this temperature was then maintained to within ± 1 K. The vapour was entrained in a flow of N_2 and carried into the central chamber, where it mixed with the flow of O_3 in N_2 before being photolysed with an ArF excimer laser at 193.3 nm (Questek, Model 2110; pulse energy, 10 mJ; pulse rate, 4 Hz). The excimer beam was shaped by a system of lenses and a pin-hole, and then loosely focussed through the chamber.

The resulting $Mg(^1S)$ atoms were probed at 285.2 nm [$Mg(3\,^1P_1{-}3\,^1S_0)$] using a nitrogen-pumped dye laser (Laser Photonics, Models LN1000/LN107; laser dye Rhodamine 590; pulse energy, 200 µJ; bandwidth, 0.03 nm), frequency-doubled with a BBO crystal (Inrad Corp.) to give a pulse energy at 285.2 nm of about 10 µJ. The diameter of the dye laser was set to be about 80% that of the excimer, and the two lasers crossed

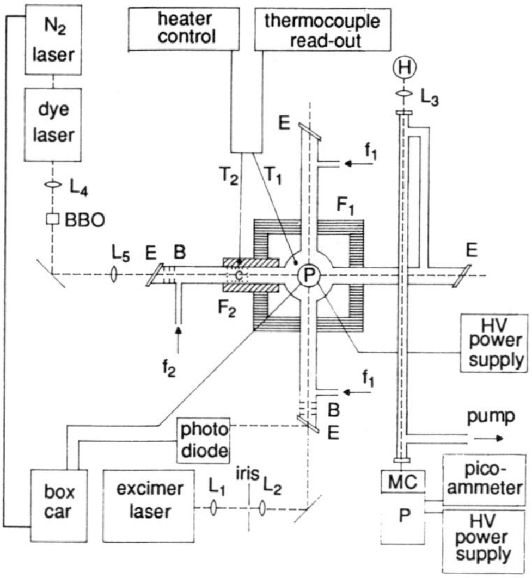

Fig. 1 Block diagram of the pulsed laser photolysis system with time-resolved monitoring of the reaction by LIF. Components are identified as follows: B, baffle to exclude scattered laser light; BBO, beta–barium–borate frequency-doubling crystal; E, Suprasil end-window at the Brewster angle; F, furnace around the central chamber of the reactor; f_1, flow of O_3 diluted in N_2 [50 standard cm^3 min^{-1} (sccm)]; f_2, flow of pure N_2 (100 sccm); H, low-pressure mercury lamp; L_1, Suprasil lens, ($f = 50$ cm); L_2, Suprasil lens ($f = 30$ cm); L_3, Suprasil lens ($f = 10$ cm); L_4, Suprasil lens ($f = 30$ cm); L_5, Suprasil lens ($f = 5$ cm); MC, monochromator; P, photomultiplier tube; T, thermocouple in the central chamber of the reactor.

orthogonally in the centre of the chamber (Fig. 1). The time-resolved LIF signal was measured by a photomultiplier tube (PMT) (Thorn EMI Gencom Inc., Model 9816QB) after passing through an interference filter centred at 285 nm [Microcoatings, Model ML3-285; full width at half maximum (FWHM), 10 nm], and was then recorded using a gated integrator (Stanford Research Systems, Model SR250) interfaced to a microcomputer.

Reaction (3) was studied by the pulsed photolysis of a higher Mg oxide to yield MgO, followed by time-resolved LIF of MgO. The Mg oxide precursor was made *in situ* by setting up a flow of Mg vapour from the heat-pipe, which reacted with the O_3 in the central chamber through reactions (2) and (3), and a probable further reaction with O_3, to yield MgO_3.[23] The Mg vapour flow was created by placing Mg metal turnings in the boat in the heat-pipe, which was then heated to between 853 and 923 K. The MgO produced upon photolysis was probed at 499.4 nm [MgO($B\,^1\Sigma^+$–$X\,^1\Sigma^+$, $\Delta v = 0$)] using the same dye laser (laser dye Coumarin 500; pulse energy, 240 μJ), and the non-resonant fluorescence signal [MgO($B\,^1\Sigma^+$–$A\,^1\Sigma^+$), $\Delta v \leqslant 0$] at wavelengths >600 nm was monitored through a long-pass filter with a cut-on at 570 nm.

Both reactions (1) and (2) were studied at low total pressures (3–18 Torr) in order to minimise the turnover time of the gas mixture in the central chamber to between 0.4 and 1.6 s, thereby preventing excessive loss of O_3, MgAcAc or MgO_3 vapour on the chamber walls. The O_3 concentration was monitored downstream of the reactor by optical absorption of the 253.7 nm line from a low-pressure mercury lamp. This was measured in a 1 m long absorption cell with a 0.9 nm resolution monochromator (Optometrics Corp., Model MC1-02) coupled to a photomultiplier tube. The ozone absorption cross-section at 298 K was taken to be 1.143×10^{-17} cm^2 from an average of several recent determinations.[26]

Materials

N_2 (Air Products, 99.999% pure) was used without further purification. O_3 was made by passing O_2 (Air Products, 99.999% pure) through a commercial ozonizer (Clearwater Tech, model M-1500). The resulting 5–8% O_3–O_2 mixture was collected on silica gel at 156 K, and the O_2 then pumped off at 195 K until a >50% O_3–O_2 mixture degassed from the gel. MgAcAc (Johnson Matthey, 98% pure) was purified by heating the sample in the heat-pipe to about 370 K under vacuum for several hours prior to kinetic experiments. Mg metal turnings (Aldrich, 99.95% pure) were used without further purification.

Experimental Results

The yield of Mg atoms from the photolysis of MgAcAc at 193.3 nm was proportional to approximately the square of the excimer laser fluence, indicating that the process is a multiphoton dissociation. A large signal was observed through the interference filter (285 ± 5 nm) at zero time after the excimer pulse, even with the dye laser switched off. This signal was proportional to the MgAcAc concentration in the central chamber, and is consistent with one of the photofragments being $Mg(3\,^1P_1)$, which then fluoresced spontaneously [$Mg(3\,^1P_1$–$3\,^1S)$, $\lambda = 285.2$ nm, $\tau_e = 33$ ns].[27]

The excimer laser energy was limited in order to keep the photolysis of O_3 to <1%. This is necessary because, by analogy with calcium,[28] the reaction

$$MgO + O \rightarrow Mg + O_2 \tag{6}$$

probably proceeds close to the collision number [$\Delta H^\circ_0(6) = -229$ kJ mol^{-1}].[22] Thus, if the ratio [O]:[O_3] were large enough, k_2 would have been underestimated.

Under the conditions of the present study, where the concentration of O_3 was always well in excess of the concentration of Mg atoms resulting from the pulsed photolysis of MgAcAc vapour, the loss of Mg atoms should be described by the pseudo-first-order decay coefficient, k', where

$$k' = k_{diff,\,Mg} + k_2[O_3] + k_1[O_2][N_2] \tag{I}$$

The term $k_{diff,\,Mg}$ describes diffusion of the Mg atoms out of the volume defined by the dye laser beam and within the field of view of the PMT. The third term of the right-hand side accounts for the loss of Mg by reaction (1). The ratio of [O_3] : [O_2] in the central chamber varied from 1 : 1 at reactor temperatures below 320 K, down to 1 : 3 at the highest temperature of 383 K. Since $k_1(T)$ is extremely slow at the low temperatures and bath-gas pressures employed here [$k_1(300\text{ K}) = 4.5 \times 10^{-34}$ cm^6 molecule^{-2} s^{-1}],[21] this term makes a negligible contribution to k'.

For the study of reaction (3), an analogous expression to eqn. (I) may be applied to interpret the pseudo-first-order decay of MgO in the presence of O_3 and O_2. The rate coefficient for the recombination reaction

$$MgO + O_2 + N_2 \rightarrow MgO_3 + N_2 \tag{7}$$

has not been measured. However, when we added a large concentration of O_2 (>10^{16} molecule cm^{-3}) to the reaction mixture, the increase in the first-order rate of removal of MgO indicated that k_7 is only about 10^{-31} cm^6 molecule^{-2} s^{-1} at 300 K (*vide infra*). Thus, this reaction would have contributed much less than 1% to the measured loss of MgO in the O_3–O_2 mixture used to measure k_3.

In the case of both reactions (2) and (3), the observed decays of the LIF signal were of a single exponential form and were well fitted to the expression $A\exp(-k't)$. This is shown in Fig. 2 which illustrates the decay of MgO in the presence of two different concentrations of O_3. These time-profiles of the LIF signal do not exhibit significant growth at early times after the excimer laser pulse, which indicates that the photolysis of

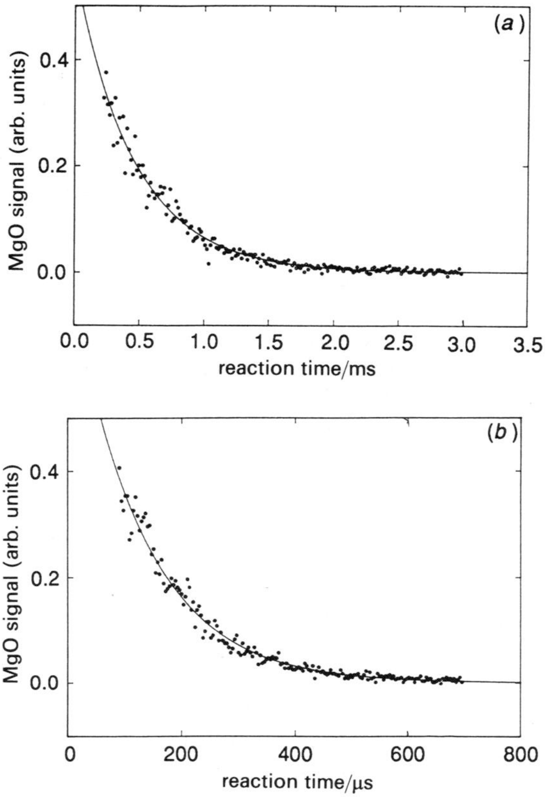

Fig. 2 Time-resolved profiles of the LIF signal obtained by pumping the MgO(B $^1\Sigma^+$–X $^1\Sigma^+$, $\Delta v = 0$) transition at 499.4 nm and monitoring emission from the MgO(B $^1\Sigma^+$–A $^1\Sigma^+$, $\Delta v < 0$) transition at wavelengths >600 nm, following the pulsed photolysis at 193.3 nm of MgO_3 in the presence of O_3 and N_2. $[N_2] = 2.5 \times 10^{17}$ molecule cm^{-3}, $T = 366$ K, excimer laser fluence at 193.3 nm = *ca.* 1×10^{16} photons cm^{-2} pulse^{-1}. (*a*), $[O_3] = 2.6 \times 10^{13}$ molecule cm^{-3}, (*b*) $[O_3] = 1.6 \times 10^{14}$ molecule cm^{-3}. The solid lines are fits to the form $A \exp(-k't)$.

MgO_3 at 193.3 nm produces MgO rather than Mg, and that the low-lying metastable states of MgO (a $^3\Pi$ and b $^3\Sigma^+$) do not play a major role in the kinetics of MgO(X $^1\Sigma^+$) in this system.

One potential source of error in these experiments could arise from an alternative pathway for reaction (3),

$$MgO + O_3 \rightarrow Mg + 2O_2 \qquad (3a)$$

which would recycle MgO back to atomic Mg and cause both k_2 and k_3 to be underestimated. The analogous reaction of NaO does indeed give rise to a significant artifact when studying the reaction between Na and O_3.[29] However, the decays of Mg were always single exponentials, with no evidence of an approach to a quasi-steady state between Mg and MgO at long reaction times, so that reaction (3*a*) appears to be a minor channel. Analysis of these decays using a time-resolved chemical model indicates an upper limit to k_{3a} (200 K) of 2×10^{-12} cm^3 molecule^{-1} s^{-1}.

Fig. 3 contains a selection of plots of k' *vs.* $[O_3]$ over the temperature ranges at which reactions (2) and (3) were studied. In both cases k' exhibits a clear linear dependence on $[O_3]$, the slopes of these plots yielding k_2 or k_3 as a function of temperature.

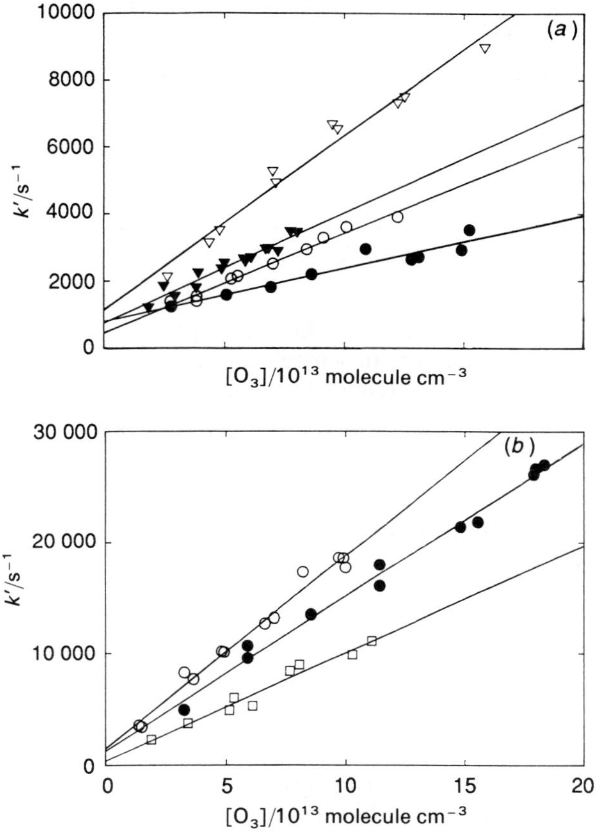

Fig. 3 Selected plots of k' against $[O_3]$ over the experimental temperature range for the reactions: (*a*) $MgO + O_3 \rightarrow MgO_2 + O_2$; ▽, 366; ▼, 313; ○, 246 and ●, 217 K: (*b*) $Mg + O_3 \rightarrow MgO + O_2$; □, 200; ●, 233 and ○, 309 K

Table 1 Experimental determination of k_2 ($Mg + O_3 \rightarrow MgO + O_2$) as a function of temperature

T/K	$k_2/10^{-10}$ cm^3 molecule^{-1} s^{-1}	$[O_2]$/molecule cm^{-3}	pressure/Torr
196	1.22 ± 0.14	5.8×10^{16}	8.8
200	0.97 ± 0.14	$<10^{14}$	9.0
231	1.38 ± 0.16	9.6×10^{16}	17.3
233	1.39 ± 0.10	8.8×10^{16}	16.0
242	1.12 ± 0.20	$<10^{14}$	9.2
300	1.22 ± 0.10	$<10^{14}$	12.9
309	1.73 ± 0.14	6.3×10^{16}	15.2
351	1.58 ± 0.26	4.8×10^{16}	13.1
361	1.51 ± 0.18	$<10^{14}$	15.2
362	1.48 ± 0.26	5.0×10^{16}	14.2
368	1.56 ± 0.10	5.4×10^{16}	15.4

The quoted uncertainty is 2σ.

Table 2 Experimental determination of k_3 ($MgO + O_3 \rightarrow MgO_2 + O_2$) as a function of temperature

T/K	$k_3/10^{-11}$ cm^3 molecule^{-1} s^{-1}	pressure/Torr
217	1.59 ± 0.38	8.8
248	2.83 ± 0.36	3.0
313	3.27 ± 0.44	3.5
366	5.18 ± 0.60	9.5

The quoted uncertainty is 2σ.

The results are listed in Tables 1 and 2. The intercepts of these plots correspond to values of $k_{\text{diff, Mg}}$ or $k_{\text{diff, MgO}}$ which are in sensible accord with the diffusional loss of Mg or MgO out of the volume of the excimer laser beam under our experimental conditions.[30] The upper temperature limit at which both reactions could be studied was defined by the decomposition of O_3 in the central chamber becoming too rapid, compared with the residence time of the gas mixture in the reactor. The lower limit was prescribed by the use of solid CO_2 as a refrigerant.

Fig. 4 illustrates the Arrhenius plots of both reactions, from which we obtain (at the 2σ level of uncertainty)

$$k_2(196 < T/\text{K} < 368) = (2.28 \pm 0.74) \times 10^{-10} \exp[-(139 \pm 84\ \text{K})/T]\ \text{cm}^3\ \text{molecule}^{-1}\ \text{s}^{-1} \quad \text{(II)}$$

$$k_3(217 < T/\text{K} < 366) = (2.19 \pm 1.8) \times 10^{-10} \exp[-(548 \pm 271\ \text{K})/T]\ \text{cm}^3\ \text{molecule}^{-1}\ \text{s}^{-1} \quad \text{(III)}$$

We consider that systematic errors, particularly those associated with determining the concentration of O_3, are contained within the quoted error limits. When studying reaction (2), a large concentration of O_2 was sometimes added in order to remove the MgO product by reaction (7) (Table 1). Inspection of Fig. 5 indicates that this did not signifi-

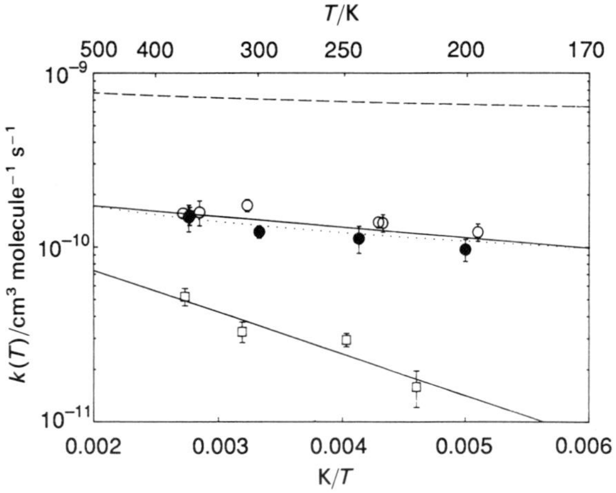

Fig. 4 Arrhenius plots for the reactions $Mg + O_3 \rightarrow MgO + O_2$ (●, $[O_2] < 10^{14}$ molecule cm^{-3}; ○, $[O_2] > 10^{16}$ molecule cm^{-3}) and $MgO + O_3 \rightarrow MgO_2 + O_2$ (□). The solid lines through the experimental points are linear regression fits. Also shown are the theoretical predictions of the harpoon mechanism (···, ref. 36) and the modified harpoon mechanism (---, ref. 38) for the reaction between Mg and O_3.

cantly increase the measured values of k_2, providing additional evidence that reaction (3*a*) does not compete with reaction (3) under these conditions.

Discussion of Experimental Results

This appears to be the first experimental determination of k_2 and k_3. In the case of reaction (2), the products ($MgO + O_2$) cannot both form in their ground electronic states unless the reaction switches from a singlet to a triplet potential surface. Since this is unlikely,[31] the formation of $MgO(A\,^1\Pi$ or $X\,^1\Sigma^+)$ must be accompained by $O_2(a\,^1\Delta_g$ or $b\,^1\Sigma_g^+)$, and $MgO(a\,^3\Pi$ or $b\,^3\Sigma^+)$ would be expected to form together with $O_2(X\,^3\Sigma_g^-)$. All of these combinations of product states arise from exothermic reaction channels.[22,23] However, some of these excited states of MgO appear to be dark {*e.g.* τ_e $[MgO(a\,^3\Pi)] = 0.1$ s}.[33] Also, $MgO(A\,^1\Pi)$ is only 0.37 eV above the ground state, and the formation of $MgO(B\,^1\Sigma^+)$ is endothermic by 74 kJ mol^{-1},[22,23] so that reaction (2) is not chemiluminescent at wavelengths <3000 nm, in contrast to the analogous reaction of Ca.[31]

Reaction (2) has two other possible reaction channels:

$$Mg + O_3 \rightarrow MgO_2 + O \tag{2a}$$

$$Mg + O_3 + N_2 \rightarrow MgO_3 + N_2 \tag{2b}$$

However, the $MgO_2(^3A_2)$ ground state has a bond energy, $D_0(Mg—O_2)$, of only 95 kJ mol^{-1},[24] so that reaction (2*a*) is probably slightly endothermic. Inspection of Table 1 reveals that there is no significant pressure dependence in k_2, so that reaction (2*b*) cannot be the dominant reaction pathway. However, using the relevant molecular parameters for MgO_3 from the *ab initio* quantum calculations described in the following section and applying the Troe formalism,[25] we estimate that $k_{2b}(T) = 3.5 \times 10^{-29}$ $(T/300\ K)^{-2.2}$ cm^6 molecule^{-2} s^{-1}. This indicates that the contribution of reaction (2*b*) to the removal of atomic Mg could have ranged from 5% at 386 K to 22% at 196 K under the conditions of the present study. Unfortunately, the slight pressure dependence of k_2 that this might have produced would not have been measurable over the restricted pressure range in this experimental system (*vide supra*). In any case, the calculation of $k_2(T)$ is almost certainly an upper limit because it assumes that only recombination can occur on the singlet $Mg + O_3$ potential energy surface, and that the highly exothermic exit channel to $MgO + O_2$ [*i.e.* reaction (2)] is closed. We therefore conclude that the role of reaction (2*b*) in this study was probably not significant. Nevertheless, the presence of a relatively deep energy minimum on the reaction potential energy surface [$D_0(Mg—O_3) = 204$ kJ mol^{-1}, *vide infra*], is probably one reason why reaction (2) is about seven orders of magnitude faster at room temperature than the apparently similar reaction $Mg + N_2O \rightarrow MgO + N_2$.[34]

Eqn. (II) demonstrates that reaction (2) proceeds essentially at the collision number, with a very small *T*-dependence. Similar behaviour has been observed for the analogous reactions of Na,[29] K,[2] Li,[2] Ca[31] and Fe.[35] The rate coefficients of these reactions vary inversely with the ionization energies of the metal atoms,[2] in accord with the original proposal by Kolb and Elgin[36] that these types of reaction proceed *via* the harpoon or electron jump mechanism. The rate coefficient for the simple harpoon mechanism is given by

$$k(T) = \pi R_c^2 (8k_B T/\pi\mu)^{1/2} \tag{IV}$$

where μ is the reduced mass of the reactants and k_B is the Boltzmann constant. The charge-transfer distance, R_c, is determined from the simple relationship

$$R_c = e^2/[4\pi\varepsilon_0(E_i - E_{ea})] \tag{V}$$

Table 3 Molecular parameters for the harpoon model calculations

	Mg	O_3
dipole moment/D	—	0.53[a]
polarizability/10^{-24} cm^3	10.6[a]	3.2[a]
ionization potential/eV	7.65[b]	12.4[b]
electron affinity/eV	—	2.10[b]

[a] Ref. 27. [b] Ref. 37.

where e is the electronic charge and, in the case of reaction (2), E_i is the ionisation energy of atomic Mg and E_{ea} is the electron affinity of O_3. This yields $R_c = 2.60$ Å if the adiabatic electron affinity of O_3, which has been accurately measured (Table 3), is adopted. As shown in Fig. 4, there is excellent agreement between the prediction of the simple harpoon mechanism and the experimental results: the $T^{1/2}$ dependence from eqn. (IV) is in good agreement with the experimental result of $T^{(0.51 \pm 0.32)}$. Also shown in Fig. 4 is the theoretical upper limit to k_2, determined by the significant attractive dispersion force between Mg and O_3 at longer range than R_c. This was calculated using the modified harpoon treatment of Gislason,[38] with the polarizability data listed in Table 3.

Reaction (3) is spin-allowed if $MgO_2(^3A_2)$ and $O_3(X\,^3\Sigma_g^-)$ are formed [in any case, $MgO_2(^1A_1)$ is only slightly bound[24]]. Furthermore, the alternative channel, $MgO + O_3 \rightarrow MgO_3 + O$, is endothermic by *ca.* 73 kJ mol^{-1}. Consideration of eqn. (III) indicates that this reaction probably has a small energy barrier. The pre-exponential factor is essentially at the collision number, in part because the large dipole moment of MgO (6.2 D† [39]) will induce a significant long-range interaction with O_3.[29]

A Model of Magnesium in the Upper Atmosphere

The importance of reaction (2) in the upper atmosphere can most easily be explored by comparing the first-order rates of loss of atomic Mg by this reaction and three other pathways. These are recombination with O_2 [reaction (1)] and charge exchange with the two most abundant postive ions in the mesosphere and thermosphere,[18,20]

$$Mg + NO^+ \rightarrow Mg^+ + NO \tag{8}$$

and

$$Mg + O_2^+ \rightarrow Mg^+ + O_2 \tag{9}$$

The rates of reactions (1), (2), (8) and (9) are plotted in Fig. 5 as a function of altitude between 65 and 110 km. In the daytime, the rates of reactions (7) and (8) increase by a factor of *ca.* 10 because of the enhancement in the concentrations of NO^+ and O_2^+ due to photo-ionization.[1,18] Atomic Mg is also photo-ionized in the mesosphere with a coefficient calculated by Swider[40] to be about 4×10^{-7} s^{-1}, although as shown in Fig. 5 this is not a competitive process. During night-time, the rate of reaction (2) increases because of the build-up of O_3 in the absence of photolysis.

Reaction (2) is clearly the most rapid removal process up to 95 km. However, MgO is probably not a stable sink for magnesium because reaction (6) cycles MgO back to Mg, and [O] is much greater than [O_3] above 80 km.[2,18] Likewise, although reaction (3) is relatively fast [$k_3(200\ K) = 1.4 \times 10^{-11}$ cm^3 molecule^{-1} s^{-1}], MgO_2 is not a strongly bound molecule[21,22] and is probably reduced efficiently back to MgO by atomic O, as in the case of NaO_2.[19]

† 1 D ≈ $3.335\,64 \times 10^{-30}$ C m.

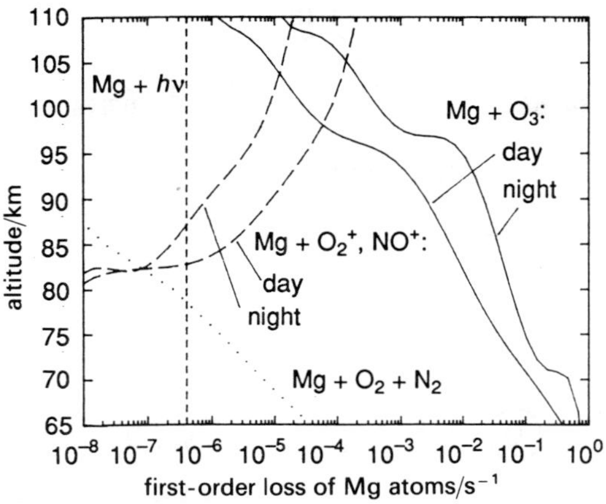

Fig. 5 Vertical profile in the earth's atmosphere between 65 and 110 km, depicting the first-order rates of loss of Mg atoms due to reaction with O_3 (this study), O_2 (ref. 21) and charge exchange with positive ions (ref. 18). The atmopsheric concentrations of the trace species O_3, NO^+ and O_2^+ are taken from a recent model (ref. 19), and the dependence of the pressure and temperature with altitude are from ref. 48. The conditions are for December, 40°N.

It is therefore likely that MgO eventually recombines with trace atmospheric species such as H_2O, CO_2 or O_2 [reactions (4), (5) and (7)], and we now examine the product molecules $Mg(OH)_2$, $MgCO_3$ and MgO_3 through *ab initio* quantum calculations. The geometries of these three molecules were first optimized, using the GAUSSIAN92 set of programs,[41] at the Hartree–Fock level with the 6−31+G basis set. We have established previously that this standard split-valence basis set, with added diffuse functions on the atoms, provides a satisfactory representation of these types of highly ionic species.[42] Fourth-order Møller–Plesset perturbation theory (MP4)[43] was then employed to determine the correlation energy corrections to these molecules at their HF geometries.

The optimized geometries, together with the Mulliken charges on the atoms, are illustrated in Fig. 6. $Mg(OH)_2$ is linear rather than bent. Interestingly, there is conflicting experimental and theoretical evidence[32] about whether MgF_2, which is isoelectronic with $Mg(OH)_2$, is linear or bent with an F—Mg—F angle of 158°. $MgCO_3$ and MgO_3 are both planar with C_{2v} symmetry. In the case of MgO_3, this geometry is in accord with the available experimental evidence from matrix isolation studies.[23] Both molecules have singlet rather than triplet ground electronic states: $MgCO_3(^1A_1)$ is 0.52 eV below $MgCO_3(^3B_2)$, and $MgO_3(^1A_1)$ is 0.65 eV below $MgO_3(^3B_1)$. In the case of all three molecules in Fig. 6, the charge on the Mg atom is between +1 and +2, and the structures of MgO_3 and $MgCO_3$, where the Mg atom is equidistant between two oxygen atoms, are more stable than the other geometric isomers of these molecules. The structure of MgOH was also optimized at this level of theory, yielding a geometry and value for D_0(Mg—OH) in good agreement with earlier theory and experiment.[44] The dipole moments, rotational constants and vibrational frequencies of all four molecules are listed in Table 4.

In order to determine the enthalpy changes of reactions (4), (5) and (7), the geometries of the molecules H_2O, CO_2 and O_2 were also optimized at the MP4/6−31+G//HF/6−31+G level. Following Langhoff and Bauschlicher[22] the absolute electronic energy of $MgO(X\,^1\Sigma^+)$ was computed by first determining the energies of the low-lying triplet states, $MgO(a\,^3\Pi$ and $b\,^3\Sigma^+)$. At the MP4/6−31+G//HF/6−31+G level we find that the a and b states are separated by 0.75 eV, in good accord with other recent theoretical estimates of 0.73 eV[22] and 0.80 eV.[33] $MgO(X\,^1\Sigma^+)$ then lies 0.326 eV below

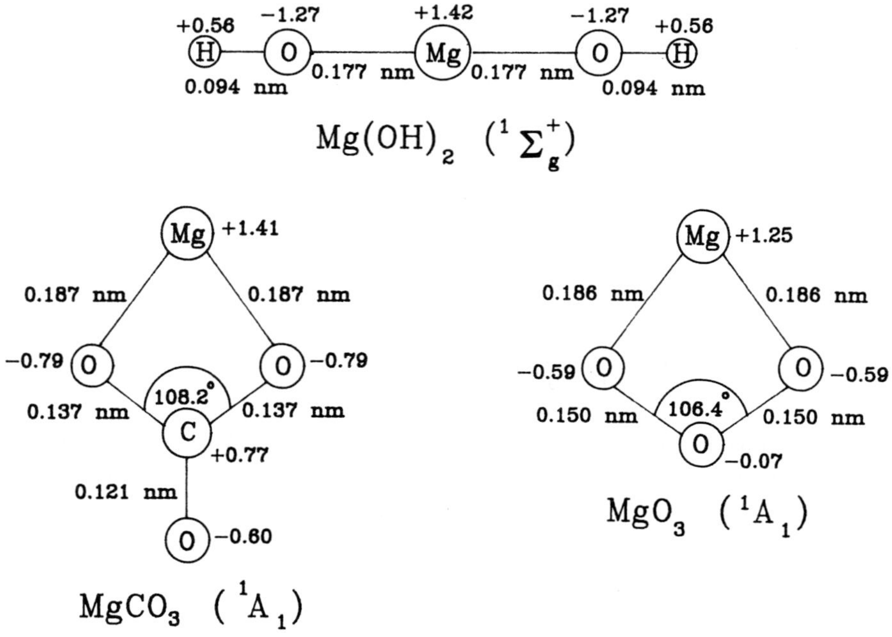

Fig. 6 Optimized geometries at the HF/6−31+G level of theory of the ground-state molecules $Mg(OH)_2(^1\Sigma^+)$, $MgCO_3(^1A_1)$ and $MgO_3(^1A_1)$. The Mulliken electron populations are indicated as a signed number adjacent to each atomic centre.

$MgO(a\ ^3\Pi)$.[45] The resulting enthalpy changes for these reactions are:

	ΔH°_0/kJ mol^{-1}	
$MgO(^1\Sigma^+) + H_2O \rightarrow Mg(OH)_2(^1\Sigma_g{}^+)$	−420	(4)
$MgO(^1\Sigma^+) + CO_2 \rightarrow MgCO_3(^1A_1)$	−94	(5)
$MgO(^1\Sigma^+) + O_2 \rightarrow MgO_3(^1A_1)$	−39	(7)
$\rightarrow MgO_3(^3B_1)$	+16	(7′)

$\Delta H^\circ_0(4)$ is in good agreement with an estimate of -451 ± 62 kJ mol^{-1} from the study of magnesium in flames,[46] if the most recent bond energy D_0(Mg—O) = 265 kJ mol^{-1} from Langhoff and Bauschlicher[22] is adopted. The rate coefficients for these reactions can now be calculated by means of the Troe formalism.[25] Note that if

Table 4 Molecular parameters of MgOH, $Mg(OH)_2$, MgO_3 and $MgCO_3$ at the HF/6−31+G level

	dipole moment/D	rotational constants/cm^{-1}	vibrational frequencies/cm^{-1}
MgOH	1.3	0.486	524 (×2), 773, 4245
$Mg(OH)_2$	0	0.146	153 (×2), 493 (×2), 511 (×2), 627, 941, 4260 (×2)
MgO_3	11.2	0.363, 0.297, 0.163	146, 430, 657, 776, 893, 927
$MgCO_3$	14.2	0.425, 0.133, 0.101	178, 496, 557, 716, 857, 864, 972, 1139, 1876

spin is conserved in the recombination of MgO and O_2 [reaction (7′)], then the reaction is slightly endothermic. In fact, even formation of $MgO_3(^1A_1)$ is not exothermic enough for the calculated k_7(300 K) to approach our experimental estimate of about 10^{-31} cm^6 molecule^{-2} s^{-1}. Of course, there is some uncertainty in the theoretical enthalpy changes at this level of *ab initio* theory:[42] the experimental result would be accounted for with $\Delta H^\circ_0(7) = -80$ kJ mol^{-1}. The application of the Troe formalism[25] to reactions (4) and (5) is summarised in Table 5, and yields:

$$k_4(T) = 1.7 \times 10^{-25}(T/200)^{-3.7} \text{ cm}^6 \text{ molecule}^{-2} \text{ s}^{-1}$$

$$k_5(T) = 1.8 \times 10^{-31}(T/200)^{-2.8} \text{ cm}^6 \text{ molecule}^{-2} \text{ s}^{-1}$$

The large value of k_4 arises because this reaction is very exothermic and $Mg(OH)_2$ has a high density of vibrational states. This result is in sensible accord with a study of the analogous recombination of CaO and H_2O in flames, where the reaction has been shown to be a very fast second-order process at atmospheric pressure.[47]

The stabilities of these neutral magnesium compounds with respect to reaction with the atmospheric trace species H and O can now be evaluated. The exothermic reaction channels, which are the only feasible pathways at the low temperatures of the upper atmosphere, are listed in Table 6. Inspection of this table reveals that there are several highly exothermic reaction pathways for the removal of $MgCO_3$ and MgO_3, whereas the much less exothermic reaction between $Mg(OH)_2$ and H appears to be the only reaction in the upper atmosphere that could remove the dihydroxide. Taking this into account, and noting that the recombination reaction between MgO and H_2O is much faster than that between MgO and CO_2 or O_2, we have concluded that $Mg(OH)_2$ is the major neutral sink for magnesium, and, for the purpose of this first atmospheric model of the metal, ignored the role of $MgCO_3$ and MgO_3.

The 1D atmospheric model that we now describe has been used successfully to model the mesospheric sodium layer.[1,19] We have assumed that magnesium is partitioned among eight constituent species [Mg, Mg^+, MgO^+, $MgO_2{}^+$, MgO, MgO_2,

Table 5 Calculation of the rate coefficients for the recombination of MgO with H_2O, CO_2 and O_2 at 300 K

	$MgO + H_2O + N_2$	$MgO + CO_2 + N_2$
$\sigma(MgX^a{-}N_2)$/Å	4.0	4.0
$[\varepsilon(MgX^a{-}N_2)/k]$/K	400	400
Z_{LJ}/cm^3 molecule^{-1} s^{-1}	5.16×10^{-10}	4.91×10^{-10}
s	10	9
m	6	4
r	0	0
Q_{vib}	5.74	2.23
a	0.940	0.939
$\rho(E_0)$	4.69×10^6	7.49
F_E	1.048	1.176
F_{anh}	1.41	1.29
F_{rot}	19.64	5.79
F_{corr}	1	1
K_{eq}/cm^3 molecule^{-1}	8.55×10^{46}	1.79×10^{-10}
k_0^{sc}/cm^3 molecule^{-1} s^{-1}	2.26×10^{-72}	1.55×10^{-21}
$k_{rec,0}^{sc}$/cm^6 molecule^{-2} s^{-1}	1.93×10^{-25}	2.77×10^{-31}
β_c	0.20	0.20
ΔE/kJ mol^{-1}	$-0.95\ (T/300)^{0.0}$	$-1.06\ (T/300)^{0.0}$

For the definition of symbols not in the present text, see ref. 25. [a] X = $(OH)_2$ or CO_3.

Table 6 Enthalpy changes[a] for the exothermic bimolecular reaction channels between magnesium compounds and the minor atmospheric species H and O

	ΔH°_0/kJ mol^{-1}
$Mg(OH)_2 + H \rightarrow MgOH + H_2O$	−58
$MgCO_3 + H \rightarrow MgOH + CO_2$	−384
$\rightarrow Mg + OH + CO_2$	−65
$MgCO_3 + O \rightarrow MgO_2 + CO_2$	−229
$\rightarrow Mg + O_2 + CO_2$	−134
$MgO_3 + O \rightarrow MgO_2 + O_2$	−283
$\rightarrow Mg + 2O_2$	−189
$\rightarrow MgO + O_3$	−61
$MgO_3 + H \rightarrow MgOH + O_2$	−439
$\rightarrow MgO_2 + OH$	−215
$\rightarrow MgO + HO_2$	−154
$MgO_2 + O \rightarrow MgO + O_2$	−170
$\rightarrow Mg + O_3$	−5
$MgO_2 + H \rightarrow MgOH + O$	−155
$\rightarrow MgO + OH$	−101
$\rightarrow Mg + HO_2$	−98
$\rightarrow MgH + O_2$	−98
$MgOH + H \rightarrow Mg + H_2O$	−175
$MgO + O \rightarrow Mg + O_2$	−228
$MgO + H \rightarrow Mg + OH$	−159

[a] Calculated using the estimates of $\Delta H^\circ_0(4)$, $\Delta H^\circ_0(5)$ and $\Delta H^\circ_0(7)$ given in the text, together with $D_0(\text{Mg—O}) = 265$ kJ mol^{-1},[22] $D_0(\text{Mg—O}_2) = 95$ kJ mol^{-1},[24] $D_0(\text{Mg—OH}) = 319$ kJ mol^{-1},[44] and $D_0(\text{Mg—H}) = 192$ kJ mol^{-1}.[32]

MgOH and $Mg(OH)_2$], and that the partitioning is rapid on the timescale of vertical mixing between 65 and 110 km. Since the transport of all the magnesium constituents is then governed by the same eddy diffusion coefficient up to the turbopause at about 105 km, the continuity equation for total magnesium can be solved as a function of altitude.[1] We have adopted the meteoric ablation profile from the model of Hunten *et al.*[14] in order to obtain the ablation rate as a function of altitude. The seasonal and latitudinal dependences of the temperature, pressure and eddy diffusion coefficient in the model are taken from the COSPAR data-set,[48] and the vertical profiles of the minor species O, H and O_3 are adapted from Garcia and Solomon[49] and Thomas.[50] The concentrations of the major ions NO^+ and $O_2{}^+$ and the electron density are taken from the model of Thomas *et al.*[51]

As in the case of our sodium model,[19] we have assumed that the partitioning of magnesium is governed by the gas-phase chemistry only, and does not include heterogeneous reactions on meteoric dust (or smoke[14]) particles. The gas-phase chemistry in the model is illustrated in Fig. 7, and the temperature-dependent rate coefficients are listed in Table 7. It will be seen that the ionic chemistry of magnesium is comparatively well understood.[20] In fact, since the radiative recombination of an atomic metal ion with an electron [reaction (21)] is known to be slow,[52] the only significant uncertainty in the ionic reaction scheme is k_{20}, the rate coefficient for the dissociative recombination reactions of MgO^+ and $MgO_2{}^+$ with electrons. This has been assigned a value based on the measured rate coefficients for the dissociative recombination reactions of several small oxygen-containing species, which tend to be very similar in magnitude.[53]

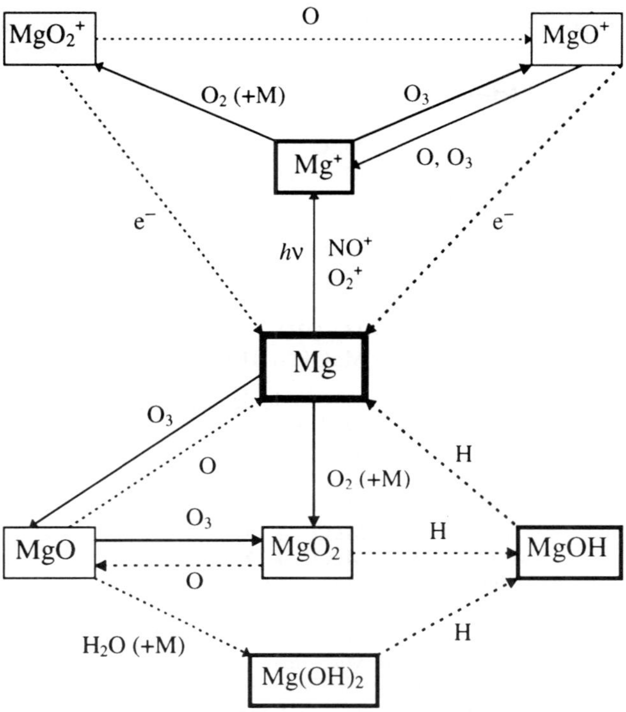

Fig. 7 Schematic diagram of the gas-phase chemistry of meteor-ablated magnesium in the upper atmosphere (Table 7). Solid lines indicate reactions for which rate coefficents have been measured.

Apart from the new results reported in this paper, very little is known about the neutral chemistry of magnesium. We have therefore had to estimate k_6, k_{10}, k_{11}, k_{12} and k_{13}. However, the thermochemical calculations described above indicate that $Mg(OH)_2$ is by far the most stable of the neutral magnesium species. Furthermore, studies of magnesium seeded into fuel-rich hydrogen flames show that the concentrations of the metal-containing species are in the order $Mg(OH)_2 > MgOH > MgO$.[46] This implies that k_{13} should be the smallest of these unknown rate coefficients, and that the model should not be particularly sensitive to the other five. By choosing to assign k_6 and k_{10-12} to the rate coefficients of their analogous sodium reactions,[19] this condition is fulfilled.

The model therefore contains two adjustable parameters: k_{13} and the Mg/Na ratio of the input flux from meteoric ablation. Whilst there appear to have been no reported observations of atomic Mg, the Mg^+ profile has been measured at a range of latitudes by rocket-borne mass spectrometry. The review by Swider[7] finds that the peak height of the Mg^+ profile lies above 90 km, which places an upper limit on k_{13}(200 K) of 5×10^{-15} cm^3 $molecule^{-1}$ s^{-1} in the model. The peak concentration of Mg^+ under quiescent conditions is about 10^3 cm^{-3}, which indicates that the Mg/Na ablation ratio is less than unity. We have therefore chosen the Arrhenius expression listed in Table 7 for k_{13}, together with a Mg/Na ablation ratio of 0.5, *i.e.* the input flux of Mg is 7×10^3 atoms cm^{-2} s^{-1}.

The resulting profiles of the major Mg constituents during daytime and night-time are illustrated in Fig. 8. Mg^+ is the dominant species above the atomic Mg layer, and $Mg(OH)_2$ is the major sink for atomic Mg below the layer, as expected. The characteristic properties of the Mg and Mg^+ layers at midday and midnight are listed in Table 8.

Table 7 Neutral and ionic gas-phase reactions of Mg species in the mesosphere

reaction number	reaction	rate coefficient
	neutral chemistry	
(2)	$Mg + O_3 \rightarrow MgO + O_2$	$2.3 \times 10^{-10} \exp(-139/T)^a$
(6)	$MgO + O \rightarrow Mg + O_2$	$2.2 \times 10^{-10} (T/200)^{1/2\ b}$
(1)	$Mg + O_2 + N_2 \rightarrow MgO_2 + N_2$	$4.9 \times 10^{-30} \exp(-2790/T)^c$
(3)	$MgO + O_3 \rightarrow MgO_2 + O_2$	$2.2 \times 10^{-10} \exp(-548/T)^a$
(10)	$MgO_2 + O \rightarrow MgO + O_2$	$5 \times 10^{-10} \exp(-940/T)^b$
(11)	$MgO_2 + H \rightarrow MgOH + O$	$1 \times 10^{-9} \exp(-1000/T)^b$
(12)	$MgOH + H \rightarrow Mg + H_2O$	$4 \times 10^{-11} \exp(-550/T)^b$
(4)	$MgO + H_2O + N_2 \rightarrow Mg(OH)_2 + N_2$	$1.7 \times 10^{-25} (T/200)^{-3.7\ d}$
(13)	$Mg(OH)_2 + H \rightarrow MgOH + H_2O$	$5 \times 10^{-10} \exp(-2500/T)^e$
	ionic chemistry	
(14)	$Mg + h\nu \rightarrow Mg^+ + e^-$	$4 \times 10^{-7\ f}$
(8)	$Mg + NO^+ \rightarrow Mg^+ + NO$	$8.1 \times 10^{-10\ g}$
(9)	$Mg + O_2^+ \rightarrow Mg^+ + O_2$	$1.2 \times 10^{-9\ g}$
(15)	$Mg^+ + O_2 + N_2 \rightarrow MgO_2^+ + N_2$	$4 \times 10^{-30} (T/200)^{-1\ h}$
(16)	$Mg^+ + O_3 \rightarrow MgO^+ + O_2$	$7 \times 10^{-10\ h}$
(17)	$MgO_2^+ + O \rightarrow MgO^+ + O_2$	$1 \times 10^{-10\ i}$
(18)	$MgO^+ + O \rightarrow Mg^+ + O_2$	$1 \times 10^{-10\ h}$
(19)	$MgO^+ + O_3 \rightarrow Mg^+ + 2O_2$	$8 \times 10^{-10\ h}$
(20)	$Mg \cdot X^+ + e^- \rightarrow Mg + X$ ($X = O$ or O_2)	$2 \times 10^{-7} (T/200)^{-1/2\ j}$
(21)	$Mg^+ + e^- \rightarrow Mg + h\nu$	$1 \times 10^{-12\ k}$

[a] Measured (this study). [b] Adapted from ref. 19. [c] Measured (ref. 21). [d] Calculated (this study). [e] Fitted model parameter. [f] Calculated (ref. 40). [g] Measured (ref. 18). [h] Measured (ref. 20). [i] Estimated (ref. 18). [j] Adapted from ref. 53. [h] Estimated (ref. 52).

Note that the model predicts no significant diurnal variation in the peak concentration, column density or FWHM of the Mg layer. However, both the peak height and the centroid height of the layer are nearly 4 km lower during the day. This is because atomic Mg is removed on the top-side of the layer through photo-ionization [reaction (14)] and the increased rates of charge transfer with NO^+ and O_2^+ [reactions (8) and (9)] during daytime. However, this removal is balanced by a release of atomic Mg from $Mg(OH)_2$ on the underside of the layer, because of the higher daytime concentration of atomic H below 85 km.[50]

By contrast with the Mg layer, the Mg^+ layer shows a nearly two-fold increase in the peak concentration and column density during the day, and the peak height is 4 km lower. This is caused by the increased rate of production of Mg^+ from Mg during the

Table 8 Characteristic properties of Mg and Na in the upper mesosphere at 40°N during winter

property	Mg (day)	Mg (night)	Na (night)
metal-atom peak concentration/cm^{-3}	1.2×10^3	1.1×10^3	5.6×10^3
metal-atom peak height/km	87.0	90.5	90.0
metal-atom column density/cm^{-2}	7.8×10^8	7.5×10^8	6.2×10^9
centroid height of metal-atom layer/km	86.6	90.4	91.5
FWHM of metal-atom layer/km	6.5	6.5	10.5
metal-ion peak concentration/cm^{-3}	2.7×10^3	1.4×10^3	3.2×10^2
metal-ion peak height/km	89.5	93.5	105.5
metal-ion column density/cm^{-2}	3.1×10^9	1.7×10^9	3.1×10^8

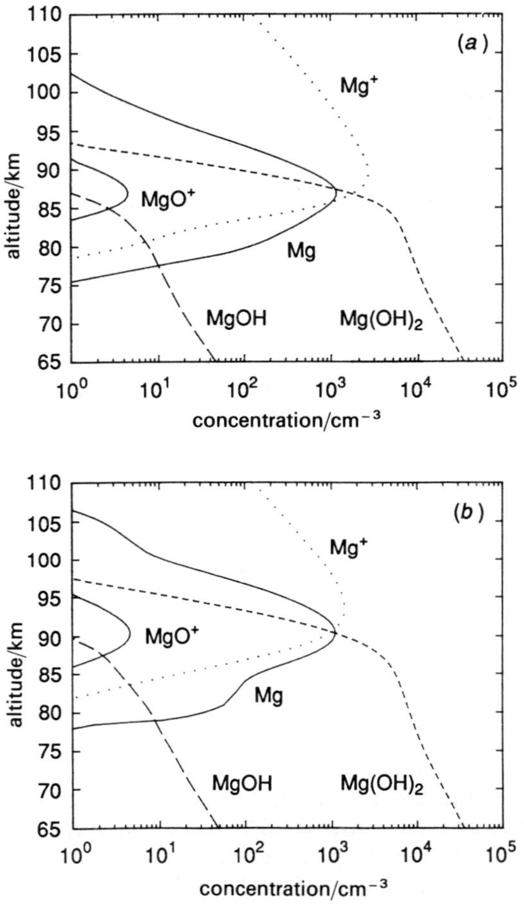

Fig. 8 Predicted profiles of the major magnesium species in the upper atmosphere for the conditions of mid-winter at 40°N: (*a*) daytime, (*b*) night-time

day, and by the higher daytime concentration of atomic O which reduces MgO^+ and $MgO_2{}^+$ to form Mg^+ [reactions (17) and (18)], and therefore more effectively competes with dissociative recombination to Mg [reaction (20)]. When Anderson and Barth[16] studied a sporadic-E layer above 100 km by using a rocket-borne UV spectrometer to observe the dayglow, they estimated that the ratio [Mg^+]/[Mg] was at least 22. Gadsden[54] subsequently reinterpreted their observations and concluded that this ratio was probably greater than 200, in agreement with the prediction of the present model that the daytime ratio above 100 km is about 300.

Fig. 9 illustrates a model comparison of the Mg/Mg^+ and Na/Na^+ layers during night-time. The characteristic properties of the layers are also listed in Table 8. It will be seen that whereas the Na layer has a much greater peak concentration and column density than the Na^+ layer, the reverse is the case for the Mg and Mg^+ layers. Furthermore, although both atomic-metal layers peak at about 90 km, the Na layer has a much broader FWHM of 10.5 km and its centroid height is 1 km higher. These differences arise principally from the ion chemistry of each metal. Na^+ is a closed-shell species and very unreactive compared with Mg^+ (which is isoelectronic with Na). Thus, Na^+ does not undergo reactions analogous to reactions (16)–(19), and so the formation of a complex between Na^+ and a ligand such as N_2, H_2O or CO_2 inevitably results in

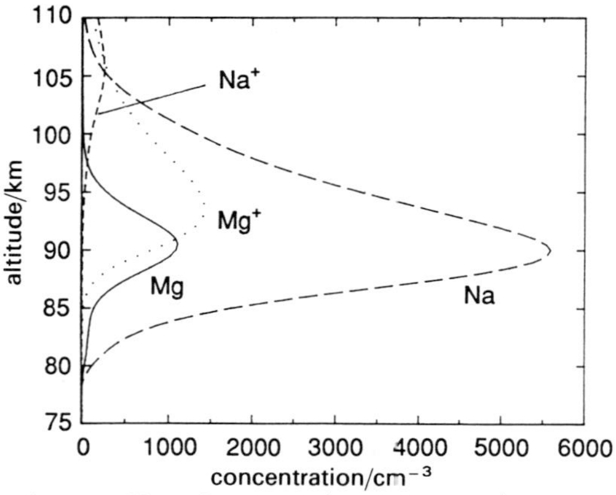

Fig. 9 Predicted night-time profiles of Mg, Mg^+, Na and Na^+ in the upper atmosphere for the conditions of mid-winter at 40°N

dissociative recombination with an electron.[19] As shown in Fig. 9, the ratio $[Mg^+]/[Na^+]$ is >100 at 93 km but <10 above 100 km, again in accord with rocket-borne observations.[7–9]

Conclusions

The experimental work reported here has established that the reaction between Mg and O_3 to form MgO is very rapid at mesospheric temperatures. *Ab initio* quantum calculations have indicated that the most stable form of magnesium in the upper atmosphere is $Mg(OH)_2$, formed from the fast recombination reaction between MgO and H_2O. A 1D model of magnesium was then constructed by using observations of the Mg^+ profile above 85 km to constrain both the chemical reaction scheme and the meteoric input flux of the metal. The atomic Mg layer is predicted to occur at the same height as the Na layer, but with a peak concentration that is smaller by a factor of *ca.* 5.

One of the most interesting findings from the model is that the ratio of the Mg : Na flux from meteoric ablation appears to be about 1 : 2, whereas the elemental abundance ratio in meteorites is about 20 : 1.[15] We consider that this is good evidence for the incomplete ablation of meteoroids, with the residual (unablated) fraction enriched in Mg. Two studies support this conclusion. First, the ablation model of Hunten *et al.*[14] predicts that only about two-thirds of the available mass of meteoroids should evaporate completely. This is because a fraction of the infalling particles are either too small, or have a relative velocity that is too low, to generate sufficient heating through atmospheric drag. Secondly, a thermodynamic model[55] has been developed to study the vaporization of a molten silicate magma of chondritic composition, heated to between 2500 and 3000 K. It was found that when *ca.* 70% of the original material had vaporized, the residue was entirely depleted in the alkali elements and largely depleted in Fe and Si, but enriched in Mg, Ca, Al and Ti. These two studies[14,55] together provide an explanation for the small ablation flux of Mg, compared with Na, that is required in our model. If this explanation is correct, then an even smaller fraction of Ca should ablate from meteoroids,[55] and it is interesting to note that lidar studies of the atomic Ca layer have shown that there is a >100-fold depletion of this metal relative to Na, when compared with their meteoritic abundances.[56,57]

In terms of future work, the most important reactions that should be studied in the laboratory are the formation and removal of $Mg(OH)_2$ [reactions (4) and (13)]. However, the real challenge must be to observe the atomic Mg layer in the upper atmosphere. The density profiles of the layer predicted by the model should be observable either by rocket-borne photometry at twilight, or by a downward-looking lidar on a platform in space.

This work was supported by the Natural Environment Research Council. We thank the Royal Society for an equipment grant, and the School of Environmental Sciences, University of East Anglia, for a research studentship (M.H.).

References

1 J. M. C. Plane, *Int. Rev. Phys. Chem.*, 1991, **10**, 55.
2 J. M. C. Plane and M. Helmer, in *Research in Chemical Kinetics*, ed. R. G. Compton and G. Hancock, Elsevier, Amsterdam, 1994, pp. 313–367.
3 D. M. Hunten, *Space Sci. Rev.*, 1967, **6**, 493.
4 M. R. Bowman, A. J. Gibson and M. C. W. Sandford, *Nature (London)*, 1969, **221**, 456.
5 T. J. Kane and C. S. Gardner, *Science*, 1993, **259**, 1297.
6 R. E. Bills, C. S. Gardner and S. J. Franke, *J. Geophys. Res.*, 1991, **96**, 22 701.
7 W. Swider, *Planet. Space. Sci.*, 1984, **32**, 307.
8 E. Kopp, P. Eberhardt, U. Hermann and L. G. Björn, *J. Geophys. Res.*, 1985, **90**, 13 041.
9 U. von Zahn, R. A. Goldberg, J. Stegman and G. Witt, *Planet. Space. Sci.*, 1989, **37**, 657.
10 G. Megie and J. E. Blamont, *Planet. Space. Sci.*, 1977, **25**, 1093.
11 M. Uchiumi, C. Nagasawa, M. Hirono, M. Fujiwara and M. Maea, *J. Geomag. Geoelectr.*, 1993, **45**, 393.
12 C. E. Junge, O. Oldenberg and J. T. Wasson, *J. Geophys. Res.*, 1962, **67**, 1027.
13 D. W. Hughes, in *Cosmic Dust*, ed. J. A. M. McDonnell, Wiley, New York, 1978, ch. 3.
14 D. M. Hunten, R. P. Turco and O. B. Toon, *J. Atmos. Sci.*, 1980, **37**, 1342.
15 B. Mason, *Handbook of Elemental Abundances in Meteorites*, Gordon and Breach, New York, 1971.
16 J. G. Anderson and C. A. Barth, *J. Geophys. Res.*, 1971, **76**, 3723.
17 J-C. Gerard and A. Monfils, *J. Geophys. Res.*, 1974, **79**, 2544.
18 M. J. McEwan and L. F. Phillips, *Chemistry of the Atmosphere*, Arnold, London, 1975.
19 M. Helmer and J. M. C. Plane, *J. Geophys. Res.*, 1993, **98**, 23 207.
20 B. R. Rowe, D. W. Fahey, E. E. Ferguson and F. C. Fehsenfeld, *J. Chem. Phys.*, 1981, **75**, 3325.
21 C-F. Nien, B. Rajasekhar and J. M. C. Plane, *J. Phys. Chem.*, 1993, **97**, 6449.
22 S. R. Langhoff and C. W. Bauschlicher, *J. Chem. Phys.*, 1986, **84**, 4474.
23 L. Andrews, E. S. Prochaska and B. S. Ault, *J. Chem. Phys.*, 1978, **69**, 556.
24 C. W. Bauschlicher, H. Partridge, M. Sodupe and S. R. Langhoff, *J. Phys. Chem.*, 1992, **6**, 9259.
25 J. Troe, *J. Chem. Phys.*, 1977, **6**, 4745; 4758; *J. Phys. Chem.*, 1979, **83**, 114.
26 W. B. DeMore, S. P. Sander, M. J. Molina, R. F. Hampson, M. J. Kurylo, C. J. Howard and A. R. Ravishankara, *Chemical Kinetics and Photochemical Data for use in Stratospheric Modeling*, JPL Publication 90-1, Jet Propulsion Laboratory: Pasadena, 1990.
27 *Handbook of Physics and Chemistry*, ed. D. R. Lide, CRC Press, Boca Raton, 73rd edn., 1992.
28 J. M. C. Plane and C-F. Nien, *J. Phys. Chem.*, 1990, **94**, 5255.
29 J. M. C. Plane, C-F. Nien, M. R. Allen and M. Helmer, *J. Phys. Chem.*, 1993, **97**, 4459.
30 J. M. C. Plane, *J. Phys. Chem.*, 1987, **91**, 6552.
31 M. Helmer, J. M. C. Plane and M. R. Allen, *J. Chem. Soc., Faraday Trans.*, 1993, **89**, 763.
32 *JANAF Thermochemical Tables*, ed. M. W. Chase Jr., C. A. Davies, J. R. Downey Jr., D. J. Frurip, R. A. McDonald and A. N. Syverud, *J. Phys. Chem. Ref. Data*, 3rd edn., 1985, 14.
33 H. Thümmel, R. Klotz and S. D. Peyerimhoff, *Chem. Phys.*, 1989, **129**, 417.
34 J. M. C. Plane, C-F. Nien and B. Rajasekhar, *J. Phys. Chem.*, 1992, **96**, 1296.
35 M. Helmer and J. M. C. Plane, *J. Chem. Soc., Faraday Trans.*, 1994, **90**, 31.
36 C. E. Kolb and J. B. Elgin, *Nature (London)*, 1976, **263**, 488.
37 S. G. Lias, J. E. Bartmess, J. F. Liebman, J. L. Holmes, R. D. Levin and W. G. Mallard, *Gas-phase Ion and Neutral Thermochemistry*, *J. Phys. Chem. Ref. Data*, 1988, **17**, suppl. 1.
38 E. A. Gislason, in *Alkali Vapors*, ed. P. Davidovits and D. L. McFadden, Academic Press, New York, 1979, p. 415.
39 H. Büsner, F. Heinrich and A. Hese, *Chem. Phys.*, 1987, **112**, 137.
40 W. Swider, *Annls. Geophys.*, 1970, **26**, 595.
41 GAUSSIAN92, Revision A. M. J. Frisch, G. W. Trucks, M. Head-Gordon, P. M. W. Gill, M. W. Wong, J. B. Foresman, B. G. Johnson, H. B. Schlegel, M. A. Robb, E. S. Replogle, R. Gomperts, J. L. Andres, K. Raghavachari, J. S. Binkley, C. Gonzalez, R. L. Martin, D. J. Fox, D. J. Defrees, J. Baker, J. J. P. Stewart and J. A. Pople, Gaussian Inc., Pittsburgh, PA, 1992.
42 B. Rajasekhar and J. M. C. Plane, *Geophys. Res. Lett.*, 1993, **20**, 21.
43 C. Moller and M. S. Plesset, *Phys. Rev.*, 1934, **46**, 618.

44 C. W. Bauschlicher, S. R. Langhoff and H. Partridge, *J. Chem. Phys.*, 1986, **84**, 901.
45 T. Ikeda, N. B. Wong, D. O. Harris and R. W. Field, *J. Mol. Spectrosc.*, 1977, **68**, 452.
46 K. Schofield, in *Gas-phase Metal Reactions*, ed. A. Fontijn, Elsevier, Amsterdam, 1992, pp. 529–571.
47 D. E. Jensen and G. A. Jones, *Combust. Flame*, 1978, **32**, 1.
48 *COSPAR International Reference Atmosphere: 1986, Part II: Middle Atmosphere Models*, ed. D. Rees, J. J. Barnett and K. Labitzke, Pergamon, Oxford, 1990.
49 R. R. Garcia and S. Solomon, *J. Geophys. Res.*, 1985, **90**, 3850.
50 R. J. Thomas, *J. Geophys. Res.*, 1990, **95**, 16457.
51 L. Thomas, M. C. Isherwood and M. R. Bowman, *J. Atmos. Terr. Phys.*, 1983, **45**, 587.
52 D. R. Bates and A. Dalgarno, in *Atomic and Molecular Processes*, ed. D. R. Bates, Academic Press, New York, 1962.
53 M. H. Rees, *Physics and Chemistry of the Upper Atmosphere*, Cambridge Univeristy Press, Cambridge, 1989.
54 M. Gadsden, *J. Geophys. Res.*, 1972, **77**, 1330.
55 B. Fegley and A. G. W. Cameron, *Earth Planet. Sci. Lett.*, 1987, **82**, 207.
56 C. Granier, J. P. Jegou and G. Megie, *Geophys. Res. Lett.*, 1985, **12**, 655.
57 J. Qian and C. S. Gardner, personal communication.

Paper 5/00360A; *Received* 20*th January*, 1995

Faraday Discuss., 1995, **100**, 431–439

Sporadic Metal Layers in the Upper Mesosphere

Chester S. Gardner
Department of Electrical and Computer Engineering, University of Illinois at Urbana-Champaign, 315 Computer and Systems Research Laboratory, 1308 West Main Street, Urbana, Illinois 61801, USA

One of the most interesting current problems in the aeronomy of the mesosphere is the origin of sporadic Na (Na_s) and sporadic Fe (Fe_s) layers. These are very thin (*ca.* 1 km), dense layers of atomic Na and Fe which often form explosively in a matter of minutes, may extend horizontally for several thousand kilometres, persist for several hours, and then disappear rapidly. While the typical concentrations of atomic Na and Fe in the upper mesosphere are respectively, 4×10^3 and 10^4 cm^{-3}, maximum densities exceeding 10^5 cm^{-3} have been observed for both species during sporadic layer events. Incoherent scatter radar and lidar observations at Arecibo have shown conclusively that many Na_s layers are linked to the formation of sporadic E (E_s) layers, while rocket and lidar observations at Andoya have shown a similar link between Fe_s and E_s. Recent lidar observations at Urbana and Hawaii have revealed the simultaneous formation of Na_s, Fe_s, and Ca^+ layers and the occurrence of substantial temperature enhancements and wind shears during some Na_s events. We present new observations of Na_s obtained with a Na wind/temperature (W/T) lidar during the recent ALOHA-93 (1993 airborne lidar and observations of the Hawaiian airglow) Campaign in Hawaii. These observations are used to characterize the wind and temperature structure associated with several of the major Na_s events observed during ALOHA-93.

The erratic formation of dense narrow layers of atomic Na and Fe in the upper mesosphere and lower thermosphere has been observed frequently at low- and high-latitude lidar sites. The thicknesses of these Na_s and Fe_s layers are typically *ca.* 1 km FWHM and their maximum densities can exceed normal species densities by a factor of 10 or more. Plotted in Fig. 1 is an Na_s layer observed by an airborn lidar northwest of Maui during ALOHA-93. While this event is unusual in that the maximum Na density is the largest that has ever been observed in the mesosphere, the profile does illustrate the remarkable physical characteristics of this puzzling phenomenon. Clemesha *et al.*[1] were the first to report observations of Na_s layers at São Paulo, Brazil (26°S, 46°W) in the late 1970s. They attributed them to meteoric ablation. It was not until von Zahn *et al.*[2] reported similar observations at Andoya, Norway (69°N, 16°E) that the unique character of Na_s was finally recognized. Strong links to E_s have been reported by several groups[3] and the simultaneous formation of sporadic Na, Ca^+, Fe, Fe^+ and E layers has also been observed.[7,8] Fig. 2 and 3 are profiles measured by the Illinois group at Arecibo (18°N, 67°W) and Urbana (40°N, 88°W) which illustrate the relationship between some Na_s, Fe_s, electron and ion layers. Airborne lidar observations during ALOHA-90[9] and ANLC-93 (1993 airborne noctilucent cloud campaign)[10] have shown that some Na_s layers may extend horizontally for several thousand km.

Various mechanisms have been proposed to explain sporadic metal layer formation. Meteoric ablation is now known to be incapable of supplying large amounts of metal

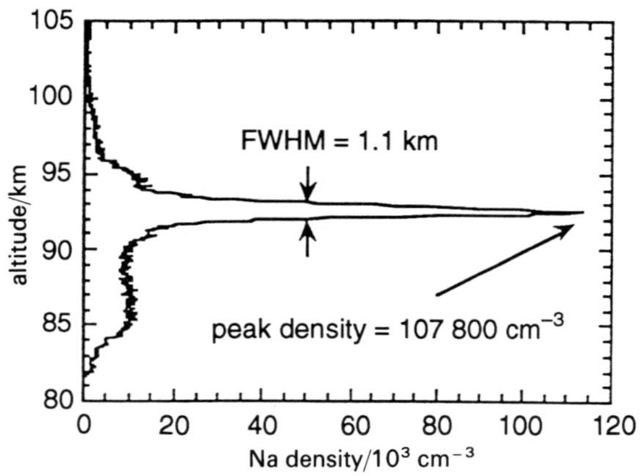

Fig. 1 Na density profile measured by an airborne lidar during the 23 October 1993 flight of ALOHA-93. The vertical resolution is 48 m and the integration period is 7 s which corresponds to a horizontal resolution of *ca.* 1 km. This Na_s layer exhibited the largest mesospheric Na density ever reported.

vapour over large geographic areas sufficient to create the sporadic layers. Von Zahn *et al.*[2] suggested that Na sputtered off the surfaces of meteoric smoke and dust particles by energetic auroral particles was the source of many of the high latitude Na_s events. Support for this mechanism was strengthened by ANLC-93 observations which showed the formation of a prominent Na_s event in the presence of a major auroral display.[10] Beatty *et al.*[11] and Kane *et al.*[6] suggested that the low-latitude events observed at Arecibo may result from the liberation of Na from dust or a chemical reservoir by the energetic ions and electrons in the associated E_s layers. Von Zahn and Murad[12] proposed $NaHCO_3$ as the chemical source while Cox *et al.*[13] suggested Na-containing silicate molecules as the reservoir. Both sources would release atomic Na through dissociative electron attachment reactions. Zhou *et al.*[14] argued that the normal temperature-dependent Na chemistry was responsible for Na_s which would form in the layers of warm air associated with the turbulent breakdown of gravity waves. Kirkwood

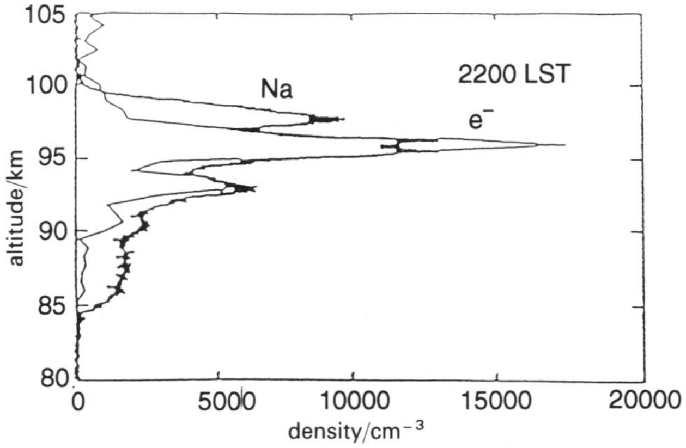

Fig. 2 Na and e^- profiles measured at Arecibo during a spectacular Na_s/E_s event observed on 30-31 March during the AIDA-89 Campaign.[9] The spatial and temporal variations in Na and e^- concentrations were highly correlated during this event. Reproduced with permission from ref. 6.

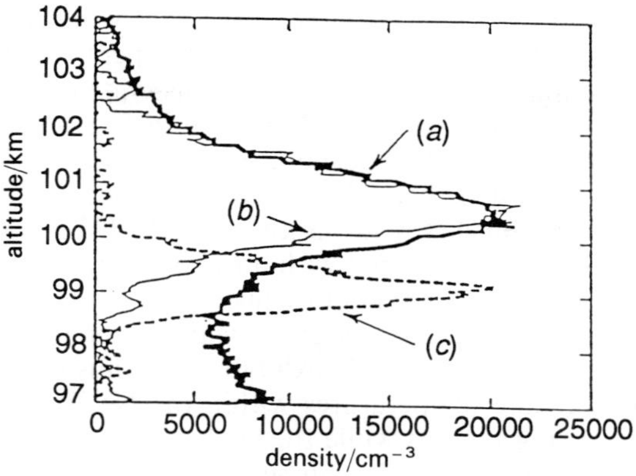

Fig. 3 Sporadic (*a*) Na(×5) (*b*) Fe and (*c*) Ca^+(×35) layers measured just after local midnight at Urbana, Il on 18 December 1991 using multiple lidars viewing a common volume of the atmosphere.[9] These sporadic layers formed simultaneously just before midnight and were observed throughout the remainder of the measurement period which ended just after 06.00 local time (LT). Reproduced with permission from ref. 9.

and Collis[15] proposed a mechanism based upon the vertical redistribution of Na associated with the passage of a large amplitude gravity wave. High-altitude (>100 km) Na_s events typically involve small amounts of Na (several hundred atoms cm^{-3}) which are believed to result from the conversion of Na^+ in the associated E_s layers to Na through clustering reactions involving N_2, CO_2 and H_2O.[16] Fe_s appears to be more common than Na_s and has been observed frequently at low-, mid- and high-latitude sites.[7,17] The neutralization of Fe^+ in E_s layers or the liberation of Fe from FeO_2^+ and FeO^+ in the presence of electrons appears capable of generating many of the observed Fe_s layers.[17]

Numerous Na_s events were observed by the airborne Na density lidar and Na W/T lidar at Haleakala (21°N, 156°W) during the ALOHA-93 Campaign. This paper focuses on the Haleakala observations which yielded temperature and wind profiles for many of these events. In particular we present measurements obtained on 6 and 21 October of two prominent Na_s events which were associated with strong vertical shears in the horizontal winds and large temperature enhancements. Possible formation mechanisms are discussed in the light of these observations.

Observations

During ALOHA-93 the University of Illinois Na W/T lidar was installed at the Air Force Maui Optical Site near the summit of Haleakala at an altitude of 3055 m. The laser was interfaced with the 0.8 m diameter beam director telescope which provided full scanning capabilities. The average power of the laser was 1 W yielding a power aperture product of 0.5 W m^2 for the lidar system. For most of the ALOHA-93 observations, the lidar was pointed sequentially at zenith and 15° off-zenith to the north and east so that the temperature and all three components of the wind vector could be measured. The data were smoothed yielding profiles with a vertical resolution of 240 m and temporal resolutions of 5 min for temperature and 15 min for winds. When smoothed to a vertical resolution of 1 km the absolute accuracies of these measurements, throughout the mesopause region, are *ca.* 1 K for temperature, 1 m s^{-1} for vertical winds, and 4 m s^{-1} for horizontal winds.

A prominent Na_s layer was observed on 6 October 1993 when measurements were made between 07.20 and 13.10 universal time (UT). Contour plots of Na density, temperature, zonal wind (u), and the vertical shear of zonal wind ($\partial u/\partial z$) are presented in Plate 1. The altitudes of the primary Na_s layer, temperature maximum and $\partial u/\partial z$ maximum are plotted *vs.* time in Fig. 4(*a*). The Na density, temperature and $\partial u/\partial z$ at the altitude of the primary Na_s layer are plotted *vs.* time in Fig. 4(*b*). The Na_s layer began forming near 93 km at *ca.* 08.00 UT, reached a maximum density of 3.5×10^4 cm^{-3} at 92 km after *ca.* 1 h and persisted throughout the remainder of the observation period. Dense secondary Na_s layers formed near 95 km at 09.20 UT and near 97 km at 11.10 UT. The temperatures were quite elevated within the primary Na_s layer exceeding 240 K at the time the Na_s layer reached maximum density. The profiles of Na density, atmospheric temperature, zonal wind and vertical shear of zonal wind at 09.05 UT are plotted in Fig. 5. The vertical wind profile exhibited no unusual features in the vicinity of the Na_s layer. However, the temperatures within the Na_s layer are more than 40 K warmer than the surrounding regions while the peak of the Na_s layer coincides with the altitude of zero zonal wind and is *ca.* 1 km higher than the altitude of maximum $\partial u/\partial z$.

Another prominent Na_s layer formed between *ca.* 08.00 and 11.00 UT on 21 October. Contour plots of the Na density, temperature, zonal wind, meridional wind (v), vertical shear of zonal wind and vertical shear of meridional wind ($\partial v/\partial z$) are plotted in Plate 2. This Na_s layer was much thicker than usual. It also formed near 93 km, moved

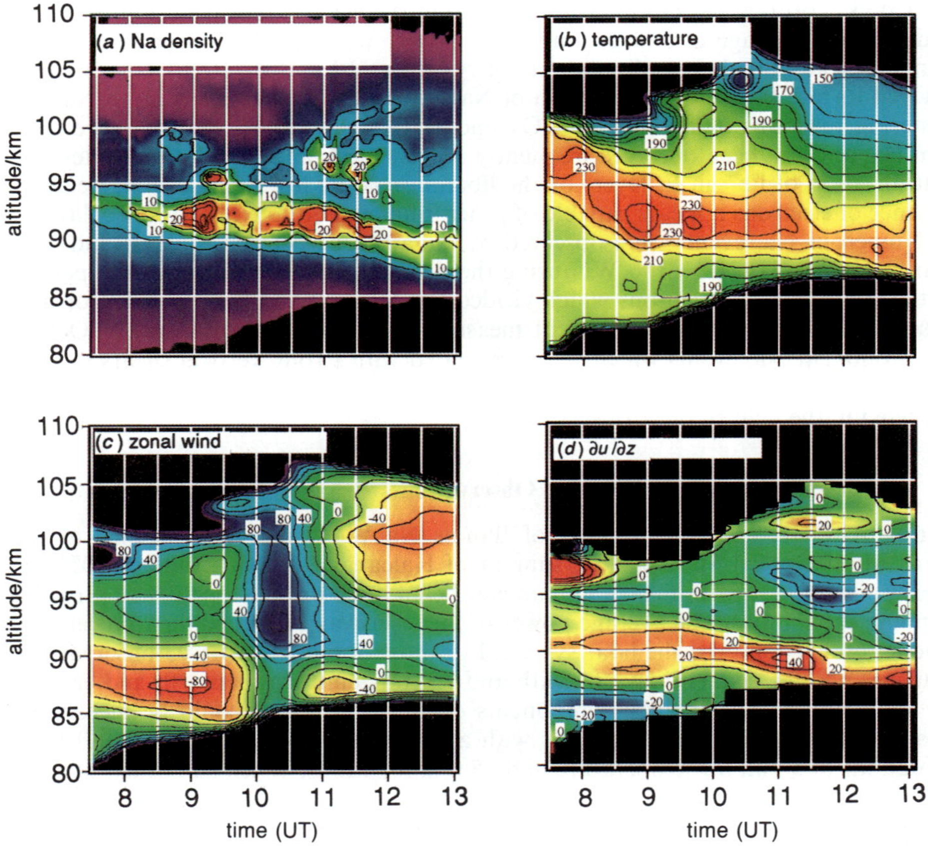

Plate 1 Contour plots of (*a*) Na density/10^3 cm^{-3}, (*b*) atmospheric temperature/K, (*c*) zonal wind/m s^{-1} and (*d*) vertical shear of zonal wind ($\partial u/\partial z$)/m s^{-1} km^{-1} observed at Haleakala on 6 October 1993

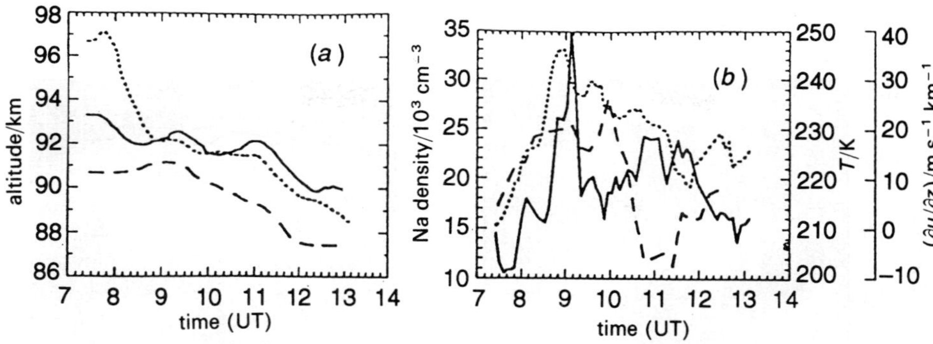

Fig. 4 (*a*) Altitudes of the primary 6 October (——) Na_s layer, (···) temperature maximum and (— —) $\partial u/\partial z$ maximum *vs.* time. (*b*) (——) Na density, (···) atmospheric temperature and (– –) vertical shear of zonal wind ($\partial u/\partial z$) at the altitude of the primary 6 October Na_s layer *vs.* time.

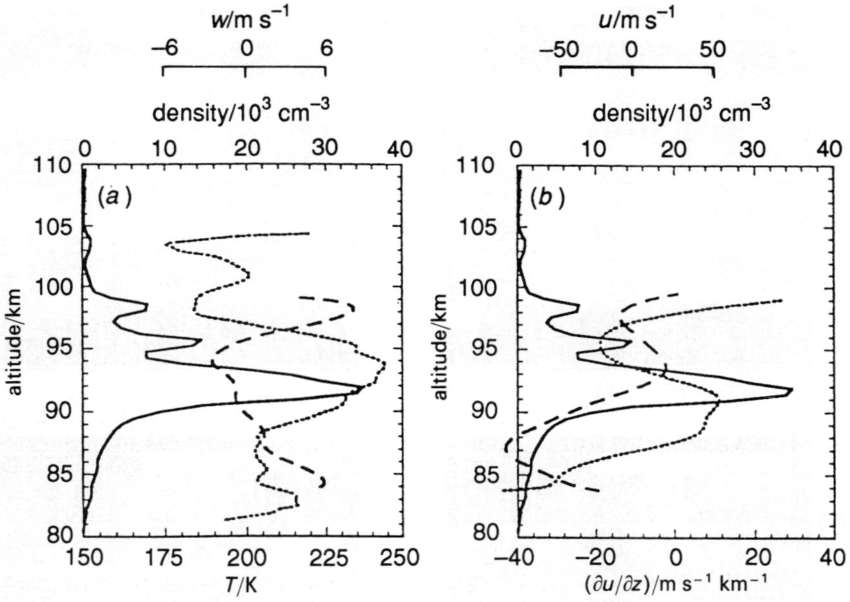

Fig. 5 (*a*) Profiles of (——) Na density, (···) atmospheric temperature and (– –) vertical wind measured at 09.05 UT on 6 October 1993. (*b*) Profiles of (——) Na density, (— —) zonal wind and (···) vertical shear of zonal wind ($\partial u/\partial z$) measured at 09.05 UT on 6 October 1993.

rapidly downward reaching maximum density near 88 km at *ca.* 10.30 UT, and then dissipated quickly. The layer appeared to be associated with a region of strong vertical shear in the meridional wind profile as well as a substantial enhancement in the temperature profile. The temperature at 88 km increased from 200 K at 9.30 UT to 235 K at 10.30 UT when the Na_s layer reached maximum density. The altitudes of the Na_s layer, temperature maximum and $\partial v/\partial z$ maximum are plotted *vs.* time in Fig. 6(*a*). The Na density, temperature and $\partial v/\partial z$ at the altitude of the Na_s layer are plotted *vs.* time in Fig. 6(*b*). The profiles of Na density, temperature, and vertical wind, meridional wind and vertical shear of meridional wind at 10.30 UT are plotted in Fig. 7.

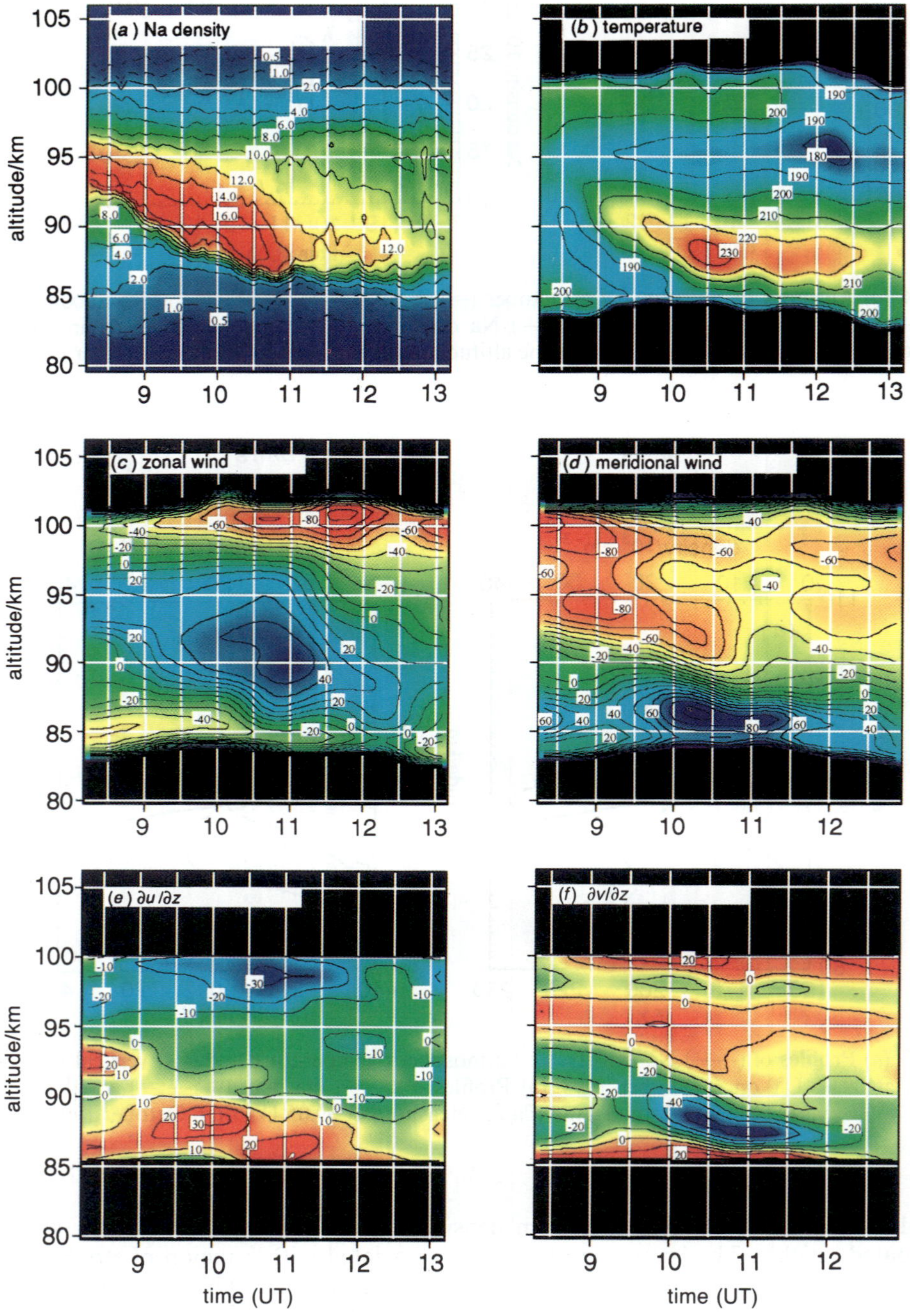

Plate 2 Contour plots of (*a*) Na density/10^3 cm^{-3}, (*b*) atmospheric temperature/K, (*c*) zonal wind/m s^{-1}, (*d*) meridional wind/m s^{-1}, (*e*) vertical shear of zonal wind ($\partial u/\partial z$)/m s^{-1} km^{-1}, and (*f*) vertical shear of meridional wind ($\partial v/\partial z$)/m s^{-1} km^{-1} observed at Haleakala on 21 October 1993.

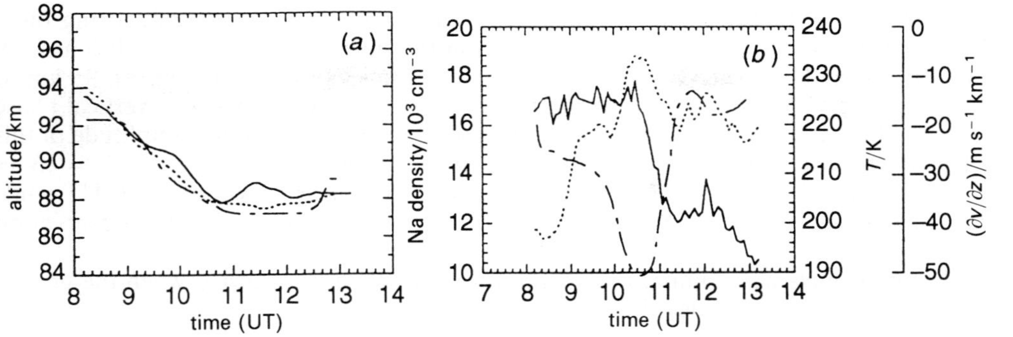

Fig. 6 (*a*) Altitudes of the 21 October (——) Na_s layer, (···) temperature maximum, and (— —) $\partial v/\partial z$ maximum plotted *vs.* time. (*b*) (——) Na density, (···) atmospheric temperature and (— —) vertical shear of meridional wind ($\partial v/\partial z$) at the altitude of the 21 October Na_s layer *vs.* time.

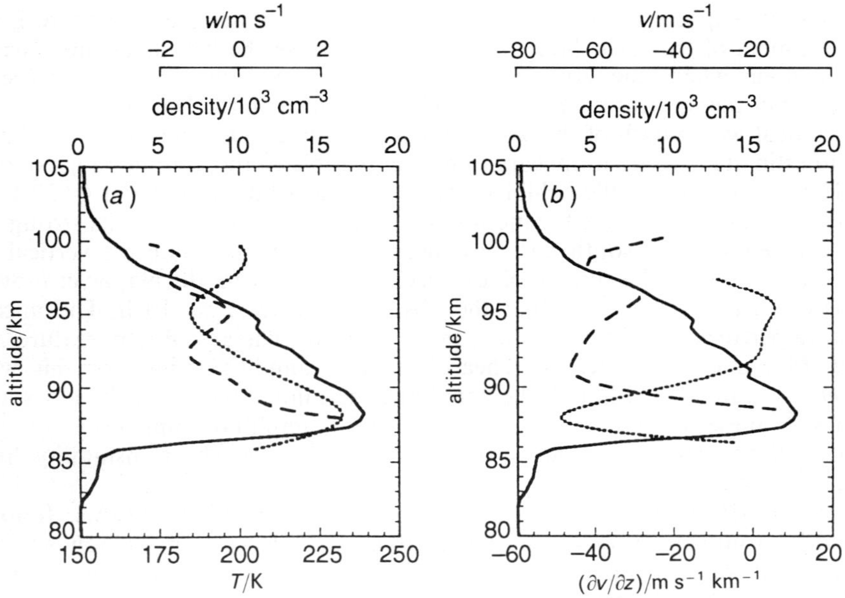

Fig. 7 (*a*) Profiles of (——) Na density, (···) atmospheric temperature and (— —) vertical wind measured at 10.30 UT on 21 October 1993. (*b*) Profiles of (——) Na density, (-–) zonal wind and (···) vertical shear of meridional wind ($\partial v/\partial z$) measured at 10.30 UT on 21 October 1993.

Discussion

The characteristics of the 6 and 21 October events are very similar to those of the rare mid-latitude event observed on 8 June 1991 at Urbana, IL.[8] After formation, the Na_s layers moved downward to *ca.* 90 km where they reached maximum densities. All three events were accompanied by substantial temperature enhancements with maximum temperatures exceeding 220 K. The Na_s densities were well correlated with the temperature variations.

The Urbana event was associated with a short vertical wavelength (*ca.* 4 km) gravity wave. The Haleakala events also appear to be associated with waves. Coherent downward motions are clearly evident in the Na density, temperature, *u*, *v*, $\partial u/\partial z$ and $\partial v/\partial z$

contour plots in Plates 1 and 2. The vertical wavelength of the wave influencing the 6 October Na_s layer is *ca.* 12 km as deduced from the shear maxima in Plate 1(*d*) near 89 and 101 km at 11.30 UT. The zero-shear contour moves downward from *ca.* 95 km at the beginning of the observation period to *ca.* 90 km at the end yielding a vertical phase velocity of *ca.* 1 km h^{-1} and an observed period of *ca.* 12 h. The zonal amplitude of this wave is very near the shear instability limit of $u = 2N/m = 80$ m s^{-1} or $\partial u/\partial z = 2N =$ 42 m s^{-1} km^{-1}. The maximum shear reaches this limit near 89 km at 11.00 UT while the zonal winds reach the 80 m s^{-1} limit in several regions during the observation period.

The maximum wave-induced temperature perturbation is achieved when the gravity wave amplitude is at the shear instability limit and is given by

$$T' = 2\Gamma/m \tag{1}$$

where $\Gamma \approx -10$ K km^{-1} is the adiabatic lapse rate of dry air. The maximum temperature fluctuation for the 12 km wave predicted by eqn. (1) is 38 K. This value is consistent with the temperature fluctuations measured at the peak of the Na_s layer [Fig. 4(*b*)]. Just below the Na_s layer, where the shear is strongest, the temperature structure suggests the atmosphere is convectively very stable. The lapse rate exceeds 10 K km^{-1} throughout most of the shear layer. However, just above the Na_s layer there are a few isolated regions where the atmosphere is unstable and the lapse rate exceeds *ca.* 10 K km^{-1}, but these features have no apparent influence on the Na_s event.

The vertical wavelength of the wave influencing the 21 October event is a bit more difficult to estimate because it appears to be comparable to the altitude range of observations. The zonal wind profile exhibits strong westward flow near 85 at 08.30 UT and near 101 km at 11.00 UT while the meridional profile at 08.30 UT exhibits strong northward flow at 86 km and southward flow near 96 km. We estimate the vertical wavelength to be *ca.* 20 km. The vertical phase speed estimated from the Na_s layer movement in Fig. 6(*a*) is *ca.* 2 km h^{-1} so that the observed period is *ca.* 10 h. The maximum temperature perturbation according to eqn. (1) is 64 K. The actual temperature at perturbation 88 km is *ca.* 40 K. The shear instability amplitude limit for this wave is $u = 2N/m = 133$ m s^{-1}. Although the zonal and meridional winds reach values as high as 80 m s^{-1}, these levels are well below the shear amplitude limit even though the meridional shear approaches and even briefly exceeds the shear instability limit of $2N = 42$ m s^{-1} km^{-1}.

During ALOHA-93 an ionosonde was operated on Maui by NOAA. Ionograms were recorded routinely every 15 min but there is no evidence from the ionograms that either Na_s event was accompanied by E_s. This is surprising since it has long been recognized that many Na_s layers are associated with E_s.[4,5,16] Discussions with NOAA revealed that the archived ionogram data were not processed to detect typical E_s events. Lidar and radar observations at Arecibo have revealed remarkable similarities in the vertical and the temporal structure of some Na_s and E_s layers.[6,11,18] These authors characterized the ion layer motions observed by the Arecibo radar during the AIDA-89 Campaign and showed that a majority of the events are associated with tidal motions. The tidal ion layers are dumped regularly by the diurnal tide between 90 and 95 km near local midnight. Both events reported here occurred near local midnight but appear to be associated with gravity waves, not tides. Interestingly, the strong zonal shears observed on both 6 and 21 October below the Na_s layers are the wrong polarity for the classical wind shear mechanism to create ion layers. Furthermore, the strong meridional shear observed on 21 October is clearly associated with the Na_s layer but would not be effective in layering ions because the wind shear mechanism requires the shear to be normal to the magnetic field lines.

Even so, the ALOHA-93 observations support the conjecture that E_s ions or electrons liberate atomic Na from a reservoir, perhaps facilitated by the high temperatures in the Na_s layer. However, it is clear from the ANLC-93 event described by Gu *et al.*,[10] that high temperatures are not essential for Na_s. In fact, Hansen and von Zahn[16] reported observing temperature minima near the height of 34 Na_s layers observed between 92 and 100 km at Andoya (69°N). We believe the large temperature enhancements reported here and observed at Urbana[8] are merely by-products of the waves responsible for dragging the E_s ions and electrons down to the Na source. Whether the source is a chemical reservoir or Na absorbed on the surfaces of meteoric smoke and dust[2] or perhaps even ice crystals, is yet to be determined.

The author gratefully acknowledges the excellent support provided by Jack Albetski and his staff at the Air Force Maui Optical Site. The author thanks Drs. Xin Tao and Yiyun Gu for their help in processing the ALOHA-93 lidar data. This work was supported in part by National Science Foundation grants ATM90-24367 and ATM94-03036.

References

1 B. R. Clemesha, V. W. J. H. Kirchhoff, D. M. Simonich and H. Takahashi, *Geophys. Res. Lett.*, 1978, **5**, 873.
2 U. von Zahn, P. von der Gathen and G. Hansen, *Geophys. Res. Lett.*, 1987, **14**, 76.
3 U. von Zahn and T. L. Hansen, *J. Atmos. Terr. Phys.*, 1987, **50**, 93.
4 K. H. Kwon, D. C. Senft and C. S. Gardner, *J. Geophys. Res.*, 1988, **93**, 14 199.
5 P. P. Batista, B. R. Clemesha, I. S. Batista and D. M. Simonich, *J. Geophys. Res.*, 1989, **94**, 15 349.
6 T. J. Kane, C. S. Gardner, Q. Zhou, J. D. Mathews and C. A. Tepley, *J. Atmos. Terr. Phys.*, 1993, **55**, 499.
7 M. Alpers, T. Blix, S. Kirkwood, D. Krankowsky, F-J. Lubken, S. Lutz and U. von Zahn, *J. Geophys. Res.*, 1993, **98**, 275.
8 C. S. Gardner, T. J. Kane, D. C. Senft, J. Qian and G. C. Papen, *J. Geophys. Res.*, 1993, **98**, 16 865.
9 T. J. Kane, C. A. Hostetler and C. S. Gardner, *Geophys. Res. Lett.*, 1991, **18**, 1365.
10 Y. Gu, J. Qian, G. C. Papen, G. R. Swenson and P. J. Espy, *Geophys. Res. Lett.*, 1995, **22**, 2805.
11 T. J. Beatty, R. L. Collins, C. S. Gardner, C. A. Hostetler, C. F. Sechrist Jr. and C. A. Tepley, *Geophys. Res. Lett.*, 1989, **16**, 1019.
12 U. von Zahn and E. Murad, *Geophys. Res. Lett.*, 1990, **17**, 147.
13 R. M. Cox, J. M. C. Plane and J. S. A. Green, *Geophys. Res. Lett.*, 1993, **20**, 2841.
14 Q. Zhou, J. D. Mathews and C. A. Tepley, *J. Atmos. Terr. Phys.*, 1993, **55**, 513.
15 S. Kirkwood and P. N. Collis, *J. Atmos. Terr. Phys.*, 1989, **51**, 259.
16 G. Hansen and U. von Zahn, *J. Atmos. Terr. Phys.*, 1990, **52**, 585.
17 T. J. Kane and C. S. Gardner, *J. Geophys. Res.*, 1993, **98**, 16 875.
18 J. D. Mathews, Y. T. Morton and Q. Zhou, *J. Atmos. Terr. Phys.*, 1993, **55**, 447.

Paper 5/00174I; *Received* 4*th January*, 1995

Faraday Discuss., 1995, **100**, 441–457

GENERAL DISCUSSION

Dr. Peter opened the discussion of Prof. Taylor's paper: Plate 9 shows the development of an extensive PSC during the end of December, 1991, and in the oral presentation it was stated that satellite measurements can now be used for much more than for mere PSC detection. How can satellite instruments measure chemical composition of PSC particles, or their physical state?

Prof. Taylor replied: Instruments like ISAMS detect the PSC at a range of wavelengths, in this case from *ca.* 4–16 μm, and the spectral dependence of the cloud's emissivity reveals its composition. For instance, the spectral signatures of water ice, nitric acid trihydrate, and sulfuric acid are quite distinct. In the case of volcanic aerosol, the concentration of the H_2SO_4–H_2O mixture can also be determined. A specific example has been discussed in the paper by Taylor *et al.*[1]

1 F. W. Taylor, A. Lambert, R. G. Grainger and C. D. Rodgers, *J. Atmos. Sci.*, 1994, **51**, 3019.

Dr. Roscoe asked: Plate 2 in your paper shows radiances of NO_2 across the terminator. I have four questions about the radiances shown: Are they wide-band or PMR radiances? What was the latitude? Are they am, pm, or the sum of the two? When can we expect to see retrievals of NO_2 mixing ratios across the terminator?

Plate 4 in your paper shows H_2O mixing ratios. The value at 50 km is 8 ppmv, much greater than the expected value of 6.3 ppmv. Do you have an explanation?

Prof. Taylor responded: All the NO_2 radiances I have shown are PMR (pressure modulated radiometry) radiances. Wide-band radiances cannot be used to detect NO_2 because there is too much interference from other species, especially water vapour.

The latitude in the example I have used was about 34°S, and for the evening transition, but we have similar data for many latitudes and for both the morning and evening terminator crossings.

NO_2 retrievals are available now from the ISAMS databases at Oxford, RAL or NASA.

The ISAMS water vapour results seem to be consistently too high by 1–2 ppm, when compared with other kinds of data. The reasons for this are still under investigation. A discussion may be found in ref. 1.

1 M. Goss-Custard, J. J. Remedios, A. Lambert, F. W. Taylor and C. D. Rodgers, *J. Geophys. Res.*, in the press.

Dr. Doughty said: Satellite measurements, such as from the UARS and ISAMS, are currently made of the IR-emission intensity from $NO_2(001)$, and used in the calculation of NO_2 concentrations. If $NO_2(001)$ is in local thermodynamic equilibrium (LTE) with $NO_2(000)$, conversion of emission intensities to concentrations of NO_2 is facile. This is not the case if 'non-thermal' processes of formation and removal of $NO_2(001)$ dominate the rates of thermal processes. In this context 'non-thermal' processes are defined to include chemical reaction, absorption or emission of radiation, and mass transfer. Under non-LTE conditions, the rates of all formation and removal processes need to be known to calculate NO_2 concentrations, including the rates of thermal formation and removal reactions. Since the thermal processes are pressure dependent, as altitude is increased it becomes less likely that the condition of LTE will be met.

At the Physical and Theoretical Chemistry Laboratory, Oxford University, 308 nm laser photolysis of NO_2 is being studied with the aim of understanding energy-transfer processes important in the calculation of NO_2 concentrations from satellite-measured $NO_2(001)$ emission intensities. Time-resolved Fourier-transform IR (FTIR) fluorescence measurements are being used to observe the temporal behaviour of vibrationally excited species. Thus energy transfer in the products of photolysis is being studied.

The photolysis of NO_2 at 308 nm proceeds by the reaction:

$$NO_2 + h\nu \rightarrow O(^3P) + NO(v = 0, 1, 2, 3)$$

The FTIR fluorescence spectra included in Fig. 1 were obtained using the FTIR fluorescence technique, 15–400 μs after dissociation (308 nm) of the NO_2 reactant. As can be seen, IR emission from vibrationally excited NO is observed initially. As the NO emission decays, emission from $NO_2(001)$ can be seen to grow in intensity. Kinetic modelling indicates that the following processes cannot explain the process by which $NO_2(001)$ is formed:

Mechanism A:

$$NO(v) + NO_2(000) \rightarrow NO(v - 1) + NO_2(001)$$

Mechanism B:

$$O(^3P) + NO_2 \rightarrow O_2(v = 1\text{–}11) + NO$$

$$O_2(v) + NO_2(000) \rightarrow O_2(v - 1) + NO_2(001)$$

In conclusion, vibrational energy transfer to and from NO_2 does not appear to occur through simple steps where a single quantum of vibrational energy is transferred. A

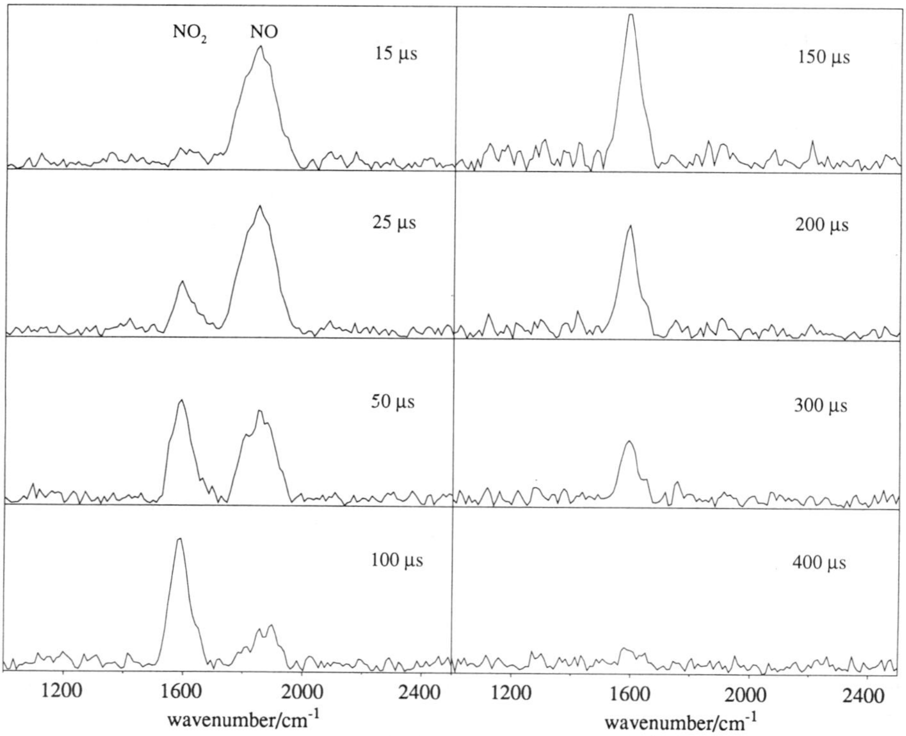

Fig. 1 FTIR fluorescence spectra subsequent to NO_2 photolysis at 308 nm

possible explanation could be that the energy transfer from NO(v) to NO_2 and $O_2(v)$ to NO_2 occurs through long-lived collisional complexes, allowing randomization of energy between colliding molecules, and resulting in inefficient v–v transfer. This is contrary to previous assumptions made in the non-LTE calculation of NO_2 concentrations (*e.g.* ref. 1). In addition, we also need to assess quantitatively the extent of the contribution of processes such as:

$$NO_2 + h\nu \rightarrow NO_2{}^* \rightarrow \rightarrow NO_2(001)$$

$$NO_2 + h\nu \rightarrow NO^{\ddagger} + O$$

$$O_3 + NO \rightarrow NO_2{}^{\ddagger} + O_2$$

to the emission observed from $NO_2(001)$ and NO($v = 1$).

To what extent could the emissions from $NO_2(001)$ observed by UARS and ISAMS be affected by NLTE? Under what circumstances would concentrations of NO_2, determined from satellite measurements of the $NO_2(001)$ emission intensity, be most likely to be in error?

1 B. J. Kerridge and E. E. Remsberg, *J. Geophys. Res.*, 1989, **94**, 16323.

Prof. Taylor responded: The NO_2 retrievals which I described (and which are referred to in the reply by Roscoe, above) were obtained assuming that the (001) band is in LTE at the levels where measurements were made, since most of the nitrogen dioxide is in the lower stratosphere where the pressures are probably high enough for collisions to maintain a Boltzmann distribution of the vibrational levels. However, we cannot be certain that this assumption is correct for NO_2 over the whole range, and we have certainly found it necessary to correct the CO, NO and H_2O radiances for non-LTE effects. Laboratory measurements such as those you describe will be valuable for use in radiative-transfer models to check on the actual behaviour of NO_2.

Dr. Chipperfield asked: When was the 2nd PSC event?

I remember that the TOVS (IR) data detected the December 27th event. TOVS PSC events are inferred to be type II events. According to ISAMS, was the December 27th event type I or type II?

Does the comment in the paper about PSC events in 91/92 being rare apply to type I events (and not simply type II events)?

Prof. Taylor responded: The two major PSC events observed by ISAMS in the northern hemisphere occurred around 28th December 1991 and 9th January 1992, both over northern Europe. The spectral properties of the PSCs on both occasions were consistent with NAT and not water ice. These were the only two distinct, dense PSCs of any type which were seen. ISAMS did not observe on every day, but when it was switched off, the NMC temperatures were too high for PSC formation.

Prof. McKenna opened the discussion of Dr. Pyle's paper: Your paper contains many impressive examples of model simulations which appear to be capturing the essence of the composition fields as indicated by observations.

Plates 1 and 2 show airborne LIDAR observations near the edge of the polar vortex and a model simulation valid at roughly the same time as the aircraft observations. The aircraft observations appear to show a portion of the polar vortex which has become 'disconnected' from the main vortex, while the high-resolution simulation indicates a filament of vortex air being ejected/extruded from the vortex and a complementary filament of mid-latitude air which is now surrounded by vortex air. This comparison can be reasonably interpreted as a successful model simulation. However, two questions which spring to mind are:

Are these filaments a ubiquitous feature of the model, so that the agreement between model and observation is coincidental, or is the model also successful in predicting that filaments will not be observed on other flights which do not exhibit such interesting structure?

If the model has skill in predicting both the presence and absence of filaments, then by implication it can give useful information on whether air is being mixed into the vortex or out of the vortex. The air which is extruded out of the vortex will clearly end up mixing with mid-latitude air. However the fate of the mid-latitude air looks less clear. Does it get deep into the vortex, does it form a new, structureless, vortex edge or does it become imbedded in a complex, multiply folded, vortex edge?

Fig. 9 of your paper shows a comparison between the ClO observed along an ER-2 flight track and a model simulation of the same feature. This comparison shows that the model is doing a good job. In fact it is doing such a good job that it permits more specific questions to be asked about what the comparison tell us about model formulation. What aspects of the model are most critically tested by the level of agreement? And conversely should the residual disagreement be interpreted as due to deficiencies in models or model input data which are of no significance or do they suggest certain aspects of the model which can be improved?

Dr. Pyle replied: The question raises important issues about the skill with which the current generation of chemical transport models can reproduce observed features, especially small-scale features. It must be said immediately that integrations of the kind shown have only recently been run at such high spatial resolution. In answering the question, we do not have a large body of information, and a large number of integrations, to fall back on. However, we can say that, based on high-resolution integrations for the EASOE, SESAME and ASHOE campaigns, the appearance of large filamentary structures is not ubiquitous. The coincidence between observed structures and modelled filaments should therefore be taken as an indication of model skill and definitely not as a statistical coincidence. Our integrations show periods when large-scale filaments are present and other periods when they are absent.

We must also admit that, in comparing observations with the model, some filamentary structures appear to be modelled better than others. Thus the depth of the structures seen, for example, in Fig. 9 and 10 is modelled well. Our simulations do less well in reproducing the depth of the feature discussed in Plates 1 and 2 at the precise location of the ClO 'blob' seen in Fig. 2 and Plate 3. Nevertheless, we are encouraged by the model's performance so far and agree with the questioner that this approach should lead to useful information about the extent of mixing of vortex air into middle latitudes.

I can only provide an incomplete answer to the second part of the question. We do see evidence, in some integrations, of air from middle latitudes being mixed into the vortex. The air moved out is subsequently mixed into the 'surf zone'. We have not studied changes to the vortex edge in detail; of course, the vortex shrinks throughout the later winter and spring.

I believe that the primary test possible here is of the transport in the model. When features, observed and modelled, coincide but show difference in magnitude, this can, of course, then provide a test of the chemistry. For example, we mention that our chemical models seem to lead to too rapid a deactivation of chlorine to $ClONO_2$. Disagreement similarly can arise from problems with the model transport or chemistry. Many of the problems outlined above are due, we believe, to problems with the specification of the meteorological analyses, their temporal resolution and their accuracies.

Dr. Tuck said: I would like to ask how optimistic you are about quantitative information being obtainable from high-resolution computations (in latitude and longitude) in view of Fig. 6 in our paper. That figure shows extensive peel-off of vortex air equatorward of the jet-stream core, as witnessed by the CH_4 trace in Fig. 6(*a*), but the numerical weather prediction model analysis seriously (by a factor of two or so) underestimated the wind speeds measured from ER2 [Fig. 6(*c*)]. There is also the question as to whether the ubiquitous small-scale variability in tracers, wind speed and direction is adequately represented by scale collapse, or whether small-scale forcing mechanisms need to be represented.

Dr. Pyle replied: As you know well, many studies are now showing that relatively fine-scale structures can be reproduced in model studies (either circulation models or contour advection studies) forced by meteorological analyses at much coarser resolution. Of course, as we indicate in our answer to Dr. McKenna, ultimately the quality of the type of model we have used here relies on having a good meteorological analysis. There will always be limitations on the analysis given, for example the uneven distribution of meteorological observations. Furthermore, some aspects of chemistry, for example, heterogeneous processing, may depend critically on very accurate resolution of small-scale temperature structures. Our confidence in reproducing these features must be limited.

Mr. Lee addressed Dr. Tuck: A model comparison on 2 June reproduces qualitatively the observed ClO structure in the observations (Fig. 4 of Dr Pyle's paper). During the dive between 83 000 and 85 000 s, the observed data indicate higher levels of processing compared with air sampled at the cruise altitude. Furthermore, the model interpretation of the two-peak feature is that the aircraft dived through a filament during the descent and ascent parts of the dive. The minimum in the two-peak feature corresponds to the bottom of the dive. Given that the scale of the model feature is smaller than the scale of the imput wind fields and that the observed data are qualitatively explained, we feel that the scale collapse within the model does represent some of the small scale variability in the tracer measurements. This is consistent with what is found in idealized studies of theoretical vortex dynamics, and their data-analytic spinoffs such as the ' contour advection' technique.[1,2]

1 W. A. Norton, *J. Atmos. Sci.*, 1994, **51**, 654.
2 D. W. Waugh and R. A. Plumb, *J. Atmos. Sci.*, 1994, **51**, 530.

Dr. Cox moved the discussion on to Dr. Tuck's paper: You presented data which showed that several periods of exposure of an air trajectory in the region of the polar vortex, to temperatures <195 K, were needed for complete conversion of HCl to active chlorine. Does this represent a limitation in chemical processing *via* the main heterogeneous reactions, for example because of limitations imposed by the initial composition, or is it because HCl is replenished by mixing with unprocessed air between the processing events?

Dr. Tuck replied: It is important to bear in mind that the vortex-edge air which was examined in early August showed no denitrification, but did exhibit about 25% loss of ozone relative to NO_y and N_2O. The detailed behaviour in Fig. 16 of my paper shows examples of all four possible air parcel trajectories: (i) processed inside staying inside, (ii) processed inside and moving outside, (iii) processed outside and moving inside, (iv) processed outside and staying outside. (i) is by far the most numerous category, but there is clear mixing across the jet core, a supposition borne out by the existence of a mean meridional mixing slope in HCl/Cl_y in the 10° latitude belt centred on the jet core in phases 3 and 4. Therefore, mixing between PSC exposures is a more likely interpretation.

Dr. R. L. Jones asked: Could you comment on the presence of 0.4 ppbv of HCl at the equatorial tropopause, in particular its relevance for the chlorine budget of the stratosphere, and for the possible transport of soluble species to the stratosphere?

Dr. Tuck responded: The measurement of HCl at 0.4 ppbv at the tropical tropopause on 29 October 1994 could have significant implications for the stratospheric chlorine budget, particularly in recently sub-tropical air as shown in Fig. 18 of the paper. Dr. Webster tells me that their error bars were larger than normal on this flight, but the measurement is 0.4 ± 0.2 ppbv. Compared to the previous flight on 26 October 1994, the HCl is several tenths of a ppbv higher throughout the flight, not just at the tropopause. According to Brock, Jonsson and Wilson, the CNC-FCAS instrument did collect what appeared to be volcanic material on these flights, and not on any others; SEM analysis does show chlorine in the particles, so it is possible that Mt. Rabaul, which erupted on 20 September 1994 in Papua New Guinea, could be responsible. However, there is still not a consistent story, because the HCl was 'normal' on 26 October 1994.

Prof. Ravishankara said: You mentioned that the level of HCl at the bottom of the tropical tropopause was, at least on the occasion when you all measured it, *ca.* 0.4 ppbv. Could you comment on the fact that the abundance of non-fluorinated chlorine-containing species is approximately of this order of magnitude? (Note: these are the halogenated compounds that may be degraded by OH reactions in the troposphere.) Could you also speculate on the implications of such high HCl at the tropopause?

Dr. Tuck replied: It is true that the schemes based upon compact, often linear tracer–tracer correlations often have non-zero intercepts on the Cl_y axis, for example in the Woodbridge *et al.* technique and in Fig. 2 of the paper. I am not sure how to interpret this; it is, however, clear that in the tropics it would be better to use short-lived tracers for this exercise, but even then the different balance between mixing and diabatic ascent may put a limit on how well we can know Cl_y without measuring it (or its components) directly. It is possible that all the chlorocarbons normally thought of as being lost in the troposphere did reach the tropopause; such behaviour was measured for ^{222}R ($\tau \approx 3.8$ days) during the STEP tropical mission from Darwin, as a result of a rapid transport in deep tropical convection. It is equally possible that tropospheric air containing up to 0.8 ppbv HCl could enter a cumulonimbus storm above the freezing level (*ca.* 5 km) and so avoid being rained out. We clearly need horizontal flight at the tropical tropopause.

Prof. Smith remarked: I wonder if I can open out the discussion by asking two rather general questions of our speakers. My first is specifically directed at Prof. Taylor. It relates to the sheer amount of data that can clearly be provided by satellite observations. Does this, in itself, cause problems, in that it can be several years before the results of a particular set of observations can be subjected to analysis and interpretation?

My second question to which I hope others might respond concerns the relationships between observers, modellers and those making laboratory measurements, and how these relationships are changing. Twenty years ago, when concern was first expressed about the effects on the stratosphere of supersonic aircraft, on the one hand, and CFCs, on the other, there was a close and valuable interaction between modellers and laboratory scientists with the former helping to identify the important chemical and photochemical processes whose rates had to be accurately known. I sense that recently there has been a shift, so that the closest relationship and interaction is now between the modellers and those observing the composition and dynamics of the atmosphere. I should be interested if others agree with this comment and whether they feel that it is inevitable, given the way in which atmospheric science has developed. Whatever the response, I would express the hope that modellers and observers will continue to see it as important that they identify crucial processes for the laboratory chemists to study.

Dr. Tuck replied: In the early days of stratospheric research, there was a very significant interaction between laboratory kineticists and atmospheric modellers. While this has continued, the modelling community has become directly involved with observational efforts, most notably by close, interactive participation in airborne field missions and on satellite science teams. The decision by NASA to adopt this approach was a wise one: in the longer term, the science is discussed more thoroughly and produced faster and better, although it requires careful handling to safeguard the interests of the experimenters. There is of course nothing, in principle, to prevent laboratory kineticists from becoming involved in airborne instrumentation.

Dr. Pyle replied: I think Prof. Smith's analysis is correct; stratospheric science in recent years has been driven by observations and the closest collaborations have been between modellers and observational scientists. I agree with the implication of his question that a balanced programme between laboratory science, field observation and modelling is the ideal. I believe that the demands to understand better heterogeneous chemistry will certainly keep laboratory scientists in the forefront of future advances in understanding.

Prof. Taylor also responded: It is certainly true that we obtained a large mass of data from ISAMS, and even though it stopped working at the end of July 1992 it will be many years yet before the results are fully analysed. I would not describe this as a problem, however. It results from the high resolution (vertically as well as horizontally) of the measurements, and the continuous, global coverage, which are particular strengths of the satellite technique. Since we are using continuously improving techniques, and are primarily interested in studying processes, and analysing events such as ozone-hole formation and development, rather than monitoring synoptic situations as our meteorological colleagues do, there is no particular hurry to complete the analysis.

I think that your observation about the emphasis having moved to the interaction between observers and modellers is consistent with my own perspective. The reason is basically that the observations are now so comprehensive that the dynamical complexity of the atmosphere has been revealed, and simple models (for example, one-dimensional chemical models) shown to be inadequate for most purposes. However, this does not mean that there is not still a great deal of important laboratory work to be done.

Dr. K. Carslaw addressed Dr. Pyle: The general agreement between large-scale numerical models and observations appears, on initial inspection, to be very good (for example for ClO and ozone concentrations). This also appeared to be the case two or so years ago after the EASOE measurement campaign. Since then, there have been a number of important developments, for example, in our understanding of heterogeneous processes leading to chlorine activation. Please could you briefly explain where the most important disagreements exist between observations and models at present.

Dr. Pyle replied: The point raised is a good one. Despite very significant advances I believe that we still have major limitations in our understanding of heterogeneous processes. This may not always prevent a reasonable diagnosis of past measurements using our current models but it certainly is a serious limitation of our ability to predict the future development of the ozone layer.

Dr. Tuck addressed Dr. Pyle: I would like to suggest that one test for the maturity of our understanding would be that time when theoretical models can predict previously unobserved atmospheric phenomena. To date, atmospheric science is still observationally driven, and models still improve by having their shortcomings revealed by new measurements—the ozone hole is an egregious example.

It is important, for example, to understand whether crystalline forms such as NAD or NAT are responsible for PSC processing, or whether it is supercooled ternary solutions—the latter should be capable of processing a larger volume of vortex, especially as a function of altitude but also as a function of radius.

Dr. Pyle agreed.

Prof. Davidovits addressed Dr. Tuck: You stated that a measure of our understanding of atmospheric chemistry would be our ability to predict the behaviour of the atmosphere, say the formation of a polar stratospheric cloud. Is this in fact a reasonable expectation? How do we know that the atmosphere, specifically its chemistry, is not chaotic?

Dr. Tuck responded: The governing equations for the physics, chemistry and fluid mechanics for the atmosphere on a rotating sphere are non-linear and coupled. They are capable of sustaining many forms of wave motion, and can exhibit many forms of quasi-periodic, irregular behaviour which, at least in numerical simulations, can be sensitive to the initial conditions. One relevant example is the demonstration a couple of years ago that the rate of stretching of material filaments at the vortex edge is described by Liapounov exponents *i.e.* it is chaotic advection. It should be remembered that molecular diffusion is presumably the ultimate dissipative mechanism at work. In the case of PSCs, my impression is that the observations are telling us that, at least in some cases, we are dealing with a non-equilibrium phenomenon, whose kinetics and phase diagrams could therefore sustain chaotic behaviour, in principle.

Dr. Sodeau addressed the audience: When I devised the initial proposal for this Discussion Meeting, I simply wanted to write: What don't we know about atmospheric chemistry? I am specifically interested today about what we do not understand about heterogeneous chemistry. We heard in Paul Crutzen's lecture about possible direct loss of ozone by heterogenous reactions. Others have also mentioned the involvement of sea-salt particles in 'sudden' ozone loss in the troposphere, but what is the current state of play? Are other particles, such as silicates and clays, able to act as active surfaces for heterogeneous chemistry? Is cold water the ingredient of prime importance in the studies to date? What don't we know?

Prof. Molina responded: We believe that a variety of atmospheric particles including sea-salt and silicates may provide sites for surface reactions. For example, metal oxide particles might be of some importance in the stratosphere. We have investigated the $HCl + ClONO_2$ reaction on alumina surfaces; although the reaction probability is about an order of magnitude smaller than on ice surfaces, it is large enough for such surfaces to play a significant role at low latitudes, where sulfate aerosols are not efficient catalysts of chlorine activation.

Dr. Cox added: Concerning the direct interaction of ozone with atmospheric particles, considerable thought has been given to this. For example, following the discovery of the ozone hole much thought was directed to the possible destruction of ozone on PSCs in the polar vortex. We now know that this is not the explanation of the phenomenon. In the troposphere available evidence does not point to a large loss of ozone in the cloud environment. However, Paul Crutzen has something to say on this subject.

Prof. Crutzen said: Direct destruction of ozone on ice particles does indeed not take place. However, one should look into the possibility that other species, with which O_3 may react, may be deposited on the surface of the ice, possibly in a quasi-liquid transition layer which may be found on the ice (the quasi-liquid layer concept actually goes back to Faraday). As an example, one may look at the role of HO_2 as follows:

$$HO_2(gas) \rightarrow HO_2(aq) \rightarrow H^+ + O_2^-$$

$$O_2^- + O_3 + H^+ \rightarrow OH + 2O_2$$

followed by volatilization of OH.

I admit that this hypothesis is very speculative, but the very low ozone concentrations measured over extended regions (*ca.* 2000 km between Fiji and Christmas Island) over the entire tropospheric column in the tropical Pacific during a field programme in March 1993 makes one think about such possibilities.

Unfortunately, it will hardly be possible to establish such O_3 loss by direct atmospheric measurements, as the chemical signals will be totally dwarfed by the dynamic variability due to heavy convection.

1 D. Kley, H. G. J. Smit, H. Vömel, S. Oltmans, H. Grassl, V. Ramanathan and P. J. Crutzen, *Science*, 1995, submitted.

Dr. Plane said: Dr. Sodeau has mentioned the possibility of other types of surface, such as minerals, providing sites for heterogeneous chemistry in the stratosphere. I would like to draw attention to the impact of meteoric debris on the stratospheric aerosol layer. At least 40 t of interplanetary dust enters the earth's atmosphere each day.[1] About 70% of this is believed to ablate completely at an altitude of about 90 km, and then recondense into 'smoke' particles before entering the stratosphere.[2] A modelling study,[3] which appears to be in broad agreement with measurements made on size-fractionated aerosols collected in the stratosphere, has shown that meteoric debris is probably the dominant natural aerosol constituent for particles larger that 1 μm throughout the stratosphere (in the form of micrometeorites), and for particles smaller than 0.01 μm above 20 km (as meteor smoke). The smoke particles also probably control the condensation nuclei abundance above 20 km, and play a role in the formation of polar stratospheric clouds.

A further important effect of the metals produced by meteoric ablation is that they form alkaline species such as carbonates and hydroxides.[1] These species will be collected by larger aerosols and provide a means of neutralizing the nitric acid that accumulates in aerosols during the denitrification of the polar stratosphere during winter. A modelling study has shown that this effect will be particularly pronounced during the polar winter when mesospheric air that is rich in meteoric debris is funnelled down to the lower stratosphere in the polar vortex.[4] Using current estimates from meteor radars of the metallic input into the atmosphere, Prather and Rodriguez[4] found that up to 1 ppb of HNO_3 would be neutralized below 25 km. Bearing in mind that the phenomenon of sporadic sodium layers discussed in Prof. Gardner's paper indicates that the meteoric input may be substantially higher, the impact of meteoric debris in the stratosphere requires further study.

1 J. M. C. Plane, *Int. Rev. Phys. Chem.*, 1991, **10**, 55.
2 D. M. Hunten, R. P. Turco and O. B. Toon, *J. Atmos. Sci.*, 1980, **37**, 1342.
3 R. P. Turco, O. B. Toon, R. C. Whitten and P. Hamill, in *Atmospheric Aerosols: their Formation, Optical Properties and Effects*, ed. A. Deepah, Spectrum Press, 1982, pp. 130–165.
4 M. J. Prather and J. M. Rodriguez, *Geophys. Res. Lett.*, 1988 **15**, 1.

Prof. Smith addressed Dr. Plane: The reaction between MgO and H_2O to form $Mg(OH)_2$ appears to be an important one in the reaction scheme proposed in your paper. The third-order rate constant for this reaction and its dependence on temperature is estimated *via* the formalism introduced by Troe.[1] However, it must be appreciated that Troe's methods were developed to estimate the rates at which two open-shell species, *i.e.* free radicals, associate. It is not at all clear that these methods should be applied when, as in the present case, the reaction involves species in singlet states and the reaction requires quite considerable reorganization of the chemical bonds. Would Dr. Plane care to respond to this comment and, in particular, indicate what effect there would be on the overall reaction mechanism if the reaction between MgO and H_2O is much slower than they have estimated?

1 J. Troe, *J. Chem. Phys.*, 1977, **66**, 4758.

Dr. Plane replied: The point you raise concerning the applicablity of the Troe formalism to the recombination reaction between two closed-shell species (MgO and H_2O) is quite valid. Because the reaction also involves some degree of rearrangement to form linear HO—Mg—OH, there is the possibility of a barrier on the potential-energy surface. Thus, the rate coefficient that we have estimated is most likely an upper limit, although, as we noted in the paper, the analogous recombination between CaO and H_2O has been shown to be very fast at atmospheric pressure.[1] However, it should be noted that the CaO + H_2O reaction was studied in a flame, from which the derivation of rate coefficients is often a complex procedure. The direct measurement of k_4 is therefore an important priority for advancing our understanding of the atmospheric chemistry of Mg.

From the point of view of the atmospheric model presented in the paper, the production of $Mg(OH)_2$ by reaction (4) is balanced by its removal through reaction with atomic H [reaction (13)]. Since k_{13} is a variable parameter in the model, both k_4 and k_{13} could be much slower than the values listed in Table 7, without significantly changing the model predictions. This would also not conflict with the two pieces of circumstantial evidence that we discuss in the paper: flame studies of Mg indicate that $k_{13} < k_{12}$, at least at about 1800 K;[2] and rocket-borne observations of Mg^+ ions in the daytime have always found the peak above 90 km,[3] which implies that k_{13}(*ca.* 200 K) $< 5 \times 10^{-15}$ cm^3 $molecule^{-1}$ s^{-1}. Of course, this evidence is indirect, and the model can only be placed on a firm foundation once the atomic Mg layer has been observed. In the meantime, this model provides a first estimate of the characteristics of the layer, which may be useful when planning field observations in the future.

1 D. E. Jensen and G. A. Jones, *Combust. Flame*, 1978, **32**, 1.
2 K. Schofield, in *Gas-phase Metal Reactions*, ed. A. Fontijn, Elsevier, Amsterdam, 1992, pp. 529–571.
3 W. Swider, *Planet. Space Sci.*, 1984, **32**, 307.

Prof. Ravishankara asked: What are the sources of metal atoms in the atmosphere? Could you tell us their relative strengths?

Dr. Plane replied: When considering the relative contributions of meteoric and terrestrial inputs of metals into the atmosphere, the picture in the mesosphere is reasonably clear. Most of the incoming meteoric mass is in the form of interplanetary dust particles in the mass range 1–1000 μg, with a radius range of 50–500 μm. The mean entry velocity of these particles lies between 14 and 45 km s^{-1}, so that frictional heating causes most of these particles to ablate in the 80–110 km region, providing a direct source of metal atoms and ions.[1] Consideration has also been given to a postulated layer of dust at about 90 km from which metal atoms are released during the day by thermal evaporation and photosputtering. The proposed origins of this dust have included sea-salt particles and volcanic debris.[2,3] However, there is now strong evidence that meteoric ablation is the major source of metals at these altitudes. This evidence has included the very good correlation between the relative abundances of the alkali metal ions observed in the lower thermosphere, and their elemental abundances in chondritic meteorites,[4] sporadic enhancements in the total column abundance of Na during meteor showers,[5] and the direct observation by a high-resolution lidar of metal deposition in short-lived meteor trails.[6]

Estimates of the daily input of meteoric material into the atmosphere have ranged from 44 to 10 000 t.[7–9] Recent models[2] have tended to employ the flux of 44 t per day derived by Hughes.[8] However, he has recently doubled this estimate,[10] in accord with a study of radio meteor rate data obtained as a function of transmitter frequency, which shows that at least 50% of meteors are not detected.[11] An estimate obtained by measuring the accumulation of iridium in sediments has indicated that the meteoric influx could be up to 250 t per day.[12]

This meteoric material eventually passes through the stratosphere, where it is presumably incorporated into aerosols. By comparison, the total mass input into the stratosphere is thought to be about 820 t per day during volcanically quiescent periods,[13] so that the meteoric source provides at least 5% of this input. In fact, the relative contributions of meteoric and terrestrial sources to stratospheric aerosols are difficult to ascertain because of the similarity of the metal isotope abundances from both sources.[13] The analysis of sulfur isotopes is also problematic because the meteoric contribution of sulfur to the stratosphere is probably very small, and there is a wide natural variability of sulfur isotope ratios.[14] Nevertheless, analyses of stratospheric aerosols have indicated a meteoric component,[15] and optical scattering measurements show that meteoric debris is probably the major aerosol component above 25 km.[16] Condensation nuclei counters flying in the stratosphere also occasionally observe the total particle mixing ratio to increase appreciably above 20 km,[17] which also implies a significant particle source from higher in the atmosphere.

1 D. M. Hunten, R. P. Turco and O. B. Toon, *J. Atmos. Sci.*, 1980, **37**, 1342.
2 J. M. C. Plane, *Int. Rev. Phys. Chem.*, 1991, **10**, 55.
3 D. M. Hunten and L. Wallace, *J. Geophys. Res.*, 1967, **72**, 69.
4 R. A. Goldberg and A. C. Aikin, *Science*, 1973, **180**, 296.
5 M. Uchiumi, C. Nagasawa, M. Hirono, M. Fujiwara and M. Maea, *J. Geomag. Geoelectr.*, 1993, **45**, 393.
6 T. J. Kane and C. S. Gardner, *Science*, 1993, **259**, 1297.
7 C. E. Junge, O. Oldenburg and J. T. Wasson, *J. Geophys. Res.*, 1962, **67**, 1027.
8 D. W. Hughes, in *Cosmic Dust*, ed. J. A. M. McDonnell, Wiley, New York, 1978.
9 D. W. R. McKinley, *Meteor Science and Engineering*, McGraw-Hill, New York, 1961.
10 D. W. Hughes, *Space Sci. Rev.*, 1992, **61**, 275.
11 D. I. Steel and W. G. Elford, *J. Atmos. Terr. Phys.*, 1991, **53**, 409.
12 J. T. Wasson and F. T. Kyte, *Geophys. Res. Lett.*, 1987, **14**, 779.
13 R. P. Turco, O. B. Toon, R. C. Whitten and P. Hamill, in *Atmospheric Aerosols. Their Formation, Optical Properties and Effects*, ed. A. Deepak, Spectrum Press, 1982, pp. 139–165.
14 A. W. Castelman Jr., H. R. Munkelwitz, B. Manowitz, *Tellus*, 1974, **26**, 222.
15 R. D. Cadle and C. S. Kiang, *Space Phys.*, 1977, **15**, 195.
16 G. Newkirk Jr. and J. A. Eddy, *J. Atmos. Sci.*, 1964, **21**, 35.
17 K. H. Käselau, P. Fabian and H. Röhrs, *Pure Appl. Geophys.*, 1974, **112**, 877.

Miss Cox opened the discussion of Prof. Gardner's paper: I would like to present some work we have been doing to investigate the nature of the source of the sporadic sodium layers (SSLs) that you have measured. We have developed a one-dimensional, time-resolved model of the sodium chemistry at this height.[1] It includes a vertical kinetic diffusion coefficient and has been used to study the creation and behaviour of a typical sporadic sodium layer. The main shortcoming of the model is that the diffusion is handled by an eddy diffusion coefficient; this is a time-averaged parametrization of the many complex forms of atmospheric turbulence at this height. As your measurements have shown, the SSLs may be linked to specific dynamical conditions. To allow for this we have also run the model with molecular diffusion as a lower limit to vertical transport in the absence of turbulence.

We have found that the chemistry does little to remove Na atoms from the layer. It is at the height of a typical SSL that the Na atom mixing ratio of the background layer reaches a maximum and practically all the Na at this height is in the atomic form. Thus, it is the dynamics that controls the layer's shape and dictates the size of the source that is needed. We have investigated three scenarios: the production of an SSL by an unspecified layer source; sustaining an SSL using a layer source; and the final decay of the SSL due to the background chemistry and dynamics. The model was used to estimate the minimum source strength required to produce a typical SSL at 95 km. This lower limit is determined by the competition between growth of the layer and removal by vertical diffusion and chemistry. The source strength needed is 3.5 Na atoms cm^{-3} s^{-1} for the case of eddy and molecular diffusion and chemical removal, and 3.1 Na atoms cm^{-3} s^{-1} with eddy and molecular diffusion alone. The time for the SSL to develop fully is 70 min, as shown in the left-hand panel of Fig. 2. For the case of molecular diffusion alone the minimum source strength is 0.2 Na atoms cm^{-3} s^{-1}. The layer source was also used to maintain the peak concentration for 3 h. The source strengths required are 1.4 Na atoms cm^{-3} s^{-1} for both eddy and molecular diffusion and chemical removal, 1.3 Na atoms cm^{-3} s^{-1} for eddy and molecular diffusion alone and 0.2 Na atoms cm^{-3} s^{-1} for molecular diffusion alone. The model was then used to observe the decay, in the absence of a source of atomic Na, of an established SSL subjected to eddy and molecular diffusion and chemistry. A sequence of the resulting density profiles is shown in the right-hand panel of Fig. 2, demonstrating a substantial broadening and reduction in peak concentration of the SSL over 3 h. In general, this peak reduction and broadening do not appear to have been reported in observations of

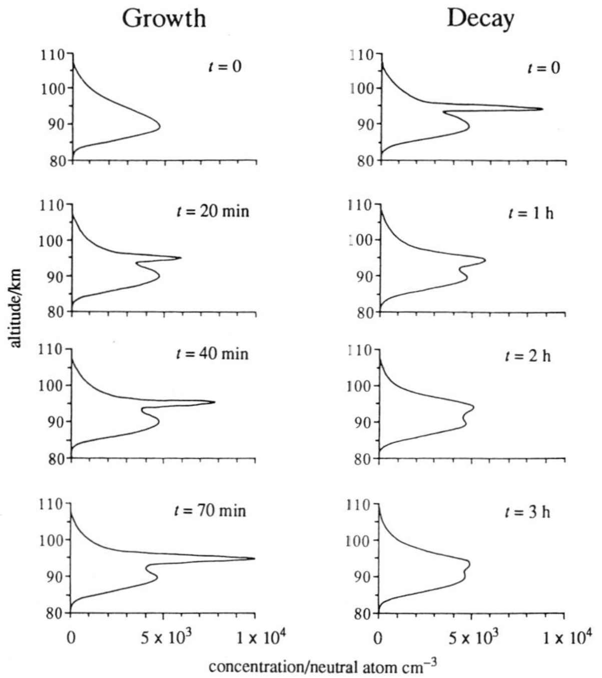

Fig. 2 A time-series illustrating the evolution of an SSL governed by eddy/molecular diffusion and chemistry. Individual plots show the concentration profiles of atomic Na as a function of time. The left-hand panels show the growth of the SSL with a source strength of 3.5 Na atoms cm^{-3} s^{-1}, acting over a 1 km region centred at 95 km. The right-hand panels illustrate the decay of a typical SSL. Reproduced from ref. 1 with permission of the American Geophysical Union.

SSLs. This may be because only the largest, most well defined layers tend to be published, because a broadened layer appearing in the lidar beam may not be recognized as an SSL, but rather as a fluctuation in the background Na atom concentration.

1 R. M. Cox, J. M. C. Plane and J. S. A. Green, *Geophys. Res. Lett.*, 1993, **20**, 2841.

Dr. Gardner replied: Your numerical calculations are important because they provide bounds on the input rates of the sources needed to produce and maintain typical sporadic Na layers in the presence of diffusion and normal Na chemistry. You are correct in noting that most published observations do not exhibit the degree of diffusive broadening of the sporadic layers seen in your models, although most of the reports have not addressed this issue quantitatively. We did compute broadening rates for 15 sporadic Na layers observed at Mauna Keal. The values varied from -9.5 to 27.5 cm s^{-1} with a mean of 8 cm s^{-1}. Only the fastest rates, which approach 1 km h^{-1}, appear to be compatible with your numerical model. One event was observed for 3 h and broadened by 2.2 km, in general agreement with your results.

1 K. H. Kwon, D. C. Senft and C. S. Gardner, *J. Geophys. Res.*, 1988, **93**, 14; 299.

Dr. Peter said: Dr. Gardner's and Dr. Plane's papers reveal considerable uncertainties in our knowledge on the mesospheric processes of meteoritic ablation and recondensation. Regardless of these uncertainties, the ablated and residual material eventually descends and passes the stratospheric aerosol layer, where it could possibly lead to (1) a partial neutralization of the acidic aerosol and (2) serve as condensation nuclei (CN) for the freezing of acidic hydrates or for new particle formation. While preliminary experiments from our laboratory indicate no influence of the meteoritic material on aerosol freezing, the neutralization issue requires answers to the questions: How much of the incoming meteoritic material is ablated? In what form does it reach the stratosphere? Does the coupled mesospheric/stratospheric advection channel the material into the polar regions?

Dr. Plane replied: My replies to Profs. Ravishankara, Davidovits and Crutzen discuss the uncertainties in the meteoritic input flux and the degree of ablation. If the average input flux of 44 t per day is correct, then our atmospheric models require that all of the alkali-metal elements and most of the Fe ablates. Immediately following ablation, the concentration of metallic species and silicates will be high enough in the meteor trail to enable some recondensation to take place. A model[1] of this process indicates that meteoric 'smoke' particles of <0.01 μm diameter are likely to form. However, the majority of the metal atoms and ions produced during ablation will diffuse out of the meteor trail and eventually be converted into hydrogen carbonate, carbonate and hydroxide molecules below the metal atom layers *i.e.* below 85 km. The fate of these compounds lower in the atmosphere is uncertain. During the approximately two-week passage down through the mesosphere, some polymerizations of these species will occur. Indeed, there is evidence for NaOH clusters of the form $NaOH_2{}^+(NaON)_m(H_2O)_n$ in the upper stratosphere.[2,3] The metal compounds also react rapidly with HCl, and the subsequent growth of large metal chloride polymers may represent a significant removal of atmospheric chlorine.[4]

Turco *et al.*[5] have carried out a modelling study of the meteoric component of stratospheric aerosols, and reached several conclusions. First, meteoric debris is probably the major natural aerosol constituent of particles larger than 1 μm (micrometeorites) throughout the stratosphere, and dominates particles smaller than 0.01 μm (meteor smoke). Second, the abundance of condensation nuclei is controlled by meteor smoke above 20 km. Third, meteoric smoke affects the stratospheric aerosol layer by increasing the total particle concentration and reducing the average particle size.

Prather and Rodriguez[6] have considered the impact of meteoric debris on polar ozone chemistry. The meridional atmospheric circulation leads to a subsidence of mesospheric air into the winter polar stratosphere, causing an accumulation of meteoric debris and providing additional CN during the early austral spring when polar stratospheric cloud formation is important. Furthermore, because the metals form alkaline compounds, their incorporation into acid stratospheric aerosols neutralizes the aerosol acidity. This may enable gas-phase HNO_3 to be more readily incorporated into the aerosol phase and enhance the denitrification process. Using the conservative meteoric input flux of 44 t per day (see my reply to Prof. Ravishankara), Prather and Rodriguez[6] have estimated that sufficient meteoric material would be present below 25 km in the early Antarctic spring to neutralize *ca.* 1 ppb of HNO_3. If the latest estimates of the meteor influx are correct, this effect could be substantially greater.

1 D. M. Hunten, R. P. Turco and O. B. Toon, *J. Atmos. Sci.*, 1980, **37**, 1342.
2 R. Arnold, D. Krankowsky and K. H. Marien, *Geophys. Res. Lett.*, 1978, **5**, 653.
3 E. E. Ferguson, *Geophys. Res. Lett.*, 1978, **5**, 1035.
4 J. J. Lamb and S. W. Benson, *J. Geophys. Res.*, 1986, **91**, 8683.
5 R. P. Turco, O. B. Toon, R. C. Whitten and P. Hamill, in *Atmospheric Aerosols. Their Formation, Optical Properties and Effects*, ed. A. Deepak, Spectrum Press, 1982, pp. 139–165.
6 M. J. Prather and J. M. Rodriguez, *Geophys. Res. Lett.*, 1988, **15**, 1.

Prof. Crutzen said: It may be worthwhile saying something about the potential role of meteoritic material in the nucleation of stratospheric sulfate aerosol. Experiments in our laboratory[1] have shown that heterogeneous nucleation by such material is not efficient.

1 T. Koop, U. M. Biermann, W. Raber, B. P. Luo, P. J. Crutzen and Th. Peter, *Geophys. Res. Lett.*, 1995, **22**, 917.

Dr. Peter said and **Dr. Biermann** communicated: We performed heterogeneous freezing experiments of stratospheric binary and ternary solution mixtures on surfaces typical for meteoritic material and space debris. As representatives for meteoritic material we have chosen powdered samples of a CM2-chondrite (Murchison, fallen 1969), the silicate fraction of an H3-chondrite (Dimmit, found 1942) and the iron components Fe_2O_3 and Fe_3O_4 (magnetite). Other freezing nuclei tested were Al_2O_3, AgI and $(NH_4)_2SO_4$.

By means of calorimetric measurements of the freezing of supercooled binary H_2SO_4–H_2O and ternary H_2SO_4–HNO_3–H_2O bulk solutions we showed that homogeneous freezing of these droplets as a possible pathway for the formation of solid polar stratospheric clouds (PSCs) at temperatures above the frost point may be excluded. In the present experiments we found that none of the tested species enhanced freezing of the binary or ternary solutions. We conclude that the heterogeneous nucleation rates leading to hydrate formation are much too small and thus not only homogeneous, but also heterogeneous freezing on undissolved nuclei appears to be an inefficient mechanism for the formation of solid PSC particles at temperatures above the frost point.

Dr. Plane added: In my reply to Prof. Ravishankara I indicated that the most quoted meteoritic input from meteor radar observations is 44 t per day. If we assume that the Na completely ablates into the gas phase as atoms or ions, this corresponds to an average input of 1.3×10^4 atoms $cm^{-2}\ s^{-1}$. Our current Na model[1] is able to reproduce in satisfactory detail the seasonal variation of the Na layer at 70 °N by using this input flux. The model contains rather few adjustable parameters, since a good deal of laboratory work on the atmospheric chemistry of Na has been carried out in the past 12 years.[2] Thus, the chemistry of the Na layer seems to be reasonably well understood (although one crucial laboratory study that remains to be carried out[1] is on the reaction $NaHCO_3 + H \rightarrow Na + H_2CO_3$).

Unfortunately, the phenomenon of sporadic Na layers (Na_s), graphically illustrated in Prof. Gardner's paper, undermines this view. There must be a large reservoir for atomic Na, which is not accounted for by our current model, and the mechanism which converts this reservoir into atomic Na remains to be satisfactorily explained.[3] These problems could be resolved if the input flux was considerably larger, with the Na being sequestered into various esoteric reservoirs. Suggestions have included the metal ablating in the form of stable silicate molecules from which Na is later released by dissociative recombination with energetic electrons;[3] Na species being adsorbed onto meteoric smoke particles, from which the Na is later released by charged particle precipitation;[4,5] or Na ablating as ions above 105 km, where Na^+ is very long-lived, and then a large concentration of the ions being swept down below 100 km and neutralized, which I will discuss later.

Prof. Crutzen has also asked about the role of meteoric material in the nucleation of stratospheric sulfate aerosol. In my reply to Dr. Peter and Prof. Ravishankara I have referred to evidence that meteoric debris is a significant constituent of stratospheric aerosol, and controls the abundance of condensation nuclei for producing sulfate aerosol above 20 km. However, Prof. Crutzen and his colleagues are particularly concerned with finding nuclei to promote the heterogenerous freezing of H_2SO_4–HNO_3–H_2O aerosol droplets at temperatures above the ice frost point. They have demonstrated recently[6] that the suspension of materials such as Fe_2O_3 and Fe_3O_4 in these supercooled solutions failed to promote freezing. However, the iron oxides would only be present in the stratosphere as the major constituents of micrometeorites *i.e.* meteoroids which were too small to ablate. Most meteoric iron appears to ablate as Fe of Fe^+, and then forms gas-phase species such as $Fe(OH)_2$ and $FeCO_3$ in the lower mesosphere (as we show in our paper for the case of magnesium, the other dominant meteoric metal). These compounds probably also react with HCl in the upper stratosphere, producing metal chlorides, and some degree of polymerization will also occur. I would rather not speculate on whether these kinds of species are effective nucleating agents, but experiments in which such metal-containing vapours were condensed into supercooled solutions might prove interesting.

1 M. Helmer and J. M. C. Plane, *J. Geophys. Res.*, 1993, **98**, 23207.
2 J. M. C. Plane and M. Helmer, in *Research in Chemical Kinetics*, ed. R. G. Compton and G. Hancock, Elsevier, Amsterdam, 1994, pp. 313–367.
3 R. M. Cox, J. M. C. Plane and J. S. A. Green, *Geophys. Res. Lett.*, 1993, **20**, 2841.
4 U. von Zahn, P. von der Gathen and G. Hansen, *Geophys. Res. Lett.*, 1987, **14**, 76.
5 T. J. Beatty, R. L. Collins, C. S. Gardner, C. A. Hostetler, C. F. Sechrist and C. A. Tepley, *Geophys. Res. Lett.*, 1989, **16**, 1019.
6 T. Koop, U. M. Biermann, W. Raber, B. P. Luo, P. J. Crutzen and Th. Peter, *Geophys. Res. Lett.*, 1995, **22**, 917.

Dr. Green said: We might plot mixing ratio instead of density. It is conserved during advection over probable height ranges whereas density is not. If we define mixing ratio in terms of the number of molecules containing (say) sodium, then chemical processes do not change it. Molecular diffusion is down the gradient of mixing ratio. Thus it is more difficult to account for irregularities in mixing ratio than in density. The mixing ratio profile of Prof. Gardner's Fig. 5 shows three layers each about 1 km deep, extending from 90 to 98 km height, rather than one layer 2 km deep at 92 km. The mixing ratio profile of Gardner's Fig. 9 shows one layer 10 km deep centred at 98 km, instead of 8 km deep at 88 km. To the extent that we believe that chemical and dynamical processes that conserve mixing ratio are important this would seem to be the more reasonable profile to demand explanation, and it does focus attention on slightly shifted aspects of the surrounding medium.

I wonder what the mesosphere is like outside the spectacular events so nicely illustrated by Gardner's paper. Is there a continual background of minor 'events', a continuous background of fluctuation? Observers have spoken of the large increase in total sodium during an SST event. How does this compare with the general variability of total sodium?

Dr. Plane replied: Dr. Green is quite correct in pointing out the advantages of considering the vertical profiles of mixing ratios rather than concentrations, since species diffuse down gradients of mixing ratio. For instance, we have shown that $NaHCO_3$, which is a potential reservoir for forming Na_s, is produced below 90 km but cannot be transported to 95 km in sufficient concentration.[1] However, the rule of conservation of mixing ratio does not apply if the precursor species is charged, since ionic species are transported differently than the neutral atmosphere in magnetic and electrical fields.[2] In particular, ions can be concentrated into thin layers at wind shears, forming sporadic E layers.[3] The striking narrowness of the Na_s layer in Fig. 1 of Prof. Gardner's paper, and the coincidence of the Na_s layers with regions of high horizontal wind shear (*e.g.* Plate 2), provides evidence that the Na_s precursor is a charged species.

We have recently considered two mechanisms for generating Na_s from ions. The first involves Na^+ as the reservoir, from which atomic Na would be generated by the following sequence of reactions:[4]

$$Na^+ + N_2 + N_2 \rightarrow Na \cdot N_2{}^+ + N_2 \tag{1}$$

$$Na \cdot N_2{}^+ + X \rightarrow Na \cdot X^+ + N_2 \quad (X = H_2O,\ CO_2\ etc.) \tag{2}$$

$$Na \cdot X^+ + e^- \rightarrow Na + X \tag{3}$$

Miss Cox and I are studying reaction (1) at present, and find that $k_1(300\ K) > 5 \times 10^{-31}\ cm^6\ molecule^{-2}\ s^{-1}$. At mesospheric temperatures, $k_1(200\ K)$ is therefore likely to be in excess of $10^{-30}\ cm^6\ molecule^{-2}\ s^{-1}$, and so at 95 km where $[N_2] = 3.2 \times 10^{13}\ cm^{-3}$, the rate of production of atomic Na *via* reactions (1)–(3) is more than $10^{-3}\ [Na^+]\ s^{-1}$ [reaction (1) is rate-determining]. Thus, a sporadic E layer at 95 km containing an Na^+ concentration of $10^4\ cm^{-3}$ would produce Na atoms at more than 1 $cm^{-3}\ s^{-1}$, which should be sufficient to produce an Na_s layer.[1] The difficulty with this scheme is that this chemistry proceeds continuously in the height range 90–100 km where Na_s are mostly observed, so that the required concentration of Na^+ could not build up before an Na_s formed. However, Na^+ ions have a much longer lifetime above 100 km [because the rate of reaction (1) falls as the square of the pressure], and so perhaps there is a mechanism for rapidly sweeping large concentrations of Na^+ downwards from higher in the thermosphere to form a sporadic E layer.

The second ion mechanism involves hydrated clusters of Na^+ ions. The most stable cluster ion formed in reaction (2) is $Na \cdot H_2O^+$, which can then add further water molecules. Such clusters are still prone to dissociative recombination with electrons,

$$Na \cdot (H_2O)_n{}^+ + e^- \rightarrow Na + nH_2O \tag{4}$$

and reaction (4) is exothermic until $n > 9$.[5] However, the primary hydration shell contains four H_2O arranged tetrahedrally, with the O bound directly to the Na^+.[6] It may be the case that reaction (4) becomes much slower once this primary hydration shell is full *i.e.* when $n \geqslant 4$. We have recently constructed a thermodynamic model which demonstrates that at a typical water vapour concentration ($10^8\ cm^{-3}$) and temperature (180 K) at 95 km, almost all Na^+ clusters will have $n \geqslant 4$. Prof. Gardner shows in his paper that Na_s often occurs where there has been a very substantial local warming, by up to 50 K (see Plates 1 and 2, Fig. 4 and 5). Our thermodynamic model shows that such a warming would cause the hydrated Na^+ clusters to lose water, until the most abundant cluster was $n = 2$. If reaction (4) was much faster for this lightly hydrated cluster, this would explain the production of the Na_s upon warming. Future observations with rocket-borne mass spectrometers of the ionic constituents of this region should search for the presence of these hydrated clusters.

1 R. M. Cox, J. M. C. Plane and J. S. A. Green, *Geophys. Res. Lett.*, 1993, **30**, 2841.
2 S. Kirkwood and U. von Zahn, *J. Atmos. Terr. Phys.*, 1991, **53**, 389.
3 G. Chimonas and W. I. Axford, *J. Geophys. Res.*, 1968, **73**, 111.
4 M. Helmer and J. M. C. Plane, *J. Geophys. Res.*, 1993, **98**, 23207.
5 I. Dzidic and P. Kebarle, *J. Phys. Chem.*, 1970, **74**, 1466.
6 E. D. Glendening and D. Feller, *J. Phys. Chem.*, 1995, **99**, 3060.

Dr. Gardner replied: The Na concentration fluctuates substantially on timescales of a few tens of minutes in response to the passage of atmospheric gravity waves through the mesopause region. The fluctuations are largest on the top and bottom edges of the Na layer where they often approach 100%. It is possible that minor sporadic Na layer events are masked by these vigorous gravity wave perturbations.

Dr. Plane summarized: The chemistry of meteoric ablation is not well understood. Important points of uncertainty are the ratio of metal atoms to ions that are produced, the fraction of metals that are ejected as cations in stable silicate molecules, and the fraction of the meteoroid mass that does not ablate completely and is perhaps enriched in the less volatile metals (see my reply to Prof. Davidovits). I think that progress could be made both by laboratory experiments employing the laser ablation of meteoritic samples, and by field observations. For instance, Prof. Gardner's group have recently employed a pair of lidars to observe freshly ablated Na and Fe in the same meteor trails.[1] Although the time resolution of the instruments was different, the relative quantities of the metals are at their chondritic abundance ratio of 1 : 10, which probably indicates that Na and Fe mostly ablate as neutral atoms. A particularly interesting lidar measurement in the future would be to analyse the relative concentrations of Na to both Ca and Ca^+ in meteor trails. The atomic Ca layer in the upper mesosphere is severely depleted,[2] and the simplest explanation is that Ca does not ablate efficiently.

1 T. J. Kane and C. S. Gardner, *Science* 1993, **259**, 1297.
2 J. Qian and C. S. Gardner, *J. Geophys. Res.*, 1995, **100**, 7453.

Faraday Discuss., 1995, **100**, 459–462

Concluding Remarks

B. A. Thrush
University of Cambridge, Department of Chemistry, Lensfield Road, Cambridge, UK CB2 1EW

The organisers of this meeting and the authors of the papers are to be congratulated on stimulating lively discussions which fitted well into their allocated slots and it would be invidious of me to make many additional comments to which the authors cannot reply. It is appropriate to begin with a brief historical résumé of a subject which has grown so much in recent years, and then to point out some of the problems which a two-day meeting could not cover.

For the chemistry of the stratosphere one naturally starts with Chapman's classic paper on what we now call 'odd-oxygen' chemistry which was published 65 years ago.[1] It was 40 years before laboratory measurements of the rates of the elementary reactions involved were measured accurately enough for the presence of other ozone-destroying reactions to be recognised.

$$O_2 + h\nu(\lambda < 242\ \text{nm}) \rightarrow O + O \qquad \text{slow (1)}$$

$$O + O + M \rightarrow O_2 + M \qquad \text{slow (2)}$$

$$O + O_2 + M \rightarrow O_3 + M \qquad \text{quick (3)}$$

$$O_3 + h\nu \rightarrow O_2 + O \qquad \text{quick (4)}$$

$$O + O_3 \rightarrow O_2 + O_2 \qquad \text{slow (5)}$$

The possible role of hydrogen radicals in the destruction of odd-oxygen (ozone plus atomic oxygen) was discussed by Nicolet[2] in *Faraday Discussion* No. 37 on *Chemical Reactions in the Atmosphere* in 1964. The publication 25 years ago of Crutzen's paper[3] on the role of nitrogen oxides in the destruction of stratospheric ozone came at a crucial time, because large fleets of supersonic aircraft which would release NO and NO_2 directly into the stratosphere had been projected.[4] This produced a major effort in laboratory studies, numerical modelling and atmospheric measurement to assess the magnitude of the problem, efforts which were redoubled after the publication 21 years ago of Molina and Rowland's letter to *Nature*[5] which focussed attention on the catalytic destruction of ozone by the chlorine released by the decomposition of chlorofluoromethanes in the stratosphere. The fact that several groups agreed with their prediction of a 10% depletion of ozone cover in about 100 years time did not prove that this would occur. It merely showed that if you used the same (highly uncertain) reaction rate and transport data you would get the same (highly uncertain) answer if your arithmetic was correct. The need for precise rate coefficients down to *ca.* 200 K for the 50–100 chemical reactions that could be important for the modelling of stratospheric chemistry provided a stimulus for laboratory studies. This had not been provided by theory which lagged behind experiment. Among the unexpected features which these measurements yielded were pressure-dependent rate coefficients for hydrogen-atom transfer reactions[6,7] such as

$$HO + HNO_3 \rightarrow H_2O + NO_3 \quad (6)$$

$$HO_2 + HO_2 \rightarrow H_2O_2 + O_2 \quad (7)$$

Further examples of such behaviour are provided by the reversible addition of HO to carbon disulfide as well as to dimethylsulfide and alkenes as reported here by the groups of Ravishankara and Atkinson.

The other important question some 20 years ago was the search for 'unknown chemistry', the molecules and reactions that were not included in the existing numerical models. This led to the recognition that chlorine nitrate ($ClONO_2$), pernitric acid ($HOONO_2$) and hypochlorous acid were relatively stable in the cold highly oxidising conditions of the lower stratosphere. They are formed in reactions between radicals which destroy ozone and are partly protected from rapid photolysis by the ozone above them. These compounds are analogous to peroxyacetylnitrate (PAN) which had been identified as an important component of, and irritant in, photochemical smog in the troposphere 35 years ago.[8]

Until 10 years ago it was believed that the chemistry of stratospheric ozone could be explained in terms of homogeneous gas reactions which can be studied in the laboratory. The unexpected discovery in 1985 of the Antarctic ozone hole,[9] which forms in their spring, and the subsequent realisation that the formation in the polar stratosphere of nitric acid trihydrate and ice in the very cold winter stratosphere results in the denitrification of the lower stratosphere and the heterogeneous conversion of chlorine compounds into forms which destroy ozone have transformed research priorities. This has been well illustrated by the number of papers devoted to such heterogeneous processes at this meeting, which consider the uptake of radicals as well as the semi-stable intermediates such as chlorine nitrate.

Although the hydrolysis of chlorine nitrate is agreed to yield hypochlorous acid, the stability of the precursor H_2OCl^+ at the relevant temperatures was an unresolved issue. Many questions remain about the nature of the surface layers of ice particles, the solvation and state of HCl, *etc.*; one common assumption is that particles are nucleated by sulfate aerosol. Whilst this is generally true in the troposphere, and occurred in the lower stratosphere when Mt. Pinatubo had a well timed major eruption, the question remains as to how nuclei are formed in the stratosphere in the periods between major volcanic eruptions. Byers Brown *et al.* pointed out in their paper that sunlight can produce nucleation in a Wilson cloud chamber[10] owing to photochemical charge transfer between H_2O and O_2 in a cluster of water molecules. Unfortunately, their calculations would require implausibly high degrees of supersaturation and we need to turn to the other property of the cloud chamber for an alternative mechanism for nucleation. The tracks in a cloud chamber are nucleated by the ions produced by the passage of an energetic particle through the chamber. There are significant numbers of ions in the atmosphere which range from atomic and diatomic species in the mesosphere to cluster ions in the lower stratosphere, where their abundance is greater than that of aerosol particles in the absence of particles of volcanic origin.[11] The strength of the forces between ions and such polar molecules as water must greatly assist nucleation.

So far we have considered only chemical processes. Fortunately, the discovery of ozone holes coincided with considerable advances in the numerical modelling of the atmosphere. Comparisons of predicted and observed concentrations of species had relied on 1D (altitude only) or 2D (altitude and latitude only) models of the atmosphere, but rapid advances in computer technology have made possible 3D chemical models of the atmosphere using the global circulation models (GCM) developed for weather and climate prediction. Such models are essential if we are to understand ozone holes and particularly their effect on other regions of the atmosphere. This is well illustrated by

Fig. 1 which shows TOMS satellite measurements of the decay of ozone cover in a recent decade.[12] The deepening of polar ozone holes owing to increased chlorine in the atmosphere is well illustrated. In addition, there has been a 10% decrease of ozone cover in spring over northern temperature latitudes. Using GCM, we can track (admittedly large) parcels of air as they enter or leave the vortex which surrounds an ozone hole but we are still unsure as to how this affects ozone levels in mid-latitudes. Is it simply the transport of ozone-depleted air from the break-up of an ozone hole, or is air steadily flowing through the vortex and being processed there, or are potential ozone-depleting species activated on aerosol particles in the lower stratosphere in mid-latitudes?

An issue not discussed here was the role of bromine in stratospheric ozone depletion and how much this contributes at different latitudes. Another puzzling ozone decay which was not discussed occurs in the Arctic spring. This is the rapid disappearance of boundary-layer ozone on a timescale of a day or so.[13] Here bromine is strongly implicated; bromoform is believed to be the precursor although its photolysis would be too slow on this timescale. Bromoform very rapidly turns yellow in the laboratory, but the mechanism of its decomposition is unknown.

Understanding the chemistry of the Earth's atmosphere involves a combination of laboratory studies, mathematical modelling and observations. We now know that the stratosphere cannot be regarded as the dry, surfaceless reactor beloved of reaction kineticists which could be regarded as homogeneous laterally. The paper by Gardner reminds us that the much less dense mesosphere has major inhomogeneities which we cannot explain. The inhomogeneities of the troposphere are all too familiar to us, because we are in immediate contact with it. The message is therefore that 3D modelling has now become essential to our understanding of the atmosphere, together with very extensive measurements for verification. Satellites provide excellent coverage but for many trace species the uncertainty in their concentrations is *ca.* 20%, even with ground truth verification. Aircraft measurements are less extensive but are more easily calibrated. Models have significant uncertainties due to their descriptions of circulation and limited spatial resolution. Paul Crutzen suggests an overall limiting accuracy of 15% on the prediction of any concentration of a trace species. This must be an important guideline as to which

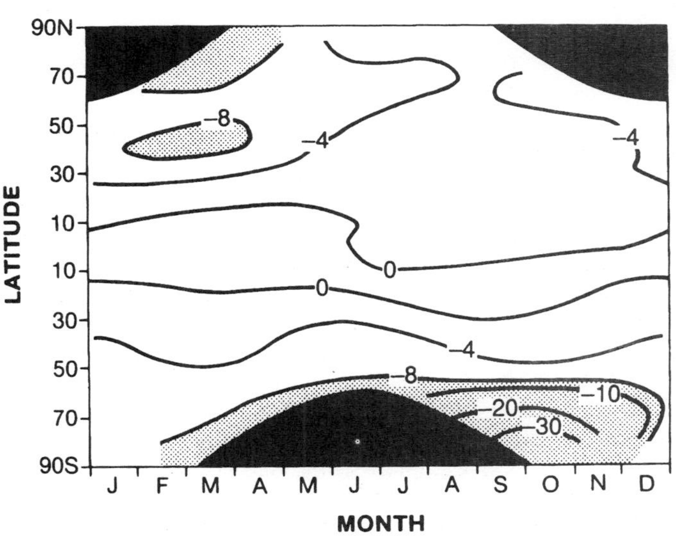

Fig. 1 Total ozone trends expressed as percentage change per decade as a function of latitude and season. The stippled areas show an ozone loss exceeding 8% per decade. The dark areas correspond to polar winters in which the TOMS satellite was unable to make measurements.

problems are worth investigating. In the case of vibrationally excited ground-state oxygen molecules (in which the author must confess to a long standing interest[14]) it follows from Wodtke's clear evidence that $O_2(X, v \geqslant 26)$ can generate odd-oxygen by the reverse of the Chapman reaction (5), that yields of odd-oxygen will be at the limit of detectability near the stratopause. The reason that the formation of only 1% of these highly excited oxygen molecules in the ozone photolysis (3) might be significant is the slowness with which odd-oxygen is generated by the photolysis of O_2 in the first step of the Chapman mechanism. This process is the closest approach to deviation from local thermodynamic equilibrium (LTE) yet found in the stratosphere or troposphere providing, of course, that $O(^1D)$ which is the main product of the UV photolysis of ozone is regarded as a separate chemical species. However, this should not discourage the search for other examples of deviations from LTE despite the large number of unreactive collisions which an energised species must undergo in these regions of the atmosphere.

References

1 S. Chapman, *Mem. R. Meteorol. Soc.*, 1930, **3**, 103.
2 M. Nicolet, *Discuss Faraday Soc.*, 1964, **37**, 7.
3 P. J. Crutzen, *Q. J. R. Meteorol. Soc.*, 1970, **96**, 320.
4 H. S. Johnston, *Science*, 1971, **173**, 517.
5 M. J. Molina and F. S. Rowland, *Nature* (*London*), 1974, **249**, 810.
6 J. J. Margitan and R. T. Watson, *J. Phys. Chem.*, 1982, **86**, 3819.
7 JPL Publication 94-26, ed. W. B. DeMore, 1994, p.43.
8 E. R. Stephens, E. F. Darley, O. C. Taylor and W. E. Scott, *Proc. Am. Petrol. Inst.*, 1960, **40**, 111.
9 J. C. Farman, B. G. Gardiner and J. D. Shanklin, *Nature* (*London*), 1985, **315**, 207.
10 C. T. R. Wilson, *Phil. Trans. R. Soc. A*, 1899, **192**, 403.
11 G. Brasseur and S. Solomon, *Aeronomy of the Middle Atmosphere*, Reidel, Dordrecht, 2nd edn., 1986.
12 R. S. Stolarski, P. Bloomfield, R. D. McPeters and J. R. Herman, *Geophys, Res. Lett.*, 1991, **18**, 1015.
13 L. A. Barrie, J. W. Bottenheim, R. C. Schnell, P. J. Crutzen and R. A. Rasmussen, *Nature* (*London*), 1988, **334**, 138.
14 F. J. Lipscomb, R. G. W. Norrish and B. A. Thrush, *Proc. R. Soc. London A*, 1956, **233**, 455.

List of Posters

Evolution of chlorine partitioning following heterogeneous processing **C. R. Webster, R. D. May, D. C. Scott** and **H. Hu,** *Jet Propulsion Laboratory, California Institute of Technology, USA*

The sticking of HCl and HBr on Type II PSC mimics **H. Ridley, H. D. Aslin** and **S. Haq,** *University of Liverpool, UK*

Time resolved studies of the integrated oxidation mechanisms of hydrocarbons in the troposphere **A. Hoffmann, W. Malms, V. Mörs, H. Sommitz** and **R. Zellner,** *Universität Essen, Germany*

High-resolution UV–VIS absorption cross-sections at stratospheric temperatures: nitrogen dioxide and chlorine dioxide **G. J. Frost, L. M. Goss** and **V. Vaida,** *University of Colorado, USA*

Mechanism reduction techniques applied to tropospheric chemistry **A. C. Heard, A. S. Tomlin, M. J. Pilling, S. Saunders** and **G. Zeng,** *University of Leeds, UK*

Uncertainties in the calculations of atmospheric emissions of industrial halocarbons **P. M. Midgley, A. McCulloch** and **D. A. Fisher,** *ICI Chemicals & Polymers, Runcorn, UK*

New kinetic model of OH(v) Meinel emission **V. M. Grigorieva, Yu. M. Gershenzon, D. V. Shalashilin** and **S. Ya. Umanskii,** *Russian Academy of Sciences, Moscow*

In situ measurements of OH production in a dark reaction **T. Pfeiffer, W. Armerding, M. Spiekermann, J. Walter, H-J. Crawack, O. Forberich** and **F. J. Comes,** *Johann Wolfgang Goethe-Universität, Frankfurt/Main, Germany*

Airborne trace gas measurements across a vegetation gradient **S. D. Chambers, N. J. Clark, J. M. Hacker** and **S. Venning,** *Flinders University Adelaide, S. Australia*

Kinetics and mechanisms for the reaction of hydroxyl radicals with organic acids under atmospheric conditions **S. Carr, J. J. Treacy, H. W. Sidebottom,** *University College, Dublin, Ireland,* **R. K. Connell, C. E. Canosa-Mas** and **R. P. Wayne,** *Physical and Theoretical Chemistry Laboratory, Oxford University, Oxford*

Detection of OCS, CH_3SOCH_3 and CH_3SCHO in the oxidation of dimethyl sulfide **I. Barnes, K. H. Becker** and **I. Patroescu,** *Bergische Universität-GH Wuppertal, Germany*

Atmospheric degradation mechanism of alkylated benzenes **I. Barnes, K. H. Becker** and **E. Weisen,** *Bergische Universität-GH Wuppertal, Germany*

REMPI spectroscopy of jet-cooled aldehydes **N. C. Shand, C. L. Ning** and **J. Pfab,** *Heriot-Watt University, Edinburgh, UK*

Kinetics and mechanism of the oxidation of dimethylsulfide by peroxo compounds in the aqueous phase **H. Elias, P. Amels** and **K. J. Wannowius,** *Edward Zintl Inst., Anorganische Chemie III, Darmstadt, FRG*

Perturbation of the NO_x/ozone photostationary state and implications for chemical modelling of the troposphere **J. N. Cape, G. G. McFadyen, K. C. Clemitshaw, L. Carpenter** and **S. A. Penkett,** *Institute of Terrestrial Ecology, Penicuik, UK and The Atmospheric Chemistry Group, UEA, Norwich, UK,* **S. Yamulki** and **R. M. Harrison,** *University of Birmingham, UK*

The state-specific formation and relaxation of vibrationally excited oxygen **M. Klatt, I. W. M. Smith, A. Symonds, R. P. Tuckett** and **G. N. Ward,** *University of Birmingham, UK*

The relative rates of F-atom reactions with hydrofluorocarbons studied by continuous photolysis/FTIR spectroscopy **C. M. Moore** and **I. W. M. Smith,** *University of Birmingham, UK*

OH radical initiated photo-oxidation of alkanes: isomerisation of 2-hexoxyl radicals **J. Eberhard, C. Miller, D. W. Stocker** and **J. A. Kerr** *EAWAG, ETH Zürich, Switzerland*

The atmospheric chemistry of iodine monoxide **B. Laszlo, R. E. Huie** and **J. J. Kurylo,** *National Institute of Standards and Technology, Gaithersburg, USA*

Environmental chemistry: complex reaction patterns in some selected sulfur compounds in aqueous solutions and gaseous mixtures **C. R. Chinake** and **R. H. Simoyi,** *West Virginia University, Morgantown, USA*

Mass accommodation of gas phase substituted aromatic compounds on aqueous droplet surfaces **M. R. Heal, M. J. Pilling, P. E. Titcombe** and **B. J. Whitaker,** *University of Leeds, UK*

The OH-initiated degradation of isoprene: Mechanism construction **M. E. Jenkin** and **G. D. Hayman,** *AEA Technology, Abingdon, UK*

Reactions of cationic and anionic water clusters with HCl and $ClONO_2$ **T. Schindler, C. Berg, G. Niedner-Schatteburg** and **V. E. Bondybey,** *Technische Universität München, Garching, Germany*

Kinetics of the self reaction of $CH_{3-n}X_nO_2$ (where X = F, Cl or Br) radicals at 298 K **C. J. Percival, J-M. Fracheboud, C. E. Canosa-Mas** and **R. P. Wayne,** *University of Oxford, UK*

The adsorption of atmospheric gases onto water clusters: a mimic for atmospheric processes **M. Ahmed, C. J. Apps, C. Hughes, N. E. Watt** and **J. C. Whitehead,** *University of Manchester, UK*

A SIMS study of some heterogeneous chemistry of simulated polar stratospheric cloud particles **H. Donsig** and **J. Vickerman,** *UMIST, Manchester, UK*

Towards comprehensive trace gas monitoring in Africa **S. O'Beirne** and **E. G. Brunke,** *CSIR, Pretoria, South Africa,* **H. E. Scheel,** *I. F. U. Garmisch-Partenkirchen*

Modelling of night-time chemistry: study of its importance on subsequent ozone formation **C. Dimitroulopoulou,** *Imperial College, London, UK*

Heterogeneous release of halogen atoms form the reaction of N_2O_5 with deliquescent sea-spray aerosols **W. Behnke, A. Frenzel, Ch. George, V. Scheer** and **C. Zetzsch,** *Fraunhofer-Institut, Hannover, Germany*

The use of emulsions to assess the physical state of H_2SO_4/H_2O aerosols at the pooles **K. Ji, J. B. Poirier** and **J. C. Petit,** *CNRS, Orléans, France*

Determination of rate constants of the sulfate radical with organic constituents of atmospheric water droplets by competition kinetic techniques **C. Brandt, W. Rolle** and **P. Warneck** *Institute for Tropospheric Research, Leipzig, Germany*

Stabilisation-dependent OH-yield in the reaction of acetylene-OH adducts with O_2 **B. Bohn, M. Siese** and **C. Zetzsch,** *Fraunhofer-Institut, Hannover, Germany*

The University of East Anglia peroxy radical chemical amplifier **K. C. Clemitshaw, L. J. Carpenter** and **S. A. Penkett,** *The Atmospheric Chemistry Group, UEA, Norwich, UK*

Photochemistry of polar stratospheric cloud mimics **T. G. Koch, C. W. Bushell, A. B. Horn** and **J. R. Sodeau,** *The Atmospheric Chemistry Group, UEA, Norwich, UK*

Study by LIF detection of temperature dependence in the reactions of the nitrate radical with monohalogenoalkenes **E. Martínez, B. Cabañas, A. Aranda** and **P. Martín** *University of Castilla-La Mancha, Ciudad Real, Spain*

Studies of the nitrate radical in the marine boundary layer **N. Smith, B. J. Allan, H. Coe, J. M. C. Plane, R. A. Burgess, L. C. Carpenter, K. C. Clemitshaw** and **S. A. Penkett,** *The Atmospheric Chemistry Group, UEA, Norwich, UK*

Determination of total hydrogen from balloon borne measurements of H_2O, H_2 and CH_4 **M. Zöger, A. Engel, C. Schiller, U. Schmidt, T. Woyke** and **D. S. McKenna,** *Forschungszentrum Jülich, Germany*

Atmospheric chemistry of dimethyl sulfide: kinetics of the $CH_3SCH_2O_2 + NO_2$ reaction in the gas phase **O. J. Nielsen, J. Sehested** and **T. J. Wallington,** *Ford Forschungszentrum, Aachen, Germany and Ford Motor Group, Dearborn, USA*

Atmospheric Chemistry of HFC-236fa and HFC-272 ca **O. J. Nielsen, T. E. Møgelberg, J. Platz, J. Sehested** and **T. J. Wallington,** *Ford Forschungszentrum Aachen, Germany and Ford Motor Group, Dearborn, USA*

Kinetic study of the reaction of Cl atoms with CH_3OCH_2Br **K. G. Kambanis, Y. G. Lazarou, I. I. Morozov** and **P. Papagiannakopoulos,** *University of Crete, Greece*

List of Participants

Ms. B. Allan *The Atmospheric Chemistry Group, UEA, Norwich, UK*
Miss C. Apps *University of Manchester, UK*
Mr. A. Aranda Rubio *University of Castilla-La Mancha, Ciudad Real, Spain*
Miss S. Ashbourn *University of Leicester, UK*
Prof. M. N. R. Ashfold *University of Bristol, UK*
Dr. S. Askey *University of Birmingham, UK*
Ms. H. D. Aslin *University of Liverpool, UK*
Dr. R. Atkinson *University of California, Riverside, USA*
Mr. S. Ball *University of Oxford, UK*
Ms. B. Bandy *The Atmospheric Chemistry Group, UEA, Norwich, UK*
Miss S. Banham *The Atmospheric Chemistry Group, UEA, Norwich, UK*
Dr. I. Barnes *University of Wuppertal, Germany*
Mr. S. Bauguitte *The Atmospheric Chemistry Group, UEA, Norwich, UK*
Prof. D. L. Baulch *University of Leeds, UK*
Ms. D. Bird *University of York, UK*
Dr. B. Bohn *Fraunhofer Institut, Hannover, Germany*
Dr. P. Borrell *Eurotrac Iss, Garnisch-Parteukirchen, Germany*
Mr. D. Bowden *The Atmospheric Chemistry Group, UEA, Norwich, UK*
Dr. C. Brandt *Institute for Tropospheric Research, Leipzig, Germany*
Miss W. Broadgate *The Atmospheric Chemistry Group, UEA, Norwich, UK*
Dr. H. Bunz *Forschungszentrum/imk, Karlsruhe, Germany*
Ms. R. Burgess *University of East Anglia, UEA, Norwich, UK*
Mr. A. M. Bush *University of Southampton, UK*
Dr. G. V. Buxton *University of Leeds, UK*
Prof. W. Byers Brown *University of Manchester, UK*
Dr. C. E. Canosa-Mas *University of Oxford, UK*
Dr. J. N. Cape *Institute of Terrestrial Ecology, Scotland, UK*
Miss L. M. Cardenas *The Atmospheric Chemistry Group, UEA, Norwich, UK*
Ms. L. Carpenter *The Atmospheric Chemistry Group, UEA, Norwich, UK*
Miss S. Carr *University College Dublin, Ireland*
Dr. D. Carslaw *British Gas PLC, Loughborough, UK*
Dr. K. Carslaw *Max-Planck Institute, Mainz, Germany*
Dr. N. J. Clark *Flinders University, Adelaide, South Australia*
Dr. D. C. Clary *University of Cambridge, UK*
Miss N. Clegg *Imperial College, London, UK*
Dr. S. L. Clegg *The Atmospheric Chemistry Group, UEA, Norwich, UK*
Dr. K. C. Clemitshaw *The Atmospheric Chemistry Group, UEA, Norwich, UK*
Dr. H. Coe *The Atmospheric Chemistry Group, UEA, Norwich, UK*
Mr. G. Collins *University of Birmingham, UK*
Ms. R. Connell *University of Oxford, UK*
Miss R. Cox *The Atmospheric Chemistry Group, UEA, Norwich, UK*
Dr. R. A. Cox *University of Cambridge, UK*
Prof. P. J. Crutzen *Max-Planck-Institute, Mainz, Germany*
Prof. P. Davidovits *Boston College, Chestnut Hill, USA*
Prof. P. T. Dawson *McMaster University, Ontario, Canada*
Prof. A. S. Dickinson *University of Newcastle, UK*
Mrs. C. Dimitroulopoulou *Imperial College, London, UK*
Dr. H. Donsig *UMIST, Manchester, UK*
Dr. D. H. V. Dos Santos *Department of Chemistry, University of Manchester, UK*
Dr. A. Doughty *University of Oxford, UK*
Prof. H. Elias *Technische Hochschule Darmstadt, Germany*
Mr. D. Firman *The Open University, UK*

Dr. D. J. Fish *University of Cambridge, UK*
Mr. O. Forberich *Johann Wolfgang Goethe Universität, Germany*
Dr. J. G. Frey *University of Southampton, UK*
Dr. G. J. Frost *NOAA Aeronomy Laboratory, Boulder, USA*
Prof. C. S. Gardner *University of Illinois, Urbana, USA*
Dr. C. George *Fraunhofer Institute Hannover, Germany*
Prof. Y. M. Gershenzon *Russian Academy of Sciences, Moscow*
Miss W. Glasspool *University of Southampton, UK*
Dr. K. Grant *Defence Research Agency, Poole, UK*
Prof. W. Hack *Max-Planck-Institute, Gottingen, Germany*
Ms. J. S. Hall *British Antarctic Survey, Cambridge, UK*
Mr. A. Hamilton *The Atmospheric Chemistry Group, UEA, Norwich, UK*
Dr. G. Hancock *Oxford University, UK*
Mr. T. Hancock *University of Bristol, UK*
Dr. M. Hand *University of Bath, UK*
Mr. M. Harwood *University of Cambridge, UK*
Dr. G. D. Hayman *National Environmental Technology Centre, Abingdon, UK*
Dr. M. Heal *University of Edinburgh, UK*
Dr. A. C. Heard *University of Leeds, UK*
Dr. D. Heard *University of Leeds, UK*
Mr. A. Heathfield *University of York, UK*
Ms. M. Helmer *The Atmospheric Chemistry Group, UEA, Norwich, UK*
Dr. R. Hernandez *University of Cambridge, UK*
Dr. H. Herrmann *University of Essen, Germany*
Prof. I. H. Hillier *University of Manchester, UK*
Prof. R. Hillman *University of Leicester, UK*
Mr. N. Holmes *The Atmospheric Chemistry Group, UEA, Norwich, UK*
Dr. O. Horie *Max-Planck-Institute, Mainz, Germany*
Dr. A. B. Horn *The Atmospheric Chemistry Group, UEA, Norwich, UK*
Dr. A. Hudson *University of Sussex, Brighton, UK*
Dr. A. J. Hynes *University of Miami, Florida, USA*
Miss A. Jackson *Lancaster University, UK*
Mr. H-W. Jacobi *University of Essen, Germany*
Dr. D. Jackson *The Open University, UK*
Dr. M. Jenkin *AEA Technology, Netcen, Abingdon, UK*
Dr. K. Ji *CNRS/LCSR ORLEANS, France*
Dr. A. Jones *British Antarctic Survey, Cambridge, UK*
Dr. R. L. Jones *University of Cambridge, UK*
Prof. J. A. Kerr *E.T.H. Zurich, Switzerland*
Mr. J. Kleffmann *University of Wuppertal, Germany*
Mr. G. Knight *University of York, UK*
Dr. T. G. Koch *University of East Anglia, UK*
Mr. P. Kunzli *Middlesex University, London, UK*
Dr. M. J. Kurylo *NIST & NASA, Gaithersburg, USA*
Dr. S. Langer *GU/CTH Gotheburg, Sweden*
Dr. B. Laszlo *National Institute of Standards & Technology, Gaithersburg, USA*
Mr. A. M. Lee *University of Cambridge, UK*
Dr. P. Lightman *Imperial College, Ascot, UK*
Dr. R. Lopez *University of Oxford, UK*
Dr. B. Luo *Max-Planck-Institute, Mainz, Germany*
Miss T. Malone *University of Leeds, UK*
Dr. G. Marston *University of Leicester, UK*
Dr. N. A. Martin *National Physical Laboratory, Teddington, UK*
Prof. E. Martinez Ataz *University of Castilla-La Mancha, Ciudad Real, Spain*
Dr. M. Massucci *The Atmospheric Chemistry Group, UEA, Norwich, UK*
Dr. M. R. S. McCoustra *University of Nottingham, UK*
Mr. A. McCulloch *ICI Chemicals & Polymers, Runcorn, UK*
Prof. D. S. McKenna *Institut für Chemie und Dynamikder Geosphare, Julich, Germany*

Dr. P. M. Midgley *M&D Consulting, Leinfelden, Germany*
Prof. T. K. Minton *Montana States University, Bozeman, USA*
Prof. M. J. Molina *Massachusetts Institute of Technology, Cambridge, USA*
Mr. C. Moore *University of Birmingham, UK*
Dr. G. K. Moortgat *Max-Planck-Institute, Mainz, Germany*
Prof. V. Mors *University of Essen, Germany*
Miss S. Murray *University of Leeds, UK*
Dr. K-H. Naumann *Forschungszentrum/IMK, Karlsruhe, Germany*
Dr. D. A. Newnham *Rutherford Appleton Laboratory, Didcot, UK*
Dr. O. J. Nielsen *Risø National Laboratory, Roskilde, Denmark*
Mr. A. Notario Molina *University of Castilla-La Mancha, Ciudad Real, Spain*
Mr. S. O'Beirne *CSIR, Pretoria, South Africa*
Prof. M. Okumura *Caltech, Pasadena, USA*
Mr. D. Oram *The Atmospheric Chemistry Group, UEA. Norwich, UK*
Mrs. S. Owen *Lancaster University, UK*
Dr. I. J. Palmer *University of Manchester, UK*
Prof. P. Papagiannakopoulos *University of Crete, Greece*
Ms. J. Parlor *The Open University, UK*
Ms. E. C. Pasteur *British Antarctic Survey, Cambridge, UK*
Miss R. Pathirage *The Atmospheric Chemistry Group, UEA, Norwich, UK*
Miss I. Patroescu *University of Wuppertal, Germany*
Prof. S. A. Penkett *UEA, Norwich, UK*
Mr. C. J. Percival *University of Oxford, UK*
Dr. T. Peter *Max-Planck-Institute, Mainz, Germany*
Dr. J. C. Petit *CNRS/LCSR Orleans, France*
Prof. J. Pfab *Heriot-Watt University, Edinburgh, UK*
Prof. M. J. Pilling *University of Leeds, UK*
Mr. J. C. Pinot de Moira *University of Oxford, UK*
Dr. J. M. C. Plane *The Atmospheric Chemistry Group, UEA, Norwich, UK*
Prof. U. Platt *University of Heidelburg, Germany*
Dr. T. Pryce-Jones *University of Hertfordshire, Hatfield, UK*
Dr. J. A. Pyle *University of Cambridge, UK*
Dr. O. U. Rattigan *University of Cambridge, UK*
Prof. A. R. Ravishankara *NOAA, Aeronomy Laboratory, Boulder, USA*
Dr. G. W. Rayner-Canham *Memorial University of Newfoundland, Canada*
Ms. C. Reed *University of Bristol, UK*
Dr. C. Reeves *The Atmospheric Chemistry Group, UEA, Norwich, UK*
Dr. S. J. Reid *Nilu, Kjeller, Norway*
Dr. H. Rieley *University of Liverpool, UK*
Dr. S. H. Robertson *University of Leeds, UK*
Mr. T. Roddis *The Atmospheric Chemistry Group, UEA, Norwich, UK*
Mr. P. Roden *University of York, UK*
Dr. H. K. Roscoe *British Antarctic Survey, Cambridge, UK*
Dr. D. M. Rowley *University of Cambridge, UK*
Dr. Y. Rudich *NOAA Aeronomy Laboratory, Boulder, USA*
Dr. C. M. Sadowski *York University, Canada*
Dr. G. A. Salmon *University of Leeds, UK*
Dr. A. Sarkissian *British Antarctic Survey, Cambridge, UK*
Mr. N. Savage *The Atmospheric Chemistry Group, UEA, Norwich, UK*
Mr. V. Scheer *Fraunhofer Institute, Hannover, Germany*
Mr. T. Schindler *Universität Munchen, Garching, Germany*
Mr. J. Sessler *University of Cambridge, UK*
Dr. D. E. Shallcross *University of Cambridge, UK*
Mr. N. C. Shand *Heriot-Watt University, Edinburgh, UK*
Dr. P. G. Simmonds *Bristol University, UK*
Prof. J. P. Simons *University of Oxford, UK*
Prof. R. H. Simoyi *West Virginia University, Morgantown, USA*
Dr. V. J. Simpson *Department of Environment, London, UK*

Dr. I. Sims *University of Birmingham, UK*
Dr. F. Slemr *Fraunhofer Institute, Garmisch-Partenkirchen, Germany*
Dr. J. Slemr *Fraunhofer Institute, Garmisch-Partenkirchen, Germany*
Dr. J. R. Slysser *University of Cambridge, UK*
Prof. I. W. M. Smith *University of Birmingham, UK*
Mr. J. Smith *University of Leeds, UK*
Miss N. Smith *The Atmospheric Chemistry Group, UEA, Norwich, UK*
Dr. H. Soathoff *Forschungszentrum/IMK, Karlsruhe, Germany*
Dr. J. Sodeau *The Atmospheric Chemistry Group, UEA, Norwich, UK*
Mr. H. Sommitz *University of Essen, Germany*
Mr. A. M. South *University of Cambridge, UK*
Mr. M. Spiekermann *Johann Wolfgang Goethe Universität, Germany*
Dr. M. Stark *University of York, UK*
Mr. D. W. A. Stewart *University of Birmingham, UK*
Dr. D. W. Stocker *University of Leeds, UK*
Miss D. Z. Stockwell *University of Cambridge, UK*
Mr. W. Scott *The Open University, UK*
Dr. W. T. Sturges *The Atmospheric Chemistry Group, UEA, Norwich, UK*
Miss J. Sully *The Atmospheric Chemistry Group, UEA, Norwich, UK*
Dr. A. C. Symonds *University of Birmingham, UK*
Prof. F. W. Taylor *Oxford University, UK*
Dr. F. Temps *Max-Planck-Institute, Gottingen, Germany*
Prof. B. A. Thrush *University of Cambridge, UK*
Ms. P. E. Titcombe *University of Leeds, UK*
Dr. A. F. Tuck *US Department of Commerce, Boulder, USA*
Dr. R. P. Tuckett *University of Birmingham, UK*
Dr. J. Uppenbrink *Science International, Cambridge, UK*
Prof. V. Vaida *University of Colorado, Boulder, USA*
Dr. M. P. van Albada *Amolf, Amsterdam, The Netherlands*
Dr. J. Vickerman *UMIST, Manchester, UK*
Dr. T. J. Wallington *Ford Motor Company, Michigan, USA*
Mr. G. Ward *University of Birmingham, UK*
Dr. K. Warr *The Open University, UK*
Prof. C. H. J. Wells *Kingston University, Surrey, UK*
Dr. J. Wenger *University College Dublin, Ireland*
Dr. C. M. Western *University of Bristol, UK*
Dr. J. B. C. Whitaker *University of Leeds, UK*
Dr. J. C. Whitehead *University of Manchester, UK*
Mrs. E. Wiesen *University of Wuppertal, Germany*
Dr. P. Wiesen *University of Wuppertal, Germany*
Mr. O. Wild *University of Cambridge, UK*
Mr. J. Williams *University of Leeds, UK*
Prof. A. M. Wodtke *University of Caifornia at Santa Barbara, USA*
Dr. E. Wolff *British Antarctic Survey, Cambridge, UK*
Prof. R. Zellner *Universität GH Essen, Germany*
Miss G. Zeng *University of Leeds, UK*
Mr. M. Zoeger *ICG-Forschungszentrum, Juelich, Germany*

Index of Contributors*

* The page numbers in heavy type indicate papers submitted for discussion.

General Discussions of the Faraday Society/Faraday Discussions of the Chemical Society

Date	*Subject*	*Volume*
1907	Osmotic Pressure	Trans. 3*
1907	Hydrates in Solution	3*
1910	The Constitution of Water	6*
1911	High Temperature Work	7*
1912	Magnetic Properties of Alloys	8*
1913	Colloids and their Viscosity	9*
1913	The Corrosion of Iron and Steel	9*
1913	The Passivity of Metals	9*
1914	Optical Rotary Power	10*
1914	The Hardening of Metals	10*
1915	The Transformation of Pure Iron	11*
1916	Methods and Appliances for the Attainment of High Temperatures in a Laboratory	12*
1916	Refractory Materials	12*
1917	Training and Work of the Chemical Engineer	13*
1917	Osmotic Pressure	13*
1917	Pyrometers and Pyrometry	13*
1918	The Setting of Cements and Plasters	14*
1918	Electric Furnaces	14*
1918	Co-ordination of Scientific Publication	14*
1918	The Occlusion of Gases by Metals	14*
1919	The Present Position of the Theory of Ionization	15*
1919	The Examination of Materials by X-Rays	15*
1920	The Microscope: Its Design, Construction and Applications	16*
1920	Basic Slags: Their Production and Utilization in Agriculture	16*
1920	Physics and Chemistry of Colloids	16*
1920	Electrodeposition and Electroplating	16*
1921	Capillarity	17*
1921	The Failure of Metals under Internal and Prolonged Stress	17*
1921	Physico-Chemical Problems Relating to the Soil	17*
1921	Catalysis with special reference to Newer Theories of Chemical Action	17*
1922	Some Properties of Powders with special reference to Grading by Elutriation	18*
1922	The Generation and Utilization of Cold	18*
1923	Alloys Resistant to Corrosion	19*
1923	The Physical Chemistry of the Photographic Process	19*
1923	The Electronic Theory of Valency	19*
1923	Electrode Reactions and Equilibria	19*
1923	Atmospheric Corrosion. First Report	19*
1924	Investigation on Oppau Ammonium Sulphate-Nitrate	20*
1924	Fluxes and Slags in Metal Melting and Working	20*
1924	Physical and Physico-Chemical Problems relating to Textile Fibres	20*
1924	The Physical Chemistry of Igneous Rock Formation	20*
1924	Base Exchange in Soils	20*
1925	The Physical Chemistry of Steel-Making Processes	21*
1925	Photochemical Reactions of Liquids and Gases	21*
1926	Explosive Reactions in Gaseous Media	22*
1926	Physical Phenomena at Interfaces, with special reference to Molecular Orientation	22*
1927	Atmospheric Corrosion, Second Report	23*
1927	The Theory of Strong Electrolytes	23*
1927	Cohesion and Related Problems	24*
1928	Homogeneous Catalysis	24*
1929	Crystal Structure and Chemical Constitution	25*
1929	Atmospheric Corrosion of Metals, Third Report	25*
1929	Molecular Spectra and Molecular Structure	26*
1930	Colloid Science Applied to Biology	26*
1931	Photochemical Processes	27*
1932	The Adsorption of Gases by Solids	28*
1932	The Colloid Aspect of Textile Materials	29
1933	Liquid Crystals and Anisotropic Melts	29*

Faraday Discussions

Date	*Subject*	*Volume*
1933	Free Radicals	30*
1934	Dipole Moments	30*
1934	Colloidal Electrolytes	31*
1935	The Structure of Metallic Coatings, Films and Surfaces	31*
1935	The Phenomena of Polymerization and Condensation	32*
1936	Disperse Systems in Gases: Dust, Smoke and Fog	32*
1936	Structure and Molecular Forces in (*a*) Pure Liquids, and (*b*) Solutions	33*
1937	The Properties and Function of Membranes, Natural and Artificial	33*
1937	Reaction Kinetics	34*
1938	Chemical Reactions Involving Solids	34*
1938	Luminescence	35*
1939	Hydrocarbon Chemistry	35*
1939	The Electrical Double Layer (owing to the outbreak of the war the meeting was abandoned, but the papers were printed in the *Transactions*)	35*
1940	The Hydrogen Bond	36*
1941	The Oil-Water Interface	37*
1941	The Mechanism and Chemical Kinetics of Organic Reactions in Liquid Systems	37*
1942	The Structure and Reactions of Rubber	38
1943	Modes of Drug Action	39*
1944	Molecular Weight and Molecular Weight Distribution in High Polymers (Joint Meeting with the Plastics Group, Society of Chemical Industry)	40*
1945	The Application of Infra-red Spectra to Chemical Problems	41*
1945	Oxidation	42*
1946	Dielectrics	42 A*
1946	Swelling and Shrinking	42 B*
1947	Electrode Processes	Disc. 1*
1947	The Labile Molecule	2
1947	Surface Chemistry (Jointly with the Société de Chimie Physique at Bordeaux Published by Butterworths Scientific Publications Ltd	
1947	Colloidal Electrolytes and Solutions	Trans. 43*
1948	The Interaction of Water and Porous Materials	Disc. 3
1948	The Physical Chemistry of Process Metallurgy	4*
1949	Crystal Growth	5*
1949	Lipo-proteins	6
1949	Chromatographic Analysis	7*
1950	Heterogeneous Catalysis	8*
1950	Physico-chemical Properties and Behaviour of Nuclear Acids	Trans. 46*
1950	Spectroscopy and Molecular Structure and Optical Methods of Investigating Cell Structure	Disc. 9*
1950	Electrical Double Layer	Trans. 47*
1951	Hydrocarbons	Disc. 10*
1951	The Size and Shape Factor in Colloidal Systems	11*
1952	Radiation Chemistry	12*
1952	The Physical Chemistry of Proteins	13
1952	The Reactivity of Free Radicals	14
1953	The Equilibrium Properties of Solutions on Non-electrolytes	15*
1953	The Physical Chemistry of Dyeing and Tanning	16*
1954	The Study of Fast Reactions	17*
1954	Coagulation and Flocculation	18*
1955	Microwave and Radio-frequency Spectroscopy	19
1955	Physical Chemistry of Enzymes	20
1956	Membrane Phenomena	21
1956	Physical Chemistry of Processes at High Pressures	22
1957	Molecular Mechanism of Rate Processes in Solids	23
1957	Interactions in Ionic Solutions	24
1958	Configurations and Interactions of Macromolecules and Liquid Crystals	25
1958	Ions of the Transition Elements	26
1959	Energy Transfer with special reference to Biological Systems	27
1959	Crystal Imperfections and the Chemical Reactivity of Solids	28
1960	Oxidation-Reduction Reactions in Ionizing Solvents	29
1960	The Physical Chemistry of Aerosols	30
1961	Radiation Effects in Inorganic Solids	31
1961	The Structure and Properties of Ionic Melts	32
1962	Inelastic Collisions of Atoms and Simple Molecules	33*
1962	High Resolution Nuclear Magnetic Resonance	34
1963	The Structure of Electronically Excited Species in the Gas Phase	35
1963	Fundamental Processes in Radiation Chemistry	36
1964	Chemical Reactions in the Atmosphere	37
1964	Dislocations in Solids	38

Faraday Discussions

Date	*Subject*	*Volume*
1965	The Kinetics of Proton Transfer Processes	39
1965	Intermolecular Forces	40
1966	The Role of the Absorbed State in Heterogeneous Catalysis	41*
1966	Colloid Stability in Aqueous and Non-aqueous Media	42*
1967	The Structure and Properties of Liquids	43
1967	Molecular Dynamics of the Chemical Reactions of Gases	44
1968	Electrode Reactions of Organic Compounds	45
1968	Homogeneous Catalysis with Special Reference to Hydrogenation and Oxidation	46
1969	Bonding in Metallo-organic Compounds	47
1969	Motions in Molecular Crystals	48
1970	Polymer Solutions	49*
1970	The Vitreous State	50*
1971	Electrical Conduction in Organic Solids	51
1971	Surface Chemistry of Oxides	52
1972	Reactions of Small Molecules in Excited States	53
1972	The Photoelectron Spectroscopy of Molecules	54
1973	Molecular Beam Scattering	55
1973	Intermediates in Electrochemical Reactions	56
1974	Gels and Gelling Processes	57
1974	Photo-effects in Adsorbed Species	58
1975	Physical Adsorption in Condensed Phases	59
1975	Electron Spectroscopy of Solids and Surfaces	60
1976	Precipitation	61
1977	Potential Energy Surfaces	62
1977	Radiation Effects in Liquids and Solids	63
1977	Ion–Ion and Ion–Solvent Interactions	64
1978	Colloid Stability	65*
1978	Structures and Motion in Molecular Liquids	66
1979	Kinetics of State Selected Species	67
1979	Organization of Macromolecules in the Condensed Phase	68
1980	Phase Transitions in Molecular Solids	69
1980	Photoelectrochemistry	70
1981	High Resolution Spectroscopy	71
1981	Selectivity in Heterogeneous Catalysis	72
1982	Van der Waals Molecules	73
1982	Electron and Proton Transfer	74
1983	Intramolecular Kinetics	75
1983	Concentrated Colloidal Dispersions	76
1984	Interfacial Kinetics in Solution	77
1984	Radicals in Condensed Phases	78
1985	Polymer Liquid Crystals	79
1985	Physical Interactions and Energy Exchange at the Gas-Solid Interface	80
1986	Lipid Vesicles and Membranes	81
1986	Dynamics of Molecular Photofragmentation	82
1987	Brownian Motion	83
1987	Dynamics of Elementary Gas-phase Reactions	84
1988	Solvation	85
1988	Spectroscopy at Low Temperatures	86
1989	Catalysis by Well Characterised Materials	87
1989	Charge Transfer in Polymeric Systems	88*
1990	Structure of Surfaces and Interfaces as studied using Synchrotron Radiation	89
1990	Colloidal Dispersions	90
1991	Structure and Dynamics of Reactive Transition States	91
1991	The Chemistry and Physics of Small Metallic Particles	92
1992	Structure and Activity of Enzymes	93
1992	The Liquid/Solid Interface at High Resolution	94
1993	Crystal Growth	95
1993	Dynamics at the Gas/Solid Interface	96
1994	Structure and Dynamics of Van der Waals Complexes	97
1994	Polymers at Surfaces and Interfaces	98
1994	Vibrational Optical Activity: From Fundamentals to Biological Applications	99

The Royal Society of Chemistry

FARADAY DIVISION

A Celebration of Physical Chemistry

To Mark the 100th Faraday General Discussion

The Royal Institution, London
24 April 1995

A Celebration of Physical Chemistry

24th April, 1995

Contents

Faraday Discuss., 1995, **100**, C1

A Celebration of Physical Chemistry

Introduction

J. P. Simons
Physical and Theoretical Chemistry Laboratory, South Parks Road, Oxford, UK OX1 3QZ

In introducing the Celebration, Professor J. P. Simons, FRS, President of the Faraday Division, said: It is my great honour, as President of the Faraday Division of the Royal Society of Chemistry, to welcome so many distinguished guests from industry, government and the broad scientific community to this Celebration of a very special centenary.

The Faraday Discussions are a unique, international scientific forum. They began almost a century ago and the series of published Discussion volumes provides an unparalleled record of scientific debate and discovery, particularly at the interfaces between chemistry and the neighbouring sciences. The 100th General Discussion, *Atmospheric Chemistry: Measurements, Mechanisms and Models*, has provided a typical (and topical) example of the genre. This Celebration, held in the Royal Institution—Michael Faraday's Workshop—marks the Centenary, and allows a moment's reflection at a time of rapid change. It is especially appropriate for our celebratory party to be held in the famous lecture theatre that is featured, along with Michael Faraday's boyish portrait, on our £20 bank notes—thereby acknowledging and symbolising the interlinking of scientific, cultural and economic wealth. The Celebration will explore the connections between curiosity and discovery, basic and applied science, imagination and application, academe and industry, the past and the future. I now invite the Right Honourable John Horam, MP, Parliamentary Secretary at the Office of Public Service and Science, to open the proceedings.

The lectures which follow provide a record of those proceedings, which were accompanied by many other supporting activities, particularly a Poster Programme highlighting current research represented by each of the 15 Subject Groups affiliated to the Faraday Division, and concluding remarks from the immediate past-President of the Society, Professor Charles Rees, FRS. The day ended with a banquet in the Society's Library at Burlington House attended by many distinguished guests from Industry, Government and Academe.

Faraday Discuss., 1995, **100**, C3–C7

The Labile Molecule since 1947

George Porter
Imperial College, London

We are here to celebrate our favourite science of Physical Chemistry, and it is right that the celebration should be here in the Royal Institution, the House of Michael Faraday, where he made all his discoveries, where he lectured, in this theatre, for over 40 years, and where he lived most of his life as the Resident professor, in the rooms on the floor above. I count it as the greatest privilege of my life to have been, for 20 years, one of Faraday's successors as Resident professor, working in his laboratories, living in his rooms, and lecturing in his theatre.

It was in this theatre, on 30th June 1903, that the Faraday Society held its foundation meeting for the purpose of 'promoting the study of electrochemistry, electrometallurgy, chemical physics, metallography and kindred subjects'. It was a society of physical chemists, named after one who never described himself as such, or even as a physicist—a species unrecognised in his time—but who liked to be called a chemist or, better still, a natural philosopher.

The first General Discussion of the new society was held in 1907 and there were about one hundred of them before they began to be numbered as separate publications with No. 1 in 1948 and the 100th in the new series held this month. The Faraday Discussions have been the seed beds from which Physical Chemistry has developed during this century. Many of the early discussions had obvious practical applications and the early presidents were pioneers in the industrial applications of physical chemistry: Sir Joseph Swann, whose electric lamp preceded Edison's by 20 years, was the first President; Lord Kelvin whose researches made possible the laying of the first transatlantic cable was second and Sir William Perkin who founded the synthetic dye industry and was present in this theatre in 1906 for the golden jubilee of his discovery of Mauvine, was the third.

Unfortunately, I was myself unable to be present at that first meeting in 1903 but I was present when the society celebrated its fiftieth anniversary, in this same room in 1953. The discussions of the Faraday Society, over the next decade or two were, to me, where everything in physical chemistry happened, where the very latest advances were published and discussed, and where one met like-minded scientists from this country and overseas who became lifelong friends. To me there was no society like it, it was part of my life.

To give a balanced, impersonal, account of the many themes of the Faraday Discussions which have taken place during the 48 years that I have been a member, even if it were possible, would be exceedingly dull. Instead, I offer a very personal account of the one area which has concerned me most and how, for me, it has been inseparable from the Faraday Discussions of the time. I am sure that many of those here could tell similar stories about the importance of Faraday Discussions to their own work.

I have called this talk 'The Labile Molecule since 1947' because it combines the title and date of the first Faraday Discussion that I attended in Oxford 48 years ago. It was only the second in the series of published discussions whose centenary we celebrate. The pattern of the Faraday Discussions is unique; the papers are written in full and distributed before the meeting; the presentation lasts only five minutes and is followed by questions and discussion, all of which are published.

I presented my first paper, on work done with Professor R. G. W. Norrish, at that 1947 meeting. There was nothing remarkable about the paper but it was important to me and its presentation was to a large Oxford audience, including Hinshelwood, Steacie, Melville, Coulson, M. G. Evans, and other names that I ranked at least equal to the old testament prophets. And there were younger prophets in waiting, like Longuet-Higgins, Rex Richards and Fred Dainton. It was a daunting occasion for a second year research student from Cambridge. I rather expected to be eaten for breakfast, along with the Oxford marmalade. But it was a friendly society and everybody was very kind.

Our paper was about the detection and measurement of free radicals in gases by a flow technique, developed in 1929 by F. Paneth, who used the removal of metal mirrors as a measure of the free radicals present after thermal decompositions. It was an unspecific and difficult method for photochemical reactions, even with the help of a large army searchlight as the light source, but it was the most direct method available.

In the whole 400 pages of this discussion there was no mention of a direct observation of any free radical, except for some emission spectra in flames and electrical discharges and, indeed, some were still sceptical about their very existence under normal conditions. Sir Harry Melville had no doubt about their existence but was sanguine about our chances of direct observation. In his opening address, he said.

> The direct physical methods of measurement simply cannot reach these magnitudes, far less make accurate measurements in a limited period of time, for example 10^{-3} sec.

This was a challenge that I heard with glee because I had, that summer, decided to try a new line of work which, if successful would do just this. I discarded my searchlight for a flash arrangement with a large capacitor bank by courtesy of my friends in the Royal Navy. The total cost was very small, or even negative, since the Navy were kind enough to say that there was five shillings due to me for every crate that I returned.

The idea, which I called flash photolysis, was very simple in principle, though less so in practice. A flash of light was used to produce chemical change in a substance and a second flash of light then recorded, spectroscopically, the products that were present a short time afterwards. By repeating the process with increasing time intervals between the flashes, a movie of the very fast chemistry of free radicals and other intermediates was recorded.

The first paper (published with Professor R. G. W. Norrish), on the use of high-energy flashes in photochemistry, appeared in *Nature* in 1949, and the second paper on the double flash, pulse and probe—the flash photolysis technique—was submitted to the Royal Society in the same year (in some haste since I was to be married two days later). I read this paper the following January, at an afternoon meeting of the Royal Society in Burlington House with the President, the redoubtable Sir Robert Robinson, in the chair. He was one of those organic chemists who had grave doubts about the existence of these labile molecules but he was gracious enough to comment after my paper 'there does now seem to be *some* experimental evidence for free radicals'.

The time was now opportune to attend my second discussion of the Faraday Society, held in Cambridge in 1950, on 'Spectroscopy and Molecular Structure' where I was able to describe in more detail the first absorption spectra of several free radicals obtained by the new method. The most interesting free radical was ClO, obtained by flash photolysis of a mixture of gaseous chlorine and oxygen.

The audience included Gerhard Herzberg and James Franck, as well as Michael Kasha; truly a contempory of mine since we were born on the same day but, owing to the transatlantic time difference, neither of us has been able to claim priority. To me, Michael's paper was the most memorable of this meeting. It dealt with types of molecular transition that were little known, singlet–triplet and n–π^* transitions. It heralded the coming of a new era in photochemistry, but received little attention at the time, least of all from photochemists.

My third Faraday Discussion, held in Toronto in 1953, was attended by 45 members of the society from Europe, including R. P. Bell, E. J. Bowen, Charles Goodeve, Alfred Egerton, R. G. W. Norrish, Harry Melville, F. S. Dainton, Peter Gray and John Polanyi along with E. W. R. Steacie, H. S. Taylor, George Kistiakowsky, Linus Pauling and the Society's secretaries, F. C. Tomkins and Beatrice Kornitzer. Many of us were shipmates on the *Empress of Britain* for the crossing. It was the first post-war overseas meeting of the Faraday Society and a quite memorable one for all of us.

The subject of this Canada meeting was 'The Reactivity of Free Radicals'. It seemed as if the Discussion subjects were being chosen especially for me because, having interpreted the spectra of some free radicals, I had now turned to the original purpose: the use of these spectra to measure reactivity.

The chemical reactions of the free radical ClO made a very interesting story but were of little practical importance, as I had to admit to visitors to the laboratory in Cambridge who might become benefactors and who, quite reasonably, were interested in possible applications. It was 30 years later that the information gained in those days became of very practical importance. It turned out that chlorine in the stratosphere, formed by photochemical decomposition of the chlorofluorocarbons used as aerosols and refrigerants, were responsible for the decomposition of ozone in the stratosphere and the recently discovered ozone hole over Antarctica. It was gratifying that, very shortly afterwards, ClO was shown, from stratospheric flights over Antarctica, to be a key intermediate in this photochemical cycle of destruction. The Faraday stories run and run, and the 100th Faraday discussion, held this month is on 'Atmospheric Chemistry'.

Within a few years, free-radical studies were everywhere, dozens of free-radical absorption spectra had been described and a new free-radical society was about to be created. We were particularly interested in the organic, and particularly the aromatic, free radicals and we recorded for the first time the spectra of benzyl, anilino, phenoxyl and phenyl and many of their derivatives. It fell to Herzberg to use flash photolysis and ultra-violet spectroscopy to discover the spectra of methylene and methyl which had been two of our original targents. About this time, in 1954, Irwin Norman and I described a complementary technique for recording free-radical spectra, by photolysis of low-temperature solid solutions.

We now turned to the flash photolysis study of another very important class of short lived intermediates, the electronically excited states of molecules. Flash photolysis was still limited by the duration of the shortest flashes available, to times of a few microseconds whilst typical allowed electronic transitions take place in much shorter times, in the nanosecond range. However, the triplet states involve much slower, forbidden transitions and their absorption spectra had just been detected in low-temperature solids by Donald McClure. All seemed set to detect these excited species in gases and in fluid solvents at normal temperatures, although it was not at all certain that their lifetimes under these conditions would be long enough for our methods. We were fortunate; the lifetimes were ten or a hundred times longer than our flashes and a whole new chemistry became available to us, the chemistry of the triplet state which, as soon became apparent, was as important to a photochemist as free radical chemistry.

These transient triplet-state molecules in solution formed the subject of a paper with Maurice Windsor in my fourth Faraday Discussion held in Birmingham in 1954, on the subject of 'The Study of Fast Reactions'. Our paper was on 'The Triplet State in Fluid Solvents' and it described the detection and kinetic study of about a dozen aromatic molecules as well as, along with Robert Livingston, chlorophyll, in their triplet states at room temperature where their lifetimes were, very conveniently for us, a few hundred microseconds.

R. G. W. Norrish and I read a paper at this same discussion on the flash photolysis technique itself. This was one of the first of many discussions on techniques for fast reaction chemistry. Other methods described included the shock tube, flow (and stopped flow) methods, and the temperature jump and relaxation techniques described by Eigen in his paper 'Study of Ionic Reactions in Solutions in Times as Short as 10^{-9} seconds." Notice that, although 'microsecond' was now in the chemical vocabulary, the 'nanosecond' had still to arrive.

We were all getting excited about fast reactions and were already talking about *very* fast reactions. There was a story going around at this meeting that Manfred Eigen sought the advice of an Oxford man, as to what, in the English language, one could call reactions that were faster than very fast. Without any hesitation Ronnie Bell replied 'Damn fast reactions Manfred and if they get faster than that, the English language will not fail you, you can call them. 'Damn fast reactions indeed!'

For chemists, biologists and physical chemists, this meeting began 'the race against time'. We awaited shorter pulses in the nanosecond region to enable us to study those very labile molecules, the excited *singlet* states of molecules. The laser, which was to transform the technique of flash photolysis and to make it the fastest of all these fast methods, appeared in 1960. It took a few years to get hold of a laser and apply it to photochemistry but this was done by Jeff Steinfeld in Sheffield by 1965, using pulses of a few nanoseconds.

In 1966 I moved from Sheffield University to the Royal Institution and a student, Micheal Topp, who moved with me, brought the precious laser and set up our first nanosecond flash apparatus here. We had to solve several further problems before we could do nanosecond flash photolysis. We had to produce a nanosecond *white light* pulse and to synchronise the two pulses within a few nanoseconds delay. We were then able to detect the singlet state absorption spectra of many molecules.

As new laser pulse techniques became available, especially the mode-locked colliding-pulse-mode laser developed by Shank, picosecond and then femtosecond flash photolysis became possible. The steps in this race against time are shown in comparison with the cosmic time scale, in Fig. 1.

In the femtosecond region of physical chemistry, there are two areas which are advancing particularly rapidly at this moment. The first, the fastest step in the course of a molecular reaction, is the crossing of the transition state. This was the subject of the Faraday Discussion of 1991 on 'Structure and Dynamics of Reactive Transition States' at which speakers included John, the second Polanyi in this field, his fellow Laureate, Dudley Hershbach and two pioneers of femtosecond transition state dynamics Dick Zare and Ahmed Zewail. The progress through the transition state of a dissociating NaI molecule, which becomes trapped in the resonant excited state for an average of 10 vibrations, was followed by Zewail using fluoresence of the product Na* and the off-resonance fluorescence of NaL*.

The second rapidly developing area of femtosecond molecule dynamics is photosynthesis. At this point I should remind you that in 1951, the Faraday Society broadened its terms of reference to read 'To promote the study of sciences lying between chemistry, physics and biology.' This was a wise move, though it is a matter for regret to me that there have been relatively very few Faraday Discussions in these interdisciplinary fields, probably because of the immense growth in the number of societies and their associated journals in the biophysico-chemical areas. Today many of the most exciting developments are in the biophysical sciences.

Particularly active at the present time are studies of the primary processes of photosynthesis. These begin with the transfer of the absorbed solar excitation energy between pigment molecules in the leaf membrane followed by the transfer of electrons and protons across the membrane. It turns out that both of these processes occur in picoseconds or less and it was fortunate that, at the time that these systems were being discovered, femtosecond flash photolysis was becoming available.

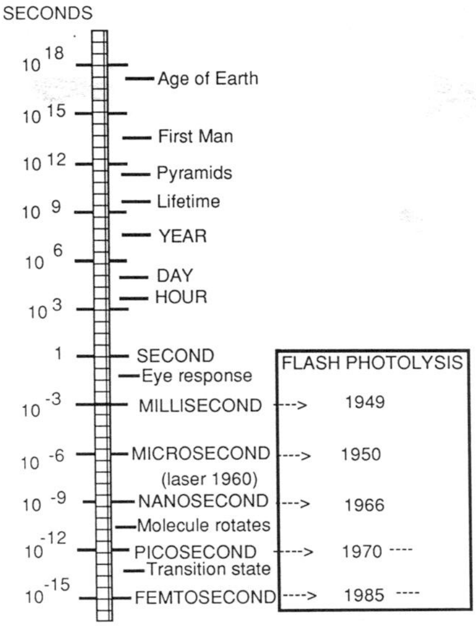

Fig. 1 Timescale of flash photolysis compared with the cosmic scale

The most recent studies of these reactions concern the reaction centre of photosystem 2 of the green leaf which is the simplest unit that still carries out both energy and electron transfer. The structure of its associated light-harvesting unit (LHC 2) is now known from the electron diffraction work of Kühlbrandt. The structure of the reaction centre (containing six chlorophylls and two pheophytins) has not yet been fully determined but much can be deduced by comparison with the structures of the reaction centres of the photosynthetic bacteria, determined by Hüber, Mlchel and Deisenhöffer, which have many close homologies with photosystem 2.

The processes of energy and electron transfer in the reaction centre occur in times that lie in the femtosecond and picosecond regions, respectively. The theory of *energy* transfer was the subject of the Spiers Memorial lecture given by Theodore Förster at the Faraday Discussion of 1959 on 'Energy Transfer with Special Reference to Biological Systems'. Theories of *electron* transfer have frequently been discussed, notably by Rudy Marcus in his R. A. Robinson Memorial lecture at the Faraday Discussion of 1982 on 'Electron and Proton Transfer'. The rates predicted for these transfers are about 100 femtoseconds and a few picoseconds, respectively, which agree very well with those measured directly by flash photolysis.

There is no part of the natural world which does not come under the scrutiny of physical chemists from time to time: the electrical, magnetic and optical properties of matter and its internal motions, rotations, vibrations and electronic transitions; the structure of molecules, new ones like buckminsterfullerene and the eternal, complex and beautiful ones of nature; chemical changes in strange places like the stratosphere and outer space, in the hottest flames and in solids near to the absolute zero. And, chosen as the subject of this talk, the changes in these structures that occur, both in the laboratory and in nature, in times so short that we are near the point where chemists, biologists and most physicists, must accept that they have reached the end of the timescale and the limits of certainty.

Faraday Discussion 100 *Celebration Paper*; *Presented* 24*th April*, 1995

Faraday Discuss., 1995, **100**, C9–C27

Tales of Tortured Ecstasy: Probing the Secrets of Solid Catalysts

John Meurig Thomas
Davy Faraday Research Laboratory, The Royal Institution of Great Britain, 21 Albemarle Street, London, UK W1X 4BS and Peterhouse, University of Cambridge, Cambridge, UK CB2 1QY

1. Introduction

In view of the fact that the first part of the title of my talk has prompted some to comment on its sado-masochistic overtones, I should point out that its origins lie deep in the poetry and insights of Shakespeare:

'... bitter torture shall winnow the truth from falsehood'

and

'... the torture of the mind (is) to be in restless ecstasy'

In common with many others, I pursue my chemistry with passion and commitment. And it seems to me to be within the bounds of poetic licence to describe one's oscillation of mood from euphoric exhilaration—when the act of probing is profitable—to saturnine gloom—when it is not—as tortured ecstasy.

Now catalysis as a phenomenon has been the subject of numerous Faraday Society Discussions. At the fiftieth, held in Cambridge in 1928, T. M. Lowry in his opening address pursued a holistic approach in which he endeavoured to encompass homogeneous, heterogeneous, biological and photo-catalysis. Looking back at that meeting, as well as an earlier one in 1921 held in London also on catalysis, we note that many of the ideas and concepts in vogue seventy years ago have fallen by the wayside, broken milestones on a vanished road. But one of the most remarkable facts is that the subdiscipline of catalysis which was arguably the least understood seventy or so years ago—biological or enzymatic catalysis—is now almost certainly the best understood.

2. Lessons from the Enzymologists and Organic Chemists

Why is it that enzymologists can nowdays successfully design new biological catalysts such as artificial enzymes that rival (even surpass) the performance of the parent natural enzyme? And why is it that the inorganic catalyst scientist and engineer are a good deal less adroit in their endeavours? The principal reason is because those working with enzymes have, for over thirty years, been engaged in *in situ* structural studies of their catalysts and have consequently gained revealing insights into mechanism, whereas those working with inorganic catalysis have not.

In the mid 1960s D. C. Phillips *et al.* elucidated the structure of the enzyme lysozyme.[1] They identified the cavity within its structures that constitutes its active site, and in particular, characterized in atomic detail the stereochemistry of the site and the manner in which reactant molecules fitted snugly within it: how, as it were, the reactant was held (Plate 1) poised for catalytic conversion. Their studies also revealed how product species and and/or inhibitors occupied the cavity of the active site (Fig. 1). The

Plate 1 Packing of lysozyme with the cleft that defines the active site occupied by the reactant (substrate)

sequence of the amino acid components lining the cavity active site was identified and the nature of the binding of reactants and inhibitors likewise elucidated.

This in turn gave the enzyme chemist and molecular biologist, equipped with the preparative and synthetic dexterity of the organic chemist, a firm basis for the design of smaller (miniature) versions of naturally occurring enzymes. It became sensible to talk of biomimetic catalysts and artificial enzymes, and with the later advent of site selective mutagenesis new dimensions of structural probing became possible. A particular amino acid close to or actually constituting the active site—which usually consists of a triad of amino acids—could be replaced, thereby enabling the catalytic performance of the enzyme to be delicately monitored as a function of subtle change in stereochemical and electronic environment within the active site. (For an elegant account of such procedural

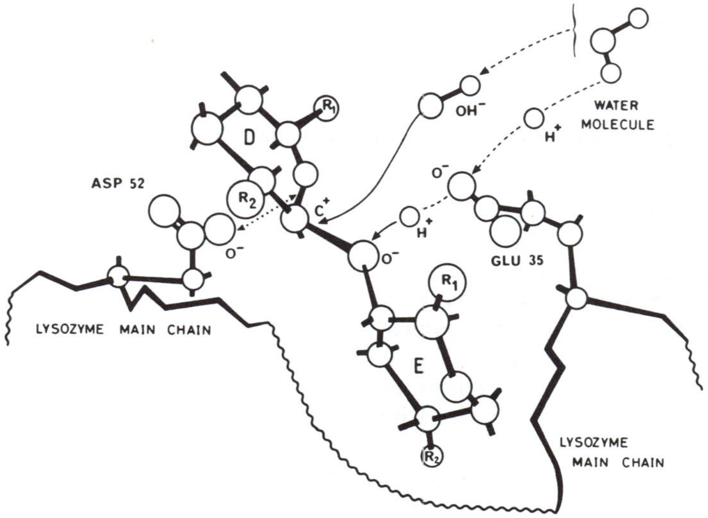

Fig. 1 Mechanism of mode of catalytic action of lysozyme (from Phillips[1])

advances, see the work of Knowles[2] on the manner in which triose phosphate isomerase catalyses the interconversion of two triose phosphates known as DHAP and GAP.)

Work on so-called artificial enzymes may be illustrated by reference to the proteolytic enzyme chymotrypsin, the structure of the active site of which was elucidated by *in situ* X-ray crystallography several decades ago. This revealed that the key components of the catalytic reaction cavity of chymotrypsin are imidazolyl, phenolic and carboxylic groups. Armed with this information a way of assembling a miniature artificial chymotrypsin presented itself to Bender and co-workers.[3] They took a naturally occurring molecule such as β-cyclodextrin as the shallow cavity and grafted on to its rim—appropriately juxtaposed just as in the chymotrypsin itself—the three components (the triad of groups) that constitute the active site. Such a creation is shown schematically in the lower half of Fig. 2 and a scalar model is shown in Plate 2. Although there is now some doubt about the validity of the claims made in Bender and co-workers' work, the notion that lies behind their overall strategy remains (Plate 2): a synthetic molecular entity of about a hundredth the size and mass of the parent enzyme can be made to function as a catalyst in such a manner as to rival the performance of the natural enzyme.

Breslow[4] has produced an even simpler piece of elegant analogous chemistry where the reality of the shape-selective catalysed chlorination of anisole is unmistakably

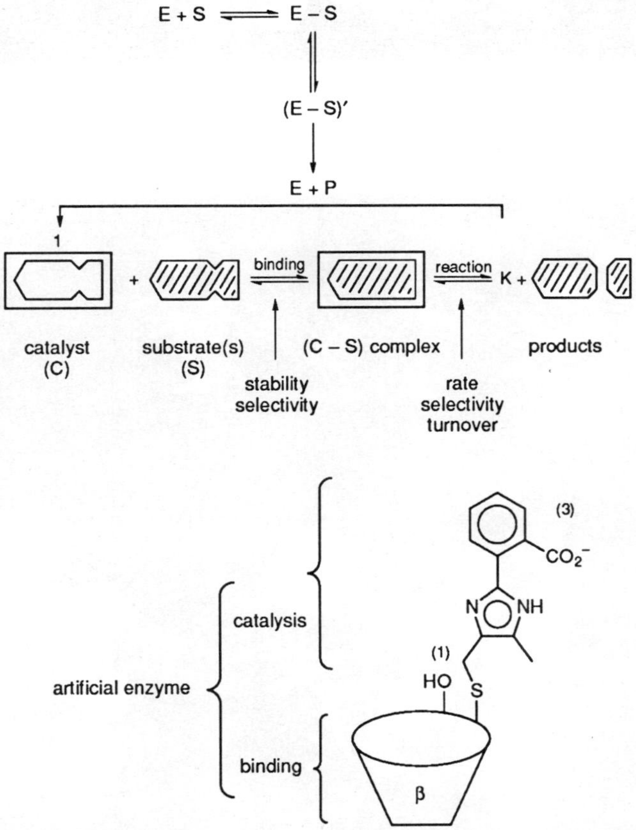

Fig. 2 (*a*) Schematic representation of the enzymatic process. The enzyme E processes the substrate (reactant) S into product P *via* the transition state E–S (after Lehn). (*b*) Strategy for the construction of an artificial chymotrypsin (miniature enzyme), based on cyclodextrin (from J. M. Thomas[6])

Plate 2 Model of artificial chymotrypsin. (after Bender and co-workers[3]). (Colour graphics by R. G. Bell and J. M. Thomas.)

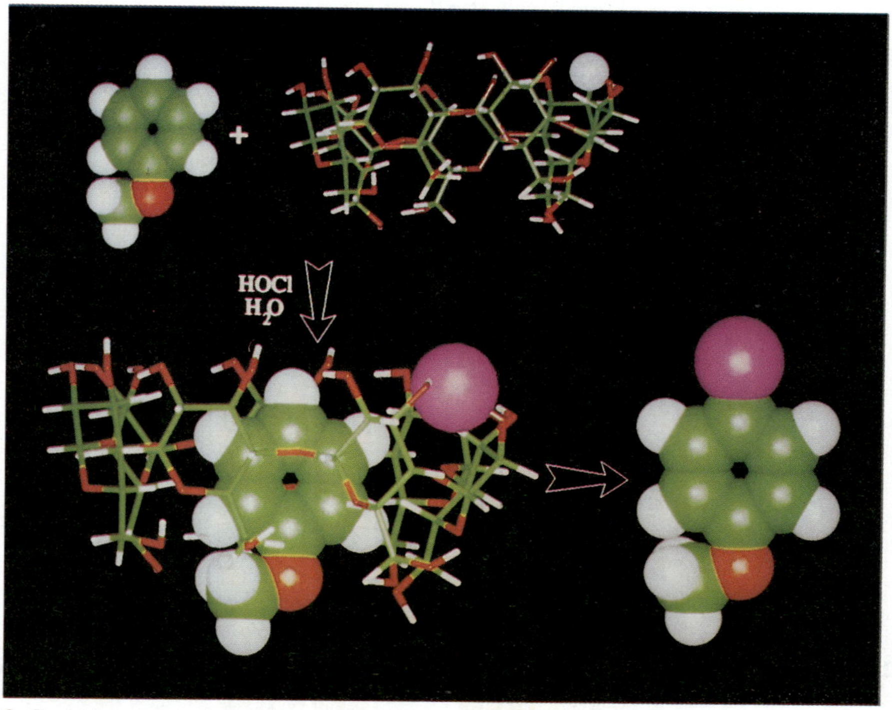

Plate 3 Cyclodextrin-catalysed shape-selective conversion of anisole to *p*-chloroanisole. (Colour graphics by Dewi Lewis based on Breslow[4].)

demonstrated (Plate 3). Only the *para* chloro isomer is produced in this cyclodextrin-steered reaction.

3. How Well can we Design Inorganic Catalysts?

Although it seems highly unlikely that we shall ever have the benefit of an inorganic equivalent to site-selective mutagenesis, great strides may be taken in designing shape-selective inorganic catalysts using molecular sieve solids typified by that shown in Plate 4. Here we see a zeolitic framework known as ZSM-5 (Zeolite Socony Mobil number 5) which is a highly siliceous aluminosilicate of general formula $H^+(Si_{1-x}Al_xO_2)^- \cdot nH_2O$, where the Si/Al ratio may range from a low of *ca.* 10 to almost infinity.

Aluminosilicate ZSM-5 is an excellent catalyst[5] for shape-selectively:

(a) converting benzene to monoethylbenzene (a precursor of styrene) by alkylation with ethene;

(b) disproportionating two molecules of toluene to one of benzene and one of *para*xylene;

(c) isomerizing a mixture of xylenes preferentially to *para*xylene; and

(d) converting methanol to petrol (gasoline)—a mixture of predominantly benzene, toluene and xylene—and water.

By introducing certain key elemental substituents into the ZSM-5 framework, numerous other highly selective catalytic reactions of profound petrochemical significance may be effected.[6] Thus, with gallium in the ZSM-5, propane and butane may be efficiently dehydrogenated to yield benzene and toluene, this being the basis of the British Petroleum cyclar process, commercialized some five years ago. With zinc in the ZSM-5 framework, dehydrogenation of isobutane to isobutene (known also as 2-methylpropene) is effected. This alkene, first prepared in this building (and discovered by Michael Faraday), is of central importance industrially as it is the precursor of MTBE (methyl tertiary butyl

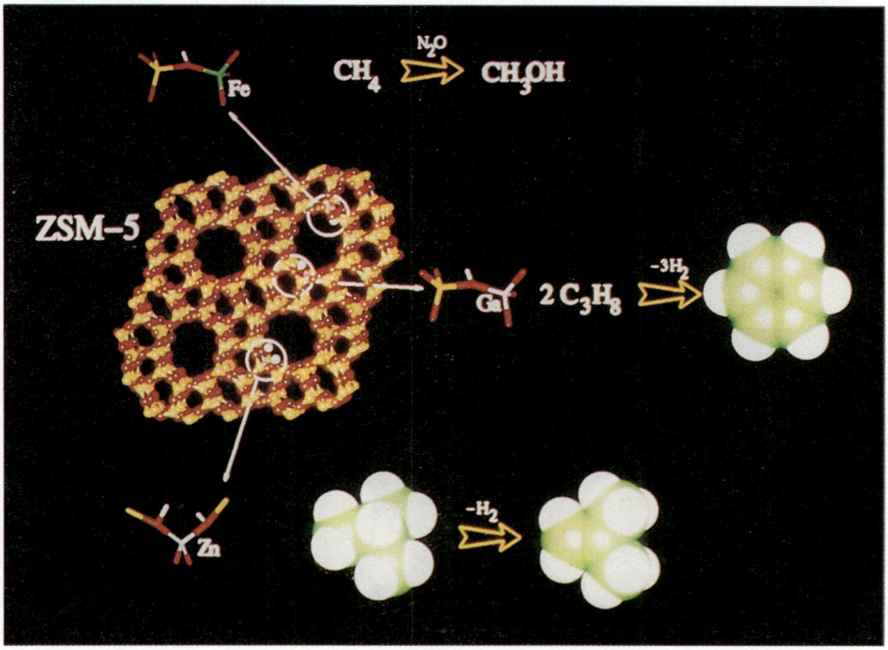

Plate 4 ZSM-5 and its use as a modified catalyst for methane oxidation to methanol;[8] propane to benzene; and isobutane to 2-methylpropene (isobutene or isobutylene)

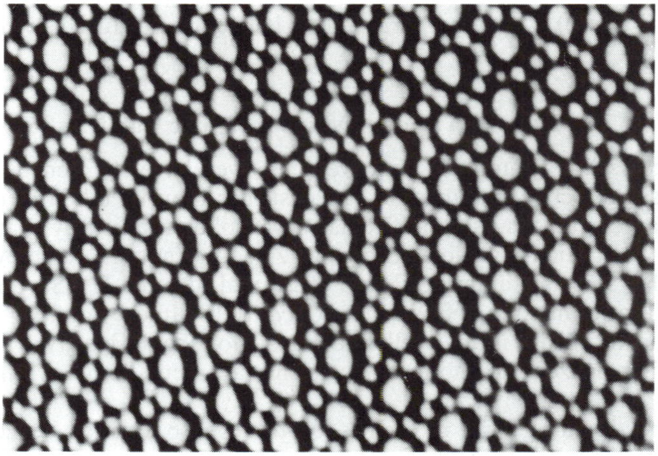

Fig. 3 High-resolution electron micrograph of a zeolitic acid catalyst showing projected structure of pores lined with active sites. Large pores are 5.5 Å in diameter [from J. M. Thomas,[6] and J. M. Thomas and P. L. Gai-Boyes, *Nature* (*London*), 1993, **364**, 478].

ether) a major constituent, nowadays, of motor car fuel since it both replaces the deleterious lead-containing additives and facilitates[7] the complete combustion of the fuel to CO_2 rather than CO. With iron in the framework, the ZSM-5 is reported to favour the oxidation of methane (with N_2O as reactant) to methanol.[8]

ZSM-5 itself (Fig. 3) was first produced (in the early 1970s) by accident. Workers at the laboratories of the Mobil Company, USA, decided to try a new organic molecule as a template for the production of synthetic zeolites using the general strategic principles first employed by the British scientist R. M. Barrer. More will be said about

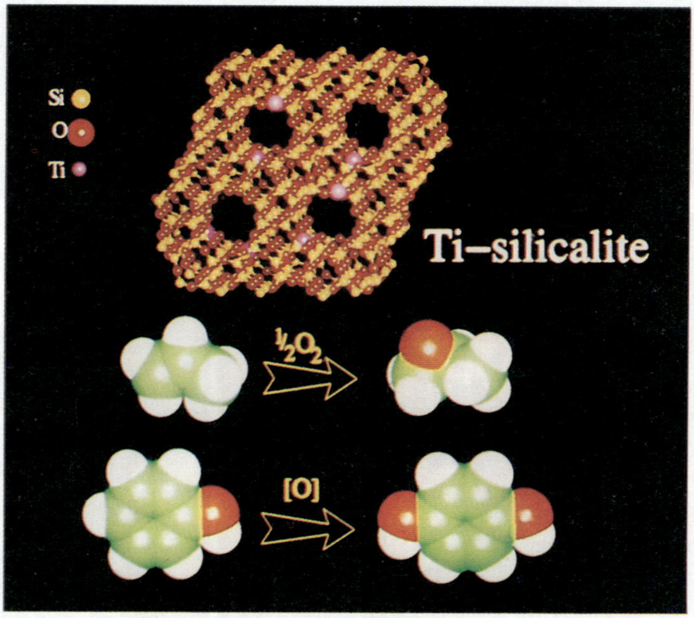

Plate 5 Graphic illustrating how microporous titanosilicalite (TS-1) converts phenol to hydroquinone using H_2O_2, and propene to propene oxide (Colour graphic by Dewi Lewis.)

templating later; what is important to note here is that it is possible to design new catalysts having a framework structure more-or-less identical to that of ZSM-5, but implanted with a 'designed' hetero-atom so as to endow the resulting solid with altogether new catalytic properties. This is precisely what researchers at the Enichem Company, Italy, succeeded in doing when they brought forth their so-called titanosilicalite-1 (TS-1). This has the same framework structure as ZSM-5 and silicalite-1 (which is the siliceous end-member of ZSM-5, *i.e.* $x = 0$ in the above-mentioned formula), but Ti^{4+} is accommodated in the framework. TS-1 is a highly selective and active oxidation catalyst, capable, amongst other things, of facilitating the epoxidation of propene and the conversion of phenol and H_2O_2 to quinol (1,4 benzene diol), an important material (as a developer) in the photographic industry (Plate 5).

4. *De Novo* Synthesis of New Inorganic Catalysts

Plate 6 shows the protonated (acidic) form of mordenite, inside the pores of which are molecules of diisopropylnaphthalene (DIIPN), a vitally important building block in the synthetic polymer industry[9,10] (DIIPN is formed in high yield when propene and naphthalene are allowed to react in the presence of an acid catalyst). Over acidic silica–alumina gel numerous other, undesirable products are formed as well as DIIPN. Over the environmentally harmful $AlCl_3$, again there is a multiplicity of products. Acidic mordenite, however, is shape-selective in that it yields predominantly 2,6 DIIPN). In Plate 6 the snugly fitting product (DIIPN) gives a clue as to the mode of action of organic templates in the synthesis of new microporous catalysts.

The degree to which templates are critical in the synthesis of a particular framework varies.[11,12] Certain templates (of which ethylenediamine is an example) may be regarded simply as void-filling species that do not contribute significantly to the preferential formation of a desired structure. We also find that a particular framework is formed by several different template molecules, a process which might be more correctly termed

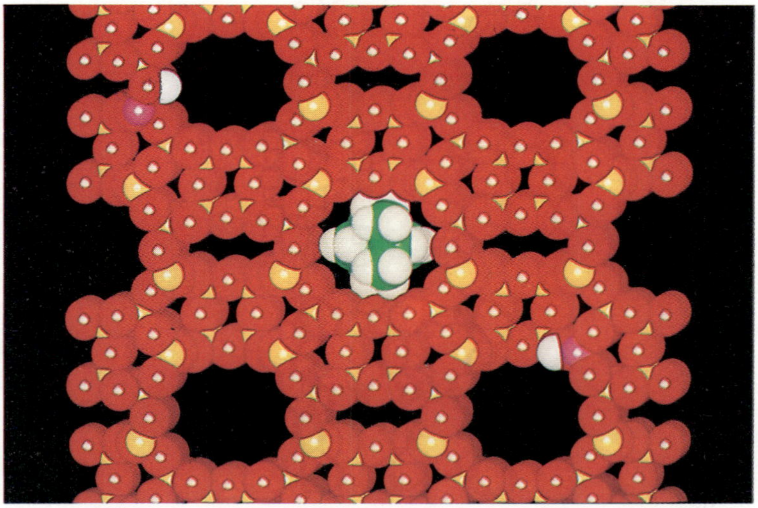

Plate 6 Colour graphic representation of protonated mordenite of high Si/Al ratio. Inside one of the large pores a molecule of diisopropylnaphthalene (DIIPN) is shown. As described in the text (see also Cusumano[9]) acidic mordenite is the catalyst of choice, especially on environmental grounds, in the conversion of naphthalene and propene to DIIPN. (After J. M. Thomas, R. G. Bell and P. A. Wright, *Bull. Chim. Soc. Fr.*, 1994, **131**, 482.)

structure-directing rather than genuine templating. In general, however, all the templates suitable for a particular framework will possess similar properties—shape, size, basicity, *etc.*—which direct the gel medium (usually under hydrothermal conditions), from which the crystals nucleate toward the formation of particular structural motifs. A great deal of accumulated experience, supplemented nowadays by computer graphics and modelling[11–13] has been acquired to enable the inorganic preparative chemist to design a wide range of zeolitic material. In particular, microporous aluminium phosphates, and metal-substituted (in the framework) aluminium phosphates may now be routinely prepared. Some of these have structures similar to those that have been characterized in the aluminosilicate zeolites (natural and synthetic). Some, however, are quite novel.

The framework structures of three entirely microporous solids, synthesized in this Laboratory or in association with collaborators in Jilin University, Peoples' Republic of China, are illustrated in Fig. 4.

It transpires that aluminium phosphates (ALPOs) may be quite readily produced as microporous analogues of naturally occurring and synthetic zeolites. Many MeALPOs (and with Me = Co, Ni, Zn, Mn, Cu, Mg and other divalent cations partially replacing the trivalent Al^{3+} of the parent ALPO) are very good solid acid and redox catalysts.[6,14]

Thanks to the efforts of many over the past decade, about a hundred distinct microporous structures, with channel apertures falling in the range from 4 to 14 Å, have been prepared. Into such structures, about a third of the elements of the Periodic Table may be incorporated at framework sites. And into the sites occupied by extra-framework, usually exchangeable, cations a further third of all the known elements may be placed. These architecturally elegant inorganic repositories have a prodigality of chemical composition that defies precise description; millions of distinct compositions are possible. Many thousands have already been prepared. The scope for delicate variation is enormous.

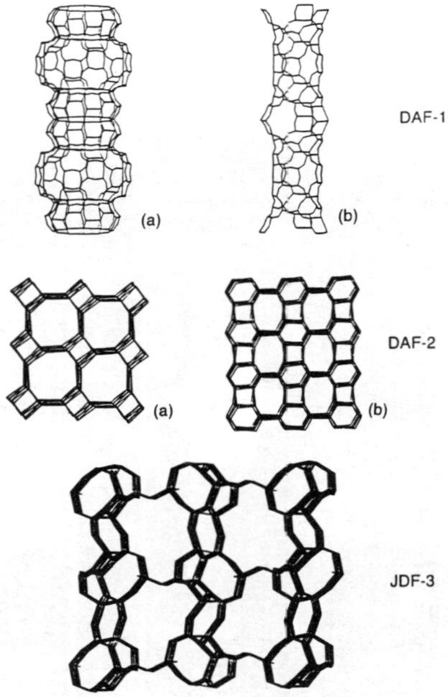

Fig. 4 Framework structures of DAF-1 (magnesium aluminium phosphate) showing (a) the wide channel containing supercages and (b) the narrower channel; DAF-2 (cobalt phosphate) containing T (T = Co, P) viewed along (a) [001] and (b) [100] axes; and JDF-3 (zinc phosphate) containing T (T = Zn, P) viewed along [100] axis. DAF stands for *D*avy *F*araday and JDF, *J*ilin *D*avy *F*araday.

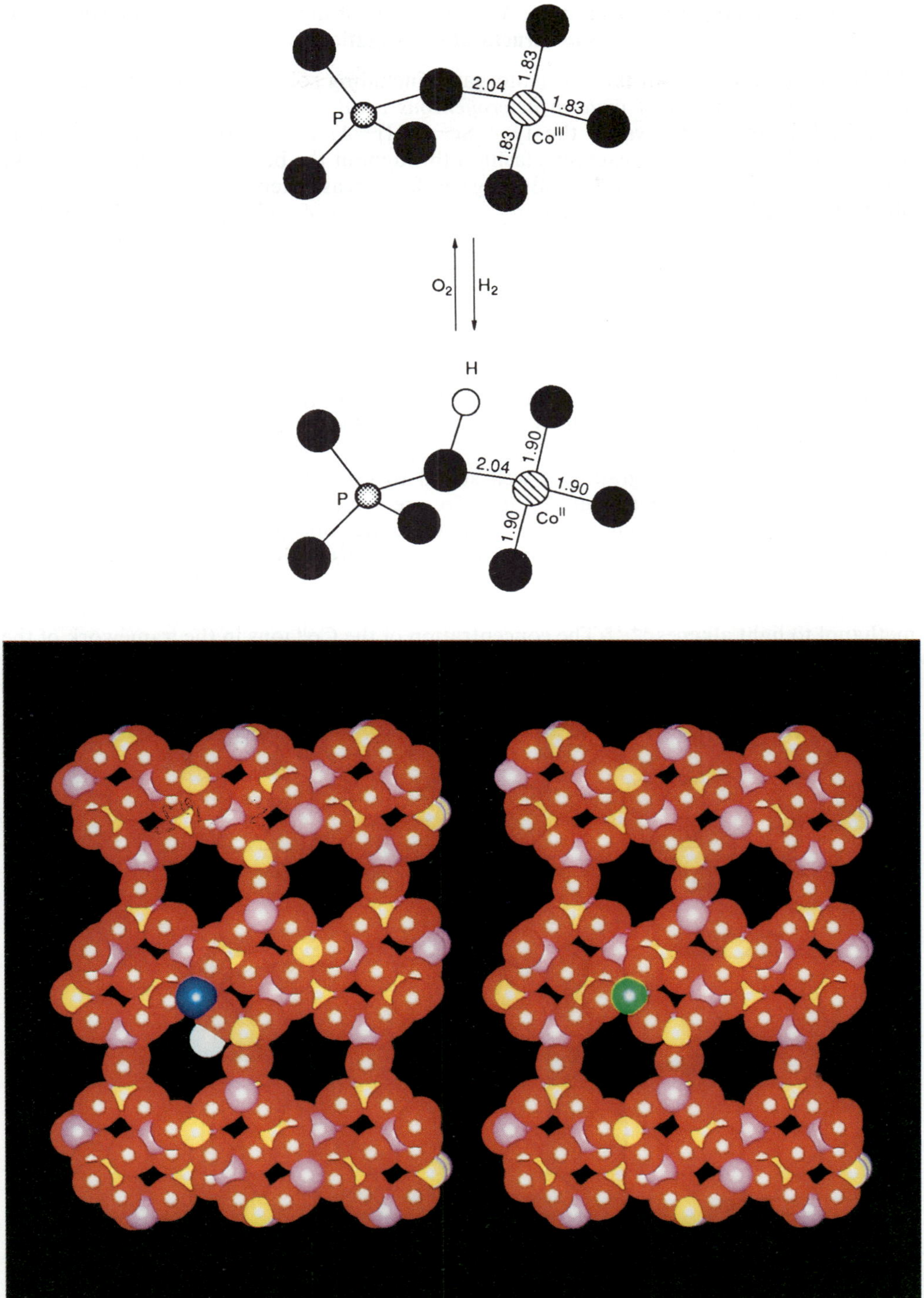

Plate 7 Quantitative description (of bond lengths and coordination) at the Co^{II} active site in the CoALPO-18 solid acid catalyst for the selective dehydration of methanol. (From J. M. Thomas and G. N. Greaves, *Sciences*, 1994, **265**, 1675.)

5. Uniform Heterogeneous Catalysts: A Vast Family of Inorganic Solids Amenable to *in situ* Structural Investigation

ZSM-5, mordenite and all the other inorganic including solid-acid, catalysts mentioned above are all examples of *uniform heterogeneous catalysts*,[15,16] where the active sites, such as bridging hydroxys of the type Si—O(H)—Al are distributed in a spatially uniform and accessible (to reactants) fashion throughout the bulk of the solid. They have other attributes, notably exceptionally large surface areas, often in excess of 600 $m^2\ g^{-1}$, ninety-five percent (or more, depending on crystal size) of which is inside the solid. In effect they are solid catalysts with three-dimensional surfaces permeating the entire material and accessible to all reactant (and product) molecules small enough to enter the apertures at their outer surfaces.

This being so, these uniform heterogeneous catalysts are amenable to the entire panoply of spectroscopic, diffraction and scattering techniques refined, over the years, by physicists and chemists. Such catalysts present unrivalled—in the experience of inorganic and physical chemists—opportunities to explore the relationship between structure and catalytic performance, much in the way that modern enzymologists pursue their investigations (described in Section 2).

In situ X-ray crystallography is well suited to such work (as it was originally for lysozyme and is for other enzymes). But there are some serious practical obstacles that have to be surmounted when, as is often the case, the concentration of the active site is low. X-Ray diffraction alone, even when imaginative use[17] is made of anomolous scattering, is not sufficiently sensitive.

Take, for example, CoALPO-18 (Plate 7), which is a good catalyst for converting methanol to light alkenes.[14,18] The concentration of the Co^{II} ions in the framework of the

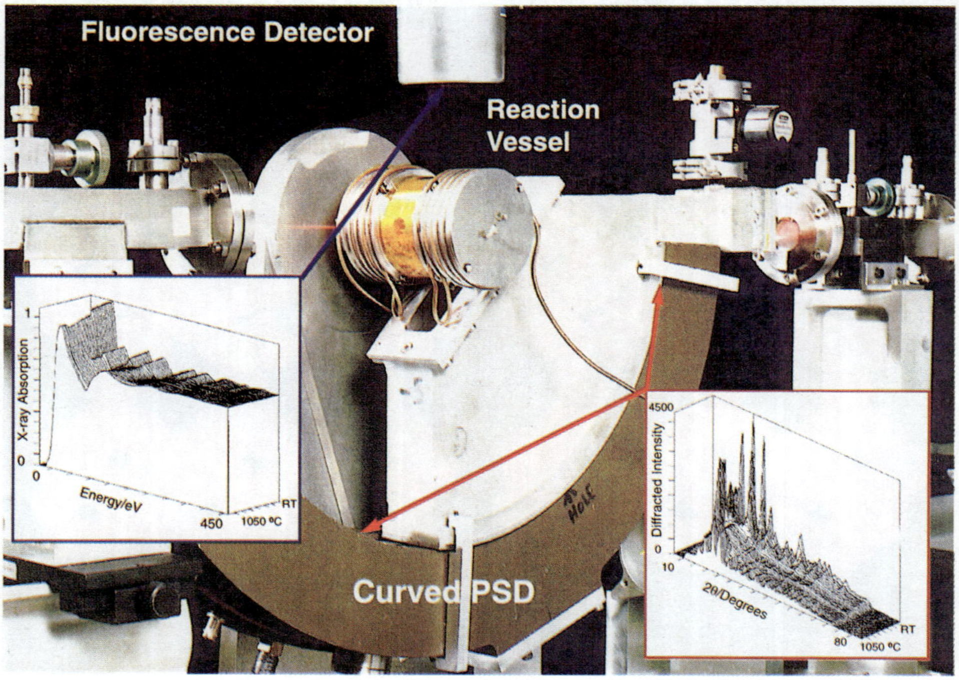

Plate 8 Experimental equipment of the combined Qu–EXAFS–XRD setup. The spectroscopy and diffraction data obtained for cordierite are shown by the two inserts relating to the two detector systems. (From J. M. Thomas *et al.*[21])

parent ALPO seldom exceeds a few percent, too low to be directly addressable by X-ray diffraction. We may, however, use X-ray absorption, thanks to the availability of synchrotron radiation. Moreover, we have evolved methods of probing catalysts of this kind under operating conditions using combined X-ray diffraction and X-ray absorption spectroscopy (XRD and XAS), recorded with the same sample under realistic conditions of catalytic use (Plate 8).[18–22] Combined studies, in parallel with infrared absorption spectroscopy, yield results such as those shown in Plate 7, where the changed bond lengths associated with conversion of the Co^{II} to Co^{III} valence states, while still retained in the framework, may be quantitatively specified. Typical results obtained by the combined *in situ* XRD and XAS techniques recorded during the calcination of the CoALPO-18 are shown in Fig. 5.

There is every reason to believe that this dual *in situ* approach to the study of open-structure (as well as other[19,23]) catalysts will provide fresh insights into the mechanisms of heterogeneous catalysis. A specific example of such new insights has recently been seen in the study of metal-ion catalysts anchored onto the interior surfaces of mesoporous supports (see below).

The dual, *in situ* XRD–XAS approach is also of great value in tracking the synthesis—the actual act of crystallization from a precursor gel—of new zeolitic catalysts, as outlined in the next section.

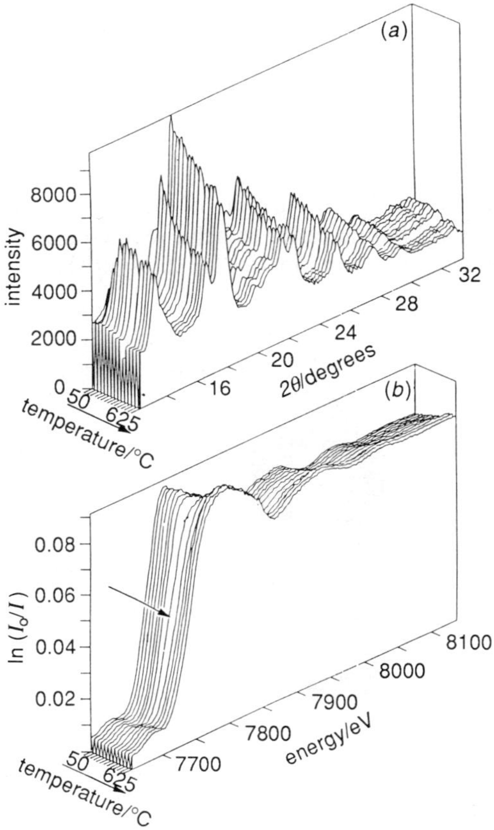

Fig. 5 Combined Qu–FLEXAFS–XRD following the calcination of CoAPO-18. (*a*) Shift in the K edge of cobalt (measured in fluorescence) during initial heating in air to burn off the template and during reduction to activate the catalyst. (*b*) XRD patterns during initial heating showing virtually no change except for a drop in background on removal of the template. (From J. M. Thomas *et al.*[21])

6. Synchrotron-based Studies for Tracking the Synthesis of Uniform Heterogeneous Catalysts

Progress in our understanding of uniform heterogeneous catalysts rests almost as much on our ability to monitor in atomic detail the formation or synthesis of such catalysts as well as the structural changes they undergo during operation. A start has recently been made by my group, working closely with Professor G. N. Greaves and his colleagues at the EPSRC Daresbury Laboratory.

Measurements that are experimentally demanding with conventional laboratory-based X-ray sources often become readily accessible when carried out with synchrotron radiation. This is so in the study of structural changes in solid catalysts under operation conditions. It is also true in the study of the nucleation and growth of crystals from solutions, melts and gels. With photon fluxes from a typical synchrotron exit port of the order of 10^{13} s^{-1} and, more importantly, with high-energy X-rays (up to 140 keV) capable of penetrating stainless-steel containers, it is relatively straightforward to study condensed phases using appropriate high-temperature and/or high-pressure cells without serious attenuation in the intensity. Moreover, taking advantage of solid-state detectors, the white radiation of synchrotrons permits the rapid buildup of time-resolved diffraction patterns (under isothermal conditions) that allows the ready tracking of structural change occurring in a cell that is held in a fixed position. Fig. 6 shows the energy-dispersive X-ray diffraction (EDXRD) geometry available on station 16.4 at the 2 GeV Daresbury Synchrotron Radiation Source. The EDXRD facility is illuminated with radiation from a 6 T superconducting wiggler. The stainless-steel autoclave shown in the figure can be operated up to temperatures of *ca.* 250 °C, permitting the *in situ* characterization during hydrothermal synthesis of most of the zeolitic materials. The solid-state detector is inclined at a fixed 2θ angle chosen so that, from the photon energies available (7–140 keV), representative crystallographic spacings for zeolite structures are readily identifiable. With SRS electron beam currents of around 200 mA energy-dispersed patterns were integrated every 2 min in order to achieve adequate statistics for phase identification and the recording of reaction kinetics for the synthesis rates encountered here.

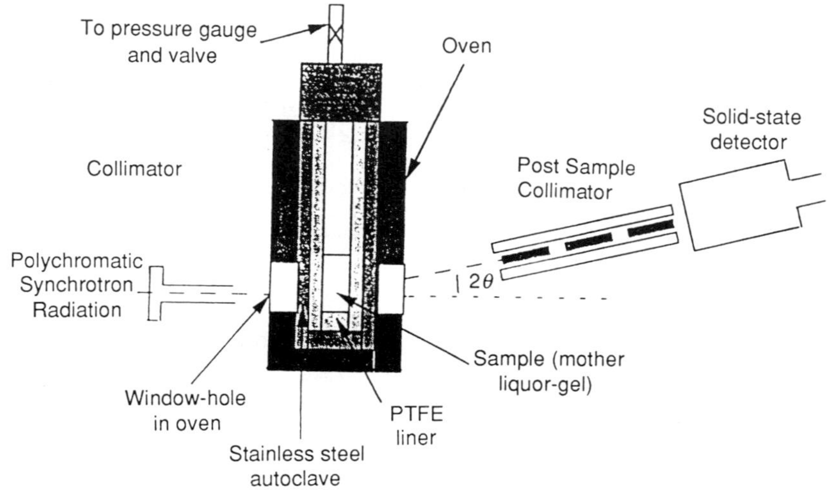

Fig. 6 Schematic diagram of the experimental set-up for energy-dispersive X-ray diffraction. Synthesis takes place in the heated stainless-steel autoclave and the crystallization is detected with a single-element (Canberra) Ge solid-state detector. This is inclined at a fixed 2θ angle of $1.4556 \pm 0.0001°$ and calibrated using NBS 640b Si. The integration time for each frame is 2 min. (From F. Rey *et al.*[24])

Time-resolved X-ray diffraction patterns of the catalyst CoALPO-5, as well as of its parent ALPO-5, have been collected (and published[24]), and these clearly show the change from structureless diffraction patterns exhibited by the mother liquor to highly structured ones corresponding to the appearance of the ALPO-5 phase along with a co-existent chabazitic equivalent of CoALPO-5.

Very recently, my colleagues and I have used the combined XRD–XAS approach, schematized in Fig. 7. With the capillary reaction vessel housed in an appropriately designed furnace, both XRD patterns (comparable to those obtained with the set-up shown in Fig. 6) as well as richly detailed X-ray absorption spectra are obtainable (see Fig. 8). The pre-edge X-ray absorption spectra, and also the XANES and EXAFS information (taken in conjunction with the XRD results) unmistakably reveal[25] that, prior to the onset of crystallization, octahedral Co^{II} ions in the precursor (templated) gel become tetrahedrally coordinated.

7. Mesoporous Catalysts

Plate 9 summarizes progress made in the last forty years in synthesizing open-structure solids with potential as catalysts. Zeolite A, produced by the Linde (Union Carbide) Company in the early 1950s, has pore dimensions around 4 Å (depending upon the exchangeable cation). These are of great commercial applicability—in detergency and water-softening especially—but they have essentially zero prospects as viable catalysts. DAF-1, (*Da*vy *F*araday One) already described above, and discovered here in 1992, is in quite a different category and has demonstrated value—but not much staying power—as an acid catalyst for the isomerization but-1-ene to isobutene (2-methylpropene), the material first synthesized by Faraday. MCM-41 (*M*obil *C*atalytic *M*aterial, number 41), first reported in 1992[26]) is of even greater potential in a new generation of shape-selective catalysts and as supports for metal-based catalytic centres.

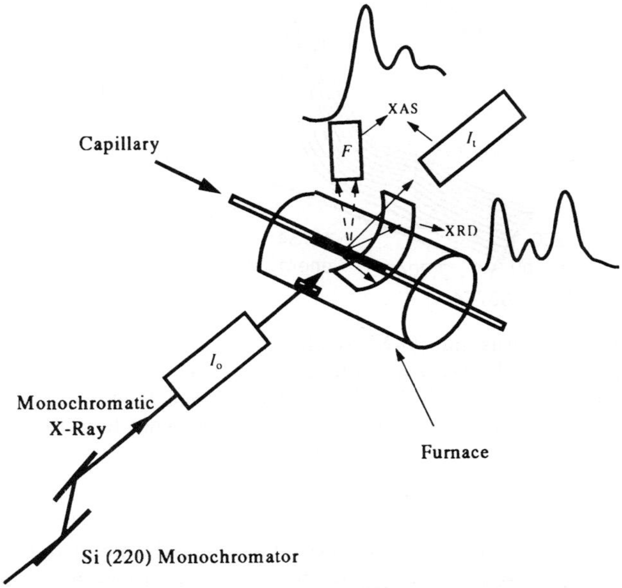

Fig. 7 Schematic representation of set-up used by Sanker, Rey, Thomas and Greaves (in ref. 24) for combined XRD–XAS *in situ* measurements to follow crystallization of molecular sieve catalysts from liquids and gels. (See also Fig. 3 of ref. 21.)

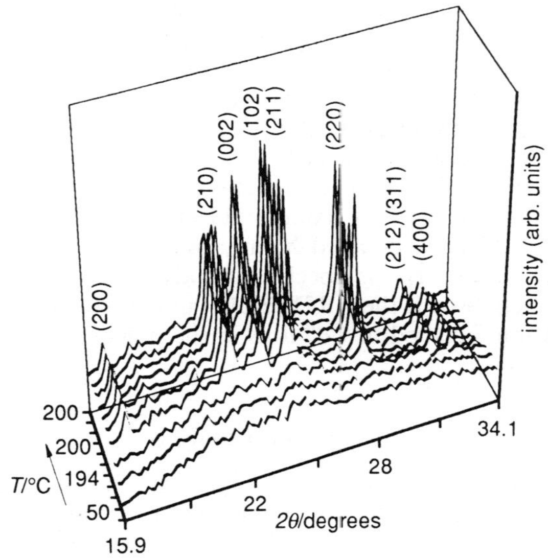

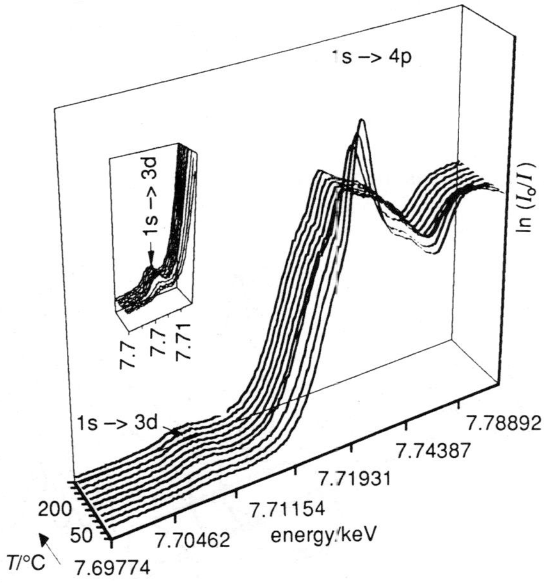

Fig. 8 Stacked (*a*) XRD patterns and (*b*) X-ray absorption spectra during the crystallization of a CoALPO-5 catalyst from the liquid phase studied using the set-up shown in Fig. 7. From the pre-edge (1s → 3d) transition, the EXAFS and the shape of the XANES peaks there is no doubt that the cobalt ion, originally in octahedral coordination, becomes tetrahedrally bonded *prior* to the onset of crystallization.

Mesoporous siliceous solids with channel apertures from 25 to 100 Å have opened up new possibilities in heterogeneous catalysis. The large diameter (*ca.* 30 Å) channels of the so-called MCM-41 mesoporous silicas that we and others have used for selective oxidation permit, in principle, the direct grafting of complete metal complexes and organometallic moieties onto the inner walls of these high-surface-area (typically >800 m^2 g^{-1}) solids. This opens routes to the preparation of novel catalysts consisting of

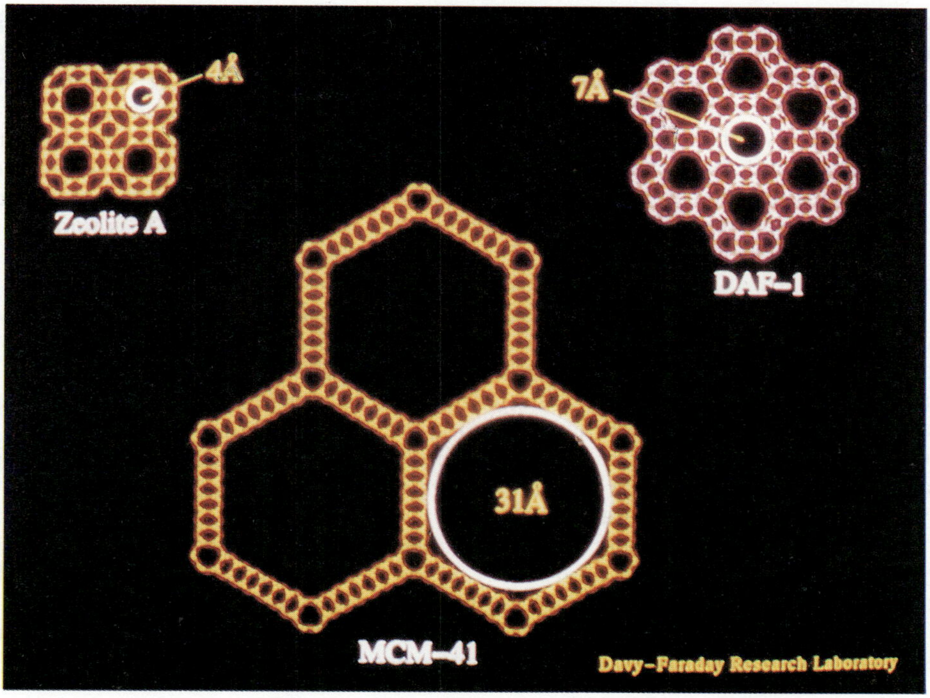

Plate 9 Zeolite A, first prepared in the mid 1950s, is a synthetic molecular sieve (*ca.* 4 Å diameter apertures) of little catalytic significance. DAF-1 (Davy Faraday one) first synthesized in 1992 (diameter of apertures >7 Å), as well as MCM-41 (Mobil catalytic material 41) a mesoporous solid, has great potential as a catalyst.

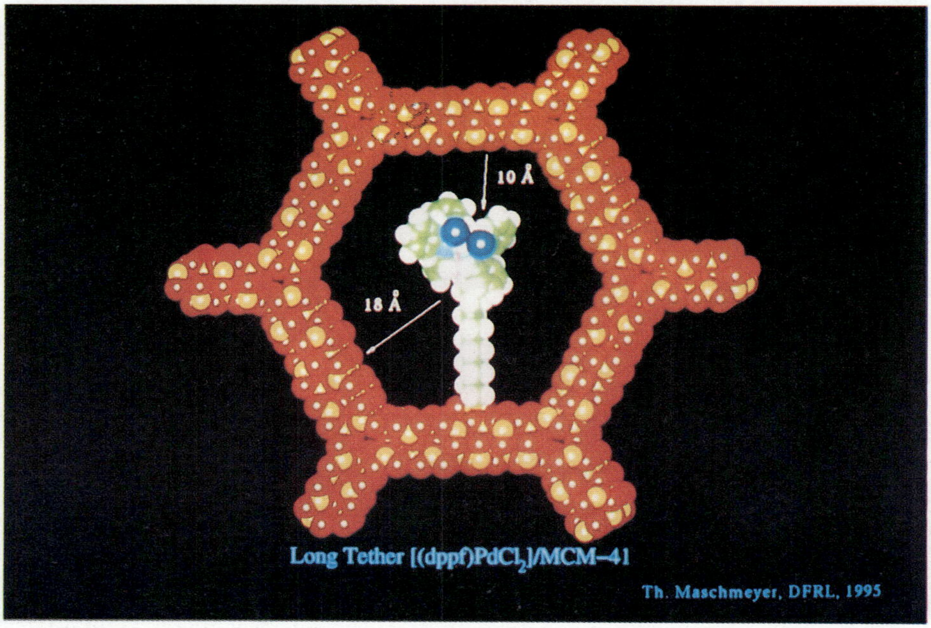

Plate 10 An example of a tethered metallic catalyst, prepared by T. Maschmeyer (unpublished), at the Davy Faraday Research Laboratory (see also ref. 27).

large concentrations of accessible, well-spaced and structurally well defined active sites. Maschmeyer *et al.*, in these laboratories, have described the production of titanocene-derived catalyst precursors anchored to the inner walls of the MCM-41, and its conversion (monitored by X-ray absorption spectroscopy) to a powerful catalyst for the epoxidation of cyclohexene.[27] Furthermore, this catalyst is also active in the oxidation of other important, bulkier reactants, such as pinene. X-Ray absorption spectroscopy not only helps (as does *in situ* FTIR) to identify key intermediates in the epoxidation catalysis, but also reveals that, in the activated state of the precursor, Ti is four coordinated with no evidence for a titanyl (Ti=O) bond. During catalytic reaction, it is six coordinated.

With MCM-41 based catalysts it is possible not only to anchor a reactive (catalytic) centre, designed to order according to the principles of organometallic chemistry, but also to tether them as schematized in Plate 10. Here the active site is situated at the extremity of the tether, and is free to flutter in the molecular breeze during the process of catalytic conversion. By deliberately restricting the spatial freedom in the vicinity of the active centre (see Plate 11) it should be possible to design highly stereo-selective (enantiometric) catalysts. Such possibilities are now under active investigation at the Davy Faraday Research Laboratory.

One other highly convenient property of MCM-41 mesoporous solids is that they may be modified in subtle, and catalytically useful ways during synthesis. Plate 12, taken from the work of my colleagues,[28] shows how hetero-atoms may be incorporated into the siliceous framework of MCM-41. When titanium is anchored in the wall of such mesoporous hosts, facile epoxidation (of cyclohexene as shown in Fig. 9) ensues.

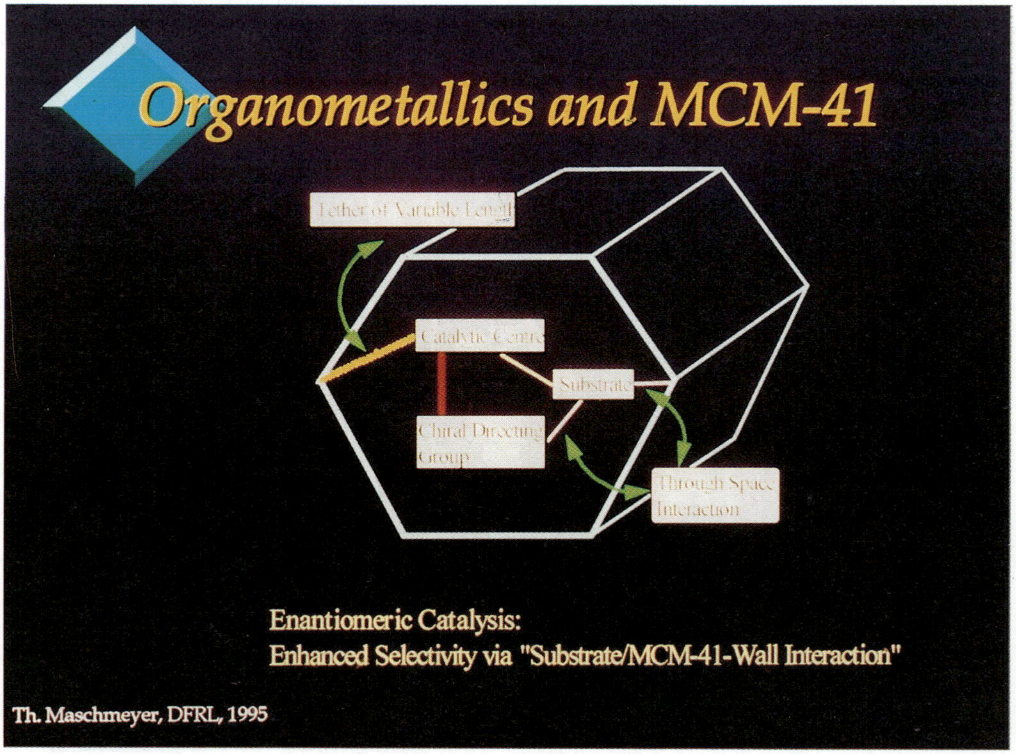

Plate 11 Deliberate restriction of spatial freedom at the active centre grafted onto the inner walls of a mesoporous silica offers scope for enantiomeric catalysis (Maschmeyer *et al.*, in preparation.)

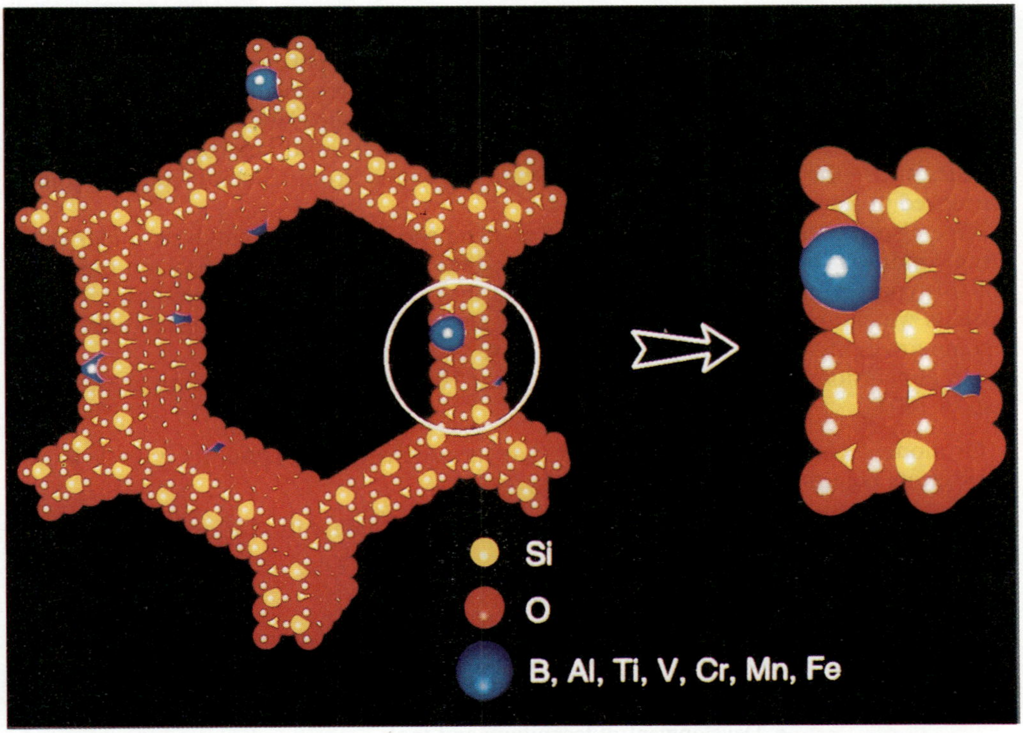

Plate 12 Idealised metal-containing MCM-41 structure generated by computer modelling methods (from Frey *et al.*[28])

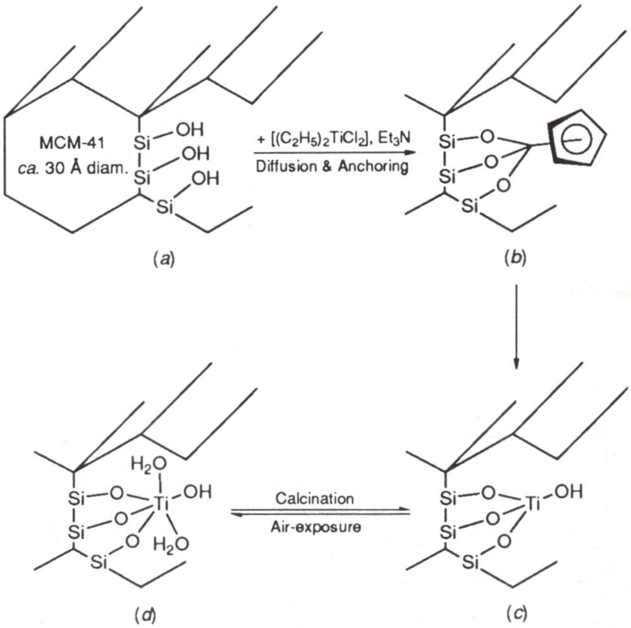

Fig. 9 Schematic representation of the preparation of the grafted catalyst: (*a*) the support, (*b*) anchoring reaction, (*c*) calcination and (*d*) reversible reaction of the organometallic-derived Ti-MCM-41 epoxidation catalyst with water upon exposure to the atmosphere

8. Epilogue

In summary, we observe that both microporous and mesoporous crystalline inorganic solids have greatly extended the scope and nature of heterogeneous catalysis. Homogeneous catalysts may be readily heterogenized so that the separate advantages of homogeneous catalysts on the one hand—their high specificity—and heterogeneous catalysts on the other—their built-in ease of separation of product from reactant—may be jointly harnessed in a synergistic mode.

We also note that *in situ* methods of probing the structure of catalysts under operating conditions are rendered feasible owing to the availability of synchrotron radiation. All this enables greater insights to be gained into the mechansim of catalytic reaction, insights which are often enriched by modern computational chemical techniques. Table 1 summarizes my own group's overall strategy.

In conclusion, I cannot help recalling what Humphrey Davy, standing at this spot, said in 1803:

> Of modern chemistry it may be said that its origins are pleasure, its growth knowledge, its objects truth, beauty and utility.

I hope that I have demonstrated that this admirable definition is as valid now as it was when it was first enunciated.

Table 1 Overall strategy of work on heterogeneous catalysis at the Davy Faraday Research Laboratory

- Synthesis of new micro- and meso-porous catalysts
- Development of techniques and tools
- Characterisation
- Performance
- Mechanism → computation
- Applications (patents?)
- Collaboration with industrial partners

References

1 D. C. Phillips, *Proc. Natl. Acad. Sci. USA*, 1967, **57**, 484.
2 J. R. Knowles, *Nature (London)*, 1991, **350**, 121.
3 V. T. D. Souza, K. Hanahusa, T. O'Leary, R. C. Gradewood and M. L. Bender, *Biochem. Biophys. Res. Commun.*, 1985, **129**, 727.
4 R. Breslow, *Acc. Chem. Res.*, 1995, **28**, 146.
5 P. B. Weisz, *Proc. 6th Int. Congr. Catalysis*, Kodansha, Tokyo, 1990, p. 1.
6 J. M. Thomas, *Angew. Chem., Int. Ed. Engl.*, 1994, **33**, 913.
7 J. M. Thomas, *Sci. American*, 1992, **266**, 85.
8 K. I. Zamaraev, *Top. Catal.*, 1996, in the press.
9 J. A. Cusumano, CHEMTECH., 1992, **22**, 482.
10 J. M. Thomas and K. I. Zamaraev, *Angew. Chem.*, 1994, **106**, 316.
11 R. G. Bell, D. W. Lewis, C. R. A. Catlow, J. M. Thomas, P. Voigt and C. M. Freeman, in *Zeolites and Related Microporous Materials: State of the Art*, ed. J. Weitkamp, H. G. Karge, H. Pfeiffer and W. Hoderich, Elsevier, 1994, vol. 84, p. 127.
12 D. W. Lewis, C. R. A. Catlow and C. M. Freeman, *J. Phys. Chem.*, 1995, **99**, 11 194.
13 D. W. Lewis, C. R. A. Catlow and J. M. Thomas, submitted.
14 J. Chen and J. M. Thomas, *J. Chem. Soc., Chem. Commun.*, 1994, 603.
15 J. M. Thomas, *Angew. Chem., Int. Ed. Engl.*, 1988, **27**, 1673.
16 J. M. Thomas, J. Chen and A. R. George, *Chem. Br.* 1992, **28**, 991.
17 A. K. Cheetham and A. P. Wilkinson, *Angew. Chem., Int. Ed. Engl.*, 1992, **31**, 1559.
18 J. M. Thomas, G. Sankar, P. A. Wright, J. Chen, L. Marchese and G. N. Greaves, *Angew. Chem., Int. Ed. Engl.*, 1994, **33**, 639.

19 J. W. Couves, J. M. Thomas, G. N. Greaves, R. H. Jones, D. Waller, A. J. Dent and G. E. Derbyshire, *Nature* (*London*), 1991, **354**, 465.
20 G. Sankar, P. A. Wright, S. Natarajan, J. M. Thomas, G. N. Greaves, A. J. Dent, B. R. Dobson, C. A. Ramsdale, R. H. Jones, *J. Phys. Chem.*, 1993, **97**, 9550.
21 J. M. Thomas, G. N. Greaves and C. R. A. Catlow, *Nucl. Instruments and Methods in Physical Research*, 1995, **B97**, 1.
22 P. A. Barrett, G. Sankar, J. M. Thomas and C. R. A. Catlow, *J. Phys. Chem. Solids*, 1995, **56**, 1395.
23 B. S. Clausen, L. Gräbek, G. Steffensøn, P. L. Hausen and H. Topsøe, *Catal. Lett.*, 1993, **20**, 23.
24 F. Rey, G. Sankar, J. M. Thomas, P. A. Barrett, D. W. Lewis, C. R. A. Catlow, S. M. Clark and G. N. Greaves, *Chem. Mater.*, 1995, **7**, 1435; see also J. S. O. Evans, R. J. Francis, D. O'Hare, S. J. Price, S. M. Clark, J. Flaherty, J. Gordon, A. Nield and C. C. Tang, *Rev. Sci. Instrum.*, 1995, **66**, 2442.
25 G. Sankar, F. Rey, J. M. Thomas and G. N. Greaves, *J. Chem. Soc., Chem. Commun.*, 1995, in the press.
26 C. T. Kresge, M. E. Leonowicz, W. J. Roth, J. C. Vartuli, J. S. Beck, *Nature* (*London*), 1992, **359**, 710.
27 T. Maschmeyer, F. Rey, G. Sankar and J. M. Thomas, *Nature* (*London*), 1995, **378**, 159.
28 F. Rey, G. Sankar, T. Maschmeyer, J. M. Thomas, R. G. Bell, G. N. Greaves, *Top. Catal.*, 1996, in the press.

Faraday Discussion 100 *Celebration Paper, Presented* 24*th April*, 1995

Faraday Discuss., 1995, **100**, C29–C45

Molecular Simulation: A View from the Bond

D. J. Tildesley
Chemistry Department, University of Southampton, Southampton, UK SO17 1BJ

The last fifty years have seen a rapid development of the molecular dynamics method for the simulation of condensed phases. The method is used to test the accuracy of effective intermolecular potentials, to validate approximate statistical mechanical theories and to explore systems at conditions and with a level of detail which would be hard to achieve by conventional experiment. In combination with data visualisation techniques, the method acts as a powerful microscope to study structure and dynamics at the molecular level. The rapid progress in the field can be attributed to the ingenuity of its practitioners in developing new algorithms and in studying imaginative applications and to the striking increases in computing speed and memory over the period.

We review the development of the technique and illustrate its use by considering three problems: the calculation of the surface tension of water; the structure of an alkyl cyanobiphenyl layer on graphite; and the testing of the wall-PRISM equation of polymer adsorption. Finally, we consider the future of these simulation methods.

1. Introduction

Traditionally science has progressed by the development of theory to explain experimental observation. The experimentalist provides the facts from which a physical model is developed and the theoretician unravels the relationship between the mathematical variables used to describe this model. For example, the motion of the planets analysed by Kepler provided a physical model of the solar system that could be described mathematically using Newtonian mechanics. In the same way, Rutherford's experiments on the scattering of α particles from gold foil indicated a basic planetary model for the atom which could then be described in precise detail by the Schrödinger and Dirac equations. Occasionally, the mathematical description precedes the observation as in the case of Einstein's general theory of relativity and Eddington's observation of the deflection of light by the sun. Regardless of the precise order of their application, theory and experiment have been the two essential pillars supporting the natural sciences.

The second half of the twentieth century has seen the development of a third methodology. The digital computer, developed during the second world war, has enabled many of the branches of natural science to simulate complicated models that could not have been solved by standard theoretical methods.[1] For example, it is now possible to model the structure and dynamics of a protein molecule buried in a lipid bilayer using molecular dynamics, to follow the propagation of a crack using finite difference methods, to develop a circulation model for weather prediction, or to solve the Euler equations of fluid flow to model a severe thunderstorm.

Apart from addressing problems of enormous complexity, the method allows the computational scientist to alter the underlying model at will. For instance, by changing the dimensionality of a problem or the structure of the model hamiltonian it is possible to test and extend approximate theories by studying systems that are not readily attainable in the laboratory.

In the discipline of Chemistry, computers have allowed us to solve the Schrödinger equation at various levels of approximation and to obtain detailed predictions of the properties of atoms, molecules and small clusters. The prediction of the energy levels in the molecule has also allowed us to understand and rationalise much experimental spectroscopy. Energy minimisation, Monte Carlo and molecular dynamics calculations, using increasingly accurate force fields, have played an essential part in the prediction of the properties of condensed phases such as liquids, liquid crystals, polymers and superconducting materials. The combination of density functional theory and molecular dynamics in the Car–Parrinello methods allows us to simulate systems where the electronic and nuclear degrees of freedom are not strongly coupled (*e.g.* chemical reactions), while detailed scattering calculations allow us to predict the rates of chemical reactions in the gas phase. This is a huge body of work, and the UK chemistry community supported by the provision of excellent central computing facilities has made important contributions in this new field over the last forty years. Since, in some sense, I represent Computational Chemistry at this celebration, I trust that my colleagues will forgive me if I concentrate on one aspect of the discipline and limit myself to some remarks on the molecular dynamics method.

2. A Brief History of the Molecular Dynamics Method

It is intriguing that the statistical mechanics of Gibbs and Boltzmann had taken such a strong hold in the early part of the twentieth century that when computers became available just after the second world war, they were used to calculate ensemble averages using the Metropolis Monte Carlo sampling method.[2] It was a further four years before trajectory calculations were used to estimate equilibrium properties by the direct route of calculating the time average.

The first molecular dynamics simulations were performed by Alder and Wainwright on the UNIVAC computer at the Livermore Laboratory in California.[3] Simulations were performed of the hard-sphere model by locating the time, collision partners and impact parameters for every collision. At each collision the new trajectories were computed from the conservation of energy and linear momentum. The properties of the system such as the pressure and the radial distribution function were calculated as time averages over the configurations of the fluid and the calculation was characterised by the conservation of the total energy and linear momentum. The effects of the small system size were mitigated by embedding the central simulation box in a lattice of replicas and using the method of periodic boundary conditions.

Even these first simple simulations provided an interesting prediction that in the hard-sphere system there would be a first-order phase transition between a solid and fluid, an observation subsequently confirmed by more detailed calculations. This is a clear example of the hamiltonian surgery discussed in the last section. A simulation of this simplified model shows that it is unnecessary to include attractive forces to induce the formation of a solid from the liquid.

Molecular dynamics of systems interacting through continuous potentials followed in a study of damage radiation by Gibson *et al.*[4] and in a study of liquid argon by Rahman[5] using central difference and predictor–corrector algorithms to solve the coupled second-order differential equations of the atomic motion. These simulations used a time-step of approximately 10^{-14} s and solved the equations of motion for a few thousand time-steps. The total simulation length of a few tens of picoseconds was sufficient to calculate equilibrium properties and to estimate single particle correlation functions such as the velocity autocorrelation function and its integral, the diffusion coefficient.

By the early 1970s the power of machines had increased dramatically and supercomputers were becoming available to academics outside the large government institutions. Although the first simulations were performed on atomic liquids and solids interacting through central forces, it became possible to simulate rigid polyatomic molecules by solving the Newton–Euler equations of motion.[6,7] This method could have been extended to study the dynamics of flexible molecules by finding the appropriate generalised coordinates and working with a time-dependent moment of inertia tensor. However, a more direct approach treated a molecule as a set of atoms with the geometry defined using stiff intramolecular potentials[8] or rigid constraints. (A typical constraint force might act to keep a bond length fixed at its equilibrium value.) These constraints can be included in the equations of motion by introducing and solving for the associated Lagrangian multipliers.[9]

The ability to simulate molecules of a completely general structure including both constraints and many-body intramolecular potentials allowed the method to be applied to the biological sciences in the simulation of the nucleic acids, proteins and polysaccharides. Although this type of simulation had become technically possible by the early 1980s, there were still significant problems concerning the accuracy of the force fields and the total time over which the molecules could be simulated. The equations of motion must be solved at a time-step that covers the fastest molecular motion (10^{-16} to 10^{-15} s), while the problem of interest might have a characteristic timescale of between 10^{-6} and 1 s. This need to simulate hundreds of millions of time-steps puts some problems beyond the range of even the fastest available computers. An example of this type of problem would be the complete simulation of protein folding where we attempt to predict the tertiary protein structure from its amino acid sequence. It is important to stress that the production of accurate force fields and the extensions to long timescales are still the key problems facing users of the method today.

The original molecular dynamics methods were performed in the microcanonical ensemble at constant *NVE*. During the 1980s, the method was extended to other ensembles such as the canonical[10] and isothermal–isoenthalpic ensemble[11] using the generalised Lagrangian method. Constant stress molecular dynamics, in which the shape and size of the central simulation box changed to balance the internal and external stresses, allowed for the direct study of solid phase transitions.[12] At the same time, the development of methods for calculating the statistical properties of systems such as the free energy difference, entropy and chemical potential[13] meant that simulation methods could be used to locate phase boundaries.

The simulations described so far have been at equilibrium. The time correlation functions can be used to calculate the transport properties of the fluid using the Green–Kubo relationships. It also proved possible to perform non-equilibrium molecular dynamics simulations by introducing a gradient that induces an appropriate current of momentum or energy. The transport coefficients can then be measured directly as the ratio of the current to the gradient. The gradient, say in velocity, can be introduced mechanically by simulating a fluid between two structured walls moving in opposite directions[14] or by using a homogeneous method.[15] In the homogeneous method an additional perturbation term of the form

$$H_p = \sum_{i=1}^{N} \boldsymbol{q}_i \boldsymbol{p}_i : \nabla \boldsymbol{u} \tag{1}$$

is added to the hamiltonian. $\nabla \boldsymbol{u}$ is the velocity gradient tensor and $\boldsymbol{q}_i$ and $\boldsymbol{p}_i$ are the atomic position and momentum of atom *i*. This kind of hamiltonian is applied using moving or Lee–Edwards periodic boundary conditions.[16]

It is clear that the last forty years has seen an enormous increase in the complexity and range of the systems that can be simulated using classical molecular dynamics. Of course, it has been necessary to develop and refine the simulation algorithms but the predictive ability of the method is most directly coupled to the significant increase in computing power over the same period.[17] The first statistical mechanical simulations were performed on the MANIAC machine at Los Alamos. In the mid-1940s a similar machine would weigh 30 tons, contain over 18 000 vacuum tubes, 6000 switches and fill a large room. A machine of this type could calculate at the rate of 400 to 1000 floating point operations per second (flops). Originally the machines were programmed by hand-setting switches and wiring plug boards. In the United Kingdom, the first stored-program computer (EDSAC) built by Maurice Wilkes became available in May 1949 and supercomputing in the UK had begun. EDSAC supported an enthusiastic community in Cambridge.

By the early 1970s the CDC 7600 at the University of London Computing Centre performed calculations at the speed of 40 megaflops. The vacuum tube had been replaced by the single transistor. In the early 1980s the Cray 1 computer became available at the Rutherford Appleton Laboratory. The clock speed of 12.5 ns obtained using integrated circuit technology gave a peak performance of 170 megaflops. This performance was sustained by the vectorisation or the pipe-lining of instructions.

The 1990s saw the introduction of massively parallel shared memory and distributed memory machines. The most recent UK supercomputer, the CRAY T3D at the University of Edinburgh, consists of 256 DEC alpha processors working in parallel to produce a peak performance of 40 gigaflops. Real scientific applications may only sustain a performance of around 7 gigaflops on this machine but it is still approximately 10^8 times faster than the first Electronic Numerical Integrator and Computer (ENIAC) used on the Aberdeen testing grounds for ballistic trajectories in 1945. This speed increase combined with a correspondingly sensational increase in size of computer memory and permanent storage has effected a revolution in Computational Chemistry.

As always, there are those who are not prepared to wait for technological developments. During an evening discourse given in this very lecture theatre in April 1900, Lord Kelvin presented his paper 'Nineteenth Century Clouds over the Dynamical Theory of Heat and Light'.[18] Unaware that a molecular vibration might not enjoy its full $k_B T$s worth of energy, he postulated that the 'anomalous' heat capacity ratio of a polyatomic gas might be a consequence of the failure of the law of equipartition of energy in vessels of an irregular shape. His assistant, one William Anderson, performed over 5000 individual trajectory calculations by geometrical construction of elastic collisions in unusually shaped containers, arguably performing the first molecular dynamics averaging. Perhaps the runs were too short or the containers too regular, but they incorrectly concluded that the Boltzmann–Maxwell doctrine of the equipartition of kinetic energy could be disproved.[19] Lord Kelvin begins the paper with a warm acknowledgement to his assistant. 'I desire to express my obligations to Mr. William Anderson, my secretary and assistant, for his mathematical tact and skill'. It is unclear whether a request of this kind today would have the graduate students reaching for their EPSRC handbooks of good practice or their workstations.

3. Molecular Dynamics at Work

The classical molecular dynamics method fits the general model shown in Fig. 1. As long as there is a strong coupling between the nuclear and electronic degrees of freedom, then a sensible input to a condensed phase simulation would be a force field consisting of a number of effective two-, three- and four-body potentials. These can be estimated from accurate quantum mechanical calculations on small clusters or isolated molecules and in some cases from the inversion of dilute gas experimental data such as the second virial coefficient or the transport properties.[20]

This force field can be used in an energy minimisation calculation to estimate the zero temperature structure of the system. Ensemble averages, at non-zero temperature,

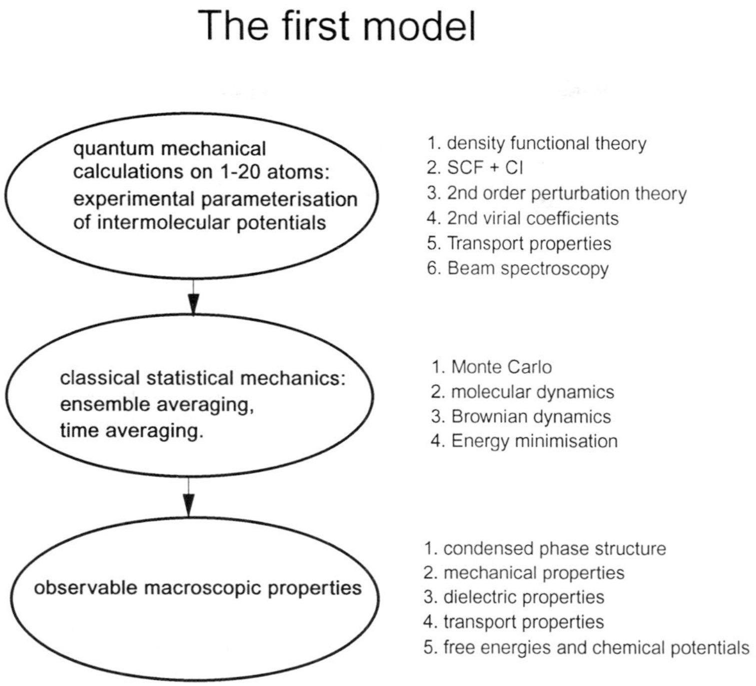

Fig. 1 Statistical mechanical simulations: the first model

can be obtained by using the Metropolis Monte Carlo method to sample the appropriate phase-space distribution. Molecular and Brownian dynamics calculations can provide additional information on the time evolution of the molecules. The outputs from the method are the measurable structural, thermodynamic and transport properties of the system.

The molecular dynamics method allows us:

(a) to adjust the intermolecular force field to reproduce experimental condensed phase data (validation);

(b) to use the force field to calculate properties that are not easily observed in the laboratory (prediction);

(c) to check the accuracy of statistical mechanical theories (testing);

(d) to obtain a detailed picture of the structure and motion at molecular level (observation).

We will illustrate some of these uses of the simulation methods by looking at three different examples.

3.1. The Surface Tension of Water

We have used the molecular dynamics study[21] to examine whether an effective pair-potential fitted to the bulk properties can reproduce an accurate estimate of the surface tension at the water liquid/vapour interface. The simulations use the SPC/E model of water that has been properly fitted to the configurational energy of the liquid by including the self polarisation term.[22] The bond length of the molecule is fixed at 1 Å and the bond angle at 109.47°. Charges of $0.4238\,|e|$ are fixed at the hydrogen nuclei and a cancelling negative charge is fixed on the oxygen nucleus. The repulsion–dispersion

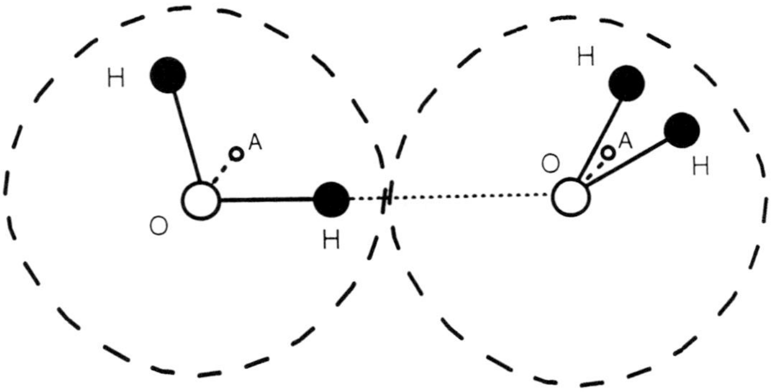

Fig. 2 The dimer configuration for the SPC/E model of water. In the SPC/E model the negative charge is placed on the oxygen nucleus. In similar models such as TIP4P and the Watanbe–Klein model, the negative charge is displaced along the bisector of the HOH bond to the position marked A.

interaction between the two water molecules is modelled by a Lennard-Jones potential centred on the oxygen atom. The total dipole moment of the model molecule is 2.351 D which is significantly higher than the gas phase value of 1.85 D. The higher dipole of the model accounts effectively for the enhancement of the gas-phase dipole by polarisation from the neighbouring molecules. Explicit polarisation is not included in this model. The dimer conformation for the SPC/E is shown in Fig. 2. There is a clear hydrogen bond induced by the interaction between the partial charges.

This particular model works well in simulations of the bulk liquid. At 298 K and 0.997 g cm^{-3} the predicted configurational energy is -41.5 kJ mol^{-1} (-41.5 kJ mol^{-1}), the predicted diffusion coefficient is 5.3×10^{-9} m s^{-1} (2.4×10^{-9} m s^{-1}) and the predicted relative permittivity is 70 (78.3).[23] The experimental measurements at the same temperature and density are included in parenthesis.

The water gas/liquid interface can be simulated using the geometry shown in Fig. 3. A three-dimensional simulation of bulk water is performed in a cubic box using full periodic boundary conditions to produce an equilibrated sample. A number of these water boxes are stacked together to form a slab of liquid and the slab is then surrounded by a number of empty boxes. The entire sandwich is reproduced periodically through space. The artificial gas/liquid interface at the start of the simulation is a step function. Water molecules evaporate quickly from the liquid slab to form a coexisting gas phase

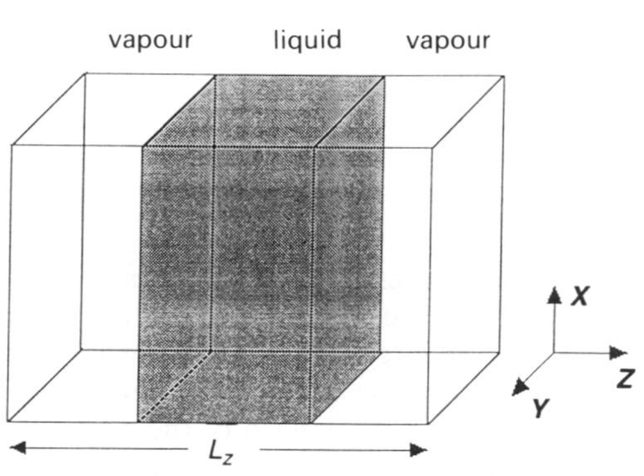

Fig. 3 The slab simulation geometry for studying the gas/liquid interface

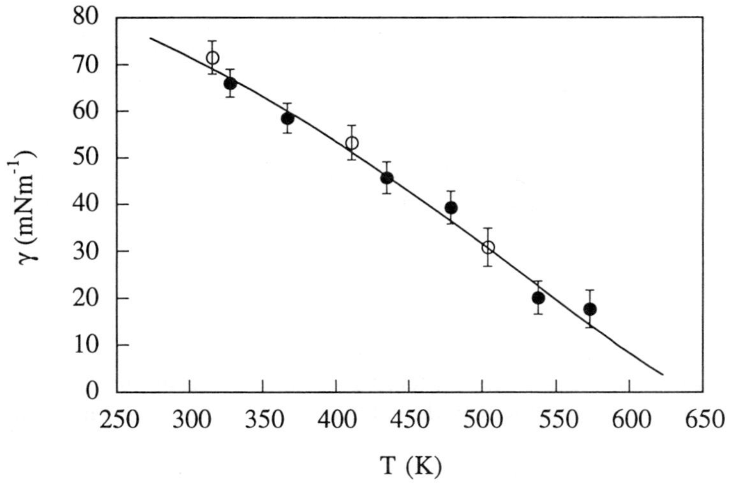

Fig. 4 The surface tension of water. The continuous line represents the experimental values from ref. 27. MD results using $N = 512$ and $N = 1000$ are shown as black and open circles, respectively. Reproduced with permission of the American Institute of Physics.

and a stable interface is formed with a density profile that decays monotonically from the bulk liquid density to that of the gas.

At a planar interface the pressure is no longer a scalar property but a tensor with a normal component ($P_N = P_{zz}$) and a tangential component $P_T = 1/2(P_{xx} + P_{yy})$. The elements of the pressure tensor can be calculated from the Irving–Kirkwood definition of the stress:[24]

$$P_{zz} = \langle \rho(z) \rangle k_B T - \frac{1}{A} \left\langle \sum_{j>i}^{N} \sum_{\alpha}^{3} \sum_{\beta}^{3} \frac{z_{ij} z_{i\alpha j\beta}}{r_{i\alpha j\beta}} \frac{du(r_{i\alpha j\beta})}{dr_{i\alpha j\beta}} \frac{1}{|z_{ij}|} \times \theta\left(\frac{z - z_i}{z_{ij}}\right) \theta\left(\frac{z_j - z}{z_{ij}}\right) \right\rangle \quad (2)$$

where $\rho(z)$ is the molecular density profile through the interface, T is the temperature, $r_{i\alpha j\beta}$ is the distance between site α on molecule i and site β on molecule j while r_{ij} is the corresponding separation of the molecular centres of mass. $u(r_{i\alpha j\beta})$ is the site–site potential function. For the oxygen–oxygen interaction this site–site potential includes the Lennard-Jones interaction and the repulsion between the partial charges. $\theta(x)$ is the unit step function and the final product of two θ functions means that the molecules on either side of a plane at z will contribute to the pressure at z. The tangential component of the pressure tensor can be calculated by replacing $z_{ij} z_{i\alpha j\beta}$ by $(x_{ij} x_{i\alpha j\beta} + y_{ij} y_{i\alpha j\beta})/2$. The normal component of the pressure tensor must be constant through the interface to ensure mechanical stability and this is an important test of equilibrium in these simulations. The tangential component changes as a function of z becoming negative around the Gibbs dividing surface. The surface tension can be calculated by integrating the difference in the elements of the pressure tensor through the interface.

$$\gamma = \frac{1}{2} \int_{-\infty}^{\infty} \left[P_{zz} - \frac{1}{2} (P_{xx} + P_{yy}) \right] dz \quad (3)$$

Here the factor of 1/2 arises from the two interfaces in the slab simulation.

We have calculated the surface tension along the orthobaric curve by performing simulations with 500 and 1000 molecules for simulations of 100 000 time-steps and calculating the surface tension from the pressure tensor profiles. The long-range interactions are included using the Ewald summation technique. The results are shown in Fig. 4. There is good agreement with the experimental surface tensions along the orthobaric curve away from the critical point. This agreement is surprising given that the model takes no account of the effective dipole moment profile that must exist in the interface. The result has encouraged us to use the SPC/E model to calculate the orientational ordering and the non-linear optical susceptibility profile for the water interface.[26] This profile cannot be measured directly but the integrated non-linear susceptibilities can be studied experimentally using the sum frequency spectroscopy technique.[26]

This work illustrates the use of the method in the validation of a model and in the prediction of a profile that is difficult to observe experimentally.

3.2. Imaging of Liquid-crystal Layers

STM images of the *n*-alkyl-cyanobiphenyls (nCBs) adsorbed on graphite suggest that the first adsorbed layer adopts a highly ordered structure extending for thousands of ångströms. The molecular arrangement within this layer depends on a fine balance between steric effects, dipolar interactions and the strong van der Waals attraction of the surface. The most commonly observed structures consist of long antiparallel rows of molecules.[28] Close examination of the STM images indicates a slight interdigitation between neighbouring rows and small periodic kinks in the arrangements within these rows. Fig. 5 is a sketch of one of the commonly observed layer structures. For 8-CB, these kinks are observed to occur after every four pairs of antiparallel molecules, indicating a unit cell of eight molecules.

The molecular model used for 8-CB was the 22-site united atom model described in ref. 29. A monolayer of adsorbed 8-CB using a system of 72 molecules was simulated with two-dimensional rhombic periodic boundary conditions. Nine separate simulations were performed, each at a different combination of head-group area per molecule, A_m, and temperature, T.[30]

In an attempt to mimic the effect of an AFM probe we have performed three additional simulations with the adsorbate sandwiched between the substrate and a smooth Lennard-Jones 9-3 half-space. This half-space, positioned 7 Å above the substrate so

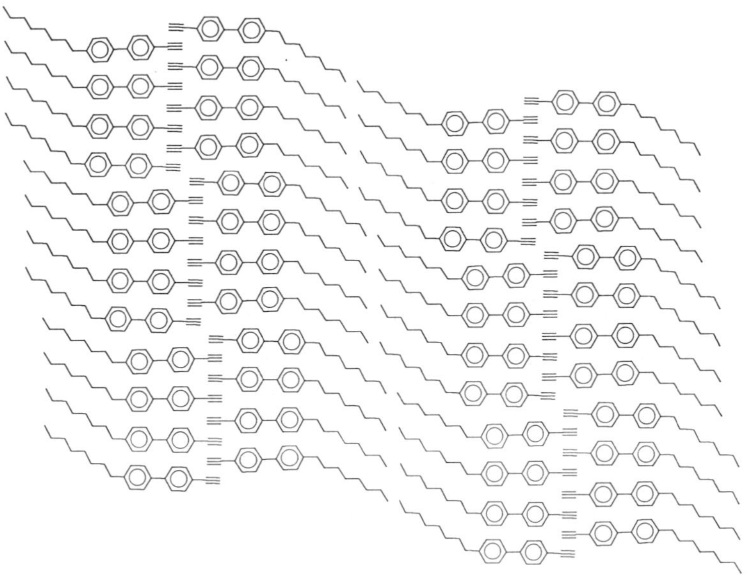

Fig. 5 *n*-Octyl-cyano biphenyl adsorbed on graphite. A schematic view of the smectic layer. (Reproduced with permission of Taylor and Francis Ltd.)

that it exerts a small net repulsive force on the monolayer, inhibits the out-of-plane motion of the molecules. In a separate run aimed at beginning our investigation of the effect of bulk liquid crystal above the monolayer, we have performed a single simulation of a bilayer of 8-CB, using an initial configuration created from two planes of the energy minimised monolayer.

Experimental STM and AFM images of adsorbed structures do not yield snapshots of static structures; rather, as the tip is raster-scanned across the surface it records a time-averaged image of the sample below it. We can, therefore, hope to make better contact with experimental results by performing a similar time-averaging of our simulation configurations. In this spirit, Plate 1 shows contour plots of $\langle \rho(x, y) \rangle$, the simulation-averaged number density of atomic sites projected onto the surface plane. These averages are taken over the entire production phase of each run. Plate 1(*a*) shows $\langle \rho(x, y) \rangle$ for a high-density system, $A_m = 118.21$ Å^2, at 298 K with planar probe above the adsorbate. When confined in this way, the system remains highly ordered throughout the run. Individual molecules are clearly visible; their locations remain virtually unchanged as the simulation progresses. This figure also displays the well defined periodic dislocations, or kinks, that define the eight molecule unit cell structure. If the probe is removed but the density and temperature are left unchanged, Plate 1(*b*), the system relaxes significantly. While individual molecules are still visible within the plot and the smectic strips remain intact, the periodic kinks are no longer visible. At low densities and high temperatures, Plate 1(*c*), the system separates into small ordered domains coexisting with fluid regions. The hole visible on the right-hand edge of this Plate was stable throughout the production cycle of 40 ps. These results support the view that the probe helps to stabilise the first layer structure by preventing out of plane motion of the mesogens.

The MD simulations described in this paper were all performed using the DL_POLY suite of routines[31] developed for Collaborative Computational Project 5 of the EPSRC. This modular package was chosen since source code is supplied in full so that problem-specific, in-house subroutines (such as the molecule–surface potential) can be easily included. Additionally, the DL_POLY offers a high degree of platform independence, since it can be run on both distributed-memory parallel computers and single-processor workstations.

This work is an example of the increasingly important use of simulation technique to support experimental studies. The prediction of adsorbate structure from experiments such as the proximal probe methods, neutron reflectivity, sum frequency spectroscopy and second harmonic generation is model dependent. Simulation can help to explore the assumptions at the heart of these models and to build confidence in the interpretation of the observations.

3.3. Polymers at a Hard Wall

Theories of molecular fluids advanced significantly in the 1970s with the development of the RISM (reference interaction site model) by Chandler and co-workers.[32] This theory treats molecular fluids as collections of atoms correlated by interactions through the fluid [described by $h_{\alpha\gamma}(r)$] and interactions within the molecule [described by $\omega_{\alpha\gamma}(r)$]. The theory introduces an Ornstein–Zernike-like equation which defines a site–site direct correlation function, $c_{\alpha\gamma}(r)$ and uses the Percus–Yevick closure to solve the equation.

The theory was extended to predict the properties of polymer melts by Schweizer and Curro.[33] The PRISM theory results in a scalar Ornstein–Zernike-like equation with an intramolecular correlation function appropriate to a freely jointed chain. This theory gives a reasonable account of the structure of a bulk polymer melt.

Theories of bulk fluids can often be extended to describe adsorption by considering a two-component mixture of atoms and allowing the radius of one of the spheres (that will

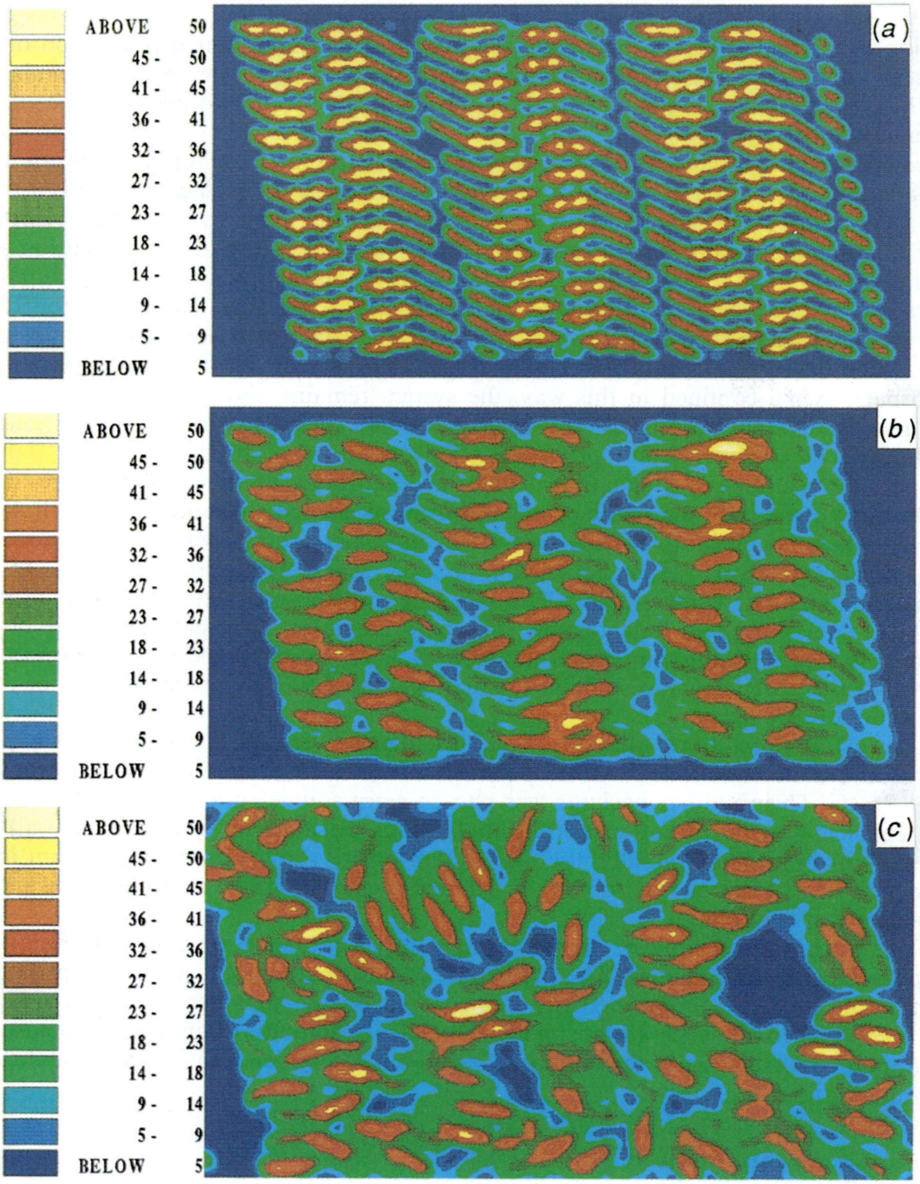

Plate 1 A contour plot of the adsorbate density $\langle \rho(x, y) \rangle$ averaged over the production phase of each simulation. The scale is the percentage probability of finding an atom: (*a*) $A_m = 118.21$ Å^2 at $T = 298$ K in the presence of the planar probe; (*b*) $A_m = 118.21$ Å^2 at $T = 298$ K for an isolated monolayer. (*c*) $A_m = 144.33$ Å^2 at $T = 350$ K for an isolated monolayer. (Reproduced with the permission of Taylor and Francis Ltd.)

form the wall) to become infinite.[34,35] The method has been extended to the polymer RISM equation by Yethiraj and Hall.[36]

The wall-PRISM equation can be written as

$$\tilde{g}_w(k) = [\hat{\omega}_b(k) + N\rho_c \hat{h}_b(k)]\tilde{C}_p(k) \tag{4}$$

where a tilde denotes a one-dimensional Fourier transform and the hat the corresponding three-dimensional transform. $h_b(r)$ is the total site–site correlation function describing the distribution of polymer sites in the bulk field. $\omega_b(r)$ is the corresponding intramolecular distribution of sites within the polymer in the bulk fluid. $g_w(z)$ is the radial distribution function describing the distribution of sites with respect to the hard surface and $C_p(r)$ is closely related to the corresponding direct correlation function, $c_w(z)$.

$$C_p(z) = c_w(z) + \frac{1}{[\hat{\omega}_b(0) + N\rho_c \hat{h}_n(0)]} \tag{5}$$

In these equations N is the number of sites in the polymer and ρ_c the density of polymers. The solution of the wall-PRISM equation requires a knowledge of the bulk fluid distributions $h_b(r)$ and $\omega_b(r)$ characterising the polymer and a closure relationship between $c_w(z)$ and $h_w(z)$. $h_b(r)$ can be obtained by solving the PRISM equation and the closure relationships for a slit-like pore of width H centred at the origin are

$$\begin{aligned} g_w(z) &= 0, \qquad |z| > \frac{H}{2} \\ c_w(z) &= 0, \qquad |z| \leqslant \frac{H}{2} \end{aligned} \tag{6}$$

This closure is analogous to the Percus–Yevick approximation for hard spheres. Three forms of the polymer intramolecular distribution function have been examined in a recent simulation study:[37]

(a) the freely jointed chain model, where

$$\hat{\omega}_{\rm FJC}(k) = \frac{1 - f^2 - 2f/N + 2f^{N+1}/N}{(1-f)^2} \tag{7}$$

where $f = \sin(k\ell/k\ell)$ and ℓ is the distance between sites in the polymer;

(b) the non-overlapping freely jointed model, where the overlaps between non-bonded monomers are excluded, and

$$\hat{\omega}_{\rm NFJC}(k) = \hat{\omega}_{\rm FJC}(k) + \frac{2}{N}\sum_{\tau=2}^{N-1}(n-\tau)\left\{\hat{\omega}_\tau(k) - \left[\frac{\sin(k\ell)}{k\ell}\right]^\tau\right\} \tag{8}$$

The detailed expression for $\hat{\omega}_\tau$ can be found elsewhere;[38]

(c) a tabulated $\hat{\omega}_b(k)$ calculated directly from simulation studies of polymer fluids close to a hard wall.

Monte Carlo simulations using a reptation algorithm have been performed to test the wall-PRISM equation.[37,39] (These simulations could have been readily performed using the molecular dynamics method.) A sketch of the simulation geometry is shown in Fig. 6. The polymers are free to move in the space between the hard walls and their periodic boundary conditions in the direction parallel to the wall.

We have studied models for butane, heptane and decane in a slit composed of hard walls. The hydrocarbon chains consist of united-atom methyl and methylene groups represented as hard spheres. The torsional potential in the model is the polynomial approximation of Ryckaert and Bellemans

$$\begin{aligned}[u_{\rm tors}(\phi)/k_{\rm B})]/K = {} & 1116 + 1462\cos\phi - 1578\cos^2\phi - 368\cos^3\phi + 3156\cos^4\phi \\ & - 3788\cos^5\phi \end{aligned} \tag{9}$$

The bond lengths and angles within the alkane chains are fixed at 1.53 Å and 109.47°, respectively, and the simulations were performed at 300 K. The temperature simply defines the torsional distribution.

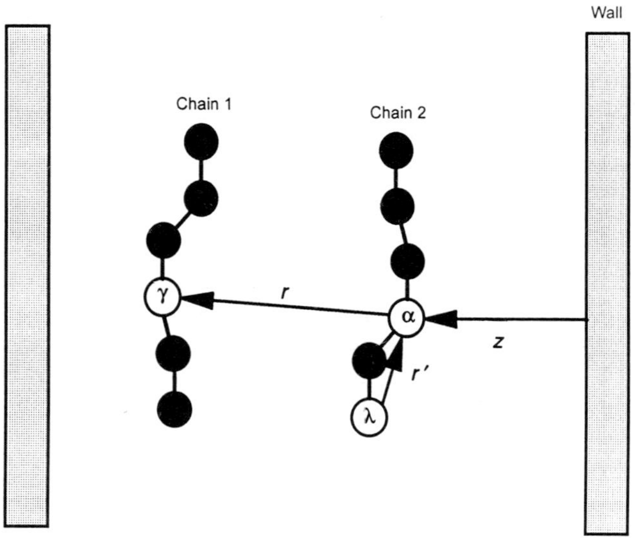

Fig. 6 A schematic view of the simulation of polymer adsorption. The structure of the polymer in the bulk phase is characterised by intermolecular and intramolecular distribution functions, $h_b(r)$ and $\omega(r')$ respectively. The distribution with respect to the surface is described by the density profile $\rho(z)$.

Fig. 7 shows a typical monomer density profile for a fluid of heptane molecules at a high bulk monomer density $\rho_0^* = \rho\sigma^3 = 1.390$. The profile shows half of the simulation cell and one of the two hard walls at $z = 0$. The three models of the intramolecular structure factor yield different results. It is encouraging that the solution using the simulated $\hat{\omega}(k)$ is in best agreement with the simulated profile. The integral equation underestimates the contact values and hence the pressure $[\beta P = \rho_b(0)]$. The theory also underestimates the depth of the first minimum and the height of the following peak although they are at correct values of z. The results with the NFJC structure factor follow those of the simulated $\hat{\omega}(k)$ quite closely at all wall–monomer separations with a

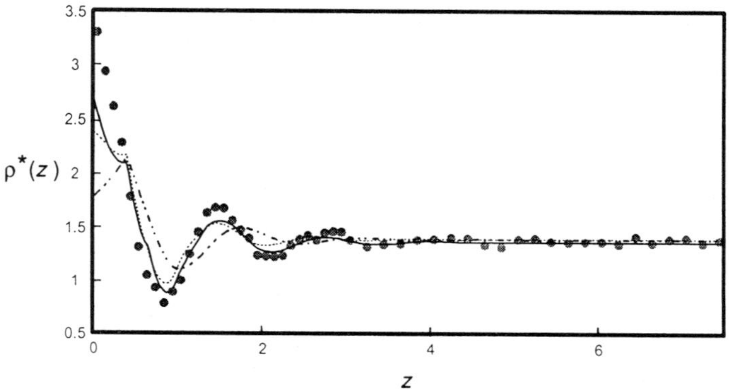

Fig. 7 The monomer density profile, $\rho^*(z)$, for athermal heptane chains between hard walls. The density profiles are plotted as a function of reduced distance from the wall, z. Simulated density profiles (●) are compared with the wall-PRISM predictions with $\hat{\omega}(k)$ from the FJC (–··–··–), from NFJC (·····) and from the simulation (——). The bulk monomer density in this simulation was $\rho_b^* = 1.390$.

slightly poorer agreement at contact. The FJC chain model is clearly poorer than the other two models. The overall agreement, regardless of the precise form of the intramolecular correlation function, is slightly poorer at lower densities and higher elongations.

This study is a good example of the way that simulation can be used to test simple theories of condensed-phase systems.

4. The Future

There are a number of directions in which we can expect significant advances in molecular simulation over the next decade. Total simulation times in molecular dynamics are limited by the time-steps of between 0.5 and 5 fs used to cover the fastest molecular motions. It seems likely that improvements in the speed of the individual processors in the massively parallel distributed memory machines will allow simulation of up to 50×10^6 time-steps and trajectories of approximately 0.1 μs. These very long timescale molecular dynamics calculations will become important as we study phenomena such as crack propagation, crystal growth and protein rearrangements, at larger distance scales and longer timescales.

The algorithms for performing molecular dynamics calculations in parallel, such as the replicated data, systolic loop and domain decomposition methods,[40] are now well understood and increasingly implemented in academic and commercial simulation codes. There is still some debate about the best way of handling long-range interactions such as the Coulomb potential, that decrease as slowly as r^{-D} where D is the dimensionality of the system. For small systems the Ewald method is well understood and can be efficiently implemented. For large systems, the Ewald methods scales as $\mathcal{O}(N^{3/2})$ and there is much active research in developing $\mathcal{O}(N)$ or $\mathcal{O}(N \ln N)$ algorithms. These general approaches follow the ideas of Greengard and Rokhlin[41] in which the electrostatic interactions between well separated parts of the system are handled through multipole expansions. We can look forward to a simplification of these algorithms and some clear statements about the systems sizes at which they really become more effective than the Ewald summation.

For some phenomena at longer times, the brute force approach of increasing the length of the simulation is inappropriate. This is the case when the observable is independent of the fast degrees of freedom and they can be integrated out of the model. The solution to this problem is to use a more coarse-grained model hamiltonian. For example, an atomic model of a poly(ethylene glycol) polymer could be replaced by a spring-bead model of a polymer in which each spherical force site represents approximately fifty monomer units. The increase in the mass of the force site allows a scaling of the time-step in a dynamics calculation and the fast vibrational motions associated with the chemical bonds can be avoided. Additional increases in time-step can be achieved by switching from a full molecular dynamics algorithm to the Brownian or Stokesian equations of motion. Using these methods we can solve the dynamics of small colloidal aggregates with hydrodynamic forces represented by a truncated multipole expansion. Finally, in some simulations we will wish to avoid an explicit representation of the particles and concentrate instead on volume elements of the fluid. The lattice Boltzmann or dissipative molecular dynamics methods are rule-based implementations of the dynamics of fluid 'particles'. Certainly the lattice Boltzmann method will reproduce the solutions to the Navier Stokes equations in certain well defined limits and this helps to make the connection between classical molecular dynamics and the methods of computational fluid dynamics.

One of the major challenges in the area of mesoscale dynamics is to relate changes in the coarse-grained hamiltonian to changes in the underlying atomic model. One interesting advance in this field has been the bond-fluctuation model described by Binder[42] in which an atomic model is mapped on to a lattice model in an attempt to predict the glass transition temperature of the polymer melt.

There are still important advances to be made in the accurate developments of force fields.[43] There is a general move towards so-called class three force fields which include a more accurate description of the intramolecular potentials. In particular, they include the cross terms, which might describe the effect of changes in bond lengths and bond angles on torsional potentials. In some cases, there is a need to use a more accurate representation of the electrostatic potential around a molecule than can be provided by the use of distributed partial charges. An accurate representation of the potential can be obtained using the distributed multipole analysis of Stone and Price but an implementation of this representation within the standard commercial and academic simulation programs is still required. Finally, in some situations, it will be necessary to include polarisation effects by allowing the the electrostatic field to induce an additional dipole at a neighbouring molecule. This can be achieved using a multipole expansion method.[44] Alternatively, it is possible to use a classical Car–Parrinello approach in which additional partial charges distributed in a molecular frame can change their magnitude under the influence of an external field, subject to constraint of total charge neutrality.[45]

All the simulation work discussed in this review has been within the framework of the first model, illustrated in Fig. 1. In this case, well defined intermolecular potentials are used with a classical simulation technique to make predictions. Chemistry normally involves the making and breaking of bonds and thus significant rearrangements of the electrons with respect to the nuclei. In these circumstances, the effective pair potential fitted between two closed shell atoms is no longer useful and we must include our quantum mechanics at the heart of the simulation technique. In other words, we must perform an electronic structure calculation for a given disposition of nuclei and use the Hellman–Feynman force to move the nuclei. This model is illustrated in Fig. 8.

Perhaps the most successful approach in this area has been the Car–Parrinello simulation technique.[46] In this method the electronic degrees of freedom are normally represented by a plane wave expansion of the single-electron wave function ψ_i. Car and

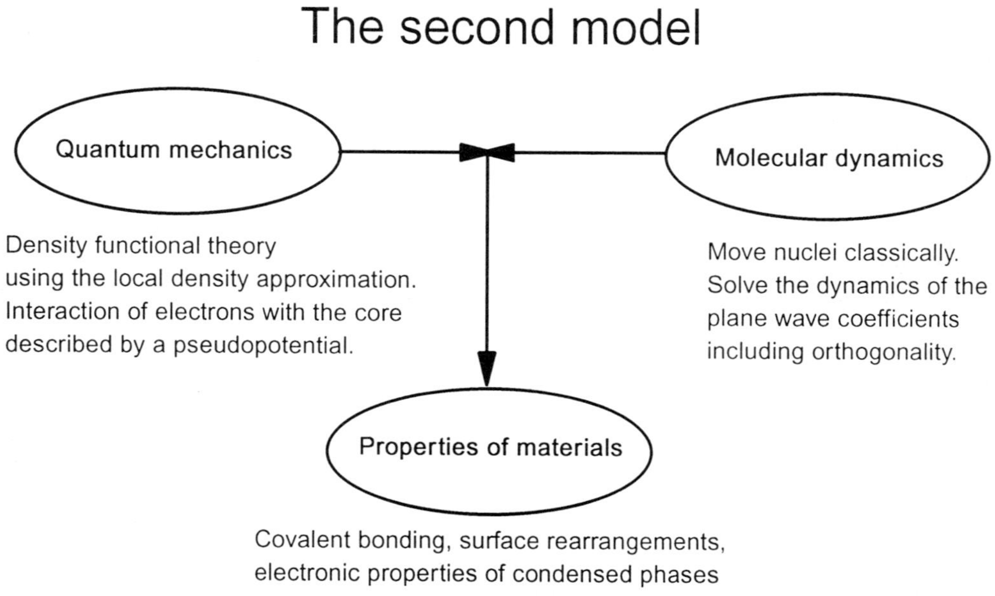

Fig. 8 The second simulation model

Parrinello set up a generalised Lagrangian

$$\mathscr{L} = \sum_i^{\text{occ}} \int d\boldsymbol{r}\mu_i |\dot{\psi}_i(\boldsymbol{r})|^2 + \frac{1}{2}\sum_I M_I \dot{\boldsymbol{R}}_I^2 - E[\{\psi_i\}, \boldsymbol{R}_I]$$
$$+ \sum_{ij} \Lambda_{ij}\left(\int d\boldsymbol{r}\psi_i^*(\boldsymbol{r})\psi_j(\boldsymbol{r}) - \delta_{ij}\right) \quad (10)$$

where the upper case variable represent the nuclei and μ_i is a fictitious mass used to define the electronic 'kinetic energy'. Λ_{ij} is a Lagrangian multiplier used to impose orthogonality on the electronic wavefunctions. This Lagrangian produces the normal Newtonian equation of motion for the nuclei and an equation of motion for the wavefunction or the plane wave coefficients.

$$\mu\ddot{\psi}_i = -\frac{\delta E}{\delta \psi_i^*} + \sum_j \Lambda_{ij}\psi_j$$
$$M_I \ddot{\boldsymbol{R}}_I = -\frac{\partial E}{\partial \boldsymbol{R}_I} \quad (11)$$

In a periodic simulation, the single particle orbitals entering the lagrangian satisfy the Bloch theorem and can be expanded in plane waves

$$\psi_i^{\boldsymbol{k}}(\boldsymbol{r}) = \exp(i\boldsymbol{k}\cdot\boldsymbol{r})\sum_{\boldsymbol{g}} c_i^{\boldsymbol{k}}(\boldsymbol{g})\exp(i\boldsymbol{g}\cdot\boldsymbol{r}) \quad (12)$$

where $\boldsymbol{g}$ is a reciprocal lattice vector of the molecular dynamics box and the wave vector $\boldsymbol{k}$ lies within the Brillouin zone of the reciprocal lattice of the box. $c_i^{\boldsymbol{k}}(\boldsymbol{g})$ are the Fourier components of the single-particle wavefunction, and are the normal dynamical variables in the calculation.

The wavefunctions at a particular time-step are used to calculate the electronic density $\rho(\boldsymbol{r})$. The total energy can be calculated from the Hohenberg–Kohn functional with the ion–ion interaction. The derivatives of this energy are used in the CP equations of motion. The interaction of the valence electrons with the nuclei and their core electrons can be handled using a transferable pseudo-potential. The orthogonality of the orbitals is maintained directly in the equations of motion using the SHAKE algorithm. Alternatively, the wavefunction dynamics can be solved by minimising the electronic energy for a particular nuclear configuration using a conjugate gradient algorithm and insuring orthogonality using the Gram Schmidt scheme.

This method has been widely applied in the UK by the CP consortium to study metal oxides, insulator metal transitions, the positions of counter-ions in zeolites, the dissociation of Cl_2 at a semiconductor surface, silicon surface reconstructions, and molecule–metal potential surfaces.

There is also active research into hybrid quantum mechanical and classical methods. In this case, a small portion or cluster is embedded in a condensed system. The cluster or region of special interest is treated quantum mechanically using self-consistent Hartree–Fock, density functional or semi-empirical methods while the region outside the cluster is treated using a classical force field. One difficulty with this technique is in the treatment of the boundary between the classical and quantum regions since the definition of the quantum region may require the breaking of a chemical bond. The problem of dangling bonds can be reduced by including additional link atoms at the boundary which conserve bond order and saturate the valencies of the cluster. This type of simulation is being used to study formamide in water. The formamide fixed in a number of different rotational conformations is treated quantum mechanically while at the same time classical Monte Carlo simulations are performed on the solvent.[47] This type of simulation method is now available in commercial modelling software.[48]

The principal advantage of the quantum models, as described, is that they avoid the use of empirical pair potentials. It is important to remind ourselves that in these cases, we are not solving the time-dependent Schrödinger equations. The dynamics of the nuclei are classical and these methods would not, for instance, provide useful information on the tunnelling of a light hydrogen atom in the active site of a protein. The methods described here allow us to generate trajectories on the ground electronic state. There are separate developments in the generalisation of the molecular dynamics method to include the possibility of transitions between electronic states driven by nuclear motion.[49]

Despite these recent advances in *ab initio* simulation, there is still considerable scientific mileage in the idea of an effective pair potential. Classical molecular dynamics can explore phase space approximately one hundred times faster than the Car–Parrinello method and will, for the present, remain the method of choice for studying large systems. There are many important problems in condensed matter physical chemistry that are adequately described by classical potentials and where the switch to a meso-scale description will be the critical development. Classical molecular dynamics will continue to provide detailed, validated information on microscopic structure and dynamics offering a grandstand view from the bond.

I would like to express my thanks to the graduate students and colleagues with whom I have worked on the areas presented in this review: José Alejandre, Gustavo Chapela, Martin Callaway, Douglas Cleaver and Giulio Galassi. I would like to thank Dr. V. Sokhan for reading the manuscript.

References

1 W. J. Kaufmann and L. L. Meyers, *Supercomputing and the Transformation of Science*, Scientific American Library, W. H. Freeman, New York, 1993.
2 N. Metropolis, A. W. Rosenbluth, M. N. Rosenbluth, A. Teller and E. Teller, *J. Chem. Phys.*, 1957, **21**, 1087.
3 B. J. Alder and T. E. Wainwright, *J. Chem. Phys.*, 1957, **27**, 1208.
4 J. B. Gibson, A. N. Goland, M. Milgram and G. H. Vinyard, *Phys. Rev.*, 1960, **120**, 1229.
5 A. Rahman, *Phys. Rev. A*, 1964, **136**, A405.
6 A. Rahman and F. H. Stillinger, *J. Chem. Phys.*, 1971, **55**, 3336.
7 D. J. Evans and S. Murad, *Mol. Phys.*, 1977, **34**, 327.
8 B. J. Berne and G. D. Harp, *Adv. Chem. Phys.*, 1970, **17**, 63.
9 J. P. Ryckaert, G. Ciccotti and H. J. C. Berendsen, *J. Comput. Phys.*, 1977, **23**, 327.
10 S. Nóse, *J. Chem. Phys.*, 1984, **81**, 511.
11 H. C. Anderson, *J. Chem. Phys.*, 1980, **72**, 2384.
12 M. Parrinello and A. Rahman, *J. Appl. Phys.*, 1981, **52**, 7182.
13 D. Frenkel and A. J. C. Ladd, *J. Chem. Phys.*, 1974, **81**, 3188.
14 W. T. Ashurst and W. G. Hoover, *Phys. Rev. Lett.*, 1973, **31**, 206.
15 W. G. Hoover, D. J. Evans, R. B. Hickman, A. J. C. Ladd, W. T. Ashurst and B. Moran, *Phys. Rev. A*, 1980, **22**, 1690.
16 A. W. Lee and S. F. Edwards, *J. Phys. C*, 1921, **5**, 172.
17 J. Palfreman and D. Swade, *The Dream Machine: Exploring the Computer Age*, BBC books, London, 1991.
18 Lord Kelvin, *Phil. Mag.*, 1901, **2**, 1.
19 W. W. Wood, *Early History of Computer Simulation in Statistical Mechanics*, Proceedings of the Enrico Fermi Summer School, Varenna, 1986, pp. 3–14.
20 G. C. Maitland, M. Rigby, E. B. Smith and W. A. Wakeham, *Intermolecular Forces*, Clarendon Press, Oxford, 1981.
21 J. Alejandre, D. J. Tildesley and G. A. Chapela, 1995, **102**, 4574.
22 A. R. van Buuren, S-J. Marrink and H. J. C. Berendsen, *J. Phys. Chem.*, 1993, **97**, 9206.
23 K. Watanabe and M. L. Klein, *Chem. Phys.*, 1989, **131**, 157.
24 J. S. Rowlinson and B. Widom, *Molecular Theory of Capillarity*, Clarendon, Oxford, 1982.
25 V. Sokhan and D. J. Tildesley, *Mol. Phys.*, manuscript in preparation.
26 Y. R. Shen, *The Principles of Nonlinear Optics*, John Wiley, 1984.
27 *International Critical Tables*, McGraw Hill, New York, 1928, vol. III.

28 J. S. Foster and J. E. Frommer, *Nature*, 1988, **333**, 542; J. K. Spong, H. A. Mizes, L. J. LaComb, M. M. Dovek, J. E. Frommer and J. S. Foster, *Nature*, 1989, **338**, 137; W. Mizutani, M. Shigeno, M. Ono and K. Kajimura, *Appl. Phys. Lett.*, 1990, **56**, 1974.
29 D. J. Cleaver and D. J. Tildesley, *Mol. Phys.*, 1994, **81**, 781.
30 D. J. Cleaver, M. J. Callaway, T. Forester, W. Smith and D. J. Tildesley, *Mol. Phys.*, 1995, **86**, 613.
31 DL_POLY is a parallel molecular dynamics simulation package developed at the Daresbury Laboratory by W. Smith and T. R. Forester under the auspices of the EPSRC for the Collaborative Computational Project for Computer Simulation of Condensed Phases (CCP5) and the Advanced Research Computing Group (ARCG) at the Daresbury Laboratory.
32 D. Chandler, in *The Liquid State of Matter: Fluids Simple and Complex*, ed. E. W. Montroll and J. L. Lebowitz, Studies in Statistical Mechanics Vol. VIII, 275–340, North Holland, Amsterdam, 1982, vol. VIII, pp. 275–340.
33 K. S. Schweizer and J. G. Curro, *J. Chem. Phys.*, 1988, **89**, 3342.
34 D. Henderson, F. F. Abraham and J. A. Barker, *Mol. Phys.*, 1976, **31**, 1291.
35 Y. Zhou and G. Stell, *Mol. Phys.*, 1989, **66**, 791.
36 A. Yethiraj and C. K. Hall, *J. Chem. Phys.*, 1991, **95**, 3749.
37 G. Galassi and D. J. Tildesley, in preparation; G. Galassi, Ph.D. Thesis, University of Southampton, 1993.
38 K. S. Schweiser and J. G. Curro, *J. Chem. Phys.*, 1988, **89**, 3350.
39 R. Dickman and C. K. Hall, *J. Chem. Phys.*, 1988, **89**, 3168.
40 P. Hilbers and K. Esselink, in *Computer Simulation in Chemical Physics*, ed. M. P. Allen and D. J. Tildesley, Kluwer, The Netherlands, 1992, pp. 473–496.
41 L. Greengard and V. Rokhlin, *J. Comput. Phys.*, 1987, **73**, 325; H. C. Petersen, D. Soelvason, J. W. Perram and E. R. Smith, *J. Chem. Phys.*, 1994, **101**, 8870.
42 K. Binder, in *Computer Simulation of Polymers*, ed. E. Colbourne, Longman, 1994, pp. 91–129.
43 W. D. Cornell, P. Cieplak, C. I. Bayly, I. R. Gould, K. M. Merz Jr., D. M. Ferguson, D. C. Spellmeyer, T. Fox, J. W. Caldwell and P. A. Kollman, *J. Am. Chem. Soc.*, 1995, **117**, 5179.
44 G. Ruocco and M. Sampoli, *Mol. Phys.*, 1994, **82**, 875.
45 M. Sprik, in ref. 40, pp. 211–296.
46 D. K. Remler and P. A. Madden, *Mol. Phys.*, 1990, **70**, 921.
47 I. Hillier, personal communication.
48 See for instance, The **Solids Embed** module available in the Catalysis program from Biosym/MSI, Biosym Technologies Inc., 9685 Scranton Road, San Diego, CA 92121-2777, USA.
49 D. Coker, in ref. 40, pp. 315–376.

Faraday Discussion 100 *Celebration Paper*; *Presented 24th April*, 1995

Faraday Discuss., 1995, **100**, C47–C59

Approaching Complexity

Stephen K. Scott
School of Chemistry, University of Leeds, Leeds, UK LS2 9JT

In 1974 a Faraday Symposium[1] was held at the Royal Institution on the subject of 'The Physical Chemistry of Oscillatory Processes'. That timely event brought together an international group of specialists in the newly developing, interdisciplinary research field of oscillations, chemical waves and other 'exotic' phenomena. In the 21 years since the Symposium, our understanding of many fundamental properties of such behaviour has deepened through the application of advances in pure and applied mathematics in collaboration with experimental studies based in many physical, biological and engineering contexts, with the physical chemist being pre-eminent, exploiting the natural 'non-linearities' in chemical kinetics. This account illustrates how these ideas have been applied and developed through two specific examples: the route to chaos in the oxidation of carbon monoxide and the experimental challenge posed by Alan Turing's predictions of chemical patterns.

1. Introduction

Many important chemical reactions exhibit the property of *feedback*. Chemical feedback arises when intermediate species formed in the reaction mechanism influence their own rate of production.[2–4] In positive feedback, such as chain branching in combustion processes or autocatalysis in solution-phase redox reactions, the intermediates accelerate their own rate of production: negative feedback or inhibition also arises and is particularly important in regulatory biochemical systems. The characteristic signature of positive feedback is clearly revealed if the overall rate of reaction (*e.g.* as defined by the rate of consumption of the initial reactant or of formation of the final product) is plotted as a function of the extent of reaction, *i.e.* the percentage consumption of the initial reactant. Fig. 1 shows such plots for (*a*) deceleratory systems typical of simple *n*th-order reaction kinetics, in which the maximum rate arises at the beginning of the reaction; (*b*) a system

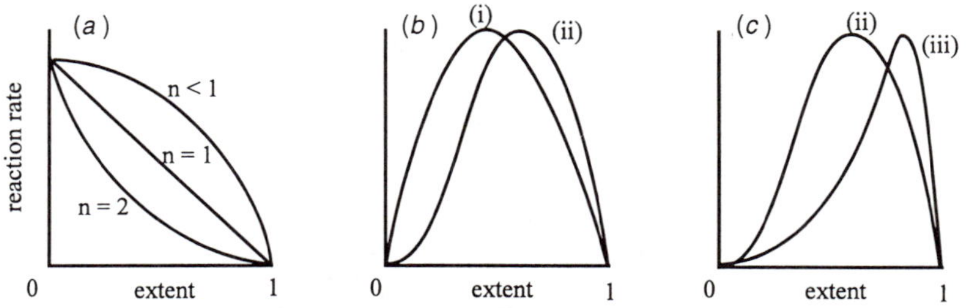

Fig. 1 Variation of reaction rate with extent of reaction; (*a*) deceleratory systems showing 1st ($n = 1$), 2nd ($n = 2$) and approximately $\frac{1}{2}$-order kinetics; (*b*) systems with feedback showing (i) quadratic and (ii) cubic autocatalysis; (*c*) comparison of (ii) cubic and (iii) thermal feedback curves

with positive feedback showing an initial acceleratory phase with the rate increasing with increasing extent of reaction until some maximum at a non-zero extent of reaction. Fig. 1(*c*) compares the shape of the rate–extent curve for a type of chemical (isothermal) acceleration and thermal feedback. The latter arises through the self-heating accompanying an exothermic reaction in the absence of infinitely fast heat transfer, feeding back through the temperature dependence of the reaction rate constant(s). It is the characteristic shape of these curves that supports the various types of *complex behaviour* that will be discussed here. This account will also indicate how the very recent advances in the mathematics of *non-linear dynamical systems* provide just the right tools for the chemical kineticist to ask the appropriate questions, and in some instances, to find the appropriate answers to those questions, when faced with 'complex behaviour'.

To provide a more concrete illustration of 'complexity' we can imagine an experiment based on the spontaneous reaction between hydrogen and oxygen. The $H_2 + O_2$ is the simplest combustion reaction: the overall stoichiometry

$$2H_2 + O_2 \rightarrow 2H_2O$$

hides a somewhat complex mechanism, involving up to 90 elementary steps and several radical and molecular intermediates including H and O atoms, OH and HO_2 radicals and H_2O_2. Its behaviour in classic, closed (static) vessels is covered in most undergraduate courses[5] and gives rise to the three ignition limits sketched in Fig. 2. To the left of the limits, there is a 'slow' reaction in which the oxidation proceeds over a timescale of minutes or longer. On the high-temperature side of the limits, reaction proceeds as a millisecond timescale ignition (perhaps after some 'induction period').

In a flow reactor, there is a similar limit structure.[6] Again, there is slow reaction at low ambient temperatures (the temperature of the oven in which the reactor is seated), with an ignition process at higher T_a. It is the nature of this 'ignition' that is of interest. Under some conditions, the reaction adjusts to a rate that simply matches the rate of inflow of fresh reactants. This gives rise to *steady-state* operation. An example of the record observed by a (coated) fine-wire thermocouple junction positioned in the reacting gas (and referenced to a similar junction fixed to the outside wall of the reactor, hence measuring the self-heating due to reaction) is shown in Fig. 3(*a*). After some initial transient behaviour, the system settles to a steady signal. Assuming that we are interested only in the long-time (post-transient) response, then the state of the system is essentially

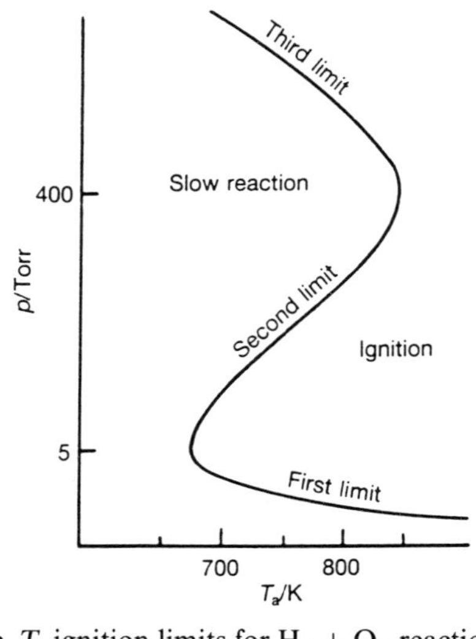

Fig. 2 The classic p–T_a ignition limits for $H_2 + O_2$ reaction in a closed vessel

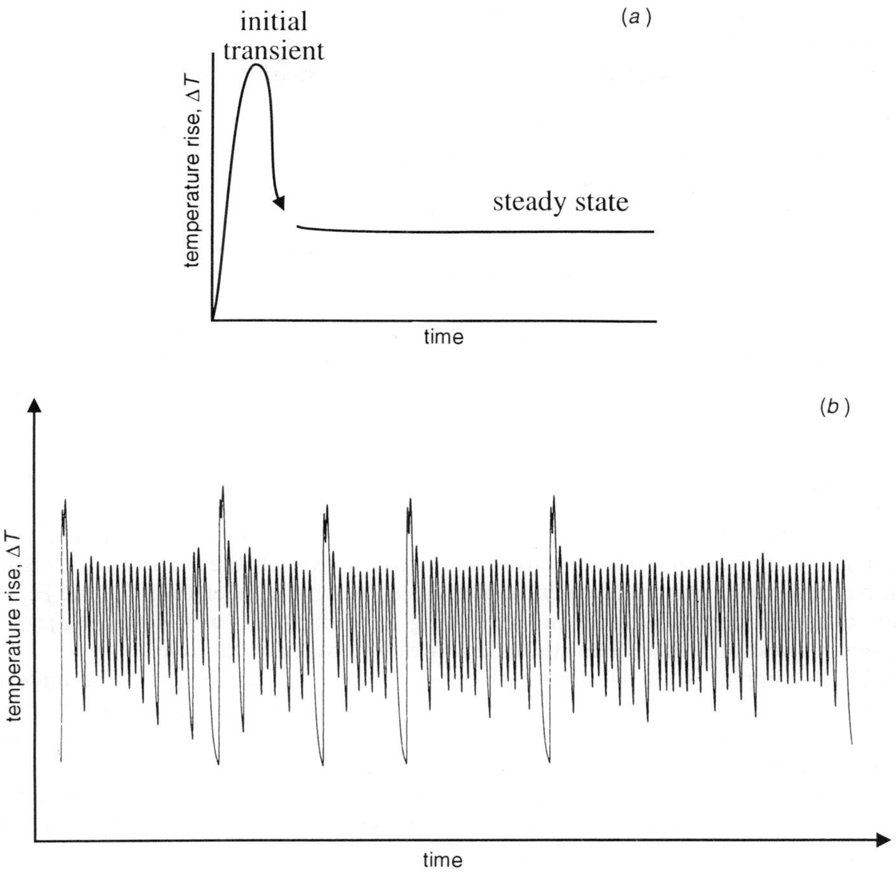

Fig. 3 Possible evolutions of combustion systems in flow reactors: (*a*) a simple, stable steady state is achieved after some initial transient from the initial conditions; (*b*) a complex oscillatory ignition response

summarised by reporting a single quantity, the steady-state temperature excess. Once this has been reported, the future behaviour can be predicted with some confidence: it will persist with the same steady-state response for as long as the experimental conditions are maintained.

If, however, the experimental conditions are changed, *e.g.* by decreasing the ambient temperature, then the response may also change. The record shown in Fig. 3(*b*) is also collected from the $H_2 + O_2$ system but is clearly much less simple than the steady-state ignition process. This time-dependent response is not 'transient' in the sense that the reaction for this set of experimental conditions will not eventually settle to a steady state. Rather this oscillatory state will persist indefinitely. This is a complex example of an oscillatory state, with no obvious period. Three fundamental questions spring to mind when such a response is encountered.

(a) How can we *describe* such a state? This question is posed in the technical sense, *i.e.* we seek to find a way such that precise information can be compactly conveyed to other scientists without having to reproduce the whole experimental record (which in the case of a non-periodic response would have to be infinitely long anyway). Clearly, reporting a single value of the temperature excess is insufficient and even a mean and a standard deviation hardly come close to conveying even the general features of Fig. 3(*b*).

(b) How did such a complex response *arise*? This question relates to the change in response between Fig. 3(*a*) and (*b*). There may have been only a 5 K change in the ambient temperature, but did the system switch suddenly from its steady-state response directly to the 'chaos' of aperiodicity, or did the complexity develop in stages, with intervening states each slightly more complex then its predecessor. Such a series of steps is known as a 'bifurcation sequence' in the language of non-linear dynamical systems theory. Do different reactions all have different bifurcation sequences, or are there some 'universal routes' from simplicity to complexity? If so, we might be able to recognise the signs that indicate that complexity is ready to emerge as we vary the experimental conditions in a particular way, allowing us either to avoid or, perhaps, move more efficiently towards such behaviour.

(c) How can we *exploit* such complexity? If complexity is just a nuisance, then recognising the early sign will perhaps allow us to move away before it develops fully. On the other hand, we may deliberately seek to maximise the complexity of a systems response if we can exploit such behaviour in a way unavailable under steady-state operation. The exploitation of chaotic systems has recently emerged as a possibility with identifiable potential applications. An introductory account of 'Controlling Chaos' can be found in ref. 7, but question (c) will not be the major focus of the present account.

Questions (a) and (b) will be addressed with reference to the oxidation of carbon monoxide, another exemplary combustion reaction for which the picture with regard to the global behaviour observed experimentally is somewhat clearer than for the $H_2 + O_2$ system (although the mechanistic understanding is, perhaps, less well developed).

A fourth question will also be addressed through a second example.

(d) Must complex systems always give rise to complex behaviour? The underlying text here is interest in whether large-scale co-operation can emerge through chemical feedback, perhaps coupled to molecular transport processes, in such a way as to form coherent spatial 'patterns' analogous to the temporal 'patterns' of periodic oscillations. This story involves a 40 year old theoretical prediction and the subsequent journey by many groups inspired by this prediction to realise these *Turing Patterns* in the chemical laboratory.

2. Complex Oscillations in the Oxidation of Carbon Monoxide

The reaction between CO and O_2 in the presence of small, but accurately controlled, concentrations of H_2 is now relatively well characterised.[8] At reduced pressures ($10 < p$/mmHg < 50) the reaction shows a second ignition limit in both closed- and continuous-flow reactors, of a very similar type to that found in the $H_2 + O_2$ system, over an ambient temperature range of approximately $750 < T_a/\mathrm{K} < 825$. We will concentrate here on the flow reactor studies as this configuration allows us to probe the behaviour on the 'ignition' side of the limit.[9] The p–T_a 'ignition diagram' for such a system is shown in Fig. 4.

At ambient temperatures lower than the ignition limit, there is a *steady slow reaction*. This is a steady-state response in which there are only small extents of reaction, with the outflow composition being mainly unreacted CO and O_2 with perhaps a few percent CO_2. Temperature excesses are less than 10 K in this region and the behaviour is definitely 'simple'. In a given experiment, it is relatively easy to vary the ambient temperature, which in the nomenclature of non-linear dynamical systems is the *primary bifurcation parameter*. Other, *secondary parameters* such as the total operating pressure, the mixture composition or the mean residence time (the average time a molecule spends in the reactor) can be varied between experiments, although in the present account only the pressure will be changed.

On crossing the ignition limit, the system displays a relatively wide region of *oscillatory ignition*. Reaction is achieved by sharp ignition events in which the fuel is effec-

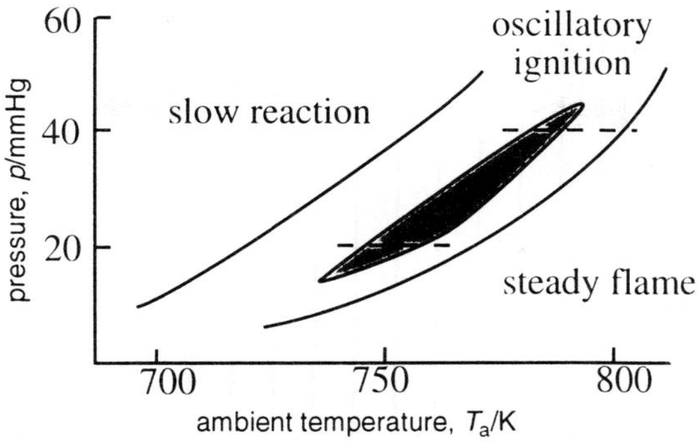

Fig. 4 The p–T_a ignition diagram for $CO + O_2$ in a flow reactor showing regions of slow reaction, steady flame and oscillatory ignition. The behaviour in the shaded region is that of complex oscillations.

tively completely consumed (the equimolar mixture is fuel-lean) and in which a large temperature excursion is detected by the fine-wire thermocouple junction: there is also a weak, pale-blue chemiluminescent emisson from electronically excited CO_2. The ignition excursions are separated by essentially quiescent periods in which the reactor is replenished by the inflow and the products are removed by the outflow. In fact, the reaction is effectively suppressed in this intervening period by 'inhibitor' species, most notably H_2O, formed as products.

Over most of the oscillatory region, the oscillations have a simple *period-1* character: provided the experimental conditions are held constant and any initial transient behaviour is allowed to die away, the system settles into oscillations for which the amplitude of each peak is the same as its predecessor and the period between any two peaks is constant throughout the sequence. An example of a period-1 ignition is shown in Fig. 5(*a*). The amplitude and period of the oscillation vary smoothly across the region of oscillatory ignition if the ambient temperature T_a or pressure are varied. As the system approaches the second main boundary on the ignition diagram, Fig. 4, the amplitude decreases smoothly to almost zero, with the oscillatory waveform then having the appearance of a small 'ripple' superimposed on an almost steady, non-zero temperature excess, Fig. 5(*b*). If T_a is increased further, so as to cross this limit into the *steady flame* region, the 'ripple' dies away and a truly steady signal is observed. The system settles onto a steady state of, now, high reactant conversion in which there is effectively a balance between the inflow of reactants and their complete conversion to CO_2 in the mean residence time. The long residence times (2–16 s) typical of such reactor studies limit the maximum reaction rate to moderately low values (maximum steady-state rate = inflow concentration/residence time, otherwise the reactant concentrations would become negative) so the steady-state temperature excesses associated with this 'steady-flame' state are typically only 10–100 K rather than those of more familiar flames.

The boundary between oscillatory ignition and steady flame represents the locus of experimental conditions at which a particular qualitative change (bifurcation) in the dynamic behaviour of the system occurs. In technical terms, this particular change from steady to oscillatory behaviour is called a *Hopf bifurcation* (and in this case it is a *supercritical Hopf bifurcation* as the oscillations grow from vanishingly small amplitude as the locus is crossed by decreasing T_a). There is an increase in the complexity from steady to simple oscillations.

Returning to the region of oscillatory ignition, there is a subregion indicated in Fig. 4 of *complex ignition*. To investigate the sort of changes in response encountered in this subregion, we can simply set the reaction system up at some ambient temperature and

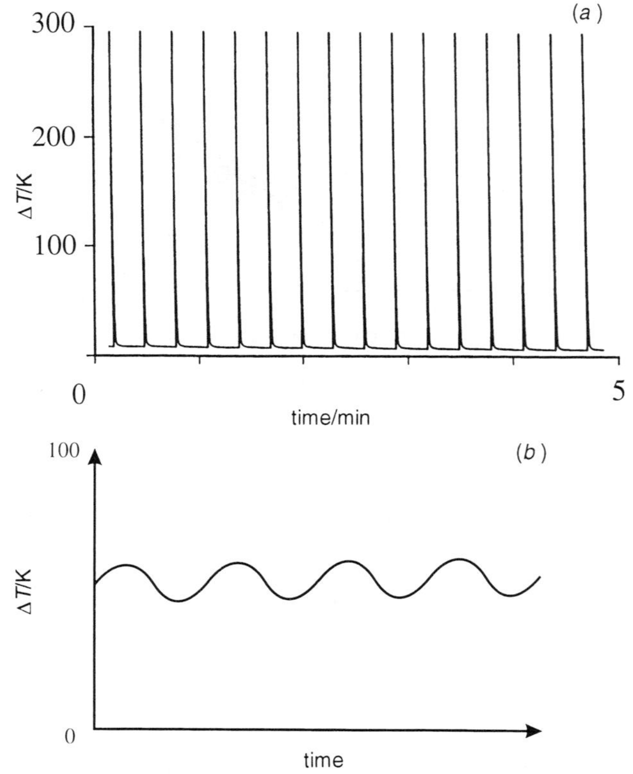

Fig. 5 Examples of simple oscillatory waveforms: (*a*) period-1 relaxation ignitions with large amplitude and long period found near the ignition limit; (*b*) small amplitude, approximately sinusoidal oscillations found near the boundary with the steady-flame state

pressure so as to lie just outside this subregion (*e.g.* at lower T_a) and then allow the oven temperature to increase slowly, so that the system drifts through the complex ignition region. Two such 'scans' are shown in Fig. 6 corresponding to the paths indicated in Fig. 4.

At a relatively high pressure, the region of complexity is rather narrow. The response of the system on entering this region is for the simple period-1 oscillation to give way to a waveform in which there is an alternation between large- and small-amplitude peaks as indicated in Fig. 6(*a*). Just inside the region the difference in amplitude between large and small is itself only small, although the large and small character is established more clearly as T_a is increased. The period between each individual peak is approximately the same: this means that the repeating unit is twice that of the period-1 response. This qualitative change in behaviour is known as a *period-doubling bifurcation* and produces a period-2 response. A further increase in the ambient temperature takes the system out of the complex oscillation region. Simple period-1 behaviour is regained by a period-halving, *i.e.* the inverse response of the way in which the increased complexity was established.

At a lower operating pressure, the region of complexity is somewhat more extensive. This allows 'more room' for a greater development of complexity as indicated by the scan in Fig. 6(*b*). The first stage is again a period-doubling to establish a period-2 waveform, but this then shows an additional period-doubling event leading to a period-4 and so on. There is clearly a suggestion of some very complex behaviour in Fig. 6(*b*), but to investigate this in a rigorous manner we really need to obtain 'transient-free' informa-

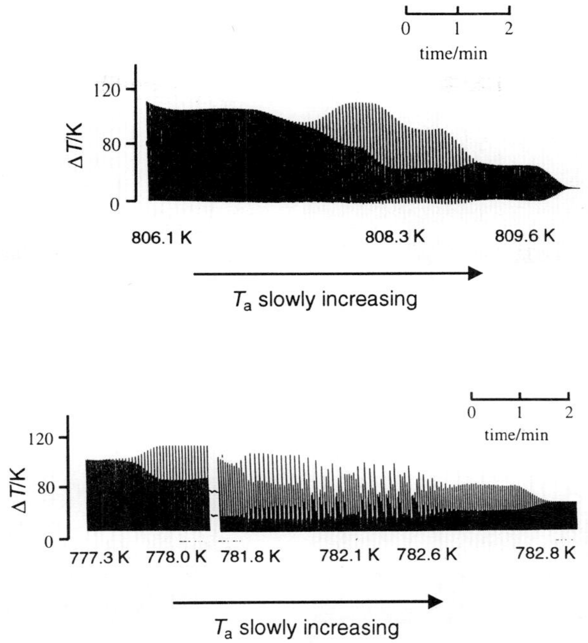

Fig. 6 Evolution of the oscillatory waveform during a scan of ambient temperature through the region of complex oscillations at different pressures

tion. This is achieved by allowing the experimental conditions to settle out completely at various points within the complex ignition region and collecting data only once this settling-out process has happened. Examples of such *asymptotic states* are shown in Fig. 7(*a*)–(*f*) which are arranged in order of increasing ambient temperature.

At the lowest T_a, just outside the complex ignition region, there is a period-1 response. Increasing the ambient temperature sees emergence of period-2 as described above and then at a slightly higher T_a, period-4 [Fig. 7(*a*)–(*c*)]. This reaction also gives strong suggestions of period-8 and period-16, but these occur over very narrow ranges of experimental conditions and it has not proved possible to stabilise the system sufficiently to collect entirely convincing time series of these higher complexity states. What does emerge at slightly higher ambient temperature is a modestly wide region of very complicated behaviour of the character shown in Fig. 7(*d*). There is no obvious repeating unit or periodicity: we will return to the analysis of this response below. If the range of the aperiodic behaviour is examined in detail, it is observed to contain very narrow 'windows' in which a periodicity can be identified: two such examples are shown in Fig. 7(*e*) and (*f*) corresponding to periods 5 and 3, respectively.

The emergence of a region of aperiodic responses has some interesting implications. Other evidence and experience indicates that systems that behave with this response are unpredictable in a specific context: small differences in initial condition do not, as is typical of periodic or steady-state behaviour, decay in magnitude; instead they grow exponentially until two systems which started 'close together' become totally uncorrelated. Thus a repeat experiment would eventually evolve in a different pattern and one system cannot be used to forecast the behaviour in another or, equivalently, the past behaviour of a system cannot be used to predict the future evolution beyond some 'threshold of acceptable error'. [It is important to appreciate a distinction between two systems subjected to the same experimental condition both behaving in the same way, *i.e.* chaotically, but whilst they both have this property they will be evolving in time in

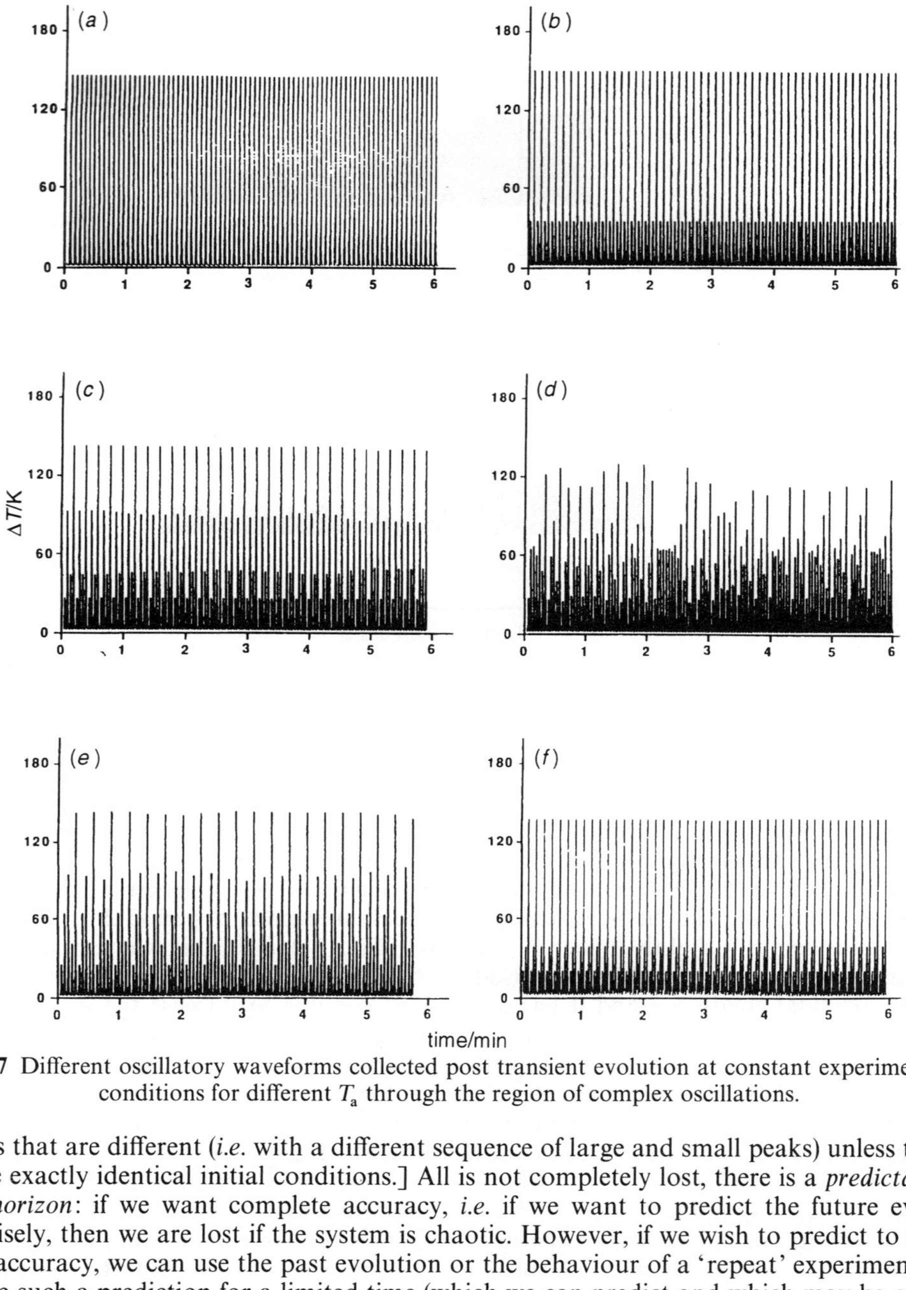

Fig. 7 Different oscillatory waveforms collected post transient evolution at constant experimental conditions for different T_a through the region of complex oscillations.

ways that are different (*i.e.* with a different sequence of large and small peaks) unless they have exactly identical initial conditions.] All is not completely lost, there is a *predictability horizon*: if we want complete accuracy, *i.e.* if we want to predict the future event precisely, then we are lost if the system is chaotic. However, if we wish to predict to say, 5% accuracy, we can use the past evolution or the behaviour of a 'repeat' experiment to make such a prediction for a limited time (which we can predict and which may be quite long if the system is only mildly chaotic). If we accept 10% precision, we can predict for longer (but not twice as long; there is a logarithmic relationship between error and time).

How does this inherent unpredictability arise? For the p–T_a conditions surrounding the narrow region of chaotic behaviour, the reaction evolves in moderately exotic ways, but always with an identifiable pattern. We would, as physical chemists, expect the evolution to be determined by some reaction mechanism comprising elementary steps each with definite rate laws and Arrhenius parameters *etc.* The resulting reaction rate equations, even if they are not known precisely, would be expected to be calling the tune, so that the behaviour should in principle be predictable and regulated. So why do we have a narrow range of experimental conditions within this range of experimental conditions for which the system becomes essentially unpredictable? Does the reaction mechanism simply give up for the 'special' set of conditions? This interpretation would be uncomfortable, but is anyway undermined by the observation of the periodic windows that are embedded within the chaotic region: does the mechanism keep switching in and out, failing and then reviving? Logic demands a different answer, with the same kinetic mechanism operating over the whole p–T_a region in Fig. 4 but capable of rationalising unpredictability.

To show that the aperiodic response of Fig. 7(*d*) is not random we can perform a simple analysis of the thermocouple record. Fig. 8(*a*) reproduces the chaotic time series and Fig. 8(*b*) shows the result of plotting the temperature maximum of one peak against the maximum of the next peak successively through the series (*i.e.* we plot the maximum of peak 1 against peak 2, of peak 2 against peak 3 *etc.*). This *next maximum map* would be a simple random array of dots if the time series were random. However, the form shown in Fig. 8(*b*) is clearly not that: it has a clear structure indicating that there is a definite relationship governing the maximum in one ignition in terms of the maxima of previous peaks. The system is thus fully deterministic. So how do we square this claim with the previous statement that the system is behaving in an essentially unpredictable manner? The answer to this is fairly involved but relates to the steepness of the slope of the map in the vicinity of its intersection with a line of unit slope drawn on the same graph. This identity line intersects with the map at a point for which the amplitude of the next maximum is equal to the amplitude of the current maximum, *i.e.* it identifies the point at which the system would sit if every oscillation had the same amplitude, which is the simple period-1 response. Because the actual map is steep in the vicinity of this point, we know that the period-1 solution is *unstable*. This means that if we have a system that has a maximum close to this point, successive iterations of the map will have maxima that move further away from the intersection point. In mathematical terms, the map has a slope of magnitude greater than 1 (in fact, we have slope < -1 as the gradient is negative) at the period-1 intersection point. Furthermore, the steepness of the slope means that two systems that have maxima that are similar but not exactly

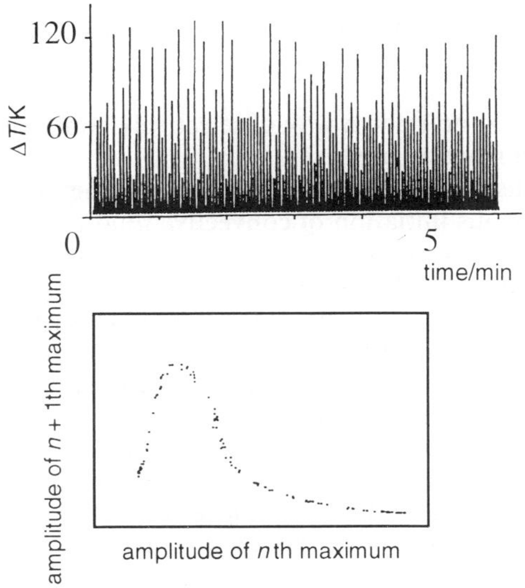

Fig. 8 (*a*) Chaotic time series from Fig. 7(*d*); (*b*) next maximum map contructed from ignition data

the same will have their differences amplified when they visit part of the map with such a steep slope. Similar comments apply to the intersections of higher-order maximum maps, such as the next-but-one maximum map where we plot the amplitude of the nth peak against that of the $n + 2$th peak. For that map, the intersections with a line of unit slope correspond to the oscillations that repeat every other ignition, *i.e.* to the period-2 solutions. If the map has slope of magnitude greater than 1 in the vicinity of these intersections, then the period-2 solution will also be unstable and small differences in the evolution of similar systems will be amplified in the vicinity of such steep segments of the map. Chaos, or aperiodicity, arises when all nth maximum maps have steep intersections with the line of unit slope. The next maximum maps convey much important quantitative information in terms of the slope at the intersection point. This information allows us to quantify just how chaotic a particular system is and how long we can predict the future evolution if we are prepared to accept certain tolerance levels for the accumulating error. The maps are also important for the new techniques of stabilising and tracking periodic states from within the natural chaos.[7,10]

3. Self-organisation and Pattern Formation in Non-linear Systems

In 1952, the mathematician Alan Turing published a paper[11] entitled 'The Chemical Basis for Morphogenesis' in which he speculated on the evolution of chemical concentrations in systems in which chemical feedback was coupled with molecular transport by diffusion. In 'typical', *e.g.* unreactive, cases, diffusive processes serve to smooth out any initial spatial dependence of concentration and the system tends towards spatial uniformity. Even in simple reacting systems, the expected tendency is for an increasingly uniform spatial distribution as diffusion operates. Turing, however, proposed that if there is a sufficient feedback mechanism and if the diffusivity of the feedback species is sufficiently reduced compared with that of the other participants, then for some experimental conditions, systems may actually evolve in the opposite way, with the development of significant and well-organised spatial dependences of species concentrations. Such *Turing Patterns* became of great theoretical interest in the 1960s to 1980s, not least because they were perceived as potential causes of spatial form in biology[12,13] as well as chemistry (articles on Turing Patterns frequently have titles that are variations on the theme of 'How Did the Leopard Gain its Spots').[14]

The obstacles to experimental observation of Turing Patterns in chemistry laboratories that presented themselves in the early years included the following:

(i) there was a general lack of known chemical reactions with clearly established feedback processes;

(ii) there appeared to be no known way of selectively controlling diffusion coefficients of individual species in a reacting mixture;

(iii) chemical reactions often have associated self-heating or changes in molar volume that give rise to spontaneous initiation of convective fluid motion in the reaction system which, in turn, readily dominates the relatively weak role played by diffusion.

The last point, relating to the relegation of diffusion processes when bulk fluid motion operates, is particularly relevant if one wishes to sustain any potential pattern development. As the reaction must be maintained away from its state of chemical equilibrium, there is a need for a continuous supply of fresh reactants and a removal of final products. This would argue for a continuous flow system much like the flow reactors described in the previous section. However, such supply and removal must be provided in a way that does not disrupt the delicate diffusion-driven patterns of interest here.

These 'technical difficulties' have been addressed by many groups over the intervening period and the final pieces of the jigsaw have recently been put together with the first experimental observations of Turing Patterns. The family of known reactions with chemical feedback has grown dramatically[15] since the acceptance of oscillatory pheno-

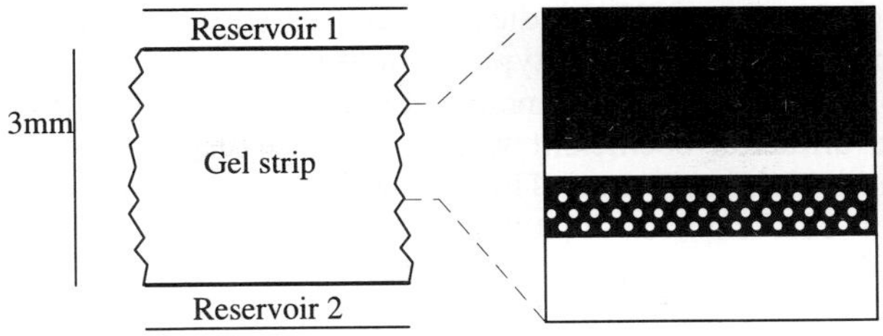

Fig. 9 Schematic representation of apparatus design and Turing Pattern obtained by Castets *et al.* (ref. 29). Separate reactant reservoirs are separated by a gel strip in which there are, therefore, natural concentration gradients from top to bottom of strip as shown, the Turing Pattern is evident parallel to the reservoirs along which direction there are no natural chemical gradients. The size of the strip is indicated.

mena by chemists mainly in response to the undeniable claims based around the Belousov–Zhabotinsky reaction. This is the oxidation of organic species such as malonic or (originally) citric acid by acidified bromate ions catalysed by an inorganic redox couple (typically Ce^{3+}/Ce^{4+} or the ferroin/ferrin couples) studied intially as an inorganic analogue of the citric acid cycle. In fact, Belousov's observations made in 1951 were effectively contemporary with Turing's work, but there was stubborn resistance to accepting that oscillations in chemical systems were allowed by the Second Law of Thermodynamics and which meant that Belousov's work was published in obscurity[16] until taken up by Zhabotinsky[17] in the early 1960's. The Belousov–Zhabotinsky system does indeed show spatial 'patterns', but these *targets* and *spiral* waves[18,19] are not true Turing Patterns as they rely in inhomogeneous initiation of reaction and reflect the underlying distribution of such *pacemaker sites.* (Spiral waves are common in natural non-linear systems including the heart,[20] where they are implicated in cardiac arrhythmias, brain tissue,[12] population dynamics as well as having recently been observed in surface-catalysed reactions under ultra-high vacuum conditions.[21])

Many new oscillatory reactions have been devised based on variations on the Belousov–Zhabotinsky reaction and on related oxyhalogen reactions, using an algorithmic appoach[22–24] pioneered by the groups led by Epstein and by Boissonade and De Kepper. One of these reactions, the CIMA (chlorite–iodide–malonic acid) reaction has proven to be suitable for observation of Turing Patterns. (This has a pleasant resonance as this system is closely related to the so-called Landolt or iodate + reductant reaction which is commonly studied as a typical *clock reaction* in undergraduate laboratories: Alan Turing had studied some aspects of the Landolt reaction at his school and been awarded a prize based on that work.[25])

The identification of a suitable reaction goes part way to meeting the demands of a Turing Pattern. The remaining obstacles were overcome by the combined ingenuity of several groups. First, a whole class of new chemical reactors[26,27] have been designed which can be termed *continuous-flow, unstirred reactors.* These allow a continuous supply of fresh reactants but in a way that the inevitable fluid flow does not influence the diffusion processes. In many cases this is achieved by conducting the actual reaction in a gel or membrane around which the reactant supplies flow. The role of I_2 in the CIMA reaction suggested that starch or some other indicator could usefully be incorporated into the gel to improve the spatial resolution of spatially-resolved spectrophotometric devices, commonly based now on video camera systems[28] and computerised frame-grabbing data acquisition. In fact, this use of I_2 indicators turned out somewhat fortuitously to solve the remaining, and perhaps most vexing, problem of controlling the species diffusion coefficients.

The high molecular mass of an indicator such as starch means that it will be relatively immobile in a gel, even though typical small ions such as I^- will have diffusivities essential that appropriate to aqueous media. If I_2 or I_3^- bind to the immobilised indicator, their effective diffusion coefficient will be reduced in a manner depending on the complexation equilibrium constant. Thus by varying the loading of the complexing agent, the diffusion coefficient of the autocatalytic species in the CIMA reaction can be effectively controlled. This observation, coupled with the other advantages of gelled media, provided the final piece in the Turing jigsaw and allowed the Bordeaux group led by De Kepper to make the first laboratory-designed observations of a Turing Pattern[29] in work developed further by Swinney and co-workers in Texas.[30] The Turing Pattern is shown in Fig. 9 and, whilst not perhaps as impressive as a leopard, is remarkable as it represents the culmination of combined (and competitive) science at its best.

4. Conclusions and Perspectives

Chemical kinetics has both much to offer and much to gain from the interdisciplinary interaction with those involved in unravelling the secrets of non-linear systems. At this stage there are perhaps more implications than direct applications, but remarkably, new ideas for exploitation are already emerging at a rapid rate. We have already identified many appropriate questions to ask of a complicated response of a chemical system so that quantitative measures can be obtained. The pursuit of Turing Patterns, an intellectual end posed by a pure mathematician, has brought direct, but perhaps more importantly, great indirect benefit to science. In addition to learning through collective efforts to this goal how to control diffusion coefficients, how to understand oscillatory reactions sufficiently well to allow the design of further systems and how to suppress natural convective effects, we now find ourselves with a whole array of both new theoretical challenges and new experimental reactors in which to study the effects of coupling molecular transport processes and complex chemistry. The latter has great implications for modelling of atmospheric and environmental chemistry systems as well as for biology and medical situations.

References

1 *Physical Chemistry of Oscillatory Phenomena*, *Faraday Symp. Chem. Soc.*, 1974, vol. 9.
2 S. K. Scott, *Oscillations, Waves and Chaos in Chemical Kinetics*, Oxford University Press, London, 1994.
3 P. Gray and S. K. Scott, *Chemical Oscillations and Instabilities*, Oxford University Press, London, 1990.
4 S. K. Scott, *Chemical Chaos*, Oxford University Press, London, 1991.
5 R. R. Baldwin and R. W. Walker, in *Essays in Chemistry*, ed. J. A. Barnard, R. D. Gillard and R. F. Hudson, Academic Press, New York, 1972, vol. 33, pp. 1–37.
6 P. Gray, J. F. Griffiths and S. K. Scott, *Proc. R. Soc. London.*, *A*, 1984, **394**, 243.
7 K. Showalter, *Chem. Bri.*, 1995, **31**, 202.
8 P. Gray, J. F. Griffiths and S. K. Scott, *Proc. R. Soc. London.*, *A*, 1985, **397**, 21.
9 B. Johnson and S. K. Scott, *J. Chem. Soc.*, *Faraday Trans.*, 1990, **86**, 3701; *Chaos*, 1991, **1**, 387.
10 V. Petrov, E. Mihaliuk, S. K. Scott and K. Showalter, *Phys. Rev. E*, 1995, **51**, 3988.
11 A. M. Turing, *Phil. Trans. R. Soc. London*, *B*, 1952, **237**, 3395.
12 J. D. Murray, *Mathematical Biology*, Springer, Berlin, 1989.
13 *Chemical Waves and Patterns*, ed. R. Kapral and K. Showalter, Kluwer, Dordrecht, 1994.
14 J. D. Murray, *Scientific American*, 1988, **258**, 80.
15 *Oscillations and Traveling Waves in Chemical Systems*, ed. R. J. Field and M. Burger, Wiley, New York, 1985.
16 B. P. Belousov, *Sb. Ref. Radiat. Med.*, 1958, 145 (Russian); see also translation in ref. 15.
17 A. M. Zhabotinsky, *Biofizika*, 1964, **9**, 306.
18 A. T. Winfree, *Science*, 1974, **175**, 634.
19 J. P. Keener and J. J. Tyson, *Physica D*, 1936, **21**, 307.
20 L. Glass and M. C. Mackey, *From Clocks to Chaos*, Princeton University Press, Princeton, 1988.
21 S. Nettesheim, A. von Oertzen, H. H. Rotermund and G. Ertl, *J. Chem. Phys.*, 1993, **98**, 9977.

22 I. R. Epstein, K. Kustin, P. De Kepper and M. Orban, *Sci. Amer.* 1983, **248**, 96.
23 P. De Kepper and J. Boissonade, in ref. 15, ch. 7, p. 223.
24 I. R. Epstein, *Chemical and Engineering News*, 1987, **65**, 30.
25 A. Hodges, *Alan Turing: the Enigma*, Burnett Books, London, 1983.
26 J. Boissonade, E. Dulos and P. De Kepper, in ref. 13, ch. 7.
27 Q. Ouyang and H. L. Swinney, in ref. 13, ch. 8.
28 S. C. Müller, T. Plesser and B. Hess, *Science*, 1985, **230**, 661.
29 V. Castets, E. Dulos, J. Boissonade and P. De Kepper, *Phys. Rev. Lett.*, 1990, **64**, 2953.
30 Q. Ouyang and H. L. Swinney, *Chaos*, 1991, **1**, 411.

Faraday Discussion 100 *Celebration Paper; Presented 24th April,* 1995

Faraday Discuss., 1995, **100**, C61–C65

How Discoveries are Made, and Why it Matters

John C. Polanyi
Department of Chemistry, University of Toronto, Canada M5S 1A1

I thought that it would be appropriate for the only North American speaker to offer an expansive title. How are discoveries made, and why does it matter?

Michael Faraday admonished us to stick to a single idea per presentation, so I answer my question in one breath by saying that nobody knows how discoveries are made, and this matters because it dictates that we not attempt to govern science as if we did know.

Science has a vocabulary of supposed facts, true to varying degrees, and a grammer of accepted practices, applicable in various circumstances. But science is more than this grammar applied to this vocabulary, since it attempts not merely to say things, but to say things that are true. Computer science has, once again, come to our aid in making this point. A program called 'Racter' ('computerese' for 'Raconteur') applying impeccable syntax to an extensive vocabulary, wrote, on command, an essay on love, of which I quote only the opening:

> We will commence with a question: does steak love lettuce? This question is implacably hard and inevitably difficult...

There is a freshness of view here. It is marred, however, by a total lack of meaning. The computer, quite evidently, has not experienced love. Just as well, perhaps.

Michael Faraday, a penniless bookbinder's apprentice, was drawn to the door of this Institution in the year 1810 out of love for science. It is a passion that many in this audience share. A decade later, Faraday was living with his young wife in the attic over our heads and shortly after was lecturing to enraptured audiences in the lecture theatre where we meet.

His encounter with the reality of science was triumphant; he ranks with the greatest scientists of all time. But the course of his love affair with science was as turbulent and taxing as were many of the great loves of history. Though himself gentle and pious, he had quarrels with his principal scientific patron and model, Sir Humphrey Davy, he was accused of plagiarism, and he was for a while thrown out of his church. This last was only indirectly owing to science; as an eminent scientist he had accepted an invitation to dine with the Queen one Sunday, and when rebuked by his church for this frivolity he refused to repent.

These dramas were merely the outward evidence of an inward collision between his ideals and the reality of science. For scientific discovery, since it cannot be performed by computers, is pursued by human beings, with their normal strengths and weaknesses.

This is not, of course, said to belittle science. Quite the contrary. In the greatest adventures, of which science is one, weaknesses contribute to strengths and vice versa. Is conservatism, to give one example, a weakness or a strength? It is a strength because it permits the explorer to steer a steady course in the midst of confusion, but a weakness since it blocks the path to acceptance for the Michael Faradays of this world. Is opportunism a strength or a weakness? It is a weakness when it causes the navigator to vacillate between goals, and a strength when it makes him quick to move the tiller if disaster threatens.

It is the balance between these countervailing influences that determines success or failure in the making of scientific discoveries, and the optimal balance is achieved by being alert to as broad a range of life's experiences as possible.

Michael Faraday came to science with a different view. This great iconoclast was naively convinced that the best scientist was the most detached from the world. He expected others to be as dispassionate, as value-free in their judgments, as he supposed himself to be. All the while he was systematically placing high explosives beneath the statue of Isaac Newton.

Matter, he believed, could have—in addition to the mass that Newton had given it—both charge and magnetism. Space could transmit the resulting forces, whether it was full or empty. With exceeding modesty he proceeded on this basis to replace the Newtonian universe by the Faradayan. He did not insist on his view. He did not need to, since he knew he was right. Others, rather to his surprise, disagreed. But not all others, and not even the most influential of others. For, had this been the case we would not have heard of Michael Faraday, and science would not be the power in the world that it is.

Faraday proceeded in this fashion for over twenty years, as a scientist should and as only a great scientist can. Understandably, he became dispirited. Gradually he became a recluse. He had overreached himself scientifically, attempting things that would not in fact be possible for decades to come. But so sure was his insight into nature's symmetries, that this premature science was as close as he came to being wrong.

With one exception. In the matter of understanding the nature of the pursuit of which he was a master—science—he was surely at times mistaken. The scientific method, is not, however, one that we shall fully understand until the day when human consciousness, focusing on itself, reveals its own nature. Meanwhile we should be content to admit that science is a skill, a way of knowing, not very different from other ways of knowing developed over the millennia and transmitted imperfectly in the course of a lifetime by way of emulation.

Those we emulate are those who, in the course of history, have exhibited the skill we seek. Among those, happily, are some who are living. Which brings us to the Faraday Discussions of which the hundredth has recently taken place, and to the role of discussion in general as a means to knowing and therefore to discovery.

The word 'discussion' comes from the Latin *discussio*, to shake apart. Unprompted, one thinks of discussion, reasonably enough, as the systematic weighing of arguments against counter-arguments. And it is this element in the Faraday Discussions that tends to find its way into the bound volumes on scientists' shelves. But it is not predominantly for this that scientists come from around the world whenever a Faraday Discussion is announced. They come to be participants or witnesses at the central mystery of 'shaking apart'.

This is not such a solemn rite as it may appear, but more akin to an act of play. Kittens do the same with balls of wool. The function of this teasing is the great mystery.

A paradox that has been debated off and on for over two thousand years, is the way in which a problem solver proceeds purposefully and with a sense of direction toward an unseen goal. In my childhood this would have been referred to as the Indian Rope-Trick. The Indian holy man was alleged to throw a rope into the sky where it remained as a result of his faith, long enough for him to climb it if he hurried.

A more revealing description of this mystery is to say that the knotty problem is 'shaken apart' by being worried this way and that until it takes on a recognisable shape. We depend upon the breadth of experience of the onlooker as a means to recognising the shape, and then on his or her deductive skills in order to relate that shape back to its original form.

If this is, as appears likely, a crucial step in determining whether discoveries are made, it is not difficult to see, in the words of my title, 'why it matters'. It matters for two immediate reasons. In the first place this scenario illuminates the function of the *group*, and in the second it exposes to view the crucial role of the *individual.*

'Shaking apart' is a labour that can indeed be shared, for example between shaker and shook. Pattern recognition, by contrast, is indivisible. It takes place initially in a single prepared mind, culturally attuned to the pattern that is unfolding. Meanwhile the group can be expected to continue, if not redouble, its efforts to shake the problem and also its purported solution apart.

How are discoveries made? In these brief remarks, I have been stressing the anguish of the process, its subtlety, its dependence on a community of truth-seekers, and equally on a lonely believer who must be permitted to defy that community. If the subtlety of this process does not give pause to would-be administrators, there are two short answers that should be given to the question of 'how discoveries are made'; imperceptibly, and infrequently. I comment on these in turn.

Scientific discoveries occur as the result of the accumulation of scientific evidence. Only the trained eye can see them gradually taking shape. There is no set level of evidence at which the bell of proof rings. It follows that governments seeking to maximise the return in terms of discoveries per research expenditure, will be dependent on the judgment of scientists in applying the accountancy that good business practice requires. The only escape from this 'peer review' is into a never-never land of waste.

Discoveries are made gradually and seldom. There is no guarantee that this gradual process once begun, will continue. Lord Kelvin made the definitive statement on the prevalence of false pregnancies in science. He was asked what, in his long life in science, stood out most clearly. He thought for only a moment before replying; 'Failure'.

High-gain activities tend to be high risk, otherwise the world would be even more unstable than it is. But high-risk activities pose a problem for administrators, whose job it is to reduce waste and therefore risk.

Josef Stalin once summoned the Soviet Union's leading film-maker to the Kremlin to inquire how many films he had made in the past year. 'Eighteen' was the impressive response. 'And how many of those would you regard as having fully succeeded?' inquired the great one, ominously. 'Three' replied the film-maker (exaggerating, surely). 'In that case, next year you will make three films' Stalin said, terminating the interview.

So, which three? In the case of science it would, in my country, be those few discoveries that had most clearly contributed to what in previous years we called 'socio-economic well-being' and today, more desperate or grasping, we simply call 'wealth'. Governments have, however, had much more success applying this type of criterion in retrospect than in prospect.

In experiments extending over three-quarters of a century, a number of European nations proved to their people's satisfaction that they were quite unable to pick out in advance the businesses that would be most profitable. It is a curious fact, however, that if to the difficulty of predicting what businesses will yield profits, one adds the further imponderable of predicting what scientific discoveries will benefit what businesses, the whole undertaking appears once more to be, if not feasible at least saleable. Basic science must, after all, be accountable, and here is a method to make it seem so. The fact that it will replace real criteria of worth, namely scientific achievement, by what must needs be a judgment based on the seat of a bureaucrat's pants, tends to be overlooked.

The problem is more serious than I have yet suggested, for science today is threatened not only by misapplied *post hoc* judgments, but also by the *pre hoc* provisions of stylistic guidelines. We may not be alone in Canada in having Centres of Excellence at our universities, in which the criterion of excellence is clearly stated to count for twenty percent in the assessment. It is outweighed four to one by ill-defined (because indefinable) criteria having to do with the qualities of management, team-work and relevance.

And yet, world-wide, the lesson has been learnt, where commerce is concerned, that the risk-taker and not the legislator is the best decision-maker. Why is that not also thought to be true in the highly competitive market for new ideas?

Perhaps because governments have yet to be persuaded that the self-regulating system of the market-place for goods has relevance to that for ideas. Does science actually have a 'bottom line'? Are international prizes (to give an example) perhaps distributed by lot? Some would be loath to admit it. Is it the case?

Consider the record of twentieth century science, guided so far almost exclusively by scientists on the basis of scientific merit, in producing scientific advances of importance. The record is so spectacularly good as to raise fears that we cannot cope with the tidal wave of discovery. It would appear that the criteria by which the scientific community have been determining the profitability of scientific investments, are real and effective.

Is there nothing, then, in the management of science that needs improvement? Do we merely need to set our scientists free so that they can make discoveries where nature permits, rather than where governments propose?

No, there is more that we can do. We can further improve the communication between the basic scientist and the applied, the academic laboratory and the market-place for goods.

Much has, however, already been done on these lines in recent years. So much that we must now be careful not to let the applied sector set the agenda for the basic, or we shall find that we have spent the bulk of our intellectual capital in pursuit of near-term profit. The objective must be to use what we know, not to restrict what we know to what we know how to use.

Beyond using what we know lies a further objective; to use it wisely. This will require that we levy a tax of citizenship on knowledge. There must be an obligation on scientists (not all of them always, but many from time to time) to participate in the public debate by which our future is determined.

This is not because scientists are possessed of special wisdom, but because they have a special form of literacy, and every form of literacy carries with it obligations. Those who can read, for example, are required to bring warning signs to the attention of those who cannot.

I said that scientists do not have a special wisdom. They do, however, share distinctive values. These values are not so much those of scientists as, more broadly, of scholars. Nonetheless, they constitute a proud element in what we term the scientific method.

Science requires that no nation, religion, ethnic group, class or gender be regarded as having a monopoly on the truth. This remains an exceptionally civilised concept, viewed in the light of today's world.

A further principle, which flows from the first, is that of civility. Since the truth is a shared concern, there should be no disrespect implied in disagreement. Not only must that be acknowledged, it must be made evident in the way we treat one another.

John Tyndall and Michael Faraday, major scientific competitors, attacked one another's views forcefully in the *Philosophical Magazine* of 1855. Strikingly, the date-line for their conflicting letters was the same; The Royal Institution, London. Conscious of their differing viewpoints, Michael Faraday had invited Tyndall to join him at the Royal Institution some years earlier. There they shared a domicile and a lifelong friendship.

Commenting on such altercations among colleagues, Faraday wrote:

> Each one gives views and ideas new to the rest. When science is a republic, then it gains; and though I am no republican in other matters, I am in that...

If this were all we knew of how discoveries are made, it would be enough to explain 'why it matters'. The pursuit of science embodies a central requirement that we note with pride on the occasion of the hundredth Faraday Discussion: 'There must be discussion to show how experience is to be interpreted'.

The quotation this time is not, however, from Michael Faraday but from Faraday's contemporary John Stuart Mill writing 'On Liberty'. What we know about how discoveries are made, matters. It matters to science, and it matters still more beyond science.

Faraday Discussion 100 *Celebration paper; Presented 24th April,* 1995

Quotations from the writings of Michael Faraday are from *Faraday as a Natural Philosopher*, Joseph Agassi, Chicago University Press, 1971.

Faraday Discuss., 1995, **100**, C67–C81

Rearranging Atoms and Trapping Electrons

F. S. Dainton
House of Lords, London UK SW1A OPW

General

According to his biographer Boswell, the great lexicographer Samuel Johnson had 'a great love of chymistry which never forsook him' and he had 'an apparatus for chymical experiments' in the attic of his house. Johnson himself, in a self-portrait, records that sitting watching a distillation was 'the solace of his life'. That he was not just idly passing the time can be inferred from his 1755 definition of 'chymistry' as—

> An art whereby sensible bodies, contained in vessels ... are so *changed* by means of certain instruments, and principally fire, that their several powers and virtues are discovered, with a view to philosophy or medicine.

Johnson was of course quite right. Chemistry *is* entirely about change and it is studied both for the intrinsic knowledge it yields about these processes and for its useful application. Without such change there would be no chemistry. The change always involves the rearrangement of atoms or groups of atoms either within molecules, radicals or molecular ions as in unimolecular structural or stereoisomeric change, or decomposition into two or more fragments; or it occurs between two reagents in bimolecular processes which may involve combination (where in the case of atoms a third body is necessary to consummate the union) or disproportionation. Such reactions may occur in gases and liquids, at a gas/solid interface; they may be accelerated by catalysts and retarded by inhibitors (especially when chain reactions are involved). In general the rate of a reaction depends on the rate at which the reagent or reagents can receive any energy necessary for the change to occur. Moreover, the group of atoms so energised and called the *activated complex* exists for some time before changing to products. Of course, if during that period the activated complex can lose even a fraction of that energy it can thereby be prevented from reaction. This is an especially important consideration in gaseous reactions where intervals between collisions can, by simple changes of pressure, be adjusted to bracket the lifetime of the complex. If the activation energy is provided by the normal collisional processes, the rate of the individual chemical reactions will increase with increasing temperature according to the Arrhenius Law. But energy can also be supplied by absorption of light in mono- or multi-photonic processes, or by ionising radiation, ultrasound, electrical discharge or even, in the case of solids, friction. These are Johnsons's 'instruments' alternative to his 'fire'.

Given the central importance of an understanding of the rate and mechanism of chemical reactions to chemistry and to biology, and in optimising old and inventing new and useful chemical processes, it is not surprising that such a high proportion of Faraday Society Discussions before 1947, and in the 100 Discussions since that date, which we are celebrating today, should have been devoted to aspects of this general theme, to which I was to become irreversibly attached 62 years ago.

In the first week of the Michaelmas Term 1933 my tutor at Oxford, the late Sir Harold Thompson, whose Ph.D. thesis work was undertaken under the supervision of Haber in Berlin, that great nursery of physical chemistry, set me an essay on the kinetics of gaseous reactions to be produced within seven days, but gave me no advice as to how

I should set about the task. Aided only by the books in the College Library I did my best but 'Tommy' as he was universally known, made plain his disapproval of my offering. Only when I was stung to ask him 'How would you do it then?', did his criticism cease and he told me how an essay should be constructed and, most importantly, recommended reading books including Hinshelwood's, which I already knew, and also gave me a string of references to original papers of which the first one was to Volume 17 of the Transactions of the Faraday Society which contains, between pages 345 and 675, a record of the 1921 General Discussion entitled, 'Catalysis with special reference to Newer Theories of Chemical Action'. Reading it was a turning point for me because chemical kinetics came alive, the comments in the discussion had all the excitement and drama of a good novel, particularly when F. A. Lindemann (later Lord Cherwell) demolished the idea separately propounded by W. C. McC Lewis of Liverpool and Jean Perrin of Paris who claimed that first-order reactions were really photochemical, *i.e.* they gained the necessary activation energy from the ambient radiation. With typical acerbity Lindemann demolished this notion with two observations on pages 596 and 606 of the discussion. These are reproduced below;

> [p. 596] My difficulty (with Perrin's theory) may be recapitulated as follows. If reaction velocity is proportional to radiation density of a given wavelength then the inversion of sucrose (a case studied by Professor Lewis, which has a temperature coefficient of 4.13 for 10^0) must be determined by the radiation density of 1.05 μ. Irradiation density of this wavelength is 5×10^{13} times greater in sunlight than in the dark yet the reaction proceeds at approximately the same rate.
>
> [p. 606] I am very much interested to hear that Professor Lewis thinks we might get a reaction velocity 50 billion times as high as that usually observed if the solution is exposed to sunlight in a thin film [because of the high absorption coefficient for light of 1.05 μ wavelength; FSD comment]—I hope he will try the experiment. Sucrose should be inverted, if we were right, within an explosive velocity when passed in a fine spray out of an ordinary watering can.†

Later in the discussion he made the prediction that unimolecular reactions would display first-order kinetics at high pressures, changing to second-order at low pressures, because of the finite lifetime of the activated complex before reaction, although, of course, he did not use that terminology. The exact form of his prophetic words was

> [p. 599] Taking the simplest possible case, a dissociation, we may assume that dissociation can only take place if the centrifugal force due to rotation is above a certain value ... it leads to Arrhenius' expression for the temperature coefficient ... if we assume that a molecule even under a large centrifugal force does not dissociate instantly but on average only after a time greater than that necessary to re-establish the Maxwell distribution.
>
> If this view is true, of course, there must be a limit beyond which the experimental law that reaction velocity [presumably expressed as a percentage decomposition per unit time; FSD] is independent of concentration, ceases to hold, namely at concentrations at which the time between collisions is of the same order as the life of the molecule after attaining its critical velocity. At very low pressures, especially in simple slow reactions with low temperature coefficients, such deviations may perhaps be observed and would form an interesting check on the theory. In liquids, of course, no such effect would occur, since in these the Maxwell distribution must be rapidly re-established with the solvent molecules whatever the concentration of the solute.

† T. W. J. Taylor demonstrated experimentally that sucrose inversion proceeded at the same rate in a fine spray as in bulk solution.[2]

The full airing of this concept, its treatment by Eyring and Polanyi, had to wait 16 years until the General Discussion at Manchester was reported in Transactions of the Faraday Society, vol. 34, pp. 1–268. It has always been a source of great regret to me that I was unable to be present at this immensely exciting meeting. Instead I was in Paris where I had been celebrating being accepted by and actually paid to do research in Cambridge. I was totally unaware that the meeting was to be held and perhaps that is a comment on the attitude of chemists in both Oxford and Cambridge where transition state theory was to my knowledge never mentioned in lectures. Indeed, I remember introducing the matter for the first time in my Part II lectures in Cambridge about 1940, much encouraged by Dan Eley, then a young Manchester graduate and post-doc in the Colloid Science Laboratory and who was well versed in the theory. A year later when the classic book by Eyring, Glasstone and Laidler[3] was published and the theory gained world currency, that Manchester Discussion and the Discussion on Free Radicals[4] which had taken place four years earlier, and on Polymerisation[5] the same year, were to influence my own research just as much as Semenov's book on Chain Reactions[6] which was published in 1935.

Tommy also directed me to many other seminal ideas often debated in Faraday Society General Discussions; and always when I read the printed proceedings I could sense the enthusiasm, excitement sometimes amounting to passion, as the discussants debated, criticising ideas and rebutting the counter-arguments of their opponents. Real drama is to be found in the pages of the Discussions and to read them the *dramatis personae* come alive as personalities. In some cases, *e.g.* the Photochemical Processes Discussion in Liverpool in 1931,[7] the Society published a photograph of those attending which enable physiognomy and personality to be conjoined.

The first General Discussion I attended was whilst a research student in Cambridge. My supervisor R. G. W. Norrish drove me to the Oxford Meeting on Luminescence. Sad to relate, this fell short of my expectations but whether this was actually less scientifically or chemically significant or whether I was preoccupied with the hazards inherent in being a passenger in the car driven back to Cambridge by my supervisor, I do not know, but certainly I did not give my full attention to the meeting.

The first meeting I attended in my own right (and from the record I seem to have been very talkative!) was in September 1945 at University College London.[8] The subject was 'Oxidation'. In some ways it was more a celebration of the end of the war, a reunion of British chemists with their European colleagues after six years separation. I remember a very friendly atmosphere which is almost evident in the published photograph. It was also the last meeting before the New Series of which this meeting today celebrates the 100th.

From Reaction Kinetics to Reaction Dynamics: The New Series

Of the 84 General Discussions in the years 1903–46 inclusive, about 14% were devoted to reaction kinetics whereas some 20% of the New Series Discussions fall into this category. These percentages are the highest of any conventional sub-division of physical chemistry and, bearing in mind that in making these counts I have excluded Discussions dealing with reactions at electrodes, at certain surfaces (including corrosion), and in the atmosphere, they testify to chemists' perception throughout the 20th century of the importance of gaining an understanding of exactly what happens to constituent atoms as the reactant molecule(s) passes into the transition state and so on to one of what may be several possible final states; and how this process in influenced by the state of aggregation of the reacting substance(s), how much excess energy is needed to ensure any particular outcome and the means by which it is acquired.

The New Series Discussions show clearly the enormous progress towards this goal which has been achieved by experimentalists and theorists since 1947. Discussion 2 'The

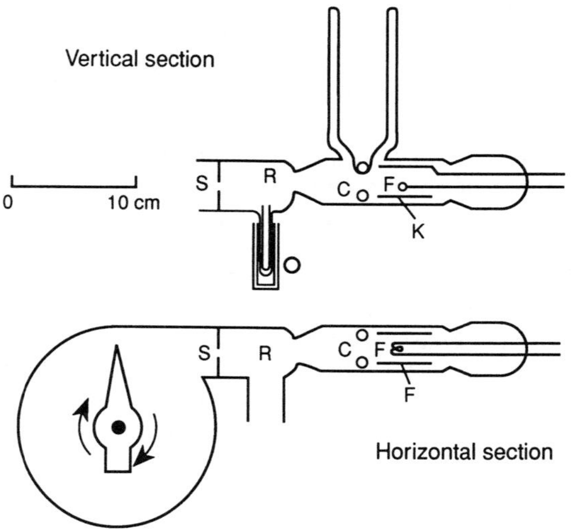

Fig. 1 Sketch of the apparatus used by Bull and Moon[11] to accelerate carbon tetrachloride molecules mechanically by a fast rotor blade to velocities at which they would react with caesium atoms in chamber R after passage through the slit S

Labile Molecule'[9] took place in September of that year and my clear recollection, borne out by the record, is that, with the possible exception of Page and Denbigh's paper (p. 263) on a stirred flow reactor for solution reactions and Robb and Melville's description (p. 132) of a variable pathlength reactor for studying rates of reaction of H atoms in the gas phase, all the other papers relied on entirely conventional and long-established experimental methods. But $6\frac{1}{2}$ years later in the Discussion entitled 'Study of Fast Reactions' held at Birmingham,[10] the situation was transformed. Not only was the time resolution enormously increased by studying the relaxation of chemical systems momentarily perturbed by a short light flash, shock wave, or temperature jump *etc.* using fast electronic detection and recording, but there is a clear pointer to a future in which the kinetic energy of reactant molecules could be precisely controlled. I refer to the brief but seminal paper of Bull and Moon[11] (p. 54) in which, to quote their words, 'gas molecules at low pressure are swept up by the tip of a rotating blade and flung off with the speed of the tip added to their thermal velocities, ... for a molecular weight of 200, 10^5 cm s^{-1} corresponds to 10^5 J = 24 kcal mol^{-1}. Fig. 1 is a diagram of their simple apparatus.

The germ of this simple idea is to be found in the letter Philip Moon wrote to three Birmingham colleagues on 7 February 1945 and which is reproduced in Fig. 2. At that time all four were involved in the work of the Directorate of Tube Alloys, the UK end of the A-Bomb project and like many others they were interested in possible ways of separating uranium isotopes including electromagnetic, diffusional (normal and thermal) and centrifugal methods. It is not inappropriate to regard Moon's idea as a 'spin off' (in more than one sense) of the last method!

Forty years after this Discussion, in memorable Closing Remarks (p. 465) to the 84th New Series Discussion entitled 'Dynamics of Elementary Gas Phase Reactions' also held in Birmingham,[12] Dudley Hirschbach (in the presence of Philip Moon) very appropriately commemorated in verse Bull and Moon's experiment.

Hey, diddle, diddle: the carbon tet riddle
The Bull rolled round with the Moon!
The little gas puffed to move so fast
But the alkali speared it soon.

Alpha and beta, E-prime coupled theta
Reaction urged by a harpoon
A rainbow† was left to show for it all
And the salt ran away with the tune.

As I look back over 58 years to my own first published research[13] on the photochemistry of the alkyl nitrites, I marvel at the progress which has been made in both the understanding and the useful application of reaction kinetics. To take the latter point first, my own work, exceptionally crude by modern standards, did establish that the NO moiety of the nitrite was detached in the primary act and abstracted a hydrogen atom either inter- or intra-molecularly. Sir Derek Barton wrote to me about ten years ago stating that this finding was a powerful influence in causing him to invent his eponymous reaction which has found such wide application in synthetic organic chemistry. To learn how much of the detailed nature of the primary photochemical act in nitrites is now known one need only look to Professor R. N. Dixon's Liversedge Lecture[14] and to the paper of Hippler and Pfab.[15] Even the rate of the reverse reaction of the photolysis of nitrous acid ($NO + OH \rightarrow HONO$) can now be measured down to 23 K[16] and the results compared with those calculated by Troe's general semi-empirical but remarkably useful calculations.[17] Progress in the experimental investigation of gaseous bimolecular abstraction reactions, *e.g.* $H + HOD \rightarrow H_2 + OD$ or $HD + HO$ has been equally dramatic. It is now possible to see the effect of selective vibrational excitation of the OH and OD bonds in water molecules on the rate of bond breaking in this reaction, a process wittily and evocatively called by Dick Zare 'choreographing chemical reactions'.[18]

Someone of my age, introduced to gas kinetics *via* the 1921 Faraday Discussion, finds these technically brilliant experiments featuring in so many New Series Discussions quite fascinating and I do wish to congratulate those younger colleagues most warmly on what they have accomplished. By extraordinary extensions of time resolution in experimental technique‡ and by theoretical and computational advances they have exposed the detailed dynamics and quantum states of atoms and molecules during reaction to a degree of precision unimaginable when the New Series began. Would that my duty ended there! But alas, Professor Simons actually invited me in his letter of 27 May 1994 to talk 'with special reference to the influence on the development of the research areas to which you have contributed'. Hence the title of this talk, 'Rearranging Atoms and Tracking Electrons', which represent two of the major themes in my research.

Rearranging Atoms

I have time to consider only two types of reaction under this heading. Although I have selected them from entirely different fields and apparently more complex than the simple reactions studied by sophisticated laser techniques to which I have just referred, I did so because the central issue in each concerns the transition state. In both cases conclusions of considerable practical value have ensued.

The first set of reactions concerns the photochemical exothermic addition of chlorine to an unsaturated organic compound, A, which may be CO or an ethene. As might be

† Here *rainbow* is a reference to Professor Hirschbach's own work on the rainbow structure in the product angle and energy distribution of the $M + CCl_4$ reaction.

‡ Made possible primarily by the invention of lasers which deliver monochromatic radiation of any desired wavelength, thereby enabling the absorbing molecules to be brought to specific quantum levels in single or multiphoton absorption processes, which can also emit light of such high intensity that very low pressures of absorbing gases can be used thereby avoiding collisional deactivation and of pulse lengths so short as to be comparable to the time taken by an atom in a molecule to execute a single step of intramolecular motion so that such motions can be 'photographed'. In addition, time-resolved laser-induced fluorescence has proved to be a very effective probe of some of these processes.

Prof. M. L. Oliphant
Prof. W. N. Haworth
Prof. R. E. Peierls.

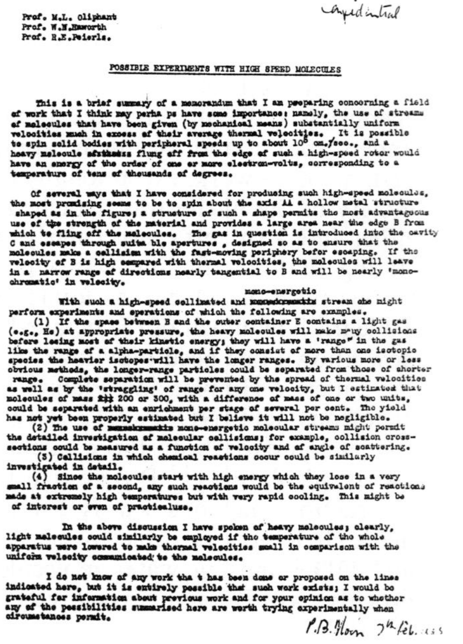

POSSIBLE EXPERIMENTS WITH HIGH SPEED MOLECULES

This is a brief summary of a memorandum that I am preparing concerning a field of work that I think may perhaps have some importance, namely, the use of streams of molecules that have been given (by mechanical means) substantially uniform velocities much in excess of their average thermal velocities. It is possible to spin solid bodies with poripheral speeds up to about 10^5 cm/sec., and a heavy molecule flung off from the edge of such a high-speed rotor would have an energy of the order of one or more electron-volts, corresponding to a temperature of tens of thousands of degrees.

Of several ways that I have considered for producing such high-speed molecules, the most promising seems to be to spin about the axis AA a hollow metal structure shaped as in the figure, a structure of such a shape permits the most advantageous use of the strength of the material and provides a large area near the edge B from which to fling off the molecules. The gas in question is introduced into the oavity C and escapes through suitable aperatures, designed so as to ensure that the molecules make a collision with the fast-moving poriphery before escaping. If the velocity of B is high compared with thermal velocities, the molecules will leave in a narrow range of directions nearly tangential to B and will be nearly 'monochromatic' in velocity.

With such a high-speed collimated and mono-energetic stream one might perform experiments and operations of which the following are examples.

(1) If the space between B and the outer container E contains a light gas (e.g., He) at appropriate pressure, the heavy molecules will make many collisions before losing most of their kinetic energy, they will have a 'range' in the gas like the range of a alpha-particle, and if they consist of more than one isotopic species the heavier isotopes will have the longer ranges. By various more or less obvious methods, the longer-range particles could be separated from those of shorter range. Complete separation will be prevented by the spread of thermal velocities as well as by the 'straggling' of range for any one velocity, but I estimated that molecules of mass 200 or 300, with a difference of mass of one or two units, could be separated with an enrichment per stage of several per cent. The yield has not yet been properly estimated but I believe it will not be negligible.

(2) The use of mono-energetic molecular streams might permit the detailed investigation of molecular collisions, for example, collision cross-sections could be measured as a function of velocity and of angle of scattering.

(3) Collisions in which chemical reactions occur could be similarly investigated in detail.

(4) Since the molecules start with high energy which they lose in a very small fraction of a second, any such reactions would be the equivalent of reactions made at extremely high temperatures but with very rapid cooling. This might be of interest or even of practical use.

In the above discussion I have spoken of heavy molecules, clearly, light molecules could similarly be employed if the temperature of the whole apparatus were lowered to make thermal velocities small in comparison with the uniform velocity communicated to the molecules.

I do not know of any work that has been done or proposed on the lines indicated here, but it is entirely possible that such work exists; I would be grateful for information about previous work and for your opinion as to whether any of the possibilities summarised here are worth trying experimentally when circumstances permit.

[signed; P. B. Moon 7th Feb. 1945]

Fig. 2 Copy of a letter from Philip Moon (see inset) to his colleagues Professors Marcus Oliphant and Rudi Peierls of Birmingham University Physics Department and Professor Sir W. N. Haworth of the Chemistry Department on 7 February 1945. I am indebted to Professor J. P. Simons, currently President of the Faraday Division, for supplying me with this copy.

expected, the quantum yield is very high and inversely proportional to $\sqrt{I_{abs}}$. They are therefore unbranched chain reactions terminated quadratically and one chain carrier must be a chlorine atom. The essential steps must therefore be:

Initiation $Cl_2 + h\nu \rightarrow 2Cl$ (1)

$Cl_2 + A \leftrightarrow ACl$ (2)

Propagation

$ACl + Cl_2 \rightarrow ACl_2 + Cl$ (3)

Quad. Term^n by 2 ACl, 2Cl + M or ACl + Cl

In principle, all the rate constants of propagation and termination can be measured by studying the effect of intermittent illumination using a rotating sector to chop the beam.[19] However, in the case of A = CO and C_2Cl_4 there were deviations from the expected simple kinetics which could only be explained by taking into account the reversibility of the first propagation step and in the case of the phosgene reaction the equilibrium constant was readily evaluated as [CO][Cl]/[COCl] = $10^{2.8}$ exp(-6.3 kcal/RT). Because in the cases of A = $C_2H_{4-X}Cl_X$ the strength of the C—Cl bond is much larger than in COCl, no reversal of reaction (2) is detectable at room temperature. However, this bond strength diminishes the more heavily chlorinated the radical and has become sufficiently small for the reverse of reaction (2) to cause the rate of photochlorination of tetrachloroethene to pass through a maximum at 203 °C (see Fig. 3).[20] For less chlorinated ethenes this inversion temperature must be higher still. However, to our surprise and great puzzlement we found that the room temperature photochlorination of either *cis* or *trans* $(CHCl)_2$, where the thermochemistry predicts that $E_{-2} \geqslant 22$ kcal, was accompanied by rapid geometrical isomerisation (Fig. 4). At the chlorine pressures used the isomerisation rate was between 20 and 80% of the photochlorination rate and this ratio was independent of temperature.

It took considerable time before we realised that the answer to this riddle lies in the fact that the lifetime of the transition state $(CHCl)_2Cl^{\ddagger}$ is comparable to the collision interval at the gas pressures in the working range from 0.01 to 1.0 atmosphere and that the only possible fate of this transition state if it is not deactivated by or reacts with on collision any other gas molecule present, is to revert to a chlorine atom and the *cis* or

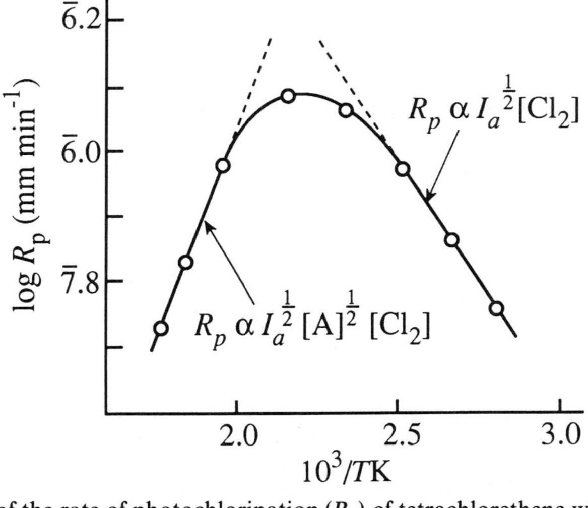

Fig. 3 Variation of the rate of photochlorination (R_p) of tetrachlorethene with temperature[20]

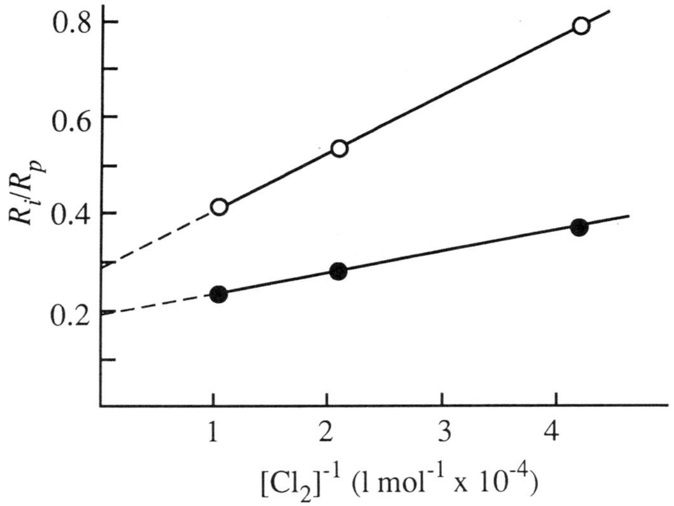

Fig. 4 Variation of the ratio of the rate of geometrical isomerisation (R_i) to the rate of photochlorination (R_p) for 1,2-dichloroethene as a function of chlorine pressure at 65 °C. Open circle denotes *cis* isomer and closed circle *trans* isomer.[21]

trans form of $(CHCl)_2$.[21] The potential-energy diagrams for the propagation steps are illustrated in Fig. 5 and make clear the possibilities for $ACl^{\ddagger}$ and ACl^{0}.

Fates of $ACl^{\ddagger}$

$ACl^{\ddagger} \rightarrow Cl + $ *cis* or *trans* A: no activation necessary

$ACl^{\ddagger} + M \rightarrow ACl^{0}$, *i.e.* quench. $M = He, CO_2, SF_6$

$ACl^{\ddagger} + Cl_2 \rightarrow$ quench or $Cl_3 +$ *cis* or *trans* A: no activation necessary

$\rightarrow ACl_2 + Cl$: no activation necessary

$ACl^{0} + Cl_2 \rightarrow ACl_2 + Cl$: requires activation

The corollaries to this mechanism are (a) that the isomerisation rate during photochlorination in the gas phase is unaffected by change of temperature but diminished as $[Cl_2]$ increases[22] or gases such as He, CO_2 or SF_6 are added[21] and (b) that such isomerisation cannot occur in the liquid phase. Those predictions are true[23] and illustrate both Lindemann's ideas of 42 years earlier as well as the later ideas of alternative product outcomes from the 'agitated dance of the atoms' in the transition state.

It is surprising that this explanation did not occur to us earlier because Professor Ivin and I and our colleagues had already studied the liquid phase addition polymerisation of various unsaturated compounds proceeding *via* both free radical and ionic mechanisms. In this case the propagation step is

$$m_j^* + m_1 \rightarrow m_{j+1}^*, \; k_p$$

where m_1 denotes monomer and m_j^* a thermally equilibrated radical or ion containing j units of monomer. For $j \geqslant 10$ the increments in thermodynamic functions for the reaction

$$m_1 \rightarrow 1/jP_j$$

are constant and equal to those for propagation. This simple fact together with the recognition that the propagation reaction is reversible but in condensed systems the activated complex is quenched rapidly, has consequences of great practical utility as well as elegance in linking the kinetics of polymerisation with the thermodynamic properties

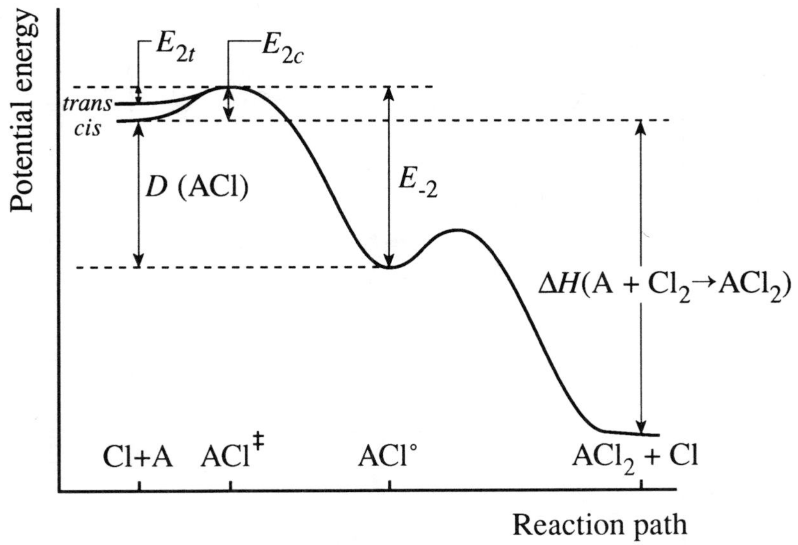

Fig. 5 Schematic potential-energy changes for the successive propagation steps in the photochlorination of $(CHCl)_2$. The first maximum denotes $ACl^‡$.

of the overall change. For example, because $E_d = E_p - \Delta H^0$ for exothermic, exentropic polymerisations there should be a sharply defined *ceiling* temperature, $T_c = \Delta H^0/(\Delta S^0 + R \ln[m_1])$, above which polymerisation cannot occur and which is independent of the chemical nature whether free radical, cation or anion, of m_j^*. For endothermic, endentropic systems there is of course a *floor* temperature. Conditions for achieving polymerisations at desired rates can therefore be predicted from measured (calorimetric)

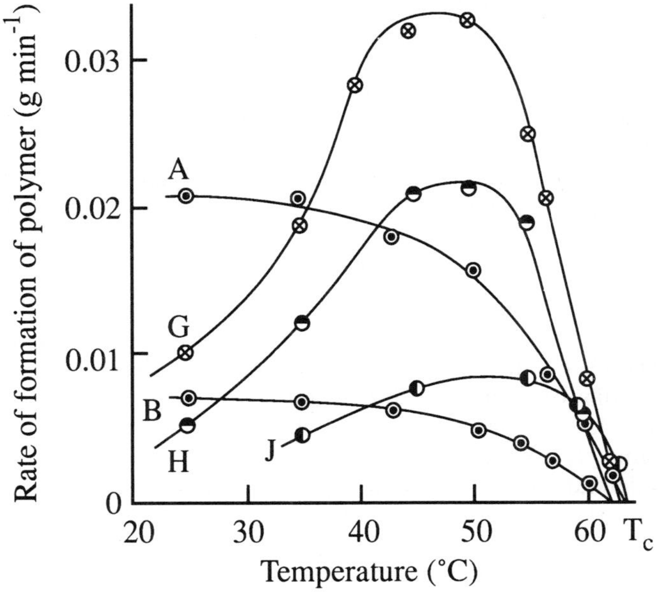

Fig. 6 The rate of formation of polysulfone from 9.1 mol% of but-1-ene in liquid sulfur dioxide as a function of temperature and for which $T_c = 61\,°C$. Curves A and B refer to photochemical initiation at different intensities, G and H to catalytic initiation by silver nitrate at different concentrations and J to initiation by benzoyl peroxide.[24]

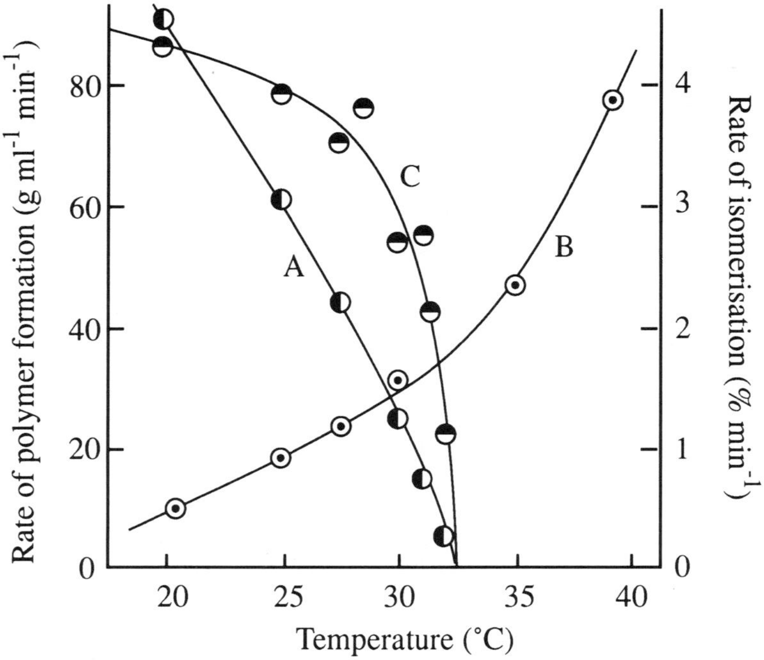

Fig. 7 The variation with temperature of the rate of polymerisation of *trans* but-2-ene in liquid sulfur dioxide (curve A photochemical and curve C catalysed by benxoyl peroxide) and the variation of the rate of geometrical isomerisation of the same monomer[24] (curve B).

or *a priori* calculated values of ΔH^0 and ΔS^0. Furthermore, when m_1 is a geometrical isomer the polymerisation is accompanied by isomerisation, the rate of which provides a direct measure of the depropagation rate and, of course, has no ceiling temperature. These two points are illustrated in Fig. 6 and 7. There are many other ramifications to this simple idea which were exploited by Professor Ivin and myself in a much quoted review.[24] Perhaps the most striking example known to every school pupil is that of liquid sulfur which, when heated to 159 °C, suddenly changes to plastic sulfur. Because the polymerisation of S_8 to S_{8n} has positive entropy (7.4 cal deg^{-1} mol^{-1}) and enthalpy (7.4 kcal mol^{-1}) changes the *ceiling* temperature is replaced by a *floor* temperature below which polymerisation will not occur.

Trapping Electrons

This story has its origins nearly 50 years ago in a need to know whether the NRX reactor at Chalk River Canada, the first to be built outside the USA, would suffer any deleterious effects when water in contact with other materials was subjected to a high flux of ionising radiation. One very important aspect of this, and an added personal motivation, was to know what happens to liquid water under radiation, because living cells are approximately 85% aqueous, and therefore such knowledge has obvious relevance to the understanding of radiotherapy and radiation-induced disease. It was, of course, well known that whether the radiation is corpuscular, *e.g.* a neutron flux or α or β rays, or high-energy quanta, *e.g.* X- or γ-rays, it always results in a loss of energy by decelerating fast charged particles which leave in their wake ionised and electronically excited molecules and that these intermediate species or their rapidly formed derivatives are the agents of chemical change in any solute. Although the effects of such radiation on aqueous systems were evident before 1900, it was common practice for the first half of this century to designate such intermediates collectively as 'activated' water.[25] In

1944 the late J. Weiss[26] produced convincing inferential evidence that this was in fact the radical pair H + OH, an idea first suggested by O. Risse[27] in 1929 but largely ignored.

If this were true, then an aqueous solution of a vinyl compound when irradiated should yield a polymer in large quantities (which it always does) and if the solvent is D_2O the end groups should be C—D and C—OD links (which we demonstrated by infrared spectroscopy). Furthermore, thermally or photochemically generated OH radicals or H atoms in water should each mimic in some degree half the effects of ionising radiation. This too is broadly the case and therefore it is not surprising that at the Leeds 1952 General Discussion 'Radiation Chemistry'[28] it was not questioned that the primary radical species in the radiation chemistry of aqueous systems were H and OH.

We therefore decided to investigate the reactions of OH and H in water (Fig. 8). The photochemical sources we used to generate H atoms in water were divalent cations like Cr^{2+}, V^{2+}, Fe^{2+}, Mn^{2+}, Co^{2+} or anions such as I^-, $Fe(CN)^{4-}$ which we assumed involved transfer of an electron to an adjacent water molecule which immediately dissociated to form $OH^-(aq) + H$, as Franck and Haber[29] had suggested 30 years earlier. Using the ideas of Evans and Polanyi to which I have already referred we drew probable potential-energy diagrams to represent these changes and predicted that the divalent cations listed above will have a minimum quantum for the process

$$M^{2+}(aq) + h\nu \rightarrow M^{3+}(aq) + OH^-(aq) + H$$

which, because the ions involved have similar radii and charge and therefore thermodynamic functions, should increase linearly with the redox potential of the relevant couple, which the experimental data in Fig. 9 show to be the case.[30]

All seemed well in the best of all scientific worlds and we happily pursued this kind of idea. However, given the existence of blue stable solutions of alkali metals in liquid ammonia, amines and cyclic ethers attributed in part to solvated electrons, the possibility could not be totally dismissed that, as had been suggested, the electron produced in water by radiation might be solvated for a time long enough before changing to the

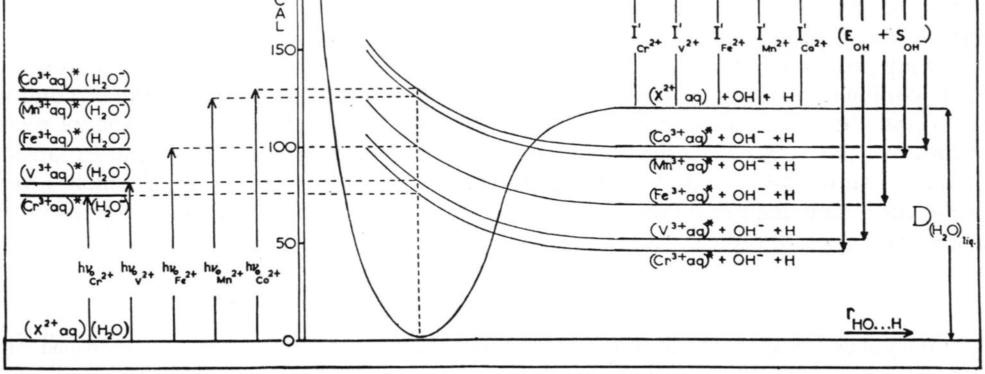

Fig. 8 The role of the ionisation potential in the photochemical reaction $M^{2+}HOH + h \rightarrow M^{3+}OH^- + H$[32]

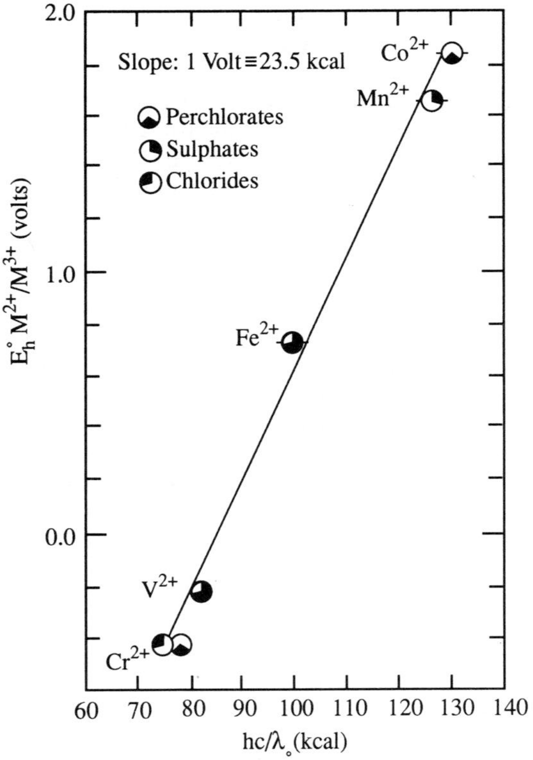

Fig. 9 The relationship of the long-wavelength limits, λ_0, of the electron-transfer spectra of the divalent ions of the first long period to their redox potential

product OH^-(aq) + H to be justifiably regarded as a distinct chemical entity. As I shall never forget, in mid October 1952, in the course of a colloquium at the Argonne National Laboratory, I mentioned this possibility only to be told in no uncertain terms by Professors James Franck and Bob Platzman, sitting in the front row, that such an aquated electron could not live longer than 10^{-11} s, the dielectric relaxation time of water. I was in no position to contradict such eminent theoreticians!

But within the next two decades as the General Discussion 'Fundamental Processes in Radiation Chemistry'[31] illustrates, the first to be held in the USA, the existence of e_s^- was never questioned because it was proved that however generated, by ionising or ultraviolet radiation, an electron could find a range of 'holes' of suitably oriented clusters of water molecules and survive for about 0.5 milliseconds, long enough for the following properties to be determined: optical absorption, thermodynamic functions, p*K* of its conjugate acid the hydrogen atom, its reduction potential, its equivalent conductivity and therefore diffusion coefficient and Stokes radius, the ionic atmosphere relaxation time, the average coefficients of isothermal compressibility and thermal expansion of the solvation holes and the rates of numerous important reactions of which a very important one is the rapid conversion by reaction with dissolved nitrous oxide to nitrogen and a hydroxyl radical.†

$$e^-(aq) + N_2O \rightarrow N_2 + OH + OH^-(aq)$$

† Many of these reactions are diffusion controlled. This is true when a water soluble vinyl compound is the solute. It first captures the hydrated electron to form the radical anion m^- and later reacts with a water molecule to form $Hm^\bullet + OH^-$. It was satisfyingly ironic to be able to show 20 years later that our evidence for H atom formation in radiolysis was in fact not a proof.[33]

Furthermore the electron can be trapped in cavities of varying depth in aqueous glasses thus enabling measurement of the electron spin resonance absorption and also the formation of trapped dielectrons. Perhaps the most striking property of such trapped electrons is that by absorption of light they can become mobile so that the absorption spectrum of an irradiated aqueous glass can be manipulated towards the blue by illuminating with red light and towards the red by illuminating with a monochromatic blue light and both these processes can be associated with electrical conductivity.[32]

Envoi

So far in this paper I have tried to illustrate the vast progress which has been made in understanding the intimate details of the 'dance of the atoms' which takes place in any chemical change, a process which Samuel Johnson correctly regarded as being at the very heart of his 'philosophy' of 'chymistry'. This achievement has been made possible primarily because, almost exclusively, academic chemists wanted to satisfy their curiosity about this aspect of 'natural knowledge'. Those chemists in the United Kingdom who contributed to this increase in knowledge received their financial support for this work largely from the public purse, through agencies such as the research councils and the University Grants Committee and its successor bodies. That they enjoyed their work, as exciting as any pursuit of the unknown, and that they are grateful for that financial support goes without saying. But in the last quarter century pressure from government on scientists to direct their energies and talents more towards the solution of practical problems has become evident and the 'customer–contractor relationship' became part of the jargon of science policy and politics. Its principal advocate in 1971 was the Lord (Victor) Rothschild and a popular conception of what this meant was neatly captured in the cartoon which appeared in the Times Higher Educational Supplement on the 19th of July 1972, reproduced here as Fig. 10. Here the customer is the state enbodied by the then Prime Minister, Edward Heath, whilst the contractor is represented by myself as Chairman of the body responsible for advising the allocation of the resources to the research councils and to that extent charged to deliver the goods to the benefit of the public.

Of course this customer–contractor relationship is entirely appropriate to the kind of enquiry in which there is a real customer for knowledge which is needed quickly to achieve a practical end, usually in the shortest possible time. Such work relies on the application of existing knowledge or some slight extension of that knowledge which it is known can be obtained by well understood methods. It is what I have called *tactical work* to distinguish from *basic* or purely curiosity-oriented work and from the intermediate kind of work called *strategic*.

Making these distinctions is not easy and unsurprisingly the pendulum of opinion as to which is the most important frequently has too wide an amplitude in its swings. Before the 1970s public money spent on any science was generally thought to be self-justifying. Now we have to be more discriminating with an eye on goals of 'wealth creation' and 'improvement of the quality of life', and we have to learn to use tools such as 'technology foresight' to guide decisions. At the end of the day spent celebrating physical chemistry I believe it would be salutary to look back and see what practical applications may have followed from increases in fundamental, basic knowledge in the areas in which each of us has been concerned. I think we should be very pleasantly surprised at their extent.

In my own case I can point to many practical gains which were unforeseen when my work was undertaken. I have already mentioned Sir Derek Barton's use of knowledge of the photolysis of the alkyl nitrites. Our curiosity to know whether the ceiling temperature in addition polymerisation was real, led us quickly to the thermodynamic criteria to predict the conditions under which unsaturated or cyclic compounds would

Fig. 10 Original of cartoon by Horner which appeared on front page of *Times Higher Educational Supplement* two days after the publication of the White Paper 'Framework for Government Research and Development' (Cmnd 5046) on 19th July 1972. Given to F.S.D. by the *Times* who stated 'Horner drew it in half an hour after the briefest explanation of what the White Paper is all about'. Horner had only seen press photographs of Rothschild and F.S.D.

polymerise and the manner and mechanism of degradation by various possible treatments of any polymers so formed.

Other benefits have accrued from curiosity driven work on free radical reactions in solution initiated by heat, light or radiation. I mention only a few, *e.g.* (a) the improvement of radiotherapy by enhancing curative power and minimising side effects, for example by the use of radiosensitisers and the development of photodynamic therapy; (b) knowledge in this area has contributed significantly to a major revolution in thought about the now widely acknowledged role of free radicals in metabolic processes during both health and disease; (c) radiosterilisation of medical equipment such as syringes, bandages *etc.* is now commonplace as is the protection of the public against anthrax by irradiation of goat hair and sterilisation of many plant and animal products; and (d) as a final example in this very brief list which combines the polymer chemistry of the first section with the radiation chemistry of the second, let me simply show you a useful gizmo for the would-be electrician or handyman. It is based on the simple principle of being able to cross-link by radiation, and therefore make more rigid, an expanded form of the polymer and then allowing it to relax under very gentle heat to make a tight-fitting insulator.

The other speakers today could similarly testify to the utility of their areas of work. I think these examples should suffice to illustrate the maxim that at the time research is undertaken or a new idea developed, future benefits are difficult to predict but one can be absolutely confident that a means will be found by government, through taxation, of recovering its investment! This is the message of encouragment which an octogenarian bequeaths equally to young enthusiastic scientists and to your governmental paymasters.

References

1 F. A. Lindemann, *Trans. Faraday Soc.*, 1921, **17**, 590; 599; 606.
2 T. W. J. Taylor, *Nature (London)*, 1921, **108**, 210.
3 S. Glasstone, K. J. Laidler and H. Eyring, *Theory of Rate Processes*, McGraw Hill, New York, 1941.
4 *Trans. Faraday Soc.*, 1934, **30**, 1–248.
5 *Trans. Faraday Soc.*, 1936, **32**, 1–422.
6 N. Semenoff, *Chemical Kinetics and Chain Reactions*, Clarendon Press, Oxford, 1935.
7 *Trans. Faraday Soc.*, 1931, **27**, 357–572.
8 *Trans. Faraday Soc.*, 1946, **42**, 99–398.
9 *Discuss. Faraday Soc.*, 1947, **2**, 1–409.
10 *Discuss. Faraday Soc.*, 1954, **17**, 1–236.
11 T. H. Bull and P. B. Moon, *Discuss. Faraday Soc.*, 1954, **17**, 54.
12 *Faraday Discuss. Chem. Soc.*, 1987, **84**, 465.
13 H. W. Thompson and F. S. Dainton, *Trans. Faraday Soc.*, 1937, **33**, 1546.
14 R. N. Dixon, *Chem. Soc. Rev.*, 1994, 375.
15 M. Hippler and J. Pfab, *J. Chem. Soc., Faraday Trans.*, 1992, **88**, 2109.
16 P. Sharkey I. R. Sims, I. W. M. Smith, P. Bocheral and B. R. Rowe, *J. Chem. Soc., Faraday Trans.*, 1994, **90**, 3609.
17 J. Troe, *J. Chem. Soc., Faraday Trans.*, 1994, **90**, 2303.
18 R. N. Zare, *Jahrbuch der Akademie der Wissenschaften in Göttingen*, 1993, 52.
19 For a summary of values for $A = C_2H_{4-n}Cl_n$, $n = 0$–4, see P. B. Ayscough and F. S. Dainton, in *Photochemistry and Reaction Kinetics*, ed. P. G. Ashmore, F. S. Dainton and T. M. Sugden, Cambridge University Press. 1967, pp. 71–82.
20 Ref. 19, Fig. 2, p. 80.
21 See F. S. Dainton *et al.*, ref. 30 in ref. 19 above.
22 P. B. Ayscough, F. S. Dainton and B. Fleischfresser, *Trans. Faraday Soc.*, 1966, **62**, 1838.
23 M. L. Poutsma and H. L. Hinman, *J. Am. Chem. Soc.*, 1964, **86**, 3807.
24 F. S. Dainton and K. J. Ivin, *Quart. Rev.*, 1958, **12**, 61.
25 C. B. Allsopp, *Trans. Faraday Soc.*, 1944, **40**, 79.
26 J. J. Weiss, *Nature (London)*, 1944, **153**, 748.
27 O. Risse, *Strahlenstherapie*, 1929, **34**, 581.
28 *Discuss. Faraday Soc.*, 1952, **12**, 1–319.
29 J. Franck and F. Haber, *Sitz. Preuss. Akad. Wiss.*, 1931, 250.
30 F. S. Dainton and D. G. L. James, *J. Chim. Phys.*, 1951, **48**, 1.
31 *Discuss. Faraday Soc.*, 1963, **36**, 1–325.
32 For a summary see F. S. Dainton, *Faraday Lecture; Chem. Soc. Rev.*, 1975, **4**, 323.
33 Compare E. Collinson and F. S. Dainton, *Discuss. Faraday Soc.*, 1952, **12**, 222 with K. W. Chambers, E. Collinson and F. S. Dainton, *Trans. Faraday Soc.*, 1970, **66**, 141.

Faraday Discussion 100 *Celebration Paper*; *Presented 24th April*, 1995

List of Posters

Laser-like emission from dying start **E. Humphreys** and **D. Field,** *University of Bristol, UK*

Physicochemical perspectives in biology—from bonds to beer! **D. Clark,** *Institute of Food Research, Norwich, UK,* **L. Fisher,** *University of Bristol, UK*

The development of buckminsterfullerene and its derivatives **P. Birkett** and **J. Hare,** *University of Sussex, UK*

Shear induced light scattering of charged stabilised polymer latex **H. Sakabe** and **J. W. Goodwin,** *University of Bristol, UK*

Scanning electrochemical microscopy: A novel probe of interfacial activity **J. V. Macpherson, M. A. Beeston, R. D. Martin** and **P. R. Unwin,** *University of Warwick, UK*

Product channels in the near UV photodissociation of ozone **S. M. Ball, G. Hancock, J. C. Pinot, C. M. Sadowski** and **F. Winterbottom,** *University of Oxford, UK*

The longest standing problem in astrophysical chemistry: the origin of the diffuse interstellar bands **J. R. Miles, T. H. Kerr, R. E. Hibbins** and **P. J. Sarre,** *University of Nottingham, UK*

Laser spectroscopy of organometallic radicals prepared in supersonic jets using pulsed discharged methods **C. K. Corlett, A. J. Bezant, S. J. Pooley** and **A. M. Ellis,** *University of Leicester, UK*

A comparison of the crystal structures of polymer and electrolyte complexes elucidated by powder X-ray diffraction **J. Thomson, P. Lightfoot** and **P. G. Bruce,** *University of St Andrews, UK*

Statistical mechanics of condensation, freezing and the formation of liquid crystals **R. P. Sear** and **G. Jackson,** *University of Sheffield, UK*

Hot oxygen and precursor states: Their role in low energy catalytic pathways **A. F. Carley, P. R. Davies** and **M. W. Roberts,** *University of Wales, Cardiff, UK*

List of Participants

Prof. E. W. Abel *University of Exeter, UK*
Mr. M. A. Adams *CCLRC, UK*
Miss I. K. Ahmed *Queen Mary & Westfield College, University of London, UK*
Sir Geoffrey Allen *Kobe Steel, London, UK*
Dr. J. A. Altmann *ULCC, UK*
Dr. A. D. Ashmore *The Royal Society of Chemistry, London, UK*
Dr. P. W. Atkins *University of Oxford, UK*
Prof. C. M. Baldwin *Union University Jackson, Tennessee, USA*
Mr. S. Ball *University of Oxford, UK*
Dr. A. D. Bangham *Cambridge, UK*
Dr. M. J. Bearpark *King's College London, UK*
Prof. H. Berg *Harvard University, Cambridge, USA*
Mr. M. P. Berry *Chislehurst & Sidcup Grammar School, UK*
Dr. J. S. Best *Retired, Bedford, UK*
Dr. J. Betts *Kingston University, Surrey, UK*
Dr. C. F. Blakely *Thames Ditton, Surrey, UK*
Dr. A. R. Blythe *BICC, Chester, UK*
Dr. G. F. Boden *University of Wolverhampton, UK*
Prof. G. C. Bond *University of Brunel, Uxbridge, UK*
Dr. G. Boocock *Retired, Maidstone, Kent, UK*
Dr. D. Bott *Courtaulds PLC, Coventry, UK*
Prof. M. Bowker *University of Reading, UK*
Dr. M. Breuer *Gillette Company, USA*
Dr. J. Brophy *BP Chemicals/RSC, UK*
Dr. C. Brown *SmithKline Beecham, Herts, UK*
Prof. P. G. Bruce *University of St Andrews, UK*
Prof. A. D. Buckingham *University of Cambridge, UK*
Sir John Cadogan *Office of Science & Technology, London, UK*
Prof. R. W. Cahn *University of Cambridge, UK*
Prof. J. A. Cahn *University of Dundee, UK*
Dr. A. Calder *Zeneca Specialities, Manchester, UK*
A. F. Carley *University of Cardiff, Wales, UK*
Dr. J. H. Carter *Royal Naval College, Greenwich, Kent, UK*
Dr. F. Cassidy *SmithKline Beecham Pharm., Harlow, UK*
Dr. P. Celani *King's College London, UK*
Prof. M. Chesters *University of Nottingham, UK*
Prof. M. S. Child *University of Oxford, UK*
Prof. R. J. H. Clark *University College, London, UK*
Dr. D. Clark *Institute of Food Research, Norwich, UK*
Prof. J. H. R. Clarke *UMIST, Manchester, UK*
Mr. S. Clifford *King's College London, UK*
Prof. J. N. L. Connor *University of Manchester, UK*
Mr. D. S. Coombes *University College, London, UK*
Mr. G. C. Corlett *University of Leicester, UK*
Dr. B. G. Cox *Zeneca FCMO Huddersfield, UK*
Mr. J. C. L. Cox *Chemical Industries Association, London, UK*
Dr. J. D. Coyle *Cookson Group PLC, Oxford, UK*
Dr. I. M. Cunningham *Glaxo Research & Development, Stevenage, UK*
Prof. J. Cunningham *University College, Cork, Ireland*
Lord Dainton *Oxford, UK*
Dr. P. R. Davies *University of Cardiff, Wales, UK*
Miss D. M. De Jesus *Imperial College, London, UK*
Prof. P. Derrick *University of Warwick, UK*

Dr. A. Dicks *British Gas PLC, UK*
Prof. R. N. Dixon *University of Bristol, UK*
Prof. G. Dixon-Lewis *University of Leeds, UK*
Prof. R. J. Donovan *University of Edinburgh, UK*
Dr. J. R. Durrant *Molecular Dynamics Group, Imperial College, London, UK*
Prof. D. D. Eley *University of Nottingham, UK*
Dr. A. M. Ellis *University of Leicester, UK*
Dr. J. W. Emsley *University of Southampton, UK*
Dr. N. Evans *British Gas PLC, Loughborough, UK*
Prof. D. Everett *University of Bristol, UK*
Dr. M. Fettweis *Molecular Dynamics Group, Imperial College, London, UK*
Mrs. Y. A. Fish *The Royal Society of Chemistry, London, UK*
Dr. L. Fisher *University of the West of England, Bristol, UK*
Prof. H. M. Frey *University of Reading, UK*
Dr. J. C. Frost *Johnson Matthey Tech. Centre, Reading, UK*
Dr. P. H. Gamlen *ICI Chemicals & Polymers, Cleveland, UK*
Dr. M. H. George *Imperial College, London, UK*
Dr. J. Gerratt *University of Bristol, UK*
Dr. D. Giachardi *Courtaulds PLC, London, UK*
Dr. J. F. Gibson *The Royal Society of Chemistry, London, UK*
Dr. D. W. Gillings *University College, London, UK*
Dr. J. S. Gow *Harpenden, UK*
Sir Frances Graham-Smith *University of Manchester, UK*
Prof. P. Gray *Gonville & Caius College, Cambridge, UK*
Dr. G. Hancock *University of Oxford, UK*
Dr. T. M. Hardman *University of Reading, UK*
Dr. A. S. Hare *Herts, UK*
Dr. K. D. M. Harris *University College, London, UK*
Dr. P. Hawker *Johnson Matthey, Royston, Herts, UK*
Dr. A. Herod *Imperial College, London, UK*
Dr. J. Hiddleston *ISIS Innovation Ltd, Oxford, UK*
Prof. J. S. Higgins *Imperial College, London, UK*
Prof. A. R. Hillman *University of Leicester, Leicester, UK*
Prof. C. L. Honeybourne *University of the West of England, Bristol, UK*
Mr. Z. Huang *Imperial College, London, UK*
Ms. E. Humpheries *University of Bristol, UK*
Prof. I. Ikemoto *Tokyo Metropolitan University, Japan*
Prof. K. J. Ivin *South Petherton, UK (ex Queens University, Belfast)*
Dr. R. A. Jackson *University of Keele, UK*
Dr. R. A. Jeffreys *Hon. Treasurer, RSC, Middlesex, UK*
Prof. B. R. Jennings *English China Clays PLC, St Austell, UK*
Prof. B. F. G. Johnson *University of Edinburgh, UK*
Prof. T. M. Jones *ABPQ, London, UK*
Dr. W. Jones *University of Cambridge, UK*
Prof. C. Kemball *University of Edinburgh, UK*
Dr. L. J. Kirsch *Shell Research Ltd, Chester, UK*
Mr. S. Kumazaki *Molecular Dynamics Group, Imperial College, London, UK*
Prof. P. J. Lachmann *University of Cambridge, UK*
Dr. G. Lake *The Malaysian Rubber Prod. Research Assoc., Hertford, UK*
Mr. A. O. Laughton *British Gas PLC, Loughborough, UK*
Dr. S. Leach *Observation De Paris-Meudon, France*
Prof. A. C. Legon *University of Exeter, UK*
Miss M. S. Levitt *SMJ Books Editorial, Oxford, UK*
Prof. T. J. Lewis *University of Wales, Bangor, UK*
Prof. J. C. Lindon *Wellcome Research Laboratory, Kent, UK*
Miss J. MacPherson *University of Warwick, UK*
Dr. G. C. Maitland *Schlumberger Cambridge Research, UK*
Sir John Mason *Chancellor of UMIST, London, UK*

Dr. J. C. Maynard *Amersham International, Bucks, UK*
Dr. J. J. McGarvey *Queens University of Belfast, UK*
Mr. S. Merry *Molecular Dynamics Group, Imperial College, London, UK*
Dr. J. R. Miles *University of Nottingham, UK*
Prof. I. M. Mills *University of Reading, Reading, UK*
Dr. R. H. Moss *BT Laboratories, Ipswich, UK*
Mr. R. G. Neal *Retired, Newbury, UK*
Dr. R. F. Newton *Glaxo Research and Development, Stevenage, UK*
Dr. J. W. Nicholson *King's College, London, UK*
Prof. J. W. Nicholson *King's College, London, UK*
Prof. R. J. Ottewill *University of Bristol, UK*
Dr. G. S. Park *Retired, University of Wales, Cardiff, UK*
Dr. R. J. Parker *Editorial Manager, The Royal Society of Chemistry, Cambridge, UK*
Dr. D. A. Parkes *Shell Research Ltd, Chester, UK*
Prof. R. Parsons *University of Southampton, UK*
Dr. J. Peto *Bromley, Kent, UK*
Dr. A. Peto *Bromley, Kent, UK*
Lord Phillips *London, UK*
Prof. D. Phillips *Imperial College, London, UK*
Prof. M. J. Pilling *University of Leeds, UK*
Prof. P. H. Plesch *University of Keele, UK*
Prof. J. C. Polanyi *University of Toronto, Canada*
Lord Porter *Imperial College, London, UK*
Dr. S. L. Price *University College, London, UK*
Dr. T. Pryce-Jones *University of Hertfordshire, UK*
Dr. F. Rappaport *Molecular Dynamics Group International*
Dr. N. V. Reed *The Royal Society of Chemistry, London, UK*
Prof. C. W. Rees *Imperial College, London, UK*
Prof. L. V. Rees *University of Edinburgh, UK*
Ms. S. Riaz *The Royal Society of Chemistry, London, UK*
Sir Rex Richards *Chancellor of Exeter University, UK*
Prof. M. A. Robb *King's College, London, UK*
Prof. M. W. Roberts *University of Wales, Cardiff, UK*
Mr. S. C. F. Robinson *British Gas PLC, Loughborough, UK*
Dr. A. P. Rood *The Royal Institution, London, UK*
Dr. W. J. Resenfelder *London, UK*
Prof. J. S. Rowlinson *University of Oxford, UK*
Mr. N. V. Russell *Imperial College, London, UK*
Prof. C. M. Sadowski *University of Oxford, UK*
Mr. H. Sakabe *University of Bristol, UK*
Prof. P. J. Sarre *University of Nottingham, UK*
Dr. P. Schoffield *Oxford, UK*
Dr. S. K. Scott *University of Leeds, UK*
Mr. R. Sear *University of Sheffield, UK*
Prof. P. A. Sermon *Brunel University, UK*
Dr. A. Shaw *University of Southampton, UK*
Prof. N. Sheppard *University of East Anglia, Norwich, UK*
Prof. J. P. Simons *University of Oxford, UK*
Prof. K. S. W. Sing *University of Bristol, UK*
Prof. E. B. Smith *University of Wales, College of Cardiff, UK*
Dr. D. J. H. Smith *Whatman PLC, Kent, UK*
Dr. A. Smith *Laporte, Luton, UK*
Mr. B. R. Smith *King's College, London, UK*
Prof. I. W. M. Smith *University of Birmingham, UK*
Prof. M. Spiro *Imperial College, London, UK*
Dr. M. Stapleton *Molecular Simulations, Cambridge, UK*
Mr. R. Stevenson *Editor, Chemistry in Britain, The Royal Society of Chemistry, London, UK*
Sir William Stewart *Cabinet Office, London, UK*

Prof. F. S. Stone *University of Bath, UK*
Prof. L. H. Sutcliffe *University of Surrey, UK*
Dr. S. Suzuki *Kobe Steel Ltd, London, UK*
Prof. S. Sykes *Queen Mary & Westfield College, University of London, UK*
Mr. Y. Tachibana *Molecular Dynamics Group, Imperial College, London, UK*
Dr. J. B. Taylor *Nottingham, UK*
Miss S. Taylor *University of Southampton, UK*
Prof. J. M. Thomas *The Royal Institution, London, UK*
Mr. J. B. Thompson *University of St Andrews, UK*
Dr. J. C. Thorn *Assistant Editor, Faraday, The Royal Society of Chemistry, Cambridge, UK*
Prof. B. A. Thrush *University of Cambridge, UK*
Prof. D. Tildesley *University of Southampton, UK*
Mr. Tompkins *Molecular Dynamics Group, Imperial College, London, UK*
Prof. R. P. Townsend *Unilever Research, Port Sunlight Lab., Wirral, UK*
Dr. C. Troyanowsky *Division De Chemie Physique, Société Francaise De Chemie, France*
Prof. J. J. Turner *University of Nottingham, UK*
Dr. M. Y. Twigg *Johnson Matthey PLC, Herts, UK*
Dr. P. Volans *Gwent, UK*
Mr. T. Vreven
Prof. R. Walsh *University of Reading, UK*
Prof. P. Wardman *Grays Laboratory, Cancer Research Trust, Middlesex, UK*
Dr. P. B. Warren *Unilever Research, Wirral, UK*
Dr. G. A. Webb *University of Surrey, UK*
Prof. P. B. Wells *University of Hull, UK*
Prof. R. E. Wetton *Wetton Associates, Leicester, UK*
Prof. D. H. Whiffen *Retired, Bridgewater, Somerset, UK*
Dr. R. A. Whitelock *Staff Editor, Faraday, The Royal Society of Chemistry, Cambridge, UK*
Dr. K. Wibley *University College, London, UK*
Dr. A. J. Wickham *Nuclear Electric, Berkeley, UK*
Prof. F. Wilkinson *University of Loughborough, UK*
Dr. A. H. Willbourn *Formerly ICI (Plastics Division), Aston, UK*
Dr. G. Williams *University of Wales, Swansea, UK*
Miss S. Wilsey *The Oxford Instruments Group, Oxford, UK*
Miss N. Yamamoto *King's College, London, UK*
Prof. J. Yarwood *Sheffield Hallam University, Sheffield, UK*